TBM Design and Construction

Kui Chen · Shengjun Jiao · Jiangka Wang

TBM Design and Construction

Translated By Yixin Chen

人民交通出版社股份有限公司
China Communications Press Co.,Ltd.

Springer

Kui Chen
China Railway Group Limited
Zhengzhou, Henan, China

Shenzhen University
Zhengzhou, Henan, China

Jiangka Wang
China Association of Construction
Enterprise Management
Zhengzhou, China

Shengjun Jiao
Engineering Committee of China Railway
Society
Shaanxi, China

CPC Committee of Shaanxi Railway
Institute
Shaanxi, China

Translated by
Yixin Chen
China Railway Engineering Equipment
Group Co., Ltd.
Zhengzhou, China

ISBN 978-981-99-0058-9 ISBN 978-981-99-0059-6 (eBook)
https://doi.org/10.1007/978-981-99-0059-6

Jointly published with China Communications Press Co., Ltd
The print edition is not for sale in China (Mainland). Customers from China (Mainland) please order the print book from: China Communications Press Co., Ltd.

This Springer imprint is published by the registered company Springer Nature Singapore Pte Ltd.
The registered company address is: 152 Beach Road, #21-01/04 Gateway East, Singapore 189721, Singapore

Editorial Board of TBM Design and Construction

Translator

Prepared by

Foreword

"Only by always thinking actively, accepting all kinds of different idea with an open mind and absorbing fresh knowledge, can we achieve success." It is glad to see that the work *TBM Design and Construction* is about to be sent to the press, which is like a clear spring trickling into the field of TBM design and construction.

It is necessary to have effective tools to do good work. Full-face rock tunnel boring machine (TBM) is a piece of large special equipment for modern tunnel construction, which is integrated with mechanical, electrical, hydraulic, induction and information technologies to complete the operations like continuous rock breaking, muck discharge and support, and is characterized by fast tunneling speed, high efficiency, safety, environment protection, automation, high informationization degree, etc.

Market globalization brings severe international competition to equipment manufacturers in various countries, and governments and enterprises are actively developing measures and adopting advanced design and manufacturing technologies to improve the product competitiveness. For China, the comprehensive competitiveness of equipment manufacturing industry is the engine to promote the industrial transformation and upgrade and the important embodiment of our national core manufacturing capacity. The "12th Five Year" Development Plan of High-End Equipment Manufacturing Industry clearly indicates that vigorous cultivation and development of high-end equipment manufacturing industry is the necessary requirement to promote our industrial core competitiveness and the strategic option to take the high ground in the future economy and technology development, and it is of important strategic significance to facilitating the transformation of economy development mode and achieving the transform from a manufacturing country to a manufacturing power.

In the face of new historic opportunity, China's tunneling machine industry has ushered in enormous development space and opportunity. In the recent years, with the expansion of construction scale of infrastructure like transportation and water diversion in China, there are more and more tunnel engineering projects, and the application of tunnel boring machine (TBM) has become more and more extensive. The solution of many key technologies in design and construction of TBM system is of significant influence to improving our enterprises' competitiveness in the world's equipment manufacturing industry. Design of TBM system shall consider both the high bearing capacity and the high reliability, as well as the design requirements of high safety and long life, and non-standard customized design must be carried out oriented to the geological adaptability according to the full-process geological conditions. A reasonable TBM system has an important effect on improving cutterhead and cutter life, increasing TBM efficiency and safety, reducing vibration and noise of tunneling machine, etc. Therefore, how to deeply study the key technologies of TBM system design and construction in the shortest time, break the foreign technical blockade and promote the independent research and development level of our tunneling equipment manufacturing industry is undoubtedly a research subject having academic value, important practical significance and extensive application foreground.

This book focuses on the key issues in TBM system design and construction to comprehensively and thoroughly describe the latest technologies regarding the structure and working principle, selection and adaptability design, cutterhead and cutter design, construction organization and risk control of TBM in a profundity and easy-to-understand way and describes the representative major TBM project examples at home and abroad in different times. And this is a very good book well combining theories with practice. This book, written in simple language, clear thought, solid data, accurate charts and a large number of pictures, contains a large number of technical information. All the data are from the engineering practice, so they are highly reliable and highly operable. This book will play an important guiding role in our TBM design and construction.

This book features novel, systematic and comprehensive contents, with close combination of theories and engineering practice. The concepts, ideas and methods are based on the author and its team's long-term engineering practice and reflect the author's achievements in theoretical research and engineering practice. I believe that the publication of this book will certainly benefit the readers and will play an important facilitation role in the future development of TBM design and construction.

Zhengzhou, China
November 2017

Hui Liu
Vice Present and Chief Engineer
of China Railway Engineering Group
Limited

Preface

TBM is a complete set of tunnel construction equipment integrating mechanical, electrical, hydraulic, induction and information technologies to complete the operations like rock breaking, muck discharge and support during continuous tunneling, with features like factorial construction, quick tunneling speed and high efficiency. The great development of tunnel engineering has promoted the progress of TBM construction technology. The history of TBM construction began with the world's first TBM developed by Maus, a Belgian engineer, in 1846, and developed from the first modern TBM successfully made by Robbins, an American engineer, in 1953.

China's first TBM is 120 years later than abroad. Then, 1966–1984 is the dawn period of our TBM technology. China's first TBM developed in 1966, with diameter of ϕ3.5 m, was used in the construction of the water tunnel of Yunnan Xierhe Hydropower Plant. Also, 1985–2012 is the digestion period of our introduced TBM technology; in 1985, the water tunnel project of Tianshengqiao II Hydropower Plant introduced the open-type TBM of ϕ10.8 m diameter from the Robbins in America. From 2013, China has entered into the independent innovation period of TBM technology and started to design and manufacture TBM with fully independent intellectual property rights.

Although the tunnel construction by TBM has started late in China, attention has been paid to absorbing and learning advanced technologies and new techniques as well as success and failure experience from foreign countries, the development of TBM is fast, but there is still a large gap compared with foreign countries. Compared with the drilling and blasting method, TBM method has the advantages like rapidness, high quality, efficiency, safety, environment protection and high degree of automation and informatization; however, due to the extremely complex geological conditions in tunnel project, engineering accidents often occur during construction by TBM method, and the construction efficiency is not ideal, thus restricting the development of TBM technology. Risks in TBM construction always look for or utilize the vulnerabilities such as geological complexity, TBM inadaptability, human perception limit and irrationality of proposal and measures as the breakthrough to cause TBM engineering accidents or affect the construction period. To reflect the new technologies and new methods in TBM construction and to summarize the experience in

TBM design and construction as soon as possible, in order to fully exploit the technical advantages of TBM construction method, the State Key Laboratory of Shield Machine and Boring Technology of China Railway Tunnel Group has organized many national experts of TBM technology to compile the book *TBM Design and Construction*, on the basis of the 6th sub-topic "Design Philosophy and Assessment Decision Making System for TBM System Adaptability to Deep Compound Stratum" (2014CB046906) of the National "973" Program Project "Interaction Mechanism Between Surrounding Rock in Deep Compound Stratum and TBM and the Safety Control" (2014CB046900).

For this book, Liu Hui, Vice Present and Chief Engineer of China Railway Engineering Group Limited; Yang Huayong, Academician of Chinese Academy of Engineering; Li Jianbin, General Manager of China Railway Hi-Tech Industry Co., Ltd.; and other well-known experts are invited as the consultants of editorial board. Chen Kui, Sun Zhenchuan, Professor-Level Senior Engineers of the State Key Laboratory of Shield Machine and Boring Technology; Li Tao, Professor of Beijing Jiaotong University; and et al. are Authors of this book. Hei Fei, Yang Yandong, Wang Jiangka, Ji Yuguo and Liao Xiaochun et al. are Associate Editors-in-Chief of this book. Tan Shunhui, Professor-Level Senior Engineer of China Railway Engineering Equipment Group Co., Ltd.; Liu Quansheng, Professor of Wuhan University; and Du Lijie, Professor of Shijiazhuang Tiedao University, are Chief Reviewers of this book. This book is of important guiding significance to improving the independent design capacity of TBM as well as the technical level of TBM construction.

The book contains 11 chapters and strives to thoroughly describe the TBM structure and working principle, adaptability design, construction organization and the latest construction technology in depth with easy-to-understand language, with focus on representative national and abroad major TBM project examples in different periods. This book has mainly introduced the use cases of open-type TBM, such as Water Tunnel of Tianshengqiao Hydropower Plant, Tunnel of Qinling Rail Line I, Mogouling Railway Tunnel, West Qinling Tunnel, Zhongtianshan Tunnel, Vereina Railway Tunnel, Gotthard Basis Tunnel, Songhuajiang River Diversion Water Supply Project and Dali-Ruili Railway Gaoligong Mountain Tunnel. This book also introduces the use cases of double-shield TBM in projects such as Datong River-Qinwangchuan Water Tunnel, Yellow River-Shanxi Water Diversion Project, Datong River-Huangshui Water Tunnel, Honryan River-Shitou River Water Diversion Project, Taiwan Hsuehshan Tunnel and South Africa Lesotho Plateau Water Diversion Project, as well as the use cases of single-shield TBM in projects such as Taohe River Diversion Water Supply Project in Gansu and Rail Transit Line 5 in Chongqing.

During compilation of this book, experts from Shijiazhuang Tiedao University, China Railway Engineering Equipment Croup Co., Ltd., Urban Rail Transit Engineering Co., Ltd. of China Railway First Group Co., Ltd., Herrenknecht AG, China Railway Tunnel Stock Co., Ltd., etc. have participated in the compilation and provided guidance and also provided many pertinent comments or suggestions. Here I would like to express my heartfelt thanks to them.

Despite our best efforts, there may be still some shortages or even errors in this book because of its wide range of involvement; the readers are welcome to criticize and correct this book.

Zhengzhou, China
November 2017

Kui Chen

Contents

Introduction of the First Author

Kui Chen was born in November 1963, male, Doctor of Engineering, Postgraduate Tutor, Professor-Level Senior Engineer and National Primary Registered Construction Engineer; enjoyed special government allowance from the state council; and awarded the May 1st Labor Medal of Henan Province, Expert of China Torch Program Expert Database, Assessment Expert of National Major Special Projects, Assessment and Acceptance Expert of State Key Laboratories, Member of the Academic Board of State Key Laboratory of Aircraft Precision Bearing, Member of the Academic Board of State Key Laboratory of High-End Engineering Machinery Intelligent Manufacturing, Member of the Technical Committee for Standardization of National Construction Machinery and Equipment, Expert of the Technological Achievement Appraisal Expert Database of Henan Province, Assessment Expert of Science and Technology Progress Award of Hebei Province, Technical Consultant Expert of Technological Business Management Sunshine Government Platform of Guangdong Province, Assessment Expert of Natural Science Fund of Zhejiang Province, the top-notch personnel of the 12th Batch of Professional Talents of Zhengzhou City, one of the 8th Batch of Excellent Experts of Luoyang City, expert of China Railway, Level 1 Expert of China Railway Tunnel Group and Secretary of Party Working Committee/Executive Deputy Director of the State Key Laboratory of Shield Machine and Boring Technology.

In the academic sector, he is also Secretary of China Strategic Alliance of Shield Machine Remanufacturing and Innovation; Vice Director of Tunneling Machine Professional Board of the Tunnel and Underground Engineering Subcommittee of China Civil Engineering Society; Deputy Director of the Academic Board Henan Province Collaborative Innovation Center of Mechanical Equipment Advanced Manufacturing; Director of Tunneling Machinery Subcommittee of China Construction Machinery Association; Editor-in-Chief of *Modern Tunnel Technology*, *Tunnel Construction*, *Shield Machine and Tunneling*; Part-Time Professor of universities like Shijiazhuang Tiedao University, North China University of Water Resources and Electric Power, Henan University of Science and Technology and Xianyang Administrative Cadres Institute; Facilitator of 3 projects of the national 973 program, 3 projects of the 863 program and 1 national/international special cooperation project; winner of 1 first prize of National Prize for Progress in Science and Technology, 1 first prize and 3 second prizes of Henan Province Prize for Progress in Science and Technology, 1 first prize of the Science and Technology Prize of China Railway Society, 5 first prizes of the Science and Technology Progress Prize of China Building Enterprise Management Association and 1 special award and 5 first prizes of the Science and Technology Progress Prize of China Railway; Owner of 16 national invention patents and over 140 academic papers on core academic journals; Facilitator in compilation of the national standard—*Code for Construction and Acceptance of Shield Tunneling Method*, the industrial standard—*Technical Guide for Rock Boring Technique in Railway Tunnel* and the group standard—*Double Shield Rock Tunnel Boring Machine*; and Author of academic monographs like *Shield Construction Technique*, *Shield Construction Technique (Second*

Edition), Construction Examples of Shield Tunneling at Home and Abroad, Structure Operation and Maintenance of Shield, Shield Driver Training Course, China Shield, etc.

Chapter 1
Introduction

This chapter focuses on the concept and classification of tunnel boring machine (TBM), its origin, development, and application in the world, advantages and disadvantages, characteristics of construction organization and design, and typical factors restricting TBM construction performance.

1.1 Concept and Classification of TBM

1.1.1 Concept

TBM is the acronym of tunnel boring machine. In Japan and China, the machine used for full-face tunneling in soft soil stratum is called a shield, while the machine used for full-face tunneling in rock stratum is called a TBM. TBM introduced in this book is the machine used for full-face tunneling in rock stratum.

TBM is a set of large equipment used especially for breaking rocks and advancing in rock stratum by rotating cutter heads. It is able to offer support to tunnel formation and discharge muck simultaneously and form the tunnel in full face just in one take. Generally, TBM refers to the full-face rock tunneling machine. It is designed to tunnel in rock stratum, and if compared with the shield, it has no tunnel face stabilization functions coming from slurry or mud pressure. Shield construction features tunnel face stabilization, tunneling and muck discharge, and segment lining and backfill grouting. Tunnel face stabilization is the core working principle of the shield, as well as the main difference from the TBM.

Applying high-tech achievements in mechanical, electrical and hydraulic fields, modern TBM has become a complete set of equipment integrating tunneling, supporting, muck discharge, and transportation under computer control and CCTV monitoring for factory-specific operation. It has made a qualitative leap over the traditional drill and blast construction in terms of one-take tunneling, construction

K. Chen et al., *TBM Design and Construction*,
https://doi.org/10.1007/978-981-99-0059-6_1

progress, construction safety, construction environment, project quality and human resource allocation.

TBM has four basic functions, i.e., tunneling, mucking, guiding and supporting. It is also equipped with advance geological forecast equipment for complex strata. Tunneling is completed by the cutter head which rotates and drives the disc cutter to break the rock on the tunnel face, together with the drive system and the thrust system which powers the TBM. Mucking generally consists of muck guiding, shoveling, conveying, and transportation. The guiding functions mainly include direction determination and adjustment, and deflection correction. The supporting functions include the pre-treatment of the stratum not excavated before tunneling, local support of the tunnel wall after excavation, and the lining or segment assembly for the whole tunnel wall. The advance geological forecast system generally consists of probe drilling machine and its material detection system.

1.1.2 Classification

TBMs are mainly classified as gripper TBMs, double-shield TBMs, and single-shield TBMs, which are used for different geological conditions.

(1) Gripper TBM: It is commonly used for tunneling in hard rocks. It is equipped with auxiliary equipment such as a ring beam erector and shotcrete bolts to be adapted to geological changes. With effective support means, it can also be used for tunneling in soft rocks.
(2) Double-shield TBM: It is an adaptable TBM, which is suitable for both weak and hard rocks, as well as weak-hard interbedding.
(3) Single-shield TBM: It is generally used for tunneling in soft rocks. During tunneling, the single-shield TBM uses segments for supporting. But compared with the double-shield TBM, it is unable to tunnel and install segments simultaneously, and it is thus slower in construction. Its working principles are similar to those of the shield. However, the single-shield TBM differs from the shield in two major aspects. First, the single-shield TBM discharges muck with a belt conveyor, while the shield discharges muck with a screw conveyor or a slurry pump through a pipeline. Second, the single-shield TBM has no tunnel face balancing function, while the shield balances the water and soil pressure on the tunnel face by using soil chamber or slurry pressure.

With the progress of TBM technology and the need for TBM to be adapted to complex geology, there are other types of TBMs, including general compact TBMs, double-shield multi-functional TBMs, and double-mode TBMs.

1.2 Origin and Development of TBM

1.2.1 Origin and Development in Countries Outside China

In 1846, Belgian engineer Henri-Joseph Maus invented the first TBM in the world.

In the middle of the nineteenth century, the Western was enthusiastic in railway construction. However, human dreams were hindered by mountains. It was ideal to go ahead and dig tunnels, but it was too expensive to afford. The working face of a tunnel project was small, and most of the time was wasted on process connection. Time was money, while the tunnel project was time-consuming and laborious. The most effective solution was to build a big machine with power increased to bring the industrial revolution underground.

Belgian engineer Henri-Joseph Maus became a pioneer. In 1845, he was authorized by the King of Sardinia to build a railway connecting France and Italy. Maus was famous in mining industry in the world, and he was super confident. He dismissed the option of climbing the pass and insisted on building a straight railway, especially insisted on passing through the Frejus Mountain near the famous Cenis Mountain by tunnel.

It was a huge challenge. In Maus' scheme, it was required to dig a 12 km long tunnel. It was almost impossible under the technical conditions at that time. The tunneling process at that time was as follows (Fig. 1.1): Drill holes on the tunnel face, fill the holes with explosives, ignite the lead, and keep workers hiding in a corner to wait for detonation. Then require workers quickly run back to the newly opened tunnel surface with supporting wood, hoping to set up the supporting wood before the cave top collapses, and transport the broken rock out with a shovel.

The problem was that detonation would produce a large amount of toxic gas and the confined space of long tunnel must be ventilated before subsequent work. With the ventilation technology at that time, it took a long time to ventilate the tunnel. Maus knew it of caused, so he planned to design and manufacture the first tunneling machine.

In 1846, his mountain-slicer was assembled and came into being in a military factory near Turin. It was complex and huger than a locomotive. It had more than one hundred drill bits. It was a cluster of cams, rods, pistons and springs. Whether it was practical, it was indeed a product of meditation. When this mountain-slicer was constructed, visitors from all over the world flocked to it and took it as a monument to history. In those days, people regarded it more as a work of art than a machine.

But visitors were full of doubts. Huge thrust force was required by the tunneling machine, but it was generated outside the tunnel and transmitted to the working face through complex machines. The longer the tunnel and connection, the greater the energy loss during transmission. As a result, the mountain-slicer (Fig. 1.2) would "die" in tunnel because of lack of power. But confident Maus believed that things would eventually sort themselves out, but those with doubts were not convinced. The political turmoil in Europe in 1848 more or less weakened the prevailing optimism, and Maus' funding was cut short.

Fig. 1.1 Process of earliest drill and blast tunneling

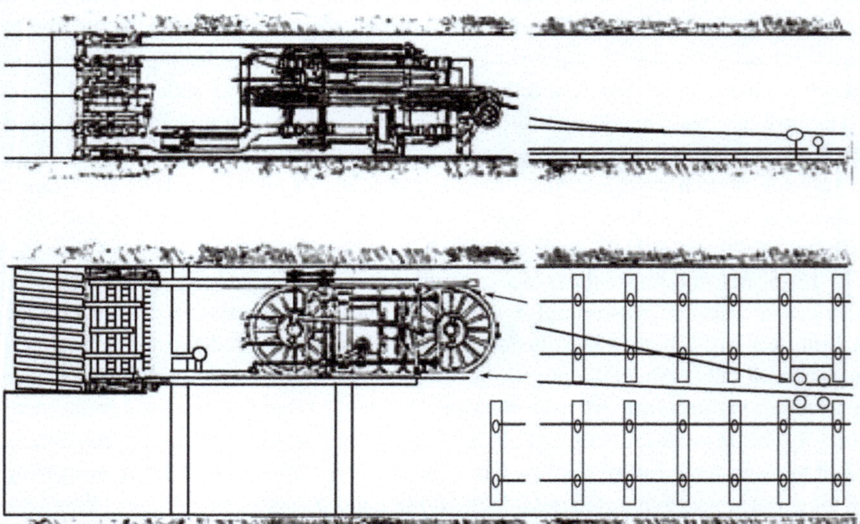

Fig. 1.2 Maus' Mountain-slicer

A decade later, thanks to the significantly improved tunnel ventilation technology, a tunnel of about 14 km long was constructed by drill and blast technique immediately adjacent to the Maus' route between 1857 and 1871 (Mont Cenis Tunnel). Maus' mountain-slicer has been recognized as the first TBM in the world although it has not been tested in practice.

In 1851, Charles Wilson from America developed a 75-ton steam engine-driven TBM for Richard Moon Corporation in southern Boston to excavate granite formation for Hoosac Tunnel in northwest Massachusetts. But this TBM stopped working after tunneling for only 10 ft (1 ft = 0.3048 m, the same below).

In 1856, Herman Haupt, a famous engineer in America, announced that he would save the Hoosac Tunnel Project with another TBM. He was confident and even paid for it out of his own pocket. However, his machine stopped working after tunneling for less than 1 ft, and Haupt went bankrupt in frustration.

In the next 30 years, engineers designed and trial-manufactured 13 TBMs of different types, and made progress, but none of them was successful. Among them, the compressed air TBM developed by Pomont in 1881 made some success. It was used to excavate the φ2.1 m exploration pilot tunnel for the Channel Tunnel, and it made an advance of 3 miles (1 mile = 1609.344 m, the same below).

From 1881 to 1926, some countries designed and manufactured 21 TBMs. But later, the development of TBMs came to a standstill as limited by technical conditions at that time. Around 1930, people were about to give up TBM. Barbara Stark, a tunneling expert, predicted pessimistically: "In the next two decades, patents on rock machinery will be limited and no similar machinery will be built."

James Robbins graduated from the then Michigan Mining School in 1930 when he knew nothing about the history of the TBM. He ran around and tried everything new in the next 15 years. He once mined in hard rocks in California, as well as washed gold in Alaska. After World War II, he became a mining consulting engineer in Illinois.

In 1952, he met a tunnel contractor, F. K. Mittry, who just won a bid of water tunnel project in Pierre, South Dakota.

The bedrock near Peel was full of cracks and faults, and geologists even named such bedrock Pierre Shale.

Pierre Shale was fragile, making blasting extremely difficult. Mittry bid for the project without hesitation even he knew that it would take a lot of time and effort to excavate the Pierre Shale, but he believed he had to solve the problem after he won the bid. Now he won the bid, and he began to visit and consult with engineers.

Robbins gave him an idea. At that time, a non-blasting excavation technology just began to develop in the mining industry. Its basic idea was to cut a batch of metal teeth or picks into the coal mining working face, and then rotate them to form a cut there. The rotating wheels suspended between the metal teeth undercut the loose coal seams on the surface.

Of course, the tunnel environment was quite different. Tunneling required much higher accuracy than coal mine excavation, and the section size was also quite different. Being uniquely insightful, Mittry selected Robbins' solution and entrusted Robbins to construct such a machine.

In 1953, the machine was successfully developed with a diameter of 7.85 m. It is the first modern weak rock TBM in the world. Unlike those TBMs developed before, this TBM performed surprisingly. The large turntable crushed the rock like a peanut shell, making an advance at 160 ft per day. It was astonishing. Such speed was almost ten times the contemporary drill-and-blast construction. Robbins built the first TBM to work efficiently in weak rock, although not the first TBM in the world.

The success of the Pierre Tunnel Project put the TBM in fashion. But this TBM was effective only when tunneling in weak rocks, and it failed in all hard rocks. Recognizing the development promise of TBM, Robbins founded S. Robbins and Associates (the Robbins Company now), the first company specializing in TBM manufacturing in the world.

In 1955, Robbins manufactured three φ2.44 m TBMs for the construction of a dam, which could not be considered successful because they presented some weaknesses quickly in the tunneling of shale, limestone interbedded rock mass and hard limestone rock mass. The biggest problem was that tungsten carbide drill bits frequently fell off when encountering hard rock, forcing the TBM to stop for repairs. Repeated shutdown of the TBM was a huge waste of time, letting alone the cost of the drill bits.

Robbins racked his brains to find a solution. One day in 1956, he decided to remove all the drill bits. The idea seemed unreasonable, because, theoretically, the drill bit was mainly used for cutting into rocks, while the disc was used for collecting broken rocks. But his intuition was correct. After all the drill bits were removed, the slightly modified discs (i.e., disc cutters, called crushing wheels by Robbins) were able to complete their work more independent and efficient, and able to effectively crush rocks with a unconfined compressive strength of 140 MPa.

In the same year, Robbins manufactured a φ3.28 m medium-hard rock TBM, which passed the industrial test. The application of disc cutter is an important symbol of full-face hard rock TBM and a major turning point in the development of hard rock TBM. During this period, Robbins manufactured a φ9m full-face TBM for a water tunnel of a dam in another country.

In 1960, the development of TBM entered a new stage. The Robbins Company constructed a φ4.89 m TBM for the Tasmania Tunnel Project, and integrated the support mechanism and thrust mechanism into a full floating system in the structural design for the first time. He used a spherical hinge structure pressed and fixed on the tunnel wall through the support shoe plate to obtain a reaction force of TBM during advancing.

This was the first Robbins TBM with a record advance of 229 m in 6 days.

The Robbins Company was also a pioneer in the development of dual-shield TBMs. As of 2001, the Robbins Company had constructed 383 gripper TBMs, 35 double-shield TBMs and 15 single-shield TBMs. In August 2006, the Robbins Company manufactured a φ14.44 m hard rock TBM (the TBM with the largest diameter in the world) (Fig. 1.3) for the Niagara Tunnel Project in Canada.

WIRTH GmbH of Germany began to construct TBMs in 1967. Its first TBM had a diameter of 2.14 m and was constructed for the 263 m long Ginzling Tunnel in Austria. In January 1982, it constructed a φ3.4 m TBM for a gold mine in South America

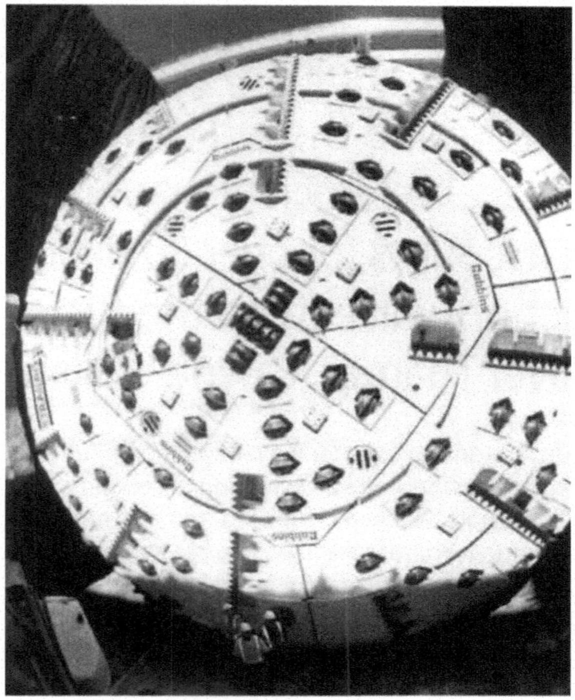

Fig. 1.3 φ14.44 m TBM from the Robbins Company

successfully, and its own disc cutter also had high bearing capacity. In 1983, Wirth began to construct double-shield TBM. In 1994, it constructed a φ7.64 m gripper TBM for the 19.062 km long VereinaTunnel in Switzerland. It also constructed two φ8.8 m gripper TBMs for Qinling Tunnel in China. In 1999, Wirth constructed a φ12.5 m gripper TBM for Spain. As of 2001, Wirth had constructed 112 gripper TBMs, 10 double-shield TBMs and 8 reaming TBMs.

At present, there are more than 30 TBM manufacturers in the world and more than 700 TBMs have been produced. Robbins of the United States, and Wirth and Herrenknecht of Germany ranked the top. The hard rock TBMs technology overseas has been proved in terms of technology and they are continuously improved in structure. There are gripper TBMs, single-shield TBMs and double-shield TBMs to adapt to different geological conditions. It is common to use TBM for tunnel construction abroad. Especially for tunnels longer than 3 km, the construction organization will usually clearly specify in the bidding documents that the bidder must use TBM for tunneling.

The following are typical engineering cases of TBM application abroad (Table 1.1).

(1) Lesotho Project

Lesotho Project in South Africa includes 4 water diversion tunnels and 2 water tunnels of about 200 km long in total, 6 dams, 2 hydropower stations and other hydraulic

Table 1.1 List of major TBM application cases in countries other than China

Project time (year)	Name	Application	TBM tunneling distance (km)	Geology	TBM diameter (m)	TBM type	Country
1990–present	Lesotho Project	Water tunnel	200	Basalt, sandstone, weak mudstone, claystone	4.95	Gripper TBM	South Africa
1995–1999	Vereina Railway Tunnel	Railway tunnel	19	Sedimentary rock, magmatic rock, granite	7.64	Gripper TBM	Switzerland
2002–2003	Khadori Hydropower Station Water Diversion Tunnel Project	Water diversion tunnel	6.5	Sandstone, shale, quartzite, quartz sandstone	3.00	Single-shield TBM	Georgia
2002–2007	Guadarrama High-speed Railway Tunnel Project	Railway tunnel	28.4	Gneiss, sedimentary rock, metamorphic sedimentary rock	9.45	Double-shield TBM	Spain
2003–2016	Gotthard Railway Tunnel Project	Railway tunnel	56.3	Hard rock, fractured breccia rock mass at both ends	8.83–9.53	Gripper TBM	Switzerland
2005–2010	Pajares Railway Tunnel	Railway tunnel	30.4	Hard rock, sandstone and slate	φ10.16 double-shield TBM φ9.90 single-shield TBM	Double-shield TBM Single-shield TBM	Spain

(continued)

Table 1.1 (continued)

Project time (year)	Name	Application	TBM tunneling distance (km)	Geology	TBM diameter (m)	TBM type	Country
2006–2013	Tunnel Works of Niagara Falls Hydroelectric Power Plant Project	Water diversion tunnel	10.4	Hard rock, and basalt	14.4	Gripper TBM	Canada
2008–2009	Choindez Safety Tunnel	Safe escape tunnel	3.2	Hard rock, molasse limestone	3.63	Gripper TBM	Switzerland
2008–2016	N.D. (Nant de Drance) Pumped Storage Power Station	Water diversion tunnel	5.6	Gneiss, hard sandstone, granite	9.4	Gripper TBM	Switzerland
2009–2011	South Cobb Tunnel	Blowdown tunnel	8.7	Hard rock, gneiss, granite	8.3	Gripper TBM	America
2013	Water Diversion Tunnel of Bärenwerk Hydropower Station	Water diversion tunnel	2.8	Hard rock, phyllitic slate, quartzite	3.8	Gripper TBM	Austria

structures. It was commenced in the late 1980s, and the total construction period was more than 30 years as scheduled. The tunnels were constructed with TBMs. The IA Tunnel Project (Phase I) under construction is 82 km long, including the 45 km long φ4.95 m water diversion tunnel from the intake of the Kaze Reservoir to the underground powerhouse of the Mola HPP, featuring basalt in Lesotho stratum. The south section of the water tunnel was surrounded by massive sandstone of Clarence stratum, while the north section featured complicated geological conditions and was surrounded mainly by weak mudstone, sandstone, claystone and dolerite dyke. A total of 5 TBMs were used for the construction.

(2) Vereina Railway Tunnel in Switzerland

The 19 km long Vereina Railway Tunnel was commenced in 1995 and completed in March 1997. The φ7.64 m diameter gripper TBM manufactured by Wirth was used to tunnel across the Alps mainly through granite, sedimentary rock and magmatic rock. Sedimentary rocks and magmatic rocks in the Alps were severely cut in tectonic movement, forming a large number of tectonic fracture zones. The tunnel overburden in these fracture zone sections reached 1200 m. It took 4 years to complete the tunnel, with a daily advance of 13.6–22.9 m. The key to successful TBM construction of the tunnel is that the geological conditions and rock characteristics that may be encountered and corresponding measures to be taken in TBM construction were fully taken into account in the design stage, and relevant measures were integrated into the selection of TBM, the determination of main technical parameters and the design.

(3) Khadori Hydropower Station Water Diversion Tunnel Project in Georgia

Khadori Hydropower Station, which is 198 km away from Tbilisi, the capital of Georgia, is a diversion type hydropower station with an installed capacity of 24 MW. The water diversion tunnel was 6491.3 m long in total and its diameter was 3.0 m. The tunnel area is dominated by sandstone, shale, quartzite and quartz sandstone. During TBM construction, adverse ground conditions such as crushed and weak rock mass, fault, groundwater and rock burst were encountered. The water diversion tunnel was commenced in September 2002 and completed in June 2003.

(4) Guadarrama High-speed Railway Tunnel Project in Spain

The Guadarrama High-speed Railway Tunnel in Spain consists of two single-track tunnels with a total length of 28.4 km. Two φ9.45 m double-shield TBMs manufactured by Herrenknecht and two double-shield TBMs manufactured by Wirth were used respectively. The construction of this project commenced in 2002 with an average daily advance of 16 m and a maximum monthly advance of 1000 m.

(5) Gotthard Railway Tunnel Project in Switzerland

The Gotthard-Basis tunnel in Switzerland, also known as the Gotthard Baseline Tunnel, is one of the most important channels crossing the Alps on the north–south axis of Europe. It took 17 years and SwF 12 billion (about EUR 11 billion) to construct this tunnel. Passing through the bottom of the Alps in Switzerland, this tunnel is about 35 mile (about 56.3 km) long and 8000 ft (about 2438.4 m) above the ground. It is longer than the Seikan Tunnel (about 53.9 km) in Japan and is the longest and deepest traffic tunnel (including railway tunnel and highway tunnel) in the world. This tunnel was constructed with 4 gripper TBMs with variable cross sections and diameters of 8.83–9.53 m.

(6) Pajares Railway Tunnel in Spain

The Pajares Railway passes through Cantabrian Mountains and the Picos de Europa along the coastline the Atlantic from Madrid to Gijon via Leon. It is as high as 2648 m. The train must wind through the valley and across the high pass. The new railway line has 24.5 km long Pajares Twin Tunnels, which were constructed with 1 single-shield TBM and 1 double-shield TBM from July and September 2005 respectively. The single-shield TBM advanced for 233 m weekly from the south towards the mountains. The first TBM advanced for more than 15 km and completed the tunnel in August 2008. Starting from the transition adit at Follo-Viadangos, the second TBM dug a 5 km long ramp first, and then two 3.9 km parallel tunnels. The tunnel was completed in November 2007. After being refurbished and customized, the single-shield TBM was used for tunneling in the northern Pajares-Sotiello section. A 5.9 km railway tunnel was built from July 2009 to March 2010.

(7) Tunnel Works of Niagara Falls Hydroelectric Power Plant Project in Canada

Running approximately in the same direction as original SAD2 water diversion tunnel, the Niagara Tunnel is 10.4 km long in total with an excavation diameter of 14.44 m, a lined diameter of 12.5 m, a maximum overburden of 140 m, and a diversion flow of 500 m^3/s. It was commenced in 2006 and constructed with an open-type TBM from the Robbins Company. It is the hard rock TBM project with the largest diameter in the world at present.

(8) Choindez Safety Tunnel in Switzerland

In consideration of the mountainous conditions in Switzerland, it is a matter of course to construct tunnels. But highway A16 (Transjurane) is a little special. It is 84 km long and has 11 tunnels, each at least 1 km long. In this project, it was required to build a 3.2 km long tunnel to bypass the Choindez Gorge. Before excavating the highway tunnel, a gripper TBM dug a safe tunnel for future escape. In addition, the TBM tunneled through strata with different mountain geological and hydrological characteristics from September 2008 to November 2009, accelerating the construction of subsequent highway tunnel.

(9) N.D. (Nant de Drance) Pumped Storage Power Station in Switzerland

The facilities of N.D. Pumped Storage Power Station are under construction underground. A gripper TBM from Herrenknecht was used to excavate a 5.6 km long entrance tunnel. After completing other projects, this hard rock TBM was specially modified for this construction project. For example, a back-up system was specially configured for this advance to adapt to the tunnel gradient. The TBM needs to tunnel on a slope up to 12% and at a turning radius of 500 m. An advanced drilling system was equipped to make detection in disturbed areas. In areas of high water pressure, the TBM poured concreted to stabilize the rock to be drilled through. Two vertical pressure shafts (the pressure water diversion tunnel) connect the Vieux Emosson Reservoir to the powerhouse in the Alps.

(10) South Cobb Tunnel in the United States

The South Cobb Tunnel Project in the United States is the largest project in the history of Atlanta, the capital of Georgia. The water recovery facilities of South Cobb were arranged close to the Chattahoochee River to collect and treat sewage from South Cobb County. The South Cobb Tunnel was built to replace the 26 km long old sewers to adapt to the population structure change. The tunnel was constructed in 2009–2011 and completed 1 year ahead of schedule. A gripper (Herrenknecht S-503) TBM was used to dig the 8.5 km long tunnel featuring gneiss and granite for the project from September 2009. It achieved a maximum daily advance of 50 m. The TBM finished the tunnel and connected it to the receiving shaft 18 months later on March 16, 2011, 11 months in advance.

(11) Water Diversion Tunnel of Bärenwerk Hydropower Station in Austria

The Bärenwerk Hydropower Station of Fusch city was first put into use in 1924, and now needs to be modernized. In this project, a 2.8 km longer water diversion tunnel was to be constructed with a slope of 4.9%. A professional construction team and a high-performance gripper TBM with a cutter head power of 1400 kW ensured the rapid progress of the project. This tunnel was constructed since the end of March 2013 with a maximum daily advance of 22 m and a maximum weekly advance of 77 m in the hard rock of the Alps. The construction team tunneled for 2.8 km under challenging geological conditions and reached the destination on September 10, 2013, less than 6 months.

1.2.2 Development and Application of TBM in China

The development of TBM covers three periods in China. The period from 1966 to 1984 is the germination period of TBM technology in China. In 1966, China researched and developed its first TBM with a diameter of 3.5 m for construction of the water diversion tunnel of Yunnan Xi'er River Hydropower Station. The period from 1985 to 2012 is the integration period of TBM technology in China. In 1985, China introduced a ϕ10.8 m gripper TBM from the Robbins Company for the Water Diversion Tunnel Project of Tianshengqiao II Hydropower Station. Since 2013, TBM development in China has entered an independent innovation period in China and Chinese enterprises have designed and manufactured TBMs with completely independent intellectual property rights.

1. Stage I: germination period of TBM technology in China (1966–1984)

Research and development of TBM in China began in 1964 when the Mechanical Design Office of Shanghai Investigation and Design Institute and the Mechanical and Electrical Department of Beijing Hydropower School had a scheme design of TBM. In 1965, the research and development of TBM was listed as a national key scientific research project of China. The former Ministry of Water Resources and Hydropower of the People's Republic of China mobilized technical forces, mainly from the Mechanical Design Office of Shanghai Investigation and Design Institute, to design a TBM in Shanghai Hydraulic Machinery Factory. In 1966, the first TBM (φ3.5 m) was manufactured in China, 120 years later than that abroad. This TBM underwent an industrial test in the water diversion tunnel of Xi'er River Hydropower Station in Xiaguan, Yunnan. It tunneled at a maximum monthly advance of 48.5 m in granitic gneiss and limestone with a compressive strength of 100–240 MPa.

In 1969, Guangzhou Electrical and Mechanical Industry Bureau built one ϕ4 m TBM which tunneled for 245 m at a maximum monthly advance of 20 m in granite and limestone with a compressive strength of 30–240 MPa.

In 1970, the machine repair workshop of Pingxiang Mining Bureau manufactured a ϕ2.6 m TBM to dig a tunnel for Pingxiang Qingshan Coal Mine. It tunneled for 622 m at a maximum monthly advance of 252.6 m in limestone and malmstone with a compressive strength of 20–120 MPa. In the same year, Xi'an Coal Mine Machinery Factory also trial-manufactured a φ3.5 m TBM, and Tongchuan Mining Bureau carried out an industrial test for it and used it to dig a 669 m long tunnel at a maximum monthly advance of 179 m in limestone with a compressive strength of 56–160 MPa.

In 1971, TBMs with a diameter of trial-manufactured in are φ2.5 m, φ5.5 m, φ3.8 m and φ5.9 m respectively; The manufacturers are Guangzhou Electrical and Mechanical Industry Bureau, Railway Corps, Fushun Mining Bureau and Shanghai Hydraulic Machinery Factory respectively for Guizhou Second Railway Bureau, Jingxi Coal Mine, Fushun Laohutai Mine and Second Engineering Bureau of Beijing Luopoling Hydropower Department. They were used to tunnel in dolomitic limestone, siliceous limestone, granitic gneiss and limestone at a maximum monthly advance of 123 m.

Later, Shanghai Hydraulic Machinery Factory manufactured a ϕ5.8 m SJ-58 TBM, which tunneled for 247.3 m in total in the industrial test in a hydraulic tunnel of Yunnan Xi'er River Hydropower Station from April 1977 to April 1978. SJ-58 TBM was optimized in design in 1981 and then used in construction of Gurenzhuang Tunnel of Water Diversion Project from the Luan River to Tangshan on November 25 of the same year. It tunneled for 2747.2 m in total at a maximum daily advance of 19.85 m and a maximum monthly advance of 201.5 m in dolomitic and siliceous limestones. This tunnel was completed on March 15, 1983, and it is the first medium section tunnel constructed with a China-made TBM in China.

During this period, Shanghai Heavy Machinery Plant manufactured EJ50 TBM and put it in the industrial test in Dongqu Coal Mine of Gujiao, Shanxi. In the test, this TBM dug a 3600 m long φ3 m tunnel at an average monthly advance of 78 m, a maximum monthly advance of 202 m, and a maximum daily advance of 12.7 m in limestone and sand shale.

From 1965 to 1984, more than 10 China-made TBMs were put into use in more than 20 engineering projects, tunneling for about 20 km in total. However, there was still a large gap in tunneling capacity compared with similar TBMs from other countries.

2. Stage II: Introduction and Integration Period of TBM Technology (1985–2012)

During this period, the main representative TBM projects in China were as follows:

(1) Tianshengqiao II Hydropower Station Water Diversion Tunnel Project

In China, water conservancy and hydropower engineering is the earliest application field of TBM construction. From 1985 to 1992, a φ10.8 m open Robbins TBM manufactured by the Robbins Company was selected for the Tianshengqiao II Hydropower Station Water Diversion Tunnel Project. But as the TBM selected could not adapt to the geological conditions and it was of high equipment failure rate, it tunneled for only 7.5 km in total at an average monthly advance of only 65 m.

(2) The Water Diversion Project from Datong River to Qinwangchuan District

This project was a large-scale inter-basin water 86.9 km to draw water from the Datong River to Qinwangchuan District. From 1991 to 1992, water tunnels 30A and 38 with a total length of about 17 km were constructed with a φ5.53 m double-shield Robbins TBM in sequence with an average monthly advance of 980 m and a maximum monthly advance of 1400 m. It was a successful TBM application. Water tunnel 30A was 11,649 m long, passing through from the inlet to outlet the pre-Sinian crystalline limestone, slate mixed with phyllite, conglomerate, glutenite, argillaceous siltstone, and sandstone in sequence and a 150 m long loess section at the outlet. Water tunnel 38 was 5400 m long surrounded by sandstone.

(3) Shanxi Wanjiazhai Yellow River Diversion Project

From 1993 to 2000, 5 Robbins TBMs and 1 NFM double-shield TBM manufactured were used in Shanxi Wanjiazhai Yellow River Diversion Project, tunneling for 122 km in total and setting a record of 113 m daily advance and 1,637 m monthly advance. Tunnels #6, #7 and #8 on the general trunk were dug in a diameter of 6.125 m with 1 double-shield TBM from Robbins, while tunnels #4, #5, #6 and #7 of southern trunk were dug in diameters of 4.82–4.94 m with 4 double-shield TBMs, including 3 from Robbins and 1 from NFM.

(4) Qinling Line I Tunnel

Qinling Line I Tunnel, which was 18.456 km in total, was constructed with TBM method and drill and blast method. In 1996, the former Ministry of Railways of the People's Republic of China introduced two φ8.8 m gripper TBMs, which tunneled for 5.244 km from the north tunnel portal and 5.621 km south tunnel portal at an average monthly advance of 252.36 m in 1998, and an average monthly advance of 290.64 m and a maximum monthly advance of 528.48 m in 1999.

(5) Taohuapu Tunnel #1

Taohuapu Tunnel #1 of Xi'an-Nanjing Railway was 7.234 km long in total, of which 6.2 km was dug at a maximum monthly advance of 551.82 m with a ϕ8.8 m gripper TBM by China Railway 18th Bureau Group Co., Ltd. (This TBM was used to tunnel the outlet section of Qinling Line I Tunnel).

(6) Mogouling Tunnel

Mogouling Tunnel of Xi'an-Nanjing Railway was 6114 m long in total, and it was surrounded by weak rock by 70.5%. It was constructed with a φ8.8 m gripper TBM by China Railway Tunnel Group and completed in November 2002. 4.653 km of this tunnel was dug with a TBM at a maximum monthly advance of 573.9 m. In this project, China Railway Tunnel Group mastered the tunneling technology of gripper TBM in soft ground.

(7) Taiwan New Wujie Water Diversion Tunnel

This was a cross-basin water diversion tunnel from Wujie Reservoir to Sun Moon Lake Reservoir, of which 7.1 km was constructed with a second hand 222 m long ϕ6.2 m gripper TBM manufactured by the Robbins Company in 1994. It was excavated from July 14, 2000 and has now been cut through. It features laccolite of 89% Jiayang terrane slate and 11% Tachien sandstone. Modified and maintained by the contractor in Hong Kong, this TBM was transported to the construction site in April 2000, assembled in May, and pushed to the predetermined excavation position and put into trial operation in late June. It began initial excavation in middle July, and full-function tunneling in early September. During the initial excavation, the TBM advanced 154.2 m only from July to August as affected by high failure rate of oil pressure and power systems, time-consuming adjustment and calibration, and insufficient spare parts. But in the full-function tunneling since September, the failure rate of machines and tools was greatly reduced, and the utilization rate was increased to more than 30%. In October 19, the TBM tunneled into Jiayang terrane slate section and progressed well. It advanced up to 44.3 m in a day, and up to 659.3 m in a month.

(8) Datong Tashan Mine Project

In this project, one ϕ4.82 m double-shield TBM was used to dig a 3.5 km long tunnel surrounded by limestone and granite mixed coal seams.

(9) Yunnan Jiu River Water Diversion Tunnel

This tunnel was constructed to relieve the serious shortage of water resources in Kunming, alleviate the backflow water pollution of Dianchi Lake and promote the benign development of water ecology. It was 22 km long surrounded by sandy slate, argillaceous slate, dolomite, etc. The construction of this tunnel began in 2003. A double-shielded φ3.65 m TBM jointly manufactured by China Erzhong and The Robbins Company of the United States was used. The planned TBM tunneling length was 13.77 km. The faults along the tunnel were basically explored during initial design, but intense small faults were encountered during actual construction, and the TBM was stuck at the head frequently. In addition, no pre-drilling and pre-grouting equipment was configured, disabling pre-drilling and pre-grouting and making TBM tunneling difficult. The advantage of TBM in fast construction cannot be brought into full play, so the drill and blast method was used later.

(10) Dahuofang Water Diversion Tunnel

Dahuofang Water Diversion Project in Liaoning was designed to provide industrial, domestic and environmental water for the central area of Liaoning Province. This tunnel was constructed from 2005 to 2009. It was 85.308 km long. The first 24.58 km of the tunnel was constructed by drill and blast method, and the rest 60.73 km, except 2 km constructed by drill and blast method, was constructed with three φ8.03 m gripper TBMs (2 from Robbins of the United States and 1 from Wirth of Germany). In this tunnel, the following technologies were firstly introduced to China: 19 cutters, variable frequency drive, continuous belt muck discharge system, steel pillow transportation for back-up system and transport vehicle, and 300 m low air leakage hose for ventilation system. The TBM construction made dozens of achievements, mainly including: a maximum monthly advance of 1208 m, a maximum daily advance of 63.5 m, an average monthly advance of 750 m (for 6 consecutive months), a maximum monthly machine-hour utilization ratio of 60.3% (over 40% on average), and successful continuous belt muck conveyor at low failure rate.

(11) Xinjiang Bashiyi Daban Diversion Tunnel

Xinjiang Daban Diversion Tunnel is 31.9 km long. Its diversion discharge is 65 m^3/s and its increased discharge is 74 m^3/s. Its excavated diameter is 6.79 m and its finished diameter is 6.0 m. 90% of this tunnel is surrounded by Class V and Class VI weak rock strata. It was constructed with a Herrenknecht double-shield TBM from February 2006 to June 2010 at a maximum monthly advance of 1,003 m. The TBM was shut down 48 times (379 days in total) before this tunnel was cut through.

(12) Qinghai Datong River-Huangshui Diversion Project

The water diversion tunnel of this project starts from the Datong River at the location basically parallel to and about 25–30 km from the Huangshui River. This tunnel is surrounded by argillaceous sandstone, argillaceous siltstone, granodiorite, quartzite, etc. It has an average buried depth of 480 m and a maximum buried depth of 1020 m. It was dug with a double-shield TBM for 20.0 km. Its excavated diameter is 5.93 m and its finished diameter is 5.0 m. This project was commenced in 2007 and completed on June 30, 2015. The tunnel is 24.17 km (19.9 km constructed with TBM) with a lined outer diameter of 5.7 m. It was constructed with a double-shield TBM manufactured by Northern Heavy Industries Group Co., Ltd. (NHI for short) under the technical support of Wirth. As affected by complicated geological condition and soft soil sections, TBM construction was not suitable and the TBM was in shutdown for more than 5 years. Later, the TBM equipment was modified, and put into operation with another TBM from Yao River Water Diversion Project to dig the tunnel from the other end.

(13) Middle Tianshan Mountains Tunnel

The Middle Tianshan Mountains Tunnel is 22.467 km long in total with a maximum buried depth of 1728 m. It was constructed with two ϕ8.8 m gripper TBMs and drill and blast method in May 2007. It featured unfavorable geological conditions such as high ground stress, mud burst, rock burst and tunnel water burst.

(14) Yao River Water Diversion Project in Gansu

This tunnel has a diameter of φ5.75 m and a length of 35 km (18 km + 17 km). One double-shield TBM and one single-shield TBM manufactured by NHI under the technical support of NFM were used. The shield was launched in 2008. The double-shield TBM arrived at the site in October 2011. The single-shield TBM was shut down for re-manufacturing of cutter head and shield and adjustment of tunneling direction as affected by loose sand and groundwater. The last 100 m of the tunnel was constructed with the freezing construction method. The tunnel was cut through in 2014, with a maximum monthly advance of 1868 m.

(15) Hongyan River-Shitou River Water Diversion Project

The tunnel was constructed from 2008 to 2016. It is 11 km long with a diameter of 3.65 m. It was constructed at a maximum monthly advance of 420 m, a maximum daily advance of 30 m with a Robbins double-shield TBM. As affected by weak rock sections and groundwater, the TBM was in shutdown for 3 years and stuck 66 times, and it resumed construction after modification.

(16) Jinping II Hydropower Station

The Jinping II Hydropower Station Project in Sichuan was commenced in 2008. Four main tunnels were constructed, of which two 16.7 km long tunnels were constructed with two ϕ12.4 m gripper TBMs (1 from Herrenknecht and 1 from Robbins). More than 90% of the engineering geology was marble, mainly Class II and Class III surrounding rocks with a compressive strength of 50–100 MPa. The maximum buried depth was 2525 m This project faced worldwide technical challenges such as extremely strong rock burst and high-pressure large water ingress. The tunnel was launched in November 2008 and broke through the tunnel in October 2011 at a maximum monthly advance of 683 m.

(17) West Qinling Tunnels Project

The West Qinling Tunnels of Lanzhou-Chongqing Railway, which are 28.236 km long, were commenced in August 2008. Located in Wudu District, Longnan City, Gansu Province, they are single-track twin tunnels and were constructed by China Railway Tunnel Group Co., Ltd. (CRTG) and China Railway 18th Bureau Group Co., Ltd. with two φ10.2 m Robbins gripper TBMs at a maximum monthly advance of 842.5 m, except that the 2113 m long preparatory tunnel at the portal section and the 4715 m long preparatory tunnel between 2 tunneling sections were constructed by drill and blast method. The whole tunnels were completed on July 19, 2014.

(18) Yunnan Nabang Hydropower Station Project

Nabang Hydropower Station was built from 2009 to 2012. Its tunnel is 4.5 m in diameter, and 10 km of this tunnel was constructed with a TBM at a maximum monthly advance of 581 m in extremely hard rock stratum with a compressive strength exceeding 220 MPa mixed with soil. It is a case of gripper TBM with minimum diameter in China.

(19) Chongqing Rail Transit Line 6

The TBM test section of Chongqing Rail Transit Line 6 (Phase I) was 12.122 km long Wulidian-Yanyanggou Reservoir Section surrounded by argillaceous sandstone and glutenite with a compressive strength of about 30 MPa. Two φ6.36 m Robbins gripper TBMs were used. This is the first time that TBM is used to construct urban subways in China. The maximum daily advance was 46.8 m and the monthly advance was 862 m.

(20) Northwest Liaoning Water Supply Project

The tunnel of this project has been one of the two water tunnels constructed with TBM in the northeast of China at present. This project was commenced in 2010. The tunnel has a diameter of 8.5 m. It consists of a 100 km long section, a 130 km long section and a 40 km long section. Eight φ8.53 m gripper TBMs were used at a maximum monthly advance of 1078 m. In this project, 20″ disc cutter was used in China for the first time.

(21) Han River-to-Wei River Water Diversion Project

This is the preliminary project of South-to-North Water Diversion Project in Shaanxi Province. It was commenced in 2012. The tunnel is 98.3 km in total. The 16 km of the section in north Qinling of the Project was constructed by China Railway 18th Bureau Group Co., Ltd. with a gripper TBM manufactured by Herrenknecht. The trial tunneling was carried out on June 15, 2014, and the first 7272 m long section was completed on August 11, 2015. The 18 km of the section in south Qinling was constructed by China Railway Tunnel Group Co., Ltd. (CRTG) with two Robbins gripper TBMs manufactured. The trial tunneling was carried out on February 17, 2015. The section was mainly surrounded by granite and quartzite with a compressive strength of 100–240 MPa, wearing cutter seriously. In addition, rock burst and sudden water burst were encountered. The excavation diameter of both TBMs was 8.02 m.

3. Stage III Independent Innovation Period TBM Technology (2013-present)

Since 2013, China has entered the period of independent innovation of TBM technology and started to design and manufactured TBM with completely independent intellectual property rights.

(1) Dual Mode TBM Manufactured by CRCHI (Fig. 1.4)

Fig. 1.4 Dual-mode TBM (earth pressure/single shield) independently developed by CRCHI

The Xinjie Taigemiao Coal Mine Inclined Shaft Project is located in Xinjie Taigemiao Mine Area, Ordos, Inner Mongolia. The inclined shaft is mainly surrounded by sandy mudstone and siltstone with a strength of 40–60 MPa. It was constructed with a dual-mode TBM jointly developed by China Railway Construction Heavy Industry Corporation Ltd. (CRCHI), China Shenhua and China Railway 13th Bureau Group Co., Ltd. On August 3, 2013, CRCHI developed the first long-distance large-gradient inclined shaft TBM for coal mine in the world. The TBM has an excavation diameter of 7.62 m, and two modes: earth pressure balanced shield and single-shield.

(2) Small-diameter TBM Manufactured by CITIC Heavy Industries (Fig. 1.5)

Fig. 1.5 Small-diameter gripper TBM Independently developed by CITIC Heavy Industries

In August 2013, the first φ5m gripper TBM in China was finished at CITIC Heavy Industry Co., Ltd. and put into operation in the construction of tunnel #1 of Luoyang Guxian Town Water Diversion Project in 2015 (Fig. 1.6).

(3) Acquisition of Intellectual Properties from Wirth by CREG (Fig. 1.7)

On November 26, 2013, CREG (China Railway Engineering Equipment Group Co., Ltd.) acquired the TBM and shaft-boring intellectual property rights from Wirth.

(4) Double Shield TBM of CSIC (Fig. 1.8)

Fig. 1.6 Application of TBM of CITIC heavy industries in Luoyang Guxian Town water diversion project

Fig. 1.7 Acquisition of TBM intellectual properties from Wirth by CREG

Fig. 1.8 DSUC double-shield TBM manufactured by CSSC

Qingdao Metro Line 2 (Phase I) is 25.2 km in total. It was constructed with TBMs and drill and blast method. Four DSUC φ6.3 m double-shield TBMs jointly manufactured by SELI of Italy and China Ship Heavy Industry (Qingdao) Rail Transportation Equipment Co., Ltd. were used in the Project. This was the first time that this type of TBM was used in metro construction projects in China. On December 22, 2014, DSUC double-shield TBM "Kunlong" manufactured by CSSC was finished, and transported to Hai'an Road Station of Qingdao Metro Line 2 for assembly and commissioning on January 23, 2015. It was launched to tunnel on March 21, 2015.

(5) Gripper TBM of CRCHI (Fig. 1.9)

Fig. 1.9 φ7.93 m gripper TBM of CRCHI

Fig. 1.10 φ8.03 m gripper TBM independently developed by CREG

The tunnel of Songhua River Water Diversion Project is one of the two water tunnels constructed with TBMs in the northeast of China at present. It is 70 km long, and it was constructed with two gripper TBMs independently developed by CREG and CRCHI. This project demonstrated the localization of gripper TBMs in China. On December 27, 2014, CRCHI developed the first φ7.93 m gripper TBM in China as supported by 863 Program. It overcame the geological characteristics and technical difficulties of "long distance, large burial depth, high ground stress, high water pressure, high ground temperature, large water burst and rock burst" and put the TBM into operation in Songhua River Water Diversion Project in Jilin, which took the lead in completing the 6000 m long tunnel in the first stage at maximum monthly advance of 1209.8 m.

(6) Gripper TBM Independently Developed by CREG (Fig. 1.10)

In 2015, φ8.03 m gripper TBM "Yongji" independently developed by CREG was used in Songhua River Water Diversion Project in Jilin, and it boasted a maximum monthly advance of 1226 m.

(7) Single Shield TBM of CRCHI (Fig. 1.11)

In March 2015, the φ6.88 m single-shield TBM "Linghang 1" independently developed by CRCHI was used for Chongqing Rail Transit Loop Line Project, and it boasted a maximum daily advance of 24 m.

(8) Dual-mode TBM independently developed by CREG (Fig. 1.12)

On April 16, 2015, the first φ6.85 m dual-mode TBM for urban metro construction in China was launched at the construction site of Chongqing Rail Circular Line Sports Park-Ranjiaba Section. It was jointly developed by CREG and Chongqing Construction Engineering Group Corporation Limited (CCEGC).

(9) Single Shield TBM Independently Developed by CRCHI (Fig. 1.13)

Fig. 1.11 φ6.88 m single-shield TBM independently developed by CRCHI

Fig. 1.12 φ6.85 m dual mode TBM jointly developed by CREG and CCEGC

Fig. 1.13 φ7.6 m single-shield TBM independently developed by CRCHI

Auxiliary shaft #2 of Shendong Bulianta coal mine is located in Bulianta Mine Area, Ordos City, Inner Mongolia. TBM construction faced six major technical problems such as long distance (2733 m), continuous downslope, overpassing and undercrossing existing tunnels, tunneling in argillaceous sandstone, repeated crossing of coal seams, and high-pressure water-rich stratums. In June 2015, a φ7.6 m single-shield TBM for coal mine inclined shaft, which was the first TBM with independent intellectual property rights in China developed by CRCHI was put it in use in auxiliary shaft #2 of Shenhua Shendong Bulianta Mine (Fig. 1.14). This TBM was about 165 m long in total and it weighed more than 1560t. The TBM passed through special strata of weak rock, hard rock and composite strata at a maximum monthly advance of 639 m, and overcame the difficulty of tunneling in vertical curve with attitude adjusted.

(10) Double Shield TBM Independently Developed by CRCHI (Fig. 1.15)

On December 24, 2015, the φ5.49 m double-shield TBM developed by CRCHI was used in Lanzhou Water Source Construction Project.

Fig. 1.14 Application of φ7.6 m single shield TBM developed by CRCHI in inclined shaft of coal mine

Fig. 1.15 φ5.49 m double shield TBM independently developed by CRCHI

Fig. 1.16 φ5.48 m double shield TBM independently developed by CREG

(11) Double Shield TBM Independently Developed by CREG (Fig. 1.16)

The Lanzhou Water Source Project was designed to draw water from "Liujiaxia Reservoir" to Lanzhou City. It consisted of water intake, main water tunnel, diversion well, branch water tunnel, etc. Extending about 14 km as planned, it passed through the topography and landform of middle and lower mountains transiting to the loess tablelands, ridges and hills in turn. It was surrounded by Class III and IV rocks at a compressive strength of 15–75 MPa. Two φ5.49 m double-shield TBMs were used for the construction of this project. On January 10, 2016, the ϕ5.48 m double-shield TBM developed by CREG was finished in Chengdu. It was used for the construction project of water source in Lanzhou City. The main water tunnel of this project is a 31.57 km long pressure diversion tunnel passing through hard rocks and different strata, of which 24.4 km was constructed with a TBM at a maximum buried depth of 918 m.

(12) World's minimum-diameter TBM independently developed by CREG (Fig. 1.17)

Fig. 1.17 World's minimum-diameter TBM (φ3.53 m) independently developed by CREG for Lebanon

Fig. 1.18 Variable-diameter TBM (φ6.53 m–φ6.83 m) developed by CRCHI

In January 2016, two ϕ3.53 m TBMs developed by CREG were finished in Zhengzhou. They were used in the construction of water supply and conveyance tunnels for Great Beirut Water Supply Project in Lebanon.

(13) Variable-diameter TBM developed by CRCHI (Fig. 1.18)

On May 5, 2016, CRCHI developed a TBM for deep-buried tunnels with variable-diameter from 6.53 m–6.83 m. This TBM was used in a major water conveyance tunnel in Xinjiang, the most challenging tunnel for TBM construction at present about 42 km long in total with an excavation diameter of φ6.53 m. TBM construction faced world-class engineering geological problems. For instance, the tunnel was buried deeply across large fault fracture zones and rock mass alteration fracture zones with strong rock burst and high ground temperature, and the surrounding rock was susceptible to deformation.

(14) Merger and Acquisition of Robbins by NHI (Fig. 1.19)

Fig. 1.19 Merger and acquisition of Robbins by NHI

On June 28, 2016, Northern Heavy Industries Group Co., Ltd. (NHI for short) of China acquired the Robbins Company of the United States. This was another major merger and acquisition effort made by NHI after the merger and acquisition of NFM of France in 2007, marking a historic leap for NHI in reshaping globalized industry pattern of the full-face TBMs.

(15) Double-shield TBM developed by CREG for Construction of Shenzhen Metro (Fig. 1.20)

The section from Meilin East Station to Innovation Park Station of Shenzhen Metro Line 10 is 3869.5 m in total, of which 2688 m was normally tunneled and 463 m was stepped without load by a TBM, while 718.5 m was constructed with a mining method.

Maximum diameter TBM (ϕ9.03 m)

This is the first section of subway construction project constructed with a double-shield TBM in Shenzhen. 95% of the tunnel was surrounded by slightly weathered rocks (mainly slightly weathered granite and migmatite) at a uniaxial compressive strength of 33.4–127.3 MPa mixed with local moderately weathered granite at a uniaxial compressive strength of 20.8–54 MPa. According to preliminary statistics, the strata crossed by TBM feature Class III and Class II surrounding rocks accounting for about 40% respectively, and Class IV and Class V surrounding rocks accounting for about 10% respectively. The double-shield TBM was independently developed by CREG, with an excavation diameter of φ6.5 m.

(16) Large-diameter TBM jointly developed by CRTG and CREG (Fig. 1.21)

Gaoligong Mountain Rail Tunnel is the most important control project of Dali-Ruili Railway. It is 34.5 km long in total. It passes through the steep Gaoligong Mountain in extremely complicated geological structures and conditions featuring high-temperature thermal damage, large deformation of weak rock, water burst, fault

Fig. 1.20 φ6.5 m double-shield TBM independently developed by CREG

Fig. 1.21 China's maximum-diameter TBM (φ9.03 m) jointly developed by China Railway Tunnel Group and CREG

fracture zone and high-intensity seismic zone. The main tunnel in the exit section of the project was constructed with the φ9.03 m gripper TBM jointly developed by CRTG and CREG.

See Table 1.2 for major TBM engineering applications in China.

Table 1.2 List of major TBM engineering applications in China

Start time (year)	Project name	TBM type	Qty. set	Diameter (m)	Tunneling length (km)	Engineering geology and construction
1985	Tianshengqiao II Hydropower Station Project	Gripper TBM	2	10.80	7.5	Limestone and sandstone (30–50 MPa); high equipment failure rate; average monthly footage: 65 m only
1991	The Water Diversion Project from Datong River to Qinwangchuan District	Double-shield TBM	1	5.53	17	Mudstone, sandstone and glutenite (2.8–133 MPa); average monthly advance: 980 m; maximum monthly advance: 1400 m

(continued)

Table 1.2 (continued)

Start time (year)	Project name	TBM type	Qty. set	Diameter (m)	Tunneling length (km)	Engineering geology and construction
1993	Shanxi Wanjiazhai Yellow River Diversion Project	Double-shield TBM	6	4.82–6.125	122	Weak and hard rocks interbedded, mainly limestone and marl; maximum daily advance: 113 m; monthly advance: 1637 m
1995	Qinling Line I Tunnel	Gripper TBM	2	8.80	10.8	Mainly hard rocks such as migmatitic granite and migmatitic gneiss (78–325 MPa); maximum monthly advance: 528.48 m
2000	Taohuapu Tunnel #1	Gripper TBM	1	8.80	6.2	Quartzite and marble (95–200 MPa); maximum monthly advance: 551.82 m
2000	Mogouling Tunnel	Gripper TBM	1	8.80	4.7	Quartz schist and marble mixed with mica quartzite (48–90 MPa), maximum monthly advance: 573.9 m
2000	Taiwan New Wujie Water Diversion Tunnel	Gripper TBM	1	6.2	7.1	Slate and sandstone; maximum daily advance: 44.3 m; monthly advance: 659.3 m
2003	Tashan Mine Project	Double-shield TBM	1	4.82	3.5	Limestone and granite across coal seams

(continued)

Table 1.2 (continued)

Start time (year)	Project name	TBM type	Qty. set	Diameter (m)	Tunneling length (km)	Engineering geology and construction
2003	Yunnan Jiu River Water Diversion Tunnel	Double-shield TBM	1	3.65	13.77	Sandy slate, argillaceous slate, dolomite, etc. (0.3–6.5 MPa); the TBM getting stuck frequently; difficult tunneling
2005	Dahuofang Water Diversion Project	Gripper TBM	3	8.03	59	Tuff, tuffaceous sandstone, migmatitic granite (25–109 MPa); maximum daily advance: 63.5 m, maximum monthly advance: 1208 m; average monthly advance: 750 m
2006	Xinjiang Daban Diversion Tunnel	Double-shield TBM	1	6.79	31.9	Composite stratum of sandstone, mudstone and andesite; maximum monthly advance: 1003 m; 48 shutdowns (in 379 days)
2007	Qinghai Datong River-Huangshui Diversion Project	Double-shield TBM	1	5.93	24.17	Argillaceous sandstone, argillaceous siltstone, granodiorite, quartzite, etc.; not suitable for TBM construction, resulting in shutdown for more than 5 years

(continued)

Table 1.2 (continued)

Start time (year)	Project name	TBM type	Qty. set	Diameter (m)	Tunneling length (km)	Engineering geology and construction
2007	Zhongtianshan Railway Tunnel	Gripper TBM	2	8.80	22.5	Metamorphic sandstone, metamorphic keratophyre, granite, etc.; unfavorable geological conditions featuring mud burst, rock burst and water ingress
2008	Yao River Water Diversion Project in Gansu	Single-shield TBM + Double-shield TBM	2	5.73	35.6	Mudstone and sandy mudstone (below 30 MPa); maximum monthly advance: 1,868 m after remanufacturing of the TBM
2008	Hongyan River-Shitou River Water Diversion Project in Shaanxi	Double-shield TBM	1	3.65	11	Weak and broken surrounding rock; abundant water and large deformation; maximum daily advance: 30 m; maximum monthly advance: 420 m; TBM shut down for 3 years and stuck for 66 times
2008	Jinping II Hydropower Station Water Diversion Tunnel	Gripper TBM	2	12.4	16.7	Marble, limestone, sandstone, slate, etc. (50–100 MPa); extremely strong rock burst and high-pressure water ingress; maximum monthly footage: 683 m

(continued)

Table 1.2 (continued)

Start time (year)	Project name	TBM type	Qty. set	Diameter (m)	Tunneling length (km)	Engineering geology and construction
2008	West Qinling Tunnel of Lanzhou-Chongqing Railway	Gripper TBM	2	10.2	19.8	Limestone, phyllite, metamorphic sandstone, sandy phyllite, etc.; maximum monthly advance: 842.5 m
2009	Yunnan Nabang Hydropower Station Project	Gripper TBM	1	4.5	10	Extremely hard rock stratum (220 MPa), mixed with mud geology; maximum monthly advance: 581 m
2009	Chongqing Metro Line 6 Project	Gripper TBM	2	6.36	12.1 × 2	Mudstone and sandstone (30 MPa), maximum daily advance: 46.8 m; monthly advance: 862 m
2010	Northwest Liaoning Water Supply Project	Gripper TBM	8	8.53	109.5	Migmatitic granite, monzonitic granite, quartz monzonite (40–80 MPa); maximum monthly advance: 1,078 m
2012	Han River-to-Wei River Water Diversion Project	Gripper TBM	2	8.02	34.0	Granite, quartzite and gneiss (100–240 MPa); severe cutter wear; rock burst and water burst
2013	Xinjie Taigemiao Coal Mine Inclined Shaft Project	Dual-mode TBM	2	7.62	6.4 × 2	Sandy mudstone and siltstone (40–60 MPa); TBM with soil pressure balanced shield and single shield

(continued)

Table 1.2 (continued)

Start time (year)	Project name	TBM type	Qty. set	Diameter (m)	Tunneling length (km)	Engineering geology and construction
2013	Luoyang Guxian Town Water Diversion Project	Gripper TBM	1	5.0	6.64	Water burst, collapse and fracture zones; maximum daily advance: 66 m; monthly advance: 1,010 m
2014	Qingdao Metro Line 2 (Phase I)	Double-shield TBM	4	6.3	25.2	Slightly weathered and moderately weathered granite; maximum daily advance: 22.5 m; maximum monthly advance: 350 m
2014	Songhua River Water Diversion Project	Gripper TBM	3	7.93,8.03	70	Limestone and granite; maximum monthly advance: 1,226 m
2015	Chongqing Rail Circular Line	Dual-mode TBM	1	6.85	1.3	–
2015	Shendong Bulianta Coal Mine Auxiliary Shaft Project	Single-shield TBM	1	7.6	2.7	Sandstone (4–46 MPa); maximum monthly advance: 639 m
2015	Lanzhou Water Source Project	Double-shield TBM	2	5.49	24.4	Quartz diorite, quartz schist, granite and metamorphic andesite (15–75 MPa)
2016	Great Beirut Water Supply Project in Lebanon	Gripper TBM	2	3.53	–	World's smallest-diameter TBM
2016	A major water tunnel project in Xinjiang	Gripper TBM	1	6.53	–	Excavation diameter adjustable between φ6.53 m–φ6.83 m

(continued)

Table 1.2 (continued)

Start time (year)	Project name	TBM type	Qty. set	Diameter (m)	Tunneling length (km)	Engineering geology and construction
2016	Shenzhen Metro Line 10 Project	Double-shield TBM	1	6.5	2.7	Mainly granite (33–127 MPa); first double-shield TBM used in Shenzhen
2017	Gaoligong Mountain Tunnel Project	Gripper TBM TBM	2	9.03/5.6	12.4	Granite, schist, slate, phyllite mixed with quartzite and metamorphic sandstone (4.6–65 MPa)

1.3 Characteristics of TBM Construction

1.3.1 Advantages

Compared with drill and blast method, TBM method has the following advantages:

(1) Fast. The drilling and blasting method is greatly affected by the geological conditions and its construction period is relatively long. As a complete set of tunnel construction equipment integrating mechanical, electrical, hydraulic, sensing and information technologies, the TBM is able to break rocks, discharge muck and offer support simultaneously in factory-specific construction to tunnel continuously at fast speed and high efficiency.
(2) High-quality. The drill and blast method is destructive to the surrounding rock, resulting in a large amount of non-human over-excavation. The TBM breaks rocks with a disc cutter to avoid blast so that the rock stratum around the tunnel will not be damaged by blast vibration and that the tunnel wall is complete and smooth, with a small amount of over-excavation.
(3) Efficient. The TBM can tunnel fast, shortening the construction period and greatly improving the economic and social benefits. In addition, thanks to small amount of over-excavation, lining costs can be significantly lowered. TBM construction requires less operator and less intense labor and consumes less materials.
(4) Safe. As explosive is used, the drill and blast method has potential safety hazards. In TBM construction, the working conditions of operators in the tunnel are much better, the physical labor is less intense, and the possible casualties caused by blast are avoided, greatly reducing accidents.

(5) Environment-friendly. As explosive is used in drill and blast method, a large amount of harmful gas will be produced, polluting the environment. In construction of a long tunnel, adits and access roads are needed, not favorable to environment. However, explosive blast is not required in TBM construction, reducing the pollution to construction site. In addition, TBM construction required less auxiliary pilot tunnels in long and large tunnels. It is conducive to ecological environment protection. The drill and blast method has great impact on surrounding buildings as affected by vibration, while TBM construction basically has no such impact.

(6) High automation and informatization degree. The drilling and blasting method requires a large amount of human resources and equipment, but it is of low degree of automation and informatization. However, the TBM is a complete set of tunnel construction equipment integrating computer control, sensor, laser guidance, measurement, advanced geological exploration and communication technologies, as well as mechanical, optical, electrical, pneumatic, hydraulic, sensing and information technologies. It is of high degree of automation. The TBM has the functions of acquisition, management and real-time remote transmission of construction data, and TBM attitude management to realize information construction.

1.3.2 Disadvantage

The TBM is not adaptable. Different TBMs with different auxiliary need to be designed for different geological conditions, tunnel faces and construction requirements.

(1) The TBM is not adaptive to geological conditions. It is more suitable for construction in strata with less change and complete good rock mass at medium compressive strength. The TBM is most sensitive to the strata of the tunnel. A different TBM is needed in a different stratum. In weak rocks, hard rocks and fault fracture zone, different types of TBMs configured with necessary reinforcing and supporting equipment are needed. However, TBM construction is not suitable for stratum with development of large karst underground river, large fault fracture zone, quartz sandstone with high quartz content, expansive surrounding rock, area with high stress, tunnel sections with serious rock burst, potential large-scale water burst, serious plastic deformation or soft rock susceptible to deformation, and other unfavorable geological conditions. In these circumstances, drill and blast method can give full play to its flexibility.

In general, gripper TBMs are suitable for construction of tunnels mainly surrounded by Class II and Class III rocks. Single-shield or double-shield TBMs are suitable for construction of tunnels mainly surrounded by Class III and Class IV rocks. Shields are suitable for construction of urban tunnels of high underwater level mainly surrounded by Class V rocks and not buried deep, or cross-river tunnels.

(2) The TBM is not suitable for construction of short and medium-length tunnels. As affected by the huge volume, it is difficult to transport. Because of complicated construction preparation and back-up system for auxiliary construction and long manufacturing period, it is difficult for the TBM to give full play to its advantages in short or medium-length tunnels. As shown in practices in other countries than China, TBM construction is more economical when the ratio of tunnel length to diameter is greater than 600. An ordinary single-track railway tunnel usually has an excavation diameter of 9–10 m. As calculated accordingly, TBM construction can be considered for tunnels longer than 6 km. In developed countries, the TBM method is generally preferred for construction of tunnels, and the drill and blast method will be considered only when the former is not suitable. But it is not the case in China. With labor redundance, China has always been good at drill and blast method, and has built more than 5000 km of railway tunnels with this method. The progress of this method is still accelerating year by year. In China, it is generally believed that it is difficult to give full play to the advantages of TBM for tunnels shorter than 10 km where the drill and blast method is more economical. For 10–20 km long tunnels, the two methods can be compared in cost and technology to select an appropriate construction method. For tunnels longer than 20 km, the TBM method is preferred. The shield method is preferred for tunnels crossing rivers or areas with dense urban buildings or with high groundwater level in consideration of construction safety, settlement control and other factors, regardless of the length of the tunnels.

(3) The TBM is not adaptive to section face. When the tunnel face is too small in diameter, it is difficult to arrange the back-up system and to advance. When the tunnel face if too large in diameter, construction will face insufficient power, difficult transportation, high expense and other problems. Generally, TBMs are suitable for construction of φ3–12 m tunnels, but not suitable for tunnels with a diameter greater than 15 m. For φ12–15 m tunnels, the construction method should be selected in full consideration of tunneling length, external conditions and other factors based on the surrounding rock conditions. In addition, variable-face tunnels cannot be constructed with a TBM. Generally, the TBM is not suitable for the construction of tunnels of which the face diameter changes in a large range.

(4) It is difficult to transport, and there are special requirements for the construction site. TBM is a large set of special equipment with a total weight up to thousands of tons. Its largest component may weigh hundreds of tons, and an assembled TBM may be more than 200 m long. In addition, there are many supporting facilities arranged outside the tunnel, including concrete mixing system, segment prefabrication plant, repair workshop, accessory warehouse, material warehouse, water supply, power supply and air supply systems, muck transportation and dumping system, loading and unloading and transportation systems, access road to the site area, and TBM assembly site. Therefore, TBM puts forward high requirements for the construction site and transportation scheme of the tunnel. Some tunnels are suitable for TBM construction in terms

of length and geological conditions, but the transportation road may not meet the requirements, or the site may not meet the construction conditions of TBM.

(5) It is costly to purchase and use the equipment. TBM construction requires high-load power, high-quality technicians and management teams, and high cost to purchase equipment in the early stage, directly affecting the applicability of TBM construction.

(6) It takes a long time to complete non-tunneling operations. When it is planned to construct a deeply-buried long and large tunnel with a TBM, it takes a long time for type selection demonstration, design liaison, design and manufacturing, assembly and commissioning in factory, transportation by sea or land, assembly and commissioning on site of TBM from ordering to trial tunneling. Even after the TBM begins normal tunneling, it may be shut down predictably or unpredictably for troubleshooting when being stuck or replacement of main bearing because of geological risks, TBM equipment risks, or professional technical management and control risks. All these operations are called non-tunneling operations, which consume long construction period but contribute no advance in TBM construction. Non-tunneling operations account for quite a large proportion in tunneling. For a tunnel constructed with a TBM, full consideration shall be given to non-tunneling operations in design and construction, to make scientific and reasonable construction organization design.

1.3.3 Characteristics of TBM Construction Organization Design

In TBM construction organization design, the long procurement cycle of TBM equipment shall be taken into account (11–12 months for manufacturing) so that the TBM and its back-up equipment can be procured in a planned and reasonably organized way according to the requirement of construction period.

Whether the TBM is adaptable to the geological conditions shall be taken into full account. To have a suitable TBM, the tunnel constructor shall orderly organize the selection of the TBM and its back-up equipment according to the surrounding rocks of the tunnel, and the supervision of TBM equipment manufacturing.

The tunnel constructor shall make preparations for the construction site according to the special situation of TBM application, including laying of power lines, construction and reinforcement of access roads and bridges, construction of temporary works, organization of transportation, installation and commissioning of major TBM components, tunneling preparation, technical training, and preparation of material, machine and tool requirement plans while purchasing the TBM.

It shall fully consider the characteristics of TBM construction, the design scheme and engineering measures for the TBM to pass through special geological section, and special requirements of the site, water and power for TBM.

1.3.4 *Advantages and Disadvantages of Different Types of TBMs*

TBM is a large tunnel construction machine integrating tunneling, mucking, guiding, supporting, ventilation and dust prevention. Generally, TBMs can be classified into gripper TBM, single-shield TBM and double-shield TBM. See Table 1.3 for the advantages and disadvantages of each type.

Table 1.3 Advantages and disadvantages of different TBMs

TBM type	Gripper TBM	Double-shield TBM	Single-shield TBM
Application	Hard rock and relatively intact weak rock	Hard rock and relatively intact weak rock	Medium hard rock, weak rock, broken rock stratum and soil
Curve radius	400 m under general conditions 300 m under tough conditions	500 m under general conditions 350 m under tough conditions	700 m under general conditions 500 m under tough conditions
Use risk	It needs auxiliary measures when passing through any unfavorable geological zone, affecting the progress, as there may be a certain risk to the surface settlement control in the primary support after excavation	There are not many auxiliary measures available for the TBM to tunnel in complex strata, and the TBM shield is too long and easy to be stuck. In addition, this type of TBM is not adaptable to plane curve radius as its segment will bear full water pressure when blocked by water from grouting behind the segment	It is the least adaptable to plane curve radius. It is long and easy to be stuck. There are not many auxiliary measures available for the TBM to tunnel in complex strata and the segment will bear high water pressure because of grouting behind it
Advance speed	It can tunnel fast with a general advance of 400–800 m/month. But it is sensitive to the change of surrounding rock. It tunnels slowly in fracture surrounding rock zone where advance supporting measures and reinforcement lining are needed and its speed is affected by the secondary lining following up	It can tunnel stably in favorable surrounding rocks generally at an advance of 350–450 m/month. It is less sensitive to the change of stratum than the gripper TBM. It can tunnel in single-shield mode or double-shield mode according to the change of the stratum	Its advance speed is 300–450 m/month generally as it needs reaction force from the jack to support the segment during tunneling so that tunneling and installation of segments cannot be carried out simultaneously

(continued)

Table 1.3 (continued)

TBM type	Gripper TBM	Double-shield TBM	Single-shield TBM
Construction safety	Relatively safe with primary supporting + advance supporting as necessary	Safe with long shield + segment lining protection	Safe with relatively long shield + segment lining protection
Tunnel face closing	Close the tunnel face with flat cutter head	Close the tunnel face with flat cutter head	Close the tunnel face with flat cutter head
Mucking mode	Integrated belt conveyor allowing for fast mucking and long distance tunneling	Integrated belt conveyor allowing for fast mucking and long distance tunneling	Integrated belt conveyor allowing for fast mucking and long distance tunneling
Simultaneous lining construction	Difficult, affecting advance speed	Proven technology with segment closely following	Proven technology with segment closely following
Lining quality	Cast-in-place composite lining of high quality	Segment lining of relative poor quality as segments are bolted, leaving many construction joints, and they may be staggered or crack later, requiring much effort for watertightness	Segment lining of relative poor quality as segments are bolted, leaving many construction joints, and they may be staggered or crack later, requiring much effort for watertightness
Lining effect and cost	Composite lining is at low cost at the primary supporting and secondary supporting can be adjusted any time according to the tunneling	Segment lining is at high cost in general as a large number of segment molds are required to prefabricate segments, and it provides less integral and waterproof supporting than molded concrete lining	Segment lining is at high cost in general as a large number of segment molds are required to prefabricate segments, and it provides less integral and waterproof supporting than molded concrete lining
Waterproof work and drainage	Reliable: drainage combined with plugging	Less reliable: It mainly relies on water plugging as there are many assembly joints. In development of fracture water, the water may be blocked by the grouting behind the segment and put the segment under full water pressure, or leak from the assembly joint	Less reliable: It mainly relies on water plugging as there are many assembly joints. In development of fracture water, the water may be blocked by the grouting behind the segment and put the segment under full water pressure, or leak from the assembly joint
Monitoring and measurement	Tunnel deformation is measurable	Tunnel deformation is not measurable	Tunnel deformation is not measurable
Durability	Excellent	Good	Good
Advanced support	Flexible	Not flexible	Not flexible

1.4 Typical Factors Restricting TBM Construction Performance

TBM tunneling is almost a completely industrialized process. Compared with the traditional drill and blast method and the step-by-step and partial section excavation method of weak surrounding rock, it has many advantages. However, TBM tunneling is not a simple and risk-free technique. To ensure successful implementation of TBM construction technology, it is not enough to introduce a TBM from a qualified TBM manufacturer. On the contrary, it is of great significance to keep the employer, the supervisor, the designer, the TBM manufacturer and the constructor collaborative at all stages of the project from the early geological engineering investigation and feasibility study to the final design and construction.

TBM method is not flexible, and requires large investment, but potentially allows for fast construction. If any adverse ground condition is encountered without early warning, TBM method is much more beneficial than drill and blast in terms of construction period and other aspects. An adverse ground condition can be a poor quality rock mass that makes the tunnel not stable, or good quality rock mass that cause low penetration (e.g., hard block rock mass). However, there is no absolute value of the influence of rock mass quality on TBM performance. In fact, the influence of rock mass quality on TBM performance depends on the TBM used and the tunnel diameter.

The type and characteristics of the TBM vary considerably because the rock, soil and environmental conditions that may be encountered during tunnel excavation are infinitely variable. The TBM is suitable to excavate large and long tunnels with medium to thick overburden in stable rock strata of medium to high strength. The basic problem is how to break rocks. There is no such TBM that can tunnel under any geological conditions. Therefore, whether a tunnel project can be successful depends on two factors: ① the type of TBM used; ② the design of the selected TBM and its special construction performance.

It is far from enough to order a particular type of TBM from a qualified TBM manufacturer. What is important is continuous cooperation between all parties involved in the TBM design, manufacture and construction. However, there is no "recognized standard" for the design and construction of any type of TBM so far, and TBM design and manufacture are continuous technological innovation processes. Each tunnel project has its own characteristics, each professional construction unit has its own core technology, and each TBM can be considered as a different prototype from others.

In general, TBM performs well in favorable strata. But in many cases, the actual progress of TBM construction is slower than the expected, and of course, the claimed by the TBM manufacturer, because problems in engineering geology, hydrogeology and rock mechanics are often underestimated or ignored in addition to unforeseen events (such as failure of TBM components).

On the one hand, there are many inevitable uncertainties in the design stage in terms of geology, geotechnique and hydrogeology, and different types of TBMs and different construction techniques may be used. Therefore, it is necessary to decide whether to optimize the construction method and TBM selection based on the understanding of the geological and geotechnical conditions of the site and the accuracy of forecasts of these conditions. On the other hand, it is quite complicated to optimize the tunnel construction in all aspects as affected by a large quantity of r geological, technical, environmental, economic and financial uncertainties.

The fundamental problem of tunnel construction is usually caused by the physical and geotechnical inhomogeneity of the stratum excavated by the TBM. Because such full-face mechanized excavation is inflexible, the stratum with inhomogeneous strength plays a more important role. Prior knowledge of the geological and geotechnical conditions of the construction site in an appropriate manner is extremely important for the construction of underground works. So far, generally not much fund is spent in early-stage survey. In fact, the fund used for early-stage survey will be compensated to a large extent by lowering of construction expenses and shortening of the construction period. The advance exploration in the pilot tunnel or main tunnel conducted by TBM cannot replace sufficient early-stage survey.

The relatively important or common adverse ground conditions that restrict TBM performance include high drillability limit, instable excavation face, faults and squeezing/expansive strata. Other adverse ground conditions may also be encountered during TBM tunneling, including cohesive soil, weak stratum where the TBM may sink, massive influx of groundwater and gas, rock burst, high-temperature rock stratum, high-temperature water, karst cave, etc.

1.4.1 Drillability Limit

If the TBM cannot penetrate the rock face in sufficient depth or the wear of the excavating cutter exceeds the acceptable limit, the rock stratum is considered as non-drillable. The drillability of rock stratum should not be determined in an absolute manner, but determined in a relative manner by comparing the TBM method with the drill and blast method in terms of project cost and construction period. The main index defining TBM's ability to excavate rock mass is its penetration under the maximum thrust.

Penetration, also called cutting depth, is the depth at which the cutter is cut into the rock per revolution of the cutter head. The penetration is related to the rock properties, such as rock type, unconfined compressive Strength, fissure development degree, wear resistance, porosity, etc. It is not possible to determine the limit of penetration below which the rock stratum is considered non-drillable. The penetration limit is also affected by the abrasion resistance and thickness of the excavated rock stratum and the tunnel diameter. High abrasive resistance of rock and low penetration will result in frequent tool changes, increasing the cost per 1m^3 of rock excavated in addition to the time taken to replace the tool. If the penetration is less than 2 mm/r, it is considered that there is a problem in the drillability of the rock. If the penetration is greater than 4 mm/r, the excavation efficiency of the TBM will be higher.

It is common in construction to push the cutter head forward to the maximum extent to gain sufficient penetration. If any part of the TBM is not designed and manufactured to work under this maximum thrust, the TBM will vibrate abnormally, the cutter head and gripper shoe structure will gradually crack, and the main bearing will be easily damaged. Since it is not easy to repair and replace the cutter head or main bearing in a tunnel, such incidents will lead to serious loss in TBM construction. Similarly, when an advancing cutter head is under too high thrust force, the main bearing and the gearbox transmitting power to the TBM cutter head may be damaged, resulting in more serious consequences.

1.4.2 Unstable Excavation Face

If the rock mass to be excavated is broken or severely weathered, the excavation face can be hardly stable, and large rock blocks and crushed rocks will collapse from the excavation face. Such situation will continue till new balance is reached. As a consequence, a large amount of over-excavation will be caused, and normal operation of the TBM will be affected. Even to a shield TBM, its advance may be obstructed for two basic factors. ① The cutter head cannot rotate under the action of or as stuck by collapsing and accumulating rock blocks. ② Serious over-excavation is caused because the excavation face is not stable. A cavity will be formed in front of the TBM so that TBM tunneling should be stopped to handle the cavity before it expands and finally develops beyond control.

During construction, such cavity is often grouted and backfilled with resin and foam to form an artificial solid. Drilling and grouting are usually carried out through special holes set on the cutter head. Or a bypass tunnel (preferably above the tunnel) is excavated to release the cutter head stuck by rock blocks. A section of tunnel can also be excavated with traditional methods to stabilize and reinforce the excavation face, or grouting or pipe shed advanced support can be used to reinforce the surrounding rock.

According to construction experience, the TBM shall be designed with the length of cutter head extruding out of the shied minimized so that the shield supports the tunnel as close as possible to the excavation face, to prevent the above accidents. In addition, under such conditions adverse to TBM advancing, it is helpful for normal advancing of TBM that a large cutter head torque can be generated and the head cutter speed can be adjusted during start and excavation of the TBM.

In addition, the adjustable scrapers should be eliminated from the cutter head. After these improvements to the TBM, the factors affecting the normal TBM tunneling can be reduced, although not be completely eliminated.

1.4.3 Unstable Excavation Wall

Unstable excavation wall is one of the factors affecting the normal tunneling of gripper TBM for hard rock. If the wall immediately behind the cutter head support is not stable, it will become difficult to install supports and position grippers. The influence of unstable excavation wall on the construction progress and the methods used to overcome this instability varies greatly, depending on the following factors: ① the size and type of unstable excavation wall; ② the type of TBM used (single or double grippers); ③ the design and construction characteristics of the TBM; ④ the diameter of the tunnel; ⑤ the device for installing tunnel supports used by the TBM and the type of supports.

Shield TBMs (both single-shield and double-shield) are not so sensitive to unstable excavation wall as gripper TBMs as prefabricated concrete segments can be installed under the protection of shield and the thrust force is applied through the segments. Therefore, they can advance regardless of whether the excavation tunnel wall is stable.

When the excavation wall is unstable, the gripper TBM may make 1–2 m advance per day only or even no advance. The following measures can be taken when the excavation wall is not stable during the tunneling of a gripper TBM: ① Stabilize and reinforce the excavation wall, and install steel arches, wooden supporting plates and spray concrete immediately after the cutter head supporting position; ② Excavate the tunnel with a conventional method in front of TBM, usually by top pilot tunnel; ③ Take measures such as drilling, grouting or installing umbrella-shaped arch frame above the TBM to pretreat the stratum in front of the excavation face.

1.4.4 Fault Zone

If its cutter head is stuck in a large fault zone, the normal tunneling of the TBM will be affected, reducing the TBM advance speed even that the construction period will not be delayed much. Although fault zones are locally distributed along the tunnel, accidents may happen because of insufficient forecasting during excavation, or insufficient estimation or knowledge of difficulties in advance.

In the fault zone, if the strata are completely weathered and high pressure groundwater exists, the excavation tunnel face may move like fluid, flooding the tunnel like river water.

If a gripper TBM encounters such fault zone not found in advanced drilling, it will be seriously hindered by stratum collapse and even buried by collapsed stones, and may become difficult to retreat in a disastrous situation. In addition, if the fault zone is treated with simple and rapid grouting, the TBM cutter head may also be stuck in the cement slurry.

If a shield TBM (single-shield or double-shield) encounters this fault zone, the result will not be as severe, although it is impossible for the TBM to continue excavating. The tunnel will be lined with segments during the tunneling of each shield TBM to form a natural extension of the shield so that the fault zone can be at least treated in the shield and tunnel being completely flooded by water can be avoid.

1.4.5 Squeezing Ground

When serious tunnel convergence occurs at a short distance (several meters) from the excavation face for any reason within 4–8 h at any time, the TBM will certainly get trapped. For designers and construction personnel, the squeezing ground is dangerous and will affect the normal tunneling of TBM. At present, there are no examples of long-term trapping of TBM because of tunnel convergence.

The shield TBM is sensitive to the rapid convergence of the tunnel and may be stuck by converging strata. If severe tunnel convergence which is related to tunnel stability occurs during the tunneling of a gripper TBM in a short time at any time, then the tunnel supports installed and the supports of the TBM grippers may present serious problems, affecting the advancement speed.

To overcome the above problems, over-excavation can be appropriate for most TBMs, and the clearance between the shield and the excavation contour surface can be adjusted from 6–8 cm to 14–20 cm. However, the main problem of open TBM tunneling in unstable strata with serious convergence lies in supporting measures such as steel support, wire mesh and shotcrete, and that the newly constructed support cannot immediately and effectively resist the trend of strata deformation and squeezing.

For a shield TBM, the maximum thrust of its longitudinal jack can be increased until the TBM can move forward under the action of high ground pressure (2–5 MPa), but the segment lining of the tunnel should be strong enough to provide necessary reaction force for the TBM thrust jack. Otherwise, the tunnel lining itself may collapse.

With these methods and measures in combination with over-excavation, the shield TBM can tunnel under almost any conditions (including so-called special conditions). However, if the TBM is forced to stop in the squeezing ground section for a considerable period of time because of mechanical failure, etc., it is much likely to get trapped.

It is relatively easy for a double-shield TBM to break out, for instance, through the opening of the TBM expansion zone at 4–5 m from the excavation face. However, the single-shield TBM must break out from its tail shield with one or two segments removed 8–9 m from the excavation face.

Questions

1. What is TBM?
2. What is the main difference between TBM and shield?
3. What are the main types of TBM?
4. Briefly describe the origin of TBM.
5. What are the main periods of TBM development in China?
6. Briefly describe the advantages and disadvantages of TBM construction.
7. Briefly describe the construction organization characteristics of TBM.
8. Try to describe the typical factors restricting TBM construction performance.

Chapter 2
TBM Principle and Structure

This chapter focuses on the structural characteristics, basic configuration and working principle of the gripper TBM, the single-shield TBM and the double-shield TBM, introduces the structure of key TBM parts such as cutter head, main drive, supporting and thrust systems, shield and accessory equipment, and describes the function, composition, structural form, main equipment and layout of the TBM back-up system.

2.1 Overview

Traditional TBMs are mainly categorized as gripper TBM, single-shield TBM and double-shield TBM.

As to gripper TBMs, most of their parts are exposed to tunnel space, except that some models are equipped with grid protection within about 120° above the head. The cutter head at the foremost position that is driven by the main drive cuts into rocks under the reaction force from the side support mechanism. Followed by is construction of primary supporting, such as the installation of ring beam with the help of manipulator, set-up of rock bolts and spraying of concrete. This type of TBM allows for convenient tunnel excavation and flexible supporting when the surrounding rock is roughly complete.

As to the single-shield TBM, the machine head is protected in the shield, while the head cutter rotates and cuts into rocks as driven by the main drive and its thrust force for advancing comes from the lining completed. The single-shield TBM enables the tunneling and lining to be carried out under the protection of shields, greatly improving construction safety. However, its efficiency is low as tunneling and lining installation cannot be carried out simultaneously. In addition, it is impossible to provide high thrust force as limited by the lining strength, so this type of TBM can only be used to tunnel in the surrounding rock with low strength.

K. Chen et al., *TBM Design and Construction*,
https://doi.org/10.1007/978-981-99-0059-6_2

The double-shield TBM is transformed from the single-shield TBM by dividing the single shield into front shield and rear shield which are telescoped together. There is a gripper set in the rear shield to generate thrust force by fraction against the surrounding rock for the cutter head to move forward. Therefore, lining installation and cutter head tunneling can be carried out simultaneously without interference. The double-shield TBM is adaptable to different surrounding rocks, including both hard and weak rocks. It can even tunnel in weak rocks and fault fracture zones so long as appropriate measures are taken.

2.2 Gripper TBM

It is called gripper hard rock TBM or gripper TBM in Europe. In gripper TBM construction, the rock support method is used instead of segment lining. The gripper TBM can advance fast in surrounding rocks of medium–high strength. To ensure fast and safe construction in weak surrounding rocks and fault zones, supporting equipment is needed to support the surrounding rocks and avoid collapse of loose and broken surrounding rocks. See Fig. 2.1 for the structure and composition of the gripper TBM. It excavates rocks with a head cutter, transports muck with a belt conveyor, supports the surrounding rocks with a bolt driller and a ring beam erector and, advances under the support of a gripper and a thrust cylinder.

By the structures of supporting tunnel walls, the gripper TBMs can be classified into Kelly type and main beam type. Wirth of Germany mainly produces Kelly type TBMs as shown in Fig. 2.2, while Herrenknecht of Germany, the Robbins Company of America and TBM manufacturers of China mainly produce main beam TBMs as shown in Fig. 2.3. The main beam TBMs can be further classified into single-pair and double-pair horizontally supported TBMs as shown in Fig. 2.4.

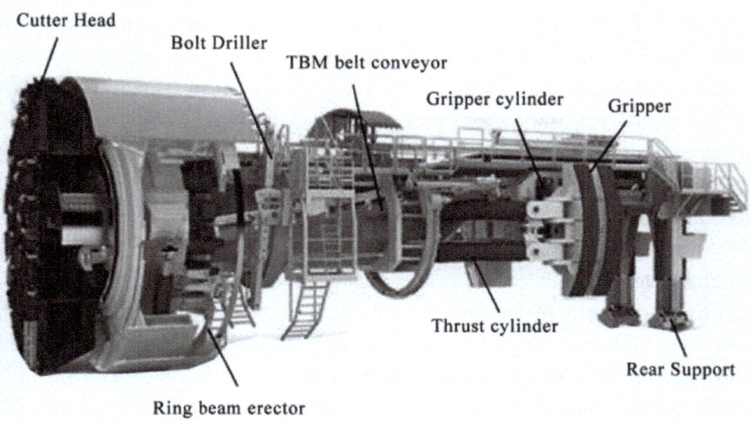

Fig. 2.1 Gripper TBM

Fig. 2.2 Kelly-type gripper TBM

Fig. 2.3 Main Beam gripper TBM

2.2.1 Kelly-Type Gripper TBM

1. Structural Features

The main feature of the Kelly-type gripper TBM is that it uses internal and external Kelly supports. In this section, the structural principle of the gripper TBM will be introduced with the example of φ8.8 m TB880E TBM manufactured by Wirth of Germany used in the tunnel of Qinling Line I of Xi'an-Ankang Railway. See Fig. 2.5 for the main structure of TB880E Kelly type gripper TBM.

The main machine of the Kelly type gripper TBM is mainly composed of cutter head, cutter head shield, cutter head main bearing, cutter head drive, auxiliary hydraulic drive, main bearing seal and lubrication system, internal and external Kelly supports, gripper, thrust cylinder, rear support, hydraulic system, electrical system, operation room, transformer, traveling device, etc. The X-shaped gripper is installed

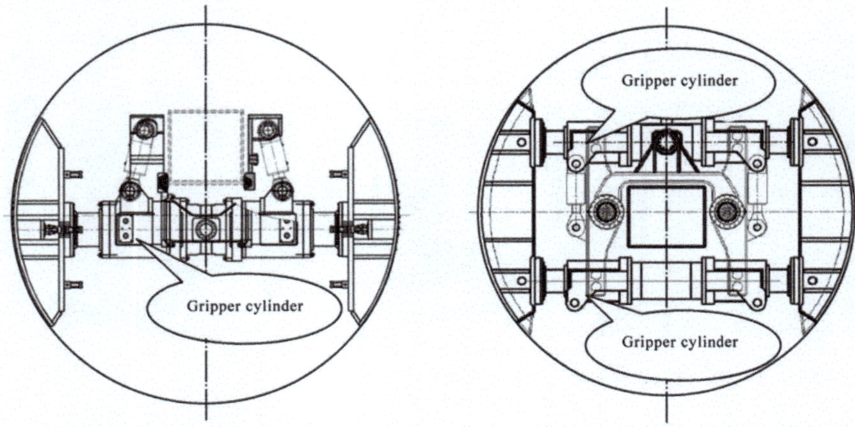

(a) Single-pair Horizontally Supported TBM

(b) Double-pair Horizontally Supported TBM

Fig. 2.4 Two types of main beam gripper TBMs

on the external Kelly support, while the main bearing and the head cutter drive are installed on the front side of and the rear support is installed on the back of the internal Kelly support. The cutter head and the cutter head drive are surrounded and supported by a floating invert shield, a retractable top shield and dust shields on both sides. The cutter head drive is installed between the front and rear grippers to provide as much space as possible behind the cutter head shield to accommodate the bolt driller and the ring beam erector. The cutter head is hollow and is equipped with disc cutter, scraper and bucket to send the muck to the belt conveyor installed in the inner Kelly support.

The back-up system, which consists of several platform trailers and an equipment bridge, is equipped with the feeding equipment and the loading system of the main machine. The back-up system is equipped with hydraulic power system, switch board, transformer, master power-off switch, cable drum, dust collector, ventilation system, operation room, belt conveying system, concrete spraying system, grouting system, water supply system, etc. Each trailer is also equipped with auxiliary equipment such as ring beam erector, invert block lifting machine, probe drilling machine, rock bolt driller, air duct box, auxiliary fan, dust collector, ventilation and cooling systems, communication system, data processing system, guidance system, gas monitor, grouting system, concrete spraying system, high-voltage cable drum, emergency generator, air compressor, water system, CCTV, etc.

2. Basic Configuration

(1) Cutter Head

See Fig. 2.6 for the cutter head structure. The cutter head is a welded steel structure in two bolted semicircle pieces to facilitate transportation and lifting in the tunnel.

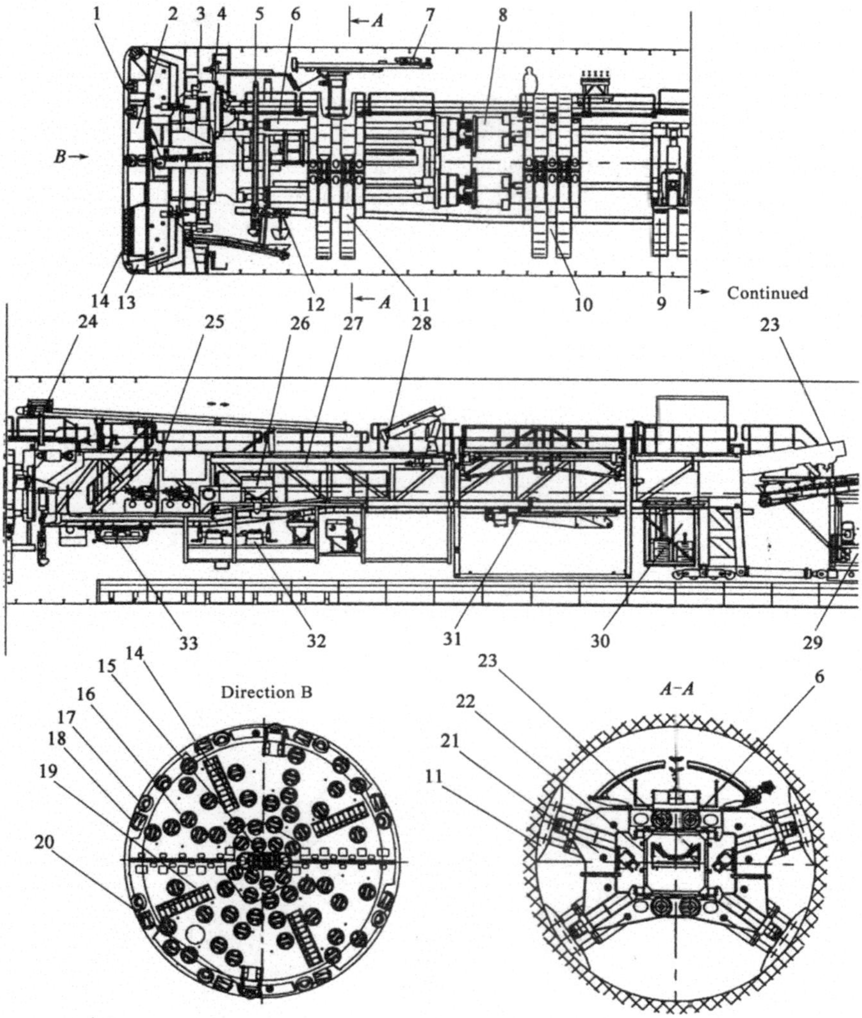

Fig. 2.5 Overall structure of TB880E Kelly type TBMs. *1*—disc cutter; *2*, *19*—cutter head; *3*—cutter head shield; *4*—ring beam erector; *5*—bolt driller; *6*—thrust cylinder; *7*—probe drilling machine; *8*—cutter head drive; *9*—rear support; *10*—X-shaped rear support; *11*—X-shaped front support; *12*—cutter crane; *13*—bucket; *14*—scraper; *15*—center cutter; *16*—face disc cutter; *17*—edge cutter; *18*—bucket; *20*—reaming cutter; *21*—front external Kelly support; *22*—internal Kelly support; *23*—belt conveyor; *24*—transport trolley; *25*—water pump; *26*—dust collector; *27*—belt bridge; *28*—crane 1; *29*—flatbed truck; *30*—operation room; *31*—crane 2; *32*—grouting machine; *33*—invert crane

The cutter head features back-loading disc cutters and concave cutter seats to allow for easy installation of cutters, and minimize its distance from the tunnel face to effectively avoid being stuck in fault fracture zones.

The scrapers and buckets installed along the circumference of the cutter head are used to convey the cut muck from the bottom to the top, and enable it to fall into the muck bucket of the conveyor along the muck chute. The open buckets and the scrapers extend a certain distance toward the center of the cutter head, so that a large amount of muck enters the cutter head before falling to the bottom, reducing the secondary extrusion of muck and the extra wear of the buckets and the scrapers. The scrapers used are replaceable wear-resistant blade connected with bolts.

The cutter head is supported on the main bearing and connected with the rotating part of the bearing with hydraulic expansion bolts. As supported by the rigidly positioned inner Kelly support and hydraulically preloaded invert shield, the cutter head will not fall or swing when encountering rock stratum changes so that its axis position will not change, the disc cutters are in their respective slits, and that vibration and wear of head cutter can be reduced during operation. The cutter head is equipped with a water spray system for preliminary control of dust on the tunnel face and as well as for the cooling of disc cutters. There is a manhole on the Kelly support for accessing to the internal of the head cutter and a manhole on the head cutter for accessing to the tunnel face.

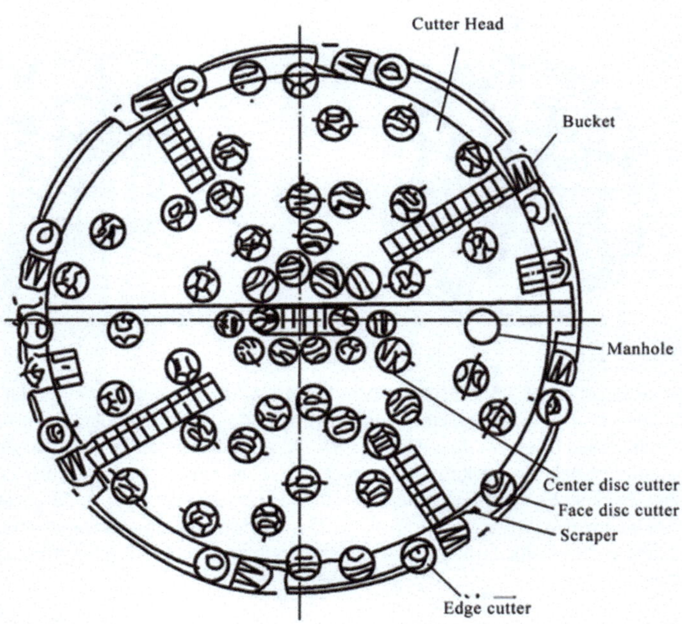

Fig. 2.6 Cutter head

(2) Cutter Head Shield

See Fig. 2.7 for the cutter head shield. The cutter head shield consists of a hydraulically preloaded invert shield and 3 retractable arch shields. The cutter head shield covers the cutter head from the scrapers to the partition, providing safety protection during the installation of the ring beam to prevent large rocks from blocking the cutter head. It supports the front part of the TBM during tunneling or regripping at the end of tunneling. All 3 telescopic arched shields can be bolted grid shields, and rock bolts can be installed when the shields are supported at the top. The shield drives the bulkhead through the cylinder connection and floats with the cutter head. The preload cylinder on the shield bears the mass of the cutter head and the drive device, keeps the shield in contact with the tunnel invert, and pushes muck forward for removal.

(3) Main bearing and cutter head drive

The main bearing is a double-axial and radial three-dimensional roller bearing with axial preload and rotating inner ring. See Fig. 2.8 for the composition of main bearing.

See Fig. 2.9 for the cutter head drive. The inner ring gear on the bearing inner ring, an integral part of the bearing, is connected to the cutter head with hydraulic expansion bolts. The cutter head is driven by 8 sets of the same cutter head drive devices through the inner ring gear. The drive pinion is supported by two bearings, and its transmission shaft 2 is connected to the water-cooled two-stage planetary

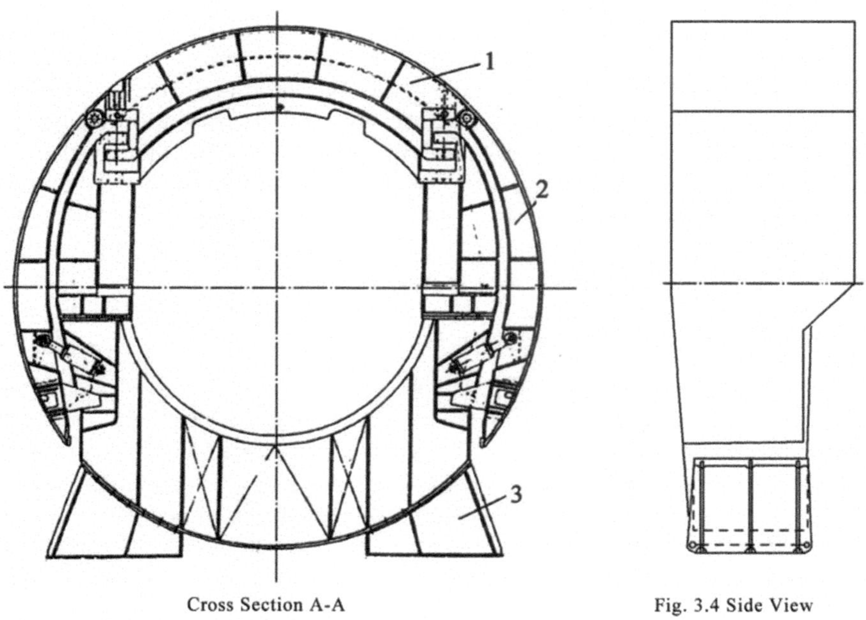

Fig. 2.7 Schematic diagram of gripper TBM cutter head shield. *1*—top shield; *2*—wing shield; *3*—temporary support

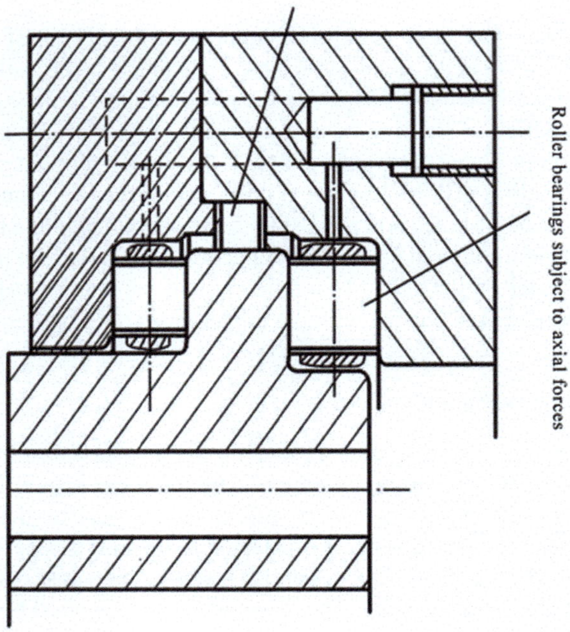

Fig. 2.8 Composition of main bearing

reducer 8 through a toothed coupling 3 and then to the drive motor 7 through a friction clutch 6. During normal operation, the cutter head is driven by a double-speed water-cooled motor 7, which is installed between the two outer Kelly supports. The double-speed reversible motor allows half-speed drive of the cutter head under unstable weak surrounding rock geological conditions and motor reversion under unfavorable conditions for breakout of the cutter head. During inching, the cutter head is driven by a hydraulic motor 4 to rotate the cutter head to the cutter change position for replacement of disc cutter or maintenance.

(4) Main bearing sealing and lubricating

The main bearing seal consists of 3 lip seals protected by labyrinth seals. The labyrinth seals are purged by an automatic grease system as shown in Fig. 2.10. The main bearing ring gear and drive pinion are mechanically lubricated in a forced manner with such equipment as lubrication pump, filter and electronic monitoring system. The grease lubrication system, oil lubrication system, and cutter head drive system are interlocked with each other. If the lubrication system fails, the cutter head will stop rotating automatically. The planetary reducer is injected with lubricating oil in a splash manner.

(5) Inner Kelly support

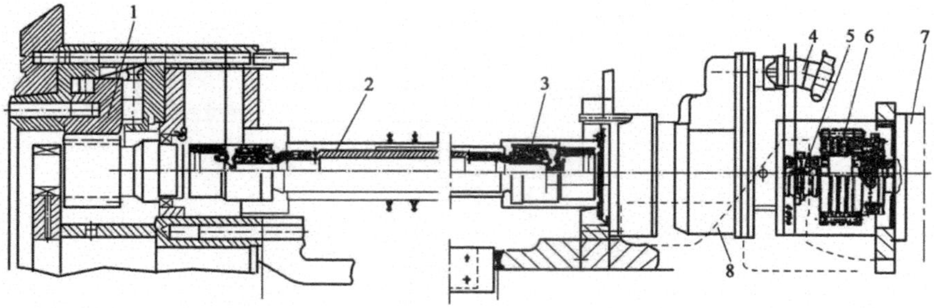

Fig. 2.9 Schematic diagram of TBM cutter head drive. *1*—inner ring gear; *2*—transmission shaft; *3*—toothed coupling; *4*—hydraulic motor; *5*—toothed coupling; *6*—friction clutch; *7*—double-speed water-cooled drive motor; *8*—two-stage water-cooled planetary reducer

Fig. 2.10 Main bearing seal

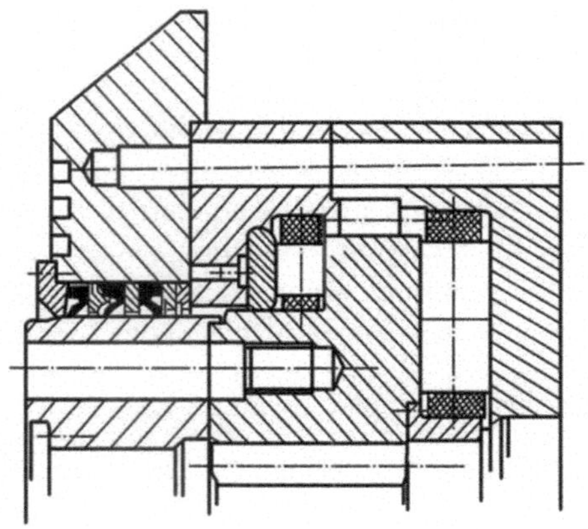

See Fig. 2.11 for the inner Kelly support. The inner Kelly support is of a welded box-section structure, on which there is a hardened slide for the bearing seat of the outer Kelly support to slide on.

The front and rear outer Kelly supports are slid by the thrust cylinder. The inner Kelly support is guided by the cutter head to transmit the thrust and torque generated during TBM operation to the outer Kelly support. The inner Kelly support is connected to the cutter head bearing, drive device, and rear support. The tail part of inner Kelly support is connected to the rear support, and its front is connected to the main bearing seat. The front end of the inner Kelly support is provided with a manhole, through which the cutter head is accessible. There is enough space inside the inner Kelly support to accommodate the belt conveyor.

(6) Outer Kelly support and grippers

As shown in Fig. 2.12, the outer Kelly support together with the grippers slides longitudinally along the inner Kelly support. The grippers are driven by 32 hydraulic

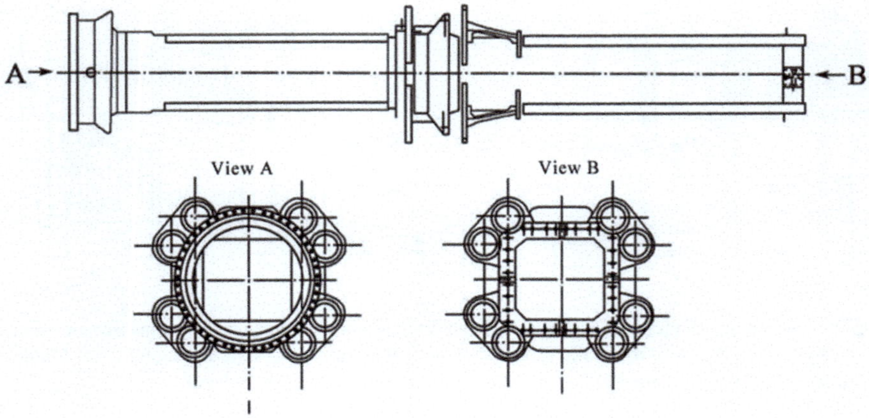

Fig. 2.11 Inner Kelly support of gripper TBM

cylinders and they are divided into two groups, each with 8 grippers, and distributed in an X shape on the outer Kelly support. The front and rear outer Kelly supports are provided with a group of grippers respectively. The 16 grippers securely fasten the outer Kelly support to the tunnel inner wall to withstand the cutter head torque and the reaction force generated during TBM advancement. The front and rear grippers can move independently to adapt to different ring beam spacing.

(7) Thrust cylinder

The thrust acting on the cutter head is transmitted to the surrounding rock through the inner and outer Kelly supports. The outer Kelly support includes two separate

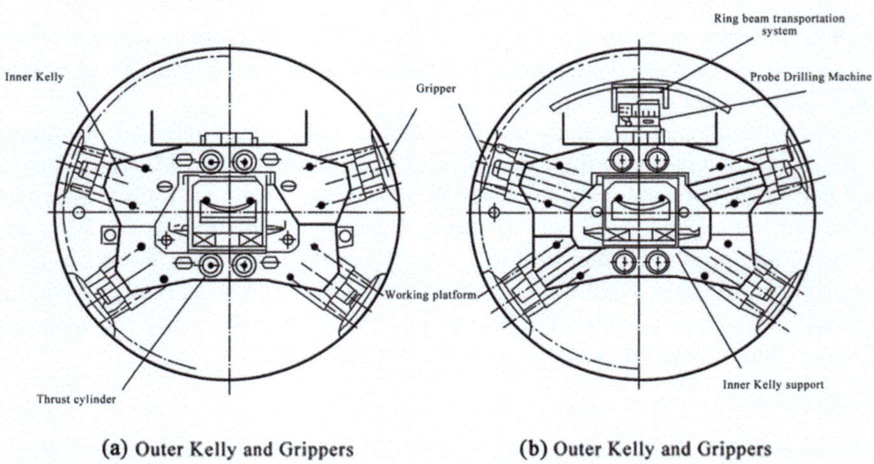

Fig. 2.12 Outer Kelly support and grippers

assemblies, each with an independent thrust cylinder. The front and rear outer Kelly supports are provided with 4 thrust cylinders respectively. The thrust cylinders of the rear outer Kelly support transmits the thrust to the inner Kelly support, while the front outer Kelly support directly transmits the thrust to the shell of the cutter head drive device. At the end of the tunneling cycle, the rear support of the inner Kelly support extends to the bottom of the tunnel, while grippers of the outer Kelly support retract, and the thrust cylinder pushes the outer Kelly support forward to prepare for the next cycle of tunneling.

(8) Rear Support

The rear support (Fig. 2.13) is installed on the inner Kelly support and located behind the rear outer Kelly. The rear support controls the expansion and retraction through the hydraulic cylinder, which can also be used for horizontal adjustment. When the rear support is retracted, the position of the inner Kelly support can be adjusted horizontally and vertically to adjust the tunnel centerline of the TBM.

(9) Connection Bridge

The connection bridge is directly articulated behind the TBM main machine and supported on the platform gantry. It is bridged upwards to enlarge the working space below for the installation of inverted blocks and laying of rails. The connection bridge is equipped with a belt conveyor system, ventilation system, invert block crane, and material lifting system.

(10) Back-up Gantry

The back-up gantry is towed on the steel rail. The gantry is equipped with TBM hydraulic power system, switch board, transformer, master power failure switch, cable reel, deduster, ventilation system, console, belt conveying system, shotcrete system, grouting system, water supply system, and other auxiliary equipment.

(11) Guidance System

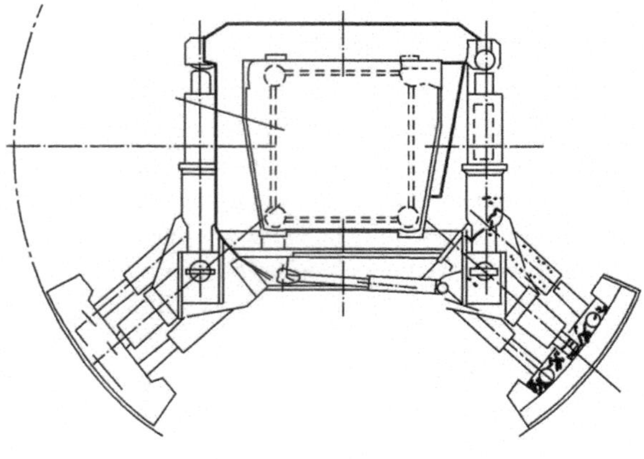

Fig. 2.13 Rear support

The guidance system consists of two laser targets mounted on the TBM and a laser mounted on the tunnel wall. The laser target is mounted behind the cutter head shield and monitored by an industrial TV monitor, which transmits the position of the TBM relative to the laser beam to the display in the operation room. The operator adjusts the support system of the TBM according to this information during the machine regripping.

(12) Data Processing System

The data processing system monitors and records the following data: date and time, tunneling length, advance rate, stroke length and duration of each cycle, current and engagement times of drive motor, thrust cylinder pressure, and support cylinder pressure. The above data can be used for monitoring and storage and can be printed for retrieval at any time. With the advancement of TBM, the recorded and stored data can be made into different tables or bar charts, and pie charts.

When a fault occurs, the warning light will remind the operator to find the fault type and cause from different screens.

(13) Dedusting System

The dedusting system is installed at the front end of the back-up gantry. The dust suction pipe is connected with the inner Kelly and the cutter head shield to form a negative pressure between the cutter head and the tunnel face so that about 40% of the fresh air in front of the TBM enters between the cutter head and the tunnel face to prevent the air containing dust from entering the tunnel. The dust-containing air sucked by the axial flow fan of the deduster passes through space with several water spraying nozzles. After the wet dust is blown to the water collecting blade of the deduster, the dust is separated highly and flows to the dust collecting box equipped with a circulation water pump for sedimentation.

(14) Bolt Driller

The bolt driller is installed next to the inner Kelly support behind the cutter head shield and next to the inner Kelly support behind the rear support gripper. The bolt driller can install the rock bolt during machine tunneling.

(15) Probe Drilling Machine

The probe drilling machine is used to drill detection holes in front of the TBM. When drilling detection holes, the TBM must stop tunneling. The probe drilling machine is installed on the outer Kelly and between the front and rear support grippers. During drilling, it is moved to the outer edge of the TBM shield and drilled in front of the TBM at a slight pitch angle. Cone guiding can adapt to the whole cutter head shield guiding and stabilizing drill rod. The power of the probe drilling machine is provided by one of the power packs of the bolt driller.

(16) Ring Beam Erector

The ring beam erector can be operated during TBM tunneling, and the pre-assembly and installation of the ring beam can be carried out behind the cutter head. The ring beam erector consists of a pre-assembly groove behind the cutter head shield, a hydraulic drive traction chain, a platform moving longitudinally over the inner Kelly, and a hydraulic cylinder for lifting and extending the ring beam. The ring beam erector is directly operated by the control desk installed behind the cutter head shield and powered by the hydraulic system of the TBM.

(17) Invert Block Crane

The invert block crane moves along the double track under the belt bridge, to lift the invert block to the installation position. The invert block crane can move horizontally and vertically. The way to move is chain drive.

(18) Grouting System

The grouting system is installed on the back-up system of TBM. The grouting system uses the mixing mortar for grouting of invert blocks, concrete bonding, pressure grouting of rocks, and fault stabilization.

(19) Shotcrete System

The shotcrete system, consisting of a wet spraying machine, a liquid metering pump, and a shotcrete manipulator, is installed on the back-up system of TBM.

2.2.2 Main Beam Gripper TBM

1. Structural Features

Utilization of horizontal grippers to provide thrust for TBM is the main feature of main beam gripper TBM. Figure 2.14 shows main beam gripper TBM for the Gotthard Base Tunnel of Herrenknecht, Germany.

The structural characteristics of the main beam gripper TBM designed and manufactured by Herrenknecht Germany can be viewed by scanning the QR code on the right.

The main beam gripper TBM is mainly composed of a cutter head, muck hopper, shield, finger shield, bolt driller, main beam, rotary joint, belt conveyor, main drive, front support, ring beam erector, working platform, probe drilling machine, gripper saddle, gripper cylinder, gripper, air duct, belt conveyor, material crane, thrust cylinder, rear support, back-up system, etc. The overall structure is shown in Fig. 2.15, and its main characteristic is that the main beam gripper system is adopted.

Fig. 2.14 Exterior of main beam gripper TBM

2. Basic Configuration

The ϕ9030 mm double-pair horizontal supporting main beam TBM manufactured by CREC engineering equipment used in the Gaoligongshan Tunnel of Dali-Ruili Railway is taken as an example in this section to introduce the basic configuration of main beam gripper TBM.

The main beam gripper TBM is composed of main machine and back-up system. The main machine is mainly composed of a cutter head, main drive system (including the main bearing), shield, thrust and support system, rear support, TBM belt conveyor, auxiliary support system, and other parts. It is the core part of the TBM system and completes the main tunneling and partial support work. The back-up system, connected with the main machine, consists of a series of steel structure jumbo connected with each other, which is used for the arrangement of the hydraulic power system, power supply and control system, water supply and drainage system, ventilation and dedusting system, muck discharge system, support system, and etc.

(1) Cutter Head

The cutter head consists of the main body of the cutter head steel structure, the disc cutter, the muck inlet, and the water spraying device, as shown in Fig. 2.16.

The cutter head is in a plane shape conducive to stabilizing the tunnel face. Considering the working conditions, processing and manufacturing, transportation, assembly, maintenance, and other comprehensive factors, the cutter head is designed as a block structure, with block form of 1 + 4.

The structure of the cutter head is a heavy-duty welded steel structure, with 1 center section and 4 side sections connected by pins and high-strength bolts. Large groove welds are designed around to ensure overall strength and reliability. The cutter head is designed with 12 uniformly distributed muck inlets, with sufficient

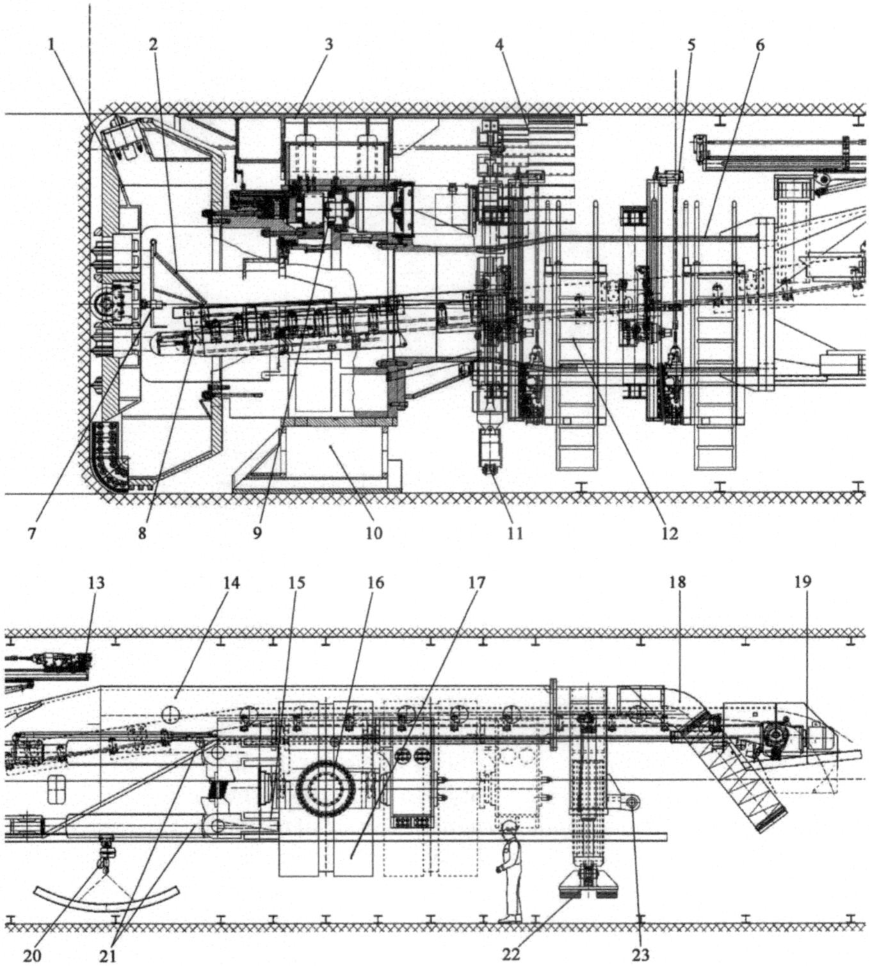

Fig. 2.15 Overall structure of main beam gripper TBM. *1*—cutter head; *2*—muck hopper; *3*—shield; *4*—finger shield; *5*—bolt driller; *6*—front part of the main beam; *7*—rotary joint; *8*—belt conveyor; *9*—main drive; *10*—front support; *11*—ring beam erector; *12*—working platform; *13*—probe drilling machine; *14*—rear part of the main beam; *15*—gripper saddle; *16*—gripper cylinder; *17*—gripper; *18*—air duct; *19*—muck discharge outlet of belt conveyor; *20*—material crane; *21*—thrust cylinder; *22*—rear support; *23*—back-up system traction point

muck discharge, which effectively reduces the secondary wear of the peripheral disc and cutter. Each muck outlet is designed with grids, which effectively limits the maximum particle size of the muck.

(2) Main Drive System

The cutter head drive is of a gripper central ring drive as shown in Fig. 2.17. The main drive consists of a drive box steel structure, main bearing, ring gear, inner and

Fig. 2.16 Cutter head

outer seals, drive pinion, planetary reducer, variable-frequency motor, and 1 set of ring parts.

The cutter head is supported by the main bearing in the drive box steel structure of the main drive, which is a heavy-duty rigid (anti-torsion) welded steel structure. The main bearing is a typical three-row roller bearing of the TBM. The main bearing diameter accounts for about 2/3 of the diameter of the cutter head, with a large

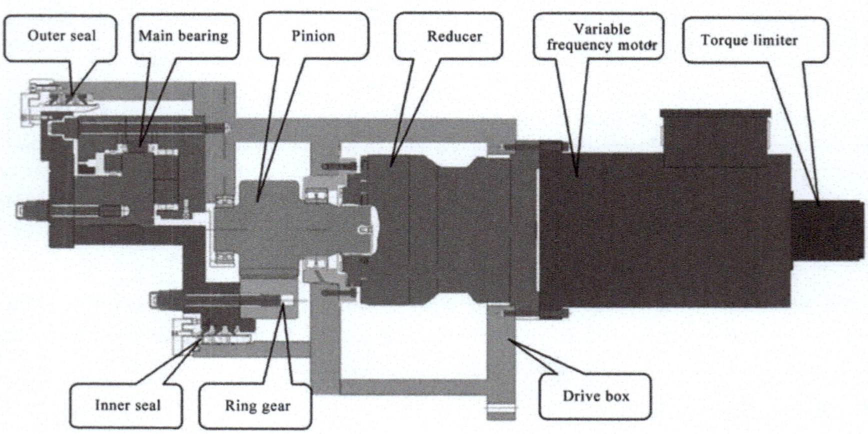

Fig. 2.17 Main drive system

capacity reserve. The layout of the main bearing and ring gear components can meet the required spatial layout and design service life. To facilitate manufacture and replacement, the ring gear from the main bearing shall be separate. The outer ring gear is driven by the pinion and is connected together with the drive adapter ring and the main bearing inner ring. The variable frequency control electric drive unit includes the following equipment: variable frequency drive and water-cooled motor, with the protection grade of IP67 and the insulation grade of H; torque limiter is adopted for mechanical overload protection.

(3) Shield

The main body of the shield is of a steel structure welding part. It is connected around the periphery of the drive box steel structure. It is used to stabilize the cutter head tightly on the tunnel wall during TBM tunneling and prevent large rock chips from falling on the back of the cutter head and at the main drive motor. The whole shield is divided into the bottom shield, wing shield, and top shield (Fig. 2.18). The bottom shield is fixed below the drive box steel structure and bears the gravity of the front part of the TBM and serves as the fulcrum for TBM steering. The wing shield and top shield are driven and telescoped by the hydraulic cylinder, and the wing shield is in tunneling. The cutter head is stabilized and the top shield structure extends behind the cutter head to provide protection for workers performing auxiliary support.

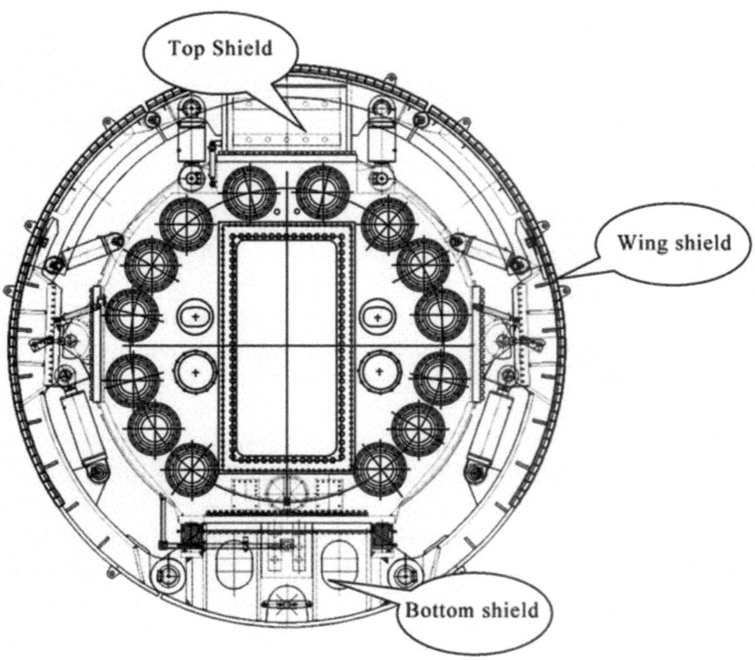

Fig. 2.18 Shield

Fig. 2.19 Thrust and support system

(4) Thrust and Support System

The thrust and support system is mainly composed of the main beam, saddle, thrust hydraulic cylinder, gripper hydraulic cylinder, torque hydraulic cylinder, and gripper, as shown in Fig. 2.19. The main function is to provide the required thrust force for cutter head tunneling through the thrust hydraulic cylinder, and the gripper cylinder will hold the gripper tightly on the tunnel wall to bear the reaction force and moment during tunneling.

① Main Beam

The main girder consisting of two sections, as shown in Fig. 2.20, are connected with each other by high-strength bolts. Its main function is to transmit cutter head thrust force and reverse torque. When the cutter head rotates, the torque generated is transmitted to the main beam through the drive box steel structure, through the main beam to the saddle, the saddle to the gripper device through the torque cylinder, and finally to the excavated rock wall.

The thrust cylinder installed on the main beam provides the thrust force of the main machine. One end of the thrust cylinder is installed in the front section of the main beam, and the other end is connected with the gripper device. The thrust reaction is provided by a gripper device holding the rock wall. When the main machine is pushed forward, the gripper is closely attached to the surface of the rock wall, and the thrust cylinder extends to drive the main beam, drive box steel structure, and cutter head to move forward along the slide rail of the saddle.

② Saddle

As shown in Fig. 2.21, the saddle installed on the main beam is a component connecting the gripper cylinder and the main beam.

Fig. 2.20 Main beam

Fig. 2.21 Saddle

Both sides of the saddle are connected to the slide rail of the main beam through coppern sleeves. The upper part of the saddle is equipped with a cross articulation device, an important part of the saddle system, which plays an important role in the horizontal steering of the main machine. And a regulating plate is used between the saddle and the cross articulation device for adjustment. The upper and lower sides of the saddle are equipped with a buffer spring and a supporting plate, and the supporting plate always fits with the customized plane on both sides of the gripper cylinder. A dustproof seal is installed on the outer side of the carrier coppern sleeve. During the forward and backward movement of the saddle along the main beam slide rail, sundries and muck particles on the slide rail surface are removed. The dustproof seal is easy to remove and can be replaced after being worn to a certain extent.

③ Grippers

The grippers, welded steel structural components with larger gripper area, can obtain proper ground pressure in the soft surrounding rock. With outer arc surface grooved, grippers can cross the ring beam without crushing. On the gripper surface, there are several antiskid nails installed in order to obtain greater friction.

The end of the grippers and the gripper cylinders are bolted in contact with spherical pairs. The outer spherical sleeves are bolted on the end of the piston rod of the gripper cylinders. The inner convex spherical surface is in contact with the concave spherical surface of the spherical sleeve, and the flange matching with the outer spherical sleeve is bolted with the gripper. This spherical auxiliary structure allows the gripper to have a certain swing angle in the 360° direction to meet the needs of TBM attitude and tunnel wall irregularity. In order to prevent excessive swing and adjust the position of the gripper, a stabilizer cylinder is set inside the gripper, and the pressure is maintained against the spherical sleeve to ensure the stability of the gripper during regripping.

④ Gripper Cylinder, Torque Cylinder and Thrust Cylinder

There are 4 gripper cylinders which are hydraulic cylinders with special designs. The cylinder barrel ends are connected by a connecting frame of appropriate size and kept on the same axis. The piston rod ends are connected with the gripper device and all of them face the rock wall.

One end of the torque cylinder is connected to the saddle, and the other end is connected to the lug support welded on the cylinder barrel of the gripper cylinder. Its main function is to transmit the reverse torque generated by the excavation of the main machine and provide power for the vertical steering of the main machine. The torque cylinders are divided into two groups, two for each group, and installed on both sides of the main beam.

The thrust cylinder provides the thrust required for main engine excavation. It is a double-acting cylinder with stroke sensor. Its one end is connected to the lug support welded on the main beam and the other end is connected to the gripper device, so that the thrust can be directly transmitted to the rock wall through the gripper device, reducing the risk of lateral force on the gripper cylinder.

(5) Rear Support System

The rear support is installed at the tail of the main beam, as shown in Fig. 2.22. During TBM regripping, the rear support cylinder extends, and the rear support gripper closely contacts the tunnel wall and supports the gravity of the main machine together with the bottom shield; During TBM tunneling, the rear support cylinder retracts and moves forward with the main beam under the action of the thrust cylinder to complete the tunneling cycle.

(6) Auxiliary Support System of Main Machine

The auxiliary system of the machine is shown in Fig. 2.23, mainly including steel slat/mesh installation, ring beam installation system, bolt driller system, transportation channel, working platform, wet spray in the L1 area, probe drilling machine, etc. The support speed of each system matches the TBM advance speed.

① Storage and Installation of Steel Slat/Wire Mesh

In order to stabilize and protect the equipment, a steel slat/wire mesh trough shield is installed at the tail end of the TBM top support, and can play a safety protection role when installing rock bolts and ring beams. The steel slat/wire mesh is stored on the connection bridge, and can be transported to the top shield through the material transportation system at any time when necessary, and then manually installed in the groove of the top shield to play the role of preliminary support.

② Ring Beam Erector

The ring beam erector is arranged under the top shield in front of the main beam, so as to timely erect the ring beam under the protection of the top shield. The ring beam is erected from multi-section steel sheets made of profile steel. The erector needs to complete rotary erection, top, and lateral bracing, bottom opening tensioning and sealing, etc. The erector can install a ring-shaped ring beam in several segments for lifting, rotating, positioning, longitudinal moving, and radial contraction movement. All the operations of the ring beam erector are completed directly by hydraulic control on the control desk behind the support shield (Fig. 2.24).

Fig. 2.22 Rear support

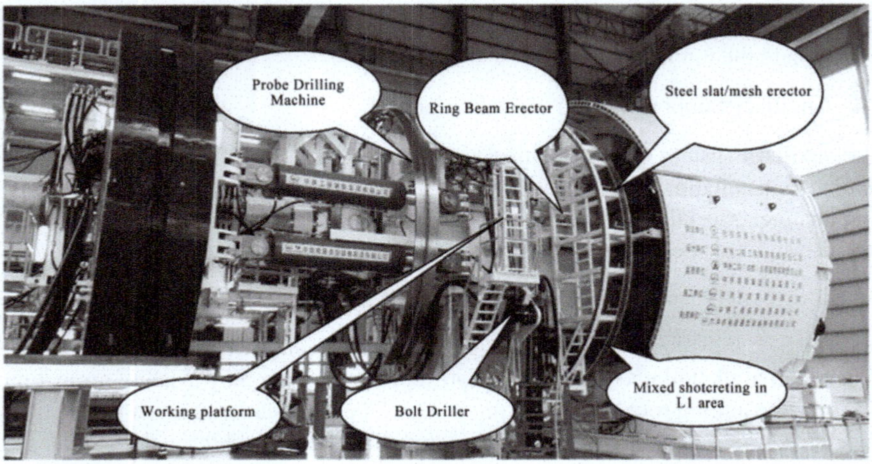

Fig. 2.23 Auxiliary support system of main machine

Fig. 2.24 Ring beam operation platform

③ Bolt Driller System

Two bolt drillers are equipped in the L1 area of the TBM, and located on both sides of the main beam, as shown in Fig. 2.25.

They can carry out drilling operations independently and synchronously during TBM tunneling to assist in the construction of rock reinforcement, wire mesh fixation, and installation of partial section ring beams. The bolt driller is fixedly installed on the ring gear beam through the base holder, and the ring gear beam can be removed. Driven by the hydraulic motor reducer assembly, it can move in the circumferential position of the ring gear beam.

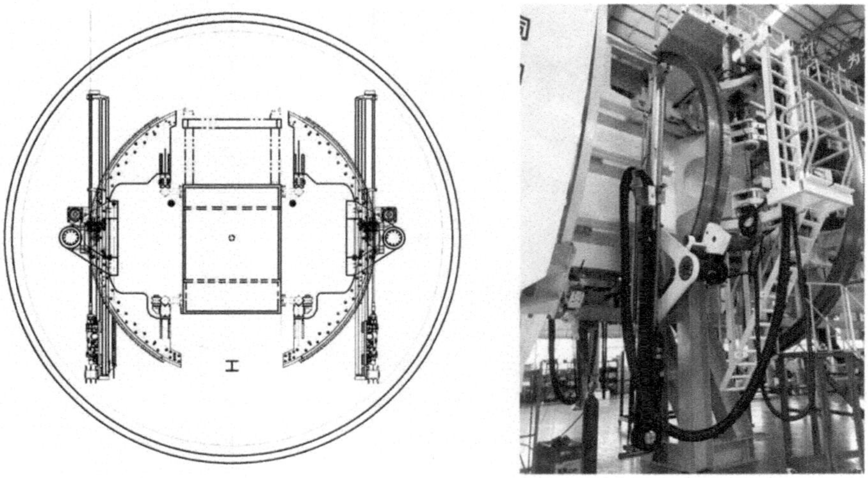

Fig. 2.25 Bolt driller

The ring gear beam is designed with a traveling device, and a traveling rail is installed on the main beam. Under the action of the horizontal moving cylinder, the bolt driller and the ring gear beam can move forward and backward in the direction of the tunnel diameter axis. The axial moving stroke of the bolt driller is greater than one tunneling stroke of the TBM, and a synchronization mechanism is installed to allow bolt drilling operation during tunneling. Equipped with a swinging cylinder to realize the radial inclination, it can cover the drilling operation of the tunnel wall at about 180° of the top of the ring beam, and the drilling depth is not less than 3.5 m each time.

④ Zone L1 Shotcrete

The shotcrete system is mainly composed of the concrete transfer pump and its control cabinet, spraying manipulator, admixture transfer pump, manipulator remote controller, transfer pipeline, etc. (Fig. 2.26). The concrete transfer pump has the functions of vibration and pumping. The power pack and control cabinet are arranged on the right side of the connection bridge, and the shotcrete is relayed by the concrete pump in the L2 area. The wet spray pipeline in the L1 area can be removed if it is without been used for a long time.

⑤ Emergency Shotcrete Equipment

The emergency spraying machine (Fig. 2.27) is independently equipped. It is usually placed outside the tunnel. When spraying is required, the spraying machine and spraying materials are transported to the bottom of the connection bridge by a flat car, and the materials are loaded manually. The spraying operation can be carried out manually within a certain range.

Fig. 2.26 Shotcrete equipment in L1 area

Fig. 2.27 Emergency shotcrete equipment

⑥ Probe Drilling Machine System

The probe drilling machine (Fig. 2.28) is installed at the tail of the main beam. It has an independent ring gear beam, a rotary mechanism, a control system, etc., which can be used for drilling operations of probe bolts, etc.; it can cooperate with pre-grouting equipment to assist in advance support of the rock in front of the cutter head. In a non-working state, it can be hidden on the lower side of the platform at the top of the main beam. When a probe drilling operation is required, the probe drilling machine can be lifted to the working position by the lifting cylinder, and the operation can

Fig. 2.28 Probe drilling machine

be carried out after the pipeline is connected. When the probe drilling machine is operating, the TBM must stop tunneling.

⑦ Working Platform of Main Machine

A large amount of support work of TBM needs to be completed at the upper part, and maintenance work is required at the gripper of the saddle. Therefore, an operation platform is designed above the main beam (Fig. 2.29), and personnel access is designed at the rear support.

Fig. 2.29 Operation platform

(7) Muck Discharge System

① TBM Belt Conveyor

The main function of the TBM belt conveyor is to convey the rock chips excavated by the cutter head to the back-up belt conveyor to realize continuous mucking and ensure the rapid tunneling of the TBM. Based on the structural characteristics of gripper TBM main machine, the TBM belt conveyor generally uses the TBM main beam as the support. Its tail height can be adjusted to meet the needs of personnel entering the cutter head for maintenance. The maintenance or replacement of the roller set can be carried out through the reserved holes on the main beam. In order to prevent the belt conveyor from deviating, the pulley and guide roll structure is adopted in the design. The muck discharge capacity of the TBM belt conveyor matches the maximum advance speed of the TBM, with some allowance reserved. In order to facilitate the muck removal inside the main beam, a simple muck removal device is set at the lower part of the TBM belt conveyor.

The belt width of the TBM belt conveyor is 1200 mm, the belt speed is 2.5 m/s, and the maximum transport capacity is 1050t/h. Its main design functions are as follows:

a. Hydraulic drive, compact structure, easy speed regulation and large torque;
b. The tail can be lifted (hydraulic drive) for easy access to the cutter head for cutter change;
c. It has the functions of anti-deviation, deviation alarm, and emergency stop;
d. The head of the belt conveyor is provided with a spray device.

② Back-up Belt Conveyor

The function of the back-up belt conveyor is to transfer the muck unloaded by the TBM belt conveyor to the tunnel belt conveyor (Fig. 2.30). The capacity of the belt conveyor can meet the requirements of the maximum advance speed of the TBM. In order to ensure safety, each key position of the belt conveyor is equipped with an emergency stop switch. The belt width of the back-up belt conveyor is selected as 1200 mm, with the belt speed of 2.5 m/s, and the maximum transport capacity of 1050t/h. The main design functions are as follows:

a. Variable frequency drive for easy speed regulation;
b. Directly connect the TBM belt conveyor and the tunnel belt conveyor, eliminate the secondary connection and transfer of materials, and reduce the risk of muck falling and leakage;
c. It has the functions of anti-deviation, deviation alarm, and emergency stop;
d. The head of the belt conveyor is provided with a spray device.

③ Tunnel Belt Conveyor

Fig. 2.30 Back-up belt conveyor

The design of the tunnel belt conveyor system is based on the principles of high reliability, universality, and reduction of the intermediate drive. In the specific design, the power margin is increased by a belt width of 914 mm, the rotation speed of 3.15 m/s, and the installed power of 4×315 kW. The models of motors and reducers are unified as far as possible to ensure the interchangeability of structural components and key parts.

The tunnel belt conveyor is mainly composed of a head-fixed part [drive device (Fig. 2.31), unloading device, belt storage cassette, tension device, vulcanization platform, etc.], body part (steel structure, upper and lower roller groups, wall-mounted bracket, etc.), tail part (tail device, installation window, buffer bed, etc.), tape, electric control system (control, communication, protection, etc.), auxiliary part (rolling belt device, vulcanization device, cleaner, water cleaning tank, etc.) and muck transfer part.

(8) Electrical System, Control System, and Data Acquisition System

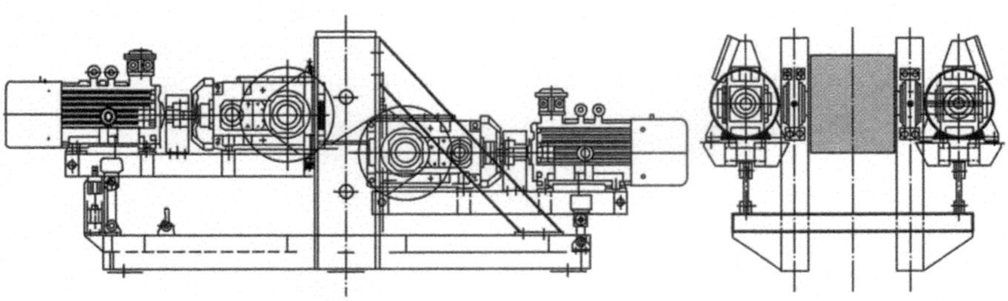

Fig. 2.31 Drive part

Fig. 2.32 Electrical control components

① Electrical System

The back-up electrical system includes the power and control circuits required by the back-up equipment (Fig. 2.32), mainly including a belt conveyor, lighting, deduster, fresh air supply system, air silo lifting mechanism, shotcrete pump, spraying mechanism in the rear support area, air compressor, industrial water system, operator rest facilities, and other power and control circuits.

② Control System and Data Acquisition System

The main component of the control system is a programmable logic controller (PLC) located in the control cabinet of the operator cabinet. The input signal of PLC comes from the operator's control and the sensor installed on the equipment. The PLC output signal is output to the operation display console of TBM (light, advance rate, chart, etc.) and to many control equipment according to the system program. The system can display and control (by program control or operator's command) hydraulic, lubrication, and electrical equipment, such as pressure, flow, limit switch, machine position, etc. The display on the control console can display most information in the form of digital or chart curves, and also display fault information (Fig. 2.33).

(9) Hydraulic and Lubrication System

① Hydraulic System

The hydraulic system is the power source for completing various actions of TBM, and the configured hydraulic components are all famous brands at home and abroad. The main circuits of the hydraulic system include the support, thrust, torque cylinder,

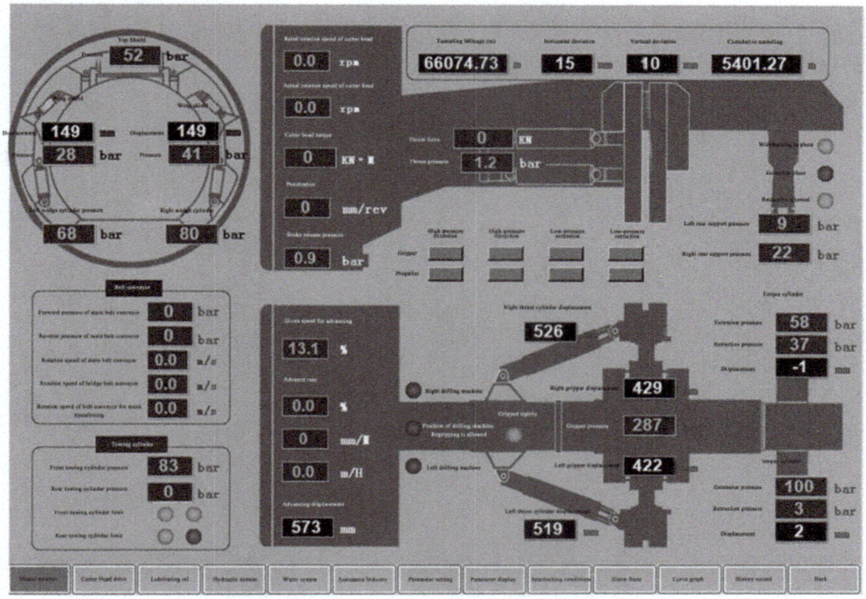

Fig. 2.33 Equipment parameter monitoring interface

and TBM belt conveyor drive system (Fig. 2.34). The gripper system and thrust system are divided into two sets of circuits, i.e. high-pressure small flow circuit and low-pressure large flow circuit, to meet the requirements of equipment tunneling performance and rapid regripping. According to the configuration requirements, the system reasonably arranges the efficiency of the hydraulic pump to improve the equipment utilization rate and reliability.

② Lubrication System

Good lubrication is the premise to ensure the normal operation of the equipment, and famous products at home and abroad are selected to ensure the reliability of the lubrication system. The main bearing seal system is lubricated by a multi-point pump and a distribution valve to ensure that the main bearing sealing grease can be uniformly and continuously extruded, which can effectively prevent external muck from entering the main drive. The main frame is automatically lubricated by multiple distribution valves and is equipped with a filter unit; The grease and lubrication system is equipped with a monitoring interlocking device; The grease lubrication system of other equipment shall also adopt automatic centralized lubrication as far as possible (Fig. 2.35).

(10) Compressed Air System

The compressed air system provides compressed air to all pneumatic tools, shotcrete equipment, and grease systems. There are compressed air pipes arranged in the main

Fig. 2.34 Hydraulic pump station

Fig. 2.35 Lubrication oil pump

machine and on the back-up flat jumbo (Fig. 2.36). The compressed air system includes an air tank, screw air compressor (water-cooled), safety valve, check valve, pressure gauge, and two-stage filtration on the air tank. A certain reserve has been considered for the capacity of the air compressor.

(11) Water Circulation Cooling System

Fig. 2.36 Air compressor

The water circulation system configured on the equipment can be divided into four parts: internal circulation cooling system, external circulation cooling system, cutter head water spray system, and cleaning water system. Note: It is necessary to control the inlet water temperature of TBM not higher than 25 °C. The water supply volume can be appropriately increased or decreased on site according to the water temperature to ensure a good cooling effect for the equipment.

The water supply line of the main tunnel is connected to the back-up cold water tank, where a cooling water pump is installed, and it is mainly used to cool the internal circulating water; The internal water circulation system, an independent closed circulating cooling system mainly cools the main drive motor, reducer, gear oil, hydraulic pump station, shotcrete manipulator pump station, frequency converter, and transformer, etc. The internal water circulation system requires distilled water to prevent the generation of scale and ensure the long-term stable operation of the cooling equipment; The hot water after the external circulation water cools the equipment is returned to the hot water tank, which is equipped with a cutter head spray pump and a cleaning water pump. The cutter head spray pump is mainly used for cutter head spraying, dedusting, belt conveyor dedusting, etc., and the cleaning water pump is mainly used for equipment cleaning and water for bolt driller, etc. (Fig. 2.37). The above four systems are equipped with filters to ensure the stable operation of the equipment.

The industrial water tank is equipped with a liquid level control device, and the operation of all water pumps can be remotely controlled in the main machine operation room.

During TBM construction, waste water may accumulate at the front end of the equipment. Therefore, a submersible pump is designed at the main machine to pump water into the waste water tank, where the wastewater pump is installed will discharge

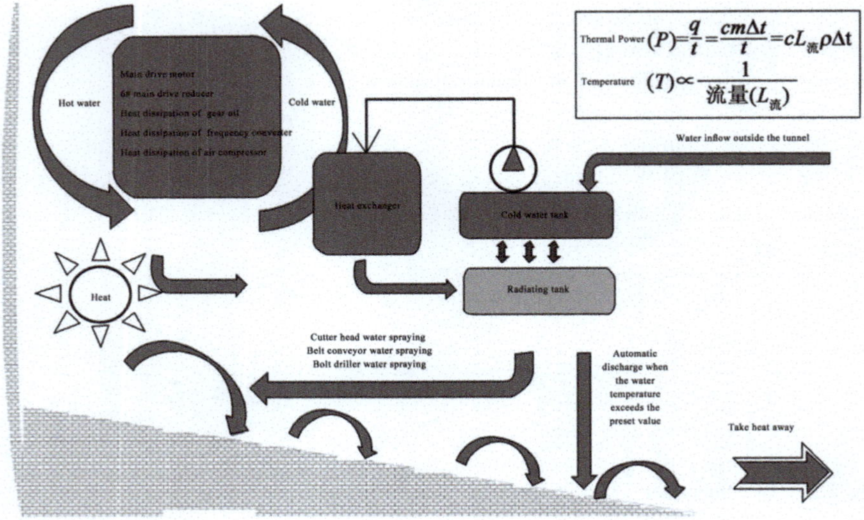

Fig. 2.37 Cooling circulation system

water outside the back-up (Fig. 2.38). The capacity of the dewatering pump equipped for the equipment meets the dewatering requirements.

(12) Ventilation, Dedusting, and Cooling System

① Ventilation System

The ventilation system mainly consists of air duct cassette, secondary fan, and air duct. The air duct cassette is arranged at the tail of the upper layer of the back-up system, and its front is close to the fan and the side of the rigid air duct along the side of the upper layer of the back-up system from the booster fan to the TBM.

After the fresh air reaches the shotcrete bridge, it is divided into two paths, one path to the tail of the gantry 1 and the other path to the rear support to ensure sufficient

Fig. 2.38 Cold and hot water tanks and internal water pump station

fresh air in the main machine area. A lifting device for replacing the air duct cassette is provided at the end of the gantry. As the TBM advances, the air duct is continuously released, and another air duct cassette is replaced after release.

② Dedusting System

The dedusting system mainly consists of a deduster, a metal air duct, an industrial exhaust fan, etc. (Fig. 2.39). The inner cavity of the TBM main beam used as an air duct is connected with a metal-air duct from the rear of the main beam and transported backward through the connection bridge. The air containing dust in the cutter head is sucked out through the industrial exhaust fan, and the filtered air after dedusting by the deduster is discharged to the back-up tail through the pipeline and industrial exhaust fan, and then flows out of the tunnel at a certain air flow rate through the tunnel.

The gap between the TBM shield and drive box steel structure of the main drive is designed with a dust baffle plate to prevent dust from directly flowing out from the tunnel face. During construction, the cutter head generally needs to be sprayed with water to reduce dust. At the same time, a dedusting system is set up to suck air from the inner cavity of the main beam for dedusting.

③ Cooling System

In order to cope with the possible high geothermal environment during tunnel construction and reduce the impact of high rock temperature on equipment and personnel. Firstly, the ventilation volume of the tunnel shall be increased; Secondly, the TBM equipped with 3 refrigeration units and air coolers (Figs. 2.40 and 2.41) can transfer the heat in the air to the water; By increasing the water inflow, the heat in the TBM construction working area is transferred to the water and discharged outside the tunnel; Thirdly, water cooling is adopted for TBM drive reducer, drive motor and main drive frequency converter.

(13) Back-up System

Fig. 2.39 Dedusting system

Fig. 2.40 Refrigeration unit

Fig. 2.41 Air cooler

The back-up system, mainly composed of a connection bridge, a shotcrete bridge, and a back-up gantry, can provide maximum convenience for material access with a compact and reasonable layout. The back-up steel structure has sufficient strength and rigidity, and the length of the gantry meets the conditions of site assembly and construction. Sufficient internal space is reserved for the back-up gantry, and the equipment layout is humanized to facilitate train access. The connection bridge is designed to facilitate the installation of relevant auxiliary equipment and facilities and is provided with a platform with guardrails for personnel to walk on. Staff can easily reach safety zones from any area. The back-up equipment is arranged with a reserved guide window to ensure the normal operation of the total station of the guidance system without hindrance.

① Connection Bridge

a. Connection Bridge

The connection bridge is connected to the tail of the main machine through the No.1 towing cylinder, with the front end supported on the rear support bracket and the rear end supported on the rail through the wheelset. The connection bridge is of a heavy-duty box structure. The interior of the box can be ventilated. A platform is set at the upper part to store certain materials. The connection bridge can bear the gravity of dead weight and reasonable materials.

The connection bridge is installed with a folding boom crane on the left side, and its right side is used to store materials. Its center of gravity is close to the right cross beam. Therefore, the heavy objects on its upper part shall be distributed as uniformly as possible, and they shall not be stored in a centralized manner (Fig. 2.42).

The equipment arranged on the connection bridge includes: the back-up belt conveyor, accelerator pump station, main control cabinet, concrete transfer pump in the L1 area, transfer pump control cabinet, lifting platform, gear oil pump station, lubrication grease pump station, folding arm crane, and rail hoist, and invert block crane.

b. Invert Block Crane

The invert block installation system, as the TBM supporting equipment, is applied to install the invert block in place, with efficient, safe, and reliable operation process by hydraulic and mechanical drive mechanisms and a PLC control system to realize its movement and adjustment of 6 degrees of freedom (Fig. 2.43).

c. Lifting Platform

Fig. 2.42 Connection bridge

Fig. 2.43 Invert block crane

The lifting platform consists of a motor reducer, coupling, synchronous transmission shaft, chain wheel, etc. The lifting platform is equipped with a crane for loading and unloading materials (Fig. 2.44).

d. Folding Boom Crane

The folding boom crane installed on the left side of the connection bridge, is mainly used for lifting steel rails and supporting materials. The system adopts PLC controller for centralized control. The control box and remote control receiver are installed on the traveling equipment, and the cable chute is used for tracking the power supply. There are two flexible and convenient operation ways, one is through a cord handle and the other is through wireless remote controller, both with efficient, safe, and reliable operation process (Fig. 2.45).

② Shotcrete Bridge

Fig. 2.44 Lifting platform

Fig. 2.45 Folding boom hoist

As shown in Fig. 2.46, the shotcrete bridge is equipped with two sets of shotcrete manipulators, with relevant control equipment, and 2 sets of nozzles can meet the shotcrete requirements with a stroke of 10 m and a range of 300°.

③ Rail Laying Operation

There are two-rail laying hoists under the connection bridge for unloading the tunnel rails from the material supply vehicle and installing rails under the connection bridge. There are also areas on the underside of the connection bridge for storage rails (Figs. 2.47 and 2.48).

Fig. 2.46 Shotcrete bridge

Fig. 2.47 Rail laying in temporary rail storage area

Fig. 2.48 Hoist

④ Back-up Gantry

The back-up gantry, a welding part of the steel structure and an important part of TBM, is the carrier of power equipment, auxiliary equipment, back-up belt conveyor, ventilation equipment, and cable reel. The gantry runs on the rail laid in the tunnel,

Fig. 2.49 Back-up gantry

and the gantry is connected by articulated connecting shafts. Each gantry is provided with a pedestrian passage and a protective fence to ensure the safety of relevant personnel. Gantry and back-up structures are equipped with inclined ladders, step ladders, or working platforms for easy manual operation (Fig. 2.49).

⑤ Passing Platform and Ramp Rail

The rear of the back-up gantry is connected with three passing platforms, and the rear of the passing platform is connected with a ramp rail on which is laid with a double-track rail and california switch. The train passes through the ramp rail section from the tunnel rail to the gantry rail. When one train of muck car is about to be loaded, the other train of muck car is in place and waiting, greatly reducing the time of mucking for delaying the tunneling operation. The passing platform is equipped with a toilet, shower room, water hose reel, and other components.

⑥ Extension Platform

When the towing cylinder is telescoped, the distance between the No.1 gantry and the shotcrete bridge will increase to avoid interruption of the lower rail. Multi-section extended platforms are arranged under the shotcrete bridge, No. 1 gantry, and connection bridge to facilitate the passing of tunnel trains.

14. Harmful Gas Detection System

Detection of harmful gas and oxygen content is an important safety guarantee measure in tunnel construction. The gas detection system deployed on the TBM can detect the concentration of CH_4, H_2S, CO_2, CO, and O_2 in the air. CH_4 detection sensor is arranged under the front shield of the main machine; H_2S detection sensor can be arranged at the muck discharge point of the TBM belt conveyor; CO_2, CO, and O_2 detection sensors are arranged near the rest room and the operator cabinet.

Fig. 2.50 Gas detector

All gas detection sensors are integrated into the PLC control system to set the content values of alarm and automatic shutdown, display the corresponding information in the operation room and give an alarm (Fig. 2.50).

(15) Guidance System

The tunnel construction requires high breakthrough accuracy of the tunnel. TBM tunneling must be carried out in strict accordance with the designed route. The guidance system is the key system that can guide TBM to advance according to the designed axis. The configured guidance system (Fig. 2.51) can complete automatic measurement and form a visual measurement result on the interface, displaying the deviation between the tunnel and the design axis and to guide the next TBM alignment operation.

(16) Industrial Monitoring System

The TBM and its back-up system are relatively long, and various auxiliary operation process and their equipment are distributed within the whole equipment range. In

Fig. 2.51 VMT guidance system

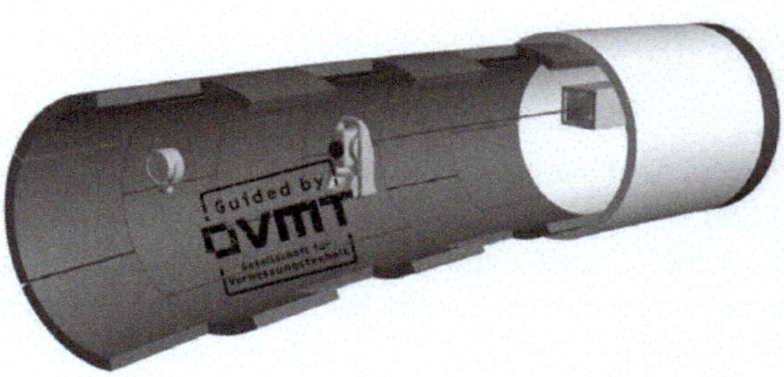

Fig. 2.52 Monitor

order to make the main driver of TBM understand the operation status and progress of other process equipment in the operator cabinet and ensure the safety and coordination of the whole operating system, a CCTV is configured on the TBM. The monitor is placed in the operator cabinet, and the cameras are distributed in the important parts to be monitored (Fig. 2.52). This equipment is equipped with CCTV monitor at 13 positions.

(17) Communication System

Wired telephones (Fig. 2.53) are installed at key parts and main operation points of TBM and its back-up systems connected with the telephones in the operator cabinet, which can be connected with the dispatching room outside the tunnel and any other telephones. In addition, the project management office is also required to configure hand-held interphones as auxiliary communication between operations at each point, and mobile phone signals can be introduced into the tunnel as one of the communication means according to the requirements of the Buyer.

(18) Lighting System

Sufficient lighting fixtures are provided on the TBM and back-up system to ensure the work safety of operators and the operation, maintenance, and inspection needs. The emergency lighting system is provided to ensure the safe evacuation of personnel in case of power supply problems, and the emergency lighting can still be maintained for not less than 2 h without power supply. The emergency lights are explosion-proof.

(19) Material Handling System

The logistics organization of the back-up material system mainly includes the transportation and storage of surrounding rock support materials, TBM energy consumption materials (by transport vehicles in the tunnel), and the transportation and transfer of excavated muck materials (by tunnel belt conveyor).

Distribution of cutters: After the transport vehicles arrive at the connection bridge, they are transported to the rear support of the main machine through the folding arm

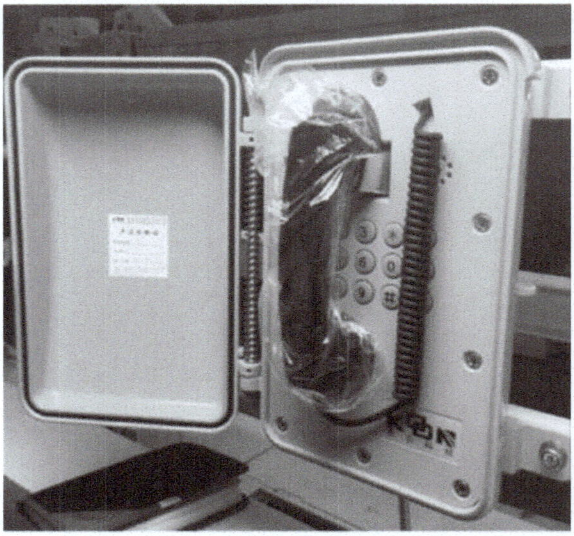

Fig. 2.53 Communication telephone

crane of the connection bridge, then to the vicinity of the manhole under the main beam of the first section through the monorail crane directly below the main machine, and then to the inside of the cutter head through the crane inside the cutter head.

Transportation of ring beams, meshes and support rock bolts: The materials are transported to the connection bridge by the folding arm crane on the connection bridge, then to the temporary storage device by the crane, and to the front end of the main machine by the ring beam transfer trolley, and then to the installation position of the ring beams one by one with a vertical lifting platform. The meshes and rock bolts are relatively lightweight to be transported manually.

Accelerator tank and concrete tank: the accelerator tank is directly transported by the crane inside the gantry. The concrete tank vehicle directly transports the tank to the designated location by a moving device.

Transportation of installation materials and tracks of tunnel belt conveyor: materials are directly transported to the material storage area of tunnel belt conveyor by the material platform truck through the locomotive to facilitate the installation by workers. The rail is also transported to the connection bridge by platform truck, and then transported to the rail installation position by crane (Fig. 2.54).

(20) HV Cable System

HV cable reels and HV cables winding in cable reels are equipped with. The back-up system is fed with main power through a cable pulley. The high voltage cable reel adopts automatic cable arrangement. It can automatically lay cable as the back-up system advances. The high voltage cable uses imported Prysmian special high flexibility cable. The type of high voltage cable is selected depending on the number of layers wound on the reel. The electric cable reel(Fig. 2.55) with a capacity of 420 m, together with 420 m long flexible cable, is an option.

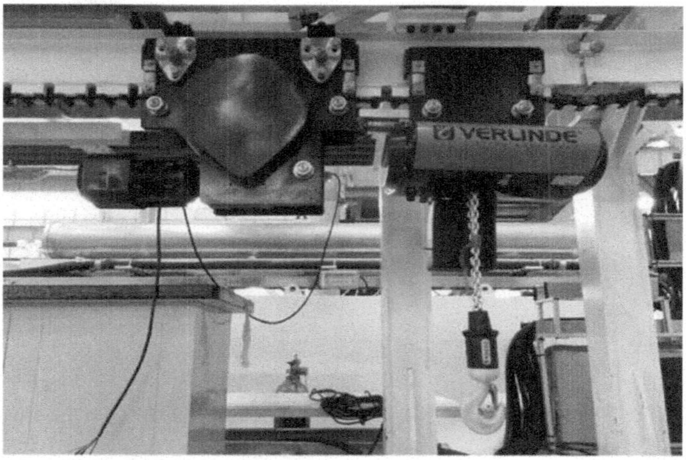

Fig. 2.54 Material hoist

Fig. 2.55 High Voltage cable reel

Generally, the back-up system has a high voltage cabinet, through which the high voltage flows to each transformer.

(21) Emergency Generator System

The emergency generator on TBM back-up system, with low noise and pollution, can support ventilation, drainage, illumination and high-voltage cable reel, as well as ensure the safety of TBM. The pollution-free diesel generator with noise enclosure, installed at the rear of back-up system, has manual and automatic functions. The emergency generator has the function of quick start when the power supply system fails (Fig. 2.56).

Fig. 2.56 Emergency generator

(22) Workshop, Storehouse, Lunch Room, Medical Room, Toilet

The workshop and storehouse of the back-up system share the same space with material storage cabinet and necessary repair tools. Lunch room is equipped with air conditioner to make the environment more comfortable for the workers. Medical room is provided with stretcher and first-aid kit for emergency relief. Toilet is environmentally friendly, which can directly move outside human feces from the tunnel. Rooms are provided with illumination system and emergency lighting system.

(23) Escape and Emergency Measures

TBM and its back-up system shall be so designed that escape and necessary emergency procedure are available. The walkway shall be kept spacious and clear as much as possible for staff escape in case of emergency. The following measures shall be added: Emergency Shut-down Button and Loudspeaker installed at key positions of equipment (Fig. 2.57), Emergency Lighting along the escape route, Pullswitch along the belt conveyor, and Emergency Stop Button for driver's cab and drive box steel structure.

(24) Fire Extinguishing System

All equipment shall be provided with portable fire extinguisher, including CO_2 extinguisher and dry powder fire extinguisher. Portable fire extinguisher is positioned along TBM main part and at key parts at the rear of back-up system.

Meanwhile, the water curtain extinguishing system is provided at the rear of gantry, and automatic fire alarm and extinguishing system are provided at such places vulnerable to fire as hydraulic pump units.

Fig. 2.57 Loudspeaker for emergency

2.2.3 Gripper TBM Operating Principle

Gripper TBM has the function of rock excavation, muck transfer and temporary support. When using gripper TBM for tunnelling, the rotating cutter head with cutters, under strong pushing force and torque, props up the tunnel face. The disc cutter revolves around the cutter head, and rotates around its center at the same time. The cutter rolls on tunnel face. When the pushing force is greater than rock compressive strength, rocks are crushed immediately and the cutter penetrates into rock. The tunnel face is squeezed and crushed by the cutter and a groove with a shape of concentric circle is formed. As groove depth increases, rock surface cracks become deep and wide. When beyond rock shear strength and tensile strength, rocks between adjacent grooves peel off and become muck (also known as lithic fragment). The scraper of cutter head scoops up rock ballasts, and the ballasts slide to the center of machine via muck chute, and fall to the conveyer of main machine via muck ring with a shape of funnel. At the end of the conveyor, the lithic fragments are conveyed to belt conveyer of back-up system or truck, and moved from the tunnel. For adverse geological conditions such as broken formation, it is necessary to provide temporary support for the rock behind cutter head. Generally, rock bolt, steel mesh and ring beam are used for temporary support. The geotechnical conditions before the machine can be explored by using additional driller, and if necessary, geological improvement shall be done. The drainage system at the bottom of TBM can be used for pumping water seepage from rock. In the area of back-up system, reinforced shotcrete is used for supporting the wall of tunnel already cut. Except reinforced shotconcrete equipment, all necessary supply facilities shall be installed in the area of back-up system, often including arch block. The gripper serves as a structure for supporting and regripping gripper TBM, and also directing TBM.

In the course of construction, the position of TBM should be continuously monitored, and according to the position of TBM in tunnelling, decide whether direction adjustment is necessary for TBM. Kelly TBM and Main Beam TBM differ in TBM direction control. For Kelly TBM, its rear support is installed on inner Kelly support and behind outer rear Kelly support. The rear support is controlled by a hydraulic cylinder, and can also be adjusted horizontally by it. When the rear support is retracted, the inner Kelly support can be adjusted horizontally and vertically to adjust the tunnel centerline where TBM is located. Main Beam TBM moves the main beam vertically and horizontally via a hydraulic cylinder in the gripper to accurately control tunnelling direction.

Gripper TBM tunnelling is realized by alternating tunnelling and regripping. In the course of tunnelling, TBM cutter head has a spiral movement composed of linear motion along the tunnel axis and rotational motion in one direction around the axis. The broken rocks fall to belt conveyor via cutter head bucket, and then are moved away.

Regripping uses supporting system. In the course of tunnelling, the gripper works to support the tunnel wall and thrust cylinder pushed cutter head to break rock. The broken rocks fall to muck transfer system via cutter head bucket, and then are moved away.

Gripper TBM is used when self-stabilization of the tunnel wall is realized and can withstand the huge horizontal support force. In the course of tunneling, stretch out horizontal support to prop up tunnel wall, retract front and rear support, start belt conveyor, and cutter head rotates to start tunnelling. after one circulation of tunnelling, switch to regripping.

The tunnelling procedure of gripper TBM is as follows (Fig. 2.58):

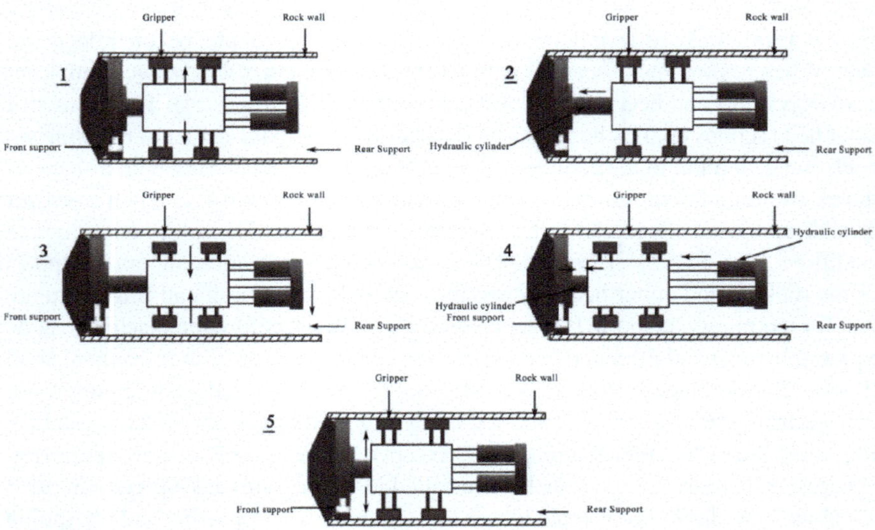

Fig. 2.58 Gripper TBM tunnelling principle

(1) Hydraulic cylinder of gripper extends, gripper props up rock face, retracting front and rear support, ready for tunnelling;
(2) Thrust cylinder extends, thrust cutter head stops after completing one circulation;
(3) Front and rear support extend to support equipment, and gripper retracts;
(4) Hydraulic Cylinder 1 becomes short and Hydraulic Cylinder 2 extends, external Kelly slides forward by a length of one stroke.
(5) The grippers of front and rear external Kelly prop up tunnel wall again, and front and rear support retracts to start a new tunnelling circulation.

2.2.4 Scope of Application of Gripper TBM

Gripper TBM, known as Hard Rock Support TBM or Gripper TBM in Europe, is mainly suitable for medium-hard complete rock strata with good self-stabilization.

Gripper TBM is equipped with auxiliary equipment such as ring beam erector and shotcrete bolt to adapt to geological changes. With effective support means, it can also be used for tunneling in soft rocks. With just a top shield, safe construction by gripper TBM can be started. In case of local unstable rock, auxiliary equipment attached to TBM should be used to strengthen rock by means of anchor rod, steel mesh, shotcrete ring, beam frame, etc., so as to maintain the stability of the wall. In the case of local soft rock and broken zone, TBM can pre-consolidate a circle of rocks around the upper part of the front by the probe drilling unit and grouting equipment attached to TBM. After the strength of surrounding rock reaches self-stability, safe excavation can be carried out. With gripper TBM, the lithology change of the tunnel wall can be directly observed during the excavation process, which is convenient for geological mapping. If using gripper TBM, the permanent secondary concrete linings are generally implemented concentratedly after the completion of the whole line.

Gripper TBM support mechanism props up the tunnel wall to withstand the counter-acting force and reaction torque that pushes ahead. The cutter head rotates, thrust cylinder pushes cutter head, and disc cutter penetrates into rock and makes a motion with a shape of concentric circles to break rock. Rock muck falls to the bottom of the hole by itself. The bucket scoops up muck, and muck falls to belt conveyer by itself via muck chute. All these repeat to complete the tunnel.

Gripper TBM is mainly used in tunnels with good rock integrity and certain self-stable surrounding rock, especially in hard and medium-hard rock tunneling, where the powerful support system provides sufficient thrust for the cutter head. With gripper TBM construction, the excavated rock face can be directly observed, making it easy to describe the excavated tunnel in terms of geology. As excavation and support are carried out separately, there is enough space near the gripper TBM cutter head to install some temporary or early-stage support equipment such as ring beam installers, bolt drills, probe drilling machine, shotcrete equipment, etc. In case of local unstable rock, temporary support can be carried out behind TBM cutter head, such as anchor rod, shotcrete, steel mesh, ring beam, to maintain the stability of the tunnel wall, or drilling hole by means of probe drilling hole and grouting to consolidate the surrounding rock ahead, and then start tunneling. Therefore, if gripper TBM adopts timely and effective support measures, it can be suitable for the tunneling of weak surrounding rock and uncertain geological tunnel.

In practical application, TBM often encounters complex geological conditions such as faults, fracture zones, local soft rock or karst caves, etc., and it may be difficult to stabilize the surrounding rock only by using these temporary support. In the soft broken rock, the excavation has to be stopped because the compressive strength of the wall rock was lower than the minimum ground pressure of the TBM support plate, leading to the failure of support for TBM.

In general, compared with the shield TBM (single-shield TBM/double-shield TBM), gripper TBM has relatively low cost, small turning radius, independent of reinforced concrete segment lining, unlikely to get stuck in bad strata, flexible configuration of support equipment according to the formation, and many other advantages. It is a type of TBM with great applicability. Therefore, more and more mountain tunnels adopts gripper TBM.

2.3 Double-Shield TBM

Double-shield TBM (Fig. 2.59) is also known as telescopic shield TBM. It has two shields, can prevent excavation face from collapsing and carry out curve excavation. It can realize telescopic sleeve and parallel operation function. The double-shield TBM has the function of hard rock boring machine. If adding the function of a soft rock shield, it can be used for both hard and soft rock, and its geological adaptability is very wide, especially it can safely penetrate the fault fracture zone.

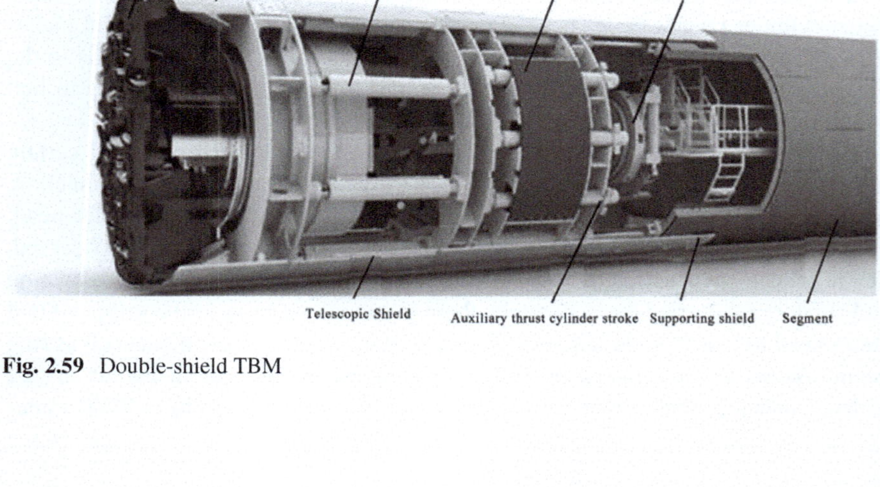

Fig. 2.59 Double-shield TBM

2.3.1 Structural Features

The double-shield TBM differs from the gripper TBM in that the former has a fully circular shield, and unlike the single-shield TBM, it can achieve both drilling and segment installation at the same time in good geological conditions, and achieve tunnelling in an open state in any cycle mode. Telescopic shield, a unique structural feature of the double-shield TBM, is the key to the transition between hard and soft rock operations.

The double-shield TBM is designed with the concept that segment assembly and tunnelling can be achieved at the same time, with tunnelling continued. According to the design requirements of rapid construction, the segment erector of TBM has the operation functions of both segment storage and transportation as well as segment assembling.

The double-shield TBM has two tunnelling modes, namely double-shield tunnelling and single-shield tunnelling. The double-shield tunnelling mode is suitable for the stratum with good stability and the stratum with small-scale spalling of surrounding rock but relative stability. The rock inside the tunnel has self-stability and can withstand the huge support force of horizontal support. In the course of tunnelling, the horizontal support is extended to prop up the tunnel wall, and the support gripper provides the driving force. The single-shield tunneling is suitable for unstable and unfavorable geological areas, and the segment provides the driving force.

The complete structure of double-shield TBM is shown in Fig. 2.60.

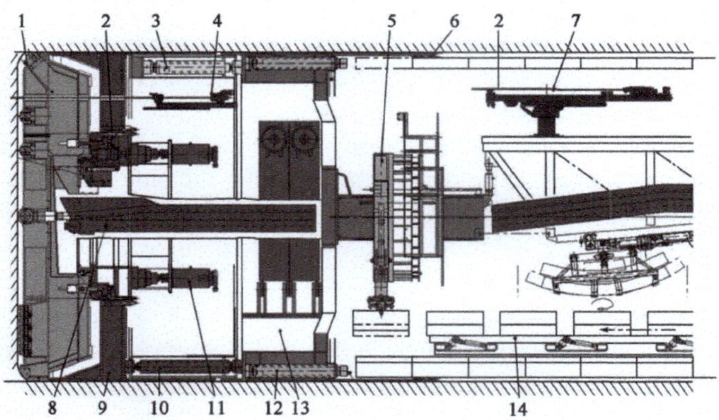

Fig. 2.60 Schematic diagram of double-shield TBM structure. *1*—Cutter head, *2*—Main bearing, *3*—Primary thrust cylinder, *4*—Multi-purpose drilling machine, *5*—Segment erector, *6*—Shield tail sealing, *7*—Probe drilling machine, *8*—Belt conveyer, *9*—Front shield, *10*—Telescopic cylinder, *11*—Cutter head drive, *12*—Auxiliary thrust cylinder, *13*—Support shield, 14—Segment feeder

Scan QR code on the right for the video of double-shield TBM manufactured by Herrenknecht AG—GERMANY.

2.3.2 *Basic Configuration*

The double-shield TBM is composed of TBM main machine, connection bridge and back-up gantry. The main machine is mainly composed of front shield with cutter, rear shield with support, telescopic device connecting front shield and rear shield, and tail shield for segment erection.

1. Cutter Head

The cutter head cuts rock only when it rotates clockwise, and its anticlockwise rotation is used to free itself from crushing zones or unstable rock formations when the cutter head is stuck. Cutter Head Design.

The hydraulic cutter wear automatic detection system can monitor cutter wear volume to ensure that the cutter head will not be severely worn or damaged due to excessive wear or damage. Cutter Head has one 600 mm manhole, through which operators can enter the tunnel face to remove obstacles ahead of cutter head. See Fig. 2.61 for cutter head structure.

Back-loading disc cutter is provided on the cutter head, and cutter can be replaced on the back of cutter head. The disc cutter seat is concave and is a part of the cutter head. Only some part of the disc cutter ring protrudes from the cutter head. This kind of cutter can prevent large rocks from blocking the cutter head in the fault fracture zone. The flat cutter head stabilizes the working face, and the shallow muck bucket and scraper narrow the distance between the cutting edge of the shield and the working surface of the tunnel.

The opening of the bucket is equipped with bucket teeth to excavate formations that may be encountered in various fault zones. A double-layer strong rubber sheet is mounted behind the rotating cutter head, forming a closed cutting chamber with the back side of the non-rotating muck funnel. The closed face of the cutter head has a long radial muck bucket that allow most of the muck to enter the cutter head before falling on the invert. The muck bucket and scraper can only excavate in one direction, which can improve the muck digging efficiency of the cutter head and reduce the wear of the bucket and scraper. The opening of the surrounding bucket is small to prevent large rocks from blocking the cutter head.

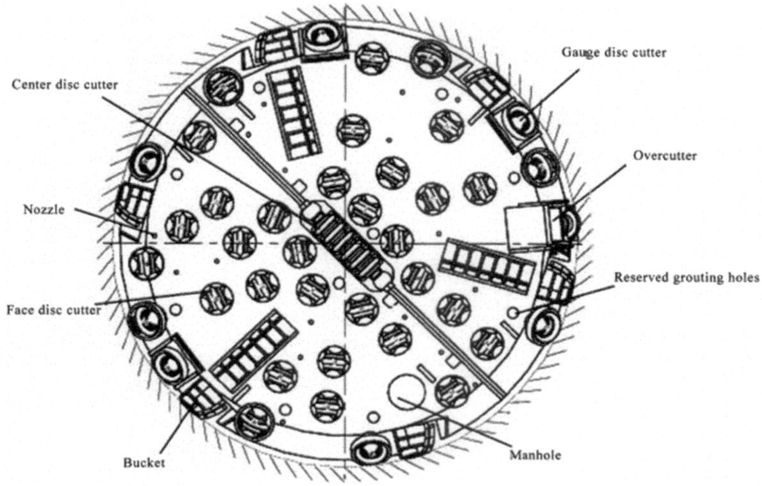

Fig. 2.61 Cutter head

During operation, the reaction force of the tunnel face to the cutter head is balanced with the thrust of the hydraulic thrust cylinder after the reaction force is transmitted to the main bearing through the periphery of the cutter head and the muck chute.

The cutter head is designed with overcutter and scraper, which can dig 200 mm outside the shield. The overcutter is extended by hydraulic control and locked mechanically, as shown in Fig. 2.62. The hydraulic oil of the overcutter is supplied by the rotary joint in the center of the cutter head. A set of fixing device is provided to make replacement of disc cutter easy. When encountering hard rock, the overcutter is used to cut a space to replace the positioning disc cutter. When replacing the rest of face disc cutter and the center disc cutter, the front shield, cutter head and driving device can be retracted.

The muck chute of the cutter head will send the much cut down to the belt conveyor placed in the center part of TBM via the muck funnel. The muck funnel is placed on much chute.

If necessary, a hydraulically-operated sliding gate is designed above the muck funnel to close the muck funnel moving towards the belt conveyor in case of water inrush. This can prevent water from flowing into the tunnel.

The cutter head is provided with a sprinkler system to suppress dust. The water is supplied to a number of nozzles mounted on the cutter head via the rotary joint in the center of the cutter head.

Double-shield TBM generally uses an eccentric cutter head design for fault or compressed formations, and a cutter head lifting device is used for expansive rock. In the design, considering the adaptability to fault or fault zone, the eccentric cutter head design is generally adopted. Generally, in the design of double-shield TBM, a 20 mm offset on the axis of the shield shell for the cutter head can be allowed, so as to achieve overcut at the upper part of the tunnel, but the bottom arch should not

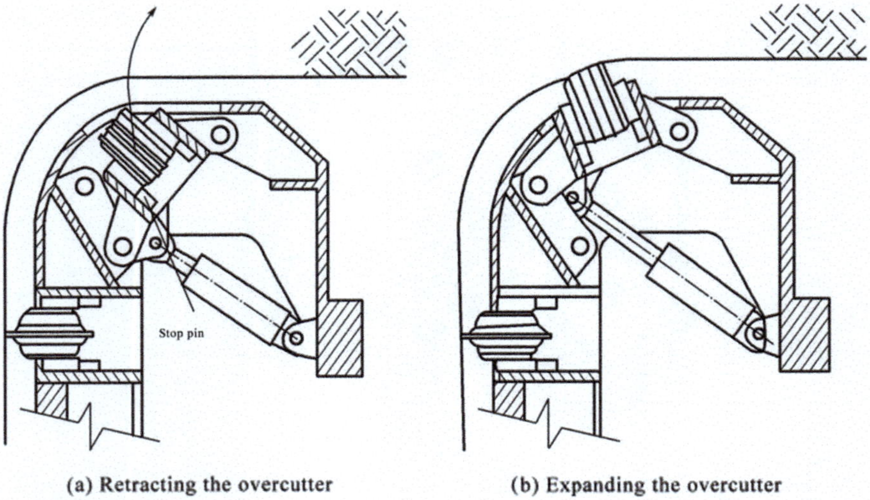

Fig. 2.62 Schematic diagram of overcutter

have excess overcut to prevent the equipment from bowing. Considering the need to further enlarge the excavation diameter if expansive rock is present, a cutter head lifting device can be provided to achieve this purpose, as shown in Fig. 2.63.

2. Cutter Head drive

Double-shield TBM generally adopts variable frequency drive unit or hydraulic drive unit.

The cutter head drive system has the following characteristics:

(1) Low speed and high torque,
(2) Breakout torque higher than the maximum torque,
(3) Enough high speed and torque to achieve maximum driving speed,
(4) The cutter head rotation speed is fully variable, and the design of the cutter head drive system has the maximum safety margin.

Cutter Head rotation speed can be changed correspondingly according to different geological conditions and drive mode supported by PLC control system. Its basic principle is to limit the maximum thrust and control the driving speed.

(1) If the thrust reaches the maximum, the driving speed will automatically decrease.
(2) If the thrust falls by a certain value, PLC will send a signal to increase the driving speed.
(3) When the operator has verified that the geological conditions of the tunnel face are even and allow for increasing speed, the operator can press the button to increase the driving speed.
(4) If the torque and driving speed change greatly, PLC will automatically change the cutter rotation speed.

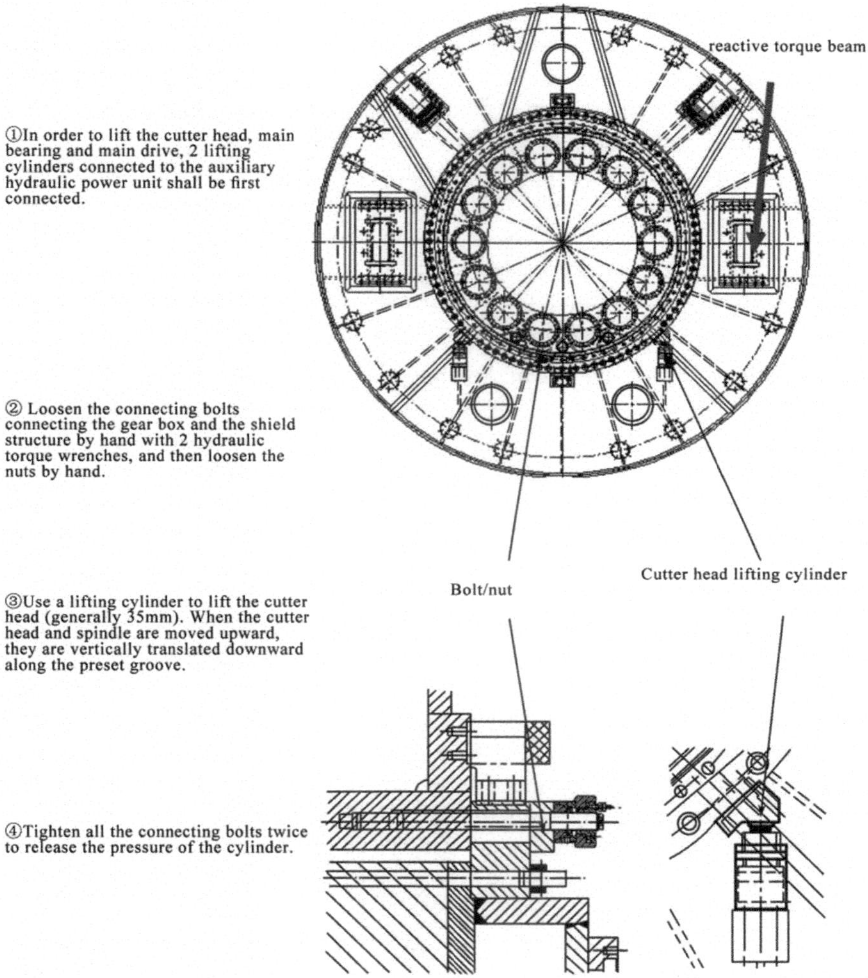

Fig. 2.63 Cutter head lifting system

(5) If geological conditions become good again, the operator can switch to full-speed operation mode again.

The cutter head drive system has the following characteristics:

(1) The advantage of hydraulic transmission is that the speed of cutter head can be adjusted to the most economical state determined by geological conditions, and the best penetration speed can be obtained. In fault zone, cutter head can operate at low speed and high torque.
(2) The pinion of the driving ring gear is supported by two bearings. A pinion, a water-cooled (or oil-cooled) planetary gearbox, an overload shear axis and a variable hydraulic pump constitutes a driving assembly.

(3) The hydraulic circuit is a closed circuit, equipped with a cleaning oil drain valve and an oil supply pump to refill the oil circuit with clean cooling oil and maintain the required oil pressure. This closed circuit has a separate tank and no hydraulic cylinder oil line is connected to this tank. The drained oil is filtered by oil filter and returned to the tank. The oil feeder itself also has a filter.
(4) Variable displacement pump and variable motor enable the cutter head drive to operate in two modes. Under normal working conditions, the hydraulic motor runs at a certain displacement. When high torque is required, the hydraulic motor runs at its maximum displacement.
(5) Hydraulically-driven cutter head can achieve anti-clockwise rotation and has a high release torque when the stratum is in unfavorable condition.

Like gripper TBM, the main bearing is a double-row axial-radial three-dimensional roller bearing with axial preloading, and the torque is transmitted from inner gear ring to the cutter head.

Large overturning moment must be considered when driving on a curve or under unstable surrounding rock conditions in fault zone. The main bearing is designed according to the maximum propulsion force, especially the maximum overturning moment.

Large diameter bearings are not only suitable for large torques, but also provide large center space and wide access to the cutter head.

The main bearing seal is used to protect the main bearing and driving device from silt, muck and water. The sealing system includes an inner seal and an outer seal, each composed of three layers of sealing rings. It is a lip seal. Grease is continuously pressed into the groove formed by sealing facing outward. The grease injection system is interlocked with the main drive. When the lubrication system fails, the cutter head automatically stops rotating.

Between the seal and the outside of "dirty", a labyrinth seal is used for additional protection. The surface of the seal seat in contact with the seal is hardened and ground to make the seal durable.

Main bearings and drives are lubricated with a forced oil circulation system. This system interlocks with the main drive and starts before the main drive. The drive system stops when it has any malfunction. The lubricating oil of the main bearing is separated from the lubricating oil of the gear transmission to prevent the abrasive particles generated by the gear transmission from entering the main bearing. All the oil flows into the bottom of the oil tank of the driving part, and then is filtered by a filter and pumped back. A number of sensors are installed to monitor the pressure of the oil supply line, the flow of oil, the degree of contamination of the filter and the oil temperature.

3. Shield

The shield consists of four main parts, namely the front shield, the rear shield (supporting shield), the telescopic part connecting the front and rear shield, and shield tail.

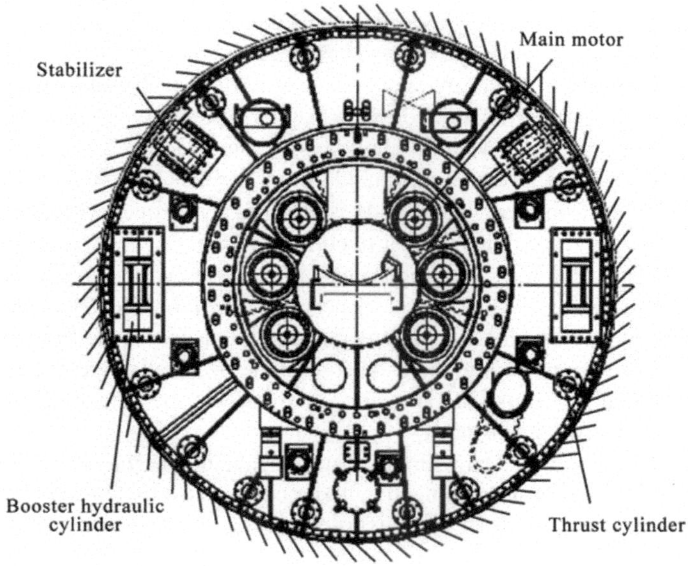

Fig. 2.64 Structural diagram of front shield

(1) Front shield

The front shield contains the cutter head and the cutter head drive device, and supports them (Fig. 2.64). The front shield is connected with the rear shield by the main hydraulic thrust cylinder.

(i.e., the telescopic hydraulic cylinder). The main hydraulic thrust cylinder is divided into four groups (Up, Down, Left and Right) to control the direction of the front shield. The position of the front shield relative to the rear shield is measured by 4 linear sensors and displayed in the operating room.

The cutting chamber is separated from the shield by the back chamber plate (sealing partition) of the cutter head. A weep hole in the back chamber leads to the bottom casing of a pump. When water enters, the gate on the conveyor belt is closed and the water can be discharged from the cutting chamber by this pump.

There are two hydraulically-operated stabilizers at the quarter of the front shield top to stabilize the front shield during tunneling into hard rock and to be helpful when the rear shield is pulled forward.

In the process of regripping, the thrust cylinder of tail shield pushes and presses the segment, while the telescopic cylinder pulls to support the shield. Because the thrust cylinder pushes the lined segment, the support shield will always be pushed forward, without pulling the front shield backward.

(2) Telescopic Shield

The telescopic section connects the front shield and the rear shield, and is designed to allow TBM drilling and segment installation simultaneously.

The main hydraulic thrust cylinder connects the front shield and the rear shield, transmitting both thrust and pull. This performance can be used to prevent the shield from tilting downwards in unstable geological conditions with heavy overburden.

Cutter Head torque is transmitted to the gripper through two heavy-duty torque beams, which effectively prevent the shield body from twisting. Two torque devices transmit cutter head torque to the rear shield from the front shield. The rolling adjustment between the front and rear shields is realized by torque hydraulic cylinder, without the need of telescopic cylinder to correct rolling.

Inspection and cleaning is available in the gap between two shells of the telescopic section. There are several windows for inspection. When the telescopic section is in retracted position and the inner shell is in contact with a seal at the front end, water or bentonite can be pumped into the gap between the two shells to remove the muck. A scraper is installed within the 120° range on the top of the outer shell to keep the space between the two shells clean.

When it is necessary to deal with the obstacles outside the shield shell or to enter the front of the cutter head, the hinged cylinder can be used to detach the telescopic inner shield and support shield, and expose the working face in contact with the surrounding rock.

(3) Rear shield

The rear shield is also called the support shield, and it is provided with an auxiliary hydraulic thrust cylinder and a support device. The rear shield bears all the thrust force of the front shield and can also pull back the front shield. Rear shield is large in size and its pressure on surrounding rock is small, which is especially important when tunneling soft surrounding rock.

The auxiliary hydraulic thrust cylinder is also divided into 4 groups to facilitate TBM steering. When tunnelling in soft surrounding rock, there is no need for support and telescopic shield. The auxiliary hydraulic thrust cylinder is connected to a common thrust boot in the corresponding 4 groups. Thrust boot is covered with polyurethane to protect the tunnel lining segment.

The total thrust of rear shield is considerable and is used to apply the required force to the cutter head and to overcome the frictional resistance of all shields.

The auxiliary hydraulic thrust cylinder has a common hydraulic power station. Each of the four hydraulic cylinders is controlled by the oil supply and monitored by TBM operator. In the course of normal tunneling, the front shield and cutter head can be pushed with a reaction force provided by the gripper, and the main hydraulic thrust cylinder can be operated by a common oil flow. Each hydraulic cylinder is fitted with measuring devices or linear sensors, enabling the operator to monitor its position. This kind of normal tunnelling is adopted provided that the surrounding rock is in good condition and can provide enough thrust force and cutting torque for the gripper. At this time, the tunneling and segment installation are carried out simultaneously. While the main hydraulic thrust cylinder completes one stroke (one step), a ring of segments is installed behind the rear shield. After that, retract the gripper, use the main and auxiliary hydraulic thrust cylinders for pulling and pushing to move the rear shield forward to realize regripping, and then support it well. Then, proceed the tunnelling and installation of segments.

When the double-shield TBM operates like a simple shield, that is to say, the telescopic section (telescopic shield) is kept in the retracted position without support, the moment of the cutter head is provided with a countermoment by the friction force between the shield and the tunnel wall, and the thrust of the cutter head is realized by the auxiliary hydraulic thrust cylinder's bearing on the segment. In case of weak surrounding rock, the tunneling and segment installation cannot be carried out at the same time.

Besides provided by the friction force of the shield housing, the countermoment of the cutter head during operation is provided by compensating the countermoment of the cutter head during operation through the oblique positioning of the auxiliary hydraulic thrust cylinder of the shield. That is to say, the two hydraulic cylinders on each thrust boot keep their piston rod ends on an adjustable fixture. This fixture can be adjusted hydraulically to tilt the auxiliary hydraulic thrust cylinder, thus generating a component force in the circumferential direction to bear the torque of the cutter head.

(4) Tail shield

Tail shield is mounted on the rear shield. It is equipped with a steel wire brush shield tail seal covered by a spring steel sheet, placed on the upper 270° round surface, and turned outward from the inside to prevent crushed stones (as concrete aggregate) from entering the tail.

4. Segment Erector

The segment erector is mounted on the erector bridge and can move longitudinally back and forth on a hardened slide. The erection bridge is also used to connect the back-up system to TBM. The segment erector is a single-unit rotary equipment, and its movement can be accurately controlled to ensure the accuracy of segment installation position. The control of segment erector includes wired control and wireless control. During construction, the wireless remote controller is mainly used to install the segment, and the wired controller is used temporarily when the wireless remote controller fails.

The segment erector can rotate 220° in both directions, and its supporting and driving devices consist of a single-seat ball bearing, an internal gear ring, two small gear rings, a planetary gear reduction box and a hydraulic motor. The drive is an infinitely variable speed system, and produces sufficient torque to erect the segments. The erector has 6 degrees of freedom, and the segment erector has self-locking ability in case of emergency to ensure safety in construction.

5. Belt Conveying System

The belt conveying system consists of TBM conveyor belt, second conveyor belt, rear supporting conveyor belt and muck discharge conveyor belt. Water nozzles are installed at the place where transfer on the conveyor belt is present to reduce dust. TBM conveyor belt is close to its centerline in the cutter head, installed in the front shield and moved with the front shield, and can be retracted with a hydraulic device for maintenance. TBM conveyor belt is hydraulically driven, with infinitely variable speed, and can operate in both directions. The bearing roller and idle pully are soft rollers.

6. Pea Gravel Filling System

The pea gravel backfilling system is arranged on the back-up machine, and the pea gravel is blown into the opening preset on the segment by compressed air for pea gravel filling. Pea gravel pump has the function of remote control operation, which can be operated remotely.

7. Slurry Mixing and Pouring System

On-site slurry mixing and pouring system is set on the back-up machine, and the cement bag treatment device is used to pour the cement into the mixer. The mortar pump pumps the mixed cement slurry into the gap behind the segment through the injection port.

8. Probe Drilling Machine

Probe drilling and advanced geological reinforcement are necessary auxiliary construction means for TBM. TBM is equipped with probe drilling machine for probe drilling for geological detection and treatment of unfavorable geology. When drilling holes, the drilling machine is fixed on the support behind the segment erector. When drilling grouting holes, the drilling machine can be installed on the segment erector to realize 360° drilling operation along the reserved holes around the shield circumference. The probe drilling machine is operated by an independent hydraulic unit.

9. Multi-function drilling machine

TBM is equipped with a multi-functional drilling machine, which can realize the function of core drilling and drilling for ground reinforcement by passing through the front shield shell and cutter head and having a conduit with a sufficient diameter. A circular support is arranged around the cutter head drive in the front shield to provide the drilling operation position passing through cutter head for the multi-functional drilling rig. The driving system is used to make the propulsion beam of the drilling machine move along the circular support and fix it at the selected position. When the drilling machine works, TBM must be shut down.

10. Dedusting System

TBM dedusting system is a dry dedusting system installed on the connecting bridge. The air suction duct is located at the front shield to pump the air in the feed inlet area of the main belt conveyor to form a negative pressure, so that some fresh air flows to the front end of the TBM and the dust-containing air is prevented from escaping into the tunnel.

A booster fan is set at the outlet of the deduster to make the air flow to the supporting area behind TBM. Clean the filter bag of the deduster regularly with compressed air. After the dust is wetted by the nozzle of the slag extractor, it is collected by the slag extractor and then discharged to the back-up belt conveyor.

11. Data Acquisition and Processing System

The data acquisition and processing system can collect, process, store, display and evaluate data related to TBM. All measurement data are continuously collected and displayed by measurement sensors controlled by clock pulses. All measurements that must be recorded are displayed graphically on PDV's monitor. The contents on the screen are displayed in pages according to functions, mainly including the following contents:

(1) Relevant parameters of tunneling: display and record of parameters such as thrust force of main thrust cylinder and auxiliary thrust cylinder, cutter head torque and over-excavation volume of over-excavation device,
(2) Display and record of injection amount and injection speed of grease, pea gravel backfilling and grouting,
(3) Temperature display and recording of key parts, hydraulic system and main bearing lubrication system,
(4) Error Display and Recording.

The operator can switch between these screen pages and obtain the required data from them. Real-time information management of TBM status can be realized through the information collected by the data acquisition and processing system. The current TBM tunneling status data can be transmitted to the relevant departments of the Employer, the Supervisor, the Designer and the Construction Contractor through the Internet, the telephone dial-up network and the computer of the data acquisition system, providing an important information source for the information management of the whole project.

12. Connection Bridge

The connection bridge (Fig. 2.65) is located behind TBM main machine and connects the main machine with the back-up gantry. Sufficient space is reserved under the connection bridge for segment hoist to transport and store segments, and lay rails. The connection bridge is equipped with belt conveyor, fresh air ventilation pipe and rail for segment hoist operation, and a power station of main machine hydraulic system is placed on the upper working platform. The length of the connection bridge meets the requirement of extending the track once.

13. Back-up Gantry

The back-up gantry adopts an open-shelf structure and runs on the special track laid. The back-up gantry is equipped with belt conveyor, dust collector, ventilation pipe, centralized grease lubrication system, pea gravel backfilling system, cement slurry mixing and injection system, electrical control cabinet, hydraulic power unit, transformer, air compressor, water system, cable drum, water hose reel and other TBM auxiliary equipment. When TBM is pushed forward, the connection bridge and these back-up gantry are towed by the rear shield to move forward on the rail.

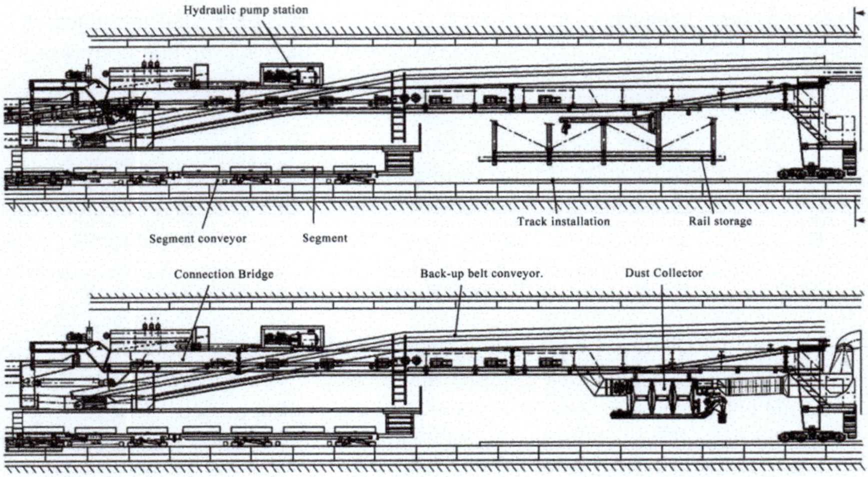

Fig. 2.65 Schematic diagram of connection bridge

14. Guidance System

The guidance system mainly consists of laser generator, laser target, inclinometer, terminal, computer, output display, printer, optical theodolite, reflective target, reflective prism, etc. The guidance system can accurately measure and display various postures of TBM during tunneling as well as the direction and position relationship of TBM tunneling. According to the information provided by the guidance system, the operator can quickly adjust the tunneling direction and posture of the TBM in a real-time manner to reduce the tunneling deviation.

The guidance system and the tunneling software continuously provide three-dimensional coordinates as well as directional and continuous dynamic information of TBM. The communication device attached to the guidance system can receive data, and the tunneling software calculates the orientation and coordinates of TBM, and accurately displays the position of the TBM in the form of charts and digital tables.

15. Grouting System

The grouting system consists of a container for mixing, a tank for mixing blades, a mortar pump, a hydraulic power unit, a hose and a shut-off valve, and a nozzle. The grouting hole shall be drilled with a rock drill installed on the segment erector, and then grouting into the hole is achieved by a grouting equipment.

16. Secondary Ventilation System

TBM air duct extends from No. 1 machine to the whole back-up system to exhaust the hot air in the gripper area and hydraulic pump area. Fresh air outside the tunnel is conveyed to the back-up system area behind TBM through the axial flow fan, and clean air can be continuously supplied to the whole TBM tunneling system through the secondary ventilator and air duct.

2.3.3 *Operating Principle*

The double-shield TBM is designed with the concept that segment assembly and tunnelling can be achieved at the same time, with tunnelling continued. According to the design requirements of rapid construction, the segment erector of TBM has the operation functions of both segment storage and transportation as well as segment assembling. Double-shield TBM can achieve tunnelling and segment erection simultaneously when the geology is good, and carry out tunneling in an open state in any cycle mode.

The double-shield TBM has two tunnelling modes, namely double-shield tunnelling and single-shield tunnelling. When TBM advances in a hard rock tunnel with relatively stable surrounding rock, the gripper prop up the tunnel wall to provide the reaction force for the primary thrust cylinder, so that TBM can advance forward. The reactive torque of the cutter head is provided by two reactive torque cylinders located in the supporting shield, and the tunneling and segment installation are carried out simultaneously. At this time, TBM operation cycle is as follows: tunnelling and segment erection → regripping → re-supporting → tunnelling and segment erection. The single-shield tunneling is suitable for unstable and unfavorable geological areas, and the segment provides the driving force.

1. Double-shield Tunnelling Mode

When TBM advances in a hard rock tunnel with relatively stable surrounding rock, the gripper props up the tunnel wall to provide the reaction force for the primary thrust cylinder, so that TBM can advance forward. The reactive torque of the cutter head is provided by two reactive torque cylinders located in the supporting shield, and the tunneling and segment installation are carried out simultaneously. At this time, TBM operation cycle is as follows: tunnelling and segment erection → regripping → re-supporting → tunnelling and segment erection. See Fig. 2.66 for specific steps.

2. Single-shield Tunnelling Mode

When TBM advances in a soft rock tunnel, the rock of the tunnel wall can't provide enough support force for the horizontal support, the support system and the primary thrust system are no longer used, and the telescopic shield is in the retracted position. The reaction force of cutter head tunnelling is provided by the friction between shield shell and surrounding rock, and the thrust force of cutter head is provided by the auxiliary thrust cylinder supported against the segment, so TBM tunneling and segment installation can't be carried out simultaneously. At this time, TBM operation cycle is: tunnelling → recovery of auxiliary oil cylinder → installation of segment → re-tunneling. See Fig. 2.67 for specific steps.

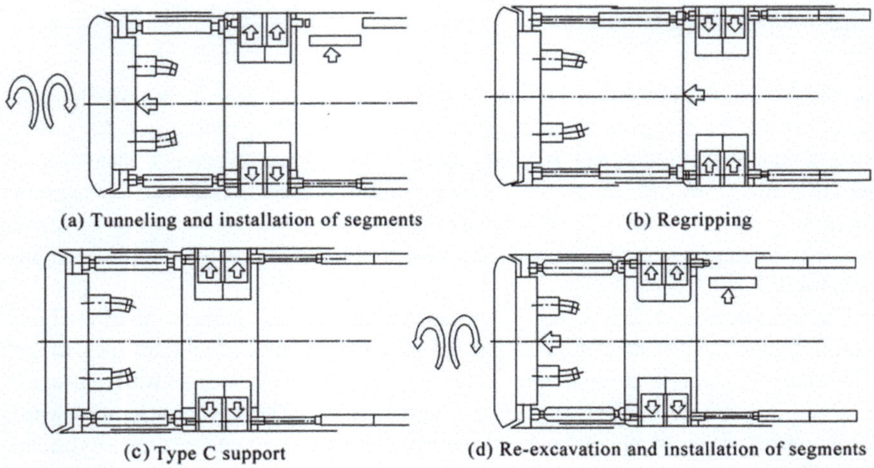

Fig. 2.66 Double-shield tunnelling mode (hard rock mode)

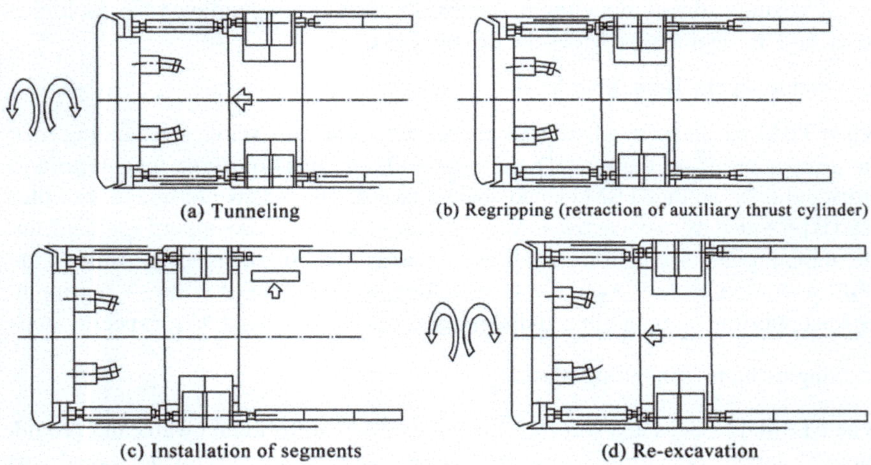

Fig. 2.67 Single-shield tunneling mode (soft rock mode)

2.3.4 Application

Double-shield TBM was developed on the basis of gripper TBM, single-shield TBM and shield structure in 1970s. Double-shield TBM is equipped with two sections of shield shells, which has the function of preventing the collapse of excavation face and is commonly used for tunneling in composite rock strata. Double-shield TBM has two tunneling modes, namely double-shield tunneling mode and single-shield tunneling mode, which are respectively applicable to strata with good stability of surrounding rock as well as strata with good stability of small-scale spalling and unfavorable geological sections. When the rock is hard and soft, and there are faults and fracture zones, the double-shield TBM can give full play to its advantages.

In case of soft rock, the soft rock cannot bear the compressive stress of the gripper, and TBM cannot achieve the support through supporting shield. The machine will work like a simple shield, and the auxiliary hydraulic thrust cylinder located at the shield tail will be supported on the assembled prefabricated lining segment to propel the cutter head to break rock.

In case of hard rock, if the rock conditions allow proper support for the tunnel wall, the gripper props up the tunnel wall, and the main hydraulic thrust cylinder pushes the cutter head to proceed the rock breaking. In the case of double-shield excavation mode, the front shield and cutter head are locked to the rear shield on the rock surface through the gripper, so the reaction force of thrust force and torque is not transmitted to the lining segment. Behind the supporting shield, under the protection of the tail shield shell and with the help of the segment erector, install the prefabricated reinforced concrete lining segments. In good rock strata, the segment assembly and excavation of the double-shield TBM can be carried out simultaneously, thus realizing high-speed and continuous tunneling. Tunnel support can also be omitted under good rock conditions.

Double-shield TBM has wide adaptability to rock stratum. It can be constructed in very hard rock, and there are engineering examples of successful implementation and completion of tunneling in 450 MPa ultra-hard rock with this type of TBM in South Africa Gold Mine Project, and it can also be constructed in unstable strata such as soft rock and fracture zone strata like ordinary single-shield TBM.

Double-shielded TBM is often used for long tunnel excavation in complex rock strata. It is generally suitable for tunnels with medium–thick buried depth, medium–high strength and basically good stability, and can adapt to various unfavorable geological conditions accounting for a part of the tunnel mileage, and has good adaptability to rock strength changes.

Double-shield TBM has good excavability when the uniaxial compressive strength of rock is 30–120 M Pa. Double-shield TBM is suitable for rock tunnels dominated by Class III and IV surrounding rocks.

2.3.5 Construction Features

1. Advantages of Double-shield TBM Construction

(1) Safe, efficient, fast

Double-shield TBM is equipped with front and rear shields, a telescopic shield is designed between the front and rear shields, and the rear shield is equipped with gripper. When the geological conditions are good, the thrush reaction force is provided by supporting the tunnel wall with gripper, and the tunneling and segment installation are carried out simultaneously, leading to a faster progress. For example, the maximum monthly advance reaches 1637 m when double-shielded TBM is used for construction in the project of diverting water from the Yellow River to Shanxi. Double-shield TBM construction, with personnel and equipment working under the protection of shield, is also safer than gripper TBM construction. Double-shield TBM construction enables tunneling, lining, mucking and transportation operations to be completed in one go under the protection of shield, realizing safe, efficient and rapid construction.

(2) Strong adaptability to unfavorable geology

For water-rich sections, a comprehensive probe geological forecast method with infrared detection as the main method and probe geological drilling as the auxiliary method is adopted for water burst forecast. The waterproof technology of combining plugging and drainage can be implemented for water burst. TBM main machine area is equipped with submersible pumps to pump water into the sewage tank located on the back-up gantry behind TBM. At the same time, TBM is equipped with a probe drilling machine, which can be used for drilling and grouting equipment can be used for advanced stratum reinforcement and water plugging.

For the fault fracture zone, double-shield TBM can adopt the single-shield tunneling mode. At the same time, probe geological forecast can be carried out for the fault fracture zone, and infrared water detector and probe drilling machine configured by TBM can be used for water detection. Probe drilling machine and grouting equipment configured by TBM are used for advanced reinforcement of the stratum. At the same time, the design of reserved grouting holes on the cutter head panel can meet the needs of tunnel face reinforcement.

For deeply-buried tunnels, due to the complicated geological structure, strong stress concentration and stress-type damage of surrounding rocks will inevitably occur. During the tunnelling of double-shield TBM, the damage to the rock mass can be reduced to a very low degree because of the smooth tunnel face, which protects the original state of the surrounding rock and stress concentration is unlikely to happen.

For rockburst sections, since TBM cutter head is equipped with a water spraying device, in the sections with high predicted ground stress and rockburst prone to occur, the probe drilling machine equipped with TBM is used to drill holes, and water is injected into the holes to wet the rocks. Water spraying can soften the rock on the tunnel face and release the stress in advance. At the same time, through segment installation, pea gravel backfilling and cement slurry grouting, TBM can quickly provide support and pass through the rockburst section.

For the karst section, stop the machine first, and then observe the karst condition through the manhole on the drive box. First, backfill the bottom with pea gravel or concrete and make it compact. When filling to the excavation diameter elevation, install the segment while moving forward, and dig manholes on both sides of the segment for filling grouting or concrete on both sides and the arch, so that the karst part is filled with concrete and compacted, and integrated with the installed segment. In order to prevent sinking of TBM drive box due to karst, double-shield TBM is equipped with probe drilling machine. For the treatment of some small karst caves, pea gravel can be backfilled between the segment and the surrounding rock after TBM passes, and then grouting for consolidation. For large-scale karst caves, as it is not easy for the segment joints to close, steel plates shall be used to connect the installed segments longitudinally.

For the plastic deformation section of swelling rock and soft rock, due to the eccentric arrangement of the cutter head of the double-shield TBM and the over-excavation cutter provided by the cutter head, the excavation diameter of TBM can be increased, and the deformation amount is reserved for TBM to quickly pass through the deformation section of the surrounding rock under the condition of small deformation amount of the surrounding rock. When the deformation of surrounding rock is large, probe drilling machine and grouting equipment configured by TBM can be used to reinforce the stratum. At the same time, the high-strength structural design and sufficient thrust reserve and torque reserve of the double-shield TBM can ensure that the TBM is not easily stuck by the deformed surrounding rock.

As for the collapse section, the closed cutter head design adopted for the double-shield TBM can effectively support the tunnel face and prevent the surrounding rock from collapse in a large area. The pressure of TBM gripper can be adjusted according to geological conditions to avoid destruction to tunnel wall rocks due to excessive supporting force. At the same time, the high-strength structural design and sufficient thrust force and torque reserve of the double-shield TBM can ensure that the TBM is not easily to get stuck by the collapsed surrounding rock.

For gas stratum, the double-shield TBM equipped with a geological forecasting instrument and an probe drilling machine can be used to detect the concentration of possible gas accumulation coal seams by probe drilling as required and to release and dilute the accumulated gas by drilling holes. The TBM is equipped with a gas monitoring system, and the data collected by the monitor is connected to the TBM data acquisition system and input into the PLC control system. When the gas concentration reaches the critical value of the first-level alarm, the gas alarm will give an alarm; when the gas concentration reaches the critical value of the second-level alarm, the TBM will automatically stop working and start the explosion-proof emergency equipment to dilute the gas through the ventilator.

(3) Industrialization Operations Realized

For the construction of double-shield TBM, the stratum are excavated by the cutter head, and the tunnel is formed by tunneling, mucking, segment erection and other operations under the protection of the shield. The auxiliary operations such as sprinkling irrigation, grouting, ventilation and power supply of pea gravel are also carried out in parallel, making full use of the space in the tunnel. Double-shield TBM construction is characterized by high mechanization degree and continuous construction procedures. The segment lining technology is adopted for the tunnel lining. The segments are prefabricated in factory and transported to the site for assembly construction. The prefabricated reinforced concrete segments are characterized by good quality and high accuracy. Compared with the traditional cast-in-place concrete tunnel lining method, the construction progress is fast and the construction period is short, and there is no need for formwork erection, binding, pouring, curing, formwork removal and other processes; wet operation is avoided for low noise on the construction site, and environmental contamination reduction. The prefabricated construction of tunnel lining not only realizes the industrialization of tunnel construction, but also even facilitates the replacement and maintenance after tunnel operation.

(4) High Automatization and Informatization Degree

The double-shield TBM adopts computer control, remote control, sensor, laser guidance, survey, probe geological drilling and communication technology. It is a complete set of tunnel construction equipment integrating mechanical, optical, electrical, pneumatic, hydraulic, and sensing and information technology, with the advantages of high automatization degree, small impact on surrounding stratum and environmental protection. There is less labor during construction, and the labor intensity and material consumption are reduced. The double-shield TBM has the functions of construction data acquisition, TBM attitude management, construction data management and real-time remote transmission of construction data, realizing informationized construction.

2. Disadvantages of Double Shield TBM Construction

(1) The price of double-shield TBM is higher. The cost of double-shield TBM with the same diameter is generally 20% higher than that of gripper TBM, and the one-time investment of double-shield TBM equipment is relatively large. At present, the factory price of TBM with a diameter of $\phi 6$ m or so is about USD 10 million, and that of TBM with a diameter of $\phi 9$ m or so is about USD 16 million.
(2) When unstable or poorly stable surrounding rocks are encountered during excavation, local loose collapse of surrounding rocks will occur and probe drilling is required to understand the stratum conditions ahead in advance and take preventive measures.
(3) During the construction of deeply buried soft rock tunnels, high ground stress may cause plastic deformity of soft rock, which is easy to jam the shield. Before construction, accurate geological exploration is required and ground stress is released first, resulting in higher construction cost.
(4) For deeply buried soft rock tunnels, the ground stress is relatively large. Since the surface of TBM tunneling is relatively smooth, the ground stress is not easy to be released, and rockburst is more likely to be induced compared with drill and blast.
(5) When it passes through the expansive rock, due to a series of unfavorable engineering characteristics such as expansion, contraction, disintegration and softening of the expansive rock, the possible load exerted by the expansion force of surrounding rock on the segment shall be fully considered in the structural design of the segment to ensure the safety of the lining structure. Attention shall be paid to the water-stop and seepage prevention of segments to prevent collapse or softening of expansive rock due to loss of water content, thus causing TBM sinking accidents.

(6) In the fault fracture zone, due to the large pressure of loose rock stratum on TBM shield, TBM jamming accidents are easy to occur; in the karst section, TBM head sinking accidents are easy to occur; corresponding countermeasures shall be taken during construction.
(7) Due to the large number of segment joints in the tunnel, its waterproofness and operation safety is still a weak link in the adverse geological tunnel section.
(8) Because the surrounding rock is isolated by the shield, it can only be known from the observation window on the side of the shield, and the construction geological description cannot be carried out systematically, and it is also difficult to measure the convergence deformity.
(9) The rock powder generated during the tunneling of the double-shield TBM is easy to deposit within about 120° of the tunnel bottom, and it is so compacted by the dead weight of the main machine that it is difficult to pour cement grouting into the rock powder layer, and it is easy to form a weak layer with lower strength than the pea gravel layer after grouting.
(10) Double-shield TBM is hard rock tunnel boring machine and is not suitable for construction in soft soil stratum. When it passes through the soft soil section, the soil mass is easy to stick to the cutter and cannot be discharged smoothly from the mucking funnel.

2.4 Single-Shield TBM

The segment lining of single-shield TBM is carried out under the protection of tail shield (Fig. 2.68). The excavation and reinforced concrete segment lining are carried out in stages, and the muck discharge system is mucking by belt conveyor. The thrust force of the single shield TBM is provided by the reaction force by hydraulic thrust cylinder acting on the installed reinforced concrete segment in the shield. During tunneling, the rear of the hydraulic thrust cylinder is supported on the already lined reinforced concrete segment, the thrush cylinder extends out to thrush the main machine forward. After the thrust stroke is completed, the TBM stops, the thrust cylinder retracts and returns to its original position, and the reinforced concrete segment is installed at the tail shield, and then, the next cycle can be entered.

2.4.1 Basic Configuration

The single shield TBM mainly consists of cutter head, drive box steel structure, main drive unit, main bearing, main bearing seal, shield structure, tail shield seal, thrust cylinder, shield structure rolling steering device, articulation cylinder, stabilizer cylinder, TBM belt conveyor, segment erector, guidance system, etc. According to geological characteristics, some single shield TBMs are equipped with main drive telescopic and lifting devices.

Fig. 2.68 Single-shield TBM

The single shield cutter head, a heavy-duty machined part is equipped with disc cutter, scraper, bucket, and nozzle, wear protection, etc. The drive box steel structure, a box-shaped structure, provides support for main bearing, main drive unit and cutter head. The thrust cylinder is arranged along the circumferential position of the shield structure. Each thrust cylinder is equipped with gripper pad, connected with the ball head and ball sleeve of the cylinder, can uniformly apply the thrust force of the cylinder on the contact surface of the segment. The shield structure rolling steering device can adjust the included angle between the thrust cylinder and the tunnel axis. The articulation cylinder is connected with the middle shield and tail shield of the TBM. The articulation mode can adjust the included angle between the tail shield and the middle shield according to the line conditions to meet the turning requirements of the TBM. The TBM belt conveyor transports the muck falling into the cutter head to the back-up belt conveyor, and it meets the needs of the maximum advance speed. The front shield is used to support the drive box steel structure and stabilizers are arranged in the front shield to reduce vibration during excavation.

2.4.2 Operating Principle

The basic principle of single-shield TBM tunneling is that during tunnel excavation, the thrust force is provided by the thrust cylinder action of the TBM. At the same time, the cutter head rotates to crush the rock in the working face, and the rock muck is transported out by the belt conveyor located in the middle of the main machine. After one stroke of tunneling, the tunneling stops. Under the protection of the tail shield, the segment is installed and its back is filled, thus to complete a working cycle. The advancement of single-shield TBM requires the lining segment to provide the thrust reverse force, so the advancement must be stopped during the installation of the lining segment, that is, the machine advancement and segment erection cannot be carried out at the same time.

During tunneling, cutter head rotation and rock breaking may cause shield structure rolling. Before tunneling, according to the rolling situation of the shield structure, the thrust cylinder is deflected by a certain angle relative to the tunnel axis by using the steering device. When the segment is tightly supported by the thrust cylinder, a tangential force generated along the circumference generates a steering torque to the front shield, to gradually steer the shield structure during the tunneling process.

Some single-shield TBMs are equipped with driving telescopic and lifting devices, and the drive box steel structure is connected to the front shield through telescopic cylinders and four sets of torque cylinders. The torque cylinder not only provides reverse torque for the rotation of the cutter head, but also realizes the up-down and left–right movement of the cutter head.

2.4.3 Structural Features

The shield structure of single shield TBM is longer than that of the open TBM, and tunneling in the stratum with convergence and deformity of weak surrounding rock may cause jamming of the shield. At present, it is mainly solved from the following five aspects: ① Eccentric design is adopted for the cutter head relative to shield structure centerline; ② The shield structure is designed to be spindle-shaped, and the diameters of front, middle and tail shields are gradually decreased; ③ The articulation force and active breakout force of the articulation cylinder shall be increased; ④ The length of the shield along the advance direction shall be as short as possible, and can also make it easier to adjust the direction of the machine; ⑤ The cutter head drive has the lifting function. The cutter head excavation diameter is expanded in the convergent stratum, and the drive lifting is carried out to increase the gap between the surrounding rock of the top of ring beam and the shield structure. In addition, the construction must be reasonably organized in combination with the characteristics of the equipment to quickly pass through the convergent geological section of the surrounding rock and avoid long-time shutdown. Even if the above measures are taken, they only reduce the risk of TBM jamming and still cannot be completely avoided.

Single Shield TBM (Fig. 2.69) During tunneling, cutter head rotation and rock breaking may cause shield structure rolling. Generally, the large reaction torque of shield structure rolling steering is obtained by the tilting adjustment of thrust cylinder. It is transmitted to the segment ring through the friction between the gripper pad and the segment after the cylinder is inclined, and the segment ring will also roll. The segment steering is more difficult. When the rolling exceeds a certain degree, the tunneling construction cannot be carried out. According to the geological adaptability, on the premise of not affecting the rock breaking efficiency, in some soft rock geologies in China, the selected single-shield TBM cutter head has the

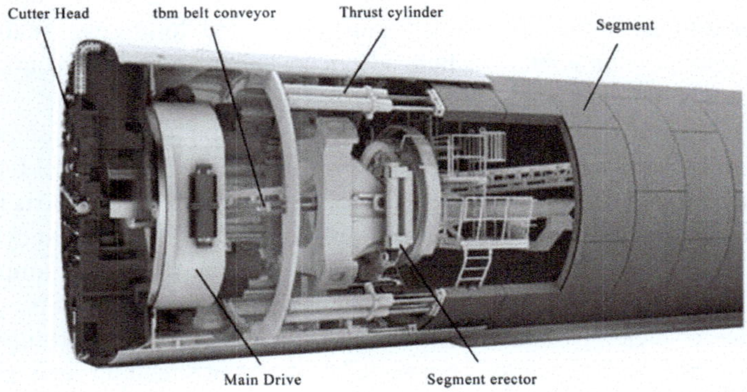

Fig. 2.69 Composition of single shield TBM

functions of two-way rotary cutting and mucking. With the help of forward and reverse rotation of the cutter head, it is simple, effective and reliable to steer the shield structure rolling, and more importantly, to avoid the segment ring rolling.

2.4.4 Application

On the premise of giving full play to the advance speed, the single-shield TBM is mainly applicable to soft rocks (5–60 MPa) with certain self-stability and shallow overburden (small ground stress). Due to the limitation of the structure, the tunneling of the single-shield TBM cannot be operated in parallel with the segment installation. Under the conditions of geological adaptation and strong production organization, the single-shield TBM will also have high tunneling efficiency.

2.5 Structure of Key Parts of Main Machine

2.5.1 Cutter Head

The cutter head of TBM consists of cutter head body and components composed of disc cutter, bucket and water spray device, as shown in Figs. 2.70 and 2.71. With the progress of TBM technology, the planar cutter head gradually adopted in the cutter head design is more conducive to stabilizing the tunnel face. The cutter head body is a heavy-duty structural member, and the large-diameter cutter head is generally assembled on the tunnel excavation site. The large-diameter cutter head with a section design structure can be divided into central split type, eccentric split type and Chinese split type according to the different cutter head structures, in order to facilitate the

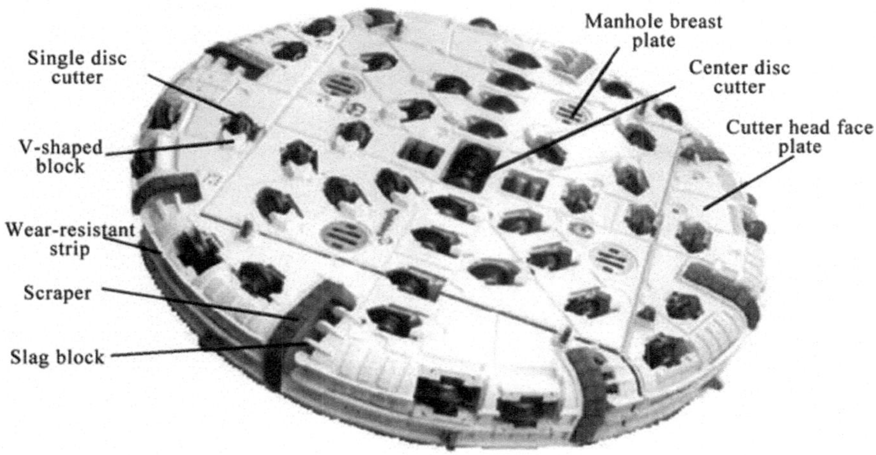

Fig. 2.70 TBM cutter head structure

transportation of the cutter head to the site. Each block is bolted on site and then welded together. Small diameter cutter head can be manufactured to be transported as a whole directly to the site.

The cutter head body structure is made from welded thick steel plates. The front and rear face plates of the cutter head features many longitudinally connected bulkheads, and the structure is complex. The connecting flange on the back face needs to be machined and connected with special bolts. The cutter seat must be precisely positioned and machined when welded. The influence of dynamic load should be considered in determination of the thickness of cutter head and the dimension of weld. Heating, heat preservation and gas shielded welding are required, and welding procedure requirements are high. The strength, rigidity and vibration characteristics of the whole cutter head need to be considered in determination of the working environment of the cutter head. In addition to sufficient strength, the weld joint also requires no cracking under impact load. The excavation diameter of the tunnel is controlled by the track of the edge cutter at the outermost edge of the cutter head. The maximum diameter of the cutter head structure is designed at the lip of the bucket. Generally, a clearance of about 25 mm is reserved from the outermost edge of the lip of the bucket to the wall of the tunnel. Too large clearance is not conducive to the flow and removal of rock chips, and too small clearance will cause the bucket to directly scrape the wall of the tunnel to wear and fail too fast. Therefore, the maximum diameter of the cutter head body structure is generally about 50 mm smaller than the theoretical excavation diameter.

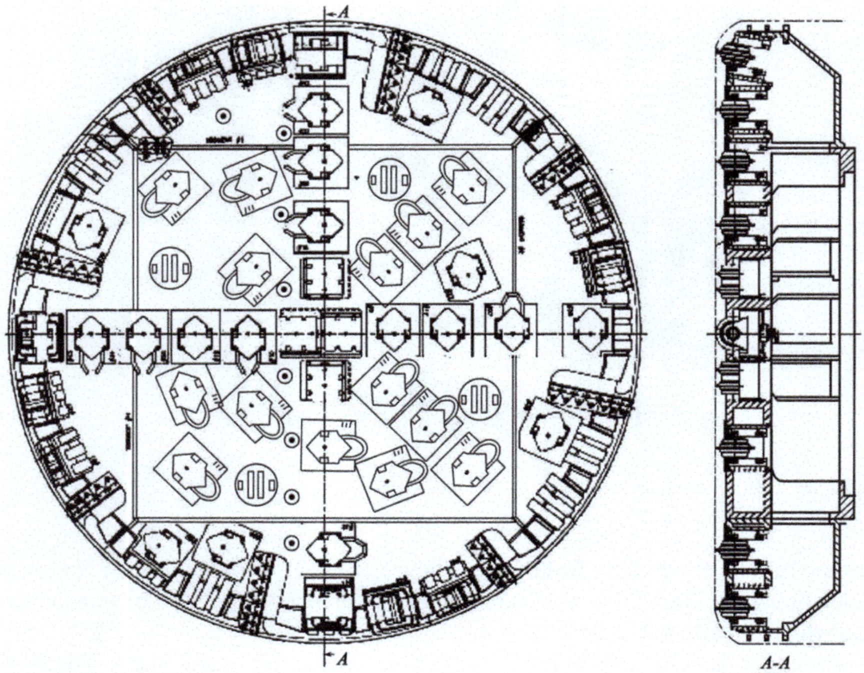

Fig. 2.71 Cutter head structure

2.5.2 Main Drive

1. Main Drive Mode

The main drive is also called cutter head drive, and the drive modes mainly include hydraulic drive, constant speed motor drive and variable frequency motor drive. Due to the development of frequency conversion technology, its reliability is greatly improved. At present, the hard rock TBM generally driven by variable frequency motor can realize stepless speed regulation in a wide range to meet the requirements of different rock tunneling. The cutter head is driven by high power and large torque. Therefore, multiple sets of motors are adopted to drive the ring gear through the reduction box and finally through the pinion to realize the function of the reducer and realize the low-speed high-torsion rotation, thus driving the cutter head to rotate. In order to reduce the overall dimension of the motor, the voltage of the drive motor is 690 V.

Due to the requirements of the unidirectional muck bucket design, the main drive rotates in one direction during tunneling. However, for the maintenance of the cutter head cutter, the cutter head drive with the inching function can be inching in two directions at a slow speed, and is equipped with a brake to stop the rotation as soon

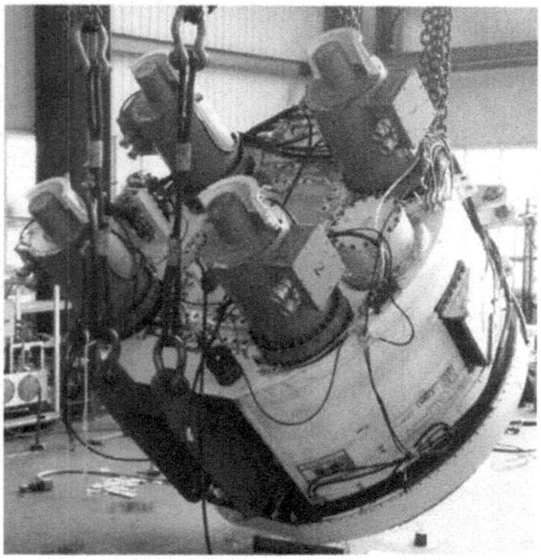

Fig. 2.72 Main drive of cutter head (rear view)

as possible after inching. But the rotation during tunneling cannot be stopped with a brake.

2. Main Drive Structure

Figures 2.72 and 2.73 shows the appearance structure of the main drive with the drive box steel structure and drive motor, the drive reducer, pinion, ring gear and main bearing are all installed in the drive box steel structure, and the whole structure is an enclosed structure by internal and outer seals. For gripper TBM, the rear middle part of the drive box steel structure will be bolted to the main beam, the upper part of the drive box steel structure will be connected with the shield by top shield cylinder, the left and right sides will be connected with the wing shield by wing shield cylinder and wedge block cylinder, and the bottom will be connected with the lower support by splines and bolts. The front of the main drive will be bolted to the cutter head.

The internal structure of the main drive is shown in Fig. 2.74. The drive box steel structure is equipped with a reduction gearbox, pinion, ring gear, main bearing, inner seal, outer seal, bearing seat sleeve, etc. Therefore, the main drive device of TBM consists of drive box steel structure, motor, reducer, pinion, ring gear, main bearing, bearing sleeve, inner seal and outer seal. The driving route is that the motor drives the secondary planetary gear reducer through the torque limiting clutch and transmission shaft at its tail, to drive the pinion outside the reducer and then the pinion drives the ring gear. Since the bull gear and the bearing pedestal sleeve are bolted, and the cutter head, the main bearing inner ring and the bearing seat sleeve are also connected by another set of bolts, the ring gear, the bearing seat sleeve, the main bearing inner ring and the cutter head will rotate together. The main bearing is a three-row spherical roller bearing, with two rows of axial rollers and one row of radial rollers. The inner ring installed on the bearing seat sleeve is a rotating part, while the outer ring is installed in the bearing sleeve of the drive box steel structure and does not rotate.

Fig. 2.73 Main drive of cutter head (front view)

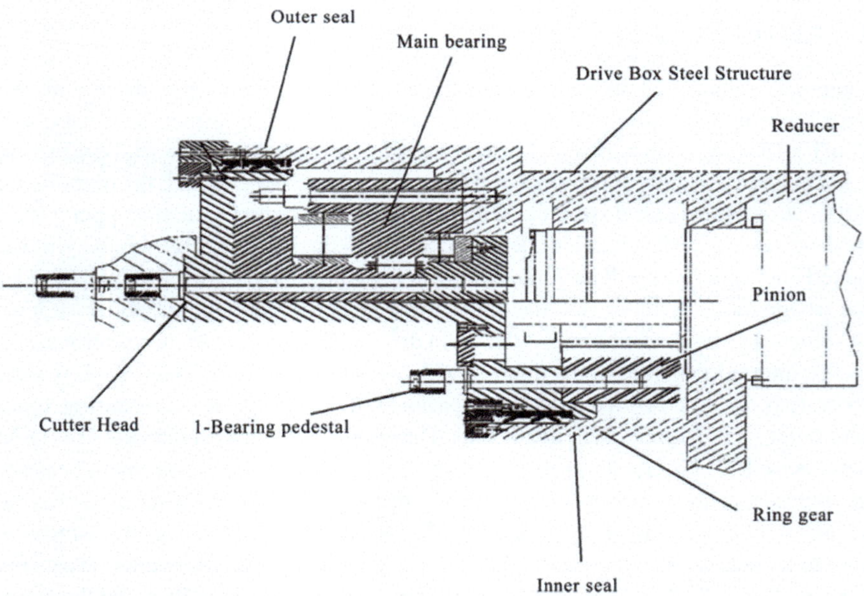

Fig. 2.74 Main drive profile

The main bearing, pinion and ring gear of the main drive are lubricated with forced circulating oil. The lubrication pump station is generally installed on the bracket fixed behind the lower support, and is filtered and cooled for circulating lubrication. Three or four lip seal structures are adopted for the inner and outer seals. The outer two lip seals need to be continuously injected with grease to prevent dust from entering, and one lip seal on the inner side prevents lubricating oil from overflowing. The planetary gear reducer is equipped with lubricating oil with a certain level in the gearbox, and splash lubrication mode is adopted. In addition, the main drive motor and the planetary gearbox are cooled by circulating cooling water. Sensors are used to monitor the oil temperature and flow of lubricating oil, the temperature of reducer and motor, and the grease injection pressure and injection amount.

3. Main Parts of Main Drive

(1) Drive Box Steel Structure

The internal structure of the drive box steel structure is shown in Fig. 2.75, and the drive box steel structure is equivalent to a large box. The upper part of the drive box steel structure is provided with a support for installing the top shield and the top shield cylinder, the lower part is a plane jointed with the lower support, and the side is provided with a wedge block joint face of the shield wedge cylinder. The front part will be connected with the inner and outer seal glands, and the rear part will be connected with the main beam. Therefore, all its upper and lower, front and rear and surrounding contact faces need to be machined. The fitting surface between the inside of the drive box steel structure and the bearing outer ring, the mounting surface of the inner and outer seal, the seat hole surface of the pinion or reducer, etc. are all important machined surfaces. The middle cavity of the drive box steel structure allows the operator to enter the cutter head position, and there are several inspection holes on the inner wall required to be well covered. There are also a large number of screw holes around the drive box steel structure to connect the lubricating oil and grease pipelines.

(2) Main Bearing, Bearing Sleeve and Ring Gear

Fig. 2.75 Internal structure of drive box steel structure of main drive

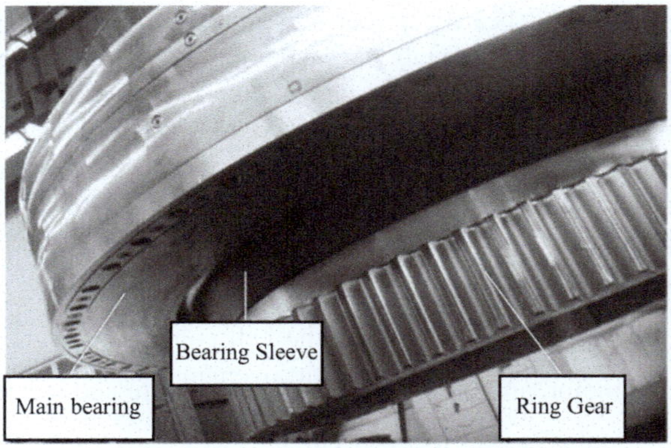

Fig. 2.76 Bearing housing, main bearing and ring gear

The main bearing generally uses a three-row spherical roller bearing, mainly composed of inner ring, outer ring, two rows of axial rollers, one row of radial rollers and retainer. The inner ring, a rotating part installed on the bearing housing, is bolted to the cutter head and the bearing housing, while the ring gear installed on the bearing housing, is bolted to the bearing housing, as shown in Fig. 2.76. The main bearing, main bearing seat and ring gear are all high-accuracy machined parts. At present, there are few manufacturers capable of producing TBM main bearings in the world, and the order period is generally more than half a year. Generally, the TBM manufacturers provide the preliminary requirements for the load spectrum, diameter size and service life of the main bearing, and the main bearing manufacturers carry out design calculation and give the size and service life of the main bearing.

(3) Motor, Torque Limiting Clutch, Reducer and Pinion

The cutter head drive drives the ring gear through the motor, secondary planetary gear reducer and pinion, thus driving the cutter head to rotate. As shown in Figs. 2.77, 2.78 and 2.79, there is a torque limiting clutch at the tail of the motor to protect the drive gear in case of overload. During normal tunneling, the motor rotates, and the torque-limiting clutch is closed to drive the transmission shaft to rotate. There are splines on both ends of the transmission shaft. The front end spline is engaged with the planetary gear reducer, and the rear end spline is engaged with the clutch, thus transmitting power to the reducer. The torque-limiting clutch has a safety valve. During installation, the clutch is injected with special oil reaching a certain oil pressure. When the overload exceeds the set torque value, the safety valve is cut off, and the high-pressure oil leaks from the safety valve, thus cutting off the power transmission between the motor rotor and the transmission shaft.

(4) Main Drive Sealing

Considering the harsh operating environment of TBM, the main drive sealing is of three or four lip seals to protect the main bearing and drive assembly. Figure 2.80 is a schematic diagram of the main drive sealing, consisting of 3 lip seals together with

Fig. 2.77 Diagram of motor, reducer and pinion

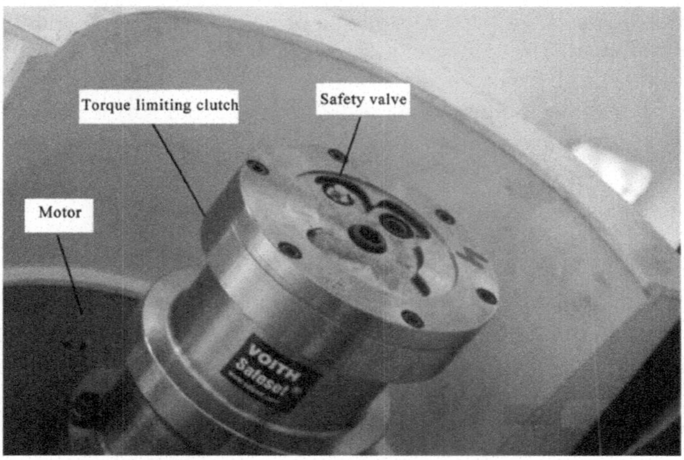

Fig. 2.78 Torque limiting clutch and its safety valve

spacer rings, of which 1 and 2 lip shapes need to be injected with grease to prevent dust and water intrusion; The 3rd seal lip faces inward to prevent the oil from flowing out of the main drive chamber, and can be used as an inspection channel to regularly check whether there is grease overflow, so as to judge the seal state. The hardening and grinding of the seal ring carried out on the contact surface of the metal ring can make the seal ring durable under normal conditions. If the contact surface is worn or the seal is replaced, it can be adjusted to a new surface without replacing the wear ring. The automatic grease system records and adjusts the total amount of grease

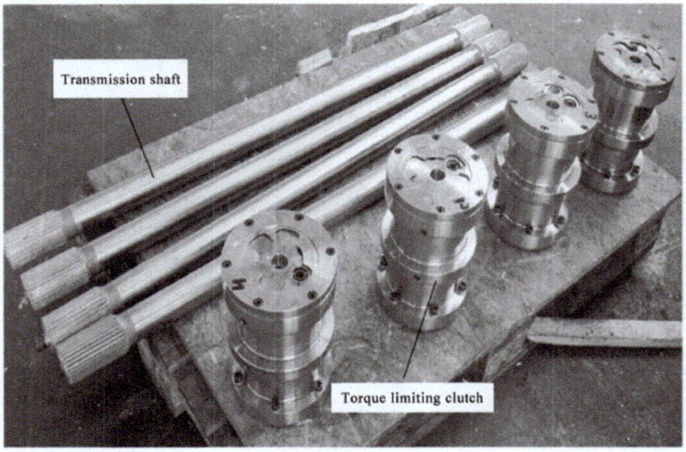

Fig. 2.79 Torque-limiting clutch and transmission shaft

entering the seal. It is interlocked with the main drive and is equally monitored. Only when it works normally can the cutter head be started.

Due to the large diameter of the seal ring, it shall be installed synchronously by multiple persons and points during installation to avoid tensile deformity of the seal ring due to distortion and asynchronization. During installation of each seal and spacer ring, the position and dimension shall be measured to determine whether they are installed in place.

4. Main Drive Lubrication

Forced pressure circulation lubrication is adopted for the main drive, and the lubricating oil is sprayed by the pump station through the filtration and cooling to the main bearing and gear engagement part in the main drive chamber through the flow valve and pipeline. The main bearing and gear are lubricated at multiple points, and the return oil also needs to be filtered by a special magnetic filter. Some TBM

Fig. 2.80 Schematic diagram of main drive sealing

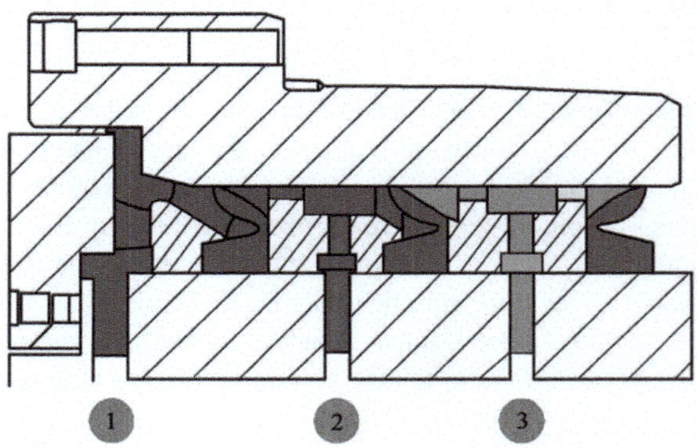

manufacturers have designed special external oil tanks, where oil is pumped back to for filtering and cooling before being pumped to the main bearing and pinion lubrication points. Some TBM manufacturers do not set a separate oil tank but use the main drive chamber as the oil tank. The lubricating oil is continuously pumped out and then continuously pumped to each lubricating point in the main drive after filtration and cooling. In order to prevent oil in the gear chamber from entering the main bearing, seal is also set between the internal main bearing and the gear. This seal allows the oil of the main bearing to flow into the gear chamber, but prevents the oil in the gear chamber from entering the main bearing.

Figure 2.81 shows a TBM main drive lubrication pump station with the main drive inner cavity as the oil tank. The lubrication oil pump is driven by hydraulic pressure. One pump supplies oil to the gear and the other pump supplies oil to the main bearing. The return oil is pumped out from the gear cavity and filtered by a magnetic filter. Before being pumped into the main drive chamber, it needs to pass through the oil supply filter element, cooling device and flow valve. The flow is monitored by sensors. The lubricating circuit is also provided with multiple pressure and temperature gauges and sensors.

Figure 2.82 shows the lubricating oil circuit and seal grease injection system of a TBM main drive. The lubrication system is externally provided with a separate pump station and oil tank. After the return oil filtration and cooling, the lubrication oil is pumped to the main bearing, pinion and lip seal on the inner side of the inner and outer seal by the lubrication oil pump through the filter element, group valve and flowmeter. The two seals with the inner and outer seal lips facing outwards are continuously injected with lubricating grease by the grease pump station, and the lubricating grease will be extruded from the labyrinth of the seal gland, thus judging whether the lubricating grease system works normally. The amount of lubricating grease consumption extrusion is very large, with high performance requirements and

Fig. 2.81 Main drive lubrication pump station

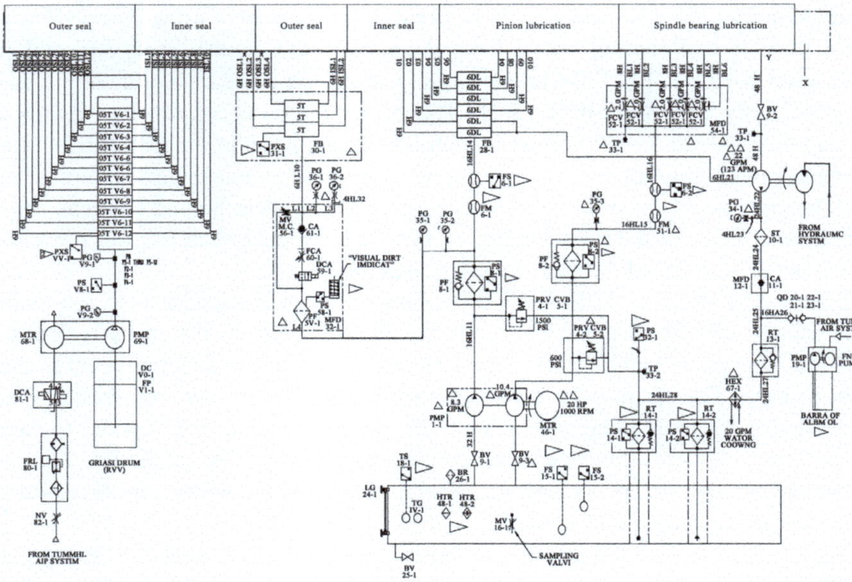

Fig. 2.82 Main drive lubrication pump station

expensive price. As long as there is tunneling, there will be consumption. Therefore, enough reserves should be on site.

2.5.3 Support and Thrust System

1. Kelly-type Gripper TBM

(1) Inner Kelly support

See Fig. 2.83 for the inner Kelly support. The inner Kelly support is a box section welded structure, with quenching hardened slideway for the bearing seat of the outer Kelly to slide on.

The front and rear outer Kelly supports are slid by the thrust cylinder. The inner Kelly support is guided by the cutter head to transmit the thrust and torque generated during TBM operation to the outer Kelly support. The inner Kelly support is connected to the cutter head bearing, drive device, and rear support. The tail part of inner Kelly support is connected to the rear support, and its front is connected to the main bearing seat. The front end of the inner Kelly support is provided with a manhole, through which the cutter head is accessible. There is enough space inside the inner Kelly support to accommodate the belt conveyor.

(2) Outer Kelly support and grippers

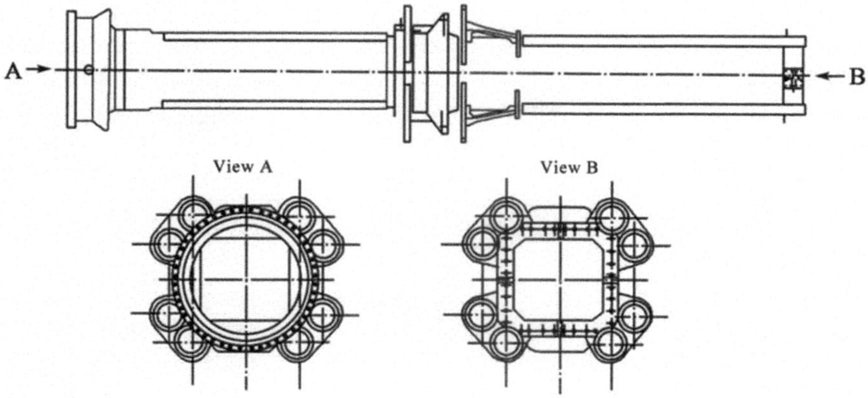

Fig. 2.83 Inner Kelly support in gripper TBM

As shown in Fig. 2.84, grippers together with the outer Kelly support slide longitudinally along the inner Kelly support. Grippers are operated by 32 hydraulic cylinders and divided into 2 groups with 8 grippers for each group in an X distribution shape on the outer Kelly support, and there is a group of grippers on the front and rear outer Kelly support respectively. The 16 grippers securely fasten the outer Kelly support to the tunnel inner wall to withstand the cutter head torque and the reaction force generated during TBM advancement. The front and rear grippers can move independently to adapt to different ring beam spacing.

(3) Thrust cylinder

The thrust acting on the cutter head is transmitted to the surrounding rock through the inner and outer Kelly supports. The outer Kelly support is of two separate assemblies,

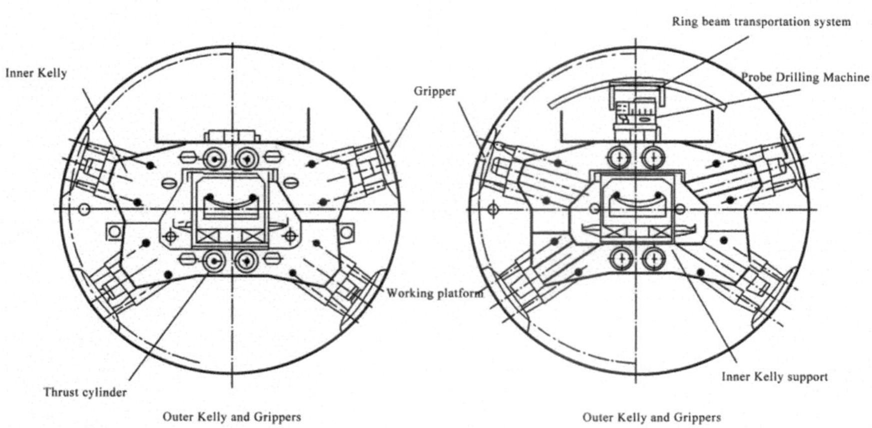

Fig. 2.84 Outer Kelly support and grippers

each with its own separate thrust cylinder. The front and rear outer Kelly supports are provided with 4 thrust cylinders respectively. The thrust cylinders of the rear outer Kelly support transmits the thrust to the inner Kelly support, while the front outer Kelly support directly transmits the thrust to the shell of the cutter head drive device. At the end of the tunneling cycle, the rear support of the inner Kelly extends to the bottom of the tunnel, while grippers of the outer Kelly retract, and the thrust cylinder pushes the outer Kelly forward to prepare for the next cycle of tunneling.

(4) Rear Support

The rear support (Fig. 2.85) is installed on the inner Kelly support and located behind the rear outer Kelly support. The rear support controls the expansion and retraction through the hydraulic cylinder, and can also be used for horizontal adjustment. When the rear support is retracted, the position of the inner Kelly support can be adjusted horizontally and vertically to adjust the tunnel centerline of the TBM.

2. Main beam gripper TBM

The functions of the thrust and support system are as follows: first, the thrust cylinder provides the thrust force required for tunneling to the cutter head; second, the gripper cylinder supports the gripper tightly on the tunnel wall to bear the reaction force and torque during tunneling. As shown in Figs. 2.86, 2.87 and 2.88, the thrust and gripper system mainly consists of saddle, thrust cylinder, gripper cylinder, torque cylinder and gripper.

(1) Saddle

As a steel structure trolley carrying the gripper system, installed on the guide rail of the main beam, the saddle can slide on the guide rail to realize the regripping after the completion of the tunneling stroke. The saddle chute inlaid with a wear-resistant plate can be replaced after being worn to a certain extent. The saddle, generally of left and right block structure, is assembled and bolted together after being installed on the guide rail.

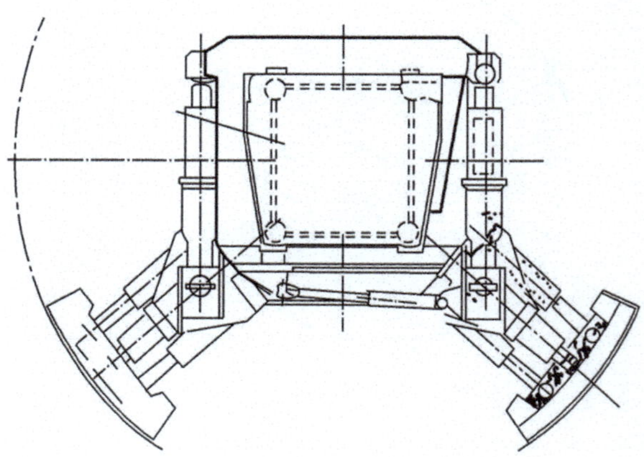

Fig. 2.85 Rear support

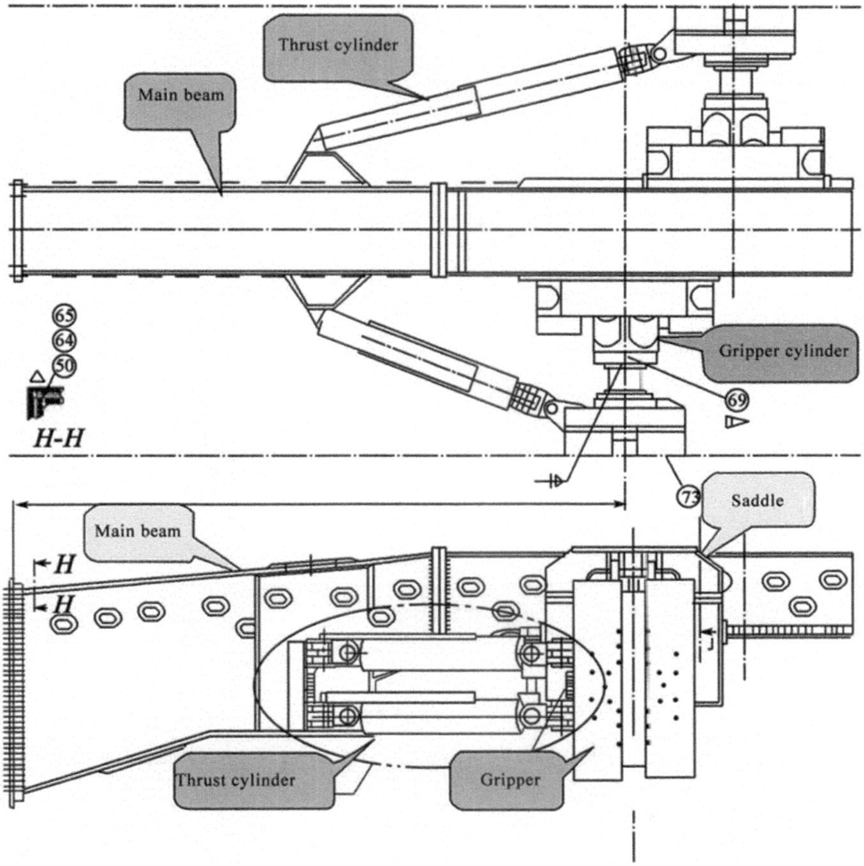

Fig. 2.86 Structure of thrust and gripper system (top and side views)

(2) Thrust cylinder

The thrust cylinder is arranged on both sides of the main beam. The front end is connected to the main beam by pin, and the rear end is connected to the gripper by pin. The front and rear ends are connected to the main beam and the gripper by two (horizontal and vertical) pins respectively through the middle adapter. In order to reduce impact, the connection between the thrust cylinder and the gripper is provided with a damping spring. During tunneling, the thrust cylinder pushes the main beam forward, together with the main drive, shield and cutter head, and the support force is transmitted to the gripper. After one tunneling is completed, the rear leg is lowered, the gripper is retracted, and the thrust cylinder is used to pull the saddle and gripper system to move forward to realize regripping. At the same time, the thrust cylinder is also retracted to prepare for the next tunneling cycle. In the design of thrust system,

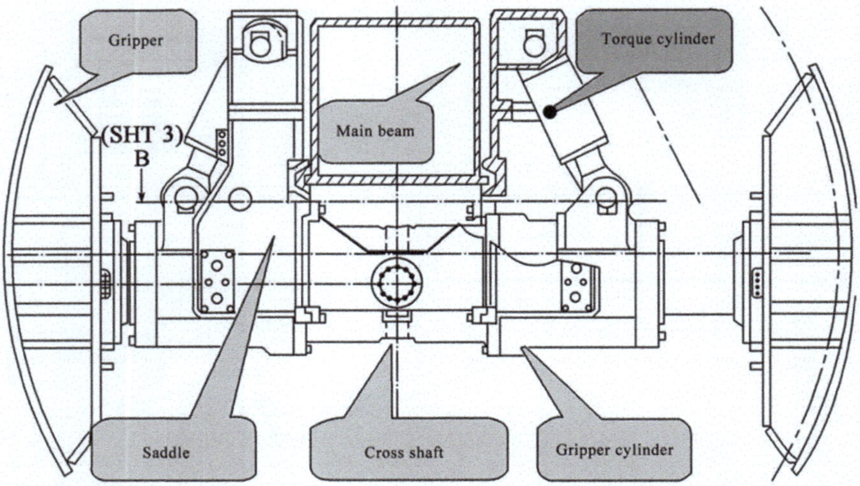

Fig. 2.87 Structure of gripper system (cross-sectional view)

Fig. 2.88 Thrust cylinder and gripper

the thrust force required by the three parts of the cutter head thrust force, shield friction and back-up towing pulling force are mainly considered.

(3) Gripper Cylinder and Torque Cylinder

The gripper system has left and right gripper cylinders connected as a whole. The middle structure is connected with the saddle through cross pin, and the gripper cylinders are suspended on the saddle through two vertical torque cylinders on the left and two vertical torque cylinders on the right, that is, the upper end of the torque cylinder is connected on the upper side of the saddle with pin, and the lower end is connected on the gripper cylinder with pin. As shown in Figs. 2.89 and 2.90, the cross pin is bolted to the saddle, and four spring-loaded top blocks are installed in the saddle at the front, back, left and right sides, respectively against the front and back sides of the left and right gripper cylinder to avoid excessive sway. The design structure of this cross pin "floating gripper system" allows the gripper cylinder to swing and tilt back and forth, up and down, left and right to a certain extent, in order to adapt to the TBM attitude change and chamber excavation tolerance. The gripper cylinder provides the supporting force during tunneling by holding the gripper tightly on the tunnel wall, and at the same time, the left and right direction adjustment of TBM can be realized by adjusting the left and right gripper cylinders. The torque cylinder transmits the torque transmitted from the main beam to the gripper, and at the same time, the rolling angle of the TBM can be adjusted through the telescoping of the left and right torque cylinders.

(4) Gripper

The gripper is of welded steel structure, and the area of the gripper shall be designed as large as possible to adapt to the tunneling of weak surrounding rock. Its middle part shall be grooved to cross the ring beam, and several conical nails shall be installed on the surface to increase its resistance on the tunnel wall. As shown in Figs. 2.91 and 2.92, the end of the gripper and the gripper cylinder are bolted in contact with spherical pair. An outer spherical sleeve is bolted on the end of the gripper cylinder. The inner convex spherical surface of the gripper is in contact with the concave spherical surface of the spherical sleeve, and the flange matching with the outer spherical sleeve is bolted to fix the grippers. This spherical auxiliary structure allows the gripper to have a certain swing angle in all directions to meet the needs of TBM attitude and tunnel wall irregularity. In order to prevent excessive swinging, several balance cylinders set inside the gripper cylinder always support the spherical sleeve after pressure maintenance.

(5) Main Beam and Rear Support

Fig. 2.89 Gripper cylinder and torque cylinder

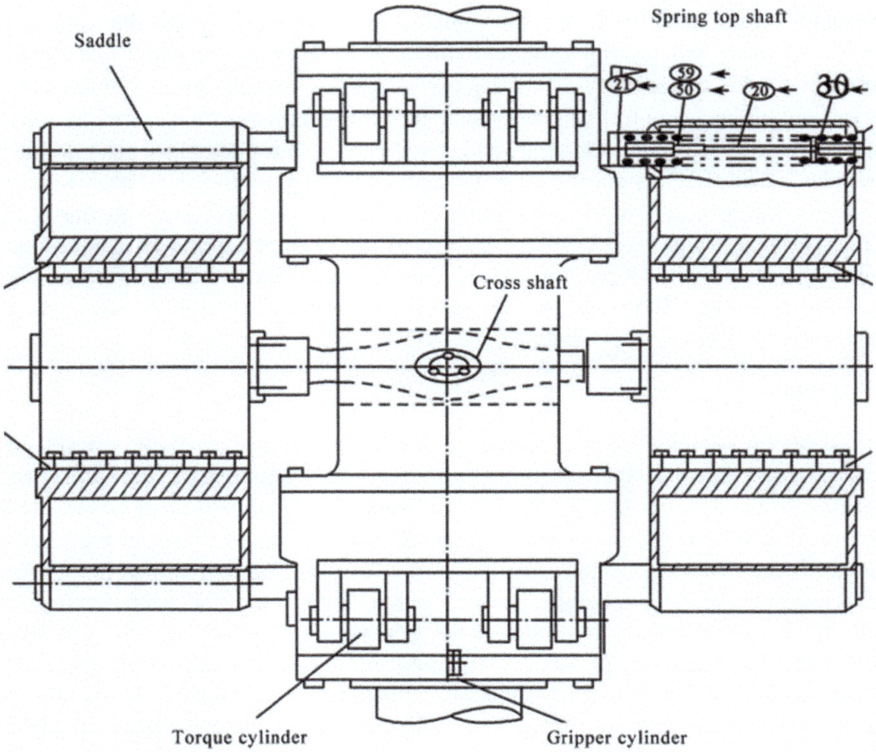

Fig. 2.90 Connection structure of cross pin of gripper cylinder

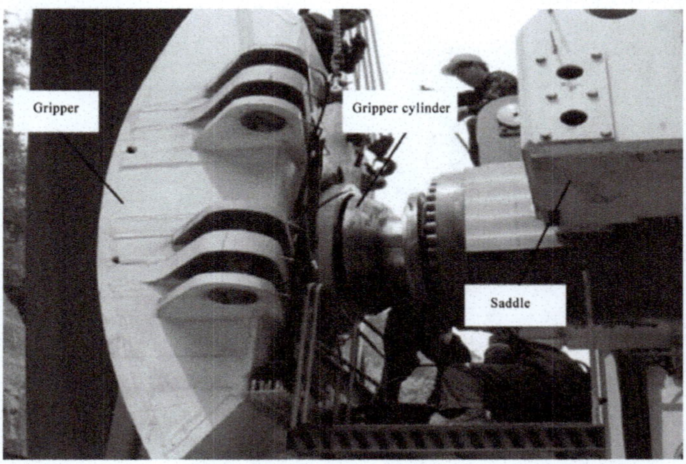

Fig. 2.91 Gripper and gripper cylinder

Fig. 2.92 Gripper inner side sphere and balance cylinder

The main beam of the gripper TBM is generally of a box-shaped steel structure with a length of about 20 m. To facilitate manufacture and transportation, it is divided into the front section, middle section and tail section, as shown in Figs. 2.93, 2.94 and 2.95. The sections will be bolted, and the front section will also be bolted with drive box steel structure of the main drive. It bears the force and action of torque transmitted by the cutter head and transmits the force and torque to the gripper acting on the tunnel wall. It shall have sufficient strength and rigidity and have the following other main structural functions:

Fig. 2.93 Front and middle sections of main beam

Fig. 2.94 Middle section of main beam

Fig. 2.95 Structural components of main beam tail section and rear leg

① Equipment such as ring beam erector, bolt driller, and probe drilling machine is installed and loaded at its front section;
② Saddle, gripper and thrust system is installed and loaded at its middle section;
③ Its tail section is integrated with the vertical steel structure of the rear leg as a whole. The rear part of the TBM is supported by the rear leg cylinder, and the vertical direction of the TBM can be adjusted up and down, and it is also connected with the back-up through the towing cylinder;
④ The main machine mucking belt conveyor, arranged in its inside, extends from its tail to transfer the rock muck to the back-up belt conveyor;

⑤ Water pipes, air pipes, cables and other pipelines leading to the main drive and cutter head are arranged on its outer side and inner wall;
⑥ The main beam shall be designed with manholes to enable operators to enter its interior and the cutter head;
⑦ Its inner cavity also serves as the air duct of the industrial exhaust fan, with an side opening connected to the external air duct leading to the deduster;
⑧ Material hoist and its slideway shall be arranged at its lower side or upper part, and cutter cranes shall be arranged in front of its inside;
⑨ The lubrication system of main drive is also arranged at the front and bottom shields of the main beam.

The rear leg is also called the rear support. There are two support grippers and cylinders in total for both the left and right. As shown in Fig. 2.96, the rear leg cylinder is installed in a vertical steel structure integral with the tail section of the main beam, and through the telescopic cylinder, the bottom supporting gripper pad is driven to move up and down. When tunneling is required, first support the gripper on the tunnel wall, and then lift the rear leg; the gripper can be withdrawn only after the rear leg is supported at the bottom of the tunnel. The gripper pad of the rear support is connected with the cylinder slide seat through a pin, with a certain degree of freedom to adapt to the possible irregular surface at the bottom of the tunnel, as shown in Fig. 2.97. The pressure and telescopic distance of the rear leg cylinder both are monitored by sensors. The telescopic speed cannot be too fast, and it is better to be adjusted. The upper and lower connection structure of the rear leg cylinder shall be firm and reliable. In practical application, there have been cases that the cylinder fell off and the gripper pad drop out.

Fig. 2.96 Rear leg cylinder

Fig. 2.97 Rear leg gripper pad

3. Single Shield TBM

Compared with gripper TBM, single shield TBM is suitable for soft surrounding rock stratum. Due to no gripper system and front and rear support, its thrust and regripping are realized by thrust hydraulic cylinder. Its thrust system typically has the following functions: enough power is provided for the shield to move forward; the forward speed of the shield is controlled to match the mucking speed and avoid accumulation; the attitude of the shield is able to control and realize the deviation correction and steering requirements of the shield; the dimension and operation requirement of segments is adapted to; from the overall perspective, the requirements for overall functional design and comprehensive construction operation of TBM shall be met.

The thrust system of single shield TBM consisting of a number of hydraulic cylinders are divided into groups for group control in order to facilitate its control. Supported on the segment, the hydraulic cylinder pushes the cutter head forward. After a stroke of tunneling, it is retracted for the segment erection.

The single shield TBM thrust system (Fig. 2.98) is composed of multiple thrust hydraulic cylinders parallel or approximately parallel to the longitudinal axis of the TBM. The front end of each hydraulic cylinder is articulated, and the rear end is jacked on the lining segment through the top gripper. With a large contact area, the top gripper can reduce the contact pressure to avoid the wear of the hydraulic cylinder and the top gripper shall be replaced if necessary. Four springs are arranged between the top gripper and the hydraulic cylinder to avoid large-angle torsion of the top gripper. The thrust system performs forward thrust and posture control functions to ensure that the TBM can advance along the design profile. There are straight lines and curves in the design profile. Therefore, during the tunneling process, the shield shall be able to move forward, upward, downward, leftward and rightward in a straight line. In order to reduce the control complexity and cost during the thrust and deviation

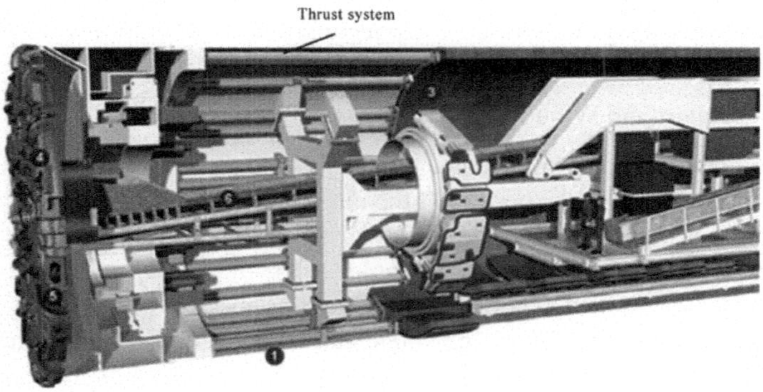

Fig. 2.98 Single shield TBM thrust system

correction operations, the thrust hydraulic cylinders are usually divided into several groups along the circumferential direction to reduce the control complexity through partitioned control.

In conclusion, the TBM thrust system mainly includes the thrust hydraulic jack and its auxiliary facilities. The thrust jack of the gripper TBM thrust system is adjacent to or parallel to the main beam and is connected with the support system through ball joint. During operation, the thrust jack extends to provide thrust force for the cutter head to propel the machine. The working principle of the single shield TBM is the same as that of the gripper TBM, and the machine is pushed forward by the extension or retraction of the thrust jack. The difference is that the thrust jack of the former is supported on the erected segment lining, while the thrust jack of the latter is supported on the supporting system of the machine itself. The Double Shield TBM has the above two thrust systems, namely the primary thrust system and the auxiliary thrust system, so it also has the above two operating modes.

2.5.4 Shield

The main body of the shield is of a steel structure welding part. The shields of double-shielded TBM and single-shielded TBM are relatively long, and are connected with the drive box steel structure by flanges. However, the shield of gripper TBM, connected with the drive box steel structure, surrounds the drive box steel structure of the main drive to stabilize the cutter head on the tunnel wall during TBM tensioning and prevent large rock chips from falling on the back of the cutter head and at the main drive motor. The whole shield is divided into bottom shield, wing shield and top shield, and the bottom shield is also called lower support. The bottom shield is connected with the bottom surface of the drive box steel structure through splines and bolts, and the wing shield and top shield are connected with the drive box steel structure through cylinders, with a large gap between them. Therefore, the gripper

TBM needs to be designed with a shield dust barrier to prevent dust from entering the back of the main machine from this gap.

1. Lower Support and Wing Shield

With reference to the previous description in this chapter, the bottom of the lower support is in contact with the bottom of the tunnel, and the upper plane is connected with the bottom plane of the drive box steel structure of the main drive by bolts and splines, with front and rear through holes in the middle, and the oil pipeline to the main drive is arranged. The lower support is connected with the left and right wing shields through pins. The wing shield can be opened or retracted through the wing shield cylinder, and wedged tightly to the tunnel wall through the wedge block cylinder during tunneling. The two inclined planes of the wedge block contact the inclined plane on the drive box steel structure on one side and the inclined plane on the wing shield on the other side. One of the inclined planes where the wedge blocks contact the wing shield and drive box steel structure should be a swinging block plane that can rotate around the pin with a degree of freedom of rotation, as shown in Fig. 2.99. There are two different designs: wing shield is connected with the wedge cylinder on the drive box steel structure and with the wedge block cylinder on the wing shield.

2. Top Shield

Figure 2.100 shows the top shield support, and it is connected with the upper machined surface of the drive box steel structure by splines and bolts. When the top shield is extended or retracted through the cylinder, the shield can slide up and down along the machining slides on the four sides of the support. As shown in Fig. 2.101, the top shield is divided into three blocks. The middle block, connected to the upper part of the top shield cylinder, can slide on the support through the top shield cylinder. The lower part of the top shield cylinder is installed on the drive box steel structure. The left and right top shield, connected with the middle block by pin, can swing around the middle block, while the other side is overlapped on the wing shield.

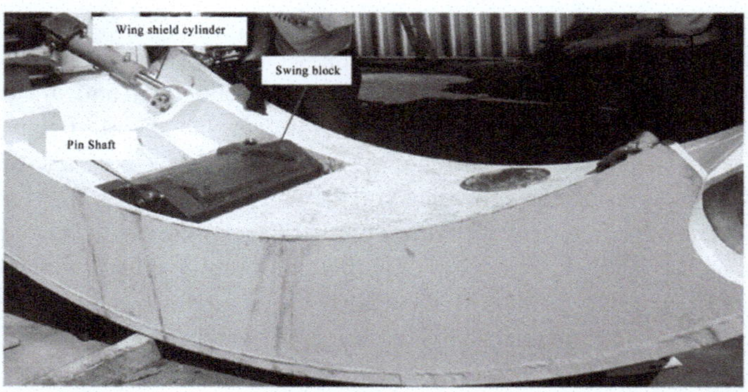

Fig. 2.99 Wing shield

Fig. 2.100 Top shield support

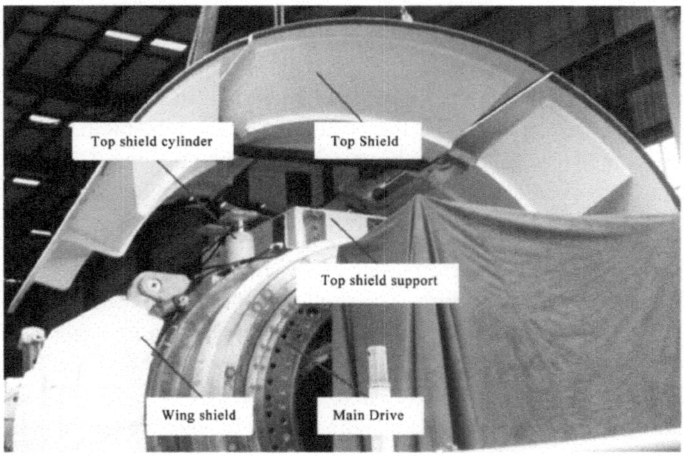

Fig. 2.101 Top shield (under installation)

2.5.5 *Main Machine Accessories*

As mentioned above, the main machine accessories of the double-shield TBM and single-shield TBM are mainly the segment erector and the mucking belt conveyor, while the accessories on the gripper TBM main machine are shown in Fig. 2.102, mainly including the ring beam erector, bolt driller, probe drilling machine and mucking belt conveyor. According to the principle of the New Austrian Tunneling Method, the ring beam erector, bolt driller and probe drilling machine are used for timely initial support operation and probe treatment, and the TBM belt conveyor for

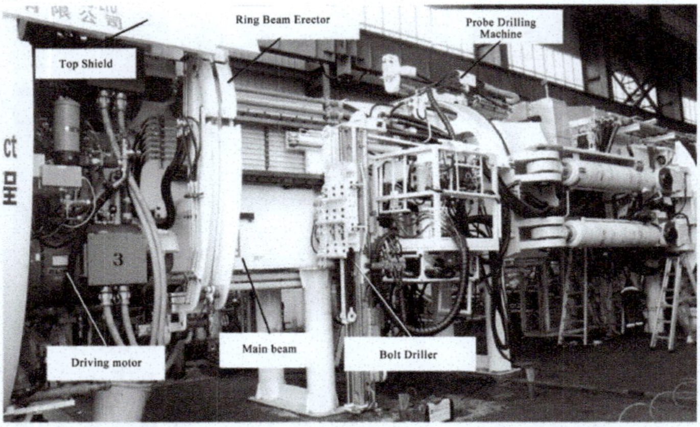

Fig. 2.102 Arrangement of ring beam erector, bolt driller and probe drilling machine

the mucking operation completion. The following will focus on the main accessories of the gripper TBM.

1. Ring Beam Erector

The ring beam erector is arranged under the top shield in front of the main beam, to timely erect the ring beam under the protection of the top shield. The ring beam is erected from multi-section steel sheets made of profile steel. The erector needs to complete actions such as rotary erection, top and lateral bracing, bottom opening tensioning and sealing, etc., and shall have the following main functions:

(1) The steel sheets of each section rotate in the slide way of the erector to complete the erection section by section. The motor winch traction scheme or gear ring drive scheme can be adopted. Figure 2.103 shows the motor winch traction.
(2) The ring beam can be perpendicular to the tunnel axis and firmly supported on the tunnel wall towards the top and side. It needs to be realized by setting a jacking cylinder and a side tension cylinder mechanism, as shown in Figs. 2.103 and 2.104.
(3) The ring beam erector shall be able to move within a certain distance in the longitudinal direction, and can be realized by arranging the towing cylinder, traveling wheel and track on the upper or side of the main beam, as shown in Figs. 2.103 and 2.104.
(4) After each section of ring beam is erected and supported tightly on the tunnel wall, a section of opening shall be reserved at the bottom. After the opening is tensioned, a section of connecting plate shall be used to connect and seal the opening. The tension device can be realized by a separate tension cylinder or screw tensioner or manual jack. The erected ring beam is shown in Fig. 2.105.

2. Bolt Driller and Probe Drilling Machine

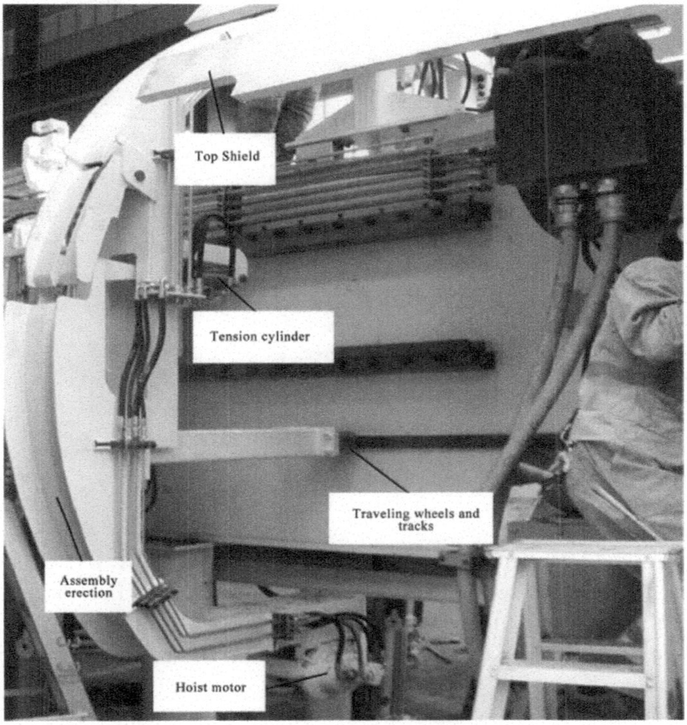

Fig. 2.103 Side tension cylinder and winch motor of ring beam erector

Fig. 2.104 Jacking cylinder and longitudinal cylinder of ring beam erector

Fig. 2.105 Ring beam support

(1) Bolt Driller

As shown in Fig. 2.106, generally, a bolt driller is arranged on the left and right sides of the main beam respectively, and the control console of the bolt driller arranged behind can move longitudinally with the bolt driller, or is fixed on both sides of the rear main beam. The hydraulic power pack of the bolt driller is arranged on the back-up jumbo.

The bolt driller shall be able to achieve circumferential rotation and longitudinal movement. The rotation within the circumferential range can be realized through the rod mechanism composed of cylinders, or through the chain wheel and chain drive arranged on the ring-shaped drilling frame of the main beam; for large-diameter tunnels, it is possible to realize the borehole within the circumferential range above 270°, while for small-diameter tunnels, the borehole is generally less than 180°.

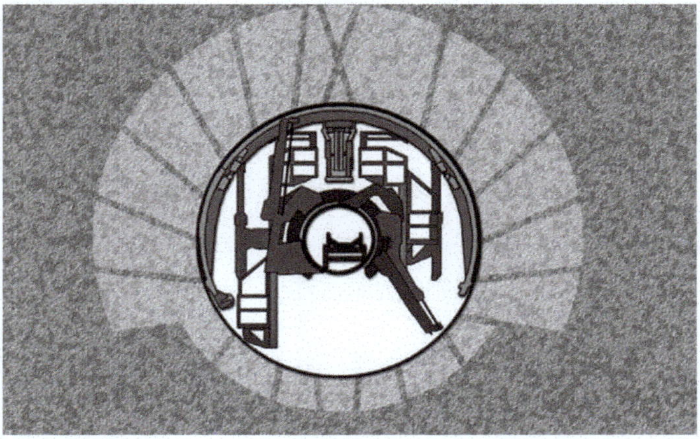

Fig. 2.106 Schematic diagram of rock bolt operation

Realized by longitudinal towing cylinder and the guide rail on the main beam, the longitudinal movement distance of the bolt driller rig shall be greater than the TBM tunneling travel. It is worth noting that, since it is expected that the bolt drilling operation can be carried out during the tunneling operation, the bolt driller system should be equipped with a synchronous mechanism to ensure that the main beam can slide freely through the driller steel structure when the TBM tunneling and drilling borehole are carried out at the same time, without broken drill rod or even damaged drilling machine. The hydraulic synchronous cylinder mechanism can be adopted to achieve this requirement.

The bolt driller has the requirements of rotation, thrust, impact, reverse rotation, recovery and other movements. Due to the limitation of operation space, the borehole depth of a tunnel with a diameter of 4.5 m is generally less than 2 m (without rod connection), while that of a tunnel with a diameter of 12.4 m can be more than 3 m. It should be noted that the rock bolt is not possible to radially pass through the center of the tunnel due to the arrangement of the drilling machine on both sides of the main beam.

(2) Probe Drilling Machine

As shown in Fig. 2.107, the probe drilling machine is generally arranged above the main beam for probe drill hole and pre-grouting operations. Due to the existence of front shield and cutter head, the probe drilling machine must be inclined with the tunnel axis for drilling at an angle of about 7° in general, and the circumferential drilling range is above 120°, and the drilling distance can reach about 30 m in front of the tunnel face.

The drilling support of the probe drilling machine is fixed on the main beam, and the chain drive mechanism of the chain wheel can make the drilling machine move along the circumferential direction of the drilling support, thus realizing the drilling operation within a certain circumference range. There shall be pilot holes

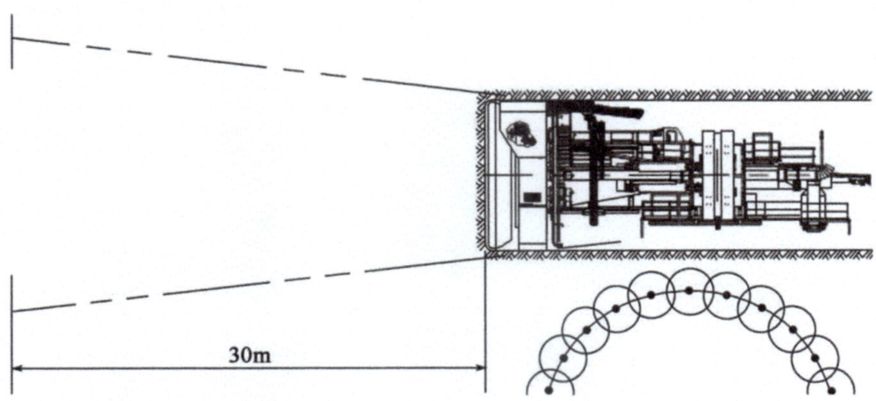

Fig. 2.107 Schematic diagram of probe drilling machine operation

at the shield through which the drill rod passes. Guide frame shall be set when the distance between the drilling machine and the shield is relatively long.

Some TBM bolt drillers can be used as probe drilling machines. At this time, there should be a special mechanism to enable the bolt driller to turn to the longitudinal direction, and it is not necessary to turn back during drilling of probe drill hole.

3. TBM belt conveyor

The TBM mucking belt conveyor arranged in the main beam, is a trough belt conveyor. Its tail extends to the inner cavity of the cutter head to receive the rock muck discharged from the muck chute of the cutter head through the muck hopper, and then transported to the head of the main machine and transported to the back-up belt conveyor. The TBM belt conveyor is mainly composed of head drive pulley, tail driven pulley, supporting frame for belt, roller, belt, scraper, muck hopper, etc. The impact-resistant roller is adopted for the tail roller of the belt conveyor in a densified arrangement, as shown in Fig. 2.108. The layout of the belt conveyor in the main beam is shown in Fig. 2.109. The small-diameter TBM needs to pull out the belt conveyor, so slides are arranged on both sides of the main beam for the supporting frame for belt to slide. Figure 2.110 shows the head of the TBM belt conveyor extending from the tail of the main beam. The drive pulley is mostly driven by a variable frequency motor within the pulley. The belt on the bearing side and the return side is equipped with a scraper and a belt tension device. The belt tension degree can be adjusted through a telescopic cylinder or a screw.

Due to the limited space in the main beam of the small-diameter TBM, the belt conveyor needs to be pulled backwards for a certain distance so that the cutter inspection and maintenance personnel can enter the cutter head for operation. Therefore, some TBM manufacturers have designed a scheme to drive a sprocket and a chain with two coaxial motors and to pull the belt conveyor out of the tail of the main beam. However, in practical application, insufficient driving force, and asynchronization of two motors, motor failure, and the jam of the supporting frame for the belt happen frequently, which takes a long time to pull out and retract the belt conveyor, seriously

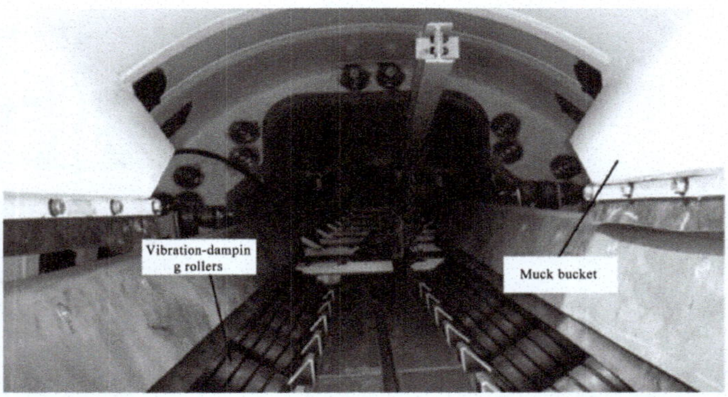

Fig. 2.108 Tail section of belt conveyor (under installation)

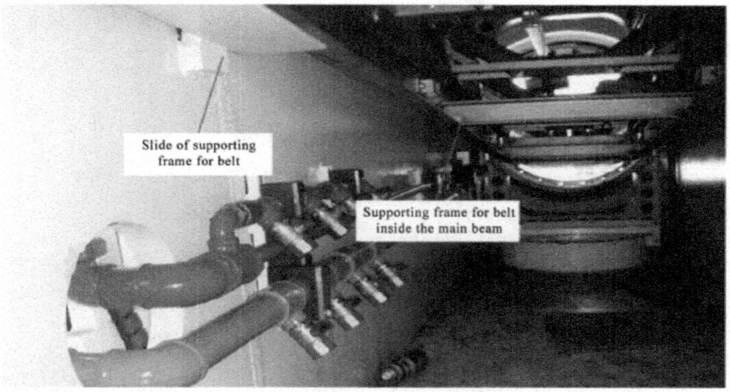

Fig. 2.109 Layout of belt conveyor in main beam

Fig. 2.110 Head of TBM belt conveyor

delaying the tunneling operation. So we have changed the motor driving scheme into a hydraulic cylinder telescopic belt conveyor scheme, as shown in Fig. 2.111. The performance in the practical application is reliable, and it only takes more than one minute to extend or retract the belt conveyor from the main beam to the required position, achieving a good improvement.

For the larger-diameter TBM, due to the larger height space in the main beam, the scheme of pulling the belt conveyor backwards may not be adopted in order for operators to enter the cutter head. Instead, a jacking cylinder is set at the tail of the belt conveyor. The tail of the belt conveyor can be lifted through the jacking cylinder to obtain a greater height space so that operators can easily enter the cutter head, as shown in Fig. 2.112.

The tail of the TBM belt conveyor is impacted by rock ballast discontinuously and unevenly. A large amount of water in the tunnel causes the muck to scatter from the belt. The working conditions are bad, and the damaged roller, the failure of the

Fig. 2.111 Scheme of changing the extending device of belt conveyor from original motor chain to cylinder

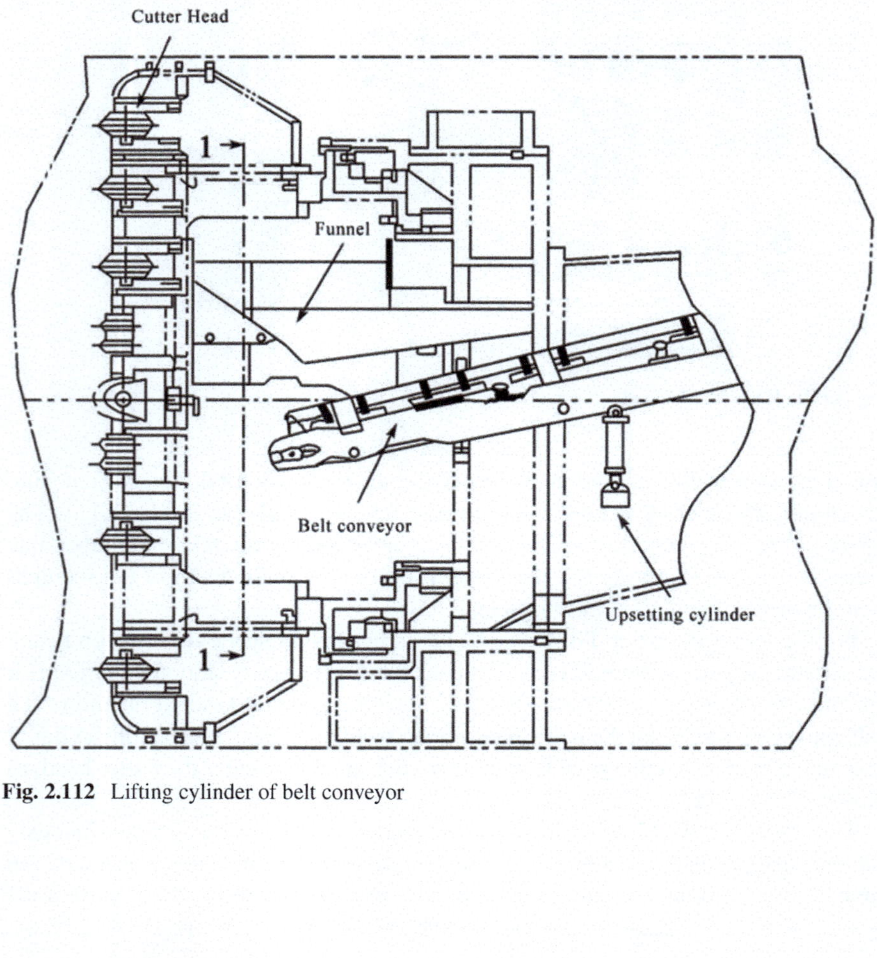

Fig. 2.112 Lifting cylinder of belt conveyor

roller bearing, the scratched belt, and the quick wear of the scraper occur frequently. Hence, reasonable tunneling measures and careful maintenance are required, and spare parts such as belt, roller, pulley, and scraper need to be stored to avoid the shutdown.

2.6 TBM Back-Up Structure

2.6.1 Back-Up System Functions

The back-up system shall be designed with the following main functions:

(1) The back-up jumbo shall be strong and rigid enough and have enough space to accommodate all auxiliary equipment and facilities, and the connection between the jumbos shall have a certain degree of freedom.
(2) The back-up system can travel together with the TBM main machine or alone.
(3) The back-up system shall be able to provide hydraulic power, electricity, water, air, compressed air, etc. for the TBM main machine or construction.
(4) The back-up system shall be able to provide equipment and space for auxiliary operations, including shotcreting, rock bolt operation, grouting, mucking, dedusting, rail laying, material transportation, muck removal, dewatering, air duct, water pipe and cable extension, etc. The capacity of the auxiliary equipment of the back-up system shall be designed to match the TBM tunneling capacity.
(5) Various pipes and lines, including air ducts, water pipes, cables, hydraulic pipes, and air pipes, shall be arranged reasonably.
(6) Personnel and equipment shall be provided with various safety and construction facilities, such as handrails, walkways, steps, step ladders, guardrails, fire extinguishing devices, waterproofing devices, lighting devices, emergency power supplies, harmful gas monitoring and alarm system, etc.
(7) It shall have necessary living functions, such as meetings, meals, toilets and first aid.

2.6.2 Back-Up System Composition

The main composition of the back-up system is described by taking the gripper TBM as an example. As shown in Fig. 2.113, the equipment and devices of the back-up system can be classified into steel-structure jumbos, supply and power systems, auxiliary operation equipment, safety facilities, living facilities, etc.

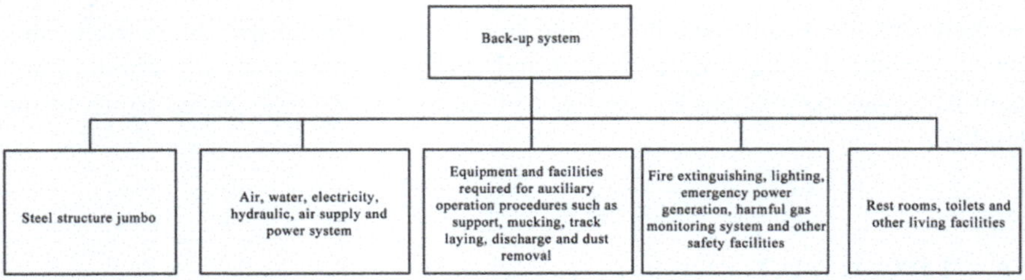

Fig. 2.113 Composition of back-up system

1. Back-up Jumbo

The back-up main body consists of several steel-structure jumbos connected by pins. Sometimes a connection bridge is added between the jumbo and the main machine. The connection bridge can also be regarded as a part of the back-up system. In addition, when a single-track muck car is adopted for mucking, a California switch is often connected behind the back-up jumbo to reduce the waiting time for the muck car. With a traveling wheel equipped at the bottom, the California switch, generally about 120 m long, runs along the tunnel rail with the back-up jumbo. It consists of a series of platform vehicles, on which there are double tracks and a switch, and a ramp at the tail for the muck car to move from the tunnel rail to its platform. The California switch may also be considered as a part of the back-up system.

2. Air, Water, Electricity, and Power Supply Systems

(1) Hydraulic Power Pack

The hydraulic power pack of the TBM main machine is generally arranged on the backup system, and sometimes on the main machine. The hydraulic power pack of the bolt driller on the main machine is also generally placed on the back-up jumbo.

(2) Power Supply and Electrical Control System

It includes HV cable reels, transformers, electrical control cabinets, and cables interfaced with cables arranged in the tunnel.

(3) Ventilation System

It includes the air duct cassette and the secondary fan interfaced with the tunnel ventilation system, metal air ducts laid from back to front on the back-up jumbo, etc.

(4) Water Supply System

It includes the water pipe reel interfaced with the water supply pipe in the tunnel, water supply pipe, water pump and water tank on the back-up system, etc.

(5) Air Supply System

It mainly refers to HP air, including compressed air for shoterete and pneumatic tools. The equipment mainly consists of air compressor, air tank, and pipeline.

(3) Auxiliary Operation Equipment

(1) Bolt Driller

The bolt driller is generally arranged at the front of the TBM main beam. However, sometimes additional bolt drillers are also required on the connection bridge or the front of the backup system, and such an arrangement is mainly used in large-diameter tunnels with a large number of rock bolts.

(2) Shotcrete System

It includes concrete transfer pump and its hydraulic power pack and control cabinet, shotcrete manipulator and its hydraulic power pack and control cabinet, concrete transfer pipeline, admixture transfer pump and storage tank, etc.

(3) Grouting System

It includes grouting pump station, agitator, and its pipeline.

(4) Muck Discharge System

Generally, the TBM belt conveyor is directly transferred to the back-up belt conveyor. Sometimes the muck discharge system consists of three belt conveyors. The rock ballast is transferred from the TBM belt conveyor to the connection bridge belt conveyor, and then to the back-up belt conveyor. If the muck is transported out of the tunnel by a muck car, the layout of the train rail, car mover, and muck unloading point on the back-up jumbo shall be considered. If the muck is discharged by a continuous belt conveyor, the space for arranging the tail of the continuous belt conveyor and installing the supporting frame for the belt on the back-up jumbo shall be considered.

(5) Dewatering System

It mainly consists of dewatering pump, waste water tank, pipeline, etc. Generally, a multi-stage dewatering pump is arranged from the main machine to the back-up tail.

The rail or invert block laying facilities include material (such as rail) hoists and slides, as well as the necessary space for rail laying. If prefabricated invert blocks are used, an invert block hoist is required.

(6) Dust Removal System

It is generally composed of deduster, industrial exhaust fan, dedusting air duct, etc.

(7) Muck Remover

During TBM construction, a muck remover is required because a large amount of muck, especially broken rocks, will be deposited at the bottom of the tunnel. Generally, the muck is manually transferred to the muck removal belt conveyor or the hoisting shovel head, and then transferred to the back-up belt conveyor. Sometimes, a screw conveyor is used at the main machine to return the rock ballast to the cutter head.

4. Safety and Living Facilities

Safety facilities include handrails, walkways, steps, step ladders, guardrails, fire extinguishing devices, lighting devices, emergency generators, and harmful gas monitoring and alarm system arranged on the back-up jumbo. Living facilities include restrooms, toilets, etc.

2.6.3 Structure of Back-Up Jumbo

According to different tunnel designs and construction process requirements, the backup structure needs different designs in many aspects.

1. Different Walking Patterns

Generally, the back-up jumbo walks through the steel idler wheels installed on both sides of the bottom of the trolley. The structure of walking on the rail is shown in Fig. 2.114. However, some trolleys adopt the form of steel pipes sliding in the tunnel or wheels rolling on the lining segment. For example, when the tunnel diameter is very small, considering the overall layout, there are still several trolleys in front of the rail laying area, and there is no rail to walk. Therefore, the trolley in front of the rail laying area is installed with a slide pipe (sled) at its lower part to slide on the tunnel wall, while the trolley behind the rail laying area walks on the rail. The trolley in front of the shield TBM rail laying area usually adopts the structural form of wheels rolling on the lining segment.

2. Multiple Trolleys Connected

Considering the design, manufacture, and transportation, the trolley needs to be designed as the one with multiple trolleys connected. In addition, considering the tunnel construction error, rail laying error, and even the design of the tunnel with curved sections, the trolley must be designed into multiple sections, and there shall be enough freedom for the connection between each section. For tunnels with very small curvature radius, special consideration is needed to design the structure of the idler wheel of each trolley. Therefore, during the TBM and its backup structure selection and design, the TBM manufacturer generally requires the user to provide the minimum curvature radius of the tunnel. In addition to the TBM main machine, the structure of the backup idler wheel is also a factor to be considered.

Fig. 2.114 Idler-wheel travelling gantry back-up jumbo

3. Multilayer Structure

As mentioned above, there are many equipment and facilities on the backup system, and there shall be enough space on the trolley. The number of the trolley sections, as well as the multilayer structure, can be considered. Therefore, the back-up jumbo generally has a multi-layer structure. For tunnels with small diameters, its radial space is limited, and generally, there are two layers at most. Only small-sized equipment like pipelines and belt conveyors can be arranged on the top layer. Therefore, it is very in the longitudinal direction, and there are many trolleys. For tunnels with large diameters, trolleys can even be arranged in three layers. Because of the large space in width and height, so the trolleys can be fewer, but their overall dimensions are larger.

4. Closed Structure and Gantry Structure

The bottom platform structure of the back-up jumbo can be a closed structure or an open gantry structure as required. In case of a closed structure, the rail material car or muck car goes up from the tunnel rail to the back-up rear ramp section and then walks to the designated position through the rail of the back-up bottom platform. Figure 2.115 shows the closed structure of the bottom platform. In case of a gantry structure, the rail material car or muck car passes through the open bottom platform of the back-up system along the tunnel rail to the front of the back-up system, while the devices are placed on the bottom platforms on both sides and the upper platform. Figure 2.116 shows the open gantry structure. The structure shall be selected according to the convenience of loading and unloading of muck and concrete materials, costs, and other factors.

Fig. 2.115 Closed back-up jumbo

Fig. 2.116 Idler-wheel travelling gantry back-up jumbo

5. Connection Bridge and California Switch

As mentioned above, for large-diameter TBMs, a steel-structure connection bridge is generally arranged between the TBM main machine and the back-up jumbo, with a connection bridge belt conveyor and a shotcrete manipulator on top and the rail laying area below. For small-diameter TBMs, the TBM main machine is generally connected directly with the back-up jumbo, the rail laying area is arranged behind the operator cabinet, hydraulic power pack and shotcrete manipulator trolley, and the back-up belt conveyor passes through the bridge above the rail laying area, as shown in Fig. 2.117. In addition, when the muck is discharged by a single-track muck car

Fig. 2.117 Back-up rail laying area with front and rear jumbos

Fig. 2.118 California switch connected behind back-up jumbo

in the tunnel, a California switch is often connected behind the back-up jumbo, as shown in Fig. 2.118.

2.6.4 Main Back-Up Equipment and Layout

1. Operator Cabinet and Hydraulic Power Pack

Since there are a large number of electrical control lines leading to the main machine, and in order to facilitate the operators and other workers to check the tunneling status of the main machine, the TBM operator cabinet is usually located on the very front back-up jumbo. The operator cabinet is equipped with a PLC control cabinet, various panels and buttons, display devices for tunneling parameters, video monitoring, and laser guidance, telephones, etc. The operator cabinet should be shock-absorbing, noise-proof, smash-proof, and equipped with air conditioner.

The hydraulic power pack mainly provides power for the hydraulic cylinders of the TBM main machine gripper system, thrust system, rear support, shield, etc. The bolt driller and probe drilling machine on the main machine also require a hydraulic power pack to provide power. Due to the large number of hydraulic pipes leading to the TBM main machine, the hydraulic power pack is also usually located on the very front back-up jumbo. The hydraulic power pack includes motor, pump, group valve, filter, heat exchanger, oil tank, control valve, and temperature, pressure, and level sensors and gauges.

Figure 2.119 shows the hydraulic power pack and the operator cabinet located on the very front back-up jumbo, with the former behind the latter. There are two hydraulic power packs, one powering the TBM main machine and the other powering the bolt driller on the main machine. Sometimes, the TBM operator cabinet and the main machine hydraulic power pack can also be located on both sides of the TBM main beam.

2. Electrical Equipment

The main equipment for power supply and electrical control system includes HV cable reel, HV switch cabinet, transformer, cutter head variable frequency drive control cabinet, electrical control cabinets of other electrical equipment, etc. During TBM tunneling, the HV cable reel continuously releases the cable. After the cable is all released, a new real of cable will be connected and wound onto the cable reel. Generally, an HV cable reel can store a 250–500 m long cable. As the operator cabinet,

Fig. 2.119 Hydraulic power pack and operator cabinet

Fig. 2.120 Electrical equipment layout

hydraulic power pack, and shotcrete system are located as forward as possible, while the ventilation and water supply systems are generally arranged backward, the electrical equipment mentioned above is generally placed in the middle of the back-up system. In a large-diameter tunnel, electrical equipment can be located on the upper or lower level. In a small-diameter tunnel, electrical equipment is generally placed on one side of the back-up system lower level, while the rail for muck cars and material cars is placed on the other side. As shown in Fig. 2.120, the ϕ4.5 m gripper TBM electrical equipment along one side of the back-up jumbo includes HV cable reel, transformer 2, power equipment electrical control cabinet, power factor correction, HV switch cabinet, transformer 1, and cutter head drive VFD control cabinet from the rear to the front in order. The VFD variable frequency control cabinet is equipped with a special water circulation cooling system.

3. Air Compressor

The air compressor system mainly consists of air compressor, air tank, and pipeline. Usually, one air compressor is used for shotcrete system, and the other provides pneumatic power for pneumatic tools and some other equipment, such as pneumatic torque wrench, pneumatic dewatering pump, high-pressure air for dry-type deduster, lubrication grease pneumatic pump, water hose reel brake system, etc. For small-diameter TBM, the air compressor is usually arranged at one side of the back-up system middle part, between the concrete transfer pump control cabinet and the electrical control cabinet. Pneumatic pipelines are arranged along the back-up system forward to the main machine and backward to the rear. Each trolley is equipped with a pneumatic interface. The air compressor for large-diameter TBM is generally located on one side of the back-up system middle part bottom. Figure 2.121 shows the main gas-consuming equipment on the TBM and its back-up system. Figure 2.122 shows

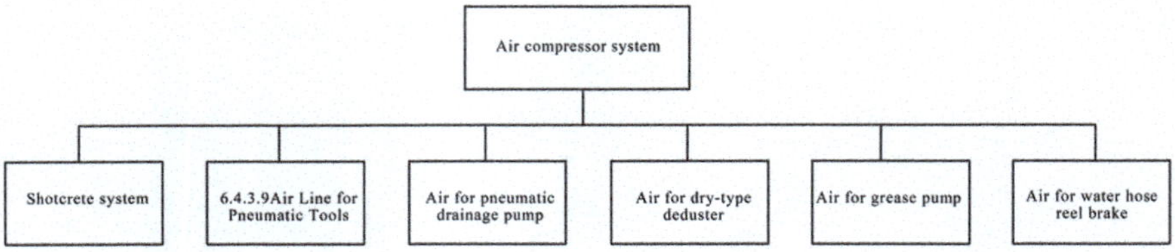

Fig. 2.121 Main gas-consuming equipment on the TBM and its back-up system

Fig. 2.122 Air compressor

two air compressors for a small-diameter TBM, arranged in sequence behind the concrete transfer pump control cabinet on one side of the back-up system.

4. Water Supply Equipment

Through the pipelines arranged on the tunnel wall, the industrial water outside the tunnel with a certain pressure, flow rate and temperature is supplied to the back-up trolley by a water pump or a height difference, and connected to the water hose reel, to complete the tunnel water supply system. The water system on the TBM and its back-up system equipment is mainly composed of water hose reels connected with the tunnel water pipelines, pumps, water tanks, filters, valves, pipelines, etc., and are used for industrial consumption water or the cooling of equipment internal circulating water. The internal circulating water cooling system for some key equipment on the TBM generally uses purified water. The industrial water supplied from outside the tunnel to the back-up system is cooled by heat exchange with the internal circulating water through the heat exchanger.

Obviously, the industrial water supplied from outside the tunnel to the back-up system, on the one hand, is used as construction consumption water, such as water spraying for the cutter head, water spraying for the belt conveyor muck-unloading

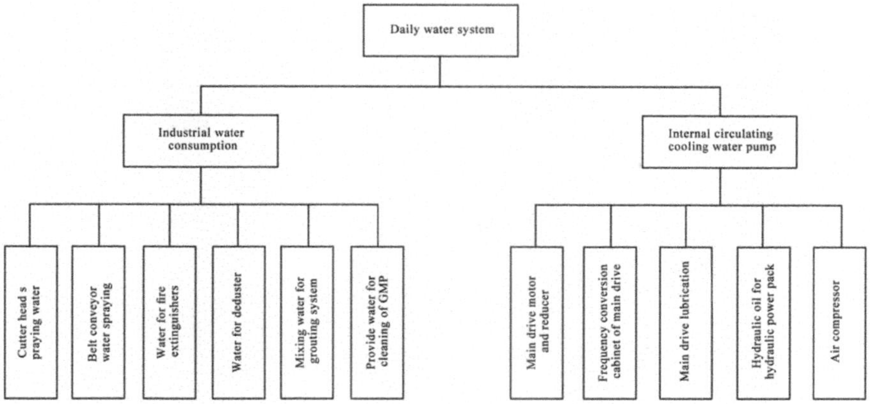

Fig. 2.123 TBM and its back-up water equipment

point, water for equipment cleaning, etc.; on the other hand, the industrial water can cool the inner circulating water by heat exchange with it, and finally cool the equipment, such as the cutter head drive motor and gearbox, cutter head drive variable frequency control cabinet, hydraulic oil for hydraulic power pack and main drive lubricant oil. The equipment for consumption water and cooling water is shown in Fig. 2.123.

As shown in Fig. 2.124, the water hose reel is usually arranged at one side of the back-up system rear bottom, which is continuously released as the TBM digs forward. After release, the steel water pipe will be extended forward along the tunnel wall, and the soft water pipe is then wound onto the water hose reel. According to the tunnel diameter, a general reel can store about 60 m of water pipes. Pump stations and water tanks are generally placed behind the electrical equipment and in front of the air duct cassette, basically in the middle and rear of the back-up system. In a large-diameter TBM, they can be placed on the lower or upper level, while placed on one side of the lower level in a small-diameter TBM. The water piping was arranged from the rear of the back-up system to the front of the TBM main machine, with several water connections along the way. Figure 2.125 shows the φ4.5 m small-diameter TBM water supply pump station arranged behind the HV cable reel and on one side of the back-up system, including pumps, heat exchangers, filters, valves, pressure and temperature sensors and gauges, industrial water tanks, etc. After the industrial water tank, there are water waste tank, dewatering pump, rest room, and toilet in order.

5. Dewatering Equipment

Generally, in the design and procurement of TBM, it is necessary to set up a dewatering system in the TBM and its back-up system, so that the tunnel gushing water and construction wastewater can be drained from the front of the TBM to the rear of the back-up system. Within the TBM and its back-up system, generally, two to three stages of drainage are generally set. To be detailed, a waste water pump is set up

Fig. 2.124 Water pipe reel

Fig. 2.125 Water pump station

under the TBM main beam and the back-up track laying area. Through the pipeline, the wastewater is discharged into the waste water tank in the middle or rear of the back-up system and then to the back-up system rear by the dewatering pump next to the waste water tank.

The waste water tank on the back-up system is generally equipped with a bulkhead for sedimentation, and the sediment should be removed during the maintenance shift. The waste water tank has a liquid level sensor, which enables to the dewatering pump to be started automatically or manually according to the water level. As shown in Fig. 2.126, the waste water tank and centrifugal dewatering pump of the ϕ4.5 m open

Fig. 2.126 Dewatering pump drains from the waste water tank to the back-up system rear

TBM are located behind the industrial water tank on one side of the back-up system, discharging the wastewater to the back-up system rear.

The dewatering from the back-up system rear to the outside of the tunnel requires an additional tunnel construction dewatering plan by the user according to the site conditions. A waste water pipe reel can be equipped to the back-up system rear and connected to the tunnel dewatering pipe for discharge to the outside of the tunnel; the wastewater can also be discharged to the outside if the tunnel without a pump, which may be determined according to the water amount and whether the tunnel is designed downhill. Depending on site conditions, the tunnel may require multi-stage dewatering pumps and water collection wells.

6. Ventilation Equipment

During TBM construction, the tunnel ventilation system delivers fresh air to the back-up system rear by tunnel entrance fans and flexible air duct. In the air duct cassette at the back-up system rear, there are 150–300 m long flexible air ducts, which are connected to the tunnel air ducts. As the TBM excavates forward, the air ducts are released continuously. After all the air ducts in the air duct cassette are released, another air duct cassette is replaced.

Since there is a long distance from the back-up system rear to the TBM main machine (usually 100–300 m), it is necessary to continue to press the air forward from the air duct cassette. Therefore, after passing the air duct cassette, the fresh air will usually be pressed to the TBM main machine rear through the booster fan and metal duct. Therefore, the ventilation facilities on the TBM and its back-up system mainly consist of an air duct cassette, a secondary booster fan and a metal air duct. The air duct cassette is arranged at the tail of the upper layer of the back-up system, just behind the booster fan, and the metal air duct is connected to the TBM main machine from the booster fan along one side of the upper layer of the back-up system.

Figure 2.127 shows the air duct cassette installed at the tail of the back-up system. Figure 2.128 shows the booster fan in front of the air duct cassette and the metal air duct supplying air to the TBM main machine.

7. Dust Removal Equipment

The gap between the TBM shield and the drive box steel structure of the main drive is designed with a dustproof plate to prevent dust from flowing directly out of the tunnel face. During construction, the cutter head generally needs to spray water to

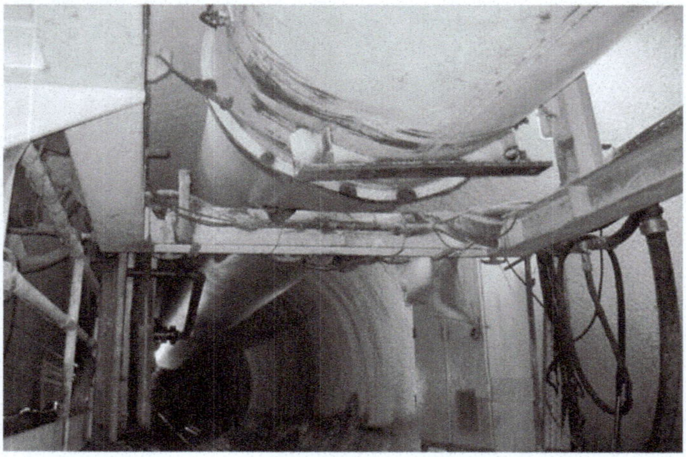

Fig. 2.127 Air duct cassette

Fig. 2.128 Booster fan and metal air duct leading to TBM main machine

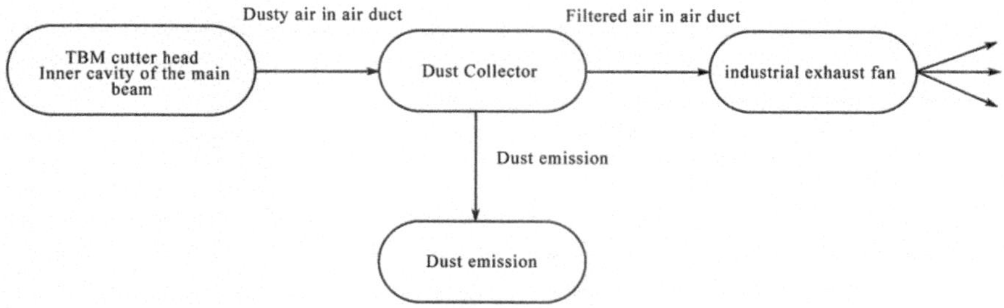

Fig. 2.129 Layout of dedusting system

reduce dust, and a dedusting system is installed for air suction and dust removal from the inner cavity of the main beam.

The dedusting system mainly consists of a deduster, a metal air duct, an industrial exhaust fan, etc. The inner cavity of the TBM main beam is used as an air duct and a metal air duct is connected behind the main beam. The air containing dust in the cutter head is sucked out by the industrial exhaust fan and then the air filtered by the deduster is discharged to the tail part of the back-up system through the air duct and the industrial exhaust fan and then out of the tunnel at certain flow rate. Generally, the deduster and industrial exhaust fan are arranged in the middle or rear part on one side of the back-up system and can be placed in the upper or lower layer of the back-up system. The industrial exhaust fan is arranged near the back of the deduster, and both are connected by a metal air duct. Figure 2.129 shows the layout of the dedusting system.

The dedusters include wet-type dedusters and dry-type dedusters. Both have achieved satisfactory results in TBM application. The wet-type deduster settles the dust via a wet-type filter element and high-pressure water mist, and then discharges the dust-containing waste water into the waste water tank for regular mud removal. The dry-type deduster settles the dust via a dry-type filter element and compressed air and then conveys the dust into the water-filled agitator through the screw conveyor. After that, the mud is discharged by the pneumatic pump to the back-up belt conveyor to transport it out of the tunnel. Wet-type dedusters should be maintained and cleaned regularly. The filter elements of dry-type dedusters should also be cleaned regularly. It is worth noting that, generally, dry-type filter elements should not be cleaned with water, hard objects or HV air. Special attention should be paid to preventing tunnel water from being sucked into the dry-type deduster from the damaged air duct, so as to prevent water and dust from mixing into mud and thus blocking the filter elements.

Both the 12.4 m large-diameter TBM of the Jinping II Hydropower Station Project and the 4.5 m small-diameter TBM of the Nabang Hydropower Station Project use dry-type dedusters manufactured by CFT GmbH. The practical application shows that the dry-type dedusters have reliable performance and good filtering and dust removal effect. Figure 2.130 shows the dry-type deduster system arranged on the back-up system, including deduster, dust screw conveyor, mud agitator and pneumatic pump.

Figure 2.131 shows the HP air valve and filter element in the dry-type deduster. Figure 2.132 shows the internal filter element and the bottom screw conveyor during maintenance and disassembly of the filter element.

8. Back-up Mucking Belt Conveyor

During TBM construction, the cutter head rotates, the muck from rock breaking is picked up by the bucket and dropped into the TBM belt conveyor through the muck chute of the cutter head, and then transferred to the back-up belt conveyor, and unloaded into the muck car or tunnel continuous belt conveyor to transport it out of the

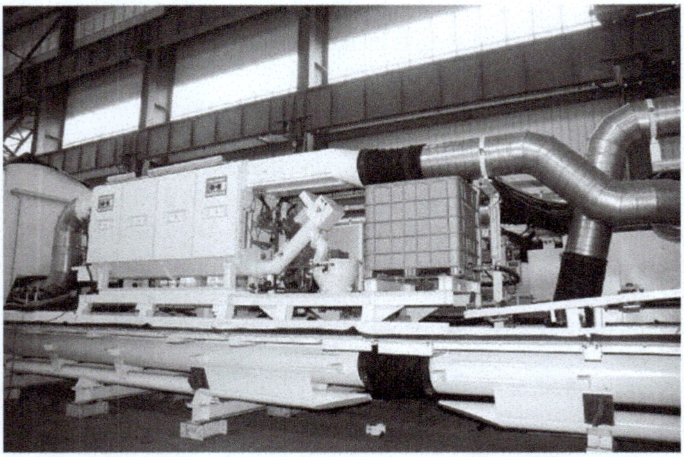

Fig. 2.130 Dry-type deduster on back-up system

Fig. 2.131 HP air valve and filter element in dry-type deduster

Fig. 2.132 Filter element and screw conveyor in dry-type deduster

tunnel. If there is a connection bridge between the TBM main machine and the back-up system, a connection bridge belt conveyor is usually added between the TBM belt conveyor and the back-up belt conveyor. Therefore, the muck transportation from the TBM main machine to the back-up system is done by belt conveyors.

For convenience of muck unloading, the back-up belt conveyor is usually arranged on the upper layer of the back-up jumbo. When the muck is transported by a muck car, the unloading point is generally close to the rear of the back-up system, so the back-up belt conveyor basically runs longitudinally through the whole back-up system. If a continuous belt conveyor is used for mucking, its tail section needs to be installed on the side of the lower layer of the back-up system, and there is enough space reserved for installation of supporting frame for belt of the continuous belt conveyor and storage of materials in front of and behind the tail section, so the muck unloading point of the back-up belt conveyor is generally located at the middle to rear part.

The back-up belt conveyor consists of belt, drive device, drive pulley, driven pulley, frame, roller, etc. A vibration-damping roller is generally used at the muck receiving position of the driven pulley. For a long back-up belt conveyor, a guide roller is added, and the supporting frame for belt can even designed to be adjustable and floating to prevent the belt from deviating. To meet the operation and maintenance needs, the back-up belt conveyor also needs to be designed with a belt tension adjustment mechanism. Emergency pull switches are arranged along the back-up belt conveyor.

During tunnel construction, the belt conveyor operates in a poor environment and muddy water often scatters from the belt, so not only the drive device and the belt must be reliable, but also the roller bearings and their seals must be of high quality and high performance. Compared with the TBM belt conveyor, the back-up belt conveyor has a low probability of belt damage, but in order to avoid TBM shutdown caused by accidental scratches, a spare belt is generally available on site. Sometimes, the belt needs to be replaced as whole or patched partially.

Figure 2.133 shows a 4.5 m small-diameter TBM back-up belt conveyor arranged along the upper layer. Due to the small diameter, the muck receiving position at the tail of the back-up belt conveyor is relatively low. Therefore, the back-up belt conveyor needs to gradually transition from the lower layer to the upper layer from the tail of the TBM main machine, as shown in Fig. 2.134. The muck unloading point of the back-up belt conveyor is shown in Fig. 2.135.

Fig. 2.133 Back-up belt conveyor on upper layer

Fig. 2.134 Belt conveyor section transitioning from lower layer to upper layer of back-up system

Fig. 2.135 Main drive and muck unloading point of back-up belt conveyor

9. Shotcrete System

For a gripper TBM, the initial supporting operations during TBM tunneling include rock bolt operation, ring beam operation, mesh hanging, shotcreting, etc. The supporting operations should be carried out as quickly as possible once the TBM shield is exposed. Therefore, the shotcrete system should be placed as forward as possible for early implementation. However, since the shotcreting at the TBM main machine is easy to cause serious contamination around the main machine and other operations cannot be carried out, generally, only an emergency shotcrete device can be arranged at the TBM main machine and connected to the conventional concrete transfer pump through a pipeline. The shotcrete system is generally arranged in front of the back-up system.

The shotcrete system is mainly composed of concrete transfer pump and its control cabinet, shotcrete manipulator and its control cabinet, admixture transfer pump, manipulator remote controller, transfer pipeline, etc. as shown in Fig. 2.136. The manipulator should be arranged as far forward as possible and such factors as circumferential shotcreting range and distance to tunnel wall should be considered. For the layout of the concrete transfer pump, the position that the rail tanker can reach should also be considered.

For a large-diameter TBM, there is generally a connection bridge between the tail of the TBM main machine and the back-up system and the shotcrete manipulator is often arranged on the connection bridge. Rail laying can generally be carried out below the connection bridge, so the concrete transfer pump is arranged on both sides of the lower layer of the first back-up jumbo.

For a small-diameter TBM, due to the small space, the manipulator needs to be arranged as low as possible to keep the shotcreting distance from the tunnel wall, so it is usually arranged directly on the jumbo. Since the TBM operator cabinet and

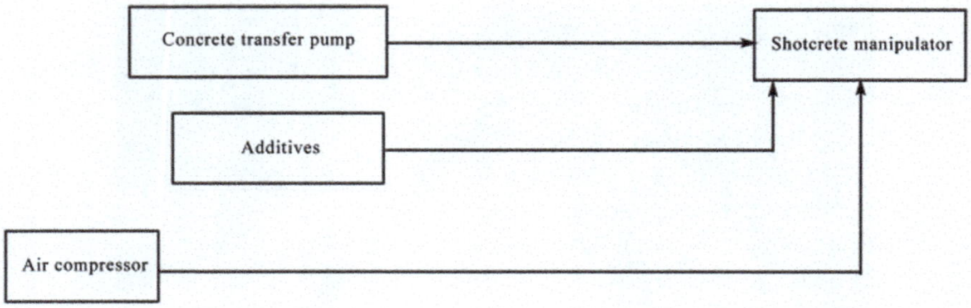

Fig. 2.136 Composition of Shotcrete System

Fig. 2.137 Concrete flowing directly from tanker into transfer pump hopper

hydraulic power pack are arranged on the first back-up jumbo, the manipulator is usually arranged on the jumbo behind the hydraulic power pack, while the concrete transfer pump on the jumbo behind the rail laying area.

The concrete transfer pump has such functions as vibration, mixing and pumping. The power pack and operator cabinet are arranged close to the pump for operation. The transfer pump should be arranged to facilitate the concrete tanker to discharge the concrete into the hopper conveniently and quickly. The relative positions between the tanker and the transfer pump, as well as whether to move the tanker and whether a concrete transfer mechanism is needed are all important considerations. The fewer processes, the better. Direct placement is the best. As shown in Fig. 2.137, the transfer pump is arranged at a low position on one side of the back-up jumbo, and the concrete tanker can discharge the concrete directly into the hopper from the side. Such an arrangement is not only convenient but also can reduce faults.

The shotcrete manipulator is operated by the operator via a remote controller for shotcreting. Such actions as circumferential rotation, longitudinal swinging, shaking and longitudinal movement are required. The longitudinal movement includes back and forth telescopic movement driven by the boom, and longitudinal movement of the boom and the manipulator driven by the dragging trolley. For a large-diameter TBM, it is possible to achieve a 360° circumferential shotcreting range, while the shotcreting range of a small-diameter TBM is generally less than 270°. A large longitudinal movement range is desired so that there is a certain buffer margin to reduce the downtime of the TBM when the tunneling and shotcreting supporting cannot be carried out synchronously. Generally, the longitudinal movement range should be above 6 m. In case of a large-diameter TBM and large amount of shotcreting, it is better to arrange two shotcreting manipulators in front and back respectively, each of which can carry out circumferential shotcreting in the left and right directions, so that circumferential shotcreting can be carried out in the left and right directions respectively when the TBM tunneling speed is fast, or when one manipulator fails, the other can still meet the circumferential shotcreting requirements. Figure 2.138 shows the arrangement and commissioning of the small-diameter TBM shotcrete manipulator.

10. Grouting Equipment

The grouting system can be used for pre-grouting or consolidation grouting of the cavity between concrete and tunnel wall. In general, its utilization rate is very low. It is mainly composed of mixing station, mixing box, grouting pump station, control box, pipeline, group valve, etc. Grouting equipment can generally be arranged in front of the concrete transfer pump. For a large-diameter TBM, it can also be arranged in the upper layer of the front back-up jumbo. The grouting pipe is laid from the pump station to the front of the TBM main beam. The grouting equipment is shown in Fig. 2.139. The TBMs are mostly equipped with grouting systems manufactured by HANY, Switzerland.

Fig. 2.138 Arrangement and commissioning of TBM Shotcrete manipulator

Fig. 2.139 Grouting equipment

11. Material Hoisting Equipment

TBM needs various materials or spare parts during construction. These materials or spare parts are generally transported into the tunnel by flat car, and then unloaded by hoist from the front of back-up system to the front of TBM main machine. Sometimes they have to be lifted from the lower level to the upper level, and the cutter has to be moved to the Cutterhead inside the main beam. Materials to be transported include those used for support, such as ring beam, rock bolt, anchoring agent and steel mesh, as well as cutter, rail, sleeper, invert (if necessary), spare parts, lubrication oil drum and lubrication grease drum.

Hoisting equipment varies with materials and positions, such as manual hoist, electric hoist and transfer trolley. The slide of hoisting equipment includes the swing arm revolving around fixed pillar, the logitudinally-fixed slide under main beam and the trolley rail. Hoisting equipment is provided all the way from a position accessible to the material flat car to the front of the TBM. Hoisting equipment is selected depending on the layout of TBM and back-up system, material quality and size, and material destination. Transfer may be required between the hoisting equipment of different level. Hoisting equipment for transporting material from flat car to back-up system is shown in Fig. 2.140. Hoisting equipment and its longitudinal slide for transporting disc cutter to the cutter inside TBM main beam is shown in Fig. 2.141.

12. Slag Remover

During TBM construction, a large amount of muck will be deposited at the bottom of the tunnel, and rocks of large quantity may collapse at fault-fracture zone of surrounding rock, and some shotconcrete may bounce back. All these make muck

Fig. 2.140 Electric swing arm hoist for handling materials

Fig. 2.141 Cutter hoist inside TBM main beam

removing necessary. In addition, it is also necessary to remove slag to lay tracks or invert, or prevent back-up system from getting stuck, as well as meet the requirements for completing the tunnelling. Slag removing looks easy but is an important link of TBM construction. The tunnelling progress is affected or delayed to some extent due to inappropriate arrangement of slag removing equipment or partial slag removing.

Slag removing is generally carried out at the lower part of TBM main beam, track-laying area under connection bridge or track-laying area between back-up system. The slag is manually removed to slag remover and then moved outside tunnel via muck conveyor. Slag removing equipment includes the lifting device with bucket,

Fig. 2.142 Slag belt conveyor

the belt conveyor and screw conveyor, etc. Figure 2.142 shows the slag removing conveyor installed at the track-laying area in back-up system. The slag is manually removed to the belt conveyor and then moved outside tunnel via cart.

13. Car Mover

In case that a low-pollution internal-combustion locomotive is used for pulling slag cart, Car Mover is generally installed for the back-up system to reduce pollution emission inside the tunnel and running time of the locomotive when slag truck arrives at the unloading point of the back-up system. The locomotive stops working once it reaches its position, and the car mover pushes slag cart one by one to fill slag carts with slag, and then the locomotive works again to tract slag carts outside the tunnel. Naturally, in case of any fault of car remover, the locomotive can be used to fill slag carts with slags.

Car movers are longitudinally arranged at the slag unloading point. Some car movers are provided at the bottom of back-up system to push the wheels forward while some are provided at the sides of slag unloading point of back-up system to push the rear wall of bucket by using toggle switch between two slag carts to move them forward. When the car movers installed at the bottom of back-up system breaks down, it is not easy to repair, while it is easy to repair for the car movers installed on the sides of slag unloading point of back-up system. The latter has been recently applied to the TBM used in Nabang Hydropower Station in China, showing great effect.

Figure 2.143 shows the car mover installed on the sides of slag unloading point of back-up system. The support of car mover is installed on the side of the parking position of slag cart, and a mobile trolley is provided at the upper and lower part of the support respectively. Mobile trolleys are pulled by a chain, and the chain is driven by a motor-driven chain wheel so that the mobile trolley moves back and

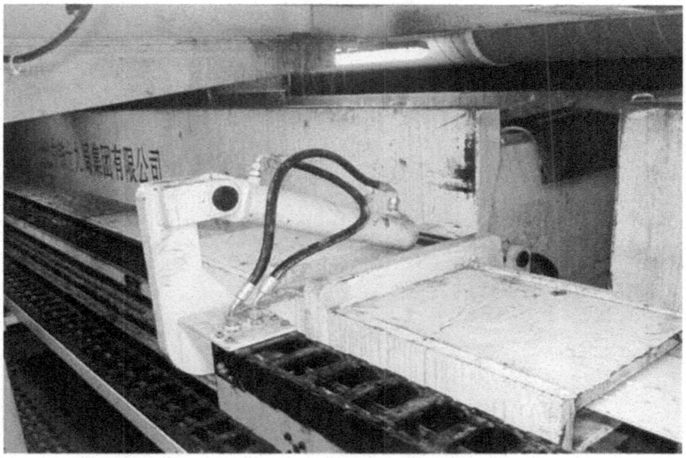

Fig. 2.143 Car mover installed on the side of back-up system

forth longitudinally. The mobile trolley device is equipped with a telescopic cylinder. Telescopic cylinder can extend a toggle switch to the position between two slag carts or retract the toggle switch. The two trolleys move back and forth under the drive of the chain to make slag carts move longitudinally so that slag carts are filled with muck.

14. Emergency Generator

In case of high-voltage power supply interruption in the tunnel, the emergency generator outside the tunnel shall ensure power supply for illumination and ventilation, and even for tunnel drainage. The emergency generator at the back-up system of TBM mainly feeds the machine, safety support equipment, DC power supply and control power supply, as well as TBM and power demand for water use safety within the back-up system. For this reason, the size of installed power should be considered in the selection of design. Emergency diesel generator is generally provided at the rear of back-up system, and it can be manually activated or activated by PLC control. Figure 2.144 shows 60 kV·an emergency generator equipped for TBM with a diameter of 4.5 m. It is provided at one side of the rear of back-up system.

15. Fire Extinguishing System

There are many devices consuming oil and electricity, as well as works such as electrogas welding and gas cutting during TBM construction. Meanwhile, the devices are expensive. Therefore, it is necessary to provide fire extinguishing system for TBM and its back-up system. The fire extinguishing system is generally composed of three parts, namely automatic foam fire extinguishing system, automatic CO_2 gas extinguishing system and portable fire extinguisher. The automatic fire extinguishing system is generally considered along with the design selection of TBM. However, customer is allowed to purchase portable fire extinguisher by themself.

Fig. 2.144 Emergency diesel generator on back-up system

Automatic foam fire extinguishing system is provided beside the hydraulic power station. Its detection sensor and nozzle are over the hydraulic power station, and the nozzle works when fire is detected, as shown in Fig. 2.145. Automatic CO_2 gas extinguishing system is provided beside the electrical control cabinet. Its detection sensor is placed inside the control cabinet, and the nozzle sprays CO_2 when fire is detected, as shown in Fig. 2.146. Several portable fire extinguishers are provided at key positions from TBM main machine to back-up system.

Fig. 2.145 Temperature sensor and nozzle of foam fire extinguishing system

Fig. 2.146 Automatic CO_2 fire extinguishing system beside electrical control cabinet

16. Gas Monitoring System

Detection of harmful gas and oxygen is an important safety guarantee measure in tunnel construction. The gas detection system provided on TBM shall be able to detect CH_4, H_2S, CO_2, CO and O_2. CH_4 sensor is provided below the front shield of main machine (Fig. 2.147). H_2S sensor can be provided at the slag discharge point of belt conveyor of main machine (Fig. 2.148). CO_2, CO and O_2 sensors are generally provided at the places where people gather, such as lounge. All gas detection sensors are integrated in the PLC control system to set the values of alarm and automatic shutdown, and display the corresponding information in the operator cabin and give an alarm.

17. Video Monitoring System

The length of TBM and its back-up system varies from 150 to 300 m. Various auxiliary operation procedures and their devices are distributed inside TBM and its back-up system. Video monitoring system is provided on TBM to enable TBM main operator in main control cabin to know equipment operation status and progress of other operation procedures in a bid to ensure the safety and coordination of the whole operating system. The monitors are set inside the main control cabin, as shown in Fig. 2.149. And the cameras are positioned at the key parts needing surveillance, generally 4–6 cameras. The overall video monitoring can be transmitted to the computer outside the tunnel via optical fiber, and even can be accessed anywhere by internet.

Fig. 2.147 CH_4 sensor below TBM top shield

Fig. 2.148 H_2S sensor near slag discharge point of belt conveyor of main machine

Figure 2.150 shows the positions needing focused monitoring, namely positions where cameras shall be provided. including: camera for the head of belt conveyor of main machine, monitoring operation of belt conveyor and slag removing, as well as any change in the geology where disc cutter is located; camera for the middle section of belt conveyor of the back-up system, monitoring the operation of longer belt conveyor of the back-up system; camera for slag loading point of slag cart, monitoring the arrival of slag cart and slag loading progress; camera for top shield over main beam, monitoring outcropped rock from shield and operation progress at

Fig. 2.149 Video monitoring in main control cabin

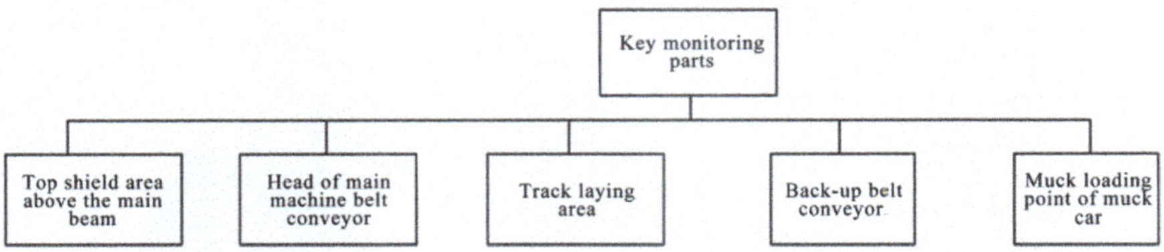

Fig. 2.150 Positions needing focused monitoring of camera system

Fig. 2.151 Pipelines from back-up system to main machine transition point

Fig. 2.152 Pipelines on side support of back-up gantry

Fig. 2.153 Pipelines on both sides at the upper part of back-up system

support; cameras for main beam bottom and rail-laying area, monitoring rail-laying progress as well as rear leg movement below main beam.

18. Communication System.

The development of communication technology makes the communication between TBM and any point inside or outside the tunnel easy and simple. Wired telephone can be installed at key parts and main operation points of TBM and its back-up systems, and connected with the telephones in the operator cabin. And the telephone in the operator cabin is connected with the dispatching room outside the tunnel and any other telephone. In addition, Walkie-Talkie can be provided as auxiliary communication between operation points. And even cellphone signal can be introduced into the tunnel as a method of communication.

19. Layout of Pipeline

The pipelines along TBM and its back-up system include high-voltage cable, utilities cable, communication cable, illumination cable, control cable, hydraulic pipe, compressed air pipe, air pipe, water pipe and grouting pipe. The entire pipeline runs from back-up system to the front of TBM main machine. The air pipe has a large diameter, generally provided on both sides of the upper part of back-up system. It is better to design other pipelines in different colors and provide them in layers on each support to avoid interference among them and facilitate pipe laying and fault inquiry. Figures 2.151, 2.152 and 2.153 show the examples of layout of pipelines.

Questions

1. Description of Gripper TBM Operating Principle.
2. What are the two types of gripper TBM? What are their respective characteristics?
3. What are the main components of TBM main drive?
4. Description of Kelly Gripper TBM Main Machine.
5. Description of Main Equipment and Layout of TBM Back-up System.
6. Description of Geology Application Range of Gripper TBM.
7. Description of Geology Application Range of Single-shield TBM.
8. Description of Geology Application Range of Double-shield TBM.
9. Description of Advantages and Disadvantages of Double-shield TBM.

Chapter 3
TBM Selection and Adaptability Design

TBM selection · TBM adaptability indicator · Adaptability evaluation

This chapter focuses on the basis, principles and influencing factors in TBM selection, describes the adaptability design of TBM from tunneling performance requirements, main parameter design, inter-system match design, etc., build a TBM adaptability evaluation method and model by using fuzzy mathematics and analytical hierarchy process, and verify it based on real project cases.

3.1 Overview

The vigorous development of major lifeline projects is the priority for rapid and sustainable economic growth, such as water conservancy, transportation and mines. The next decade will witness the construction of more than 40 super-large hydropower stations and some inter-basin water transfer projects in western China, the highway and railway network extension to the west, and mineral resources mining up to over 1000 m deep on a large scale. Deep and long tunnels are key control works in these lifeline projects. In terms of environment protection, high mountains and deep valleys in the western region make it difficult to drill and blast by multi-headed sublevel tunneling for deep and long tunnels. Given the duration, cost and techniques, TBM will be the preferred choice for long and large tunnels. In recent years, TBM is usually preferred in construction of long and large tunnels for its obvious advantages such as rapidity, high quality, safety, environmental protection and economic efficiency. But its main disadvantage is that its adaptability to geological conditions is not as good as that of conventional drill and blast, and its initial investment cost is high. A wide variety of TBMs are available, and they are classified in different ways according to different reference standards. To ensure the rational and efficient use of TBM, the following elements shall be met: clear geological conditions and necessary

K. Chen et al., *TBM Design and Construction*,
https://doi.org/10.1007/978-981-99-0059-6_3

pretreatment measures, reasonable TBM model and perfect supporting system, experienced construction team, and professional control measures and scientific management. When the geological conditions are given, TBM selection directly determines the construction cost and work efficiency of the project and even determines the success or failure of the project.

In a word, reasonable TBM selection is the key to the success of the project. From 1985 to 1992, a φ 10.8 m open Robbins TBM was selected for the water diversion tunnel works of Tianshengqiao-II Hydropower Station. Due to inadaptability between type selection and geological conditions and high equipment failure rate, project progressed so slowly that the average advance per month was only 65 m. In the main canal works for water diversion from Datong River to Huangshui River in Qinghai, a φ 5.93 m WIRTH double shield TBM was used. In October 2006, tunneling works began. By March 9, 2010, the completed length was 7258 m. On April 3, 2008, after entering the southern margin fault zone (F5, F4, F19 fault groups) of Dabanshan, the TBM underwent nine times of jamming and breakout. Therefore, half-section manual excavation and TBM tunneling methods had to be used together. In Hongyanhe-to-Shitouhe Water Diversion Project, an φ 3.655 m Robbins double shield TBM was used. In November 2008, tunneling works began. Nearly 20 times of jamming occurred during the construction. Seriously affected by frequent change of surrounding rock, high crustal stress and large surrounding rock deformation, and fault fracture zone, the segment was damaged and the shield crushed, and it was difficult to control TBM direction.

Fundamental reasons for these adverse consequences include unclear understanding of the interaction mechanism between surrounding rock and TBM, and lack of scientific and reliable TBM adaptability evaluation standards and safety control theory. The basic theoretical research on TBM adaptability and safety control has become an urgent and important subject in the engineering fields like water conservancy, transportation and deep resources exploitation in China. Since there is no comprehensive and in-depth domestic study on TBM adaptability design and evaluation in the complex geological environment, the application of TBM in complex geological strata has been seriously restricted. Therefore, it is urgent to research and explore new design theories, analyze adaptability evaluation methods, and build a new evaluation and decision-making system. Many factors affect TBM adaptability evaluation. The deep analysis of various influencing factors is essential for adaptability evaluation to determine their independent and coupled influences. Therefore, this chapter will, based on the structural characteristics and adaptability of several common TBMs, compare and demonstrate some common TBM types to consider several TBM adaptability influencing factors, and will, based on the analysis of the influencing factors, build a TBM adaptability evaluation method and mode by using fuzzy mathematics and analytic hierarchy process, and verify it by real engineering cases to guide TBM adaptability design.

3.2 TBM Selection

3.2.1 *Basis and Principles*

1. Basis

Before tunnel construction, TBM shall be selected to achieve reasonable matching, give full play to the overall efficiency of construction machinery and improve mechanized construction. TBM selection is based on the following:

(1) Engineering geological and hydrogeological conditions of the tunnel, including stratum lithology, rock strength, integrity, joint development degree, quartz content, groundwater development degree, groundwater level, water inflow in tunnel and adverse ground condition, etc.
(2) Design parameters such as shape and geometric size of the section and the length, slope, turning radius and overburden of the tunnel.
(3) Environmental conditions around the route, site conditions along the route, structural characteristics of adjacent pipelines, buildings and underground caverns, foundation forms, present conditions and possible deformation.
(4) Sufficient assembly sites at the entrance and exit of the tunnel and conditions for transport and hoisting of large parts as well as climate conditions, water and electricity supply, traffic conditions and other geographical and environmental factors of the construction site.
(5) The length of TBM continuous tunneling at a time and the maximum length of a single section.
(6) Schedule requirements of tunnel construction such as total duration, preparation period and excavation period.
(7) Deformation monitoring and measurement data of tunnels constructed by drilling and blasting method in the same area.
(8) Flexibility and economic efficiency in dealing with adverse ground conditions.
(9) Performance and technical service of TBM manufacturers.
(10) Professional technical level and management level of the construction team, etc.

2. Principles

The performance of TBM and its adaptability to geological conditions and engineering construction features are critical to the success of tunnel construction. TBM with advanced technology and reliable quality and TBM manufacturers with rich experience and professional service are the key factors. The TBM systems and components shall be selected based on the unity of their reliable performance, advanced technology, economic efficiency and safety. The environmental conditions such as outer diameter, length, overburden, and geological conditions of the tunnel, topography along the route and the conditions of the entrance shall be complied according to the geological data provided in the bidding documents. Existing TBM engineering examples at home and abroad and related technical specifications shall be referred to. The requirements of safety, quality, duration, cost and environmental protection shall be met. Back-up equipment shall be matched with the main machine, and the advance speed of the main machine shall match with the production capacity to achieve low energy consumption and high efficiency. In addition, safe construction, simple structure, reasonable layout and easy maintenance is required. Electrically powered machinery shall be preferably selected as machines entering the tunnel.

The following principles should be followed in TBM selection:

(1) Unity of safety, advanced technology and economic efficiency.

TBM shall be so selected that safety and reliability shall be ensured firstly, and then the advanced technology and economic efficiency. The technological level shall be up-to-date, reliable, appropriately advanced, conforming to the project characteristics, meeting the purpose, and featuring the unity of safety, advanced technology and economic efficiency.

(2) Conforming to environmental conditions.

TBM equipment shall be selected based on the environmental conditions such as outer diameter, length and overburden, and geological conditions of the tunnel, topography along the route, tunnel entrance conditions, and an in-depth analysis of the tunnel construction environment. Due to the strong geological pertinence of TBM, the performance of TBM depends largely on the engineering geological conditions and hydrogeological conditions both of which are important factors affecting the tunnel construction quality and important basis for the selection of TBM equipment. Geological survey data shall be comprehensive, true and accurate. In addition to detailed and accurate geological survey data, the landform and geological lithology of the tunnel site and the stability conditions of the section across the valley, the shallow overburden section near the mountain, and the slope at the entrance and exit, etc. shall be included. TBM is most sensitive to the stratum that the tunnel passes through. Different TBMs are used for different strata. Generally, open TBM is more suitable for hard rock tunnels with Class II and III surrounding rocks, while shield TBM is more suitable for tunnels with Class III and IV surrounding rocks. Shield tunneling is often used under changeable stratum, soft soil stratum, complex surface structure, and high requirements for settlement control.

(3) Meeting the requirements of safety, quality, duration and cost.

The TBM equipment shall be configured as rationalized and standardized as far as possible, and selected based on the scale, difficulty, safety, quality, duration, cost, environmental protection, civilized construction of the project after full research and investigation. The duration requirements include the whole process of TBM preparation, tunneling, lining, disassembly and turnaround. The preparatory work includes bid procurement, design, manufacturing, transport, site, installation, commissioning, stepping, etc. The total excavation period shall meet the requirements of the scheduled period. For simultaneous tunneling and lining works, the total duration for TBM tunneling should meet the requirements of the scheduled tunneling period. The disassembly and turnaround of TBM shall meet the requirements of the scheduled follow-up period.

(4) Back-up equipment matching with the main machine.

The back-up equipment shall be matched with the TBM to ensure well-matched production capacity and advance speed, conformable working state, low energy consumption, and high efficiency, and shall feature safe construction, simple structure, reasonable layout and easy maintenance. Electrically powered machinery shall be preferably selected as machines entering the tunnel, and relevant supporting equipment shall be reasonable. The production capacity shall first meet the construction period required in the construction organization design, so as to in step with the set schedule. The back-up equipment shall be selected based on the requirements of occupational health and safety such as labor protection and environmental protection, and the requirements of safe and well-organized construction. In addition, the operator's labor intensity and improved working conditions shall be met and less-polluting, energy-saving and high-efficient construction machinery shall be provided to reduce the environmental pollution in the workplace and benefit the environmental protection. At the same time, construction managers shall have a strong sense of labor protection and environmental protection, regard environmental protection as the most important content of site management from beginning to end, strengthen environmental management work and formulate environmental protection measures.

3.2.2 Selection Steps

TBM is generally selected in the following steps: Determine the type of TBM according to geological conditions, construction environment, construction period requirements, economy and other factors; Select open TBM or shield TBM; Compare the structure and parameters of similar TBMs according to the tunnel design parameters and geological conditions, and determine the main technical parameters of the TBM; Determine the technical parameters and functional configuration of the back-up equipment according to "well-matched production capacity and advance speed" principle.

After determining the TBM type, it is necessary to determine the main technical parameters of the main machine according to the tunnel design parameters, geological conditions and tunneling length of the specific project, and select the main machine with strong adaptability to stratum, reliable function of the whole machine, and operability and safety. For open TBM, special attention shall be paid to the selection and matching of auxiliary supporting equipment such as ring beam erector and shotcrete to adapt to the change of tunnel geology.

TBM equipment is composed of main machine and back-up equipment, forming a mobile mechanized tunneling line. The main machine is mainly used for rock breaking and slag loading. The technical parameters, functions and forms of the back-up equipment shall be well-matched with the main machine. The capacity of the main machine and progress shall be taken as the standard for calculation. In order to give full play to the advantages of TBM and ensure the smooth completion of the project, the capacity of matching equipment shall be appropriately expanded, with appropriate margin to meet the needs of normal construction progress and possible expansion. The back-up system is roughly classified to three types: track type, continuous belt conveyor type and trackless tire type. The continuous belt conveyor type is gradually popularized for its single structure and quick slag transportation.

TBM is classified to three types: open type, double shield type and single shield type. They are suitable for different geological conditions, and therefore shall be selected after comprehensively analyzing the engineering geological and hydrogeological conditions, construction environment, construction period requirements, economic situation and other aspects according to Table 3.1.

3.2.3 Factors Affecting TBM Selection

TBM is generally selected with the following factors considered:

(1) Design parameters of tunnel such as geometric length, plane and longitudinal section size.
(2) Geological conditions of the tunnel: the level of rocks, lithology, hardness of rocks (uniaxial saturated compressive strength Rc), number of faults, width of faults, types of fillings and physical characteristics of the tunnel; rock mass integrity degree and rock mass integrity coefficient; wear resistance and quartz content of rock; combination between occurrence of main structural planes of rock mass and tunnel axis; Initial ground stress state of rocks; hydrogeology such as water-bearing and water leakage of the tunnel; distribution of harmful, combustible gases and radioactive substances in the tunnel.
(3) Tunnel construction environment: surrounding environment, inlet and outlet construction sites, traffic conditions; climatic conditions; water and electricity supply.
(4) Total duration of tunnel construction and node construction period requirements.
(5) Economic and technical comparison.
(6) The following factors that restrict the construction performance of TBM should be considered emphatically when selecting the type: boreability of rock stability of tunneling surface, stability of tunnel wall during tunneling, width of fault zone, existence of squeezing ground.

TBM selection is determined by comprehensive analysis of tunnel geological conditions. Generally, the soft rock, hard rock and fault fracture zone can be tunnelled by different types of TBM with necessary pre-reinforcement and support measures, but TBM construction is not suitable for tunnels with large karst underground rivers, tunnels with high ground stress, tunnels with large deformation of soft rock, tunnels with special adverse ground conditions such as large-scale water and mud inrush. Under these circumstances, drill and blast method can give full play to its flexibility.

Table 3.1 Comparison between open TBM and shield TBM

Comparison items	Open TBM	Double-shield TBM	Single-shield TBM
Geological adaptability	It is generally used in good geological conditions, featuring good adaptability to hard rock tunneling. In soft rock, stratum shall be reinforced in advance. It is suitable for tunnels dominated by Class II and III rocks	Its adaptability to hard rock tunneling is the same as that of open type. In soft rock tunneling, single shield is better than open type. It is suitable for tunnels dominated by Class III rocks	It is used under relatively poor geological conditions (but the excavation face can be self-stabilized). It is suitable for tunnels dominated by Class III and IV rocks
Tunneling performance	On the premise of giving full play to the advance speed, it is mainly suitable for hard rock stratum (50–150 MPa) with relatively complete to complete rock mass and good self-stability. When effective support measures are adopted, it can also be applied to soft rock tunnels, but the advance speed is limited	On the premise of giving full play to the advance speed, it is mainly suitable for soft rock to medium hard rock stratum (30–90 MPa) with relatively complete rock mass and certain self-stability	It is suitable for soft rock (5–60 MPa) with certain self-stability in medium-length tunnels
Construction speed	Only bolting and shotcreting is needed if the geological conditions are good. In this case, the supporting workload is small and the speed is fast. However, it needs to be reinforced in advance if the geological conditions are poor. In this case, the supporting workload is large and the speed is slow	When the geological conditions are good, the tunnel wall is supported by grippers to provide the reverse thrust, and the tunneling and installation of the segment are carried out at the same time to speed up the progress. In soft ground, single shield tunneling is adopted, and tunneling and segment installation cannot be carried out at the same time, so the construction speed is limited	Tunneling and segment installation cannot be carried out at the same time, so the construction speed is limited

(continued)

Table 3.1 (continued)

Comparison items	Open TBM	Double-shield TBM	Single-shield TBM
Safety	Equipment and personnel are exposed to rocks, so protection should be strengthened	Personnel safety can be secured under the protection by shield. There is a danger of being stuck when the ground stress is high and the stratum is broken	Personnel safety can be secured under the protection by shield. There is a danger of being stuck when the ground stress is high and the stratum is broken
Advance speed	It is affected by geological conditions, but	Less affected than the open type	It is less affected by geological conditions than the open TBM
Lining mode	Secondary concrete lining can be carried out depending on specific situation	Segment lining is adopted	Segment lining is adopted
Description of geological conditions for construction	The lithological change of the tunnel wall can be directly observed during the tunneling process. It is convenient for geological map description. When the geological survey data is not detailed, it is less risky to choose open TBM for construction	It is difficult to systematically describe the construction geological conditions and measure the convergence deformity. The construction risk will be high if the geological survey data is not detailed	It is difficult to systematically describe the construction geological conditions and measure the convergence deformity. The construction risk will be high if the geological survey data is not detailed

(1) Uniaxial saturated compressive strength (Rc) of rock is one of the main factors affecting TBM selection. Generally, the lower the R_c, the higher the rock-breaking efficiency of TBM and the faster the tunneling. The higher the Rc, the lower the rock-breaking efficiency and the slower the tunneling. If the self-stabilizing time of rocks is extremely short or even unable to self-stabilize, the influence of Rc height on advance speed is not the first factor. *When Rc value is within a certain range, it is ideal that TBM tunneling can be kept on a certain speed, and the rocks of the tunnel can keep its self-stabilization within a certain period of time.*

(2) The development degree of rock discontinuities is another major factor that determines the level of rocks and TBM efficiency. Generally, TBM tunneling efficiency is higher when the joints are more developed and completely developed. When the joints are undeveloped and the rock mass is intact, it is difficult to break the rock. When Joints are well developed, rock mass is broken, and self-stabilizing ability is poor, TBM support workload will increase. At the same time, the reaction provided by rock mass to TBM grippers is low, resulting in

insufficient driving thrust, which is not conducive to the improvement of TBM efficiency. The more developed the structural plane of rock mass, the greater the density, the smaller the joint spacing, the lower the integrity coefficient and the higher the TBM advance speed. When the structural plane of rock mass is particularly developed, the density is extremely high, the joint spacing is extremely small, and the integrity coefficient of rock mass is very small, the rock mass has become fragmented or loose, and the strength of rock mass is extremely low. As a tunnel project, the rock mass has no self-stabilizing ability. When TBM method is used in this kind of rock, the advance speed will not be increased, but will be greatly reduced due to a large number of reinforcement measures for unstable rocks.

(3) The wear resistance of rock is also one of the main factors affecting the efficiency of TBM, which plays a decisive role in cutter wear. The higher the hardness and wear resistance of rock, the greater the wear of cutter and cutter head. The increase of quantity and times of TBM cutter replacement will inevitably affect the utilization rate of TBM. The failure of cutter, cutter ring and bearing has a great influence on the use cost of TBM. The hardness of rock, the size and content of mineral particles in rock, especially high-hardness mineral particles such as quartz, determine the wear resistance index of rock. Generally speaking, the higher the hardness of the rock, the better the wear resistance of the rock, the greater the wear on the cutter and the lower the tunneling efficiency.

(4) The influence of the combination between the occurrence of main structural planes of rock mass and the tunnel axis on the working efficiency of TBM is mainly manifested in the influence of the combination on the stability of rocks. This kind of influence will affect the working efficiency of TBM. When the angle between the main structural planes of rock mass and the tunnel axis is less than 45° and the dip angle of the structural planes is slower ($\leq 30°$), unstable wedges will appear in the part above the arch foot of the side wall of the tunnel and the surrounding rock of the arch due to the adverse combination of the structural planes and the free face of the tunnel excavation. Accordingly, falling blocks and collapse will occur, affecting the work of TBM, reducing the work efficiency of TBM and even endangering the safety of TBM and construction personnel.

(5) Under high ground stress, the brittle, relatively complete or complete surrounding rock is easily subject to rock burst, endangering the safety of TBM and construction personnel in severe cases, while the soft surrounding rock will be deformed greatly. Both will cause difficulties to the TBM advancement.

(6) The influence of water bearing and water leakage of tunnel on TBM work efficiency depends on the water content, the scope of surrounding rock with water bearing and water leakage and the hardness of the surrounding rock. Generally speaking, the strength of surrounding rock will be reduced to varying degrees in areas rich in water, water gushing and leakage, especially in soft rock, which will undermine the stability of surrounding rock and affect the work efficiency of TBM. In addition, a large amount of water gushing and

leaking will definitely worsen the work environment and the work efficiency of TBM.

The adverse ground conditions that restrict the construction performance of TBM may be the rock mass with poor quality that causes tunnel instability, or the rock mass with low penetration (such as the whole rock mass with high strength). However, the influence of rock mass quality on TBM performance does not have an absolute value. TBM is usually suitable for long tunnels boasting surrounding rock with good stability, medium to thick overburden and medium strength. It is extremely important to grasp the geological conditions of the tunnel before the construction of the tunnel project and very important to conduct the preliminary geological survey in the design stage. The funds used in the preliminary survey will be greatly compensated by the reduction of construction cost and the shortening of tunneling duration. The advance exploration in the pilot tunnel or main tunnel conducted by TBM can't replace the full preliminary geological survey. The relatively important or common adverse ground conditions that restrict TBM performance include the limit of boreability, the instability of excavation face and excavation wall, faults and squeezing/expansive ground. In addition, the adverse ground conditions affecting TBM construction include cohesive soils, groundwater, mine gas, rock burst, high-temperature rock stratum, high-temperature water and karst caves, etc.

The main index defining TBM's ability to excavate rock mass is the advance speed of TBM under the maximum thrust. It is related to rock type, uniaxial compressive strength of rock, crack development of surrounding rock, rock abrasivity, porosity, and so on. If TBM can't advance at a fast speed or the cutter wear exceeds the acceptable limit, the rock stratum boasts poor boreability.

If the excavated rock mass is broken, it will lead to significant instability on the excavation face. Accordingly, the cutter head can't rotate because the collapsed and accumulated stones affect the cutter head or the cutter head is stuck; or serious over break caused by unstable excavation face results in holes in front of TBM; or the excessive support workload greatly reduces the utilization ratio of TBM.

The unstable wall will affect the normal tunneling of open TBM; the relatively wide fault zone will bring difficulties to normal excavation, support and treatment; the shield TBM is very sensitive to the rapid convergence of surrounding rock.

3.2.4 Main Geological Scope of TBM Application

Giving full play to the advance speed, the main geological scope of TBM application is as follows:

(1) Open TBM is mainly suitable for hard rock stratum (50–150 MPa) with relatively complete–complete whole rock and good self-stability. When effective support measures are adopted and proved, it can also be applied to soft rock tunnels, but the advance speed should be limited.

(2) Double shield TBM is mainly suitable for soft rock to hard rock stratum (30–90 MPa) with relative completeness and certain self-stability.
(3) Single shield TBM is mainly suitable for soft rock stratum (5–60 MPa) with certain self-stability.

Open TBM is more suitable for hard rock tunnels with Class II and III surrounding rock. Open TBM can also be used in soft rock tunnel after effective support measures are adopted. Open TBM can be equipped with ring beam erector, shotcrete and other auxiliary equipment to adapt to geological changes. Open TBM with top shield and finger shield can conduct safe construction. In case of locally unstable surrounding rock, the auxiliary equipment attached to TBM can be used to reinforce and stabilize the tunnel wall by anchoring, adding wire mesh, shotcreting and erecting ring beams, etc. When encountering weak surrounding rock and fracture zones in local areas, TBM can pre-consolidate the surrounding rock around the upper front through the attached probe drilling machine and grouting equipment, and then excavate after the surrounding rock obtain self-stability. With the open TBM, the permanent linings are generally implemented in an intensive way after the breakthrough of the whole line. With the open TBM, the lithology change of the tunnel wall can be directly observed during the excavation process, which is convenient for depicting the geological map. When the hydrogeological data is not sufficient, the open TBM can give full play to its advantage of providing timely support through the NATM theory. In addition, the small-diameter open TBM can cooperate with the drill and blast method to advance the tunneling of double-track (large section) tunnels.

Shield TBM is more suitable for tunnels with Class III surrounding rock. Double shield TBM is generally suitable for tunneling in surrounding rock with medium to thick overburden, medium to high strength and basically good stability, and has good adaptability to the change of rock strength.

Double shield TBM with two excavation modes is suitable for tunneling in soft rock and hard rock, and most ideal for rock with the uniaxial saturated compressive strength being 30–90 MPa.

When TBM advances in a hard rock tunnel with relatively stable surrounding rock, the grippers tightly support the tunnel wall to provide the reaction force for the primary thrust cylinder, so that TBM can advance forward. The reactive torque of the cutter head is provided by two reactive torque cylinders located in the supporting shield, and the tunneling and segment installation are carried out simultaneously. At this time, TBM operation cycle is as follows: excavating and installing segment → regripping → re-supporting → re-excavating and re-installing segment.

When TBM advances in a soft rock tunnel, the rock of the tunnel wall can't provide enough support force for the horizontal support, the support system and the primary thrust system are no longer used, and the telescopic shield is in the contracted position. The reaction force of cutter head tunneling is provided by the friction between shield shell and surrounding rock, and the thrust of cutter head is provided by the auxiliary thrust cylinder supported against the segment, so TBM tunneling and segment installation can't be synchronized. At this time, TBM operation cycle is as follows: excavating → auxiliary thrust cylinder retracting → installing segment → re-excavating.

Single shield TBM is more suitable for tunnels with Class III–IV surrounding rock. It is often used in soft rock and medium-length tunnels with certain self-stability. When the surrounding rock level is slightly low, TBM can advance at a faster speed and require lower investment than double shield TBM.

3.2.5 TBM Adaptability Analysis

1. Selection factors

(1) Function

The selected TBM must be suitable for geological conditions, meet the construction needs of the stratum that it penetrates, conform to the tunnel characteristics and achieve the purposes of tunnel.

(2) Linear conditions

The selected TBM must be able to meet the requirements of turning radius and longitudinal gradient of plane curve of the tunnel, i.e., the advance direction of TBM should be timely adjusted and effectively controlled according to the linear conditions, and the guidance system should be able to meet the demands of the final tolerance of the tunnel breakthrough.

(3) Safety, quality, duration, cost and environmental protection requirements

The selected TBM must be able to meet the requirements of safety, quality, duration, cost and environmental protection. Among them, safety requirements include the safety of tunnel itself and the safety of surrounding buildings (structures) under TBM construction. The safety of the tunnel itself means that the initial support of the tunnel adopting composite lining according to NATM must meet the needs of structural safety during the tunneling duration, and the secondary lining must meet the requirements of structural safety and waterproofing under long-term operation conditions. Segment lining should meet the requirements of structural safety in both construction and operation periods. The duration requirements include the whole process of TBM preparation, tunneling, lining, disassembly and turnaround. Among them, the early-stage preparation, such as design, manufacturing, transportation, site, installation, commissioning, etc., should meet the predetermined requirements of tunnel excavation; the total tunneling duration of TBM excavation should meet the requirements of the scheduled tunnel excavation, and the total tunneling duration of TBM for side excavation and side lining should meet the requirements of the scheduled tunneling duration; TBM disassembly and turnaround should meet the scheduled requirements of the subsequent working period.

(4) Long distance tunneling

Continually conducting long-distance tunneling requires TBM to have good performance, long service life and sufficient spare parts and accessories. Due to the absence of the conditions of constructing branch hole, long-distance tunneling has enhanced the requirements for long-distance ventilation, power supply, transportation and long-distance water supply and drainage. It is required to have ventilation system, power supply system, water supply and drainage system with high reliability and strong ability, and transportation system with high efficiency.

(5) Dealing with adverse ground conditions in flexible ways

TBM needs to be equipped with advanced geological forecasting unit, probe drilling unit, pre-grouting drill and other equipment to prevent stratum collapse, stratum damage or TBM jamming, and the special geology should be handled in flexible methods.

2. Geological adaptability analysis

(1) Hydrogeological conditions

As an essential prerequisite, detailed and reliable hydrogeological data directly determine the success of TBM project. Hydrogeological data determine the feasibility of TBM in the project, the TBM type, main technical parameters of TBM, the selection of auxiliary equipment and the formulation of emergency plan.

The hydrogeological data must be detailed, accurate and reliable. The fundamental problem of tunnel construction is usually determined by the physical and geotechnical inhomogeneity of the stratum excavated by the TBM. Because the full-section and mechanized excavation is inflexible, the stratum with inhomogeneous strength plays a more important role. Mastering the geological and hydrological conditions of the project in advance as appropriate is very important for TBM construction. A large number of project construction examples at home and abroad have proved that the funds used in the preliminary investigation will be greatly compensated due to the reduced construction costs and the shortened construction period. Only by mastering detailed, accurate and reliable geological and hydrological data, can TBM be correctly selected and specific construction measures be formulated.

Generally, TBM excavation is not suitable for the following geological conditions. If TBM is used for construction under these geological conditions, it will encounter great difficulties in tunneling, and other corresponding technical measures must be adopted for auxiliary construction.

① Weak surrounding rock with large plastic ground pressure. Such surrounding rocks are prone to huge plastic deformation due to their low rock strength and high surrounding pressure, and TBM can easily get stuck.

② High pressure water gushing area. In case of serious water leakage, water gushing, fault zone and weak surrounding rock, the engineering geological conditions of surrounding rock will be greatly deteriorated, thus bringing difficulties to TBM construction.

③ Karst development zone. When TBM passes through the strong karst development zone, it is very likely that it will encounter underground river channels, water-filled caves or huge karst caves. It will be very difficult for TBM to tunnel or pass through. In severe cases, TBM will be trapped or buried, and the consequences will be disastrous.

④ Extremely strong rock burst. Tunnels with high in-situ stress and large overburden depth, such as those overburden depth exceeding 1500 m, are likely to encounter extremely strong rock burst, which will seriously threaten the safety of construction workers and equipment. Moreover, it will cause fatal damage to TBM and lead to a project failure.

⑤ Extreme hard rock. According to the current level of TBM technology, if the uniaxial compressive strength of rock exceeds 300 MPa, and the rock is highly abrasive and the joints are not developed, it is difficult for TBM to advance forward, and it is extremely uneconomical.

(2) Main geological factors affecting TBM tunneling

① Uniaxial compressive strength Uniaxial compressive strength is the ultimate compressive stress value of rock specimen when it is damaged under uniaxial compression. It is an important indicator reflecting TBM rock breaking as well as one of the key factors affecting TBM tunneling efficiency. The rock compressive strength suitable for TBM tunneling is 5–150 MPa, and should have a certain self-stabilizing ability. TBM is most suitable for tunneling rocks with a compressive strength of 30–150 MPa. If the uniaxial compressive strength of the rock is less than 30 MPa, the penetration of disc cutter is large, and it is not easy to produce the extrusion zone. This may lead to not achieving the expected rock breaking effect. When the uniaxial compressive strength of rock is more than 150 MPa, and the joints are not developed, the TBM advance speed will decrease significantly. At the same time, the disc cutter consumption will increase, and the cutter head wear, vibration and weld cracking will also be obviously aggravated, and the time for cutter head maintenance, tool inspection and tool change will see huge increase. That means the impact of this factor must be considered when calculating the construction period.

② Hardness and wear resistance of rock. The content and particle size of feldspar, Shi Ying and other abrasive components in the rock are the main factors that affect the hardness and wear resistance of the rock. The higher the content, the greater the hardness and wear resistance of the rock, the faster the abrasion of the hob in the tunneling process, and the higher the disc cutter consumption and construction cost. At the same time, the downtime for checking and changing the cutter increases, seriously reducing the advance speed of TBM.

③ Development degree and orientation of discontinuity of rock mass. The rock discontinuity structural plane includes schistosity, minor faults, joints, bedding, etc. Its development degree, that is, the degree of fissure or integrity of rock mass, is also one of the key factors affecting TBM tunneling efficiency. Generally, the degree of structural plane development is characterized by the volumetric joint count Jv, the integrity coefficient Kv of rock mass or the rock quality index RQD. The advance speed of TBM mainly depends on the integrity of rock mass, and the most applicable range is the relatively complete and broken state (Kv = 0.45 – 0.75).

3. Analysis and treatment measures of various risks of TBM construction and unfavorable influencing factors

(1) Main construction risks and unfavorable influencing factors of open TBM

① Only the initial support of surrounding rock can be carried out after open TBM excavation. In order not to affect the advance speed and avoid construction interference, the subsequent secondary lining can only be applied after TBM turnaround or all tunneling works are completed, which makes the excavation section of the tunnel in a state of only initial support for a long time.
② When TBM passes through the surrounding rock fracture zone or the back-fill stratum, it is necessary to take surrounding rock reinforcement measures in advance, which will increase more advanced reinforcement measures and auxiliary treatment measures, and will greatly affect the advance speed.
③ When the ground specific pressure of the weak fracture zone around the cave is insufficient, the head of TBM sinks and cannot be tunneled.

(2) Main construction risks and unfavorable influencing factors of shield TBM

① Shield TBM is a thoracotomy mode, it has a great construction risk when passing through unstable strata such as soil layers and is difficult to slag out.
② Shield TBM has poor adaptability to small curve radius.
③ Segment prefabrication requires segment yard, which is characterized by large investment, large land occupation, numerous molds and high segment cost.
④ The advancing power of the single shield TBM is obtained by pushing the subsequent segments through the cylinder. Therefore, it is required that the segment must follow closely, and the advancing and segment assembly cannot be proceeded simultaneously to avoid certain impact on the advance speed.
⑤ For its longest body, front shield, tail shield, and telescopic shield, the double shield TBM is susceptible to jamming in case of a section where the rock stratum is relatively broken, collapsed and possible deformation may occur. As a result, falling blocks may damage the jacking cylinder during construction. In severe cases, it is even impossible to advance, and the construction flexibility is poor.
⑥ In case the converging deformation around the tunnel under excavation is large, the double-shield TBM may be stuck by the convergence around the tunnel after it excavates and passes through.

(3) Engineering measures against risks and adverse factors

Engineering measures against risks and adverse effects proposed based on the analysis of the risks and adverse effects of various TBM types are shown in Table 3.2.

Table 3.2 Measures against TBM risks and adverse factors

TBM type	Risks and adverse factors	Measures
Open type	Long-time exposure of the initial support due to long-delayed secondary lining	Keep the shotcrete machine as close as possible to the cutter head. When necessary, add another artificial shotcrete machine on the platform in the operation room for shotcrete support as soon as possible. When the section with poor surrounding rock is short, implement secondary lining for the section in advance during the TBM maintenance. If this section is long, let the supply of wind, water and electricity still follow through the original tunnel, and carry out mucking in the cross passage or the transport switch located at the station switching to another tunnel, or transport out through the access of the nearest station, and complete the secondary lining in this section
	Pass partially through the fracture zone of the surrounding rock, or through the backfilled stratum section	Use the probe drilling machine of the TBM for the pipe and shed pre-grouting support, and then strengthen the initial support to ensure the stability of surrounding rock
	Insufficient ground pressure in weak fracture zone causes the head of TBM to sink, making it impossible to advance	Install rock bolts at the location of the gripper and grout to reinforce the surrounding rock, or add auxiliary measures such as sleepers and steel formwork to increase the bearing area of the gripper to avoid insufficient reaction force and deep sinking of the gripper
Single shield	Difficult to adapt to curve segments with radius less than 500 m	Shorten the regripping distance, reduce the width of segment, add overcutter, and pass through in a polygonal manner instead of curve manner by "short advance and big over breaking" gradually. As a result, TBM equipment cost increases, tunnel construction becomes more difficult, falling behind the schedule, segment types increase, and engineering investment raises greatly
	Advancing and segment erection cannot be done synchronously, affecting the advance speed	

(continued)

Table 3.2 (continued)

TBM type	Risks and adverse factors	Measures
	Long body makes it easy to get stuck and there are relatively few measures to deal with complex geological sections. Its flexibility is poor, affecting the advance speed	Implement advance reinforcement in the shield body in advance in case of the sections under adverse ground conditions such as fault fracture zone ahead based on the advance geological forecast
	The pre-set segment prefabrication yard occupies a large area and there are many molds at a large investment and cost	Set the segment yard based on specific conditions of the surrounding area and local conditions. Use domestic segment molds to reduce the cost of the segment production
Double shield	The longest body makes it easy to get stuck and there are few measures to deal with complex geological sections. Its flexibility is insufficient, affecting the advance speed	Make enough advance reinforcement measures, and pass through as fast as possible to avoid long exposure of the telescopic cylinder, damage caused by the falling rock blocks of the rock, or getting stuck in the position of the telescopic shield due to the rock convergence
	Poor ability to adapt to small radius	Shorten the regripping distance, reduce the width of segment, add overcutter, and pass through in a polygonal manner instead of curve manner by "short advance and big over breaking" gradually. However, TBM equipment cost increases, tunnel construction becomes more difficult, falling behind the schedule, segment types increase, and engineering investment raises too
	The pre-set segment prefabrication yard occupies a large area and there are many molds at a large investment and cost	Set the segment yard based on specific conditions of the surrounding area and local conditions. Use domestic segment molds to reduce the cost of the segment production

3.3 TBM Adaptability Design

3.3.1 Advance Performance Requirements in Adaptability Design for TBM Selection

In adaptability design, reasonable requirements for the tunneling performance shall be put forward against various factors such as construction period, engineering geology, engineering design and construction process. Main indicators to measure TBM advance performance include Penetration, Rate of Penetration or Advance

Speed, availability ratio of equipment, utilization ratio of advance operation, cutter consumption and others. Under certain rock conditions, TBM advance performance depends not only upon the performance of TBM and cutter itself, but also upon the operation, maintenance and construction organization and management of TBM.

1. Penetration and Pure Advance Speed

Penetration is the cutting depth of the cutter head per revolution, in mm/r. Pure advance speed is the penetration multiplied by the rotation speed of cutter head, in mm/min or m/h. If the penetration and rotation speed of cutter head are available, the pure advance speed can be calculated. Once the utilization ratio of TBM advance operation is further known, the daily advance rate, weekly advance and advance per month can be estimated. Penetration and pure advance speed are mainly influenced by machine design parameters and geological parameters. Therefore, according to the engineering geological conditions and duration requirements, reasonable machine characteristics requirements can be put forward to the TBM design manufacturer, and whether the design parameters meet the requirements of advance speed can be judged, or the advance speed can be predicted.

Depending on the design parameters and geological parameters of the machine, the penetration may be at 2–20 mm/r, and the rotation speed of the cutter head is generally at 5–12 r/min. When selecting and designing TBM, the manufacturer may be required to provide the reference value of penetration or pure advance speed of designed TBM under different rock conditions. In recent 30 years, the technical level of TBM tunneling is roughly as follows: in 1980s, the maximum daily advance of TBM was 30–40 m, and the average advance per month was 300–500 m; after 1990s, the maximum daily advance was 45–60 m, and the average advance per month was 500–700 m. In 2005, 8 m diameter TBM was selected for the Dahuofang water transfer project in Liaoning. The maximum daily advance was 63.5 m and the maximum advance per month was 1208 m.

2. Availability

The availability ratio of TBM includes the reliability of TBM system and maintenance degree. The higher the reliability, the shorter the maintenance time of fault, the higher the availability ratio of TBM system, and the larger the proportion of pure advance time, the better the performance of the machine. Therefore, this performance parameter mainly depends on TBM manufacturers. Of course, the high availability can only be obtained by the contractor according to correct use method and maintenance requirements given by the manufacturer. In the design of TBM selection, the contractor can put forward reasonable requirements to the manufacturer for the TBM availability ratio, which is usually required to reach 90% in the contract. However, it is very difficult to obtain high equipment system availability ratio because TBM is consist of a large number of subsystems and equipment integration. According to statistics, the past 25 engineering applications of Wirth's TBM show that the availability ratio of TBM equipment system is generally stabilized at 70–90%, and few projects can reach more than 90%.

TBM availability ratio can be determined according to the following formula:

$$AV = \frac{ET + RT}{ET + RT + OUT} \tag{3.1}$$

Type:

AV availability ratio;
ET tunneling time (min);
RT regripping time (min);
OUT downtime caused by fault (min).

3. Utilization Ratio of Tunneling

The utilization ratio of tunneling is the proportion of tunneling time to the total construction duration. The utilization ratio depends on both the availability and the engineering geological conditions and on-site organization and management level. The higher the utilization ratio of tunneling, the more likely it is to get a high advance. Equipment fault, rock support, mucking, material transportation and other factors often cause downtime, thus reducing the utilization ratio of tunneling. At present, the average utilization ratio of TBM tunneling is around 40%. The utilization ratio can even reach more than 50% in some months with low equipment fault ratio and good rock conditions. And the utilization ratio can even reach less than 20% with poor rock conditions. It is very difficult for the average utilization ratio of the whole project to exceed 40%.

TBM penetration, pure advance speed and utilization ratio of tunneling under different rock conditions are the important basis for planning the construction duration. For the contractor, the manufacturer can be required to improve the availability ratio of the machine. And the coordination relationship among subsystems such as tunneling, mucking and support should be considered in the type selection design, and the construction organization and management and maintenance of TBM should be improved to increase the utilization ratio of TBM tunneling.

4. Cutter Consumption

Except for TBM investment consumption, one of the most important expenses is cutter. The prediction of cutter consumption is more difficult than that of advance speed, and the increase of cutter consumption will increase downtime, thus affecting the construction duration. Those all depend more or less on the unknown rock conditions to be excavated. The compressive strength of rock, crack condition and quartz content are main factors that influence cutter consumption. Because of different rock conditions in different projects, the consumption quantity and expense of cutter may vary greatly. Some projects hardly need to change cutter in 1 km tunneling, while others may cost several million yuan of cutter in 1 km tunneling.

In addition to the normal cutter ring wear, the cutter consumption also includes the edge breakage of the cutter ring, the damage of the cutter bearing and seal, and even the cutter body damage caused by the partial wear of the cutter ring not found in time. Under certain rock conditions, the consumption of cutter depends not only on the quality of cutter and the overall arrangement of cutter on the cutter head, but also on the installation, use and maintenance. When select and design TBM, manufacturers can be required to provide reference values of the cost or quantity of cutter for excavation per cubic meter of rock, so as to judge the quality of cutter and predict the investment and consumption of cutter.

3.3.2 Adaptability Selection of Mucking Transportation and Feeding Equipment

In TBM construction, the efficiency of tunneling largely depends on whether mucking transportation and material transportation are in place in time. A large amount of muck produced by TBM excavation are transported outside, and maintenance equipment such as materials and cutters needed for tunnel support and extension are transported inside.

In the selection of mucking transportation and feeding equipment, firstly, it is necessary to consider the matching with the advance speed of TBM. Secondly, it is necessary to analyze from the technical and economic point of view, and select a technically reliable, economical and reasonable scheme. The specific dimension and quantity of the equipment are determined by the excavation diameter, tunneling cycle advance, tunnel length and gradient, etc.

Mucking transportation and feeding equipment are mainly divided into two types: rail mucking and feeding, continuous belt conveyor mucking discharging and light rail feeding.

1. Rail Mucking and Feeding

According to the tunnel advance length, excavation section, tunnel gradient, advance of each excavation cycle and loose coefficient of rock, the bucket capacity and quantity of mine train of each muck train can be calculated.

The locomotive selection should meet the requirements that not only one mine train of heavy-duty harvesters can be pulled, but also the required number of material trains and manned trains can be driven. At the same time, the gradient should be considered, and the number and dimension of locomotives are finally determined.

According to the tunneling length and the average running speed of trains, the number of muck trains needed is determined. First, the quantity and dimension of trains and locomotives contained in each muck train should be determined. To meet the requirements, the muck quantity of one tunneling cycle should be transported away once by one muck train. According to the tunneling length, the average running speed of trains and the requirement of TBM continuous muck, the quantity of muck trains needed is determined.

At the initial stage of excavation, the distance is short, and the quantity of muck trains is small. With the increase of excavation distance, the quantity of muck trains will gradually increase. In order to improve the tunneling efficiency of TBM as much as possible, at least four rolling stocks are needed in the construction. During the tunneling cycle, one train is loaded with muck at the hopper of the mucking discharging belt conveyor, one train is on standby at the side of the double track, one train is waiting to enter at the side of the exit track, and one train is unloading muck outside the tunnel. Each truck should include locomotives, mine trains, material trains, etc.

2. Belt Conveyor Mucking and Light Rail Feeding

When continuous belt conveyor is used for mucking, the track in the tunnel only undertakes the transportation of tunnel supporting materials, TBM maintenance personnel and equipment, and light rails can be used.

The belt conveyor moves with TBM, and is connected from TBM to the entrance of the tunnel for mucking. Belt conveyor is mainly composed of belt storage cassette, main drive device, auxiliary drive device, driven wheel, belt, idler and other parts. Simple in structure, high in transportation efficiency and easy to maintain and manage, the belt conveyor can reduce the quantity of transport vehicles in the tunnel and air pollution, and is conducive to forming a rapid and continuous muck discharge system.

The key to continuous mucking with belt conveyor is that the belt conveyor can be extended with each stepping of TBM, and the conveyor can turn. The tail of the belt conveyor is installed on the back-up equipment. When the back-up equipment moves forward, the adhesive tape is pulled out from the belt storage cassette by section, so that the belt conveyor can continuously complete the muck transportation. When TBM completes a cycle stroke step each time, the back-up system is pulled forward one stroke, and the belt conveyor also extends. Therefore, it is necessary to install the belt frame, roller and grooved roller in front of the tail of the belt conveyor to provide conditions for belt transportation. In order to meet the requirement that TBM can continuously extend forward within a certain distance without extending the tape at any time, a tape storage device is set up. The muck transported by the back-up belt conveyor is discharged to the mucking belt conveyor. When the adhesive tape in the belt storage cassette is used up, the mucking belt conveyor needs to stop working, and the extended adhesive tape should be vulcanized. The main drive device of continuous mucking belt conveyor consists of motor, reducer and drive wheel, and uses variable frequency speed regulation motor. The transmission between the drive wheel and the tape is friction transmission. When the horizontal conveying distance increases, another set of auxiliary drive devices with the same structure will be added, and two sets of drive devices will drive the tape. In order to coordinate the operation of the main drive devices and auxiliary drive devices and automatically adjust the tension of the tape when starting, the continuous belt conveyor is controlled by PLC, and this control system matches the control of TBM to ensure that TBM controls the starting and stopping sequence. In order to keep the tape centered, the continuous belt conveyor has the steering capacity of hydraulic drive. Under the action of hydraulic cylinder, the tail of the continuous muck belt conveyor can move in the X-axis and Y-axis directions of the tunnel section, and can rotate along the Z-axis. These movements swing with the belt and are controlled by the control console installed at the tail of the belt conveyor.

3.3.3 Main Parameter Design of TBM

In recent years, the general trend of TBM design is developing large-diameter cutter to improve the bearing capacity of cutter, increase the thrust and torque of cutter head, improve the rotation speed of cutter head, and increase the tunneling stroke. The main key components that affect the durability (service life) and performance of TBM are cutter head, cutter, main bearing, cutter head drive system, TBM main machine, thrust system and support system, etc.

1. Diameter of Cutter Head

TBM diameter tends to develop towards two directions: large diameter and miniature TBM. At present, TBM diameter is generally between 3 and 15 m. The larger the TBM diameter, the greater technical difficulty in TBM design, manufacture, transportation, assembly and construction, while the smaller TBM diameter results in narrow working space and the difficulty in equipment layout. The diameter of the cutter head shall be determined according to the diameter of the excavated tunnel, and a certain capacity of excavation is required. Comprehensive factors such as the diameter of the excavated tunnel, support requirements and the deformation of the surrounding rock shall be considered. Diameter excavation is mainly to prevent the deformation of the surrounding rock from blocking the shield. A certain degree of excavation can be achieved by adding side cutter adjustment pads, generally around 60 mm; If it is necessary to obtain greater excavation capacity, telescopic cutters with hydraulic cylinders shall be arranged on the cutter head.

In order to reduce the impact of unfavorable geology, TBM is designed as flat cutter head and concave cutter mounting seat as much as possible. This design reduces the distance between tunnel face and the cutter head structure, which can better maintain the stability of tunnel face, and reduce the risk of overloading mechanical components caused by blocking the cutter head. Generally, the cutter head is designed as a back-mounted tool changer to prevent people from appearing in the unprotected zone between tunnel face and the front end of the cutter head.

The strength, stiffness, wear resistance and welding strength of the cutter head design are key performance indicators that need to be investigated. The design and manufacture of the cutter head is very critical, especially under hard rock conditions, the vibration, wear and weld cracks of the cutter head will become salient problems, so full consideration must be given to its design.

2. Tool diameter

Hard rock TBM generally uses disc cutter, and the basic components of the cutter are: cutter ring, cutter body, cutter shaft, bearing assembly, seal, end cover, etc. The design and manufacture of tools and their arrangements on the cutter head are critical to the rock breaking effect, advance speed and tool consumption.

In order to increase the penetration and the pure advance speed of TBM tools, the technology of tool bearing and tool ring materials has been continuously developed to obtain higher bearing capacity and wear resistance, thus having higher rated thrust of

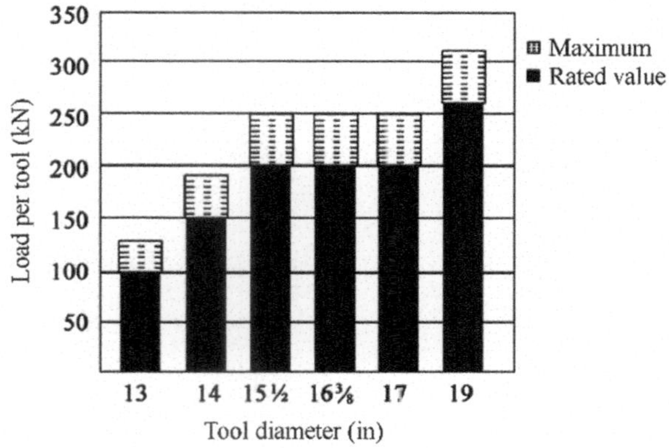

Fig. 3.1 Bearing capacity of tools with different diameters

cutters. At present, the tool manufacturer has offered a wide range of tool diameters, ranging from 300 to 480 mm. The bearing capacity of tools with different diameters is shown in Fig. 3.1. Tools with a diameter of 17 inch (432 mm) are commonly used, 19 inch (483 mm).

Diameter tools have also been widely used, and 20-inch tools have also been developed abroad. All famous manufacturers are still working on the development of larger diameter tools. The general advantages of large diameter tools are that they allow greater wear, bearing capacity and lower rolling resistance coefficient. The water delivery project of Dahuofang Reservoir selected a φ 19 in (483 mm) tool for the first time in China, which proved to be successful in practice.

The biggest difference between tools from different manufacturers lies in their materials, heat treatment process and the geometry of the cutter ring profile. The main issue is to find a balance between wear resistance and toughness. In terms of the cutter ring, its outer profile has changed from the wedge shape to quasi equal thickness, so that the contact area between the cutter ring and the rock can be maintained during a long wear period, therefore, it will not lead to the increase of required thrust or the decrease of advance speed.

3. Cutter spacing and number of tools

The thrust of each disc cutter and the cutter spacing are the most important parameters to improve the cutting depth (penetration) and thus the advance speed.

The test shows that under the given geological conditions, when the cutter spacing is reduced, the thrust required to achieve a certain advance speed will decrease, as shown in Fig. 3.2. This means that in the case of very hard rock, the thrust per cutter is kept constant, and the advance speed can be increased by reducing the cutter spacing. However, correspondingly, it is required to increase the total thrust of the machine and produce smaller crushed stone. The smaller the crushed stone, the greater the specific energy required to cut the rock, and the less economical the excavation process.

Comparing the energy consumption of excavating rocks with different cutter spacing shows that each rock has an optimal cutter spacing. This optimum cutter

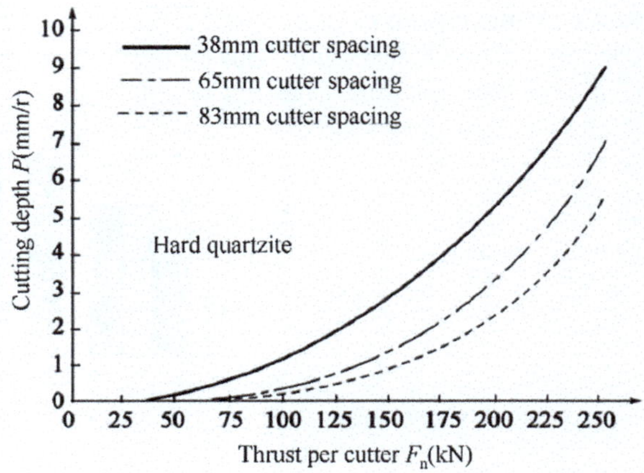

Fig. 3.2 Impact of cutter spacing on cutting depth (penetration)

spacing is mainly a function of geology. In the case of hard rock, it is 10–20 times the cutting depth, between 65 and 90 mm; Soft rock is about 100 mm. In actual design, in order to improve the adaptability of TBM, the cutter spacing should be as small as possible, so as to run through the whole tunnel and encounter the hardest rock.

Under given geological conditions and comparable cutter spacing, the number of tools is proportional to the excavation diameter. Figure 3.3 shows the relationship between the tunnel excavation diameter and the required tools parameters.

Reducing the number of tools without increasing the maximum average load per tool. The increase of the cutter spacing will lead to the decrease of advance speed. For certain rock conditions, large cutter spacing increases the wear of the tool, correspondingly reduces the life of the cutter ring and increases the downtime

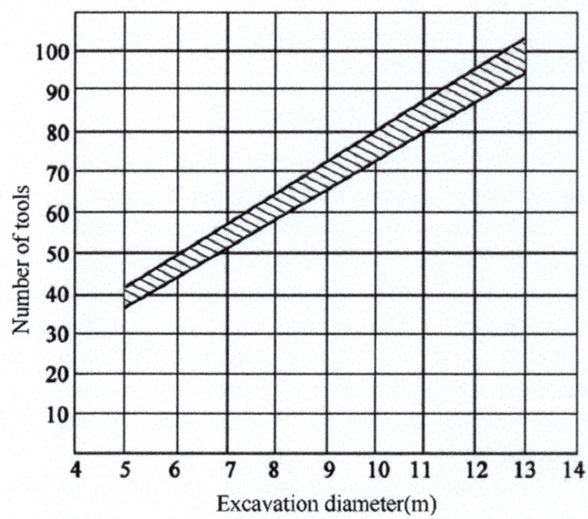

Fig. 3.3 The relationship between number of tools and excavation diameter

of TBM. If the instantaneous rock strength is too high, and the cutter head with large cutter spacing can't break the rock, in a short period of time, if TBM doesn't monitor it carefully, the single knife as well as the adjacent tools will be damaged. More tools are in contact with the rock, which can also reduce the vibration of TBM and the size of the caves in the fractured rock section, thus reducing the risk of blocking the rotation of cutter head.

4. Cutter head rotation speed

Generally, the permissible peripheral speed of cutter bearing and seal of disc cutter is about 150 m/min. Therefore, cutter head rotation speed should be inversely proportional to the cutter head diameter, namely:

$$n = \frac{X}{D} \tag{3.2}$$

where:

n cutter head rotation speed(r/min);
D excavation diameter (m);
X velocity coefficient, usually 45 or 50.

Fig. 3.4 shows the relationship between relative cutting depth, cutter head rotation speed and excavation diameter at constant advance speed. The figure is based on a velocity coefficient of about 45. With the improvement of TBM technology, the cutter head rotation speed tends to increase. With large-diameter disc tool, it is theoretically allowed to increase the cutter head rotation speed, but the velocity coefficient should not be too large, so as to maintain the slag discharge capacity of the cutter head.

5. Cutter head torque

The total rolling resistance of all tools on the cutter head determines the torque. Calculate the rolling resistance coefficient, that is, the relationship between rolling force and thrust is mainly determined by the cutter ring diameter and cutting depth.

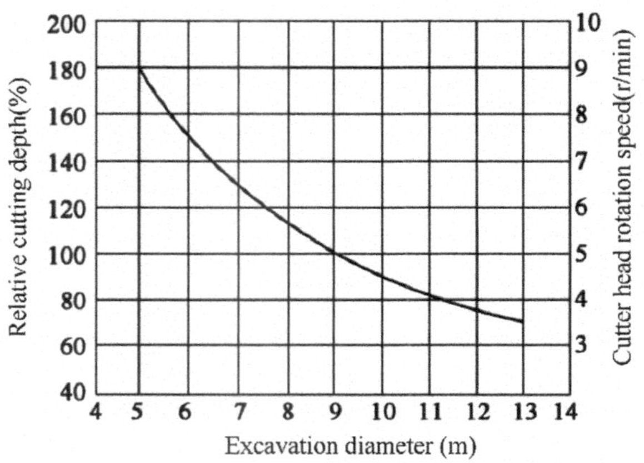

Fig. 3.4 Relationship between cutter head rotation speed and excavation diameter

The following equation provides a simple and reliable method to calculate the rolling resistance coefficient k:

$$k = \frac{4}{5}\left(\frac{p}{d}\right)^{0.5} \tag{3.3}$$

where:

d cutter ring diameter (mm);
P cutting depth (mm).

Therefore, when the cutting depth is constant, the rolling resistance coefficient decreases with the increase of cutter ring diameter. It is roughly estimated that the average percentage of rolling resistance as thrust is 10% for hard rock and 5% for easily tunneled rock.

The power requirements of TBM cutter head are determined by rotation speed and torque, while the torque is affected by geology. Under comparable cutter head thrust, the easily tunneled rocks allow higher cutting depth, and require a larger cutterhead torque due to the increase of rolling resistance. In terms of rocks with good drillability, the theoretically practicable cutting depth cannot be adopted because it is uneconomical. The required cutter head torque Md is generally considered according to the relationship that increases with the square of cutter head diameter D:

$$M_{\mathrm{d}} = YD^2 \tag{3.4}$$

where:

Y torque coefficient, usually 45 for previous TBM design.

Due to the higher tool load, the current coefficient reaches 60. Figure 3.5 shows the relationship between cutter head torque and cutter head diameter.

6. Cutter head driving power

The main forms of cutter head driving system include hydraulic drive, two-speed motor drive and variable frequency motor drive. Because of the reliability improvement and cost reduction of variable frequency control technology, variable frequency motor drive represents greater adaptability to different rocks. Variable frequency motor drive has shown strong momentum of development and become the first choice of TBM drive in recent years. Generally, the motor drives the planetary gear reducer, which in turn drives the pinion and the large gear ring to rotate, thus driving the cutter head to rotate.

Cutter head driving power depends on cutter head torque and cutter head rotation speed. At present, for a φ 4.5–15 m open TBM in practical engineering applications, the design value of the cutter head driving power is shown in Table 3.3.

In the TBM design, the user can ask the manufacturer to provide the cutter head driving curve, so as to judge the characteristics of the driving system. For example,

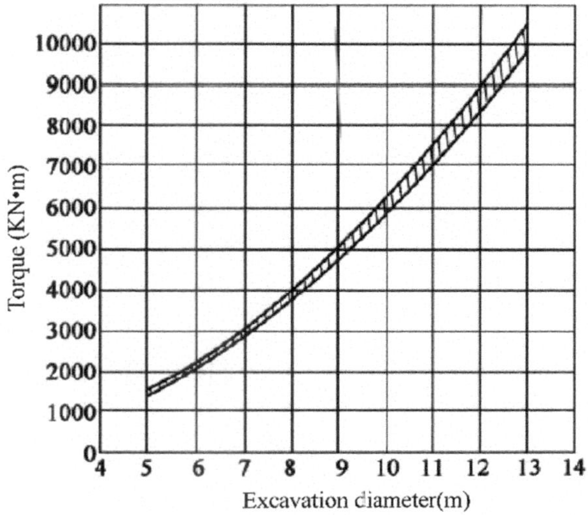

Fig. 3.5 Relationship between cutter head torque and excavation diameter

Table 3.3 of driving power of TBM cutter head with different diameters

Cutter head diameter (m)	4.5	6.36	8.03	10.2	12.4	14.4
Driving power (kw)	1400	2300	3000	3800	4500	4900

the driving characteristics curve of a φ 4.5 m open TBM in Nabang Hydropower Project is shown in Fig. 3.6. Cutter head driving curve is divided into constant torque section and constant power section. Generally, the constant power section is used for tunneling when the rock is hard and intact; the weak and broken surrounding rock section is tunneled by constant torque section.

7. Main bearing and its service life

TBM main bearing, also known as cutter head main drive bearing, is the most critical component of TBM. Its development direction is to use three-axis roller bearing (two rows of thrust and one row of radial) instead of tapered roller bearing to improve its bearing capacity and extend its service life, especially in the design of large diameter TBM. At present, the service life of cutter head bearing reaches 15,000–20,000 h, and the normal tunneling range is 15–30 km. It is certain that the operation and maintenance of TBM are also very important to the service life of TBM main bearing.

Due to the long ordering and manufacturing period (about 6 months), high cost and difficulty in disassembly and replacement in the tunnel during construction, the service life, size, quality, lubrication, sealing and monitoring facilities of the cutter head bearing are very important factors in TBM selection and design.

At present, there are only a handful of companies that can produce TBM main bearings in the world, such as SKF and HoeschRotheErde. Generally, TBM design manufacturer provides the bearing manufacturer with ordering specifications and

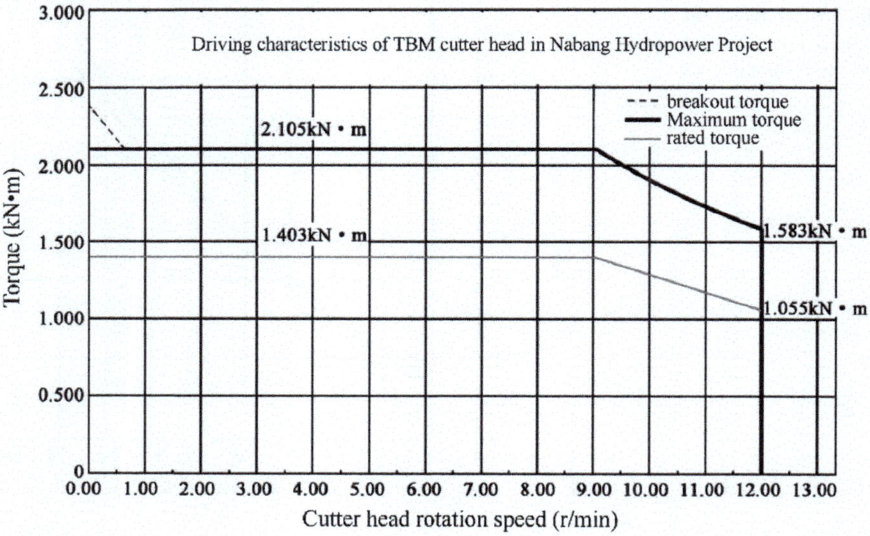

Fig. 3.6 Cutter head driving curve

load spectrum of the bearings, which shall be designed and manufactured by the bearing manufacturer, and the service life calculations are made subsequently.

Table 3.4 gives an example of the load spectrum estimation and service life calculation by a TBM manufacturer for the main bearing of φ 4.5 m TBM used in a water diversion tunnel project.

Generally, a pressure oil circulation lubrication system is used in the main bearing to keep it well-lubricated. In addition to the external labyrinth seal, more than three lip seals are used for inner and outer sealing of the main bearing respectively. An automatic grease injection system is provided for the sealing to discharge grease outward and prevent intrusion of dust and dirt during tunneling. A monitoring system

Table 3.4 Load spectrum and service life of TBM main bearing

Item	Load under working condition 1	Load under working condition 2	Load under working condition 3
Time proportion under each working condition (%)	80	15	5
Axial load (kN)	8010	8010	8010
Radial load (kN)	1100	1350	1600
Tilting moment (kN m)	2880	4530	6680
Rotation speed (r/min)	12.00	6.00	3.00
Service life of bearing (h)	23,161 (service life of bearing L10 calculated according to ISO281)		

is designed to monitor the lubricating and sealing states of the main bearing, and thus ensure the sound working state and service life of the main bearing.

8. Thrust System and its Thrust Force

The thrust system is mainly composed of hydraulic thrust cylinders arranged on both sides of the main beam and other components. It is designed to provide thrust force for the cutter head, overcome the friction resistance of the shield, and provide pulling force required by the back-up system. It is able to generate sufficient thrust force based on the conditions of rocks, excavation diameter, cutter diameter and number of cutters, to allow certain penetration in hard rock. Generally, the required thrust force is comprehensively determined according to the rated thrust of all cutters, the friction resistance between the cutter head and shield, and the traction resistance of the back-up system.

Take the φ 4.5 m open TBM as an example for design and calculation.

(1) Shield Friction Force

The shield friction force will be calculated with the following formula:

$$F_{\mu} = \mu \cdot [2\pi \cdot r \cdot l \cdot (P_v + P_h) \cdot 0.5 + G] \tag{3.5}$$

where:

F_{μ} shield friction force (kN);
μ friction coefficient, 0.2;
r shield radius, 2.25 m for this project;
L shield length, 3.5 m for this project;
P_v vertical load, 134 kPa;
P_h horizontal load, 0;
G dead weight of TBM, 3183 kN for this project.

Then: $F_{\mu} = 1299$ kN.

(2) Rated Thrust Force of Cutter Head

The rated thrust force of cutter head will be calculated with the following formula:

$$F_c = n \cdot F_{sc} \tag{3.6}$$

where:

F_c rated thrust force of cutter head (kN);
F_{sc} rated thrust force of a single cutter, 267 kN for 17-inch cutter;
n number of cutters, 30 for this project.

Then $F_c = n \cdot F_{sc} = 30 \times 267 = 8010(\text{kN})$.

(3) Pulling Force of Back-up System

The pulling force of back-up system F_d depends on the design and mass of the back-up system, and will be estimated based on experience. It takes 500 kN for this project.

(4) Thrust Force of TBM

The thrust force F_t required by the TBM is the sum of the above three forces, then:

$$F_t = F_\mu + F_c + F_d \tag{3.7}$$

where:

F_t the maximum thrust force required by TBM.

The maximum thrust force required by the TBM of this project is:

$$F_t = F_\mu + F_c + F_d = 1299 + 8010 + 500 = 9809(\text{kN})$$

(5) Dimension of Thrust Cylinder

The thrust cylinder forms a certain angle α with the axis of TBM, which is the largest at the beginning of tunneling. This angle is 22.6° for the designed TBM of this project. The required maximum thrust force of thrust cylinder is:

$$F_{\text{tmax}} = \frac{F_t}{\cos\alpha} \tag{3.8}$$

Then, $F_{\text{tmax}} = 9809/\cos 22.6^\circ = 10624(\text{kN})$.

It is assumed that totally 4 thrust cylinders are arranged on both sides, the diameter of the cylinder piston is 360 mm, and that the maximum oil pressure is 350 bar, then, the maximum thrust force of the designed thrust cylinder $F_{\text{tcy}} = 14250$ kN@350bar.

9. Supporting System and its Supporting Force
(1) Supporting Force

The supporting system consists of a horizontal supporting cylinder with grippers and a vertical torque cylinder. The grippers of the horizontal supporting cylinder will tightly support against the cave wall during tunneling. The supporting cylinder and torque cylinder should be able to resist the thrust force and torque transmitted by the cutter head. The grippers should be designed according to space size, rock not prone to collapse, crossing ring beam, and other requirements. Sufficient strength and area should be provided, and the ground pressure with the tunnel wall should be calculated.

According to the engineering practice experience, the supporting force is generally about 2.6 times of the thrust force. The rated supporting force required is 20,826 kN as calculated with formula (3.9).

$$F_g = 2.6F_c \tag{3.9}$$

The maximum supporting force required is 25,503 kN as calculated with formula (3.10).

$$F_{gmax} = 2.6F_t \tag{3.10}$$

(2) Gripper Cylinder

In the structure design of TBM, the left and right gripper cylinders are installed on the saddle of main beam. For the φ 4.5 m TBM in this project, the cylinder with φ 810 mm piston is used, and its rated thrust force is 25,765 kN @ 250 bar, and maximum thrust force is 36,071 kN @ 350 bar.

3.3.4 Matching Design Between TBM Systems

TBM is a huge factory-like operation system and an organic whole composed of many subsystems and equipment. The optimal configuration and integration of each system is the key to the success of a TBM project. When designing a TBM and its back-up system, the designer should consider the characteristics of parallel operation of tunneling, supporting and mucking, to allow optimal configuration and coordination of the system. The TBM should be designed with tunneling as the core, and the capability of other auxiliary operation systems should be configured to minimize the delay to tunneling. The back-up system should meet the needs of materials, water supply, power supply, ventilation, dust removal, etc. in tunneling, supporting and mucking, and tunnel safely, rapidly and efficiently as an integrated, optimized and well-matched operation system.

The key points of matching design of TBM system include:

(1) Matching of tunneling capacity (advance speed) with muck discharge capacity;
(2) Matching of tunneling capacity (advance speed) with supporting capacity;
(3) Matching of materials transportation, water supply, power supply, ventilation, dust removal, track laying and other systems with the main machine design.

Take TBM with an excavation diameter d = 4.5 m and an excavation stroke L_s = 1.5 m as an example for calculation and analysis. It should be noted that, to prevent the auxiliary systems and process operation capability from delaying the tunneling operation, the auxiliary system should be designed with the capacity matching the maximum tunneling capacity as much as possible. Therefore, in the following analysis, the possible maximum advance speed should be fully considered in the first place.

1. Calculation and Analysis of Advance Speed
(1) Average Advance Speed

The average advance speed V may be set at 3.0 m/h according to the rock condition of a project and the tunneling performance of φ 4.5 m TBM.

① Time of one tunneling stroke: $t = L_s/v = 1500/50 = 30$ min;
② Regripping reset time: 5 min;
③ Time of one tunneling cycle: 35 min;
④ In consideration of TBM maintenance, troubleshooting and delay by auxiliary operation, the utilization ratio of tunneling operation is calculated as 35%.
⑤ Effective tunneling time per day (including the regripping time): about 10 h;
⑥ Number of tunneling stroke per day: about 17;
⑦ Advance per day: about 25 m;
⑧ Calculated as 30 days per month, the average advance per month: 750 m.

(2) Maximum Advance Speed

For φ 4.5 m TBM in a project, the maximum advance speed can reach 7.2 m/h. But in the analysis, the maximum advance speed 6 m/h, i.e., 100 mm/min, is taken as the actual advance speed.

① Time of one tunneling stroke: $t = L_s/v = 1500/100 = 15$(min);
② Regripping reset time: 5 min;
③ Time of one tunneling cycle: 20 min;
④ In consideration of TBM maintenance, troubleshooting and delay by auxiliary operation, the utilization ratio of tunneling operation is calculated as 46% when the rock is in good conditions.
⑤ Effective tunneling time per day (including the regripping time): about 14 h;
⑥ Number of tunneling stroke per day: about 34;
⑦ Possible advance per day: about 50 m;
⑧ Possible advance per month as calculated at 30 days per month: 1500 m.

2. Calculation and Analysis of Supporting Capacity

Supporting capacity covers shotcrete speed, rock bolt installation speed, ring beam installation speed, etc. for open TBM, and lining segment installation speed, pea gravel injection speed and grouting speed for a shield TBM. The following focuses on the design, calculation and analysis of the shotcrete capacity of open TBM.

The shotcrete system mainly consists of concrete transfer pump and concrete shotcrete manipulator. For φ4.5 m tunnel and 10 cm thick shotcrete, the shotcrete capacity is calculated and analyzed as follows:

(1) Tunnel diameter: 4.5 m.
(2) Shotcrete thickness: 10 cm.
(3) Inner diameter of tunnel: 4.3 m.
(4) Concrete to be shot per 1 m advance: 1.38 m^3.
(5) As calculated at the maximum advance speed of 6 m/h and in consideration of matching the advance, a shotcrete capacity of 8.28 m^3/h is required. In addition to 20% kick-back amount, a shotcrete system with a shotcrete capacity above 10 m^3/h is needed.

(6) The stroke is 1.5 m and concrete to be shot is about 2.5 m^3 for one stroke.
(7) If the concrete for two strokes is transported with one concrete tank truck, a tank car with a capacity of 5 m^3 is required.

It should be noted that two shotcrete systems are sometimes arranged in practical projects, to provide some allowance for the theoretical value, and act as a standby in case of failure of one shotcrete system on the other hand to prevent delay to TBM excavation.

3. Calculation and Analysis of Muck Discharge Capacity

The muck from rock breaking by the disc cutter will be picked up by the cutter head bucket, unloaded to the TBM belt conveyor, continuously transferred to the back-up belt conveyor, unloaded to the muck car or the tunnel continuous belt conveyor, and sent out of the tunnel. Therefore, the muck discharge capacity of the TBM belt conveyor, the back-up belt conveyor, the muck car or the tunnel continuous belt conveyor should match the advance speed of TBM.

Taking the φ 4.5 m TBM of a project as an example, the muck discharge capacity is calculated and analyzed as follows:

(1) Excavation area: $A = \pi \cdot d^2/4 = \pi \cdot 4.5^2/4 = 15.9(m^2)$.
(2) Volume of muck excavated in one tunneling stroke (compacted): $V = A \cdot L_s = 15.9 \times 1.5 = 24(m^3)$.
(3) Muck discharge capacity per stroke (not compacted): Given the looseness coefficient $\zeta = 1.7$, then the muck discharge capacity $Vt = \zeta \cdot V = 1.7 \cdot 24 = 40(m^3)$; For 2 tunneling strokes, 80 m^3 of muck should be transported out. If 10 m^3 muck cars are used, a train of 8 muck cars is required.
(4) Given that the rock density is 2.7 t/m^3, 65 t of muck will be produced in one tunneling cycle, and about 130 t in two tunneling cycles.
(5) Given that the instantaneous maximum advance speed is 7.2 m/h, the maximum transfer capacity of the belt conveyor must be higher than 309 t/h.

3.4 Method and Model of TBM Tunneling Adaptability Assessment

3.4.1 Adaptability and Assessment Method

1. Concept of Adaptability

Adaptability refers to the capacity of things to adapt to objective conditions and different internal and external environments and needs. According to the theory of system science, a system is an organic aggregate composed of a number of elements (or components) that are different and interrelated, but interact and restrict each other, and it can achieve certain overall functions. Adaptability refers to the ability of the system subject to adapt to external environmental factors, and keep consistent and coordinated development with them. For a TBM, tunneling adaptability means that it should adapt to different influencing factors in the process of tunneling, to ensure that the whole unit can play its design role and finish the task of tunneling on time with both quality and quantity guaranteed.

2. Adaptability Assessment Indicator and System

The indicator refers to the characteristic basis that can definitely reflect some aspects of the research object according to the research object and purpose. Each assessment indicator describes a certain characteristic of the object from different aspects. The indicator system refers to a whole composed of a series of interrelated indicators. It is able to comprehensively reflect all aspects of the object according to the research object and purpose. It is not only extremely limited by the object and target of assessment, but also influenced by the values of the assessment subject.

In adaptability assessment, the determination of assessment indicator system is the basis. The favorable indicators play an important role in the research object. If too many indicators are selected, some of them may be repetitive and cause interference. If too few indicators are selected, they may not be representative enough, but one-sided. The assessment indicator system is an organic whole composed of several interrelated and interactive assessment indicators in a certain hierarchical structure. Only with a scientific and reasonable assessment indicator system can we draw a scientific and fair comprehensive assessment conclusion.

(1) Assessment Indicator

Because many factors will affect the adaptability of the system, it is necessary to first screen out the main or key influencing factors, and then describe them with some measurable indicators as much as possible. The assessment indicators should be selected based on engineering practice, scientific research findings, expert opinions, mathematical methods, etc. In addition, assessment indicators should be easily available for statistics and analysis.

(2) Assessment Indicator System

The assessment indicator system (AIS) should be constructed to not only truly reflect the influence of different indicators on adaptability, but also fully reflect the relationship between indicators. The following three principles should be observed to construct the AIS:

① Hierarchy: The hierarchy principle comes from the hierarchical system structure established by system analysis. Hierarchy refers to the multiplicity of the indicator system itself. Because the content of adaptability evaluation covers multi-levels, the indicator system is also in a multi-level structure, and it should reflect the characteristics of each level. In addition, all elements are interrelated to form an organic whole, and adaptability is the result of comprehensive influence and action of multiple levels and factors. First, in the indicator system, some indicators should be selected to present the coordination procedures of evaluation objectives from the overall level to keep the assessment comprehensive.

Second, indicators must be arranged in hierarchical progressive relationship to clearly reflect the hierarchy and fully implement the principle of hierarchical assessment as far as possible, to ensure a scientific indicator system.

② Validity: Influencing factors are not only numerous but also interrelated. If they are all included in the AIS, there will be too many indicators, and priorities will be unclear. Therefore, it is necessary to keep the assessment indicators effective, that is, of high validity, when characterizing their influence.

③ Concise: To effectively evaluate the adaptability, the assessment indicators used should be neither too detailed and numerous, nor too few and simple, and indicators with the same meaning should not be repeated.

3. Assessment Methods of Adaptability and Selection

The model is the generalization of reality. A conceptual model that can reflect the actual change process and is suitable to actual circumstances can be extracted from the actual system only when we have deep understanding of the composition and structure of the studied system, collect a large amount of actual data, and master the change rules. In other words, only by inputting high-quality parameters appropriately reflecting the characteristics of each influencing factor into the mathematical model can we obtain convincing results.

Many methods can be used to assess the adaptability, such simple methods as expert assessment method, scoring method, ranking method, average indicator method, and efficacy coefficient method, and such complex methods as principal component analysis method, factor analysis method, analytic hierarchy process, neural network method and fuzzy comprehensive assessment method. Each method has its own characteristics and its applicable scope and limitations. Table 3.5 shows the advantages, disadvantages and applicability of different assessment methods.

In the TBM adaptability assessment, many quantitative and qualitative uncertain indicators should be combined, while fuzzy comprehensive assessment method gives enough attention to uncertain factors and it is able to make a comprehensive assessment of things determined by many factors. Based on the comparison between this objective fact and the above assessment methods, the fuzzy comprehensive assessment method is used in this chapter for TBM adaptability assessment.

3.4.2 TBM Adaptability Assessment Method

When a comprehensive assessment method is introduced to assess TBM construction adaptability, the first problem to be solved is to select and determine the assessment indicators. In comprehensive assessment of TBM construction adaptability, different quantitative and uncertain influencing factors should be taken into account. There are many interrelated and closely related factors that affect TBM adaptability. If all relevant influencing factors are taken into account, the assessment system will become too cumbersome and not conducive to the assessment. To evaluate the system

Table 3.5 Adaptability assessment method

Method	Advantage	Disadvantage
Expert assessment method	This method is simple and easy. Experts will make accurate judgments on some vague and uncertain problems based on their experience	But this method is conservative, and the results are greatly influenced by the subjective factors of experts. It can be used to judge and make selection from existing schemes, but cannot be used to create a scheme
Delphi method	This method can avoid the role of authority and the situation that "the majority has the final say". In several turns of consultation for comments, experts have enough time to revise their opinions	But because of the lack of face-to-face communication, experts may insist on wrong opinions. In addition, they may lose patience to answer questions in several turns of consultation
Failure mode and effects analysis (FMEA)	This method is designed to analyze every possible failure model of each component of the system in detail, and infer its influence on the whole system, possible consequences and how to avoid or reduce losses	It can only be used for analysis on non-hazardous failure, but not on the comprehensive factors of different failures
Accident tree analysis	This method is used to make a comprehensive and concise description of inducing factors and their logical relationships. It is convenient for logical operation, qualitative and quantitative analysis and systematic analysis	But it contains too many steps. Its calculation is complicated. It is prone to omit influencing factors. In addition, a large failure trees may generate more than one solution mathematically
Event tree method	This method features clear hierarchy and obvious stages. It can be used for the process analysis of dynamic development of multi-stage and multi-factor complex events to predict the development trend of events in the system	But there is not much statistical data about tunnel accidents, and quantitative analysis still needs a lot of efforts. Therefore, this method is not suitable for analysis on the parallel consequences and discussion about conditional independence
Influence diagram method	This method can present the time sequence, information and probability relationship between variables in a decision analysis problem	But it is difficult to obtain the marginal probability of nodes and conditional probability between nodes. Therefore, the subjective probability analysis may violate probability theory

(continued)

Table 3.5 (continued)

Method	Advantage	Disadvantage
Analytic hierarchy process	This method is able to clearly and vividly present the hierarchy of the system	But it can be used to select the best scheme from the existing schemes, but cannot be used to create a scheme. In addition, the construction of hierarchy and the formation of judgment matrix are both subjective
Monte Carlo method	This method can be used for any calculation type including random variables. Random variables can be in any distribution form, and the number of variables is unlimited	The system to be simulated in practice is complex, and it is difficult to obtain the probability distribution of uncertain factors, so it is difficult to establish the model. The mutual influence between influencing factors is not taken into account
Neural network method	This method features great learning ability, resistance to failure and parallelism	In case of insufficient data or being incapable of constructing accurate training sample set, other comprehensive assessment method may be required. If the sample size is large and the number of neurons in the neural network is large, the amount of calculation will be large, and the global limit may not be found
Fuzzy comprehensive assessment method	This method gives enough attention to uncertain factors. It is able to make a comprehensive assessment of things determined by different factors	As affected by the randomness in the assessment process, the subjective uncertainty of evaluators and the fuzziness of cognition, the assessment process has a certain degree of subjectivity

adaptability, the key lies in the role of the selected indicators in the system, not the number of the indicators. The assessment indicator in the system should not only reflect the objectives and requirements of TBM adaptability assessment, but also be as comprehensive, scientific, practical and concise as possible. The technical route of the assessment method is shown in Fig. 3.7.

3.4.3 TBM Adaptability Assessment Indicator System

Many factors will affect TBM construction adaptability, which is divided into TBM design, TBM construction, tunnel design, stratum factors & problems, and construction organization & management (Fig. 3.8) based on consultation with experts. The

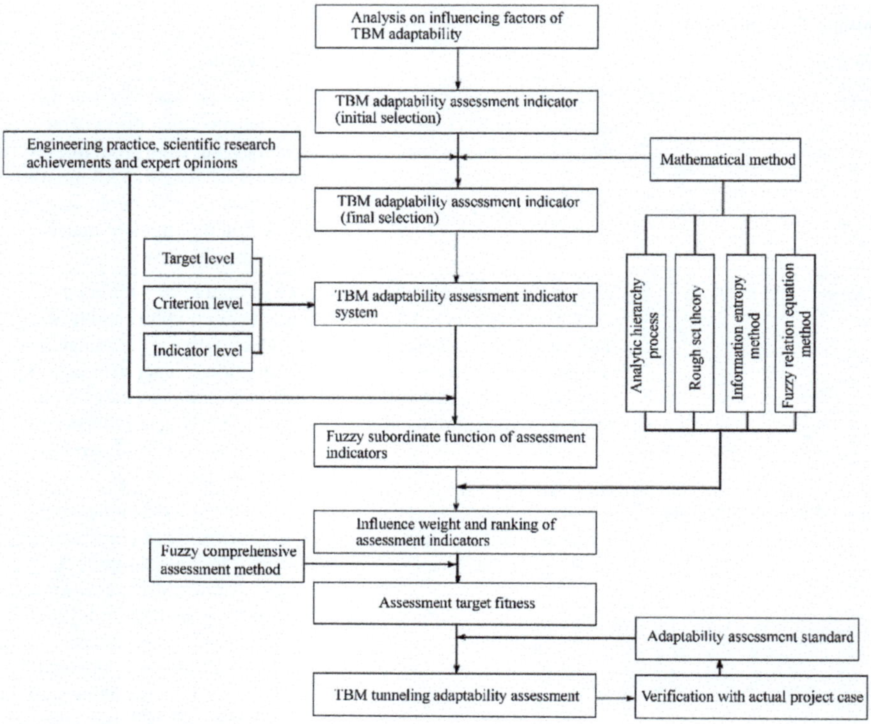

Fig. 3.7 Route of TBM tunneling adaptability assessment methods

influencing factors of TBM construction adaptability will be analyzed based on the six aspects, and the assessment indicators will be initially selected below.

To more clearly explain the different influencing factors that need to be taken into account in TBM adaptability assessment, Tables 3.6 and 3.7 generally present these factors and their influences.

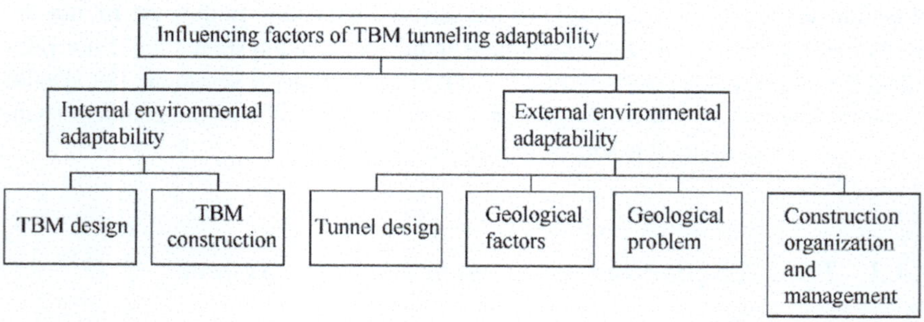

Fig. 3.8 Influencing factors of TBM tunneling adaptability

Table 3.6 Internal environmental impact factors of TBM tunneling adaptability assessment

Fitness D_1	Influencing factor	Influence consequence
TBM design P_1	TBM selection	Type selection should be determined after comprehensive analysis on engineering geological and hydrogeological conditions, tunnel design requirements, support and lining forms, segment manufacturing technology and mold cost
	Cutter head structure design	Inappropriate cutter head design may lead to instability of excavation face, rapid wear of cutter head, clogging of cutter head and screw jamming of cutter head
	Cutter design	Inappropriate selection and layout of cutters may lead to difficulties in advance, rapid wear of cutter head and clogging of cutter head
	Opening ratio design	If the opening ratio of the cutter head is too low, the opening is not set at a proper position, and the muck cannot be transferred into the excavation chamber in time, the cutter head may worn more quickly. Too low opening ratio may also result in clogging of head cutter, not only increasing the torque and thrust force of the cutter head, but also reducing the advance speed, and accelerating the wear of cutters and cutter head. Consequently, excessive ground settlement or uplift may occur
	Cutter spacing design	Inappropriate cutter spacing will reduce rock breaking efficiency, increase tool wear and affect full-section cutting and cutter life
	Tail shield clearance design	If the clearance is too small, the tail shield will interfere with the segment during the advancing of the TBM
	Muck improvement	The muck improvement device of an improper design may cause engineering problems such as blocking of excavation chamber, clogging of cutter head and gushing at screw outlet

(continued)

Table 3.6 (continued)

Fitness D_1	Influencing factor	Influence consequence
	Mucking mode	The muck discharge capacity must meet the requirements at maximum TBM advance speed
TBM construction P_2	Total thrust force	In case of excessive thrust force, the soil in front may swell, the cutter and the cutter head may be worn more seriously, and the segment may crack
	Cutter head rotation speed	The cutter head rotation speed is closely related to torque, and improper control will disturb the soil too much
	Advance speed	The advance speed depends on the thrust force. If the advance speed is too high, the soil in front may swell, and the cutter and cutter head will be worn more seriously
	Cutter head torque	In case of improper control of cutter head torque, the soil may be disturbed too much
	Pressure of excavation chamber	In case of improper control of pressure of excavation chamber, the excavation face may become unstable, and ground settlement or uplift may occur
	Rotation speed of screw conveyor	In case of improper control of the rotation speed of screw conveyor, the pressure control of the excavation chamber may be affected, leading to instability of the excavation face and excessive flow of water-soil mixture into the excavation chamber
	Supporting force of gripper	The grippers should be designed according to space size, rock not prone to collapse, crossing ring beam, and other requirements. Sufficient strength and area should be provided, and the ground pressure with the cave wall should be calculated
	Ventilation	As the tunneling distance is long and the hole diameter is small, the ventilation is very difficult. Ventilation in tunnel construction should also be specially considered

Table 3.7 External environmental impact factors for TBM tunneling adaptability assessment

Fitness D2	Influencing factor	Influence consequence
Tunnel design P_3	Tunnel section size	
	Overburden of tunnel	One of the main factors that should be considered in tunnel classification according to overburden is the deformation and failure mode of surrounding rock, which is closely related to the initial stress state of surrounding rock
	Length of tunnel	It is more economical to use a TBM to construct long tunnels longer than 3 km
	Tunnel turning radius	
	Tunnel gradient	
Geological factor P_4	Uniaxial compressive strength of rock (Rc)	The lower the Rc, the higher the advance speed of TBM, and the faster the tunneling, and vice versa
	Hardness and wear resistance of rock	Generally, the harder the rock, the better its wear resistance and the higher wear to TBM cutter
	Completeness of rock mass	The more developed the structural plane of rock mass, the greater the density, the smaller the joint spacing, the lower the intactness coefficient and the higher the TBM advance speed
	Quartz content of rock	The higher the content of quartz in rock, the harder and more resistant to wear the rock, the greater the cutter loss, and the lower the TBM advance speed (PR) and construction speed (AR)

(continued)

Table 3.7 (continued)

Fitness D2	Influencing factor	Influence consequence
	Water-bearing state of rock mass	Generally, the strength of surrounding rock is lower to varying degrees in areas rich in water, water gushing and leakage, especially in soft rock, which will undermine the stability of surrounding rock and affect the work efficiency of TBM
	Surrounding rock grade	In the specification, the tunnel surrounding rock is classified into six grades, namely I, II, III, IV, V and VI. The lower the grade, the better the surrounding rock properties
Adverse ground condition P_5	High geostress	When the tunnel overburden is large (> 250 m), the geostress will obviously increase with the increase of the tunnel overburden. The stress of the surrounding rock will be redistributed after tunnel excavation, and may be concentrated locally, easily leading to rock burst, damaging the construction equipment and endangering the personnel of constructors
	High ground temperature	As the ground temperature increases with the increase of depth, the high-temperature heat damage intensifies in underground chamber with the increase of excavation depth

(continued)

Table 3.7 (continued)

Fitness D2	Influencing factor	Influence consequence
	Collapse and water gushing of fault fracture zone and surrounding rock	Fault fracture zone is a common geological phenomenon in tunnel construction, which often causes landslides ranging from a few meters to tens of meters because of insufficient understanding, improper construction and other reasons. It makes construction more difficult, causes delay in the construction, wastes funds and brings hidden hazards to the safety of tunnel construction and operation
	Composite strata	In compound strata, the stratum conditions on the same end face are obviously different, it is impossible to establish pressure balance on the excavation face, and the tunnel is susceptible to uneven settlement, resulting in the local additional stress and stress concentration in the lining structure. In addition, it is extremely difficult to control the advance parameters. Therefore, the compound strata are extremely vulnerable to collapse, roof caving and slurry leakage
	Rock burst	In slight rock burst, only rock pieces flake without ejection. In severe rock burst, the measurable magnitude may reach 4.6. Such rock burst usually lasts for several days or months

(continued)

Table 3.7 (continued)

Fitness D2	Influencing factor	Influence consequence
	Serious deformation of soft rock	The rapid convergence of tunnel during excavation often leads to the deformation and damage of concrete segments. In severe cases, it even leads to TBM jamming accidents
	Karst water inrush and mud outburst disasters	In case of improper handling, there may be engineering problems such as integral segment sinking, joint opening, and steps between segments, even catastrophic accidents such as machine nose sinking and collapse, large-scale karst cave water inrush flooding the tunnel
	Coal-bearing strata and gas outburst	Coal-bearing strata often contain flammable and harmful gases such as CO and CH_4, which seriously threatens the health and life of construction workers in the tunnel
Construction organization and management P_6	Working arrangement	Including available working hours, working regulations, shift arrangement, personnel organization, personnel training and experience, and personnel salary and bonus system
	Logistics system	Including water and power supply, ventilation and cooling
	Safety guarantee	Including control of dust, light, vibration and noise
	Management	The authority of managers and procurement conditions
	Local actual situation	Including the conventional practice of tunnel construction, the quality of labor force, the supply of goods, and local laws and regulations

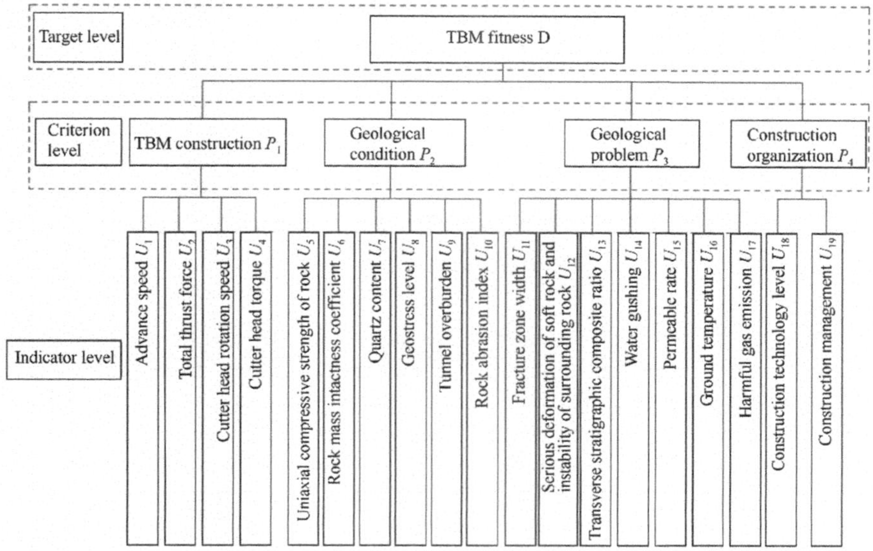

Fig. 3.9 TBM adaptability assessment indicator system

According to the construction principle of adaptability assessment system, TBM assessment indicators will be selected finally in combination with mathematical methods, specific engineering practice, scientific research achievements and expert opinions after initial selection of assessment indicator, and the TBM adaptability assessment system will be constructed, as shown in Fig. 3.9.

(1) Target level

The target level is of TBM construction fitness D. It is used to reflect the adaptability between TBM construction and related influencing factors.

(2) Criterion level

The criterion level covers TBM construction, geological conditions, geological problems and construction organization, expressed with P1, P2, P3 and P4 respectively.

(3) Indicator level

The indicator level includes 19 indicators. It is used to express 4 elements in criterion level, including TBM construction, geological conditions, geological problems and construction organization. TBM construction indicators include advance speed U_1, total thrust force U_2, cutter head rotation speed U_3 and cutter head torque U_4. Geological conditions include uniaxial compressive strength of rock U_5, rock mass intactness coefficient U_6, quartz content U_7, geostress level U_8, maximum overburden U_9 and rock abrasion index U_{10}. Geological problem indicators include fracture zone width U_{11}, serious deformation of soft rock and instability of surrounding rock U_{12},

Transverse stratigraphic composite ratio U_{13}, water gushing U_{14}, permeable rate U_{15}, ground temperature U_{16}, and Harmful gas emission U_{17}. Construction organization indicators include construction technology level U_{18} and construction management U_{19}.

3.4.4 Determination of Membership Function of TBM Adaptability Assessment Indicator

The influencing factors of TBM adaptability include several assessment indicators, but the attributes, units and importance of each indicator are different in different degrees and not commensurable. To solve this problem, we must rely on experienced experts, sufficient field survey data and modern mathematical methods. To allow the commonality of each assessment indicator, each assessment indicator is quantified, and the corresponding membership function is constructed in this chapter based on the theory and method of establishing membership function with fuzzy mathematics.

When TBM construction does not adapt to a certain influencing factor at all, its membership function is set to 0, that is, its assessment indicator value is quantified to 0. When TBM construction fully adapts to a certain influencing factor, its assessment indicator value is quantified as 1, and a certain value between [0, 1] can be used to express the different degrees between TBM adaptability and the influencing factor. When the quantitative value selection of assessment indicator is selected, the quantitative criteria should express the range and interval size of each indicator as accurately as possible. In this chapter, the membership function of specific assessment indicator is determined by linear relationship.

1. TBM construction

(1) Advance Speed

The advance speed is a comprehensive parameter of TBM construction. It is influenced not only by TBM thrust force, but also by key parameters such as rotation speeds of cutter head and screw conveyor. If the advance speed is too low, the tunneling efficiency of the TBM will be reduced. If it is too high, the cutter head may vibrate excessively, leading to abnormal loosening of the connecting bolt, plastic deformation and fracture of the main bearing, and seriously affecting the construction safety.

According to the experience of TBM construction, the advance control parameters and advance speed of TBM are different under different engineering geological conditions. The main influencing factors include surrounding rock lithology, rock strength and rock mass intactness. However, the strength of rock will directly affect the penetration depth of TBM, and then affect the advance speed. It is generally believed that the higher the strength of rock is, the lower the cutting depth of cutter head per revolution and the advance speed are. During excavation, the TBM should be operated effectively based on the specific stratum conditions, and the cutter head rotation speed and cutter head torque should be adjusted within the working load range of TBM, to increase excavation efficiency without damage to TBM equipment.

Yuan Xingze, et al. made statistics on the geological records and construction records of a TBM project. They found that when the uniaxial compressive strength of rock was in the range of 30–70 MPa, the average advance speed was about 2.43 m/h, and the advance speed remained relatively stable. When the uniaxial compressive strength was greater than 70 MPa, the advance speed obviously decreased with the increase of surrounding rock strength. When the uniaxial compressive strength was in the range of 70–120 MPa, the average driving speed was 1.81–1.24 m/h. Yan Yanbin collected and sorted out the main parameter categories of factors affecting TBM advance speed, including rock mechanics parameters, geological structure and hydrogeological environment parameters, and found that, when the uniaxial compressive strength was less than 70 MPa, the advance speed of TBM ranged from 1.92 m/h to 2.86 m/h.

Li Zhun, et al. made a classification by rock mass strength and surrounding rock stability based on engineering practice, and obtained the corresponding average daily advance speed (including standard deviation). They found that when the strength of rock mass is low, the advance speed is low as well. As the strength of rock mass increases, the advance speed gradually increases. When the strength reaches a certain value, the advance speed gradually decreases with the increase of strength (as shown in Table 3.8).

Zhao Weigang obtained the relationship curve between compressive strength of rock and advance speed based on the corresponding relationship between compressive strength of rock and advance speed in Qinling tunnel construction (Fig. 3.10), and numerically simulated the compressive strength of rock and advance speed by using the Qinling tunnel thrust force model. Both the curves of theoretical simulation and actual construction showed that the driving speed decreased with the increase of compressive strength of rock. Figure 3.11 shows the relationship between surrounding rock grades and advance speed during the construction of Qinling tunnel.

Table 3.8 Average daily advance speed V (including standard deviation) under different rock strength and stability

Uniaxial compressive strength of rock (MPa)	A compactness FS < 1.25	B fragility 1.25 < FS < 2	C stability 2 < FS < 6	Total (m)
Rc < 20	$S_{AI} = 600$ m $\phi_V = (29 \pm 11)$m/d	$S_{BI} = 350$ m $\phi_V = (32 \pm 8)$m/d	$S_{CI} = 1050$ m $\phi_V = (34 \pm 7)$m/d	$SS_I = 2000$
20 < Rc < 80	$S_{AII} = 600$ m $\phi_V = (38 \pm 14)$m/d	$S_{BII} = 1855$ m $\phi_V = (41 \pm 11)$m/d	$S_{CII} = 2375$ m $\phi_V = (43 \pm 10)$m/d	$SS_{II} = 4830$
80 < Rc < 140	$S_{AIII} = 200$ m $\phi_V = (22 \pm 10)$m/d	$S_{BIII} = 1855$ m $\phi_V = (24 \pm 7)$m/d	$S_{CIII} = 0$ m $\phi_V = (26 \pm 7)$m/d	$SS_{III} = 2050$
Total (m)	$SS_A = 1400$	$SS_B = 4055$	$SS_C = 3425$	8880

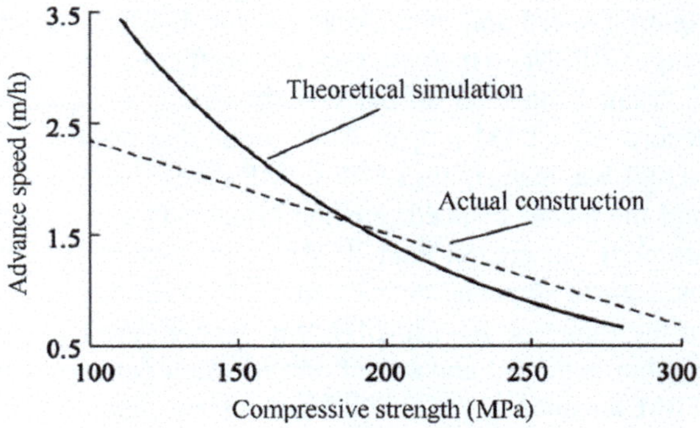

Fig. 3.10 Relation between rock compressive strength and advance speed

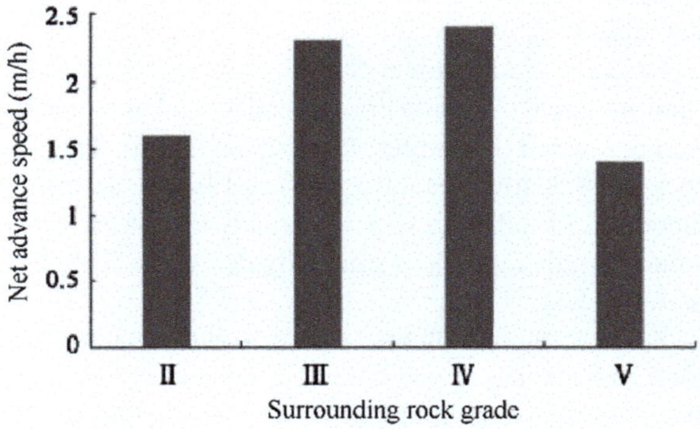

Fig. 3.11 Column of relation between surrounding rock grade and advance speed

It is shown that the advance speed of surrounding rocks of grades II and V is lower, and surrounding rocks of grades III and IV are more suitable for TBM excavation, and TBM can operate at a higher advance speed in such rocks.

In conclusion, the advance speed can be classified according to the uniaxial compressive strength of rock. In case of 15 MPa $\leq$ Rc < 40 MPa, the advance speed will gradually increase with the increase of rock compressive strength. Therefore, the advance speed within this strength range is counted in combination with engineering experience in China, and the fitting relationship between uniaxial compressive strength and advance speed within this compressive strength range is obtained. The fitting formula can be expressed as follows:

$$v_c = 0.7134e^{0.0336R_c} \tag{3.11}$$

where:

v_c advance speed related to uniaxial compressive strength.

In case of 40 MPa $\leq$ Rc < 180 MPa, the advance speed gradually decreases with the increase of compressive strength. Therefore, the advance speed within this strength range is estimated in combination with engineering experience in China, and the

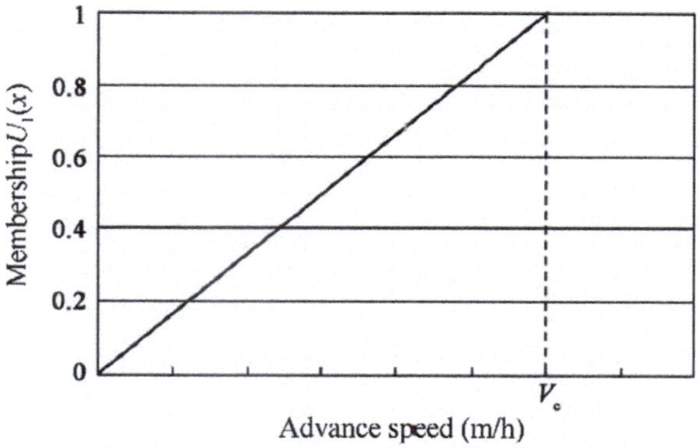

Fig. 3.12 Membership function of advance speed

fitting relationship between uniaxial compressive strength and advance speed within this compressive strength range is obtained. The fitting formula can be expressed as follows:

$$v_c = -1.036 \ln R_c + 6.8749 \tag{3.12}$$

On this basis, the membership function of advance speed is constructed as follows:

$$U_1 x = \begin{cases} \frac{x}{v_c} & (0 < x \le v_c) \\ 1 & (x > v_c) \end{cases} \tag{3.13}$$

Refer to Fig. 3.12 for the membership function.

(2) Total thrust force

The total thrust force of TBM is one of the main performance indicators. It is closely related to construction parameters such as advance speed and cutter head torque. The calculation of total thrust force of TBM is the basis of TBM type selection, as well as an important guarantee for successful tunneling of TBM. In a hard rock stratum, the diameter of the tunnel is larger than that of the shield. Therefore, the thrust force is used to thrust the cutter head and overcome the friction force at the bottom of the shield, and not much thrust force is needed. In a weak or broken rock stratum, the shield is tightly wrapped by rock, additional thrust force is required to overcome the friction of cutter head and the friction of the whole shield, and much thrust force is needed. Therefore, it is important to reasonably calculate the primary thrust force for the normal tunneling of TBM according to the known parameters, and the main thrust force is related to the normal operation of the whole tunneling project.

When the primary thrust system works, the back-up equipment will not move accordingly, but move forward after the primary thrust cylinder retracts when completing a tunneling stroke. Therefore, the primary thrust force will be calculated without the tractive force required for pulling the back-up equipment. The main machine is connected to the back-up equipment through a traction cylinder. During tunneling, the traction cylinder is in a free flexible state, and the back-up equipment does not move accordingly. Therefore, the tunneling reaction force does not include the tractive force required for pulling the back-up equipment.

① When the rock stratum is in good condition, that is, in hard rock state, the tunnel bored is larger than the shield structure in diameter and it will not contract or extrude. Then the reaction force F provided by the grippers mainly includes: the force required for cutter head advance (F_1) and the friction force between the shield structure bottom and the tunnel wall (F_{U_1}).

The force F_1 required for cutter head advance is calculated as follows:

$$F_1 = nP \tag{3.14}$$

where:

n number of cutters;
P rated bearing capacity of disc cutter (kN).

The friction force between the shield and the tunnel wall F_{U_1} is calculated as follows:

$$F_{U1} = \mu W \tag{3.15}$$

where:

μ friction coefficient;
W shield weight (kN).

Therefore, when the surrounding rock is stable, the total thrust force F is calculated as follows:

$$F = F_1 + F_{U1} \tag{3.16}$$

② If the tunnel wall is still able to provide stable support and the thrust force needed for excavation can be met when the surrounding rock is in an unstable geological section, the rock stratum will be closely attached to the shield structure in consideration of the characteristics of soft rock. Therefore, the friction force between the front shield, telescopic shield and tunnel wall is mainly caused by the soft rock pressure acting on the outer surfaces of the front shield and telescopic shield. Then the reaction force F provided by the grippers mainly includes: the force required for cutter head advance (F1) and the friction force between the shield structure and the tunnel wall (F_{U_2}). The soft rock pressure acting on the outer surface of shield structure is distributed as shown in Fig. 3.13.

where:

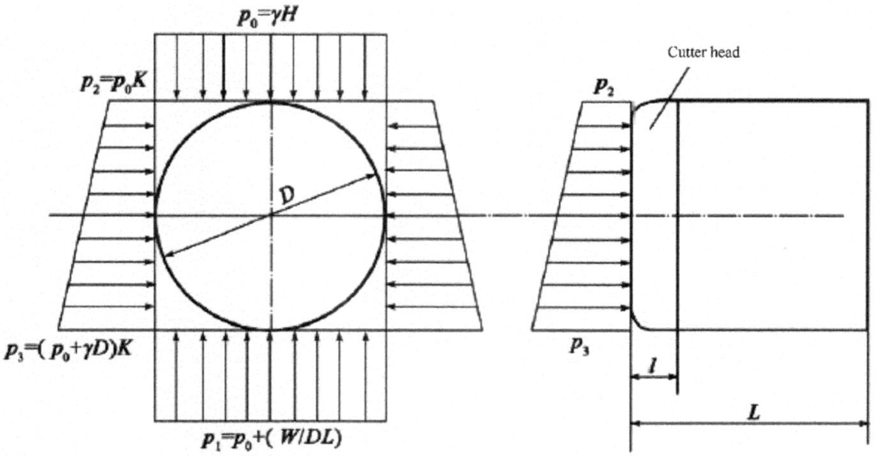

Fig. 3.13 Distribution of cutter head pressure

D	diameter of cutter head;
W	gravity of shield structure;
γ	unit weight of soft rock;
H	thickness of broken surrounding rock (height of collapsed arch), H = 0.7D;
K	lateral pressure coefficient K = (1-sinφ)/(1 + sinφ), where φ is the internal friction angle;
p_0, p_1, p_2, p_3	top pressure, bottom pressure, top lateral pressure and bottom lateral pressure of shield structure.

For a shield TBM, L is the sum of the lengths of front shield and outer telescopic shield. For an open TBM, L is the width of cutter head, that is, e.

The shield friction force F_{U_2} caused by soft rock will be calculated with the following formula:

$$F_{U2} = \mu(p_0 + p_1 + p_2 + p_3)DL \quad (3.17)$$

where:

μ friction coefficient.

Therefore, when the surrounding rock is unstable, the total thrust force F is:

$$F = F_1 + F_{U2} \quad (3.18)$$

Therefore, the adaptive range of the total thrust force of the cutter head is F_U ~ ($F_1 + F_U$). When the surrounding rock is stable, $F_U = F_{U1}$. When the surrounding rock is unstable, $F_U = F_{U2}$, given $F_{low} = F_U$, $F_{high} = F_1 + F_U$.

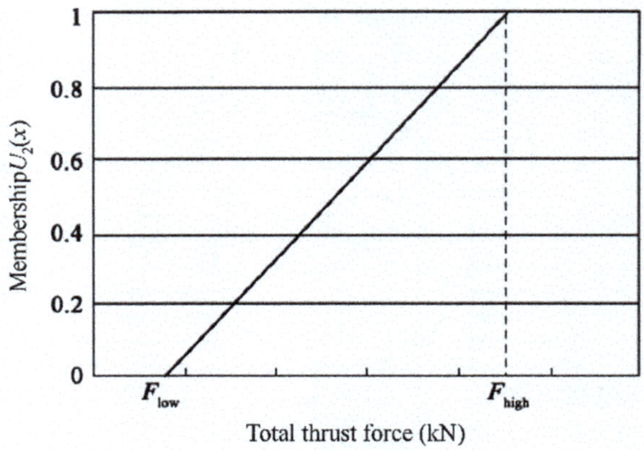

Fig. 3.14 Total thrust force membership function

On this basis, the membership function of the total thrust force is constructed as follows:

$$U_2(x) = \begin{cases} 0 & (x < F_{\text{low}}) \\ \frac{x - F_{\text{low}}}{F_{\text{high}} - F_{\text{low}}} & \left(F_{\text{low}} \leq x < F_{\text{high}}\right) \\ 1 & \left(x \geq F_{\text{high}}\right) \end{cases} \tag{3.19}$$

Refer to Fig. 3.14 for the corresponding membership function.

(3) Cutter head rotation speed

The main working parameters of TBM cutting mechanism include cutter head torque, rotation speed and thrust force, among which the cutter head rotation speed is an important parameter in TBM design. Song Kezhi, et al. made a statistical research based on cutter head rotation speed and TBM diameter involved in projects both at home and abroad, and obtained the correlation between cutter head rotation speed and diameter that the cutter head rotation speed decreases with the increase of cutter head diameter, as follows:

$$n = 22.266 D^{-0.7453} \tag{3.20}$$

In addition, if the cutter head rotation speed is too high, the temperature will rise, affecting the service life and performance of the bearing and its seals, and speeding up the aging of edge cutters. Generally, the allowable peripheral linear speed of cutter head V is 2.5 m/s. Based on this principle, the cutter head rotation speed can be determined with the following formula:

$$n = \frac{v}{\pi D} = \frac{2.5 \times 60}{\pi D} = \frac{47.7}{D} \tag{3.21}$$

Therefore, the relationship between cutter head rotation speed and its diameter can be expressed as:

$$n = \frac{x}{D} \tag{3.22}$$

where:

x rotation speed factor, generally 45–50, with corresponding cutter head peripheral linear velocity V = 2.4 – 2.6 m/s.

When the cutter head is rotating, cutting muck will be produced, generating centrifugal force. Too large centrifugal force will affect muck discharge. Therefore, in the determination of rotation speed, the centrifugal acceleration generated by muck should be kept not exceeding the gravitational acceleration G, namely:

$$\frac{v^2}{r} \leq g \tag{3.23}$$

where:

r cutter head radius.

$$v = \frac{n\pi D}{60} \leq \sqrt{\frac{gD}{2}} \tag{3.24}$$

Further derived is:

$$n \leq 42.6D^{-0.5} \tag{3.25}$$

In comparison of the calculation results of the above formulas, it can be seen that, the design rotation speed of TBM cutter head does not exceed the allowable value at present, and there is still a certain allowance. On the premise of not damaging the cutters, the tunneling efficiency of TBM gradually increases with the increase of cutter head rotation speed. When the rotation speed of TBM cutter head reaches the rated rotation speed, the TBM will work at the maximum tunneling efficiency.

However, according to the research of Wang Mengshu, the cutter head rotation speed is generally about 6 r/min in hard rock. It is adjusted to 3–4 r/min when the TBM enters the end of the transition section to weak surrounding rock, and it is kept at about 2.0 r/min when the TBM completely enters the weak surrounding rock.

Then, the membership function of cutter head rotation speed is constructed as follows when TBM is tunneling in hard rock:

$$U_3(x) = \begin{cases} \frac{x}{6} \ (0 < x \leq 6) \\ 1 \ (x > 6) \end{cases} \tag{3.26}$$

The membership function of cutter head rotation speed is constructed as follows when the TBM is tunneling in the transition section from hard rock to weak surrounding rock:

$$U_3'(x) = \begin{cases} \frac{x}{3} & (0 < x \leq 3) \\ 1 & (x > 3) \end{cases} \tag{3.27}$$

The membership function of cutter head rotation speed is constructed as follows when TBM is tunneling in weak rock:

$$U_3''(x) = \begin{cases} \frac{x}{2} & (0 < x \leq 2) \\ 1 & (x > 2) \end{cases} \tag{3.28}$$

Refer to Fig. 3.15 for the corresponding membership function.

(4) Cutter head torque

The design cutter head torque mainly depends on the type of TBM, cutter head form, and type and quantity of cutters. Under normal circumstances, the cutter head torque should be controlled within a reasonable range, reflecting the adaptability of TBM construction to stratum. Improper control of cutter head torque will not

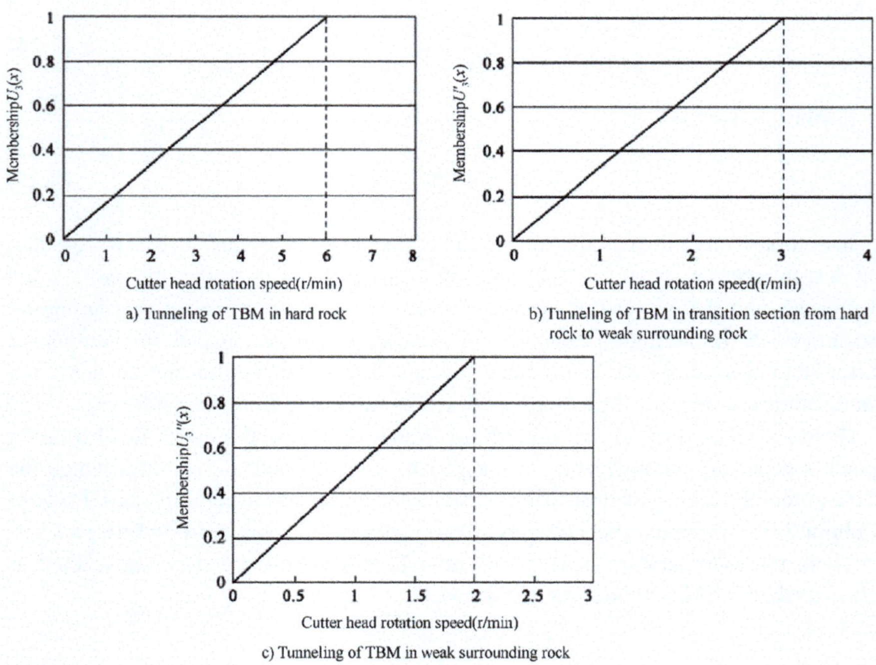

Fig. 3.15 Membership function of cutter head rotation speed

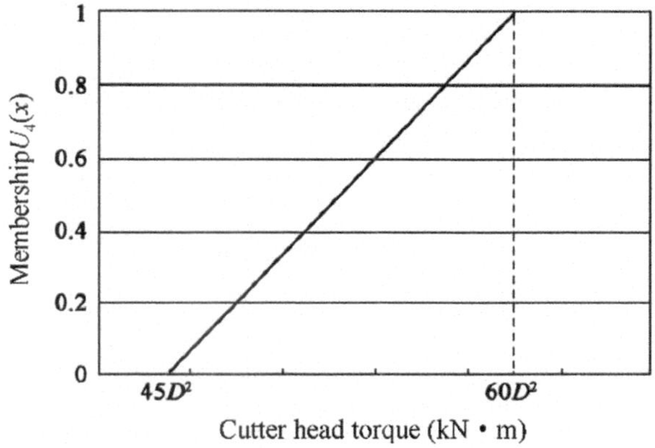

Fig. 3.16 Membership function of cutter head torque

only affect the tunneling efficiency of TBM construction, but also may cause engineering accidents in severe cases. Reasonable driving torque of cutter head allows not only high tunneling efficiency and tunneling performance under different geological conditions, but also optimized structure and total installed power of TBM.

Cutter head torque M_d is generally considered according to formula (3.4).

Therefore, the membership function of cutter head torque is constructed as follows:

$$U_4(x) = \begin{cases} 0 & (x < 3251) \\ \frac{x-45D^2}{60D^2-45D^2} & (3251 \le x < 4335) \\ 1 & (x \ge 4335) \end{cases} \tag{3.29}$$

Refer to Fig. 3.16 for the corresponding membership function.

2. Geological conditions

(1) Uniaxial Compressive Strength of Rock (R_c).

Uniaxial compressive strength of rock is an important indicator to predict TBM assessment model. The lower the R_c, the higher the advance speed of TBM, and the faster the tunneling, and vice versa. With the increase of uniaxial compressive strength of rock, the tunneling efficiency of TBM tends to decline gradually. However, if the R_c is too low, the self-stabilization time of surrounding rock after TBM excavation is extremely short. Sometimes, the surrounding rock cannot be self-stabilized. The TBM is able to keep a certain speed, and the surrounding rock of the tunnel is able to be self-stabilized within a certain period of time only when R_c is within a certain range. It is generally believed that the TBM is more suitable for tunneling in a medium hardness environment with an uniaxial compressive strength of rock ranging from 30 to 60 MPa and in good geological conditions. In addition, the TBM can tunnel faster in surrounding rock of grades III and IV. According to the relevant data at home and abroad and the actual construction situation, it can be found that the TBM is adaptable when the uniaxial compressive strength of rock (R_c) is 40–75 MPa.

However, when R_c exceeds 180 MPa or is lower than 15 MPa, it is not conducive to TBM construction.

The membership function of uniaxial compressive strength of rock (R_c) is constructed as follows:

$$U_5(x) = \begin{cases} 0 & (x < 15) \\ \frac{x-15}{40-15} & (15 \le x < 40) \\ 1 & (40 \le x < 75) \\ \frac{180-x}{180-75} & (75 \le x < 180) \\ 0 & (x \ge 180) \end{cases} \tag{3.30}$$

Refer to Fig. 3.17 for the corresponding membership function.

(2) Rock Mass Intactness Coefficient (K_v)

The rock mass intactness coefficient (K_v) is one of the main indicators to reflect the rock mass intactness. It is equal to the square of the longitudinal wave velocity ratio between rock mass and rock, and it can be measured with dynamic method. The rock mass can be classified into intact, relatively intact, relatively broken, broken and extremely broken by rock mass intactness coefficient. The development degree of structural planes (joints, bedding, schistosity, faults) in rock mass (i.e., the rock mass intactness) is an important geological factor affecting TBM tunneling efficiency. Generally, intact rock mass is not conducive to TBM tunneling. The TBM is able to work at higher advance speed and efficiency when the rock mass is less intact. However, when the structural plane is excessively developed and the rock mass is not intact, or even broken or loose with low overall strength, the rock has no self-stability as surrounding rock, and the TBM will work at very low advance speed and tunneling efficiency. Therefore, both excessively developed and undeveloped structural planes of rock mass are not conducive to TBM tunneling.

According to the tunneling situation of TBM construction in China, when $K_v \ge 0.85$, the rock mass is intact. When the compressive strength of rock mass (R_c) in the corresponding section is greater than 150 MPa, the tunneling efficiency of

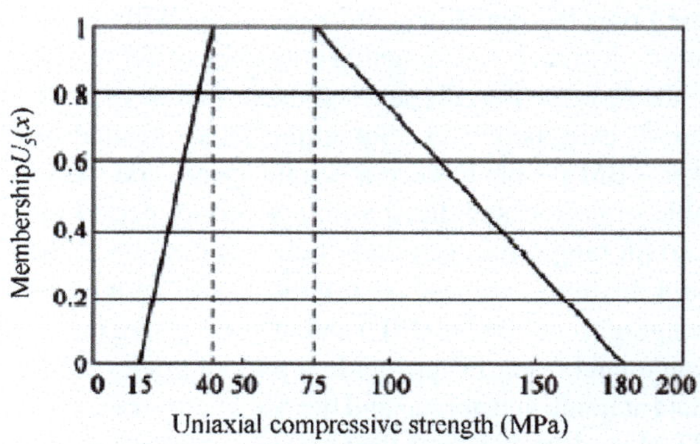

Fig. 3.17 Membership function of uniaxial compressive strength of rock

TBM is very low. When $0.45 \leq K_v \leq 0.75$, the rock mass is relatively intact. When corresponding R_c is 30–60 MPa, the tunneling efficiency of TBM is the highest. When $K_v < 0.35$, the rock mass is very broken and mostly located in the fault-affected zone, and the engineering geological conditions are not favorable. The surrounding rock of the tunnel is prone to collapse and not stable, seriously affecting the normal tunneling of TBM, and resulting in extremely low tunneling efficiency.

The membership function of rock mass intactness coefficient K_v is constructed as follows:

$$U_6(x) = \begin{cases} 0 & (x < 0.35) \\ \frac{x-0.35}{0.45-0.35} & (0.35 \leq x < 0.45) \\ 1 & (0.45 \leq x < 0.75) \\ \frac{0.85-x}{0.85-0.75} & (0.75 \leq x < 0.85) \\ 0 & (x \geq 0.85) \end{cases} \tag{3.31}$$

Refer to Fig. 3.18 for the corresponding membership function.

(3) Quartz Content

Quartz, with a hardness of 7, is a common clastic component in clastic rocks, as well as a bad mineral frequently encountered in TBM construction. Quartz of high content in rock mass will directly affect the construction efficiency and cutter service life of TBM, while cutter wear will make the construction period and cost unpredictable, and seriously restrict the adaptability of TBM construction.

Generally, the content of quartz in sandstone (including mixed sandstone) and siltstone is about 65–80%, but less than 20% in slate. The higher the content of quartz in the rock, the higher the hardness and wear resistance of the rock. Therefore, the TBM cutter will suffer from more serious wear, the advance speed will be slower, and the adaptability of TBM to geological conditions will be degraded. Current researches both at home and abroad show that the quartz has slight influence on TBM construction at a content less than 5%, has obvious influence at a content of

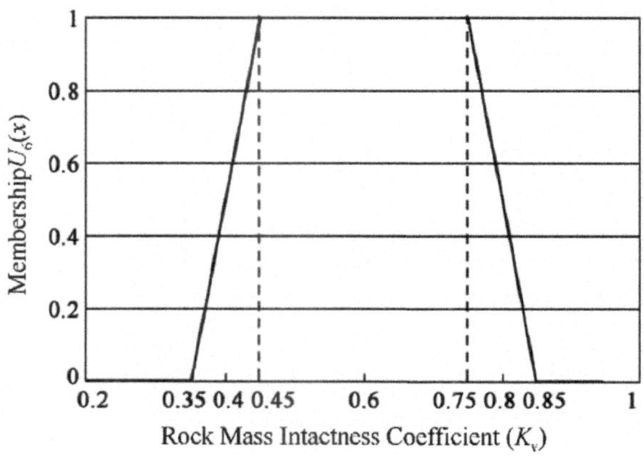

Fig. 3.18 Membership function of rock mass intactness coefficient

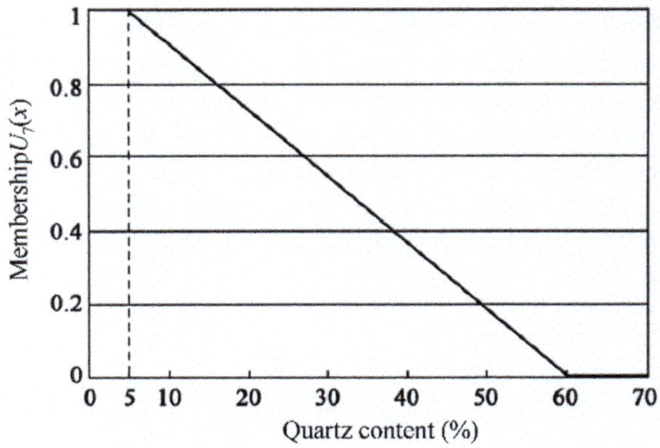

Fig. 3.19 Adaptive membership function of quartz content

5–30%, has moderate influence at a content of 30–60%, and has serious influence at a content higher than 60%.

Accordingly, the membership function of quartz content is constructed by using the descending half-step distribution function:

$$U_7(x) = \begin{cases} 1 & (x < 5) \\ \frac{60-x}{60-5} & (5 \le x < 60) \\ 0 & (x \ge 60) \end{cases} \tag{3.32}$$

Refer to Fig. 3.19 for the corresponding membership function.

(4) Geostress Level

In the TBM project for deep-buried large and long tunnels, the problems like serious deformation and rock burst caused by high geostress are very prominent. In deep-buried composite strata, high geostress is the dominant factor that endangers engineering safety. At present, it is difficult to directly measure the magnitude of geostress. According to the field data and measured data of borehole, the geostress level can be analyzed by linear regression in different depth ranges. The geostress level is generally classified by the strength-stress ratio of surrounding rock:

$$T_s = \frac{\sigma_{max}}{R_c} \tag{3.33}$$

where:

σ_{max} the maximum primary stress of the tunnel section;
R_c uniaxial compressive strength of rock.

According to researchers at home and abroad, the geostress level "σmax/Rc ≥ 0.4" has serious influence on TBM construction; the geostress level "σmax/Rc < 0.15" has rare influence on TBM construction. Now, based on the engineering experience at home and abroad, the working conditions of TBM can be classified into the following 4 grades according to the ground stress level (Table 3.9).

Accordingly, the membership function of ground geostress is constructed by using the descending half-step distribution function:

$$U_8(x) = \begin{cases} 1 & (x < 0.15) \\ \frac{0.4-x}{0.4-0.15} & (0.15 \le x < 0.4) \\ 0 & (x \ge 0.4) \end{cases} \tag{3.34}$$

Refer to Fig. 3.20 for the corresponding membership function.

(5) Tunnel Overburden

When the tunnel overburden is large, many unpredictable complex and uncertain factors may be encountered in TBM construction. When the tunnel overburden exceeds 250 m, the geostress will obviously increase with the increase of the overburden, and a series of geological problems may be caused, such as rock burst and TBM jamming accident caused by serious deformation of surrounding rock. For the tunnel with an overburden of more than 1000 m, the overburden is the key factor that restricts the construction efficiency and economic benefits. For the tunnel of shallow overburden, the surrounding rock will be less self-stable at shallower overburden. Generally, the larger the tunnel overburden, the worse the adaptability of TBM construction to geological conditions.

Benardos et al. classified the tunnel overburden into 4 grades according to the scoring criteria, as shown in Table 3.10. It can be seen that overburden less than

Table 3.9 Adaptability assessment criteria of geostress

Assessment indicator	TBM working conditions			
	Good	Moderate	Fair	Bad
Geostress level	< 0.15	0.15–0.2	0.2–0.4	≥ 0.4

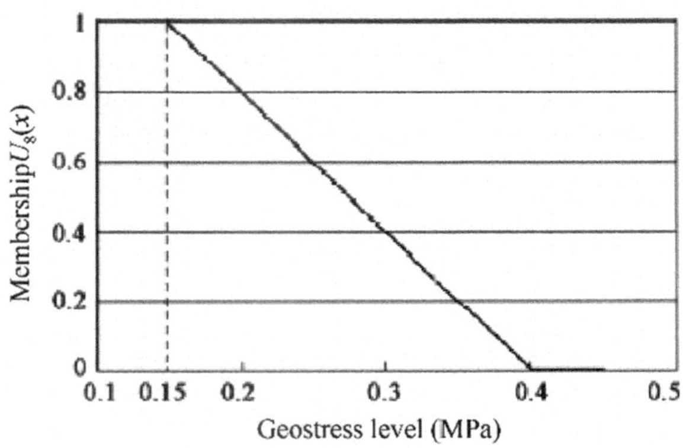

Fig. 3.20 Adaptive membership function of geostress level

7.5 m was not conducive to the TBM construction. With the increase of the tunnel overburden, the conditions for TBM construction would become more favorable. However, they did not involve the influence of deep tunnel overburden on TBM construction conditions in the research.

According to the criteria for deep and shallow overburdens specified in *Code for Design of Railway Tunnel* (TB10003-2016), shallow overburden refers the overburden lower than 2.0–2.5 times of collapsed arch height. For a single-track tunnel, the collapsed arch height is expressed as follows:

$$h_{\mathrm{a}} = 0.41 \times 1.79^{S} \tag{3.35}$$

where:

S surrounding rock grade. For surrounding rock of grade II, S = 2.

For a tunnel of two or more tracks, the collapsed arch height is expressed as follows:

$$h_{\mathrm{a}} = 0.45 \times 2^{S-1} \times \omega \tag{3.36}$$

where:

ω influence coefficient of excavation width. Given tunnel width B = 5 m, the increase or decrease rate of surrounding rock pressure is ω = 1 + i(B − 5) for every 1 m increase or decrease of B. When B < 5 m, i = 0.2; When B > 5 m, I = 0.1.

Xu Zemin, et al. classified hard rock tunnels into shallow overburden tunnels, deep overburden tunnels and ultra-deep overburden tunnels in terms of initial geostress characteristics of rock mass, surrounding rock pressure and surrounding rock stability (Table 3.11). They found that rock burst is easy to occur when the overburden is more than 500 m based on a large amount of tunnel engineering experience. Xu Linsheng found in his research that the overburden of main rock burst section is 500–760 m, and the probability of rock burst on both side walls and vaults is basically the same.

Accordingly, the adaptability assessment criteria of the tunnel overburden are constructed, as shown in Table 3.12.

The adaptive membership function of tunnel overburden is constructed by using the descending half-step distribution function;

Table 3.10 Classification criteria of tunnel overburden

Assessment indicator	Scoring criteria			
	0	1	2	3
Tunnel overburden (m)	< 7.5	7.5–12.5	12.5–17.5	> 17.5

Table 3.11 Classification criteria of tunnel overburden

Assessment indicator	Tunnel classification criteria		
	Shallow overburden tunnel	Deep overburden tunnel	Ultra deep overburden tunnel
Tunnel overburden (m)	$\leq (2\text{–}3)h_a$	$(2\text{–}3)h_a$–500	> 500

Table 3.12 Adaptability assessment criteria of tunnel overburden

Assessment Indicator	TBM working conditions			
	Unfavorable	Unfavorable–favorable	Favorable–unfavorable	Unfavorable
Tunnel overburden (m)	$< h_a$	$h_a \sim 2.5$ ha	2.5 ha ~ 600	≥ 600

$$U_9(x) = \begin{cases} 0 & (x < h_a) \\ \frac{x-h_a}{2.5h_a-h_a} & (h_a \leq x < 2.5h_a) \\ \frac{600-x}{600-2.5h_a} & (2.5h_a \leq x < 600) \\ 0 & (x \geq 600) \end{cases} \quad (3.37)$$

Refer to Fig. 3.21 for the corresponding membership function.

(6) Rock Abrasion Resistance Index

The rock abrasion resistance index is an important assessment indicator. It reflects TBM cutter wear and tunneling efficiency. Generally, with the rock of higher the abrasion resistance of rock, the wear to TBM cutters, cutter ring and bearing will be more serious, more cutters will be consumed, and more construction cost will be needed. In addition, more shutdown time is needed for more cutter replacement, affecting the normal tunneling operation of TBM. The rock abrasion resistance can be measured on an indoor CERCHAR rock abrasiveness tester. The CERCHAR

Fig. 3.21 Adaptive membership function of tunnel overburden

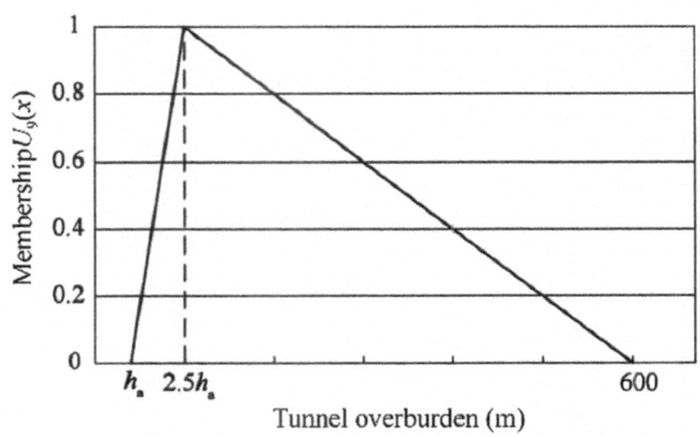

Table 3.13 Corresponding relationship between TBM working conditions and rock abrasion resistance index

Assessment indicator	TBM working conditions				
	Excellent	Good	Moderate	Fair	Poor
Abrasion resistance index of rock (0.1 mm)	< 3	3–4	4–5	5–6	≤ 6

rock abrasiveness tester consists of abrasion test device and measurement recorder. According to the calibration conversion of microscope measurement, take 0.1 mm as the basic unit, convert the measured abrasion into the abrasion of steel needle, and compare it against the empirical table of the abrasion of rock-steel needle, to judge whether the measurement is reasonable. Calculate the average value of the abrasion values measured at 3 angles, and record it as a single tested value. Test on the surface of each sample 3 times, and calculate the arithmetic average value of the 3 tested values as the final CERCHAR abrasivity index (CAI) of the sample.

According to a large number of examples of TBM construction projects at home and abroad, the TBM is able to work at high tunneling efficiency in surrounding rock with low to medium abrasion resistance, but at much lower tunneling efficiency in surrounding rock with strong to extra strong abrasion resistance. Therefore, the working conditions of TBM are classified into the following 5 grades according to the abrasion resistance index (Table 3.13).

Accordingly, the membership function of rock abrasion index is constructed by using the descending half-step distribution function:

$$U_{10}(x) = \begin{cases} 1 & (x < 3) \\ \frac{6-x}{6-3} & (3 \leq x < 6) \\ 0 & (x \geq 6) \end{cases} \tag{3.38}$$

Refer to Fig. 3.22 for the corresponding membership function.

3. Adverse Ground Condition
(1) Fracture Zone Width

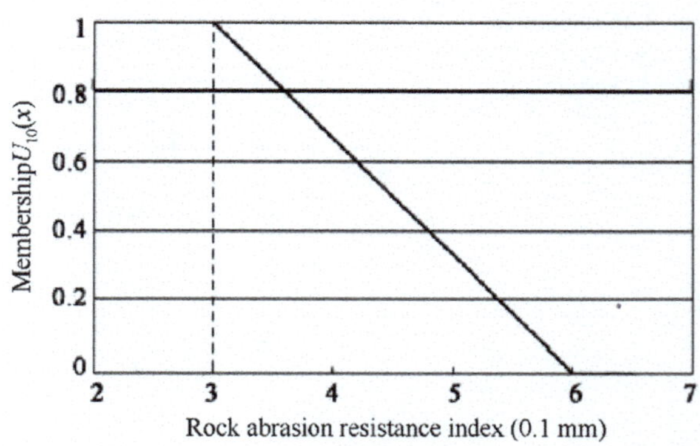

Fig. 3.22 Adaptive membership function of rock abrasion resistance index

The fracture zone, especially the large-scale fault zone, is an unfavorable engineering geological condition that most tunnel excavation will encounter. The main problems encountered by TBM when it enters the weak and broken surrounding rock section are the surrounding rock slump and collapse. The fault and fracture zones feature developed rock mass fracture, abundant groundwater, and rock and soil layer of different hardness, and are prone to causing cutter wear, cutter head damage and water gushing which may endanger construction personnel and equipment. Therefore, the fracture zone width is an important assessment indicator to characterize the geological adaptability of TBM construction.

The structural planes can be classified into the following 5 grades by their width and extension length. Refer to Table 3.14 for the scale and engineering significance of structural planes at each grade.

Liu Xuezeng, et al. classified the fault fracture zone width into 4 grades. The fracture zone with a width less than 2 m has little influence on tunnel construction. According to the research of Yin Juntao, it is preferable to directly pass the small-scale fault fracture zone (width < 5 m) at low rotation speed, small stroke and fast advance speed without shutdown or with downtime minimized to avoid TBM cutter head getting stuck. The large-scale fault fracture zone (width 10–30 m) that the TBM is not able to directly pass through should be excavated slowly after the tunnel face and the broken surrounding rock are grouted with polyurethane foam, cement slurry, water glass, etc. Larger-scale (width > 30 m) fault fracture zone that the TBM is not able to pass through should be excavated by drilling and blasting after a side adit is excavated manually, and then be stepped by the TBM under no-load condition.

Therefore, the working conditions of TBM are classified into the following 5 grades by fracture zone width (Table 3.15).

According to the assessment criteria, the membership function about the fracture zone width adapting to different assessment criteria is constructed by using the descending half-step distribution function:

$$U_{11}(x) = \begin{cases} 1 & (x < 1) \\ \frac{30-x}{30-1} & (1 \le x < 30) \\ 0 & (x \ge 30) \end{cases} \tag{3.39}$$

Refer to Fig. 3.23 for the corresponding membership function.

(2) Serious Deformation of Soft Rock and Instability of Surrounding Rock

The serious deformation of soft rock is one of the important factors that affect the normal tunneling of TBM. The rapid convergence of tunnel in the process of excavation often leads to the deformation and damage of concrete segments, and, in severe cases, TBM jamming. Extra-long traffic tunnels, such as roads and railways, are mainly distributed in complex mountain areas with developed geological structures in central and western China, with an overburden of 300–1000 m. Technical problems such as large deformation and long-term stability of high-stress soft rocks are serious. When underground works such as tunnels pass through the shallow

Table 3.14 Adaptability assessment criteria to fracture zone width

Grade	Structure plane name	Scale	Engineering and geological significance
I	Regional fault and major fault	Extend from several kilometers to tens of kilometers, and break through the rock mass, with a fracture zone width of more than 30 m	Affect the stability of regional structure and directly affect the stability of engineering rock mass
II	Large fault	Extend for thousands of meters with a fracture zone width of 10– 30 m	Affect the stability of engineering rock mass directly
III	Fault, bedding fault, and primary weak intercalated layer	Long but not wide regional geological interface with fracture zone width ranging from several centimeters to several meters	Control the stability of the mountain or rock mass in the project area, and affect the project layout. Buildings should be kept away from such places or provided with necessary treatment measures
IV	Minor fault, large joint, intercalated layer, well-extended bedding plane and bedding fault	Faults with a length of tens of meters to hundreds of meters and a width of several centimeters to 1 m	Affect or control the stability of engineering rock mass, such as the stability of surrounding rock of underground chamber and the stability of slope rock mass
V	Poorly extended joint, bedding plane, secondary interstice, schistosity and cleavage	Generally several centimeters to 30 m in length and several centimeters in width	Form the boundary plane of rock mass, destroy the integrity of rock mass, affect the physical and mechanical properties and stress distribution of rock mass, and affect the failure mode of rock mass to a great extent

Table 3.15 Adaptability assessment criteria to fracture zone width

Assessment indicator	TBM working conditions				
	Bad	Moderate	Fair	Good	Excellent
Fracture zone width (m)	≤ 30	10–30	5–10	1–5	< 1

overburden bias pressure area with high geostress and high residual tectonic stress, and weak broken surrounding rock mass, serious deformation of surrounding rock is often encountered. The serious deformation of the surrounding rock of the tunnel is a serious geological disaster. It not only causes difficulties in construction, delays

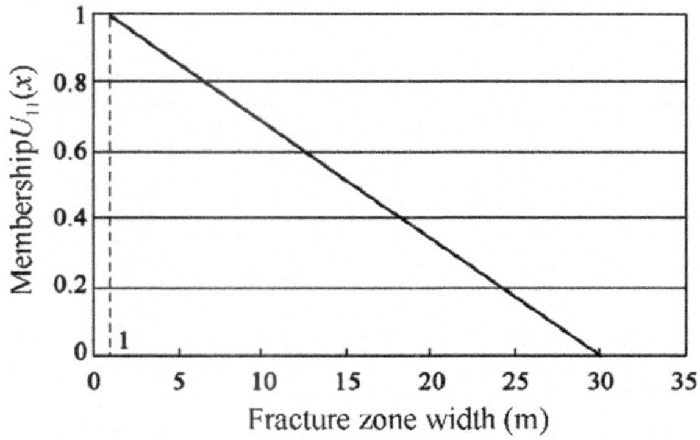

Fig. 3.23 Adaptive membership function of fracture zone width

in construction period, increases in cost, but may even lead to the termination of the project.

TBM jamming includes cutter head jamming and shield jamming. Cutter jamming mainly occurs in the fracture zone of surrounding rock, and the instantaneous breakout torque of cutter head (up to 1.7 times the rated torque) is usually used for breakout. Shield jamming means that the TBM shield is jammed due to the serious deformation of surrounding rock, and the resistance caused by the surrounding rock stress acting on the shield exceeds the breakout thrust force of TBM.

① Cutter Head Jamming

a. When the TBM is excavating hard rock, the rolling resistance coefficient of disc cutter will be calculated with the following formula (Fig. 3.24):

$$\xi = \frac{F_t}{F_d} \tag{3.40}$$

where:

F_t tangential force of disc cutter (kN);
F_d radial force of disc cutter (kN).

It can also be estimated with an empirical formula:

$$\xi = \frac{4}{5}\sqrt{\frac{P_e}{d}} \tag{3.41}$$

where:

d diameter of disc cutter ring (mm);
P_e cutting depth (mm), generally set to 7–15 mm/r.

Rolling resistance coefficient ξ of disc cutter will be influenced by disc cutter type, dimension, cutting angle and rock condition. Generally, the rolling friction

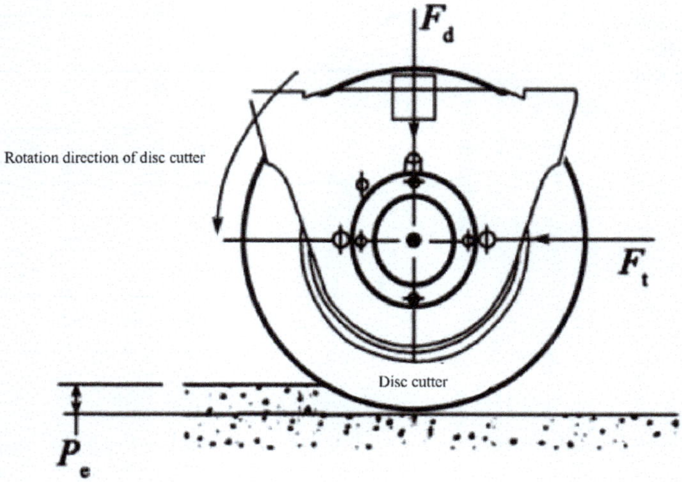

Fig. 3.24 Rolling cutting with disc cutter

coefficient will increase with the increase of cutting depth. At present, the rolling friction coefficient ξ is about 0.1 for excavation in hard rock.

Therefore, the resistance torque of TBM cutter head for excavation in hard rock is:

$$T_R = \sum_{i=1}^{n} (F_t R_i) = \sum_{i=1}^{n} (\xi F_d R_i) \approx 0.35 \xi n D F_d \tag{3.42}$$

Type:

T_R cutter head drive torque (kN m);
n number of disc cutters;
R_i the turning radius (m) of the ith disc cutter on the cutter head;
F_d the acting force of disc cutter, in the range of 210–310 kN, often set to 240 kN.

b. When TBM is excavating soft rock, the cutter head will be resisted by surrounding rock in front and around.

When the TBM cutter head rotates, it should overcome 3 moments, namely, the circumferential friction force of cutter head T_{r1}, the friction moment in front of cutter head T_{r2} and the rotation resistance moment of disc cutter T_{r3}. These 3 resistance moments will be calculated as follows respectively.

Torque required to overcome circumferential friction force of cutter head, T_{r1}:

$$T_{r1} = \frac{\pi D^2 e}{8} (p_0 + p_1 + p_2 + p_3) \mu \tag{3.43}$$

where:

e cutter head width;
D diameter of cutter head;
μ friction coefficient.

Torque required to overcome friction force in front of cutter head, T_{r2}:

$$T_{r2} = \frac{\pi D^2}{4} \cdot \frac{p_2 + p_3}{2} \cdot \frac{D}{3}\mu = \frac{\pi D^3}{24}(p_2 + p_3)\mu \tag{3.44}$$

The rotation resistance moment of disc cutter Tr3 required for TBM tunneling in broken rock can be expressed as:

$$T_{r3} = \sum_{i=1}^{n} (\xi F_d R_i) \approx 0.35\xi n D F_d \tag{3.45}$$

Therefore, when the TBM is excavating broken rock, the resistance torque of cutter head is as follows:

$$T_R = T_{r1} + T_{r2} + T_{r3} \tag{3.46}$$

To evaluate the influence of rock mass collapse on cutter head rotation, k is the safety factor, T_R is the torque generated by resistance, and T_r is the rated torque of cutter head. Usually, the rated thrust force of TBM can be exerted under certain conditions, and it should be reduced according to specific conditions under different conditions. Meanwhile, to provide a certain safety margin, $k > 1$ is taken here. If the cutter head of TBM can rotate normally, the following formula should be met:

$$\frac{T_r}{k} \geq T_R \tag{3.47}$$

Similarly, to prevent the cutter head jamming, the breakout torque T_{BT} of the cutter head should be greater than the torque T_R generated by the resistance, and the safety factor k is also introduced, as shown in the following formula:

$$\frac{T_{BT}}{k} \geq T_R \tag{3.48}$$

According to this, the membership function of cutter head jamming is constructed as follows:

$$U_{12a}(x) = \begin{cases} 1 & (x \leq T_{BT}/k) \\ \frac{T_{BT}-x}{T_{BT}-T_{BT}/k} & (T_{BT}/k < x \leq T_{BT}) \\ 0 & (x > T_{BT}) \end{cases} \tag{3.49}$$

Refer to Fig. 3.25 for the corresponding membership function.

② Shield Jamming

When the deformation of surrounding rock around the shield exceeds the reserved deformation for excavation, the surrounding rock begins to contact with the shield and squeeze the shield. As the TBM advances, a friction resistance is generated between the surrounding rock and the shield. If the TBM thrust force is unable to overcome the friction resistance of surrounding rock to the shield, the TBM shield will get jammed.

The judgment criteria for TBM jamming are made according to its mechanism as follows:

$$F_R = F_r + F_b \geq F_{I(\text{Jammed})} \tag{3.50}$$

$$F_R = F_r + F_b < F_{I(\text{Not jammed})} \tag{3.51}$$

where:

- F_r the thrust force required to overcome the friction resistance acting on the shield (kN);
- F_b thrust force (kN) in normal continuous excavation of TBM.
- F_I TBM rated thrust force (kN).

The friction resistance of tunnel surrounding rock acting on shield F_r is:

$$F_r = \mu(p_0 + p_1 + p_2 + p_3)DL \tag{3.52}$$

where:

- μ friction factor between shield and surrounding rock;
- L shield length (m);
- D shield diameter (m).

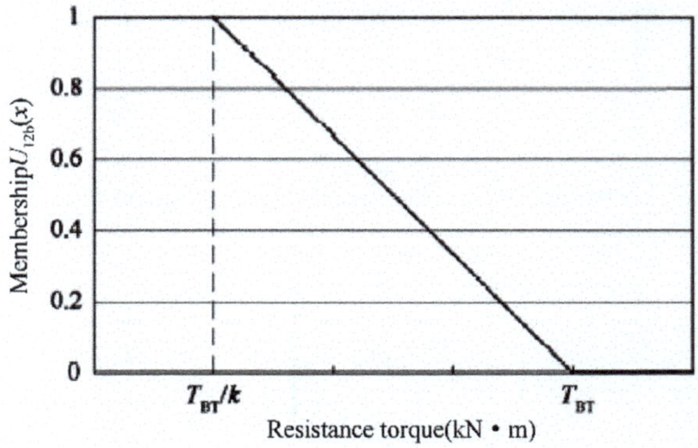

Fig. 3.25 Adaptive membership function of resistance torque

To evaluate the influence of serious deformation of rock mass on the shield, k is the safety factor. Usually, the rated thrust force of TBM can be exerted under certain conditions, and it should be reduced according to specific conditions under different conditions. Meanwhile, to provide a certain safety margin, k > 1 is taken here. If the TBM can advance normally and the shield is not jammed, the following formula should be met:

$$\frac{F_{\mathrm{I}}}{k} \geq F_{\mathrm{R}} \tag{3.53}$$

According to this, the membership function of shield jamming is constructed as follows:

$$U_{12b}(x) = \begin{cases} 1 & (x \leq F_{\mathrm{I}}/k) \\ \frac{F_{\mathrm{I}}-x}{F_{\mathrm{I}}-F_{\mathrm{I}}/k} & (F_{\mathrm{I}}/k < x \leq F_{\mathrm{I}}) \\ 0 & (x > F_{\mathrm{I}}) \end{cases} \tag{3.54}$$

Refer to Fig. 3.26 for the corresponding membership function.

(3) Transverse Composite Strata

Homogeneous strata refer to one or several strata or strata combination with similar geotechnical mechanics, engineering geology and hydrogeology, and other characteristics within the scope of excavation section and in the direction of excavation extension. In a homogeneous stratum, it is not necessary to adjust the TBM construction in operation mode and structural form.

Composite strata consists of two or more different strata of different characteristics such as geotechnical mechanics, engineering geology and hydrogeology. When TBM is tunneling in uneven soft and hard composite strata, it is necessary to take corresponding treatment measures in construction technology and construction parameters. The "upper soft and lower hard" strata are the most typical composite

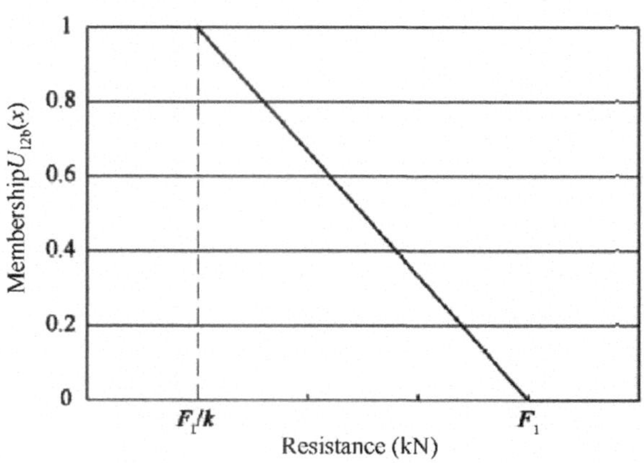

Fig. 3.26 Adaptive membership function of resistance

strata in the vertical direction. In such strata, the upper part of the tunnel section is soft soil layer and the lower part is a hard stratum. Sometimes, a soft rock layer is sandwiched between hard rock layers, or a hard rock layer is sandwiched between soft rock layers. In composite strata, large strength difference on excavation face is not conducive to tunnel excavation: ① It is difficult to control the advancing attitude of TBM; ② An eccentric stress will act on the TBM main bearing and cutter head to produce excessive overturning moment and unbalanced force, and damage the cutter head because of being eccentric and vibrating; ③ Uneven stress acting on the cutter accelerates the wear of the cutter, reduces the service life of the cutter, and causes uneven wear of disc cutter, such as eccentric wear, fracture, shift or falling off of the cutter ring, etc.

Therefore, the geological characteristics of uneven soft and hard composite strata are quantified with the transverse stratigraphic composite ratio, as shown in the following formula:

$$R_{\mathrm{c}} = \frac{H_{\mathrm{h}}}{D} \tag{3.55}$$

where:

R_c transverse stratigraphic composite ratio;
H_h the thickness of hard rock on the excavation face;
D excavation diameter.

Accordingly, the adaptability assessment criteria of the transverse stratigraphic composite ratio are constructed, as shown in Table 3.16.

According to the assessment criteria, the membership function of transverse composite stratigraphic ratio is constructed as follows:

$$U_{13}(x) = \begin{cases} 1 & (x < 30) \\ \frac{50-x}{50-30} & (30 \le x < 50) \\ 0 & (50 \le x < 60) \\ \frac{x-70}{70-60} & (60 \le x < 70) \\ 1 & (x \ge 70) \end{cases} \tag{3.56}$$

Refer to Fig. 3.27 for the corresponding membership function.

(4) Water Gushing

Table 3.16 Assessment criteria for adaptability of transverse stratigraphic composite ratio

Assessment indicator	Assessment criteria			
	Not adapted	Not adaptable–adaptable	Adaptable–not adaptable	Adaptable
Stratigraphic composite ratio (%)	50–60	60–70	30–50	< 30 or ≥ 70

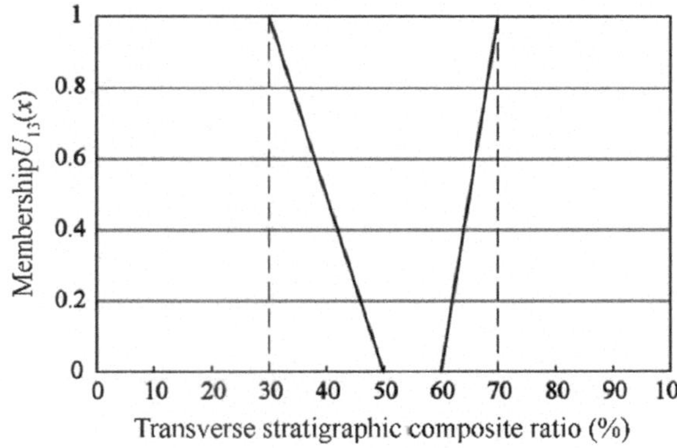

Fig. 3.27 Adaptive membership function of lateral stratigraphic composite ratio

Karst water gushing in tunnels is an adverse ground condition often encountered in TBM construction. Several factors will affect the karst water gushing in tunnels. Because of the complexity and variability of water gushing in tunnels, it is not accurate enough at present to predict the water gushing in tunnels. Tunnel water gushing rate is an important indicator to assessment the adaptability of TBM construction to karst water gushing. Water gushing is inevitable in tunnel construction. It features large amount of water, high water pressure and abruptness, and often leads to instability of surrounding rock, collapse and even submergence of the tunnel, endangering the construction personnel and equipment in the tunnel.

According to the measured data in field, the larger the unit maximum water gushing rate, the more hazardous the water gushing disaster in the tunnel, and the worse the geological adaptability of TBM construction. HamidiJK, et al. evaluated the boreability of TBM with RME assessment criteria, and classified the water gushing rate in tunnel into 5 grades, namely < 10 L/s, 10–30 L/s, 30–70 L/s, 70–100 L/s, and > 100 L/s. The water gushing rate less than 10 L/s has the highest score of 5, while the water gushing rate greater than 100 L/s has the lowest score of 0. When predicting the performance of TBM with penalty factors, Jean-Paul Dudt, et al. took the reduction coefficient 1 for water gushing rate < 20 L/s, 0.8 for water gushing rate between 20 and 50 L/s, and 0.5 for water gushing rate > 50 L/s.

Accordingly, the adaptability assessment criteria of the maximum water gushing rate are constructed, as shown in Table 3.17.

According to the assessment criteria, the membership function of water gushing rate is constructed by using the descending half-step distribution function:

Table 3.17 Adaptability assessment criteria of water gushing rate

Assessment indicator	TBM working conditions				
	Bad	Fair	Moderate	Good	Excellent
Water gushing rate (L/s)	> 100	70–100	30–70	10–30	< 10

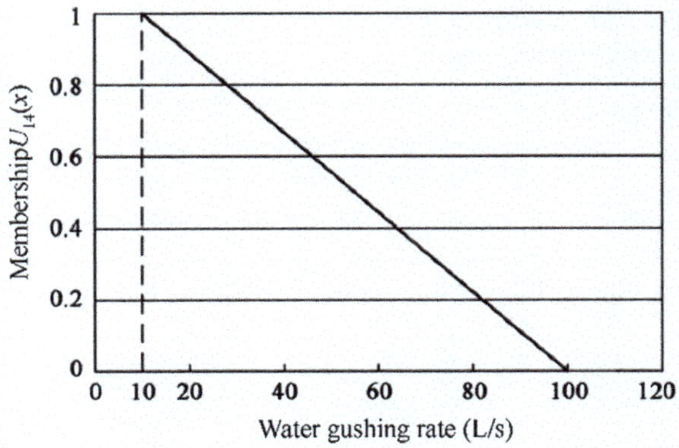

Fig. 3.28 Adaptive membership function of water gushing rate

$$U_{14}(x) = \begin{cases} 1 & (x < 10) \\ \frac{100-x}{100-10} & (10 \le x < 100) \\ 0 & (x \ge 100) \end{cases} \tag{3.57}$$

Refer to Fig. 3.28 for the corresponding membership function.

According to the *Exploration Specification of Hydrogeology and Engineering Geology in Mining Areas* (GB12719—1991) and the *Regulations on Water Prevention and Control in Coal Mines* (2009), the aquifer should be classified into not abundant, moderately abundant, abundant and extremely abundant based on unit water gushing rate in borehole, as shown in Table 3.18.

According to the assessment criteria, the membership function of unit water gushing rate is constructed by using the descending half-step distribution function:

$$U_{14}(x) = \begin{cases} 1 & (x < 0.1) \\ \frac{5-x}{5-0.1} & (0.1 \le x < 5) \\ 0 & (x \ge 5) \end{cases} \tag{3.58}$$

Refer to Fig. 3.29 for the corresponding membership function.

(5) Permeable Rate

In the construction of underground works for transportation, water conservancy and nuclear waste disposal, the permeable rate of rock mass is an essential survey indicator. The permeability coefficient of fractured rock mass is an indispensable

Table 3.18 Water abundance grade of aquifer

Assessment Indicator	Water abundance grade			
	Extremely abundant	Abundant	Moderately abundant	Not abundant
Unit water gushing rate [L/(s m)]	> 5.0	1.0–5.0	0.1–1.0	≤ 0.1

Fig. 3.29 Adaptive membership function of unit water gushing rate

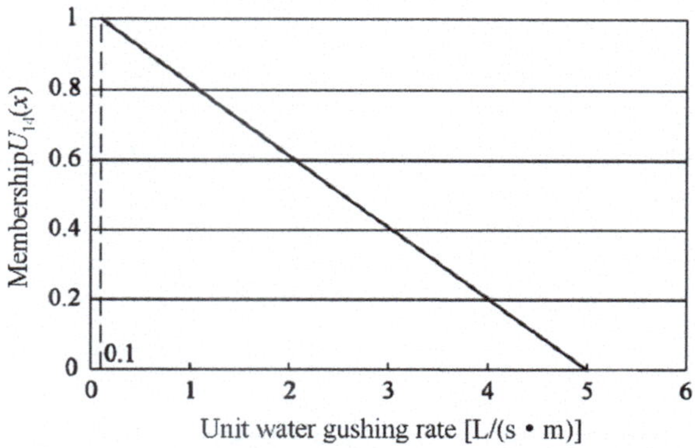

parameter to evaluate the permeable rate and simulate the seepage of fractured rock mass. It is also of great significance in practical engineering application. After the permeable rate of rock mass is determined, the permeability coefficient can be determined according to its relationship to the permeable rate. Therefore, the permeable rate of rock mass is an important parameter to calculate the permeability coefficient of rock mass. The anti-seepage criterion of rock foundation works is generally 1 LU–5 LU (Lugeon value). In the conventional water pressure test, 1 LU is generally used as the criterion for termination of anti-seepage grouting. Up to 0.5 LU or down to 10 LU may be taken in special cases. The permeable rate can be calculated with the following formula:

$$q = \frac{Q}{Lp} \tag{3.59}$$

where:

q permeable rate (LU);
Q pressurized flow in test section (L/min);
p test pressure (MPa);
L test length (m).

The *Code for Engineering Geological Investigation of Water Resources and Hydropower* (GB50487—2008) provides a comparison table for the relationship between permeable rate and permeability coefficient of rock mass (Table 3.19).

Benardos et al. classified the permeability coefficient of rock mass into 4 grades according to the scoring criteria when evaluating geological disasters for TBM. As shown in Table 3.20, when K > 10-2 cm/s, it was not conducive to TBM construction, but with the decrease of permeability coefficient, the conditions for TBM construction became more favorable, and when K < 10-6 cm/s, it was suitable for TBM construction.

Accordingly, the adaptability assessment criteria of the permeable rate are constructed, as shown in Table 3.21.

Table 3.19 Relationship between permeability coefficient and permeable rate of rock mass

Permeability coefficient K (cm/s)	Permeable rate (LU)	Permeability grade
$K < 10^{-6}$	$q < 0.1$	Rarely permeable
$10^{-6} \le K < 10^{-5}$	$0.1 \le q < 1$	Slightly permeable
$10^{-5} \le K < 10^{-4}$	$1 \le q < 10$	Weakly permeable
$10^{-4} \le K < 10^{-2}$	$10 \le q < 100$	Moderately permeable
$10^{-2} \le K < 1$	$q \ge 100$	Permeable
$K \ge 1$		Extremely permeable

Table 3.20 Classification criteria of permeability coefficient

Assessment indicator	Scoring criteria			
	0	1	2	3
Permeability coefficient (cm/s)	$> 10^{-2}$	10^{-4}–10^{-2}	10^{-6}–10^{-4}	$< 10^{-6}$

Table 3.21 Adaptability assessment criteria of permeable rate

Assessment indicator	TBM working conditions			
	Bad	Fair	Good	Excellent
Permeable rate (LU)	≥ 100	10–100	1–10	< 1

According to the assessment criteria, the membership function of permeable rate is constructed by using the descending half-step distribution function:

$$U_{15}(x) = \begin{cases} 1 & (x < 1) \\ \frac{100-x}{100-1} & (1 \le x < 100) \\ 0 & (x \ge 100) \end{cases} \tag{3.60}$$

Refer to Fig. 3.30 for the corresponding membership function.

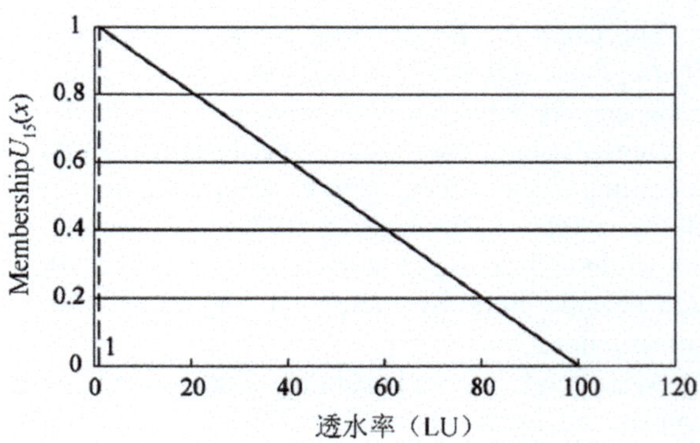

Fig. 3.30 Adaptive membership function of permeable rate

(6) Ground Temperature

When TBM is tunneling in the regional active fault zone or stratum of high geostress or deep overburden, it may encounter high temperature and high heat. High ground temperature has become a common geological disaster for underground works such as tunneling and mining. It is difficult for construction in the section with high geothermy that the tunnel needs to pass. In the tunnel at high temperature, it is difficult for the human body to dissipate heat, the body temperature will rise, the heart rate will speed up, and the pulse may accelerate to 120–150 beats /min, destroying the thermal balance of the human body, resulting in abnormal metabolism, and endangering human health. In a gold mine of South Africa, the working face temperature was up to 60 °C, many workers suffered from heat stroke or even died. In Pingdingshan Tianan No. 8 Coal Mine, the working face temperature rose to 33–34 °C, and some construction workers fainted or vomited, and all suffered from infectious eczema. In Heshanlilan Mine in Guangxi, the tunneling face once reached 32 °C. According to statistics, 415 construction workers suffered from various skin diseases, and many of them suffered from heatstroke and even fainted once. It can be seen that high geothermy greatly reduces the work efficiency, and increases the accident rate as well as the project cost.

When the original ground temperature of the tunnel reaches 28 °C, appropriate cooling measures should be taken during the construction. When the local temperature reaches 35 °C and the humidity reaches 80%, it will harm the health and safety of workers, and reduce the mechanical efficiency and labor productivity. Karadogan et al. defined the membership of temperature change from 20 to 30 °C as 0 to 1, and took the temperature of 30 °C as the boundary of high temperature. The ground temperature in the tunnel will rise by 3 °C for every 100 m increase in overburden. To ensure normal and safe construction in the tunnel, relevant departments of China have issued some sanitary standards for the working environment of tunnel construction. For example, the temperature in the tunnel should not exceed 28 °C as specified by the former Ministry of Railways, and not exceed 30 °C as specified by the Ministry of Transport of the People's Republic of China.

As mentioned above, the adaptability assessment criteria of ground temperature is constructed, as shown in Table 3.22.

According to the assessment criteria, the membership function of ground temperature is constructed by using the descending half-step distribution function:

Table 3.22 Adaptability assessment criteria of ground temperature

Assessment indicator	TBM working conditions			
	Bad	Relatively unfavorable	Moderate	Good
Ground temperature (°C)	≥ 40	30–40	20–30	< 20

$$U_{16}(x) = \begin{cases} 1 & (x < 20) \\ \frac{40-x}{40-20} & (20 \le x < 40) \\ 0 & (x \ge 40) \end{cases} \quad (3.61)$$

Refer to Fig. 3.31 for the corresponding membership function.

(7) Harmful gas emission

The main harmful gas is methane (CH4, commonly known as biogas), which emits from coal and rock during tunnel construction. There are also some other harmful gases, including carbon monoxide (CO), carbon dioxide (CO_2), hydrogen sulfide (H_2S), nitrogen dioxide (NO_2), sulfur dioxide (SO_2), ammonia (NH_3), hydrogen (H_2), radon (Rn), etc. In a narrow sense, methane is also called gas because it accounts for more than 80% in the gas composition.

According to the analysis of the measured data in the jobsite, the Harmful gas emission is a direct assessment indicator to characterize the adaptability of TBM construction to harmful gases. The more harmful gases are emitted, the greater the harm to human body, and the lower the adaptability of TBM. In Japan, a comprehensive indicator scoring system is used to classify gas-containing tunnels, but it is difficult to assign value to some indicators, so this system can be used for reference only. Yuan Yulin in China put forward a more detailed set of classification criteria in combination with this Japanese system, and classified the gas emitting rate into 4 scoring criteria. The higher score represents higher risk (Table 3.23).

To prevent outburst and harmful gas when highway tunnels pass through coal seams, the former Ministry of Communications initiated the research on the design and construction technology of gas-containing highway tunnels in western China

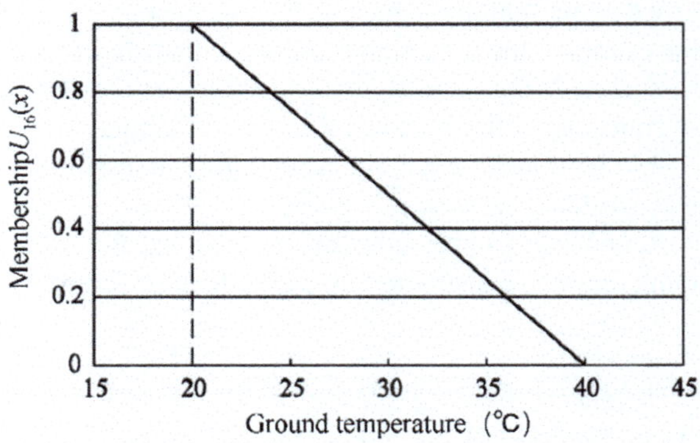

Fig. 3.31 Adaptive membership function of ground temperature

Table 3.23 Scoring criteria of harmful gas emission

Assessment indicator	Scoring			
	6	4	2	1
Harmful gas emission (m^3/min)	3	1–3	0.5–1	0.5

in 2007. In this research, the classification indicators of gas-containing highway tunnels was studied, and the concept of micro-gas-containing tunnel was put forward according to the actual construction demand. In this research, the absolute gas emitting rate was calculated back with the gas concentrations at 0.1 and 0.3%, ventilation rate calculated at the minimum wind speeds of 0.15 m/s (micro-gas-containing tunnel) and 0.5 m/s (high-gas-containing tunnel), and the cross-sectional area of 70 m^2, the typical cross-sectional area of highway tunnel in western China. The safety factor is 2, and the boundary indicator values are 0.3 m^3/min and 3 m^3/min respectively after rounding. Therefore, the tunnel a micro-gas-containing tunnel, low-gas-containing tunnel and high-gas-containing tunnel respectively when the absolute gas emitting rate in the tunnel is less than 0.3 m^3/min, not less than 0.3 m^3/min but less than 3 m^3/min, and not less than 3 m^3/min.

Kang Xiaobing et al. classified the gas-containing tunnels into 4 grades: micro-gas, low-gas, high-gas, outburst risky. In the Technical Code for Railway Tunnel with Gas (TB10120-2002) issued by the former Ministry of Railways in 2002, the tunnel gas grades are quantitatively classified. When the absolute gas emitting rate of the whole work area is less than 0.5 m^3/min, it is a low gas work area, and when it is greater than or equal to 0.5 m^3/min, it is a high gas work area. At present, the baseline of absolute gas emitting rate of 0.5 m^3/min set in the *Technical Code for Railway Tunnel with Gas* (TB10120-2002) is too low for defining low gas and high gas grades. As a result, some jobsites that should have been taken as low-gas-containing sites are treated as high-gas-containing sites, so some unnecessary measures are taken, and some construction costs are wasted.

Accordingly, the adaptability assessment criteria of the Harmful gas emission are constructed, as shown in Table 3.24.

According to the assessment criteria, the membership function of Harmful gas emission is constructed by using the descending half-step distribution function:

$$U_{17}(x) = \begin{cases} 1 & (x < 0.3) \\ \frac{3-x}{3-0.3} & (0.3 \leq x < 3) \\ 0 & (x \geq 3) \end{cases} \tag{3.62}$$

Refer to Fig. 3.32 for the corresponding membership function.

4. Construction Organization
(1) Construction Technology Level

The construction technology level is an important indicator that reflects the core competitiveness of construction organizations. Construction technology mainly

Table 3.24 Adaptability assessment criteria of harmful gas emission

Assessment indicator	TBM working conditions			
	Bad	Fair	Moderate	Good
Harmful gas emission (m^3/min)	≥ 3	1–3	0.3–1	< 0.3

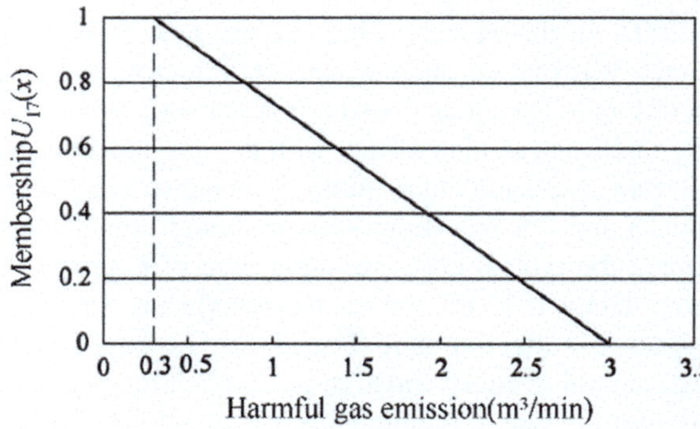

Fig. 3.32 Adaptive membership function of harmful gas emission

includes excavation and support, advanced geological prediction, and monitoring & measurement. The construction technology level can be simplified into enterprise reputation and construction experience. Refer to Table 3.25 for the adaptability assessment criteria of construction technology level.

According to the assessment criteria, the membership function of construction technology level is constructed by using the ascending half-step distribution function:

$$U_{18}(x) = \begin{cases} 0 & (x < 0.25) \\ \frac{x-0.25}{1-0.25} & (0.25 \le x < 1) \\ 1 & (x \ge 1) \end{cases} \tag{3.63}$$

Refer to Fig. 3.33 for the corresponding membership function.

(2) Construction Management

Construction management includes organization management and technical management in the process of project construction, and the management quality will directly affect the safety and efficiency of the whole project. The construction management level of tunnel project is the concrete embodiment of the business

Table 3.25 Adaptability assessment criteria of construction technology level

Assessment indicator	Assessment criteria			
	Not adapted	Basically adaptable	Relatively adaptable	Adaptable
Construction technology level	0.25	0.5	0.75	1
Specific description	Poor reputation, insufficient construction experience and technical force	Poor reputation, average construction experience and technical force	Good reputation, rich construction experience and strong technical force	Excellent reputation, rich construction experience and strong technical force

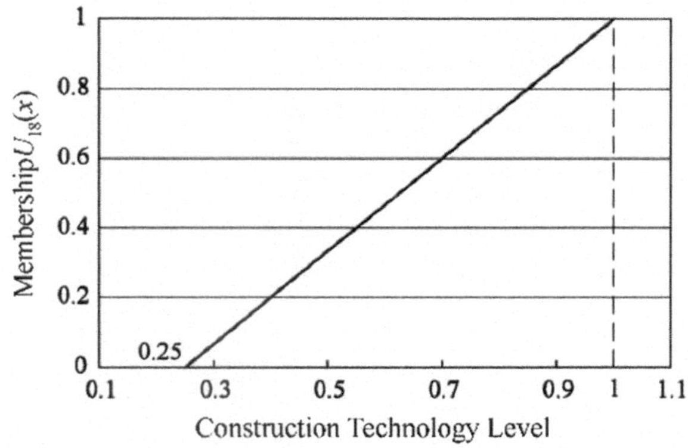

Fig. 3.33 Adaptive membership function of construction technology level

level of the construction, and an important part of the core competitiveness. Good construction management can create considerable economic benefits and social reputation for construction organizations and improve the competitiveness. The higher the level of construction management, the stronger the adaptability of TBM construction. Construction management can be simplified into control information & planning, and quality of constructors. Refer to Table 3.26 for the adaptability assessment criteria of construction management.

According to the assessment criteria, the membership function of construction technology level is constructed by using the ascending half-step distribution function:

$$U_{19}(x) = \begin{cases} 0 & (x < 0.25) \\ \frac{x-0.25}{1-0.25} & (0.25 \leq x < 1) \\ 1 & (x \geq 1) \end{cases} \tag{3.64}$$

Refer to Fig. 3.34 for the corresponding membership function.

Table 3.26 Construction management adaptability assessment criteria

Assessment indicator	Assessment criteria			
	Not adapted	Basically adaptable	Relatively adaptable	Adaptable
Construction management	0.25	0.5	0.75	1
Specific description	Lack of information and planning, and poor quality of construction personnel	Lack of information and planning, and poor quality of construction personnel	Complete information and planning, and relatively good quality of construction personnel	Complete information and planning, and good quality of construction personnel

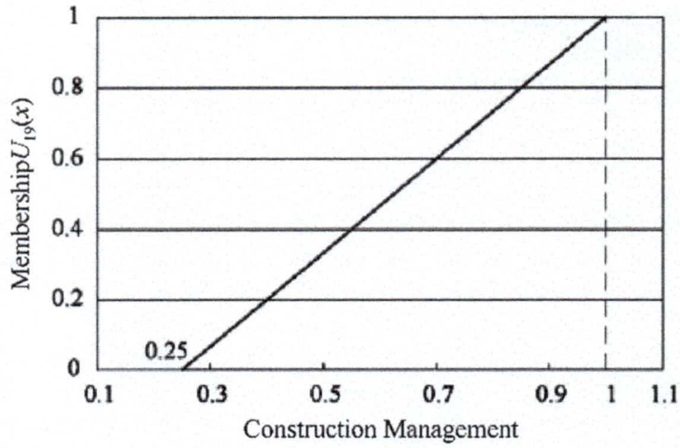

Fig. 3.34 Adaptive membership function of construction technology level

3.4.5 Determination of Weight of Assessment Indicator

In the constructed adaptability assessment indicator system, it is necessary to further determine the relative weight of each assessment indicator. Many methods are available, including interval statistics, AHP, fuzzy matrix, Delphi method, etc. Each weight definition method has its specific advantages and scope of application for specific assessment indicator system. For TBM adaptability analysis, the nine-scale method in AHP is used in this chapter to analyze the weight of each assessment indicator. For easy calculation, the comparative judgment matrix of each level of assessment indicator is constructed according to the principle and steps of analytic hierarchy process first. Then the maximum eigenvalue and corresponding eigenvector of the judgment matrix are calculated and then undergo a consistency test. Whether the consistency test is passed or not is judged according to the result CR. Finally, the total sort of the weight of each assessment indicator of TBM adaptability is calculated according to the index weight of each level.

1. Assessment in Hierarchical Single Sort

To reduce the bias caused by subjective assumptions, the value is assigned to the relative importance of each assessment indicator through expert seminars, telephone contact and consulting experts in person in this section. According to the weight of each indicator obtained by feedback, the comparative judgment matrix of each level of the assessment indicator is calculated, and the weight of the judgment matrix is calculated separately with Matlab program. The calculation results are shown in Table 3.27.

After calculation, the maximum eigenvalue λ_{max} of the judgment matrix is 4.0434, and the relative weight of the assessment indicator at this level is obtained after normalization (Fig. 3.35). The consistency test of the judgment matrix showed that $CR = 0.0162 < 0.1$, meeting the requirements.

The same method is used to calculate the judgment matrix of TBM construction, geological conditions, geological problems and construction organization at

Table 3.27 Analysis on impact weight of criteria level indicators in target level D

D-P	D	P_1	P_2	P_3	P_4	Weight	
	P_1	1	1/2	1/3	4	0.1754	$\lambda_{max} = 4.0656$ CI = 0.0218
	P_2	2	1	1/2	5	0.2903	RI = 0.8931 CR = 0.0245
	P_3	3	2	1	6	0.4756	
	P_4	1/4	1/5	1/6	1	0.0587	

Note CI is related to the maximum eigenvalue λ_{max} and the order n of the matrix, $CI = (\lambda_{max} - n)/(n-1)$; RI is the average random consistency indicator, and its value is related to the order n; CR the random consistency ratio, CR = CI/RI

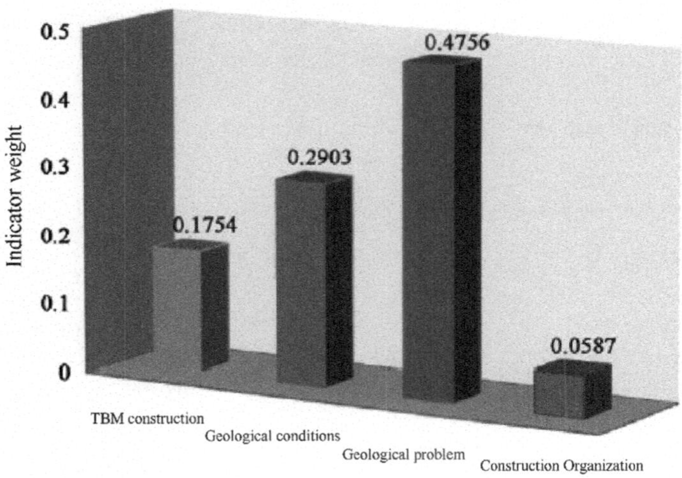

Fig. 3.35 Weight of assessment indicator at criterion level P

the criterion level respectively. Refer to Table 3.28 for weight analysis of indicators at criterion level P_1 (TBM construction conditions).

After calculation, the maximum eigenvalue λ_{max} of the judgment matrix is 4.0428, and the relative weight of the assessment indicator at this level is obtained after normalization (Fig. 3.36). The consistency test of the judgment matrix shows that $CR = 0.0103 < 0.1$, meeting the requirements.

Table 3.28 Weight analysis of indicators at criterion level P_1 (TBM construction conditions)

P_1-U	P_1	U_1	U_2	U_3	U_4	Weight	
	U_1	1	1	4	5	0.4067	$\lambda_{max} = 4.0428$ CI = 0.00923 RI = 0.8931 CR = 0.0103
	U_2	1	1	4	5	0.4067	
	U_3	1/4	1/4	1	2	0.1143	
	U_4	1/5	1/5	1/2	1	0.0723	

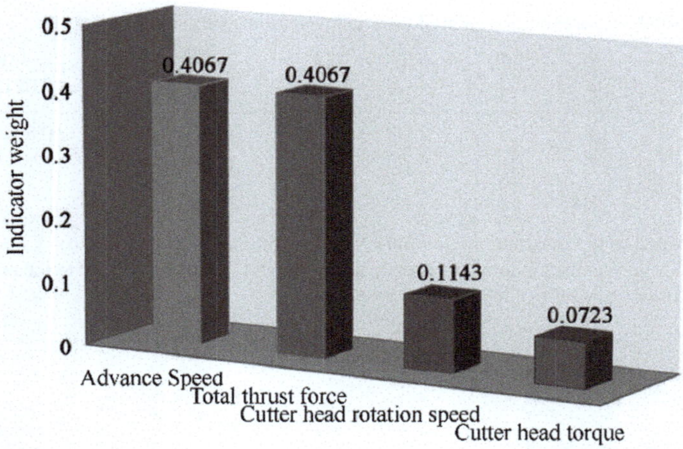

Fig. 3.36 Weight of each assessment indicator at criterion level P_1 (TBM construction conditions)

Table 3.29 Weight analysis of indicators at criterion level P_2 (TBM geological conditions)

P_2-U	P_2	U_5	U_6	U_7	U_8	U_9	U_{10}	Weight	$\lambda_{max} = 6.1165$
	U_5	1	4	7	6	6	7	0.5055	CI = 0.0233
	U_6	1/4	1	4	3	3	4	0.2097	RI = 1.2494
	U_7	1/7	1/4	1	1/2	1/2	1	0.0526	CR = 0.01865
	U_8	1/6	1/3	2	1	1	2	0.0898	
	U_9	1/6	1/3	2	1	1	2	0.0898	
	U_{10}	1/7	1/4	1	1/2	1/2	1	0.0526	

Refer to Table 3.29 for weight analysis of indicators at criterion level P_2 (TBM geological conditions).

After calculation, the maximum eigenvalue λ_{max} of the judgment matrix is 1.2493, and the relative weight of the assessment indicator at this level is obtained after normalization (3.37). The consistency test of the judgment matrix shows that CR = 0.01865 < 0.1, meeting the requirements. (Fig. 3.37)

Refer to Table 3.30 for weight analysis of indicators at criterion level P_3 (TBM geological problems).

After calculation, the maximum eigenvalue λ_{max} of the judgment matrix is 7.2294, and the relative weight of the assessment indicator at this level is obtained after normalization (3.38). The consistency test of the judgment matrix shows that CR = 0.02843 < 0.1, meeting the requirements.

Refer to Table 3.31 for weight analysis of indicators at criterion level P_4 (TBM construction organization).

After calculation, the maximum eigenvalue λ_{max} of the judgment matrix is 2, and the relative weight of the assessment indicator at this level is obtained after

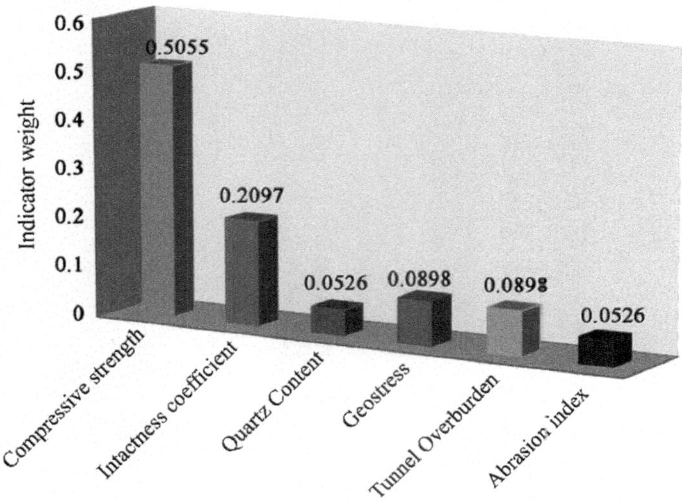

Fig. 3.37 Weight of each assessment indicator at criterion level P_2 (TBM geological conditions)

Table 3.30 Weight analysis of indicators at criterion level P_3 (TBM geological problems)

P_3-U	P_3	U_{11}	U_{12}	U_{13}	U_{14}	U_{15}	U_{16}	U_{17}	Weight	$\lambda_{max} = 7.2294$ CI = 0.03824 RI = 1.3450 CR = 0.02843
	U_{11}	1	1	2	3	4	7	8	0.2814	
	U_{12}	1	1	2	3	4	7	8	0.2814	
	U_{13}	1/2	1/2	1	2	3	6	7	0.1818	
	U_{14}	1/3	1/3	1/2	1	2	5	6	0.1194	
	U_{15}	1/4	1/4	1/3	1/2	1	4	5	0.0804	
	U_{16}	1/7	1/7	1/6	1/5	1/4	1	2	0.0321	
	U_{17}	1/8	1/8	1/7	1/6	1/5	1/2	1	0.0235	

normalization (3.39). The consistency test of the judgment matrix shows that CR = 0, meeting the requirements (Figs. 3.38 and 3.39)

2. Assessment in Hierarchical Total Sort

The total sort of TBM adaptability evaluation indicators at target level D is calculated according to the single sort of indicators at each level, as shown in Table 3.32.

Refer to Table 3.32 for the total sort calculation.

The large deformation of soft rock and the instability of surrounding rock specified in Table 3.32 can be classified into two cases: cutter head jamming and shield jamming. For an open TBM, only cutter head jamming is considered, and the weight of U_{12} is 0.1338. For a shield TBM, both cutter head jamming and shield jamming are considered, U_{12a} and U_{12b} are of the same importance, and their weights are 0.0669.

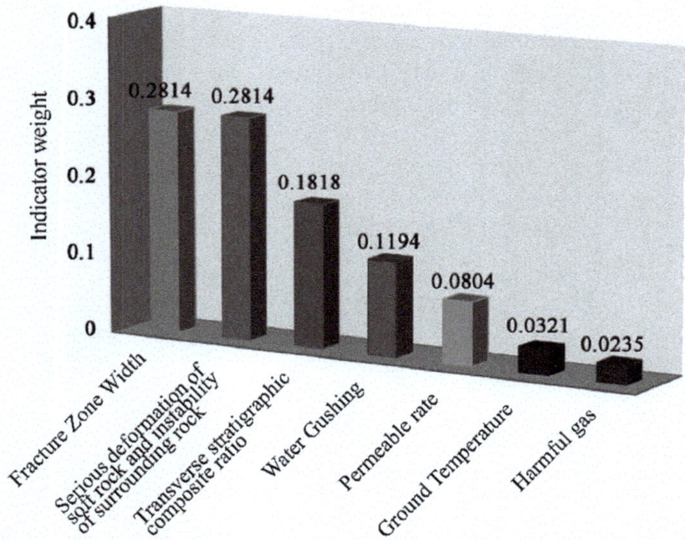

Fig. 3.38 Weight of each assessment indicator at criterion level P3 (TBM geological problem)

Table 3.31 Weight analysis of indicators at criterion level P4 (TBM construction organization)

P_4-U	P_4	U_{18}	U_{19}	Weight	$\lambda_{max} = 2$
	U_{17}	1	4	0.8000	CI = 0
	U_{18}	1/4	1	0.2000	RI = 0 CR = 0

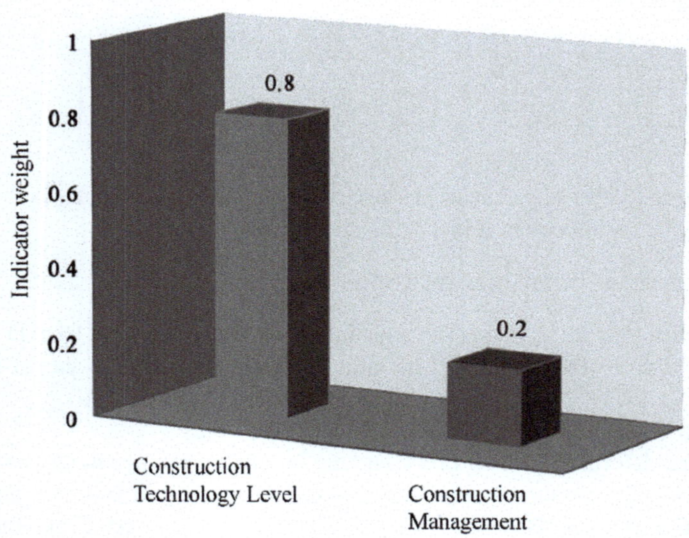

Fig. 3.39 Weight of each assessment indicator at criterion level P_4 (TBM construction organization)

Table 3.32 Total sort of TBM adaptability evaluation indicators from calculation of each indicator

Indicator level	Criterion level				Total sort weight of indicator level U to target level D	Sort
	P_1	P_2	P_3	P_4		
	0.1754	0.2903	0.4756	0.0587		
U_1 (advance speed)	0.4067				0.0713	4
U_2 (total thrust force)	0.4067				0.0713	4
U_3 (cutter head rotation speed)	0.1143				0.0200	10
U_4 (cutter head torque)	0.0723				0.0127	12
U_5 (uniaxial compressive strength of rock)		0.5055			0.1467	1
U_6 (rock mass intactness coefficient)		0.2097			0.0609	5
U_7 (quartz content)		0.0526			0.0153	11
U_8 (ground stress level)		0.0898			0.0261	9
U_9 (tunnel overburden)		0.0898			0.0261	9
U_{10} (rock abrasion index)		0.0526			0.0153	11
U_{11} (fracture zone width)			0.2814		0.1338	2
U_{12} (serious deformation of soft rock and instability of surrounding rock)			0.2814		0.1338	2
U_{13} (transverse stratigraphic composite ratio)			0.1818		0.0865	3
U_{14} (water gushing)			0.1194		0.0568	6
U_{15} (permeable rate)			0.0804		0.0382	8
U_{16} (ground temperature)			0.0321		0.0153	11
U_{17} (Harmful gas emission)			0.0234		0.0111	14
U_{18} (construction technology level)				0.8000	0.0470	7
U_{19} (construction management)				0.2000	0.0117	13

The total sort consistency ratio of indicator level U to target level D is calculated as $0.01798 < 0.1$ according to formula (3.65).

$$\mathrm{CR} = \frac{\sum_{j=1}^{m} \mathrm{CI}(j)a_j}{\sum_{j=1}^{m} \mathrm{RI}(j)a_j} \tag{3.65}$$

The overall hierarchical consistency test shows good consistency of sort. Figure 3.40 shows the total sort weight of indicator level U to target level D.

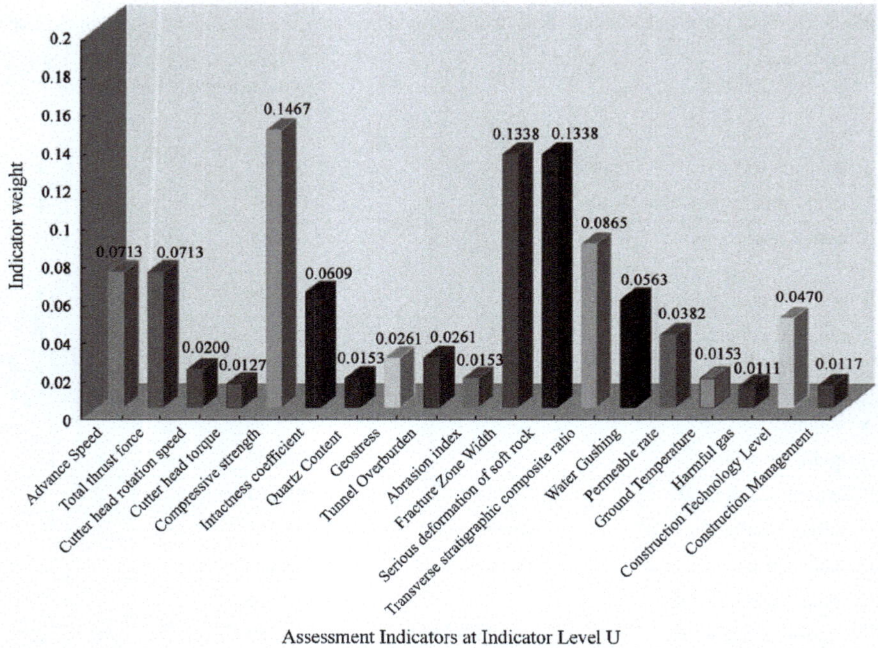

Fig. 3.40 Weights of assessment indicators at indicator level U

According to the total sort of hierarchical weights determined by AHP, the weights of TBM adaptability assessment indicators are sorted as follows: uniaxial compressive strength of rock U_5, fracture zone width U_{11}, large deformation of soft rock and instability of surrounding rock U_{12}, transverse stratigraphic composite ratio U_{13}, advance speed U_1, total thrust force U_2, rock intactness coefficient U_6, water gushing U_{14}, construction technology level U_{18}, permeable rate U_{15}, geostress level U_8, maximum overburden U_9, cutter head rotation speed U_3, quartz content U_7, rock abrasion index U_{10}, ground temperature U_{16}, cutter head torque U_4, construction management U_{19} and Harmful gas emission U_{17}. According to the total sort of weights for TBM adaptability assessment, the top five assessment indicator of the greatest influence of weights are uniaxial compressive strength of rock U_5, fracture zone width U_{11}, large deformation of soft rock and instability of surrounding rock U_{12}, transverse stratigraphic composite ratio U_{13} and advance speed U_1.

3.4.6 TBM Adaptability Assessment Method

The TBM adaptability assessment indicator system composed of 19 assessment indicators and weights of these assessment indicators were determined above through the analytic hierarchy process (AHP), and the membership function of each assessment indicator was constructed with fuzzy mathematics method. The TBM tunneling adaptability assessment method is proposed as follows:

(1) Determine the influencing factors of TBM adaptability, classify, select and reduce them, and determine the assessment indicators of TBM adaptability.
(2) Construct the TBM adaptability assessment indicator system. The above assessment indicators are classified by attributes, including target level, criterion level and indicator level, and the upper level governs all assessment indicators of the next level.
(3) Construct the membership function of TBM adaptability assessment indicator according to the proposed adaptability assessment criteria. Determine the change range of each assessment indicator of TBM adaptability, construct membership functions of different assessment grades with fuzzy mathematics method, and determine the assessment indicator matrix.
(4) Determine the weight of each assessment indicator of TBM adaptability. Construct the fuzzy judgment matrix through the AHP according to hierarchical levels, and make assessment in hierarchical single sort and hierarchical total sort respectively to obtain specific values and total sort of the weight of each assessment indicator.
(5) Comprehensively calculate the indicator parameter matrix and the weight of each assessment indicator in TBM construction with the fuzzy comprehensive judgment method to obtain the membership matrix of the construction adaptability, and determine the assessment grade and fitness of TBM adaptability assessment grade.

3.4.7 TBM Tunneling Adaptability Assessment in Actual Project

1. Data for Assessment Indicator

The data for TBM adaptability assessment indicators are listed for corresponding projects, as shown in Table 3.33, with composite average values and analysis on adverse conditions based on geologic characteristics and conditions for TBM construction and main technical parameters of TBMs used in Northwest Liaoning Water Supply Project (open TBM), Chongqing Rail Transit Line 6 Phase I Project (single shield TBM) and Project of Drawing Water from Datong River into Huangshui River (double shield TBM).

2. TBM Adaptability Assessment

The membership of each assessment indicator (Table 3.33) for TBM construction sections in three projects is obtained as shown in Table 3.34 with the above

Table 3.33 Data for TBM adaptability assessment indicators

Criterion level	Indicator code		Indicator name	Indicator value			Indicator unit
				Northwest Liaoning Water Supply Project	Chongqing Rail Transit Line 6 Phase I Project	Project of Drawing Water from Datong River into Huangshui River	
P_1, TBM construction	U_1		Average advance speed	4	9	6	m/h
	U_2		Total thrust force	20,491	12,800	25,000	kN
	U_3		Rated rotation speed of cutter head	7.98	11.97	7.4	r/min
	U_4		Rated torque of cutter head	6713	4050	2000	Kn m
P_2, geological conditions	U_5		Uniaxial compressive strength of rock (R_c)	60	48	94	MPa
	U_6		Rock mass intactness coefficient (Kv)	0.55	0.67	0.5	–
	U_7		Quartz content	25	54	40	%
	U_8		Geostress level	0.33	0.172	0.327	–
	U_9		Tunnel overburden	424	40	1082	m
	U_{10}		Rock abrasion index	7	4	6	0.1 mm
P3, geological problems	U_{11}		Fracture zone width	50	0	340	m
	U_{12}	U_{12a}	TBM jamming (cutter head jamming)	4541.04	2322.4	5630.3	kN m

(continued)

Table 3.33 (continued)

Criterion level	Indicator code		Indicator name	Indicator value			Indicator unit
				Northwest Liaoning Water Supply Project	Chongqing Rail Transit Line 6 Phase I Project	Project of Drawing Water from Datong River into Huangshui River	
		U_{12b}	TBM jamming (shield jamming)	–	–	20,783.2	kN
	U_{13}		Transverse stratigraphic composite ratio	0	0	25	%
	U_{14}		Unit water gushing rate	7.06	0.1	579	U(s × m)
	U_{15}		Permeable rate	4.6	5	100	LU
	U_{16}		Ground temperature	21.7	18.3	28	°C
	U_{17}		Harmful gas emission	0	0	0	m^3/min
P4, construction organization	U_{18}		Construction technology level	0.85	0.85	0.85	–
	U_{19}		Construction management	0.85	0.85	0.85	–

mentioned TBM adaptability assessment methods based on the membership function constructed for each assessment indicator.

The basic model of fuzzy comprehensive judgment is used to assess the TBM tunneling adaptability in the three projects based on the values of TBM adaptability assessment indicators obtained above and their corresponding fuzzy membership, and the calculation results are as follows

Table 3.34 Membership of assessment indicator corresponding to membership function

Assessment Indicator			Membership U(x)		
Indicator code		Indicator name	Northwest Liaoning water supply project	Chongqing rail transit line project	Project of drawing water from Datong River into Huangshui River
U_1		Advance speed	1	1	1
U_2		Total thrust force	1	1	1
U_3		Cutter head rotation speed	1	1	1
U_4		Cutter head torque	1	1	0.792
U_5		Uniaxial compressive strength of rock (R_c)	1	1	1
U_6		Rock mass intactness coefficient (Kv)	1	1	1
U_7		Quartz content	0.636	0.109	0.364
U_8		Geostress level	0.280	0.912	0.294
U_9		Tunnel overburden	0.296	0.949	0
U_{10}		Rock abrasion index	0	0.667	0
U_{11}		Fracture zone width	0	1	0
U_{12}	U_{12a}	TBM jamming (cutter head jamming)	1	1	0
	U_{12b}	TBM jamming (shield jamming)	–	–	0.506
U_{13}		Stratigraphic composite ratio	1	1	1
U_{14}		Unit water gushing rate	0	1	0
U_{15}		Permeable rate	0.964	0.960	0
U_{16}		Ground temperature	0.915	1	0.6
U_{17}		Harmful gas emission	1	1	1
U_{18}		Construction technology Level	0.8	0.8	0.8
U_{19}		Construction management	0.8	0.8	0.8

$$D = \overline{U} \times R = \begin{bmatrix} 0.0713 \\ 0.0713 \\ 0.0200 \\ 0.0127 \\ 0.1467 \\ 0.0609 \\ 0.0153 \\ 0.0261 \\ 0.0261 \\ 0.0153 \\ 0.1338 \\ 0.1338 \\ 0.0865 \\ 0.0568 \\ 0.0382 \\ 0.0153 \\ 0.0111 \\ 0.0470 \\ 0.0117 \end{bmatrix}^{T} \times \begin{bmatrix} 1 & 1 & 1 \\ 1 & 1 & 1 \\ 1 & 1 & 1 \\ 1 & 1 & 0.792 \\ 1 & 1 & 1 \\ 1 & 1 & 1 \\ 0.636 & 0.109 & 0.364 \\ 0.280 & 0.912 & 0.294 \\ 0.296 & 0.949 & 0 \\ 0 & 0.667 & 0 \\ 0 & 1 & 0 \\ 1 & 1 & 0.253 \\ 1 & 1 & 1 \\ 0 & 1 & 0 \\ 0.964 & 0.960 & 0 \\ 0.915 & 1 & 0.6 \\ 1 & 1 & 1 \\ 0.8 & 0.8 & 0.8 \\ 0.8 & 0.8 & 0.8 \end{bmatrix} = \begin{bmatrix} 0.737 & 0.964 & 0.581 \end{bmatrix}$$

The matrix on the right side of the equation is the comprehensive fitness value of TBM tunneling adaptability in the three projects: 0.737 in Northwest Liaoning Water Supply Project, 0.964 in Chongqing Rail Transit Line Project and 0.581 in water diversion tunnel of Project of Drawing Water from Datong River into Huangshui River. In combination with the quantitative expression of each assessment indicator grade, the adaptability grade of the project and its description are obtained (Table 3.35).

3. Analysis of Adaptability Assessment Results

According to the field measured data and project construction:

Table 3.35 Adaptability in actual projects

Project name	Comprehensive fitness	Adaptability description	Adaptability grade
Northwest liaoning water supply project	0.737	Adaptable	II
Chongqing rail transit line 6 phase I test section project	0.964	Fully adaptable	I
Water diversion tunnel of project of drawing water from datong river into Huangshui River	0.581	Slightly adaptable	III

(1) The stratum conditions in the construction section of Northwest Liaoning Water Supply Project are favorable, and the geological parameters such as uniaxial compressive strength of rock and rock mass intactness coefficient are conducive to tunneling with an open TBM. The main problem encountered by TBM is the influence of fault fracture zone on tunneling. The main faults that TBM crosses are F65 and F64, of which F65 is up to 50 m wide. During the construction, the tunneling parameters of the main machine of TBM were adjusted and controlled in time to adapt to the stratum conditions more quickly, and effective measures were taken to deal with the fault fracture zone. According to the construction report of the construction organization, Lot V of Section II of Northwest Liaoning Water Supply Project has been completed without downtime caused by geological disasters such as rock burst, TBM jamming in serious deformation and instability of surrounding rock during the construction. As shown in the engineering verification, the TBM model selected for Lot V of Section II of Northwest Liaoning Water Supply Project is appropriate and the actual construction conditions meet the mechanical needs, allowing great tunneling efficiency and good adaptability of TBM construction to Lot V. The actual construction situation on site is consistent with the results obtained by the adaptability assessment method.

(2) The planned total construction period of Chongqing Rail Transit Line 6 Phase I Test Section Project was 35 months, but this project was completed 3 months ahead of schedule, showing efficient excavation. Moreover, no adverse geological conditions such as rock burst and TBM jamming occurred during the construction, showing the high adaptability of TBM construction method in this section, and the actual construction situation on site was consistent with the results obtained by the adaptability assessment method.

(3) In construction of the diversion tunnel for the Project of Drawing Water from Datong River into Huangshui River, the TBM experienced 10 major and long-term jamming accidents (Table 3.36), including both cutter head jamming and shield jamming, in which the TBM was unable to advance. It shows that the adaptability of TBM construction method in this section is weak, and the actual construction situation on site is consistent with the results obtained by adaptability assessment method.

3.5 Summary

Reasonable selection of TBM is of great significance. State-of-the-art TBM equipment of reliable quality designed based on specific geological conditions is the key to the success of tunnel project constructed with TBM. It can effectively improve the geological adaptability of TBM, reduce the equipment failure rate and improve tunneling efficiency. In this chapter, the adaptability of common TBM forms are compared and demonstrated based on the corresponding basis and principles of

Table 3.36 Description of TBM jamming in project of drawing water from datong river into Huangshui River

Times	Time	Stake No	Cause
First TBM jamming	April 3, 2008	K17 + 140.416	Convergence of surrounding rock
Second TBM jamming	April 30, 2008	K17 + 135.284	Collapse + convergence
Third TBM jamming	June 29, 2008	K17 + 133.868	Collapse
Fourth TBM jamming	September 3, 2008	K17 + 117.984	TBM attitude out of control + convergence
Fifth TBM jamming	November 3, 2008	K17 + 102.586	Collapse + convergence
Sixth TBM jamming	December 24, 2008	K17 + 086.993	Collapse + convergence
Seventh TBM jamming	September 18, 2009	K16 + 905	Collapse
Eighth TBM jamming	October 18, 2009	K16 + 981	Collapse + convergence
Ninth TBM jamming	December 25, 2009	K16 + 876	Collapse + convergence
Tenth TBM jamming	July 23, 2010	K16 + 775	Tunnel face instability + convergence

TBM selection and the analysis of influencing factors of TBM adaptability. The TBM selection process is further detailed according to actual projects for guidance.

Reasonable adaptability design is the key to the success of a project. In this chapter, fuzzy mathematics and analytic hierarchy process are used to establish the TBM adaptability assessment method and model which are verified in actual projects. As shown in verification in several actual projects, the adaptability assessment method and model established in this chapter are reasonable and feasible, and can generate results consistent with the actual situation on site. The TBM adaptability assessment is of some guidance in TBM adaptability design.

Questions

1. Briefly describe the basis of TBM selection.
2. Briefly describe the principle of TBM selection.
3. What are the key points of matching design of TBM system?
4. What are the main assessment methods of TBM adaptability?
5. Briefly describe the main characteristics of fuzzy mathematics comprehensive assessment method used in TBM adaptability assessment.

Chapter 4
TBM Cutter Head and Cutter Design

This chapter focuses on TBM rock breaking mechanism, TBM load calculation method, arrangement law of disc cutter, and cutter head structure and cutter design.

4.1 TBM Rock Breaking Mechanism

4.1.1 Study on Rock Breaking Mechanism of Disc Cutter

The disc cutter blade is pressed into the rock surface under the thrust of the hydraulic cylinder, and the disc cutter rolls along the concentric circle track under the action of the cutter head driving system (Fig. 4.1) simultaneously. When the thrust force exceeds the compressive strength of the rock, the rock under the disc cutter tip is directly broken, and the tip penetrates into the rock, forming a crushed zone and radial cracks. When the disc cutter spacing S meets certain conditions under further pressure, the cracks in the rock between adjacent disc cutters extend and penetrate each other so that rock peels off in chips (Fig. 4.2). The rock chips, after falling into the bottom of the tunnel under the action of dead weight, are scooped up by the bucket of the rotating cutter head, and then discharged into the muck conveyor through the muck chute.

The rock breaking process of TBM disc cutter is divided into two stages: the process of pressing disc cutter into rock mass and the process of forming rock chips between adjacent disc cutters.

The loading process of pressing disc cutter into rock mass can be divided into four stages: establishing stress field, forming crushed zone, forming rock chips and forming pressing pit. With the increase of stress in the initial stage, the tiny cracks in the rock will be closed, and the rock will be in the linear elastic deformation stage, with slight damage. As the load continues to increase, conical cracks will appear at the blade edge, and then a crushed zone will be formed at the bottom of the pressed

K. Chen et al., *TBM Design and Construction*,
https://doi.org/10.1007/978-981-99-0059-6_4

Fig. 4.1 Rock breaking track of TBM disc cutter on tunnel face

Fig. 4.2 TBM rock breaking chips

blade. At this stage, the dislocation and expansion of the rock chips will be limited by the surrounding original rocks. To crush the rocks, the load will continue to increase, resulting in violent crushing of the rock chips. After that, these crushed rocks will be compacted again under the restriction of the positive pressure and lateral pressure of the disc cutter. The movement and expansion of rock chips inward and on both sides will cause radial, middle and lateral cracks in the rock. During the formation of rock cracks, the disc cutter will be gradually pressed into the rock. When the side

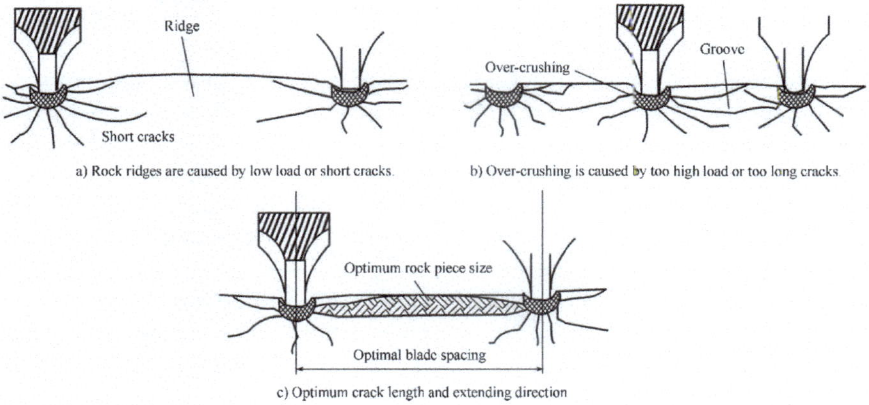

Fig. 4.3 Three rock breaking modes

crack reaches the rock surface, the rock at the bottom of and around the blade will be separated as rock chips.

For TBM excavation, when TBM disc cutter rolls on the excavation surface, the crushed zone will continuously expand, and cracks will appear and develop. One or more cracks will reach the free surface or extend to the cracks generated by adjacent disc cutter under the action of disc cutter, and be connected with these cracks to generate rock pieces.

When the distance between two adjacent disc cutters is too large or the load of disc cutters is too low, the cracks will develop toward the cutting surface, and, when they reach the free surface, triangular small rock pieces will be produced, forming uncut rock ridges between two disc cutters (Fig. 4.3a). When the cutter spacing of the disc cutter is too small or the load is too high, the ineffective long cracks under the disc cutter will develop into the rock or rock mass. A large cracking area will result in large energy loss, and a groove will be formed between the two disc cutters (Fig. 4.3b). With optimal disc cutter spacing and thrust force, the lateral cracks will expand toward the adjacent disc cutters in a roughly straight path and produce straight and flat rock pieces between adjacent disc cutters (Fig. 4.3c). The straight path is the shortest, and only half of the cutter spacing.

4.1.2 *Influencing Factors of Rock Breaking by Disc Cutter*

The rock breaking process and effect of TBM disc cutter are influenced by many factors such as construction environment, cutter head design, tunneling parameters, etc. In hard rock environment, geological factors are complicated. There are many kinds of rocks, such as granite, limestone, tuff, shale, gneiss, sandstone, etc. Their compressive strength, tensile strength, elastic modulus, Poisson's ratio and other mechanical properties are quite different. With the increase of rock strength and elastic modulus, it is more difficult to develop rock cracks. The inhomogeneity of rock itself will also affect the rock breaking process of disc cutter.

In addition to the inherent mechanical properties of rock, surrounding pressure, joints, temperature and many other factors will influence the rock breaking process of TBM disc cutter (Figs. 4.4 and 4.5). The development degree, strength and direction of the rock joint and the expanding cracks of the rock under the action of disc cutter will interact with each other to form broken rock blocks. Generally, proper joint development and direction are helpful to improve the rock breaking efficiency of disc cutters, but the requirements for supporting operation are stricter. Under high surrounding pressure, the development of rock cracks will be limited, making the rock breaking by disc cutter more difficult. Other factors, such as temperature, water content, etc., will affect the mechanical properties of rock, and then affect the rock breaking effect of disc cutter.

It is important to reasonably design TBM cutter head and select tunneling parameters for higher tunneling efficiency based on the geological conditions on jobsite. Generally, φ 17 in or 19in (1in = 2.54 cm, the same below) disc cutters are used. At present, φ 20 in and 21 in disc cutters have been developed. The diameter of disc cutters tends to increase.

The diameter of center cutter is generally 17 in, while that of face disc cutter is mostly 19 in. There are two types of disc cutter blades: wedge blade and flat blade, of which the wedge blade cutter ring can be pressed into rock easier. But with the increase of wear, the contact area between the wedge blade and the tunnel face also increases. To ensure the tunneling efficiency, the TBM needs to provide greater forward thrust force, and its stability will be affected. However, the contact area between the disc cutter of flat blade and the tunnel face will not change much before and after wear, so the service life of the disc cutter is longer. At present, the disc cutter of flat blade is more commonly used. In addition to type of disc cutter,

Fig. 4.4 Joint development

Fig. 4.5 High surrounding pressure

the other two main factors that influence the rock breaking efficiency of disc cutter most are the disc cutter spacing and penetration.

The rock is broken by several disc cutters arranged in certain spacing simultaneously. Reasonable selection of cutter spacing is an important part in TBM design. Once the disc cutter spacing is selected during design and manufacturing, it cannot be changed in the excavation. Disc cutter spacing is crucial to TBM tunneling efficiency and cutter service life. When the disc cutter spacing is too large, the rock between adjacent disc cutters cannot be completely peeled off, and will form a "rock ridge", making TBM tunneling difficult. Meanwhile, the rock breaking burden of the disc cutter will be heavier, and the disc cutter will be more susceptible to wear and fracture failure. If the disc cutter spacing is too small, the rock between adjacent disc cutters will be excessively broken, resulting in waste. As the disc cutter spacing is too small, more space must be reserved on the TBM cutter head to accommodate disc cutters. This is not conducive to cutter head structure design. Selection of reasonable disc cutter spacing according to the geological environment is of great significance for efficient rock breaking of TBM.

The penetration of disc cutter refers to the depth of the face disc cutter pressed into the rock in one revolution of the cutter head. It is a tunneling parameter that can be adjusted in the operator cabinet during TBM tunneling. Similar to the cutter spacing, the penetration will also affect the tunneling efficiency of TBM, and must be set properly. At fixed cutter spacing, rock type and other parameters, the cracked zone under rock squeezing and shearing will be larger with the increase of penetration, and it is more prone to rock over-crushing and cutter overload. However, with insufficient penetration, the rock cannot be completely broken and the TBM tunneling efficiency will be low.

4.2 TBM Load Calculation

4.2.1 Calculation of Disc Cutter Load

In rock breaking of disc cutter, forces in three directions interact with each other between the cutter blade and rock as shown in Fig. 4.6: vertical force F_V (also known as normal force or radial force), which points to the excavation face and is provided by the thrust force of cutter head; rolling force F_R (also known as tangential force or cutting force), which points to the tangential direction of disc cutter and is provided by cutter head torque; side force F_s, which points to the center of the cutter head, and is generated by the pressing force of disc cutter on rock and centrifugal force of cutter head rotation, but it is low.

The resistance torque generated by these forces to the rotation of the cutter head is balanced by the cutter head torque. The component along the rotation axis of the cutter head is balanced by the cutter head thrust force. The radial component along the cutter head and its radial effect on the cutter head will be borne by the large bearing of cutter head. This is one of the reasons for machine vibration and noise, and premature failure of the large bearing of cutter head. Force analysis of cutter head provides the basis of cutter arrangement and cutter head main parameter design. Many scholars have established the prediction model of cutting force of disc cutter

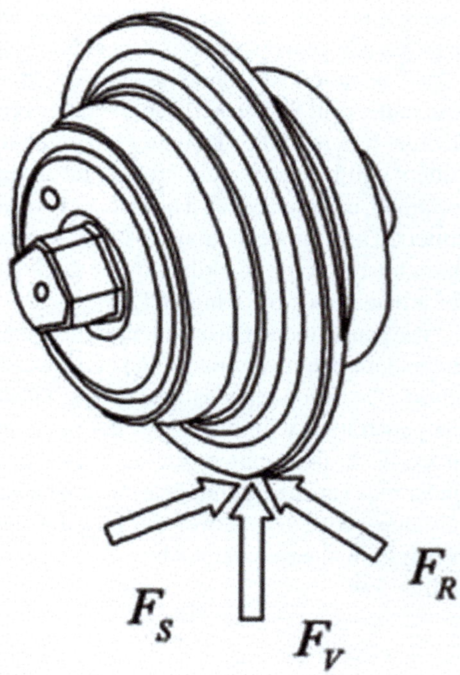

Fig. 4.6 Forces acting on disc cutter

through experimental test and analysis of jobsite test data. Literature [6] summarized the typical force models, including vertical thrust force prediction formula of Evans (1966), prediction formula of Akisanto Saburo, prediction formula of Roxborough (Australia, 1975), CSM prediction formula of America (1976, 1993), Robbins calculation model, prediction formula of former Northeast Institute of Technology Rock Fragmentation Research Institute, prediction formula of Shanghai Jiao Tong University, prediction formula of former North China Institute of Hydropower graduate department (Beijing), etc. Among them, Robbins calculation model features a clear mechanical principle, but strongly depends on cutter structure and shape. CSM model takes all factors into account and has high prediction accuracy. Both of them are widely used, especially CSM model.

CSM model is a force prediction formula of disc cutter obtained from linear cutting experiment by Colorado School of Mines. The disc cutter first crushes the rock below. It is assumed that the lateral component of the force acting on the rock by the wedge-shaped blade side will produce a shear force to the rock ridge (the rock between adjacent cutter spacing). The formula of CSM model is as follows:

$$F = C \cdot \frac{\varphi R_0 T}{1 + \psi} \left(\frac{S \sigma_c^2 \sigma_t}{\varphi \sqrt{R_0 T}} \right)^{1/3} \tag{4.1}$$

$$F_V = F \cos \frac{\varphi}{2} \tag{4.2}$$

$$F_R = F \sin \frac{\varphi}{2} \tag{4.3}$$

where

- R_0 disc cutter radius (mm);
- ψ pressure distribution coefficient of the top blade of cutter ring, generally -0.2 ~ 0.2;
- φ contact angle of disc cutter (rad), $\varphi = \arccos[(R_0\text{-}h)/R_0]$;
- h disc cutter penetration (mm);
- S cutter spacing (mm);
- C dimensionless coefficient, 2.12;
- σ_c uniaxial compressive strength of rock (MPa);
- σ_t shear strength of rock (MPa).

4.2.2 Calculation of TBM Cutter Head Load

During TBM tunneling, the cutter head is subjected to thrust resistance, rotation resistance torque and unbalanced force.

The cutter head thrust resistance consists of the component of vertical force of each disc cutter along the TBM axis, and it is balanced by the thrust force provided by the thrust system. Then,

$$F_P = \sum_{i=1}^{n} (F_{Vi} \times \cos \beta_i) \quad (4.4)$$

where

- F_p cutter head thrust resistance (kN);
- F_{Vi} the vertical force of the i-th disc cutter (kN);
- β_i the angle (rad) between the vertical force of the i-th disc cutter and the TBM axis;
- n the number of disc cutters on the cutter head.

The rotation resistance torque of the cutter head consists of the resistance torque generated by the rolling force of each disc cutter, and it is balanced by the torque provided by the drive system.

$$T_P = \sum_{i=1}^{n} (F_{Ri} \times R_i) \quad (4.5)$$

where

- T_P resistance moment of cutter head rotation (N·m);
- Fri the force of the i-th disc cutter (kN);
- R_i the installation radius of the i-th disc cutter (mm);
- n the number of disc cutters on the cutter head.

The unbalanced force of cutter head is the resultant force of disc cutter load on cutter head plane. It is easy to cause eccentricity and vibration when cutter head rotates, and adversely affects the service life of main bearing. It should be minimized. Unbalanced force is calculated as follows:

$$F_u = \sqrt{F_x^2 + F_y^2} \quad (4.6)$$

$$F_X = \sum_{i=1}^{n} (F_{Vi} \sin \beta_i \sin \theta_i - F_{Ri} \cos \theta_i) \quad (4.7)$$

$$F_y = \sum_{i=1}^{n} (F_{Vi} \sin \beta_i \cos \theta_i - F_{Ri} \sin \theta_i) \quad (4.8)$$

where

- F_u unbalance force of cutter head (kN);
- F_x unbalanced force in X axis direction (kN);
- F_y unbalanced force in Y axis direction (kN).

4.3 TBM Disc Cutter Arrangement

The arrangement of TBM disc cutters is crucial in TBM cutter head design. It is directly related to the overall stress, overall performance and service life of the cutter head. It may affect the rock breaking efficiency and advance speed of disc cutter, and will determine the tunneling performance of the cutter head. The disc cutter can be arranged in radial or circumferential direction. In radial arrangement, disc cutter spacing is crucial, while, in circumferential arrangement, the installation angle should be determined following the disc cutter spacing.

4.3.1 Arrangement Zones

Disc cutters can be arranged in the following 4 typical zone: center zone, face zone, transition zone and edge zone (Fig. 4.7). In different zones, disc cutters and their arrangement may be different.

(1) Center zone. It is the center of the cutter head, and the arrangement of disc cutter is greatly limited by space. The center disc cutters are mainly arranged in I shape or cross shape. Generally, a center disc cutter has six or eight blades. Single disc cutters, twin disc cutter or multi-disc cutter can be used,
(2) Face zone. The face zone accommodates the largest number of disc cutters. It is the main zone for the TBM cutter head to break rock. Different disc cutter arrangement types lie in the arrangement of face disc cutters.
(3) Transition zone. On some cutter heads, there is a transition zone between the face and edge zones. The transition disc cutters protrude from the cutter head face plate with an amount the same as the face disc cutter, but to reduce wear, their spacing is smaller.
(4) Edge zone. In the edge zone, the envelope of the disc cutter tip is circular arc. Because of the long cutting distance of the gauge disc cutter, the disc cutter spacing is less than or equal to the disc cutter spacing (S) in the transition zone, and gradually decreases to reduce the wear.

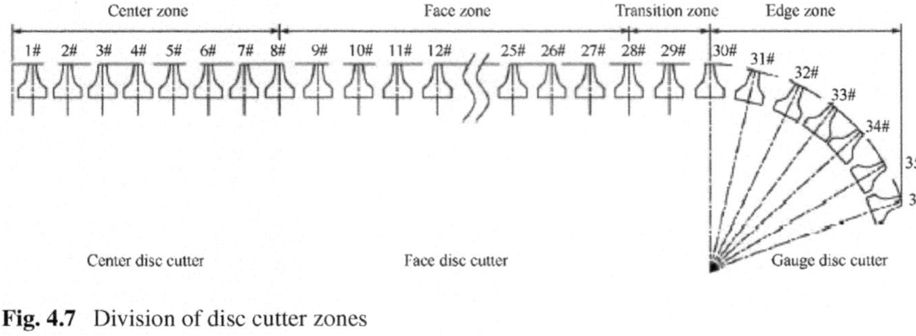

Fig. 4.7 Division of disc cutter zones

4.3.2 Arrangement Principles

The cutter head face plate is divided into three different zones: the center zone, the face zone, the transition zone and the edge zone. Because of the difference of the rock breaking volume and the stress characteristics of the cutter, the performance of the cutter is also different. The center zone is the center of the cutter head. It features small space and small turning radius of disc cutter. It bears sliding and rolling of disc cutters and large axial force. The cutter ring used in this zone must be tough. The face zone is the main zone where the cutter head breaks rock. Different disc cutter arrangement types lie in the disc cutter arrangement in the face zone. When the disc cutter in the face zone works, the disc cutter is facing the tunnel face with a certain turning radius, and pure rolling is dominant. The transition zone is located between the face zone and the edge zone. The transition disc cutters protrude from the cutter head face plate with an amount the same as the face disc cutter, but to reduce wear, their spacing is smaller The edge zone is located at the edge of the cutter head, and the disc cutters in this zone not only bear a large amount of rock breaking, but also are under a complex force. In addition, the excavation diameter should be ensured, and the cutter ring should have strong abrasion resistance.

Although different arrangement types are available, the disc cutters must be arranged according to certain design principles. The disc cutters must be arranged to achieve one of the main targets: balance the cutter head load, so as to minimize the radial load and overturning moment of cutter head, and ensure uniform load distribution of cutter head, and balanced and reliable comprehensive mechanical properties during tunneling. In addition, the main function of the cutter head is rock breaking, and the disc cutters must be arranged to meet the requirement of efficient rock breaking. The penetration of adjacent disc cutters should be kept within a certain range so that continuous rock breaking can be achieved.

In addition to above mentioned main performance indexes, the arrangement of disc cutter is also limited by the structure of the cutter head. Therefore, the wear to the disc cutter should be taken into account. A certain number of manholes, muck chute, water spray holes, etc. need to be arranged on the cutter head face plate, and the arrangement of disc cutters must be coordinated with these structures. Meanwhile, most cutter heads are designed to be assembled in blocks, and disc cutters should not be arranged on the contact face of each connecting block.

For arrangement of disc cutters, the following should be taken into account: ① minimized radial load; ② minimized overturning moment; ③ requirements for continuous and sequential rock breaking of adjacent disc cutters; ④ minimized difference of rock breaking volume; ⑤ interference-free requirements; ⑥ uniform mass distribution, and center of mass close to center of rotation; ⑦ technological requirements of cutter head assembly.

To arrange the disc cutters, reasonable spacing should be determined first according to the selected disc cutter type and engineering geology, and then the circumferential arrangement angle should be determined. After the position of the first disc cutter is selected, the position of the next one should be determined based on the spacing and arrangement angle. The disc cutter spacing is mainly designed according to the requirements of rock breaking. The arrangement angle should be so determined based on the requirements of rock breaking that comprehensive mechanical properties of the cutter head are ensured.

The following principles should be considered in cutter head design and disc cutter arrangement:

1. Reasonable Disc Cutter Spacing

If the cutter spacing is too large, the rocks between the adjacent cutters will not be completely peeled off, and then "rock ridges" will be formed, making TBM tunneling difficult and disc cutters vulnerable to damage. If the disc cutter spacing is too small, the rock will be broken into too small pieces. It is a waste of TBM energy and will reduce rock breaking efficiency. Therefore, the disc cutter spacing should be reasonably set according to geological conditions, to completely crush the rock between adjacent disc cutters, in other words, to make crushing grooves of adjacent disc cutters intersect. Because the size of crushing groove is related to disc cutter dimension, rock properties, thrust force and cutting depth, the disc cutter spacing is also related to the above factors. According to the rock breaking test results of disc cutters, the cutter spacing should be smaller to cut rock of higher strength.

Different domestic and foreign manufacturers may apply different disc cutter spacing, as each of them has their own proven ideas and design habits. For example, Robbins TBM often uses 83 mm cutter spacing, while Herrenknecht TBM often uses 90 mm cutter spacing.

In practical design, the cutter spacing is generally set from 50 to 100 mm according to different working conditions. The disc cutter spacing is different in different zones as well. Generally, the center disc cutter spacing is larger than face disc cutter spacing, and the gauge disc cutter spacing is usually small. In addition to commonly used 17-in disc cutters, 19-in disc cutters become more popular, so the cutter spacing can be appropriately increased. The optimal disc cutter spacing should be scientifically determined in combination of different factors.

2. Force Balance of Cutter Head

The rock breaking load of each disc cutter on the cutter head should be as equal as possible. That is to say, when each disc cutter breaks rock, the load and rock breaking volume of each disc cutter are as equal as possible, and the lateral reaction forces on both sides of the blade offset each other as much as possible, so the radial load acting on the major bearing is as close as possible to zero. When the forces of the cutter head are balanced, the radial load and overturning moment will be minimized.

The resultant force of horizontal cutting components (perpendicular to the axis of the cutter head) of each disc cutter on the cutter head is called unbalanced force.

The unbalanced force in design should be minimized, and the ratio of unbalanced force to thrust force should be less than 1%. Under too large unbalanced force, the cutter head is susceptible to vibration during excavation. It is unfavorable to the bearing capacity of the cutter head main bearing. Therefore, the force on the cutter head of TBM in the process of tunneling must meet the balance condition of space force system. If the xoy plane is located on the cutter head, the axis of TBM cutter head is the Z axis along the TBM advance direction. Taking the unit (detached part) composed of disc cutter and cutter head taken as the research object, the algebraic sum of each force acting on the unit projected on 3 coordinate axes and the algebraic sum of the moment of each force on the coordinate axis should be zero or as close to zero as possible.

Overturning moment produced during rock breaking by disc cutters has a great influence on the overall performance of cutter head. To avoid damage to cutter head components such as main bearing because of vibration, the overturning moment should be minimized.

The disc cutters should be arranged on the TBM cutter head to meet both the needs of engineering design and the structural characteristics of cutter head. At present, there is no unified and complete design theory for disc cutter arrangement, but the basic design principles and concepts must be followed. The ultimate goal of disc cutter arrangement is to ensure TBM cutter head to complete the tunneling operation reliably and efficiently. It is necessary to comprehensively consider different factors and select reasonable disc cutter arrangement parameters to ensure stable mechanical properties and tunneling performance of cutter head.

3. Equal Service Life of Disc Cutters

This principle is to minimize the difference of rock breaking volume of each disc cutter. It is also known as the principle of equal wear volume. The ideal disc cutter arrangement should enable the wear volume of the disc cutter at each disc cutter position to be balanced as much as possible, that is, the principle of "disc cutters arranged with equal wear", under the rock breaking conditions (the rock breaking ring of the disc cutter should cover the excavation face without any raised rock ridge ring), However, as limited by the maximum disc cutter spacing, the rotation radius and wear of the disc cutters get less at the position closer to the center of the cutter head. Therefore, disc cutters can be arranged with equal wear only in the cutter head area with the radius of rotation larger than a certain value.

(4) Other Principles

The principle of rock breaking in sequence: The former disc cutter should provide the next one with a rock breaking free surface to allow smooth muck discharge. Centroid distribution principle: The whole center of gravity of the disc cutters should be dispersed as much as possible to avoid coinciding with the center of the cutter head. Interference-free installation principle: The disc cutters should be installed at positions not interfering with any other configuration on the cutter head, and different disc cutters should not penetrate each other geometrically.

4.3.3 Determination of Disc Cutter Blade Spacing

Disc cutter spacing refers to the difference of the distance between the blades of adjacent disc cutters relative to the center of the cutter head, i.e., the distance between adjacent circular tracks left by the disc cutter blades on the rock section during TBM excavation. The disc cutter spacing directly affects the thrust force acting on the disc cutter and the power consumption of rock breaking. It is an important index of disc cutter arrangement, and an important working parameter for rock breaking of disc cutter. The disc cutter spacing should be selected so that the rock between two blades is completely broken and there is no repeated rock breaking. When rock conditions and rated thrust force of disc cutter are determined, there is a critical cutter spacing and an optimal cutter spacing. The critical cutter spacing refers to the maximum spacing when there is influence between the cutting grooves of two adjacent cutter blades, that is, the minimum spacing when the adjacent disc cutters roll on the rock section and there is no accumulated rock ridge between the cutting grooves of the two disc cutters. When the cutter spacing is greater than the critical cutter spacing, the cutting grooves will not influence each other, that is, under the action of two adjacent disc cutters.

When cracks produced on rock cannot meet and get connected, accumulated rock ridges will be produced between two cutting grooves.

To break an accumulated rock ridge between two cutting grooves, the cutting grooves should be deepened by repeated rolling of disc cutter, reducing the rock breaking efficiency and worsening the disc cutter blade wear. If the cutter spacing matches the thrust force of the disc cutter when the cutter spacing decreases gradually, more large broken rock blocks will be produced. When the energy consumption per unit volume of rock is minimized and the rock breaking efficiency is the highest, the cutter spacing is the optimal one. The optimal cutter spacing S is related to the physical properties of rock and the cutting depth h of disc cutter.

According to the intersection degree of the rock breaking volume during rock breaking by TBM disc cutter, the rock breaking zone can be classified into non-intersected, just intersected or intersected.

Fig. 4.8 shows the critical cutter spacing, S_0.

$$S_0 = B + 2h \tan \beta \tag{4.9}$$

where

B width of disc cutter top;
h cutting depth of disc cutter;
β half angle of rock breaking, see Table 4.1 for its value.

As shown in formula (4.9), the critical cutter spacing is related to the cutting depth of disc cutter and the properties of rock (breaking angle). In practice, the cutter spacing of the same TBM is constant, and different geological conditions are adapted to mainly by changing the thrust force of the cutter head. There should be

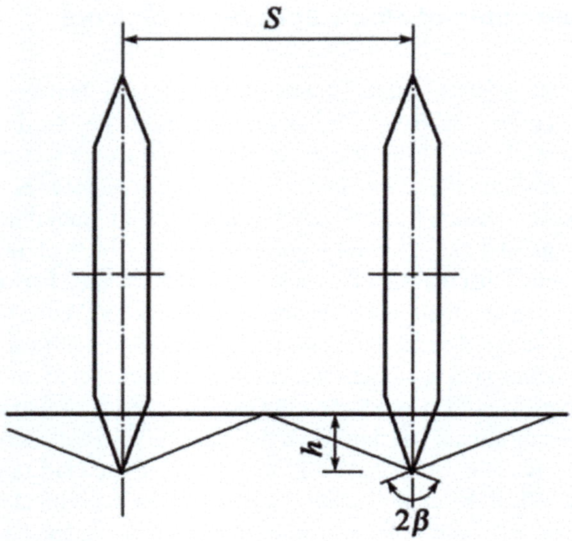

Fig. 4.8 Schematic diagram of disc cutter cutting into rock

no accumulated rock ridge between the two disc cutters, that is, there should be no unconnected rock ridge when the cutter head rotates repeatedly (two or more revolutions), and the design of the blade spacing of TBM disc cutter should meet the condition $S < S_0$.

4.3.4 Plane Arrangement of Disc Cutter

The disc cutters can be arranged on the cutter head in helix, dynamic star or random modes, as shown in Fig. 4.9. In multi-helix and dynamic star modes, the disc cutters are arranged symmetrically with respect to the rotation axis of cutter head. In multi-helix mode, if the angle of each helix is 0°, it is a star mode.

In random mode, the disc cutters seems to be arranged randomly on the cutter head. In fact, the influence of cutter head partition is considered, and the disc cutters are basically arranged symmetrically with respect to the cutter head rotation axis.

Generally, the multi-helix mode is suitable for tunneling in extremely hard and intact stratum. The star mode is suitable for tunneling in soft rock or broken rock stratum. The random mode pattern is suitable for tunneling in the stratum with properties between above two strata.

The disc cutter should be so arranged on the cutter head that the force acting on the cutter head is balanced. When TBM works, the rock breaking force of disc cutter is balanced by reducing the disc cutter spacing in different degrees along the radius direction of cutter head according to the specific operation object of TBM.

a) Multi-helix b) Dynamic Star c) Random

Fig. 4.9 Arrangement of disc cutter on TBM cutter head

Refer to Fig. 4.10 for the force acting on the disc cutters. The plane position of the disc cutter on the cutter head can be expressed with polar coordinates as (ρ_i, θ_i), where ρ_ii (0, R_t] is the mounting polar diameter of the disc cutter, and $\theta_i i \in [0, 2\pi]$ is the mounting polar angle of the disc cutter. $\alpha_i i \in [0, \pi/2]$ is introduced to describe the installation tilting angle of disc cutter, and R_S and R_f are the radius of the envelope arc of gauge disc cutter tip and the distance from the center of the arc to the center of the cutter head respectively.

(1) Minimized Radial Load

In ideal cutter arrangement, the radial unbalance force generated by the cutter head can be minimized. Given that the resultant force of radial load of disc cutter acting on the cutter head is the objective function f1(x), its expression is as follows:

$$\begin{cases} F_x = \sum\limits_{i=1}^{n} \left(\mathrm{F}_{si} \cos \alpha_\mathrm{i} \cos \theta_\mathrm{i} + \mathrm{F}_\mathrm{Ri} \sin \theta_\mathrm{i} - \mathrm{F}_\mathrm{Vi} \sin \alpha_\mathrm{i} \cos \theta_\mathrm{i} + \mathrm{m}\omega^2 \rho_\mathrm{i} \cos \theta_\mathrm{i} \right) \\ \mathrm{F_y} = \sum\limits_{i=1}^{n} \left(\mathrm{F}_{si} \cos \alpha_\mathrm{i} \sin \theta_\mathrm{i} - \mathrm{F}_\mathrm{Ri} \cos \theta_\mathrm{i} - \mathrm{F}_\mathrm{Vi} \sin \alpha_\mathrm{i} \sin \theta_\mathrm{i} + \mathrm{m}\omega^2 \rho_\mathrm{i} \sin \theta_\mathrm{i} \right) \\ F = \sqrt{\mathrm{F_x^2} + \mathrm{F_y^2}} \end{cases} \tag{4.10}$$

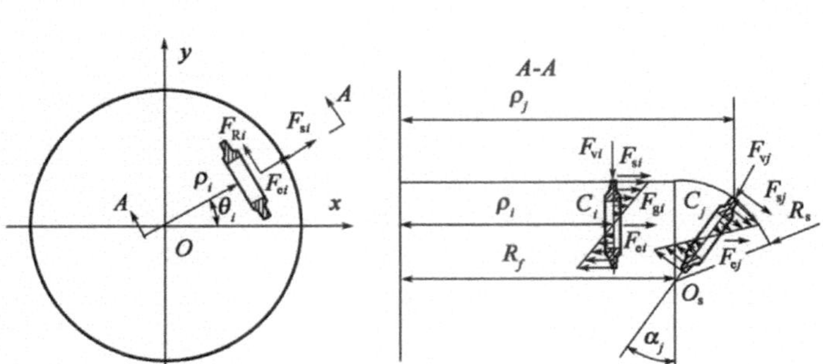

Fig. 4.10 Forces acting on disc cutter

where

F_{Ri} rolling force of the i-th disc cutter;
F_{Si} the lateral force of the i-th disc cutter. Fsi = (τ/2)(rφ)2sin(rφ/2ρi), where r is the disc cutter radius, and φ is the disc cutter contact angle. φ = arccos[(r-h)/r] if the penetration is h;
n number of disc cutters;
Fx the radial force resultant in x-axis direction of the cutter head face;
F_y radial force resultant in Y-axis direction of cutter head;
F the total radial force resultant acting on the cutter head.

(2) Minimized Overturning Moment

Overturning moment generated in hard rock breaking by disc cutter has a great influence on the overall performance of cutter head. To avoid damage to cutter head components such as main bearing by vibration, the overturning moment should be minimized. Given that the overturning moment of cutter head is the objective function $f_2(x)$, the expression is as follows:

$$\begin{cases} M_x = \sum_{i=1}^{n} (F_{Vi}\rho_i \cos\alpha_i \sin\theta_i + F_{si}\rho_i \sin\alpha_i \sin\theta_i + M_i \sin\theta_i) \\ M_y = \sum_{i=1}^{n} (F_{Vi}\rho_i \cos\alpha_i \cos\theta_i + F_{si}\rho_i \sin\alpha_i \cos\theta_i + M_i \cos\theta_i) \\ M = \sqrt{M_x^2 + M_y^2} \end{cases} \tag{4.11}$$

where

F_{Vi}, F_{Si} vertical force and lateral force of i-th disc cutter;
M_i couple moment generated by Coriolis inertia force of disc cutter movement;
M_x the moment of vertical load of disc cutter acting on x-axis of cutter head face;
M_y the moment of vertical load of disc cutter acting on y-axis of cutter head face;
M the total overturning moment acting on the cutter head.

The disc cutters should be arranged on the TBM cutter head to meet both the needs of engineering design and the structural characteristics of cutter head. At present, there is no unified and complete design theory for disc cutter arrangement, but the basic design principles and concepts must be followed. The ultimate goal of disc cutter arrangement is to ensure TBM cutter head to complete the tunneling operation reliably and efficiently. It is necessary to comprehensively consider different factors and select reasonable disc cutter arrangement parameters to ensure stable mechanical properties and tunneling performance of cutter head.

4.4 TBM Cutter Head Structure and Cutter Design

4.4.1 Cutter Head Structure Design

1. Cutter Head Structure
(1) Cutter Head Body Structure

The cutter head body structure is made from welded thick steel plates. The front and rear face plates of the cutter head features many longitudinally connected bulkheads, and the structure is complex. The connecting flange on the back face needs to be machined and connected with special bolts. The cutter seat must be precisely positioned and machined when welded. The influence of dynamic load should be considered in determination of the thickness of cutter head and the dimension of weld. Heating, heat preservation and gas shielded welding are required, and welding process requirements are high. In the overall cutter head structure, the strength, rigidity, wear resistance and vibration stability must be taken into account, and the welding seam must be of sufficient strength to prevent cracking under vibration conditions. The excavation diameter of the tunnel is controlled by the gauge cutter at the outermost edge of the cutter head, while the maximum diameter of the cutter head structure is usually designed at the lip of the bucket. Generally, there is a gap of about 25 mm from the outermost edge of the bucket lip. If this gap is too large, it is not conducive to the removal of rock chips, and if the gap is too small, the bucket is susceptible to damage because of scraping the tunnel wall directly. Therefore, the maximum diameter of the cutter head body structure is generally about 50 mm smaller than the theoretical excavation diameter.

(2) Cutter Seat and Cutter

The cutter head body structure is welded with a cutter seat for mounting the disc cutters, as shown in Fig. 4.11. The cutter seat must be strictly positioned before being welded on the cutter head, and the positioning face and mounting contact face of the cutter seat must be machined. The disc cutters are classified into center disc cutters, face disc cutters and gauge disc cutters according to the position of installation on the cutter head. Generally, double disc cutters are used as the center disc cutters, while single disc cutters are used as the face disc cutters and gauge disc cutters. Disc cutters on the cutter head are removable and replaceable. Generally, they are installed and fixed on the cutter seat with wedge bolt. The cutter head is so designed that both front mounting and back mounting of disc cutters can be realized. Back-mounted disc cutters are beneficial to ensure the safety of operators.

(3) Bucket and its Teeth

The bucket opening is arranged with teeth on one side and vertical breast plates on the other side, as shown in Fig. 4.10. The bucket teeth are bolted to the teeth seat to shovel up the muck during excavation, and can be replaced after being worn or damaged. The breast plates opposite the bucket teeth are used to prevent large muck

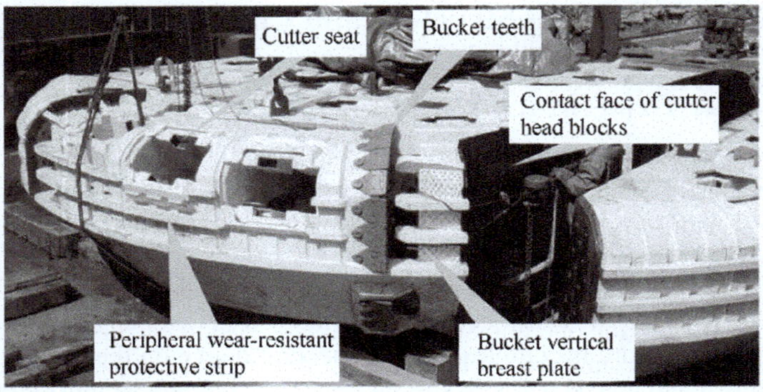

Fig. 4.11 Field erection and welding of cutter head blocks

from entering the cutter head from the bucket opening and reaching the TBM belt conveyor on the one hand, and break large rocks and protect the bucket teeth on the other hand. Buckets are also designed on the back of the cutter head to scoop up the muck dropping or left between the back of the cutter head and the lower support at the bottom of the tunnel. The cutter head rotates in one direction during excavation, and it can rotate in two directions by inching for maintenance of cutter head and disc cutter only.

(4) Nozzle and Rotary Joint

The water spray device of the cutter head is mainly used to reduce dust from the tunnel face and cool the disc cutters. The TBM water supply system supplies water to the rotary joint installed at the center of the back of the cutter head through water pipes, and then to several water spray nozzles in front of the cutter head through the pipes in the cutter head. The water spray system must work at certain water spray rate and water pressure, which may be adjusted in the operation room based on water gushing and muck discharge in the tunnel. Because the water nozzle is vulnerable to clogging, it must be inspected and maintained regularly in front of the cutter head.

(5) Manhole

Generally, the cutter head is also designed with a number of manholes for operators to enter the tunnel face to inspect and maintain the cutter head when necessary, as shown in Fig. 4.12. The manhole is of a closed or semi-closed structure to prevent large rocks from entering the cutter head, and the opening can be fixed or opened. Refer to Fig. 4.13 for the back structure of the cutter head after welding.

(6) Wear Resistance Design

To protect the cutter head body structure and the cutters, wear resistance is taken account into the design of cutter head, such as the wear-resistant plate in front of the cutter head, the peripheral wear-resistant strips and the wear-resistant column, etc., as shown in Figs. 4.14, 4.15 and 4.16. These wear-resistant structures can be

Fig. 4.12 Front structure of cutter head after erection and welding

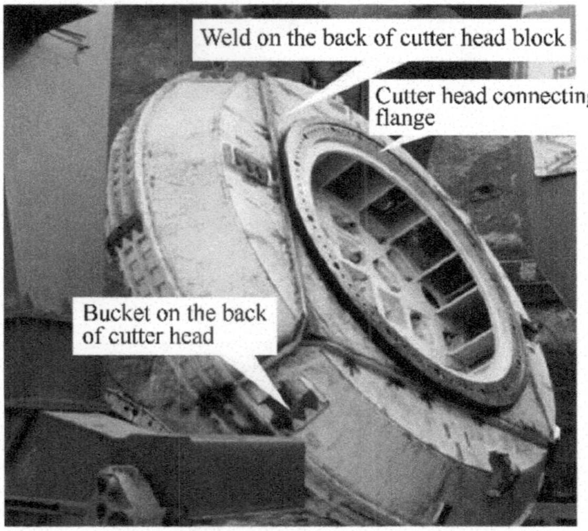

Fig. 4.13 Back structure of cutter head after welding

replaced or repaired after wear. It should be noted that, for tunnels in hard rocks, the cutter head is more prone to weld cracking and wear. It should be fully considered in design, manufacturing and maintenance of the cutter head.

(7) Expanding Excavation Design

Generally, the expanding excavation should be taken into account during design of the cutter head for excavation of larger tunnel, especially for tunnel section with surrounding rock subject to serious deformation to prevent TBM jamming. At present, adjusting washers or overcutters with telescopic cylinder are mainly used

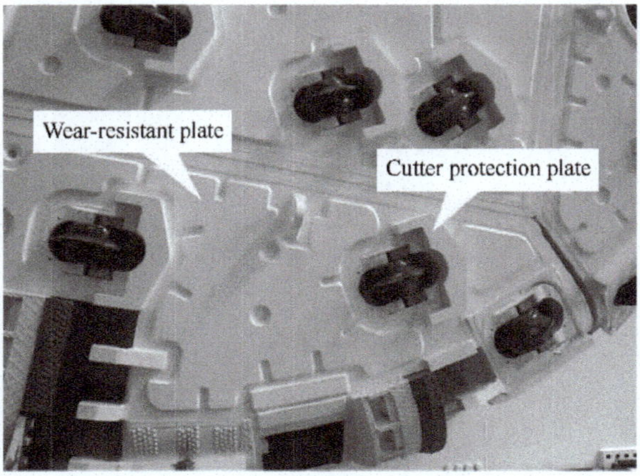

Fig. 4.14 Surface wear-resistant plate of cutter head

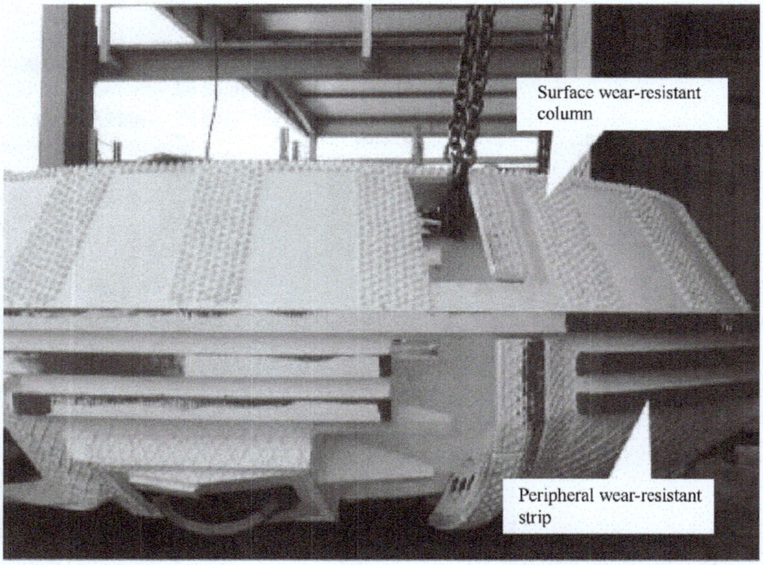

Fig. 4.15 Peripheral wear-resistant strip and surface wear-resistant column of cutter head

for expanding excavation. The adjusting washer is installed on the gauge cutter seat to extend the gauge cutter outward for expanding excavation. This structure is simple, but allows an expanding excavation diameter of about 70 mm only. The overcutter with telescopic cylinder is able to achieve larger expanding excavation, but its structure is complicated, and it will weaken the cutter head to some extent, so it is seldom used at present.

2. Cutter Head Design

Fig. 4.16 Wear-resistant structure of bucket teeth and teeth seat on cutter head

The design of cutter head is a very long process, including preliminary scheme design and detailed design.

(1) Preliminary Scheme Design

The preliminary scheme design of the cutter head is an important stage in the whole cutter head design process, and main design parameters will be determined in this stage, including cutter head diameter, cutter head expanding excavation, cutter head block arrangement, muck inlet, center cutter structure, face cutter number, cutter diameter, cutter spacing, gauge cutter number, water nozzle number, manhole number, etc.

① Cutter Head Diameter and Expanding Excavation

The cutter head diameter should be determined according to the engineering tunnel diameter and TBM model. Cutter expanding excavation scheme should be determined according to the geological conditions and the construction requirements of the owner. In general, in a tunnel with small turns, the cutter head should have the function of expanding excavation to allow smooth turning. The expanding excavation scheme should be determined according to the length of expanding excavation. Three expanding excavation schemes are available: hydraulic mechanism, cushion block and replaceable wedge, as shown in Fig. 4.17.

② Cutter Head Block Design

Cutter head block design mainly depends on construction, transportation and processing technology. Construction factors include the lifting capacity of crane on the site. Transportation capability will limit the cutter head size and mass. Cutter

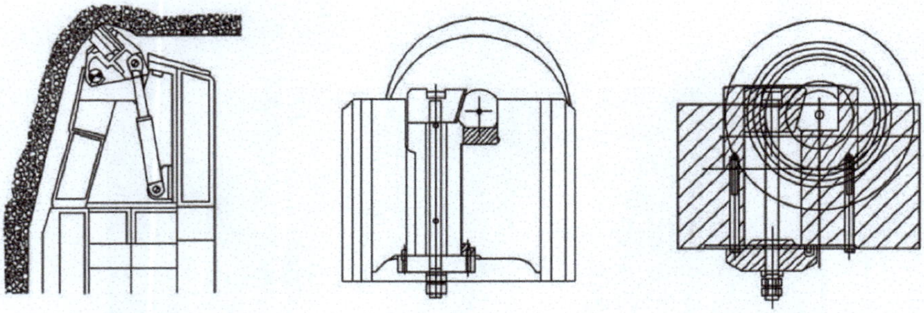

a) Hydraulic mechanism b) Cushion block c) Replaceable wedge

Fig. 4.17 Expanding excavation scheme

head blocks may be designed in 1 + 1, 1 + 2, 1 + 4 and 1 + 8, and may be adjusted according to the cutter arrangement.

③ Design of muck inlets

The design involves the quantity and radial dimensions of the cutter head muck inlets. The quantity, which generally corresponds to the diameter of the cutter head (1/m), shall be determined according to the excavation diameter. Radial dimensions of the muck inlets around the cutter head shall be determined according to the stacking height of the muck at the tunnel bottom. In accordance with their layout, the muck inlets are divided into front muck inlets which are mainly for mucking, and cone plate muck inlets which function as auxiliaries. In terms of their structures, the seats of the muck inlets are divided into integrated seats and modular seats. The integrated seats are more impact-resistant, whose structural strength is noticeably higher than that of the modular ones. As shown in Fig. 4.18, a grid is welded at the muck inlet which can limit the size of mucks entering the cutter head, and will crush large stones twice to prevent them from blocking the cutter head or damaging the belt.

④ Structure of central cutters

As shown in Fig. 4.19, the structures of central cutters can be roughly divided into the "–" shape and the " + " shape.

Central cutters of the "–" shape and the " + " shape have their advantages and disadvantages. The cutter ring of the "–" shaped central cutters is usually 17in, and the cutter spacing can be made smaller. The 1 + 1 block structure adopted in the cutter head requires a higher machining accuracy for the cutter seat whose quality is not guaranteed. Subject to the disc cutter structure and installation, the " + " shaped central cutter features a larger cutter spacing, with two cutter ring diameters of 17in and 19in. According to the use, central cutters of both sizes can meet the requirements of actual engineering construction. Besides, the installation of the "–" shaped central cutters, which is more complex than that of the " + " shaped central cutters, requires a higher process standard. Once the disc cutter is damaged, it is troublesome to replace it.

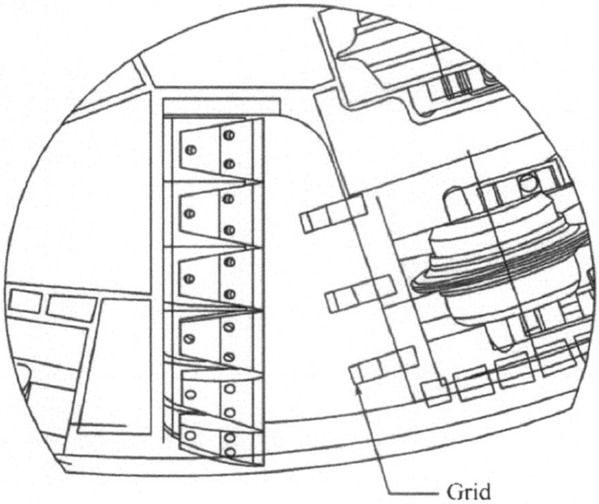

Fig. 4.18 Grid

Fig. 4.19 Structure of central cutters

⑤ Determining the cutter spacing and cutter diameter

Cutter spacing and cutter diameter are the key parameters in cutter head design. An appropriate cutter spacing plays a crucial role in the rock-breaking capability of the cutter head. Subject to the cutter head structure, the cutter spacing in the central area of the cutter head generally varies in a small range, and there is little change for different cutter heads. The cutter spacing in the positive cutter area of the cutter head shall be demonstrated and selected according to the engineering geological conditions. The selection of cutter spacing is affected by numerous geological parameters. At present, the main geological parameters for reference in the design are rock uniaxial

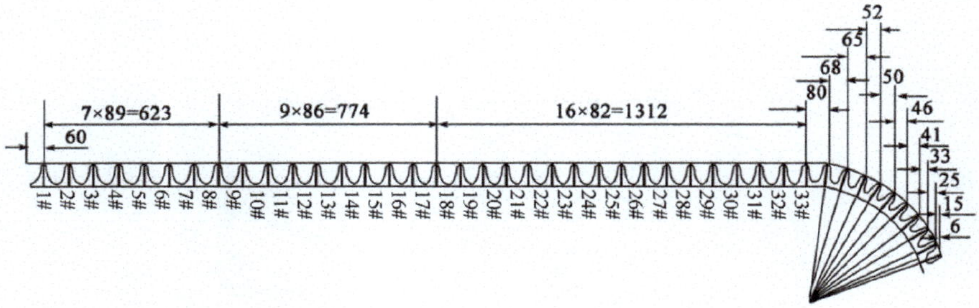

Fig. 4.20 Cutter spacing diagram (Dimension unit: mm)

compressive strength and rock integrity coefficient. Without a doubt, the selection of cutter spacing shall also be compared with that of previous similar projects, and the cutter spacing shall be adjusted appropriately according to the use of previous similar projects. The stress of the edge cutter of the cutter head is complex, therefore, under the condition that the spatial layout of the cutter head meets the requirements, the number of edge cutters shall be maximized to reduce the rock-breaking volume of a single edge cutter. The cutter spacing is shown in Fig. 4.20.

⑥ Water spray nozzles and manholes

Installed on the cutter head, the water spray system with rotary joints is to wash away the dust on the tunnel face and cool the cutter. The rear supporting water flows into the cutter head through the rotary joint of the cutter head water spray system and is distributed to each nozzle in the radial direction of the cutter head through the shunt valve block. The water spray nozzles shall be evenly arranged on the whole excavation surface of the cutter head to remove the dust removal cool the cutter. The pipeline layout between the rotary joints and the water spray nozzles of cutter heads shall be convenient for maintenance, and shall be equipped with sufficient wear-resistant protection structures. A manhole is arranged on the cutter head, through which people can enter the cutter head front to observe the tunnel face after shutdown.

(2) Detailed design of cutter heads

The detailed design of the cutter heads is also a very important part of the whole cutter head design process, mainly including determining the cutter head structures, cutter mounting, cutter seats, etc. After determining the cutter mounting and cutter seats of the cutter heads, the stiffened plates between the front face plate and the rear face plate shall be arranged, and the detailed design shall be optimized.

① Structures of cutter heads

Structures of cutter heads are divided into the curved and the plane. As the name suggests, the front face plate of the curved cutter head is curved, while that of the plane cutter head is plane. The curved cutter head structure extends the layout space where more cutters can be arranged, and the cutter spacing can be made smaller in the plane perpendicular to the TBM thrust. However, the lateral force on the curved cutter head will increase, which will shorten its service life.

The plane cutter head is commonly used at present. Its size in the tunnel axis direction is small, accordingly, the size of the exposed shield structure is scaled down, which reduces the disturbance to the surrounding rocks of the tunnel wall and prevents the cutter head from being stuck. Therefore, the plane cutter head is preferred when the geological conditions are not particularly good and landslides may occur.

② Cutter head mounting

Cutter head installation is divided into the front-mounted and the back-mounted. Personnel shall enter between the cutter head and the tunnel face to install the front-mounted cutters. In order to ensure the safety of construction personnel, only when the geological conditions are excellent can the front-mounted cutters be adopted. The back-mounted cutter head structure is a commonly used one. The cutter replacement is carried out inside the cutter head, which will ensure the safety of operators, as shown in Fig. 4.21.

③ Type of cutter seats

As the most complex structure on the cutter head, the cutter seat is also the most fault-prone. The use effect is subject to its strength index, hardness index and structure. Each manufacturer has its own structural style of the cutter seats, mainly including the inner and outer welding type, C-shaped block type and L-shaped block type, as shown in Fig. 4.22.

The inner and outer welding cutter seat is formed by welding the inner and the outer cutter seats together. After being manufactured, the inner and outer welding cutter seats are welded to the main structure of the cutter head. The inner and the

a) The front-mounted cutters b) The back-mounted cutters

Fig. 4.21 Cutter mounting

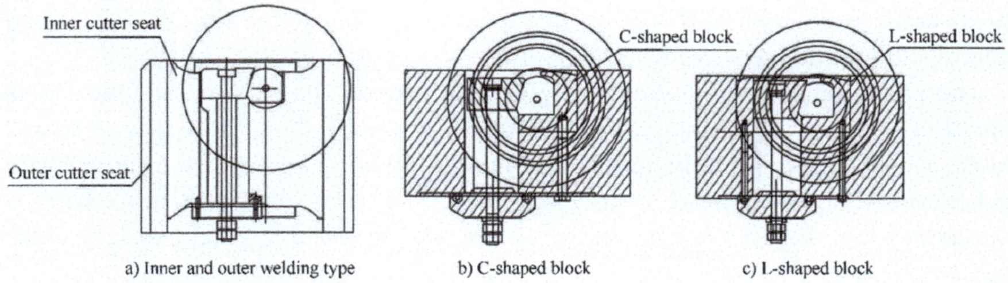

Fig. 4.22 Structures of cutter seats

outer cutter seats are made of special alloy steel and ordinary steel respectively and are connected by welding. The C-shaped block cutter seat is composed of a C-shaped block made of special alloy steel and a cutter seat base, which are connected by a bolt. The L-shaped block cutter seat is composed of an L-shaped block made of special alloy steel and a cutter seat base, which are connected by the left and right bolts.

The above-mentioned three types of tool holders have their own characteristics. The inner and outer welding cutter seats feature higher manufacturing requirements, a more complex heat-treatment process and a longer production cycle; the cutter seats feature an excellent overall stiffness and a good impact resistance if the weld quality is guaranteed, and the cutter bolts feature a firm connection and a sound effect. Cutter seat structures of the C-shaped blocks and L-shaped blocks are relatively complex but easy to manufacture; in terms of the stiffness and stability for the cutters, the C-shaped block cutter features a defective structure, and the contact surface between the C-shaped block and the cutter seat base is prone to impact deformation; initially, the contact surface between the L-shaped block and the base will also be deformed due to impact during the use of the L-shaped block cutter seat, but the assembly process adjustments give rise to higher overall stiffness and stability, which abates impact deformation of the contact surface.

④ Wear-resistant design

Wear-resistant designs are required on the cutter head face plates, rings and muck inlets, which is crucial to the service life of cutter heads. The clad wear protection plate is welded on the cutter head face plate, and the Hardox wear-resistant steel plate strip is welded in the circumferential direction of the cutter head ring. Wear-resistant nails or wear-resistant grids shall be welded at the muck inlet which is crucial to the wear-resistant design. The wear-resistant welding layer shall be buttwelded on the part where the cutter seat on the edge protrudes from the cutter head body to protect the cutter seat from wear.

⑤ Layout of cutter head stiffened plates and muck scoops

In order to endow the cutter heads with better overall strength and stiffness, appropriate stiffened plates functioning as a support for the cutter seat shall be arranged between the front face plate and the rear face plate of the cutter head according to the layout of the cutter seats; the muck scoops shall be reasonably arranged between the rear face plate of the cutter head and the rear cone ring for muck guiding, cutter head supporting and keeping the structural stiffness. The stiffened plates and cutters shall be arranged to make the cutter head weight uniform so that the eccentric load value is kept small during the rotation of the cutter head. The muck scoop layout shall consider the uniform circumferential distribution, the muck inlet positions and the interference with the cutter hole (for the back-mounted cutter head). During the main structure design of the cutter head, the operation space for welding flaw detection shall be considered when arranging the welds. The welds shall be arranged appropriately. On the premise of ensuring the cutter head quality, the number of cutter head welds shall be minimized to abate the welding deformation of the cutter head.

4.4.2 Cutter Structure Design and Materials

1. Cutter structure and parts

Figure 4.23 shows the face cutter or gauge cutter with a diameter of 19 in, and Fig. 4.24 shows the central cutter with a diameter of 19 in Fig. 4.25 shows the structural sections of the face disc cutter and the gauge disc cutter. As can be seen from the figures, the cutter is mainly composed of cutter ring, cutter body, cutter shaft, bearing, floating metal ring seal, retainer, cutter ring axial retaining ring, bearing adjusting nut, etc. The difference between the central cutter and the face cutter is that the face cutter has a cutter ring, a cutter body, a pair of bearings, two groups of seals and two retainers, while the central cutter has two cutter rings, two cutter bodies, two pairs of bearings, four groups of seals and three retainers (including the middle retainers). In addition, the wedge bolts and central cutter are used to position and fix the cutter shaft and the retainer respectively when installing the face cutter and gauge cutter on the cutter head.

Disc cutter technology is one of the core technologies of TBM, which requires high bearing capacity, good wear resistance, red hardness and good impact toughness.

Fig. 4.23 Face cutters and gauge cutters

Fig. 4.24 Central cutters

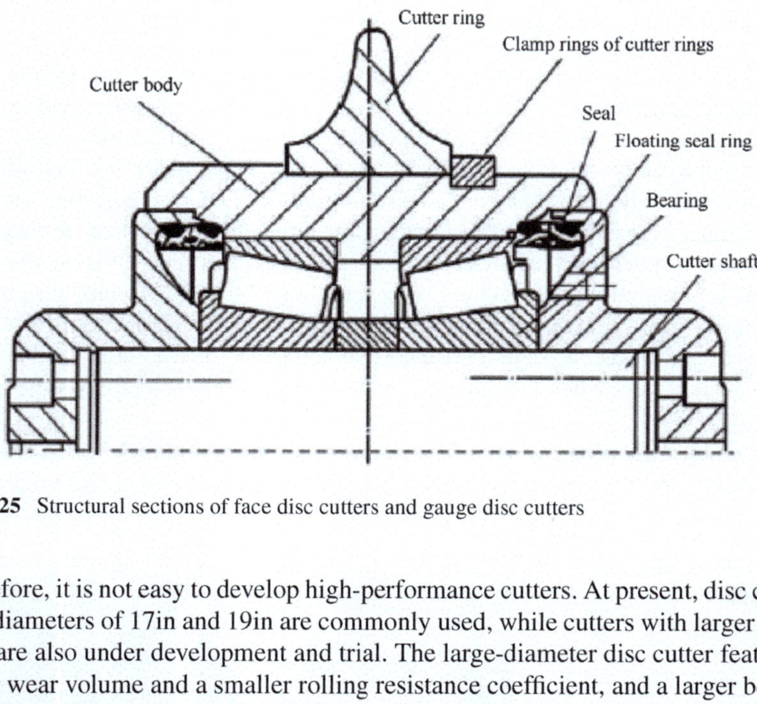

Fig. 4.25 Structural sections of face disc cutters and gauge disc cutters

Therefore, it is not easy to develop high-performance cutters. At present, disc cutters with diameters of 17in and 19in are commonly used, while cutters with larger diameters are also under development and trial. The large-diameter disc cutter features a larger wear volume and a smaller rolling resistance coefficient, and a larger bearing can be used to improve its bearing capacity. Among the disc cutter parts, cutter rings, cutter bearings and metal floating ring seals are the key parts. The material composition and heat treatment process of the cutter ring are the core technologies, which are difficult to control. TIMKEN and SKF tapered roller bearings are commonly used as the cutter bearings, and special high-temperature resistant grease or lubricating oil must be injected into the bearing cavity. Metal floating ring seal is the key to the normal operation of the cutter bearing. The cutter is often subject to uneven wear due to seal failure, therefore, the material and processing technology of metal floating ring seal also demands stringent specification. Generally, the cutter shaft, cutter body and retainer are not prone to damage and can be reused after repair minor damages. However, if the cutter is unevenly worn and cannot be stopped promptly, the cutter body may also be unevenly worn and scrapped. The cutter ring, bearing and seal are consumables. The cutter ring shall be replaced after wear. The cutter ring shall be replaced in case of uneven wear or large cracks. The bearing can be reused after the cutter is disassembled. The seal is generally disposable, or reused after inspection and grinding.

In addition to the above three key parts, the other three design and assembly technologies are also crucial to the high performance of the cutters. First, the fit between the cutter ring and the cutter body shall adopt interference fit, and the cutter ring must be heated during installation; Second, the tapered roller bearing of the cutter adopts the pre-tightening assembly design. During installation, it shall be pre-tightened through the adjusting nut on the cutter shaft to eliminate the bearing clearance and generate a certain elastic deformation. The pre-tightening degree is determined by measuring the rotating torque of the cutter during assembly; Third, after the cutter assembly, the sealing test shall be carried out. Pump air into the cutter bearing cavity and then maintain the pressure for a period to observe whether the pressure is stable, so as to determine whether the seal has leakage.

The cutter inspection is divided into two parts. One part is the cutter inspection, installation and disassembly on the cutter head inside the tunnel, while the other is the cutter disassembly, repair and assembly in the cutter maintenance workshop outside the tunnel. The new cutter or repaired cutter assembled in the cutter workshop outside the tunnel shall be transported into the tunnel and installed on the cutter seat by the inspection operators in the tunnel. Predicting and reserving the consumption of cutter spare parts is an important task of TBM construction. The timely supply of cutter spare parts shall be ensured to avoid delaying the tunneling.

2. Cutter ring materials

In order to increase the penetration and advance speed of cutters, the cutter ring material has been continuously studied and developed. The biggest difference between the cutters produced by different manufacturers is the material quality, composition, heat treatment process and geometry of the cutter ring section. It is difficult to identify the best balance between wear resistance and toughness of specially hardened alloys. In terms of the cutter ring geometry, its outline has changed from the wedge shape to quasi equal thickness, which can keep the contact area between the cutter ring and rocks a constant in a long wear period, therefore, it will not lead to the increase of required thrust or the decrease of advance speed.

The cutter ring material is one of the core technologies of TBM manufacturers, which is strictly confidential. Accordingly, there is little published research. The cutter ring manufactured by a foreign company is tested to preliminarily explain its material and mechanical properties. The results and analysis are as follows.

(1) Chemical composition

Table 4.1 Natural rock fracture angle

Rock type	Soft sandstone	Hard sandstone	Basalt	Fine grained granite	Hard quartzite
Breaking angle (°)	130	144	146	140	150

Table 4.2 TBM cutter ring material composition (%)

Chemical composition	C	Si	Mn	P	S	Cr	Mo	V
Mass fraction	0.41	1.08	0.33	0.011	0.007	5.22	1.45	0.97

Chemical composition analysis is shown in Table 4.2. Besides C, there are Cr, Mo and V(Si, Mn, S and P are permanent impurity elements) in the steel for TBM cutter ring, which belongs to hot working die steel. Cr and Mo, known as alloying elements, can improve the hardenability, high temperature strength and wear resistance of steel; Alloying element V, due to the high solubility of carbide VC in steel, begins to dissolve noticeably above 1200 °C, therefore, in addition to increasing the hardenability of steel when partially dissolved, it can also improve the coarsening temperature of austenite, generating a fine grain microstructure which is beneficial to the strength and toughness of the material. Besides, martensite containing alloying elements V and Mo also features high tempering softening resistance, which improves the red hardness of the steel.

(2) Tensile and impact tests

Conduct tensile and impact tests on the cutter ring material samples. See Table 4.3 for the test results. In accordance with Table 4.2, the cutter ring material is a medium carbon and medium alloy high-strength steel, with an average tensile strength $R_m > 2000$ MPa; simultaneously, it also ensures good impact toughness (with an average impact energy $A_{Ku} > 9$ J), indicating that the material features excellent comprehensive mechanical properties.

(3) Hardness test

The surface hardness test results of the cutter ring end face are 56.8, 56.2, 57, 57.7 and 56HRC, ranging from 56 to 57.7 HRC. Cross-section hardness uniformity test: take the full section of the cutter ring sample, test one spot at different parts from the surface to the core or at the core every 3 ~ 4 mm, test 5 spots at each part, and the hardness range of the whole section is 54.5 ~ 57HRC.

Table 4.3 Tensile and impact test results of TBM cutter ring

Sample No	R_m(MPa)	A_{Ku}(J)
1	2058.65	9.8
2	1995.92	9.5
3	2026.88	8.7

(4) Microstructure

Observe the prepared metallographic sample and transmission electron microscope (TEM) sample of the cutter ring on PMG-3 inverted optical metallographic microscope and JEM-2100 high-resolution transmission electron microscope respectively, and collect the electronic pictures. The results are shown in Figs. 4.26 and 4.27 respectively. The TBM cutter ring microstructure under the optical microscope is band-shaped or bundle-shaped, but its outline is vague and the details are difficult to distinguish. Through TEM, it can be clearly seen that the organizational unit is plate-like martensite, and a large number of dislocations are accumulated in the plate; The microstructure is uniform and basically retains the quenched martensite morphology, indicating that the material for the tool ring has high tempering stability; Twin martensite can be seen in some areas; There is no large undissolved carbide in the structure, indicating that the carbide is almost completely dissolved at a higher quenching temperature. According to the chemical composition, the material should be tempered in a high temperature. The tempered martensite structure of disc cutters abroad and the fine and evenly distributed granular carbides ensure good comprehensive mechanical properties.

(5) Fracture scanning

The cleaned fracture sample is observed on JSM-5610LV scanning electron microscope (SEM), and the results are shown in Fig. 4.28. From the fracture scanning results, it can be seen that the dimple edge is similar to a sharp one, the estimated brightness is large, the dimple bottom is relatively flat, and the image brightness is low. Some dimples have small second phase particles in the center. Due to the small particle size, the incident electron beam can excite more secondary electrons on its surface, which explains why this kind of particle is relatively bright. No matter

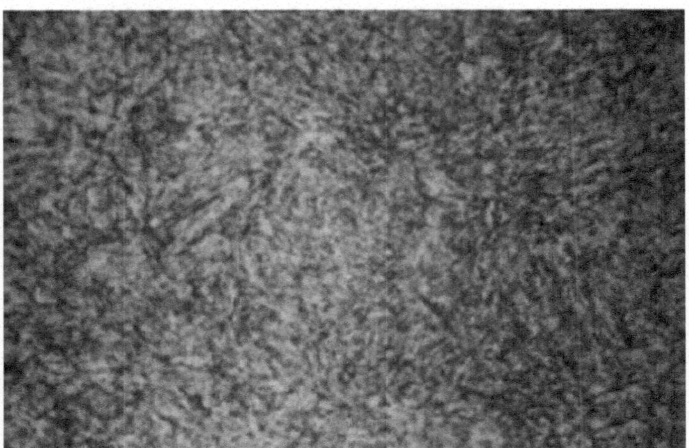

Fig. 4.26 Metallographic structure of the cutter ring (1000 times)

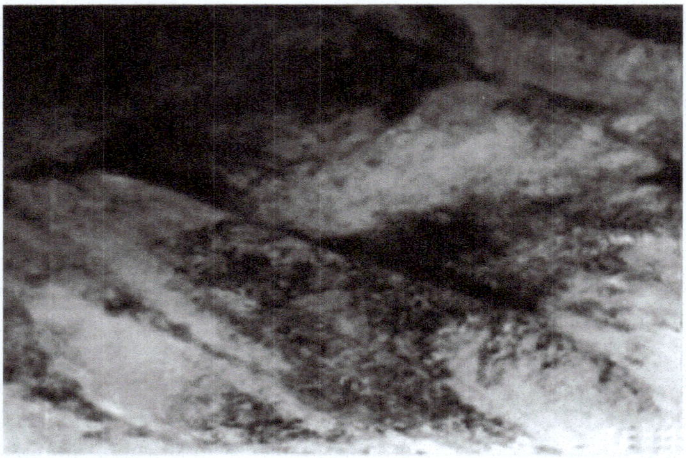

Fig. 4.27 TEM photos of cutter rings

from the macroscopic deformation of the sample or from the microscopic region of the fracture, noticeable plastic deformation can be seen. The second phase particles generally exist at the bottom of the dimple, because the interface between the second phase particles and the matrix cracks first to form a crack (dimple) source during shear deformation. With the increase of stress and deformity, the dimple is gradually torn, and protrusion tearing edges with great plasticity are formed around the dimple. The dimple fracture is transgranular toughness fracture.

(6) Heat treatment process

According to the previous research and foreign information, we recommend the heat treatment process as illustrated in Fig. 4.29. The objective of further research is to develop a high-quality cutter ring through numerous tests and adjustments based on the results of material analysis and heat treatment process curve. Due to the large size of the disc cutter ring, step heating is adopted to ensure complete heat penetration;

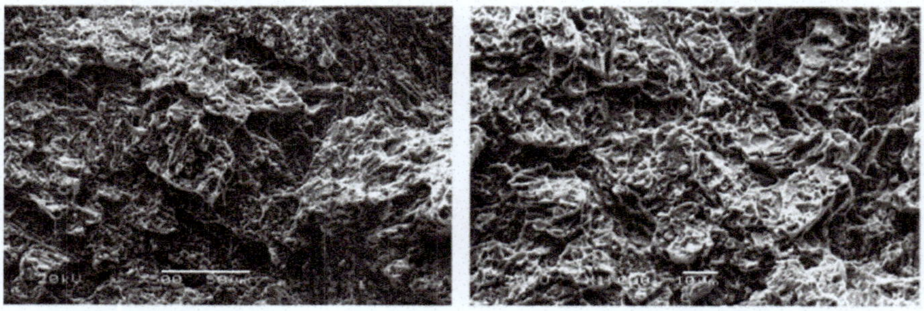

Fig. 4.28 Port scanning results

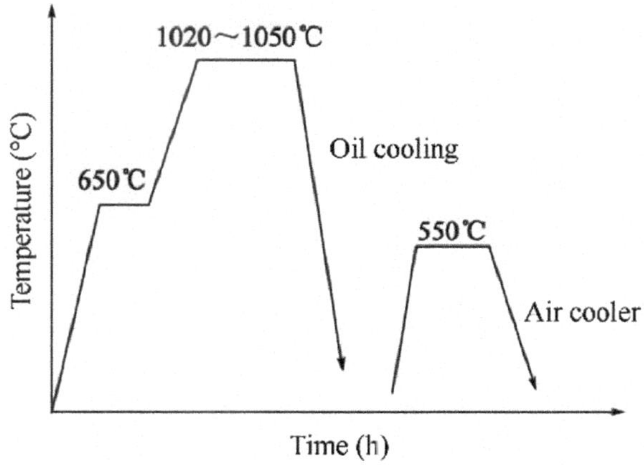

Fig. 4.29 Cutter ring processing technology

The cutter ring material contains lots of alloying elements that can be fully dissolved by heating at a temperature of 1020 ~ 1050 °C. The quenching heating shall be conducted in a controllable atmosphere furnace or salt bath furnace. Oil cooling can abate the deformation and cracks caused by quenching. After heat treatment, the microstructure shall be tempered martensite and a small amount of fine and uniformly distributed granular carbide with a hardness of 54.5 ~ 57HRC.

Questions

1. Briefly explain the mechanism of rock-breaking by TBM disc cutters.
2. Briefly explain the main factors affecting the rock-breaking by TBM disc cutters.
3. Briefly explain the area and characteristics of the TBM disc cutter layout.
4. Briefly explain the principles of the TBM disc cutter layout design.
5. Briefly explain the key points of the TBM cutter head structure design.

Chapter 5
TBM Project Construction Organization

Key points of this chapter: this chapter first analyzes the TBM construction system from the aspects of tunneling and mucking, support operation, transportation operation and energy supply, then analyzes the main TBM construction process organization from the aspects of TBM construction organization, construction preparation, testing and commissioning, tunneling operation, and finally introduces the quality and safety management of TBM.

5.1 Analysis of TBM Construction System

The fundamental objective of tunnel construction is to obtain the best progress, construction period and construction quality at an appropriate cost. To achieve this, each construction process of TBM and the relationship between them must be studied.

The advancing process of TBM is the tunneling process of construction. "Tunneling → Mucking → Support → Tunneling" is the fundamental operation flow of tunnel construction. The guarantee system of construction consists of direct and indirect parts, that is, the organization and transportation of materials and the maintenance of TBM.

The fundamental process and relationship of TBM construction are illustrated in Fig. 5.1. As can be seen from the figure, tunnel construction is based on tunneling, support and mucking, and supported by the power supply, air and water systems. The environmental protection system shall meet the basic environment required by the construction, and the monitoring system is used to ensure the smooth progress of automatic (production) construction.

K. Chen et al., *TBM Design and Construction*,
https://doi.org/10.1007/978-981-99-0059-6_5

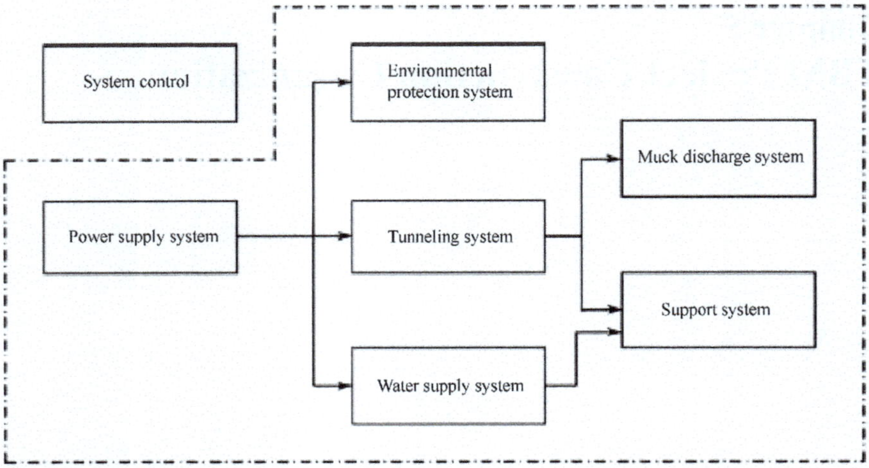

Fig. 5.1 TBM construction basic engineering and relations

5.1.1 Tunneling and Mucking System

TBM tunneling is the process of cutter head advancing → rotating → cutting. The cut-off rocks enter the cavity of the cutter head through the bucket and scraper on the cutter head, and then are transported out through the belt conveyor.

The operation process of the tunneling system and transportation system is illustrated in Fig. 5.2. This operation is the core process of TBM, which determines the advance speed of the project, the coherence of corresponding supporting processes and operation quality requirements.

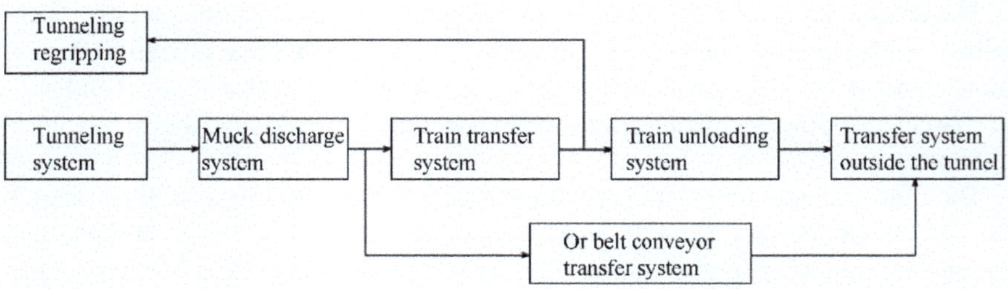

Fig. 5.2 Operation process of tunneling system and transportation system

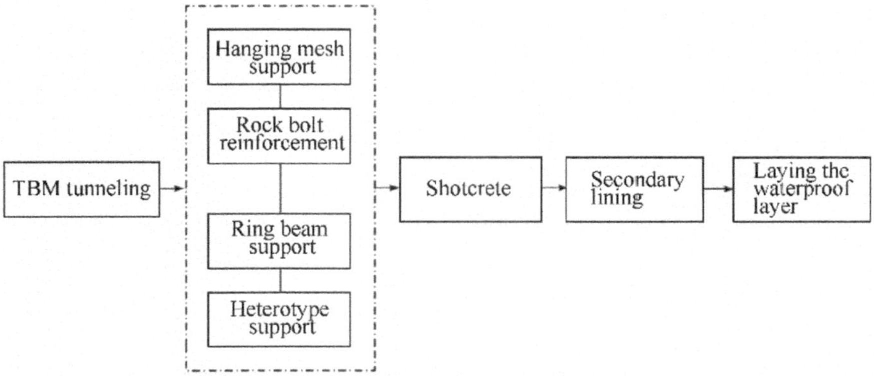

Fig. 5.3 Tunneling and supporting operation process

5.1.2 Tunneling Support Operation System

During tunnel construction, various problems involving stratum stability will inevitably be encountered and shall be solved. The stability of surrounding rocks is not only closely related to the safety of personnel and equipment, but also related to the success of the project.

The supporting mode is subject to engineering hydrogeological conditions. Its supporting structure, parameters and measures are also subject to the geological conditions. In addition to meeting the engineering quality requirements and corresponding equipment and manual operation quality requirements, the support shall also be planned and designed according to the advance speed and time sequence. It is of great importance to avoid or reduce operation interference and waiting time for support, and improve the proportion of pure tunneling time and project progress.

Tunneling support includes hanging mesh support, rock bolt reinforcement, ring beam support and other supports. Among them, other support modes include single-layer simple grid (steel slat) processed with curved shape channel steel and reinforcement, insert plate grouting, backfilling and reinforcement, etc.

See Fig. 5.3 for the tunneling and supporting operation process.

5.1.3 Transportation System

It mainly consists of the mucking transfer and the transportation of materials such as invert blocks. Materials mainly include engineering materials and equipment maintenance materials. Figure 5.4 shows the relationship between tunneling and transportation.

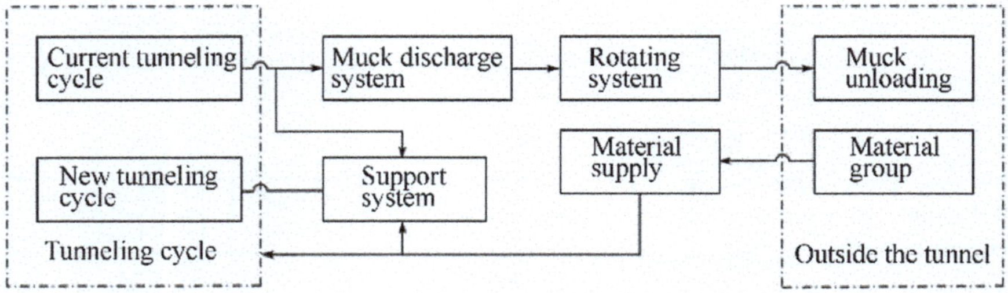

Fig. 5.4 Tunneling and supporting operation process

5.1.4 Energy Supply and Labor Environment System

Electric power is an important energy source of TBM. Therefore, the TBM system operation is subject to the quality of the power supply and distribution system. The power distribution shall be set according to the power distribution standard of electric equipment and the operating condition requirements of the tunneling system.

Wind, water and electricity are the fundamental energy consumed in both tunnel construction and TBM construction. Therefore, the supply of wind, water and electricity is the fundamental condition for normal and effective construction.

Tunnel construction may generate dust, harmful gas, noise, humidity, geothermy, rock burst, etc. which are harmful to human body, equipment and surrounding environment. Therefore, labor environment protection is necessary to avoid injuries and equipment failure.

TBM industrialized construction and production management is reflected in coordinating and organizing the operation sequence of the macro system and controlling the time and quality of a single micro-operation. System operation and process management include monitoring of hardware system, construction organization of production system, operation organization of guarantee system, troubleshooting for abnormalities, etc. The efficiency of traditional drill and blast is subject to the cycle time of the whole process, while that of TBM construction is subject to the operating time of the machine. Figure 5.5 shows the relationship of TBM construction systems.

5.2 Organization of Major Construction Processes

5.2.1 Construction Organization

The practical construction organization design shall make preparations according to the construction period requirements and the bid winning cost, arrange the construction period, construction method, TBM selection and supporting equipment, organize the mobilization, transportation, installation, commissioning and tunneling

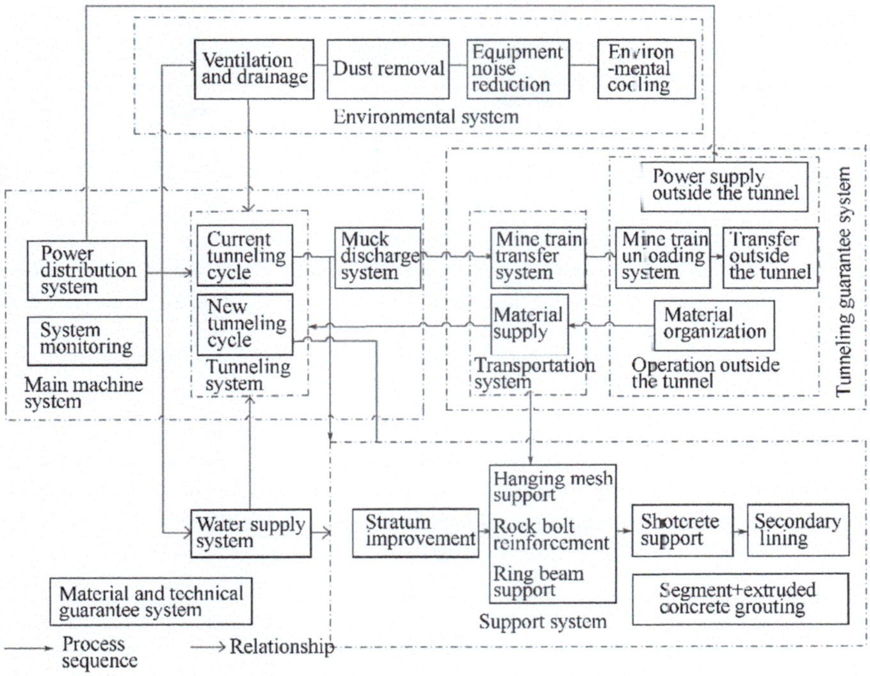

Fig. 5.5 Schematic diagram of TBM construction system relationship

preparation of TBM large pieces, and prepare the requirements of personnel, materials and machines and other tools. When preparing the construction organization schedule, planned progress, tunneling indicators, personnel dynamics, etc., the design scheme and emergency plan of TBM passing through special geological sections, site and hydropower conditions, progress indicators, construction management, quality, safety, environmental protection, etc. shall be taken into account.

(1) Support for adverse ground conditions

Due to uncertainties of tunnel geological conditions and different geologies, the capacity of TBM construction to tackle adverse ground conditions is crucial to the construction safety and period. Potential schemes, means and emergency plans such as advance drilling shall be formulated for adverse ground conditions during the construction organization design.

(2) Key technologies and difficulties in construction

Find out the potential key technologies and difficulties, conduct analysis and research, and formulate countermeasures in the light of the construction characteristics, geological conditions, construction period requirements and equipment conditions. Strengthen the advance geological prediction, compare and analyze the confined water, collapse, fault, rockburst, karst cave, etc. with the existing geological data, summarize the difficulties in construction, carry out targeted prevention and formulate corresponding solutions.

(3) TBM construction supporting equipment

The supporting equipment for TBM construction shall match the capacity and progress indexes of the TBM main machine. In order to give full play to the advantages of TBM and ensure the completion of project construction, the production capacity of supporting equipment shall be appropriately expanded. The calculated quantity of equipment shall meet the requirements of normal construction progress and its potential expansion.

(4) Main temporary facilities

The main temporary facilities include construction roads, mucking dump, substation, material factory, invert prefabrication factory, site hardening, maintenance of construction production equipment, office, staff living facilities, etc. In particular, perform well in drainage and arrange flood control facilities according to the hydrometeorological investigation to ensure safety during the flood period.

(5) Wind, water and electricity for construction

The electricity for TBM construction shall be supplied from the substation to the tunnel according to the reliability requirements of class I load, and 10 kV or 20 kV armored high-voltage cables shall be adopted. TBM transformer capacity, fan power, production and domestic power consumption of concrete mixing plant and invert prefabrication, as well as power factors, utilization factors and line loss, shall be taken into consideration when determining the capacity of the portal substation. The standby generator is used to supply power to the portal fan, concrete production, lighting and off-site living facilities when necessary.

Factors such as dust removal and cooling of cutter heads, cooling of bolt drillers, ejector and equipment shall be considered for TBM water consumption.

Tunnel ventilation is an important link in construction. In addition to the press-in fans set at the portal according to the requirements of air flow rate and air volume, there shall also be relay fans and dedusting systems at the TBM end in the TBM ventilation system.

(6) Construction site layout

The layout of construction sites outside the TBM tunnel shall consider a large enough site to meet the assembly requirements. The site shall be hardened and can bear TBM gravity to assemble medium and large gantry cranes. TBM construction is a fast operation process integrating excavation, transportation and support. The site shall be equipped with a complete transportation system, and the off-site production facilities shall meet the TBM progress requirements in different construction periods.

5.2.2 Construction Preparation

As an industrialized operation mode and one of the most advanced tunnel mechanized construction modes in the world, TBM has reached a higher level in construction progress, safety, environment and quality. However, the construction is subject to the scope of the construction site, the surrounding environment and the coherence of adjacent projects, which must be investigated and improved to meet the appropriate operating conditions. Site investigation shall be conducted before tunnel construction according to the characteristics and contents of TBM construction to keep abreast of its construction conditions, construction scope, local traffic, communication and material supply.

Engineering geological conditions have a great impact on TBM advance speed and quality, accordingly, relevant drawings, documents and geological data shall be carefully checked, and technical requirements, support methods, quality acceptance requirements and relevant technical descriptions shall be fully mastered before construction. The following data must be mastered before tunnel construction: engineering geology and hydrogeological exploration report; local meteorological, hydrological and water quality conditions; project construction contract documents, subcontract documents and supervision contract documents; design drawings and engineering technical descriptions for construction; investigation of terrain and facilities of transport road from the arrival port (bank) to the construction site of TBM.

Before TBM construction, the following main works should be completed: check the location of the portal and the coordinates of the entrance; Determine the setting-out accuracy of the portal and the elevation and coordinates after positioning; Assembly, commissioning and acceptance of TBM; Preparation of precast segments/invert blocks; Various statements for TBM construction; Connection of supporting projects; Construction organization scheme for transportation of TBM equipment components.

TBM construction operators should be equipped with complete expertise, meet the construction requirements, and they should receive professional training and work with certificates. In view of various adverse ground conditions in TBM construction, technicians shall work out detailed operation procedures and quality control requirements, so that the operation documents are issued to each operator to clarify the quality and safety standards of construction. Vocational skills should be learned in various forms such as theoretical training, on-site operation training and learning from other units, etc. Every employee is required to receive pre-job training, and work with certificates after passing the examination, so as to improve operation quality. It is strictly forbidden to work without certificates.

1. Geological investigation

Geological conditions have a great impact on TBM construction. Detailed and reliable geological and hydrological data serve as the basic conditions for the success of

TBM project and directly determine the success or failure of the project. Hydrological data determine whether it is feasible to adopt TBM, the type selection of TBM, the main parameters of TBM, the selection of auxiliary construction equipment and the formulation of emergency plan.

In order to fully understand and master the geology, rock mass characteristics and groundwater conditions along the tunnel, it is necessary to carry out sufficient geological investigation, especially the investigation into unfavorable strata, which not only affects the selection of the basic form of TBM, but also the main factor determining whether the TBM construction method can be adapted to.

The geological conditions affecting TBM adaptation mainly include: the nature, scale, distribution range, main crushed materials, fragmentation degree, water storage capacity, expansive surrounding rock, etc. of the main faults and weak fracture zones that the tunnel passes through. The hydrogeological conditions of the tunnel shall be mastered, the groundwater type and the source of replenishment shall be determined, and the sectional water inflow and possible maximum water inflow of the tunnel body shall be predicted, as well as the possible serious water inrush and water inflow points (sections). In karst areas, the scope, depth and scale of karst development in the tunnel area, and whether there is a risk of karst water or fillings suddenly gushing out should be ascertained to determine whether the TBM can pass through safely.

2. Technical Preparation

Before tunneling construction, technicians should be familiar with and recheck the design documents and construction drawings, know well with relevant technical standards, technical conditions, design principles and design specifications, and master the basic principles, assembly sequence, operation procedures and maintenance procedures of TBM and ancillary equipment.

During tunneling construction, the operative construction organization design should be compiled according to the overview, hydrogeological conditions, quality control, construction period requirements, and resource allocation of the project, the construction scheme should be demonstrated and optimized, and approved according to relevant procedures. Operative construction organization design is a technical document that directly guides the construction. It is compiled after fully investigating the project site, being familiar with the engineering design drawings and further understanding the geological data. It is different from the compilation content of the bidding period. Besides meeting the construction period requirements of the project, it also needs to meet the investment plan and the requirements of environmental protection and safety, so as to make the tunnel construction balanced and orderly.

(1) Construction investigation

Construction investigation should be conducted before and after winning the bid. The investigation contents are roughly the same, but the depth and focus are different. The post-bid investigation shall take into account the specific work arrangements, such as: TBM transportation scheme, road conditions for over-size transport and safety measures; Whether the water source meets the construction needs and the power supply; TBM storage after the arrival of the site, concrete mixing, inverted arch prefabrication, etc.; The location and capacity of the slag field; Local material supply, hydrological and meteorological data, communication, environmental protection and verification of mobilization design documents are all contents of construction investigation.

(2) Verification of design documents

The process of verifying design documents is also the process of mastering and digesting documents. The purpose of verifying documents is to prevent errors or shortages of documents. Verification includes engineering construction contract documents, subcontract documents, supervision contract documents, relevant technical standards, main technical conditions, data of vertical and horizontal sections of the tunnel, survey data, as well as design scheme, construction methods, technical measures and construction period when passing through adverse sections, etc.

(3) Compilation of operative construction organization design

The operative construction organization design is supposed to reflect the progressiveness, cost-effectiveness and reliability. The following principles should be followed in the compilation of operative construction organization design:

① Meet the contract requirements;
② On the basis of detailed investigation, technical schemes should be compared and selected, and the best scheme should be selected for compilation.
③ We should improve the construction technology and actively use new technologies and processes;
④ According to local conditions and materials, the construction should meet the requirements of environmental protection.

The compilation of practical construction organization design should meet the contract requirements. On the basis of detailed investigation, technical schemes should be compared and selected, and the best scheme should be selected for compilation. We should improve the construction technology and actively use new technologies and processes. According to local conditions and materials, the construction should meet the requirements of environmental protection and meet the technical characteristics of TBM construction.

In addition to being advanced, economical and feasible, the principle of implementing construction organization design should also emphasize the importance of environmental protection, safe production and occupational health, and should list the detailed measures for environmental protection and safe production in detail.

The main contents of the implementation construction organization design should include: project overview and technical standards, geological conditions and analysis of major difficulties, TBM assembly, commissioning, tunneling and dismantling construction organization, layout of construction site and temporary facilities, construction progress index and schedule, organization and resource allocation, measures to guarantee quality protect environmental, TBM construction treatment measures and emergency plan for complex geological conditions, main research contents of key and difficult project technologies and problems.

The implementation construction organization design is more specific and accurate than the guidance construction organization design, and it is more suitable for the actual construction. The key point is to do a good job in the construction investigation and check the design documents. If there is any variation in the construction conditions, it should be adjusted in time.

Before construction, it is necessary to formulate detailed rules for the implementation of the construction procedures, prepare work instructions and complete the technical disclosure of key procedures. After checking the technical standards used in the design, the construction unit should study the relevant construction specifications and regulations to be implemented in the construction according to the specialties involved in the project construction, and at the same time, formulate practical implementation rules for on-site construction procedures, prepare work instructions and complete technical disclosure of key procedures.

(4) Personnel Training

Training all the personnel participating in TBM construction is the main basic work, and the allocation and training of technical force is one of the tasks in the construction preparation stage. Personnel in related positions can be familiar with the structural principle of equipment, master the performance of equipment, and operate and maintain the equipment correctly through training. Before construction, it is necessary to formulate detailed rules for process implementation, work instructions, complete technical disclosure of key processes and conduct pre-job technical training.

3. Jobsite Preparation

The layout of construction site shall be designed according to the following requirements: It is beneficial to production and civilized construction, save land and protect the environment; It should realize overall planning and phased arrangement to facilitate orderly construction activities and avoid mutual interference; It should ensure the needs of excavation, mucking, lining, transfer and shunting, and meet the assembly and initial conditions of the equipment. TBM is a multi-link joint operation system, which includes rock breaking, muck loading, segment/invert blocks making, material supply, shunting mechanism and some auxiliary measures. In order to meet the assembly and initial operation conditions of the huge system, a large site is needed. Temporary settings should have the capability to prevent disaster.

The temporary construction layout of the construction site includes determine the location and scope of the muck dump and the arrangement of mucking line, material preparation line, marshalling line and other operation lines during rail transportation, the layout of automobile transportation roads and other transportation facilities, the assembly and spare parts storage site of TBM, the location of wind, water and electricity facilities, the location of segment/invert blocks prefabrication factory, the storage or stacking site of materials such as sand, stone, cement and other machinery materials, mechanical equipment accessories, the location of various houses for production and living and the layout of water supply and dewatering system in the site.

Segment/invert blocks processing area should be equipped with hoisting transportation line, mold workshop, mixing station, steel bar processing workshop, steam curing and curing pool, and outdoor storage site for segment/invert blocks to meet the requirements of rapid production schedule and storage quantity of segment/invert blocks.

The waste muck field should meet the requirements of environmental protection, and the waste muck should not clog the groove and squeeze the river channel, and the slope foot of the waste muck field should be protected by gravity retaining wall.

TBM is bulky and can only be shipped in the form of bulk containers and can be transferred from the port or assembly workshop to the construction site. There are many types and quantities of parts (components). Therefore, a large enough temporary parking (assembly) site must be set at the entrance of the tunnel, and the assembly of bulk parts must be near the entrance of the tunnel. After the TBM assembly and commissioning is completed, it will enter the main tunnel along the track stepping for construction. The assembly site is divided into two parts: main machine equipment and back-up equipment. The dimension and placement of these large pieces should be considered in the layout of the assembly site. The site should be hardened with concrete, and the strength should meet the requirements of bearing capacity. Adverse ground condition should be avoided as much as possible. Support and drainage measures should be taken to ensure the implementation safety of TBM method. The assembly site should be located near the mouth of the tunnel, and the site should be hardened with concrete, and the strength should meet the bearing capacity requirements.

(1) Outside site

The main supporting facilities outside of TBM construction include concrete mixing station, invert prefabrication yard, repair workshop, spare parts and materials warehouse, substation, muck transportation and muck dumping system outside, roads, assembly site, etc. In order to meet the assembly and initial operation conditions of the huge system, a larger site is needed. Temporary facilities supporting TBM must also meet the requirements of TBM construction in different periods. Scientific overall arrangement is the premise to ensure the smooth construction of TBM.

For example, for the site layout outside the entrance of Qinling Tunnel Line I (entrance), the key considerations are: The design capacity shall meet the requirements of the highest efficiency of TBM and the construction progress index of TBM (the maximum daily advance rate is 45 m, the average advance per month is 350 m). Temporary facilities shall meet the requirements of tunneling and lining for repeated use in different worksites in the future, and realize serialization and standardization. We shall combine permanent facilities with temporary ones to reduce waste. Temporary facilities shall be able to prevent disasters properly. Based on topographical conditions, we shall take up as little cultivated land as possible and protect land resources, which fully practices the environmental protection.

According to the characteristics and requirements of TBM construction, the arrangement of temporary facilities outside the tunnel at the entrance of Qinling Tunnel Line I is mainly based on the transportation track, which is correspondingly arranged along the transportation line in strips according to the sequence and transportation mode of the muck train. The facilities are set to meet the needs of TBM construction (mucking, feeding and maintenance), to ensure that the operation of each process is coordinated, reasonable and orderly, and the tunneling of TBM is continuous and rapid.

TBM components are heavy, bulky, resulting in higher requirements for transportation and hoisting. By controlling the length of the assembly site, we can meet the requirements of rapid assembly and initial operation. The length of the assembly site is at least equal to the length of TBM + connection bridge + the first gantry + the towing cylinder + the storage length of the main bearing + the storage length of the cutter + the maneuvering length, and the width shall at least meet the requirements of TBM assembly and the access of large pieces.

The width of the entrance site of TB880ETBM shall be greater than 26 m, and the minimum length shall not be less than 70 m, so as to meet the requirements of transportation and quick assembly. In addition to the requirements of length and width, the ground of assembly sites shall be compacted and leveled to meet the requirements of safe operation of 75t $\times$ 2 + 15t gantry crane and TBM stepping.

(2) Preparation tunnel and launching tunnel

Because of the thin overburden (30 ~ 40 m) at the tunnel entrance, the stone may be weathered, which is not suitable for TBM construction. In order to put TBM into normal construction as soon as possible, drill and blast method is taken to excavate to the tunnel section with good surrounding rock, which is called the preparation tunnel. The preparatory tunnel at the north entrance of Qinling tunnel is 300 m long and the south one is 190 m long.

Launching tunnel refers that when TBM steps to the working face to start tunneling, the gripper shall tightly support against the wall during TBM tunneling to overcome the reactive torque and reverse thrust when breaking rock with the cutter. A certain length of launching tunnel is built to serve as the auxiliary chamber for TBM to launch and tunnel. The length of the launching tunnel depends on the structural dimension of the selected main machine.

The preparation tunnel depends on the assembly site and the assembly conditions. For the assembly site which can meet the assembly conditions of main machine and back-up equipment, i.e., the length of the site is sufficient enough to accommodate the length of main machine and back-up equipment, the construction of the preparation tunnel can be cancelled, and only construct the TBM launching tunnel. For the assembly site which cannot meet the assembly conditions of tunneling main machine and back-up equipment, i.e., the assembly site does not have enough length, which can only accommodate the length of main machine or part of the back-up equipment, the preparatory tunnel and the launching tunnel must be prepared at the same time in advance to leave enough site for assembling back-up equipment outside the tunnel. The construction length of the preparation tunnel depends on the needs of assembly.

In the launching tunnel where process changes from stepping to evacuating, so it is not only required to ensure the smooth passage of TBM, so that the gripper of the cutter head can enter the launching tunnel, but also to ensure that the gripper can support against the tunnel wall, and the concrete wall after lining has enough strength, which can provide enough counterforce for TBM advance and promote TBM cutter head advance. Therefore, the requirements for the launching tunnel are relatively high.

When assembling TBM, the requirements for the ground of the assembly site and the preparation tunnel are relatively high, that is, to ensure the ground even, and the relative height difference shall be controlled within 3 mm (for the area where the stepping mechanism passes). Because the layout of the assembly site, especially the stepping mechanism, is the basis of the whole assembly, the unevenness of the ground in the preparation tunnel will have a great impact on the stepping of TBM, causing the connecting bolt of the stepping device on the lower shield to be cut off.

The ground of the assembly site and the preparation tunnel shall be of sufficient strength, otherwise the ground will crack or sink during assembly and stepping, which shall be highly valued.

(3) Design and construction of temporary power (taking the export section as an example)

The temporary power required for the tunnel construction of Qinling Line I is provided by 35/10 kV substation, which is equipped with two transformers with capacity of 5000 kV·A each, and the two transformers operate in parallel. After high-voltage distribution, 10 kV voltage on the secondary side of the transformer is divided into three circuits to be fed out, one of which is TBM dedicated circuit, and the other two ways are connected to the switching station of the tunnel, and the voltage are supplied to other auxiliary equipment outside the tunnel after distribution. Therefore, the design and construction of TBM supporting power facilities can be divided into two parts:

① Design and construction of TBM dedicated circuit. After being fed out by 35/10 kV substation, a high voltage switch cabinet is installed in the switching station at the entrance of the tunnel, in order to stop/transmit power and meet the requirements. TBM high voltage switch cabinet has the functions of over-current, quick-break, grounding protection, and can realize automatic power outage through cable joints switch-off or cable pull-off. Zero-sequence current transformer is used to monitor zero-sequence current for grounding protection.

High voltage switch cabinet supplies 10 kV high voltage to two transformers with capacity of 2700 kV·A on TBM to ensure safe and reliable power supply of TBM.

② Power supply for auxiliary equipment outside the tunnel. Two circuits passing through 35/10 kV substation are used as standby, the other one is connected to the high-voltage distribution system, and the other is powered by self-provided generator. In the generator room, there are two generators with a capacity of 256 kW operating in parallel. When the grid power fails, the generator starts to work. 400 V is converted into 10 kV high voltage by the boosting transformer and then 10 kV is sent to the whole high voltage distribution system, so as to ensure the ventilation and lighting in the tunnel after the grid power fails, and the power supply of other equipment with higher power consumption. After distribution, the high-voltage distribution system is divided into four circuits to feed out, so as to solve the problem of dispersed auxiliary equipment outside the tunnel.

Two transformers at the entrance of the tunnel are close to the switching station, so cables are used to supply power to the transformers at the entrance. After feeding out, cables used in the prefabrication factory and storage yard shall be overhead until reaching the transformers. In response to the problem of scattered electrical equipment, low voltage distribution panels are used for local power distribution of low voltage units.

(4) TBM water supply system

Domestic water is used for employee's drinking, restaurants, medical institutions, etc. Production water is used for concrete production, macadam yard, invert segment precasting plant, tunnel construction, etc.

For example, in a TBM project, two pumping stations are used for the water system, and one pumping station is equipped with two water pumps, with specifications of $100m^3/h$ flow, 126 m lift and 55 kW power. The other pumping station is equipped with two DA150 × 5 multi-stage pumps. One $200m^3$ and one $100m^3$ reinforced concrete pool were built on the hill near the entrance of the tunnel. The water pipe with a diameter of 150 mm for construction in the tunnel was introduced into the tunnel from the $200m^3$ pool, and the water pipe with a diameter of 100 mm for production and living outside the tunnel was introduced from the $100m^3$ pool. Divert the water outside the tunnel to the $20m^3$ water tank on the back-up gantry of TBM.

(5) Ventilation system

Ventilation system is an indispensable part of TBM for excavating long and large tunnels. It plays a vital role in introducing fresh natural air into the tunnel during TBM tunneling and maintenance, which is extended by the air duct outside the tunnel and connected with the air duct on TBM, pressing fresh air into the tail of the main machine, while discharging the turbid air containing rock dust and the like outside the tunnel. It is forced ventilation.

For example, in a TBM project, the main components of the ventilation system are 250 kW fans outside the tunnel, 3 in total. Usage and distribution: 1 ~ 3 km tunnel length, completed by 1 fan. 3 ~ 6 km tunnel length, completed by 2 fans. 6 ~ 9 km tunnel length, completed by 3 fans. The tail of TBM consists of relay fan and air duct.

(6) Production system of prefabricated invert segment

Generally, after excavation, concrete segments shall be installed immediately, and the lining of the full face is completed. With prefabricated production of tunnel lining in factory ahead of schedule, we can simplify the working procedure, greatly reduce the construction interference and speed up the progress. Fabrication and installation of lining precast blocks (segments) is an important back-up technology for TBM excavation. TBM tunneling inside the tunnel and precast block production outside the tunnel are equally important.

After tunneling with TB880E, an open TBM, only the bottom lining precast blocks are installed, and the tunnel lining of Qinling line I is designed with a composite lining structure. After the excavation of the whole tunnel is completed, the TBM shall be removed, and then the full-circular lining shall be completed. With the excavation of TBM in each cycle, the precast blocks of invert and the bottom backfill are completed synchronously, becoming a part of the lining structure and the foundation part of the whole track bed. In TBM construction, the precast blocks of invert are installed to facilitate the rapid laying of construction tracks that meet the transportation requirements of TBM construction and support tracks (four tracks and two lines). At the same time, the rapid laying of invert also plays an important role in stabilizing the surrounding rocks after excavation.

Preparation includes design, construction and set up an invert prefabrication factory. For example, a TBM project mainly includes a steel-structure plant with a length of 60 m, a width of 24 m and a height of 13.2 m. The plant is equipped with two production lines consisting of 17 sets of fixed molds, an indoor storage line, a steel structure processing workshop, and concrete transportation tracks, hoisting, curing and demolding facilities. There are heat sources, mixing stations, and storage yards outside the plant. The designed production capacity of the prefabrication factory is 17 blocks per day, which can meet the maximum monthly requirements of TBM.

(7) Transportation of TBM

Most components of TBM are transported in containers, and railway transportation can be preferred to solve the problem of high long-distance transportation costs and reduce transportation costs. For the large components exceeding the limit of railway transportation, the bulk transportation company can be entrusted, and the transportation of TBM large components from the factory to the construction site shall be undertaken by a professional transportation company with corresponding qualifications. Temporary reinforcement measures shall be taken to improve or reinforce the transportation road in the transportation of large components, and transportation insurance shall be handled.

5.2.3 TBM Assembly

1. General Requirements

(1) Assembly outside the tunnel

The assembly site outside the tunnel shall meet the following requirements: the surface of the assembly site shall be flat, and the flatness of the assembly site of the main machine shall be controlled within 3 mm, with the centerline basically aligned with the tunnel centerline. Support and drainage measures shall be taken. The foundation shall be tamped, the foundation of gantry crane track shall be reinforced concrete structure, and the ground compressive strength and concrete thickness of the assembly site shall meet the requirements of TBM manufacturer. The length and width of the assembly site shall at least meet the requirements of TBM assembly and access of large components, and have a certain maneuvering length. The length of the assembly site is at least equal to the sum of the length of the main machine, the length of the connection bridge, the length of the towing cylinder, the length of the first section of the gantry, the storage length of the main bearing, the storage length of the cutter head and the maneuvering length. $L_{\text{assembly site}} \geq L_{\text{main machine}} + L_{\text{connection bridge}} + L_{\text{towing cylinder}} + L_{\text{the first section of gantry}} + L_{\text{main bearing}} + L_{\text{cutter head}} + L_{\text{maneuvering}}$. The span of the gantry crane shall be greater than the diameter of the cutter head to ensure the safety during hoisting.

(2) Assembly inside the tunnel

Extra-long tunnels are constructed by multiple TBM sections, and TBM in the middle section is generally assembled in the tunnel. The site assembled in the tunnel shall meet the following requirements:

The dimension of the chamber is determined according to the space dimension that TBM large components must occupy in the assembly and the space dimension during crane installation and operation. The chamber construction shall comply with the economic principle to minimize the construction cost. The dimension of the assembly chamber shall meet the working conditions of hoisting in the chamber.

In order to meet the needs of transportation during assembly, the length of the site shall meet the length requirements of main machine and back-up equipment, and a certain safe distance is also required. The length of the assembly site in the tunnel can generally be calculated by the following formula:

$$L \geq L_1 + L_2 + L_3 + e \tag{5.1}$$

where

- L total length of the site (m)
- L_1 TBM length (m)
- L_2 total length of traction equipment and transfer equipment (m)
- L_3 length of shunting section of track (m)
- e maneuvering length (m)

Immediately after the main machine steps, concrete invert precast blocks shall be laid, and concrete cushion blocks shall be placed under the adjacent invert precast blocks. Concrete shall be poured between the bottom of the invert precast blocks and the bottom of the tunnel. Pay attention to the intersection angle, slope and section dimension of the built branch tunnel and the main tunnel, so as to deliver the components into the main tunnel safely and conveniently. Branch tunnels can be used as air ducts and power pipeline passages, and can also be used as transportation passages when suitable for transportation scheme conditions, so as to reduce transportation chainage. A large tonnage bridge crane is required to be erected in the main chamber, and the section dimension of the assembly chamber is determined by the technical requirements of TBM assembly and the technical parameters of the crane used (the type of crane depends on the technical requirements of assembly). The model and structure of the crane shall be determined by the mass and dimension of the largest components of TBM equipment, and the length, width and height of the assembly chamber shall be determined by the installation procedure of the equipment. The reference drawings of section dimension of the assembly chamber shall be offered by the manufacturer providing TBM equipment. The tonnage of the bridge crane installed in the assembly chamber shall be considered according to the quality of the largest component of TBM, and the beam of the bridge crane and track shall be anchored. In the TBM assembly yard, it is necessary to ensure that the track bed has sufficient strength for assembly of back-up track bed and layout of track, that is, the track bed must have a certain width and density to ensure the smooth passage of the TBM back-up equipment.

The width of the site shall meet the requirements of hoisting and transportation of the largest component, and the dimension requirements of the crane. The height of the chamber shall meet the hoisting requirements, and the maximum effective lifting height of the hook relative to the ground shall be 3 ~ 4 m greater than the diameter of the main machine.

After the large components of TBM arriving at the jobsite, unloading sequence and storage shall be arranged reasonably according to the assembly sequence requirements of TBM. Unloading of TBM heavy equipment shall be carried out by professional trained lifting personnel and directed by special personnel. The model and structure of gantry crane (assembled outside the tunnel) or bridge crane (assembled inside the tunnel) shall be determined according to the maximum mass and dimension of TBM equipment. The gantry crane (or bridge crane) for TBM assembly must be the products produced by qualified professional manufacturers that meet the safety requirements. After the gantry crane (or bridge crane) is assembled, it must be put into trial operation, and the local technical supervision department shall be asked to carry out quality acceptance before it can be used.

For unpacking inspection or unpacking before assembly, care shall be taken to ensure the original dimension and machining accuracy of TBM components. After confirmation of the components, we shall adopt the corresponding disassembly scheme according to the actual packaging, so as to avoid blindly disassembling and damaging the original machining accuracy of TBM components.

Bolts, contact faces, hydraulic components, etc. shall be carefully cleaned. Bolts and contact faces shall be degreased and derusted, and cleaned with cleaning agent (oil protection if necessary) to ensure their machining smoothness before installation. Painted contact faces shall be derusted and cleaned. The scratches caused by carelessness during transportation shall be treated within the original design dimension range to ensure the assembly accuracy. The hydraulic components must be cleaned with cleaning agent. It is strictly prohibited to wipe the hydraulic components with cotton yarn and they shall be wiped with lint-free cloth or wool.

During hoisting, select reasonable lifting points for each large component. The lifting and hoisting shall be stable, which shall be based on the designed hoisting position. Confirm its quality. Use the hoisting equipment whose capacity is larger than the load and hoist equipment within the safe range to lift components and ensure absolute safety. During hoisting, select reasonable lifting points for each large component, lift them in the correct way, and slowly and accurately assemble the components on the equipment.

The hoisting of large components shall be carried out according to the relevant operation safety regulations and the assembly requirements of TBM manufacturer. Before installation, the drawings and atlas shall be carefully studied, and the assembly relationship (chronological order, front and back, left and right, up and down sequence) of components shall be confirmed before correct assembly, so as to prevent rework caused by blind assembly. The accuracy and torque of bolts shall be confirmed and verified, and the materials coated on the bolt ports shall be determined (ordinary level 8.8, level 10.9 bolt ports greased, HV10.9 high-strength bolts sprayed with MoS_2), and the bolts shall be tightened with correct tools, correct tightening sequence and specified torque.

Assemble according to the correct assembly relationship, and tighten bolts according to the specified sequence and torque. Ensure fire protection work at the jobsite is in place. When welding, there must be a special personnel to protect it. After assembly, no-load commissioning of each system and the whole machine must be carried out.

2. TBM Assembly

(1) Work to be completed before TBM assembly

① The construction of assembly site, temporary storage site, preparation chamber and stepping chamber shall be completed.

② TBM is of large structure. Before assembly, the assembly drawings and technical requirements shall be studied to determine the assembly sequence. Ensuring the quality and accuracy of the assembly process has a great influence on the service performance and service life of TBM. Therefore, detailed assembly process regulations shall be formulated, assembly drawings and technical requirements shall be studied, the assembly structure, assembly characteristics and adjustment methods shall be understood, and the detailed assembly process specifications shall be developed according to TBM components and site conditions.

③ Assembly tools, measuring tools, fixtures, lifting slings and materials that meet safety requirements shall be prepared, and the assembly components shall be visually inspected. Select hoisting equipment according to the dimension of the largest component and the specifications of the heaviest component, and make preparations for the assembly site. Tools, materials, oil, hydraulic pipe joints, special tools, etc. shall be set up in a special warehouse nearby, and containers/warehouses such as tools, common materials storage, hydraulic parts, electrical components, water pipe joints, etc. shall be placed in the assembly site in turn.

④ Ventilation, water supply, power supply and electric welding equipment shall be prepared, and the power socket shall be waterproof and explosion-proof. Ventilation, water supply, power supply equipment shall be provided during assembly. Electric welding machine, oxygen, argon, carbon dioxide, acetylene, etc. shall be equipped. Pressure of high pressure air: 6 ~ 8 bar. Air duct: 3 × 60 m. Matching joint. High pressure water pipe: 2 × 60 m. Electricity: 380 V,3 ~ 50 Hz. 220v single phase, 50 Hz. Distribution box: The main machine assembly area is equipped with a 380 V power distribution box with 4 ~ 5 external outputs, including 1 ~ 2 external multi-purpose combined sockets. There is a 380 V power distribution box in the front, middle and back of the back-up equipment assembly area, each distribution box has 2 ~ 3 outputs, and there shall be 1 ~ 2 external multi-purpose combined sockets.

⑤ Provide enough firefighting equipment. A number of hand-held fire extinguishers, fire sand and other fire-fighting products shall be set in the main machine assembly area and the back-up equipment assembly area.

(2) Main machine assembly

The main machine shall be assembled in strict accordance with the assembly procedure, otherwise, it will bring difficulties to the next step, and may even cause rework. In the assembly work of the main machine, all connecting bolts must be tightened to the specified tightening torque to form the basic framework of the main machine.

The basic requirements for assembly of the main machine are as follows:

① Strictly follow the assembly procedure.

② All connecting bolts must be tightened to the specified tightening torque to form the basic framework of the main machine.

③ The horizontal tolerance of the four corners of the stepping mechanism shall be controlled within 5 mm.

④ The pre-tightening pressure of contact faces shall be uniform.

⑤ The cutter head shall be closed in blocks first, and then hoisted as a whole after connection.

⑥ After the main machine is assembled, necessary electrical and hydraulic system connections shall be carried out, and necessary commissioning and functional tests shall be conducted on some auxiliary equipment of the main machine.

The basic sequence of main machine assembly of an open TBM is as follows:

① Positioning of stepping mechanism.

② Connect the thrust cylinder to the front and rear outer frames (or main beams).

③ The lower parts of the front and rear outer frames (with thrust cylinders) are connected with the stepping mechanism.

④ Install the front and rear couplings of the transmission shaft and the front and rear protective sleeve.

⑤ A transmission shaft with a protective cover is sleeved at the lower part of the front outer frame and preliminarily fixed on the outer frame.
⑥ Install the front and rear inner racks on the front and rear outer racks, and connect the front and rear inner racks.
⑦ Install bolt driller track on both sides of the front inner frame.
⑧ Install the upper parts of front and rear outer frames (with thrust cylinder).
⑨ A transmission shaft with a protective cover is sleeved on the upper part of the front outer frame and preliminarily fixed on the outer frame.
⑩ Sleeve the ring beam erector.
⑪ Shield under the cutter head is in place.
⑫ The rear support with stepping mechanism is installed at the rear of the rear inner frame.
⑬ Install the rear of the inner frame.
⑭ Hoist the main bearing.
⑮ Install the upper shield support frame.
⑯ Install the hopper in the drive assembly.
⑰ Hoist the cutter head.
⑱ Install the flank and top of the cutter head shield.
⑲ Install the transmission shaft and the thrust cylinder in the correct position.
⑳ Install transmission mechanism, ring beam erector and drive device, main machine belt conveyor, main machine working platform and other working platforms, other auxiliary equipment and accessories in turn.

The basic sequence of main machine assembly of double shield TBM is as follows:

① Install TBM bracket on TBM sliding track.
② Install the front shield. First, install the lower shield of the front shield, then install the inner components of the front shield such as the main bearing, then install the upper shield of the front shield, and bolt the upper and lower shields of the front shield firmly.
③ Install the telescopic shield. First install the lower outer shield, weld the lower outer shield on the front shield, then install the lower inner shield and the internal components of the telescopic shield, then install the upper inner shield, and install the upper outer shield and weld the upper outer shield on the front shield.
④ Install the rear shield (support shield). Install the lower shield first, then install the internal components such as support device, support gripper and auxiliary thrust cylinder, and then install the outer shield.
⑤ Install the tail shield. Install the tail seal, weld the tail shield on the rear shield, and install the segment erector and other devices in the tail shield.
⑥ Install the cutter head. Install the cutter head on the main bearing assembly, and finally install the cutter, cutter head bucket, etc.
⑦ Install other components and connect various systems.

(3) Assembly of the connection bridge

The connection bridge is a bridge connecting the main machine and the back-up gantry. The key point of assembly is the connection and hoisting of the connection bridge. Because the connection bridge is long, it is necessary to select the appropriate lifting point to ensure the safe hoisting of the connection bridge. For the installation of invert crane, attention should be paid to the straightness of its running track.

The assembly of the connection bridge shall meet the following requirements:

① Select suitable lifting points to ensure the hoisting safety of the connection bridge; Before the connection bridge is connected with the main machine, necessary temporary support should be provided.
② The installation of invert crane (or segment installation machine) should pay attention to the straightness of its running track.
③ Clean the connecting surface to ensure the tightening torque of bolts and the pre-tightening force of the connecting surface.
④ Install all kinds of cranes, grouting equipment and industrial exhaust fan correctly.
⑤ Install data recording system, PLC system, TV monitoring system, etc., and connect with the main machine correctly.

The assembly procedure of the connection bridge is as follows:

① Select a suitable assembly site and make preparations before assembly.
② Ground connection of connection bridge.
③ Install both sides and upper auxiliary frames on the connection bridge.
④ Install the rear support of the connection bridge on the extended track.
⑤ The hoisting connection bridge is connected with the main machine.
⑥ Install the auxiliary equipment on the connection bridge.
⑦ Install the working platform on the connection bridge.
⑧ Install the operator cabinet.
⑨ Each system is connected with the main machine.
⑩ Install belt conveyor of connection bridge.

(4) Assembly of back-up gantry

The back-up system has multiple gantries, so the assembly work is relatively simple. According to different assembly conditions, there are different assembly methods, and a reasonable assembly method can speed up the assembly. In the early stage of assembly, it mainly includes the assembly of gantry frame and the installation of stereotype equipment, and the connection of hydraulic and electrical system in the later stage.

The assembly of the back-up gantry shall meet the following requirements:

① First, the assembly of each gantry and the installation of corresponding auxiliary equipment on the gantry should be carried out.
② Then connect each gantry.
③ Install the belt bracket and belt.
④ Install electrical system, hydraulic system and muck discharge machine.
⑤ Connect with the main machine and carry out necessary commissioning.

The assembly sequence of the back-up gantry is as follows:

① Assembly of each gantry and installation of corresponding auxiliary equipment on the gantry.
② Connect each gantry.
③ Install the truck hauling system.
④ Install the belt.
⑤ Install electrical system, hydraulic system and muck discharge machine.
⑥ Connect with the main machine for commissioning.

(5) Assembly of continuous belt conveyor

Continuous belt conveyor is mainly composed of belt storage cassette, main drive device, auxiliary drive device, driven wheel, belt and roller. The tail of the continuous belt conveyor is installed on the back-up gantry. When the back-up gantry moves forward, the belt is pulled out from the belt storage cassette section by section, so that the continuous belt conveyor can continuously complete rock ballast conveying. As each time the TBM completes a cyclic stroke stepping, the back-up system is pulled forward for a stroke, and the continuous belt conveyor extends accordingly. Before installation, it shall be ensured that all civil works have been completed and that the horizontal plane is accurate and the civil works are qualified. Before installation, it shall be checked to ensure that the section markings have been marked. In case of missing markings, the corresponding section must be marked.

The assembly of the continuous belt conveyor shall meet the following requirements:

① The installation position of the belt storage cassette and the site required by muck dumping equipment should be considered for the entrance site.
② Install in accordance with manufacturer's instructions. Install belt conveyor frame, roller and grooved roller in front of the tail of continuous belt conveyor to provide conditions for tape transportation; Too many apron board sealing strips should not be used on the belt; Pay attention to the quality and dimension of bolts, and all dimensions should be consistent with those indicated in drawings and parts details; The adhesion of belt should be smooth; The installation of the pulley bearing should pay special attention to the center distance and the fastening of the fastening sleeve; Avoid earthing near roller or mechanical components during welding.
③ When the tape in the belt storage cassette is used up, the extended tape should be vulcanized.

5.2.4 TBM Commissioning

The main contents of TBM equipment commissioning include visual inspection, functional test, technical performance test and adjustment.

Compile the *Equipment Test Function Table* based on the main machine, auxiliary equipment and accessories. The functional commissioning of each single equipment shall be carried out first, and then the complete machine interlocking functional commissioning of TBM equipment shall be carried out, and the test data shall be compared with the standard values in the table. After commissioning, the installed quality shall be judged based on the test results and the design performance, and the problems existing in each system shall be handled in time. The main contents of commissioning are as follows:

(1) Mechanical part: Whether design actions and noise test can be completed.
(2) Hydraulic pressure: Pressure, flow, frequency (grease system), leakage, etc.
(3) Electrical part: Test voltage, current, control voltage, frequency, power factor, PLC module function, etc.
(4) Water and gas system: Test pressure, leakage, pipeline layout, etc.
(5) Data recording system and communication system: Test function.

Testing and commissioning items mainly include: Hydraulic system, electrical system and PLC control system. The commissioning of hydraulic system mainly includes pressure setting of each system, and make matching adjustment of control components such as pumps and valves in the system, and signal test elements. The commissioning of the electrical system is mainly to confirm the input and output levels, adjust the correlation action and action time of the control elements, conduct the synchronization test of the hydraulic magnetic valve, conduct the safety test of the power supply and consumption equipment, and check the installation position and performance of the sensing elements. The commissioning of PLC control system mainly includes the commissioning of interlocking and interlocking functions. The commissioning of PLC control system is the focus of TBM commissioning, which is crucial for whether the TBM can tunnel normally and ensure the safety of the equipment. In particular, the interlocking function of the main machine, the front and rear support of cutter head, thrust cylinder, gripper, etc., as well as the interlocking function of advance state and regripping state, and the interlocking function of cutter head, thrust, etc. and belt conveyor.

See Table 5.1 for the format of *Equipment Test Function Table.*

The commissioning of open type TBM shall refer to the following, and the commissioning of TBM shall be carried out in the actual works according to the detailed commissioning items provided by the TBM manufacturer.

(1) Cutter head: Shrinkage/extension and function test of reamer.
(2) Function test of cutter head and connected shield: PLC system program test, shield clamping/loosening, cutter head lifting/falling, cutter head shield expansion/ contraction.
(3) Cutter head drive: Check the PLC system program, start/stop the motor through the program, inching function, start/run the cutter head through the clutch, interlock the brake and PLC system program, cooling water circulation and cutter head water injection, turn on/off the cutter head drive motor cooling water, turn on/off the cutter head water injection device, check the PLC system program (oil and grease lubrication), oil circulation lubrication, oil pump on/off, grease lubrication test and pneumatic grease pump control. Long-term

Table 5.1 Format of test function of TBM equipment

S/N	Test items	Parameter standard value	Test value
–	Main machine		
1	PLC system program test		
2	Variable frequency motor start/stop		
3	Cutter head start/rotation		
4	Related interlocking test		
5	Cooling water circulation and cutter head spraying water		
6	PLC control of lubricating oil (grease)		
7	Revolution test		
8	Temperature detection of each part		
…	…		
II	Auxiliary equipment		
1	…		
…	…		

Table 5.2 Reference format of TBM *Acceptance Outline*

S/N	Function description of system	Design Standard	Acceptance record	
			Factory Acceptance	Site acceptance

operation test of cutter head drive system, inspection of lubricating oil cleanliness, speed detection, temperature measurement of main drive parts, drive motors, planetary reducer, water cooling, oil lubrication, etc.

(4) Rack thrust: PLC system program test, thrust, return stroke, fast thrust, fast return stroke.

(5) Outer frame: PLC system program test, high-pressure clamping/releasing, low-pressure (fast moving) clamping/releasing, reset time test (extension/retraction time of each hydraulic cylinder).

(6) Rear support: PLC system program test, hydraulic support rising/falling, hydraulic support extending left/right.

(7) Hydraulic control system: Oil level and oil temperature function test, pressure setting function test and PLC system program check.

(8) Electrical devices: PLC input/output test, program interlock of transformer, switch cabinet and PLC system, test temperature and signal on/off of transformer controllers, safety circuit (emergency off) test, simulation test of switch

cabinet and switch circuit under power-on, switch/on of lighting circuit and function test of control desk.

(9) Mucking removal belt conveyor: Function test.
(10) Main machine belt conveyor: Running speed test, tension test, extension/retraction test of belt conveyor.
(11) Ring beam installation system: Check the movement of ring beam, function test of transport trolley, indicator and alarm signal check, and electrical interlocking check.
(12) Invert precast block crane: Function test, check the distance between test block cars, and use test blocks for dynamic test.
(13) Geological probe drilling machine: Check the movement and move to the limit position, check the borehole position and the disturbed limit.
(14) Bolt driller: Check the movement and move to the limit position, check the interference of the ring beam transport trolley, and check the surrounding working area.
(15) Ventilation system: ventilation function test of deduster, start-up test of wind machine, function test of air duct and air duct cassette, and function test of air cooling system.
(16) Communication system: Speech test of interphone, speech test of wired phone.
(17) Data recording system: Data transmission and collection test.
(18) Guidance system: Control indoor data display and data recording outside the tunnel.
(19) Harmful gas alarm system: Alarm test of harmful gas alarm test.
(20) Materials and rail crane: Function test, electrical interlocking test.
(21) Transmission crane, mortar barrel crane, auxiliary hoist: Function test, clearance test.
(22) Muck discharge machine and back-up belt conveyor: Power function, belt tension test and inspection.
(23) Truck hauling device: Chain drive function test.
(24) Grouting system, shotcrete system, cable reel, emergency generator, air compressor, water supply system, monitoring system, etc.: function test.

The commissioning of double shield TBM shall refer to the following, and the commissioning of TBM shall be carried out in the actual works according to the detailed commissioning items provided by the TBM manufacturer.

(1) Cutter head: Shrinkage/extension and function test of reamer.
(2) Cutter head drive: The contents of inspection and commission are the same as those of open type TBM.
(3) Sealing and lubrication of main drive: Interlocking test between automatic grease system and main drive, adjustment and recording of grease injection volume, pneumatic operation of grease pump.
(4) Shield: Function test of telescopic cylinder, stroke display, stabilizer function test, function test of main thrust cylinder and supporting device, and function test of auxiliary thrust cylinder.
(5) Segment installer: 6 degrees of freedom function test.

(6) Segment transportation/segment register: Function test.
(7) Main machine belt conveyor: Running speed test, tension test, extension/retraction test of belt conveyor.
(8) Muck discharge machine and back-up belt conveyor: Power function, belt tension test and inspection.
(9) Hydraulic control system: Oil level and oil temperature function test, pressure setting function test and PLC system program check.
(10) Electrical devices: PLC input/output test, program interlock of transformer, switch cabinet and PLC system, test temperature and signal on/off of transformer controllers, safety circuit (emergency off) test, simulation test of switch cabinet and switch circuit under power-on, switch/on of lighting circuit and function test of control desk.
(11) Geological probe drilling machine: Check the movement and move to the limit position, check the borehole position and the disturbed limit.
(12) Communication system: Speech test of interphone, speech test of wired phone.
(13) Data recording system: Data transmission and collection test.
(14) Guidance system: Control indoor data display and data recording outside the tunnel.
(15) Safety control system: Emergency stop switch test, function test of grounding protection device, harmful gas monitoring and alarm device, TBM interlocking function test (lubricant flow of main bearing, lubricant flow of output pinion of reduction gearbox, motor cooling water flow, motor overload alarm, belt conveyor operation, oil pressure of hydraulic motor of belt conveyor, revolution, sealed lubricant flow, pressure of support cylinder, synchronous protection of rear lower support and parallel support), function of information display (motor temperature, frequency converter temperature, gear reducer temperature, transformer temperature, oil level of lubricating oil tank, oil level and temperature of hydraulic oil, and traction cylinder pressure).
(16) Segment hoist, lifting equipment, cooling system, dedusting system, air compressor, grouting equipment, mortar equipment, pea gravel spraying equipment, cable reel, emergency generator, ventilation system, monitoring system, lighting system, etc.: Functional test.

The commissioning points of continuous belt conveyor are as follows:

(1) Horizontal tolerance and vertical tolerance of belt support should be controlled within the allowable range.
(2) After the drive system is installed, no-load commissioning of individual drive elements shall be carried out first, then linkage no-load commissioning shall be carried out, and finally the belt shall be tensioned for on-load commissioning.
(3) Adjust the belt tension device to make the belt pulling force reach the required value before starting the belt.
(4) When starting the belt for the first time, the rotation speed of the belt should be set at a lower speed. After starting, observe whether the difference and variation range of driving torques of each motor are within the allowable range, otherwise, it needs to adjust.

(5) Before the belt runs for the first time, check whether each pull switch and emergency button work normally.
(6) The belt conveyor cannot be tensioned until the strength of the belt conveyor mounting foundation reaches 80%.
(7) Adjust the pulley and roller to make the belt deviation within the allowable range. Adjustment should be carried out under no-load and on-load conditions.

Single tests and inspections, such as appearance inspection, function test of single equipment and technical performance test, can be carried out synchronously during the assembly period, and the commissioning of the system can be carried out after assembly; The equipment must be tested item by item according to the prepared equipment test function table, and corresponding records shall be made during equipment commissioning. When the data exceeds the standard value, the reason shall be found until the commissioning reaches the specified range of the tested data; The stepping of the TBM can only be started under the precondition that all performance indexes of the equipment fully meet the technical description of the TBM and the installation of each equipment is confirmed to be correct.

5.2.5 Site Acceptance of TBM

During the procurement period, the *Acceptance Outline* shall be jointly determined with the manufacturer according to the main design functions and use requirements of the TBM equipment; After the TBM assembly and commissioning are completed, the acceptance shall be carried out item by item in accordance with the subsystem of the *Acceptance Outline*. See Table 5.2 for the reference format of the *Acceptance Outline*.

The acceptance of TBM main machine must meet the following requirements:

(1) The outer diameter must meet the design requirements.
(2) The auxiliary equipment in the main machine meets the functional requirements, and shall not interfere with each other during operation.
(3) The shield must be a straight cylinder with flat surface.
(4) For shield TBM, the inner surface of the tail shield is flat without protruding weld within the range of movement of the auxiliary thrust cylinder, and the true roundness of the tail shield is within the allowable range.

The cutter head acceptance must meet the following requirements:

(1) All bolts used for connection must be configured according to the design requirements of the manufacturer, and the hydraulic torque wrench should be calibrated regularly to reach the design torque value.
(2) No-load operation of the cutter head is in the forward and reverse direction for 15 min respectively, operates smoothly, and there is no abnormal sound from each reducer and transfer system.

(3) The centralized lubrication system should be tested for flow and pressure, and the oil receiving conditions of all lubricating components must meet the design requirements.

The acceptance of shield TBM segment installation machine must meet the following requirements:

(1) During the no-load test run, the stroke, rotation limit, lifting distance, translation distance and adjustment distance of each component, and the working pressure of each system must meet the design requirements.
(2) During the on-load test run, the segment installation machine runs smoothly by turning, translating, lifting, adjusting and other actions, the installation and positioning of each idler wheel and thrust roller shall be accurate, safe and reliable, and the working pressure of each system shall be normal.

The acceptance of belt conveyor must meet the following requirements:

(1) No belt deviation is allowed during no-load test run.
(2) During the on-load test run, it runs smoothly without vibration and abnormal noise; All rollers and pulleys shall operate flexibly.

The acceptance of continuous belt conveyor must meet the following requirements:

(1) Running speed test; Tension device test; Manual function test; Electrical interlocking test; The whole signal alarm test of belt conveyor.
(2) The PLC control system is matched with the main machine PLC system to ensure that the main machine controls the start and stop of the continuous belt conveyor.

5.2.6 TBM Stepping

Stepping construction is the construction in the moving process without TBM tunneling.

Under different conditions, there are different ways of stepping, and there are roughly two ways of TBM stepping. One is to make TBM move forward through the cylinder supported on supports, split heads, segments, etc. The other is to move forward directly on the ground through the stepping mechanism of TBM.

Before TBM stepping, the clearance of the drilling and blasting section shall be measured with a profiler, and transfinite is strictly prohibited. The flatness and strength of the bottom should meet the stepping requirements. The tunnel design centerline is measured on the top surface of the preparatory tunnel pavement to facilitate the guidance construction of the TBM bottom. When stepping, the supporting cylinders of the probe drilling machine, bolt driller and ring beam erector should be locked in the minimum state. After the stepping of TBM main machine, the back-up equipment follows the main engine synchronously forward. When stepping, the operator should pay close attention to the display of relevant instruments in the operation room, strengthen stepping monitoring, and the operators should strengthen patrol work and do a good job in extending the construction track. After the stepping is completed, the TBM is supported, the step device is removed and ready for launch.

Preparations for stepping construction are as follows:

(1) Preparatory tunnel pavement

The requirements for preparatory tunnel pavement are the same as that of the TBM erection yard, and it is carried out one month in advance, and the strength and flatness meet the requirements of TBM stepping.

(2) Precast wedge block

Because the pavement of the preparatory tunnel is 15 cm lower than the invert of TBM tunneling section, a larger wedge block is needed, and the prefabrication of wedge blocks shall also be carried out in advance. In order to facilitate the rapid paving of invert blocks, wedge blocks are placed on both sides of the preparatory tunnel in advance (without affecting the stepping through of TBM main machine).

(3) Preparation for measurement

In order to facilitate the paving of TBM stepping guidance invert block, the tunnel design centerline must be surveyed and set out in advance on the top surface of the preparatory tunnel paving before TBM stepping.

(4) TBM stepping construction

① Stepping construction of erection section. When TBM passes through the erection yard, it is mainly the extension of the construction track, with sleepers and crushed stone roadbed (1.0 m thick) under the track, and the long track is extended once every 12.5 m stepping of TBM. At the entrance and before entering the preparatory tunnel, in order to meet the drainage needs, a central ditch is laid to transition to the back reinforced concrete invert block.
② Stepping construction of preparatory tunnel section. After entering the preparatory tunnel, the TBM shall lay one invert precast block every 1.8 m stepping, and grouting should be performed at the bottom of each 3 ~ 5 invert blocks; Extend the track every 7 invert precast blocks; Extend the water pipeline every 50 m advance.
③ Material supply during TBM stepping. When the back-up of TBM cannot fully meet the requirement that materials such as invert blocks can be lifted to the flatbed car from the clearance of the back-up by using the door crane at the entrance to the jobsite during material transportation.

5.2.7 TBM Tunneling

1. Overview

Advanced geological prediction and forecast shall be carried out during TBM construction; The selection of advance speed and thrust force shall be determined according to the geological conditions; The mucking amount shall be strictly controlled in the crushed section to avoid large-scale collapse in front of the tunnel face.

There are generally three operating modes of TBM: Automatic torque control, automatic thrust control and manual control mode. It should be reasonably selected according to geological conditions. Generally speaking, under the condition of homogeneous hard rock, the automatic control thrust mode is selected; Under the condition of joint development or weak surrounding rock, the automatic control torque mode is selected; The surrounding rock of the tunnel face is uneven in hardness, if the state of surrounding rock cannot be judged, the manual control mode is selected.

Automatic torque control is suitable for homogeneous soft rock; Automatic thrust control is suitable for homogeneous hard rock, and manual control mode is convenient to operate and flexible to respond.

The basic principles for selection and adjustment of tunneling parameters under different geological conditions are as follows:

(1) Operation in hard rock with undeveloped ~ developed joints

① Select cutter head to tunnel at high speed.
② Under normal circumstances, the advance rate is generally $\leq 35\%$ (setting value of potentiometer).
③ The compressive strength of surrounding rock itself is large, and it is not easy to be broken. If the advance speed is too low, it will cause a large amount of wear of the cutter and cutter ring. If the advance rate is too high, it will cause overload of the cutter, so reasonable parameters must be selected for tunneling.

(2) Operation in soft rock with developed joints

If the tunneling thrust is small, the automatic torque control mode should be selected, and the torque change should be closely observed to adjust the best advance parameters. Observe the supporting capacity of double-shield TBM gripper and determine the working mode.

(3) Operation in surrounding rock with developed joints and large hardness changes

The advance rate should be controlled below 30% of the setting value of potentiometer. Because of the uneven distribution of surrounding rock, the hardness changes greatly, and sometimes there will be large vibration, and the change of thrust and torque is large, so manual control mode must be selected to closely observe the change of thrust and torque.

(4) Operation under geological conditions such as relatively developed joints, many cracks, or fracture zones and faults

During tunneling, the tunneling parameters shall be selected and adjusted mainly in the automatic torque control mode, and the torque change, current change, thrust value and surrounding rock condition shall be closely observed. The torque change range shall be controlled below 10%, the advance rate shall be reduced, the penetration index shall be controlled, and the working mode shall be adjusted for double-shield TBM.

In the process of tunneling, observe whether the display of each instrument is normal, check whether the supply of air, water, electricity, lubrication system and hydraulic system is normal, check whether the gas alarm system is in working condition and whether the gas concentration exceeds the limit, and there will be no alarm or cut-off instruction on the error indicator. Conduct a lighting test to check the function of all indicating elements.

During tunneling, patrol shall be strengthened to ensure good operation of the equipment.

Check the tunnel face support, invert block laying, segment installation, muck car in place, continuous belt conveyor normal, operators in place, etc., to ensure normal tunneling.

Before each cycle operation, the operator should conduct steering operation according to the position data of main machine displayed by the guidance system.

Automatic guidance system is used to monitor TBM attitude. Carry out regular manual measurement and recheck the automatic guidance system.

2. TBM operation

The operator cabinet is the brain of TBM, and 90% of the instructions on the equipment come from the operator cabinet, which is internally equipped with operation panel, display instrument (including parameter display, instrument display, fault display, video display and indication, etc.), PLC system, steering display, etc. The most important one is the operation panel with more than 100 buttons and handles.

(1) Key points of operation before getting on the machine

Key points of operation before getting on the machine refer to the key points of operation before operating the machine after the main driver on duty takes over the shift.

① Observe whether all instrument indicators are normal.
② Check whether the supply of air, water and lubrication system is normal.
③ Analyze the types of surrounding rocks, and choose reasonable tunneling parameters instead of directly operating the equipment.

(2) Key points of operation during tunneling.

After the inspection before getting on the machine, the equipment is all right and can be used for tunneling. The operation steps are as follows:

① Start the main pumping station, including the water supply system.
② Start the ventilation and dedusting system.
③ Select high or low speed or auxiliary drive of the motor (only one can be selected).
④ Start the belt conveyor in turn.
⑤ After all the main motors are in normal operation, turn the cutter head.
⑥ Choose a reasonable advance speed to tunnel.

The above operation sequence cannot be changed at will, otherwise the equipment will be damaged, resulting in adverse consequences.

(3) Operating points during downtime

Under normal circumstances, the downtime should correspond to the operation during tunneling. First stop tunneling, reverse the cutter head by 2 ~ 3 cm, idle the cutter head for 1 min, stop the cutter head rotation, stop the motor rotation, stop the belt conveyor conveying, etc., adjust the advance direction, and change to the next cycle operation.

(4) Key points of operation in emergency

Master the normal operation points such as tunneling, downtime and regripping, and also know the operation in emergency.

Emergency conditions generally refer to situations such as cutter damaged and detached from belt conveyor, failure of unloading muck bucket, failure of liquid concealment system. When an emergency occurs, stop the tunneling immediately, reverse the cutter head by 2 ~ 3 cm, stop the cutter head rotation immediately, stop each system in turn, inspect the failed parts, and it is better not to press the emergency stop button to avoid damaging the equipment components. In case of very special circumstances that endanger the personnel safety of workers, the emergency button must be pressed to reduce unnecessary economic losses and casualties.

(5) TBM operation

① Regripping and steering operation. According to the design requirements, the relative position of the cutter head displacement and the last cycle should not exceed 3 cm, so as to prevent the edge cutter from being damaged due to excessive steering. How to realize a smooth transition of steering: First of all, we should ensure that the pressure of the front support is around 10 MPa, the front gripper and the rear gripper are relaxed, and the rear support is adjusted; After the front and rear supports are stable (adjust the angle so that it does not change), first tighten the frontgripper, then tighten the rear gripper, and recover the front and rear gripper at the same time to meet the smooth transition between each cycle for the next cycle of tunneling.
② Launch operation during tunneling. The launching operation is important regardless of the surrounding rock conditions. The launching operation must be carried out at a low speed. Generally, after 10 cm of tunneling, it shall be increased

slowly by 10% until the rated requirements of thrust force or torque are met. Especially the starting operation after changing the edge cutter, otherwise the edge cutter will be easily damaged.

③ Tunability of operation speed of belt conveyor. TBM muck is generally transported by multiple long and large belt conveyors through relaying, and reasonable transportation speed is selected, which is conducive to cost reduction. The selection shall be based on the output of slag charge and the advance speed. When the advance speed is greater than 50% of the set value, the speed of belt conveyor shall be adjusted between 85 and 90% of the set value; When the advance speed is between 30 and 50% of the set value, the speed of belt conveyor is adjusted between 25 and 85% of the set value.

3. Selection of excavation mode

There are three modes available in the TBM operator cabinet, namely automatic thrust mode, automatic torque mode and manual excavation mode. The working mode selected is determined by the operator according to the condition of the surrounding rock.

Under the condition of homogeneous hard rock, the automatic control thrust mode should be selected, which will neither overload nor ensure the judgment basis for selecting this working mode at the highest advance speed; If the thrust force reaches the maximum value first and the torque does not reach the rated value during tunneling, it can be judged as hard rock state,

Automatic control excavation mode can be selected. Under the condition of homogeneous soft rock, generally, the thrust force is not too large, and the change of cutter head torque is the main factor, so the automatic torque mode should be selected. Judgment basis for selecting this mode: During tunneling, if the torque reaches the rated value first, but the thrust force fails to reach the rated value or reaches the rated value at the same time, it can be judged as the soft rock state, and if the geology is relatively uniform, the automatic torque control mode can be selected.

If the state of surrounding rock is uncertain, or the hardness of surrounding rock varies unevenly, or there are many fracture zones, faults and fissures in surrounding rock joints, the manual control mode must be selected.

In the operation process of manual control mode, if the surrounding rock is hard, the thrust reaches the rated value first, and the surrounding rock is relatively complete. At this time, the operation should be based on the thrust operation mode, and the thrust pressure should be limited to not exceed the rated value. If the surrounding rock joints are well developed, there are many cracks or fracture zones and faults, etc., at this time, the operation should be based on the torque mode, and the tunneling parameters should be selected mainly by torque changes combined with thrust parameters.

4. Selection of tunneling parameters

The tunneling parameters involved in tunneling mainly include the following parameters: Revolution, cutter head torque, motor current value, thrust force, thrust cylinder pressure, penetration (per rotation) and advance rate. Among them, the current value of the motor is proportional to the cutter head torque, the pressure of the thrust cylinder and the thrust force. The actual advance speed is: Advance speed = revolution × penetration.

After the revolution has been selected, the only thing that the main driver can directly control is to select the advance rate. Due to different surrounding rock conditions, different torques and thrust force are displayed, and the actual driving speed is also different. The main driver selects the advance speed based on the torque, thrust force, cutterhead vibration and mucking conditions.

When TBM enters weak and broken surrounding rock from hard rock, the corresponding main advancing parameters, slag quantity and slag particles of belt conveyor will change obviously. Accordingly, the surrounding rock condition of cutterhead working face can be judged, and the advancing parameters can be adjusted in time.

Advance speed: In the case of hard rock, the penetration is generally 9 ~ 12 mm; When entering the transition section of weak surrounding rock, the penetration increases slightly, and when completely entering the weak surrounding rock, the penetration is generally 3 ~ 6 mm.

Thrust: In the case of hard rock, the advance speed is generally about 75% of the rated value, and the advancing pressure is in a corresponding proportion; When entering the transition section of soft rock, the advancing pressure decreases in an inverse parabola form, and the advancing speed decreases with the decrease of the advancing pressure. When fully entering the soft rock, the pressure tends to be relatively stable, and the advancing speed is generally maintained at about 40%.

Torque: In case of hard rock, it is generally about 50% of rated value; When entering the soft rock transition section, the torque rises slowly; When fully entering the soft rock, the torque is generally maintained at about 80% due to the decrease of the advance rate.

Revolution: In case of hard rock, high rotation speed is generally adopted; Adjustment is made when entering soft rock transition section; Low rotation speed is generally used when entering soft rock completely.

Gripper supporting force: In the case of hard rock, it is generally rated value; When entering the soft rock transition section, it is adjusted to about 90% of the rated value; When fully entering the soft rock, the supporting force of the gripper is adjusted to the lowest limit value.

When the hardness of surrounding rock of the tunnel face is uneven, even if the torque and thrust do not reach the rated value, some cutters may encounter local hard rock, resulting in overload failure. At the same time, the main bearing and inner Kelly will have unbalance loading. At this time, measures to reduce the advance speed shall be taken based on the changes of torque and thrust and the vibration of the cutter head, especially when the tunnel face is partially collapsed, the vibration of the cutter head is intensified, and the machine should be stopped to enter the tunnel face to confirm the surrounding rock situation, and if necessary, the low-speed excavation mode should be adopted. Reduce the revolution from 5.4r/min to 2.7r/min to reduce the disturbance to surrounding rocks, but at this time, attention should be paid to halving the advance speed, otherwise the penetration rate will be doubled, which will easily lead to the overload of cutter head and the unbalanced loading of cutter head.

In addition, besides reasonable regripping and correct selection of tunneling parameters, the speed-up process during the transition period from regripping to normal tunneling is also a factor of time saving and rapid tunneling. There is a misunderstanding here, that is, after the regripping, it is believed that we should tunnel at a very low speed until every cutter is in uniform contact with the tunnel face, otherwise high-speed tunneling at the beginning will cause a large thrust, which is entirely borne by some of cutters first contacting with the tunnel face, resulting in overload. If the speed is increased slowly in this way, the process will last for 10 ~ 15 min, which will seriously affect the advance speed. After careful analysis, a solution can be found, which can be called the "hardest rock" speed-up principle. If the hardness of the hardest rock on the tunnel face can be known, tunneling at a speed less than its corresponding maximum advance speed will not lead to overload of cutter. Because the maximum hardness of rock can't be accurately known, generally, when the tunnel face is complete, the advance speed at the end of the last cycle is reduced by 10% after regripping, which not only saves valuable tunneling time, but also reduces the probability of overloading caused by partial cutting of hard rock, thus finding the best point between speed increase and overload prevention. In addition, in order to avoid the excessive instantaneous advance rate of the cutter head contacting with the tunnel face, and the impact on motor transmission and partial load of the cutter head, the speed can be reduced appropriately before the cutter head contacting with the tunnel face (which can be judged by the cutter head displacer), and then the speed is gradually increased to enter the normal tunneling operation after the thrust rises. This operation usually takes 3 ~ 5 min, and there are few cases of overloading of cutter and oil leakage in actual tunneling, but the benefits are obvious, and each tunneling cycle time can be shortened by nearly 10 min.

5. Feeding and Mucking

The productivity of TBM under normal geological conditions depends on its matching with muck discharge capacity. The choice of transportation scheme (rail transportation, trackless transportation or belt transportation) is based on the excavation diameter, tunnel length, tunnel gradient, ventilation and investment cost. In the TBM construction of Qinling Tunnel TB880E, the belt conveyor is the key equipment in TBM construction, which transports the rock chips in the excavation to the mine car and completes the TBM excavation construction. Once the belt conveyor fails, the tunneling cannot be carried out. Belt conveyors are divided into three levels: 1, 2 and 3, all of which are driven by hydraulic motors and controlled by PLC program. Among them, No.1 and No.2 belt conveyors are driven and controlled by the same hydraulic pump station, and No.3 belt conveyor is driven and controlled by another hydraulic pump station, which can realize the functions of stepless speed regulation, automatic speed measurement and automatic control.

(1) Layout of path line in the tunnel

The transportation track in the tunnel is consistent with the back-up track behind TBM. It adopts four-rail and double-line system with a gauge of 900 mm, and the left and right lines are symmetrical with the center line of the line, which is directly laid on the arch precast block. The distance between the two inner rails between the left and right rails is 1110 mm, and the gauge of the running-through lining formwork trolley is 2980 mm. Combined with the size of the transported blocks, the mucking quantity of TBM and the equipment, 43 kg/m rail is used. The track line in the tunnel is designed as a straight-line 'longitudinal slope, which is the same as the longitudinal slope of the line, with an uphill gradient of 11‰. The right line is an empty train line (feeding).

(2) Layout of track and california switch outside the tunnel

In order to meet the requirements of TBM for quick mucking and feeding, 43 kg/m rail, crushed stone ballast bed, wooden sleeper and No.6 california switch are also used for the track outside the tunnel, and the sleepers are arranged at 1,760 pieces/km.

The total length of single track outside the tunnel is 3 km, and double tracks are laid from the entrance of the tunnel to the muck dumping device. Due to the limitation of the site topography, only 160 m straight road is set on the heavy truck line at DK63 + 627 muck dumping device to meet the requirements of mucking unloading, and other sections are arranged according to the site.

(3) Standard for track laying

The allowable tolerance of gauge is + 6 mm and -2 mm. Widening and superelevation are provided in curved sections, and gauge pull rods are added if necessary. The allowable horizontal tolerance of two rails in straight sections is ± 4 mm. Splints and sleepers are placed at rail joints. Sleeper spacing is less than 0.7 m, sleeper length is 1.5 m, and long sleepers are laid at california switch.

(4) Configuration of transportation equipment

① Rolling stock configuration. According to the technical parameters of TBM construction, each TBM is equipped with five CFL-200DCL diesel rolling stocks produced by German SCHOMA. As there is a 15% uphill slope at the outlet section during mucking, the configured locomotive is 45t tractive force.
② Vehicle Configuration. The vehicle is produced by MUHLHAUSER Company of Germany, and the vehicle parameters are shown in Table 5.3.

(5) Formation of mucking transport vehicles in TBM tunneling

① When the tunneling length is less than 1 km, two groups of trains can be used alternately due to the short transportation distance. Each group of trains is arranged in turn with one carriage, one 35t locomotive, 10 mining cars, one grout tank car, one invert block car and one material car. Among them, grout tank car can be formed as required. Since the TBM grouting equipment cannot meet the requirements of

Table 5.3 Vehicle Main Parameters

Model Parameters	Muck car	Concrete truck	Ring beam/rail	Invert block car	Carriage
Loading mass (t)	45		16	13	18 persons
Carrier capacity (m^3)	20	6			
Quantity (section)	36	4	4	4	4

actual use, after grouting with the concrete transfer pump, it is no longer necessary to specially arrange the grouting truck on the train and use the grout tank car for feeding (practice has proven that this method has fully met the construction requirements). At the initial stage of tunneling, three rolling stocks can be put into use, of which two are engaged in pulling trains and one is in shunting.

② When the tunneling length is 1 ~ 3 km, with the extension of the tunneling length, the third muck car should be put into use at the right time according to the transportation distance to load muck in turn. At this time, there should be three traction trains and 1 shunting train.

③ When the tunneling chainage exceeds 3 km, due to the extension of the transportation distance and the reduction of the sealed gas storage performance of the brake system of the muck car, at this time, a locomotive should be towed to reduce the trailer reversing, and one outside the tunnel should be used for shunting operation until the breakthrough of tunnel.

6. Precast installation and bottom treatment of invert block

(1) Fabrication and heating and curing process of invert block

The process flow of fabrication and heating and curing of invert blocks is shown in Fig. 5.6.

(2) Operating points of heating and curing process

① Invert block concrete is produced by HZS50 mixing plant, which realizes automatic metering control; The loading hopper of the rail platform truck is transported to the molding site. The 16t bridge crane lifts the hopper concrete to mold automatically (with an attached vibrator). It is vibrated and formed with the attached pneumatic vibrator on the mold. After 3 days of storage indoors after curing and demolding, the rail invert transport platform truck is used to transport it to the invert block turnover platform, and the invert block is lifted to the storage place with a 16t gantry crane.

② The heat supply and air supply of 17 sets of molds are in parallel, and the curing temperature and humidity of each set of molds are monitored centrally and separately.

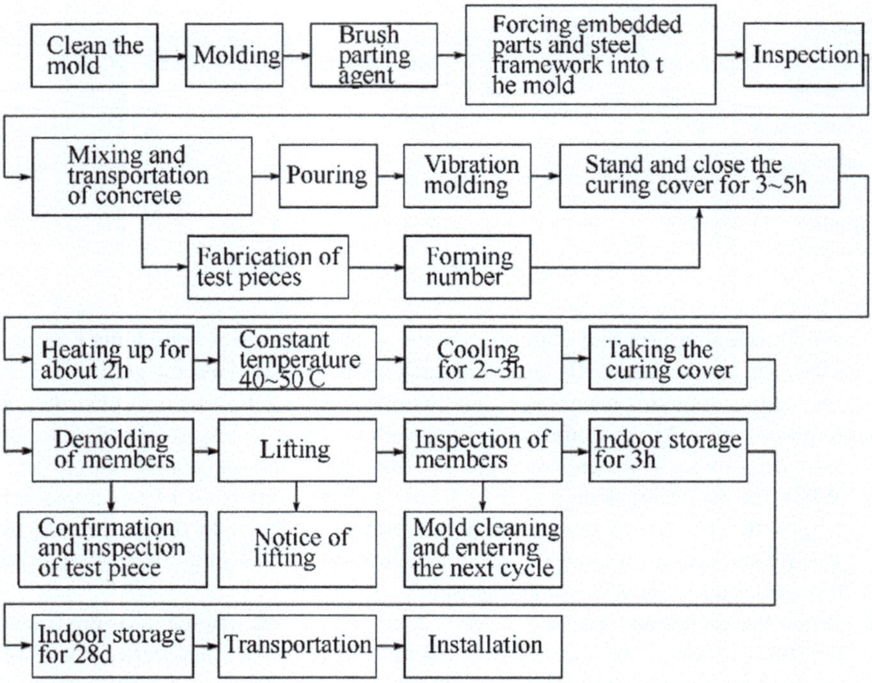

Fig. 5.6 Fabrication and heating and curing process of invert block

③ Precast blocks are cured by covering the curing cover. The steam boiler shall heat the heating radiator under the mold to maintain the ambient temperature at about 40° C. After constant temperature curing for about 18 h, the temperature shall be gradually reduced to room temperature and the curing cover shall be removed.

④ The curing constant temperature is determined to be 40 ~ 50 °C, the constant temperature time is 10 ~ 12 h, the standstill time is 3 ~ 5 h, the temperature rising and falling time is 2 ~ 4 h, and the temperature rising and falling speed is not more than 15 °C/h.

⑤ When demolding, the difference between the temperature inside the cover and the internal temperature of concrete and the room temperature is not more than 25 °C.

⑥ The demolding strength reaches more than 50% of the design strength of the top block, that is, not less than 20 MPa.

⑦ Keep the humidity in the cover above 90%, and the humidity is regulated at any time by replenishing the steam in the cover (switching the air escape valve).

(3) Laying and precise positioning of invert block

① After the invert block truck enters the loading and unloading area under the connecting bridge, it is manually rotated by 90° on the rotary table, and then the invert block is lifted from the truck by the 13t invert block crane under TBM with three chain load hooks, and transported forward to the required installation position.

② The rock chips and waste at the bottom of the invert block are manually cleaned into the muck bucket, unloaded to the belt conveyor behind the cutter head with the monorail bucket, and sent to the front of the cutter head with the belt conveyor.

③ Control the position of the invert block with laser pointer and track level installed in the central ditch of the invert block (check the position of the target in the central ditch of the invert block with laser light, and check the level of the invert block with a track level).

④ Determine, check and adjust the laser direction of the laser pointer with the ZED260 guidance system on the TBM.

(4) Inspection of backfill compactness of invert block

① The maximum particle size of backfill concrete coarse aggregate is 10 mm and slump is 15 ~ 18 cm.

② Grouting shall be carried out by grouting holes on the invert block, with a grouting pressure of 0.3 ~ 0.4 MPa.

After grouting for 14 days, the compactness of the bottom of the invert block shall be checked by geophysical prospecting technology, and the uncompacted part shall be grouted again.

7. Regripping operation of TBM and adjustment of tunneling deviation

For the first time, the GPS positioning system and national first-class leveling survey are adopted for the plane and elevation control and survey outside the Qinling Tunnel. ZED guidance system suitable for TBM rapid tunneling is adopted in the tunnel. The rapid, accurate and high-accuracy tunneling of TBM in long and large tunnels is ensured.

The control and survey and construction survey of Qinling Line I tunnel have achieved success. The breakthrough measurement results show that the transverse breakthrough tolerance is 10 mm, the longitudinal breakthrough tolerance is 13 mm, the elevation breakthrough tolerance is 4 mm, and the TBM tunneling tolerance is ± 30 mm.

(1) Control and Survey

① Construction survey will be conducted based on traverse points and elevation points. The accurate survey team of the engineering division shall review every 500 m of tunneling, and the measuring points shall be reviewed every 1000 m of tunneling.

② The surveying accuracy of the control survey in the tunnel, the horizontal control survey shall be carried out according to the requirements and accuracy of the second class conductor, the elevation control survey in the tunnel shall be carried out for the instrument inspection regularly once a year according to the second class level and GM3 optical micrometer and GPL3 steel ruler.

(2) Construction Survey

TBM is equipped with ZED260 guidance system, T2 theodolite, Grade S3 level gauge and supporting instruments due to its fast tunneling and high accuracy. When surveying, the coordinate parameters are provided for ZED system with conventional traverse survey method, and the tunneling parameters are provided for TBM by ZED260 guidance system to guide tunneling.

(3) ZED260 system

ZED260 guidance system can quickly, continuously and accurately provide TBM with the relative relationship between its axis and the design axis of the tunnel, and automatically and digitally display the deviation value of the cutter head center position from the design position to TBM operators, so that they can know the spatial position of TBM at any time and adjust the direction in time to ensure the high - accuracy tunneling of the TBM.

① The ZED system works on a laser basis, which is provided by an electronic theodolite with laser goggles.

Combined target/tilt target unit: This unit is used to measure the vertical and horizontal displacement of TBM with laser as reference, and simultaneously measure the horizontal angle of laser incidence, TBM rolling angle and tilt angle.

Control unit: Operator's main display screen and computer.

② Installation of ZED guidance system. Instrument calibration and equipment allocation; Processing and installation of various supports, installation of target combination unit; Control unit installation and cable connection; Installation and connection of laser/theodolite /EDM and interface unit assembly; Computer networking inside and outside the tunnel. Calculation and input of DTA table of tunnel design axis: The DTA table is calculated by taking the tunnel design axis as a trajectory formed by space points; The axis parameters of TBM (elevation angle, rolling angle, guiding angle) are measured by conventional measurement method, and the relationship between the axis of target unit and TBM axis is corrected by the parameters. At the same time, input the coordinates (x,y) of the center of target unit relative to TBM axis, and the theodolite coordinates of laser/theodolite /EDM combination unit are provided with its position (x,y,h) by conventional traverse measurement method. After the system is checked, it enters the automatic connection state, and the system works normally, and the interface unit assembly of the theodolite moves forward every 100 m, and the position of the theodolite (x, Y, H) is remeasured.

(4) Guidance principle and steering principle of ZED260 system

ZED260 guidance system is based on the laser emitted by the laser/theodolite and diastimeter installed in the tunnel, and the interface unit transmits the parameters to the control unit. The target unit receives the laser beam, measures the position of the light spot and the relationship between the laser incident, axis and the axis of the target unit, measures the axis of TBM, and sends them to the control unit. After considering the correction parameters, the computer system in the control unit can calculate the deviation of the current position (x,y) of the TBM axis at the target plane relative to the design axis, and calculate the deviation of the center position (x,y) of the TBM cutter head according to the included angle between the TBM and the design axis. These data are displayed on the ZED operation screen, and TBM operators can adjust the TBM direction at any time. The TBM operator mainly observe the first and fourth screens of the display screen, the first screen shows the deviation of the axis point at the TBM target plane, the fourth screen shows the deviation of the position of the TBM cutter head, and the direction can be adjusted with the angle between the TBM axis and the design axis (unit of which is mm/m). For example, the elevation angle of TBM is 15 mm/m, the current position is 50 mm lower and the cutter head is 10 mm lower.

Adjust the rear support cylinder so that the axis of the main engine continues to tunnel at 15‰, and the centerline of the lower circulating cutter head is closer to the design position. Observe the screen page for each cycle regripping of tunneling and adjust the direction at any time so that the center of cutter head is always adjusted within the design deviation.

(5) Control of Tunneling Deviation

ZED260 guidance system is a set of guidance system suitable for TBM rapid tunneling. However, due to many factors affecting TBM tunneling deviation, it is particularly necessary to establish a TBM tunneling deviation control system. In case of special circumstances, timely measures should be adopted to strictly control the driver's steering to ensure the tunneling quality. There are seven factors affecting TBM deviation:

① Measurement accuracy. Because the excavation section of TI3M is round, the system can't control the deviation between measurement and construction measurement, directly affecting the accuracy of tunneling and invert block laying.
② Heavy dust will directly result in the invalidation of the laser beam. Since the system cannot receive light signals and the system cannot work normally, the ventilation system must ensure normal operation, especially the normal operation of the industrial exhaust fan. In case of special circumstances, the operation must be shut down for ventilation.
③ High temperature and moisture will make the air uneven and refract the laser beam, which is the main reason for tolerance in ZED system.
④ Poor geological conditions will lead to the weakening of TBM support system, the main engine can't tunnel in the set direction, and the corresponding treatment measures should be adopted when the surrounding rock collapses.
⑤ Due to the dead weight of TBM cutter head and the reaction force of rotation on surrounding rock, cutter head tends to be low and left, so it is necessary to consciously tunnel to the right and up when tunneling.

⑥ It is necessary to ensure that there is no direct light on the target screen and that the laser path is unobstructed.
⑦ The steering technical level of drivers needs to be further improved. Through learning, training and strengthening management, the professional level of operators can be significantly improved.

(6) Key points of regripping in TBM tunneling

After one tunneling cycle is completed, it is transferred to the next cycle for regripping. Regripping mode has great influence on cutter and advance speed. The key points of regripping are:

① Set the cutter head back 2 ~ 3 cm.
② Idle the cutter head for 10 ~ 20 s, stop the cutter head rotation, power off the motor and stop the belt conveyor.
③ Pull out the lower support of the shield to increase the cylinder pressure to 18 MPa.
④ Use the pine folder to clamp the cylinder. At this time, the supporting pressure under the shield drops, and the pressure should continue to rise to 14 ~ 16 MPa. During this period, if you feel the edge cutter scraping the ceiling of tunnel, the support can be stopped.
⑤ Pull out the rear support and raise the pressure of the vertical cylinder of the rear support to about 15 MPa.
⑥ Relax the outer Kelly I and outer Kelly II.
⑦ Move the outer Kelly forward and make deflection adjustment for the outer Kelly and main machine attitude at the same time.
⑧ After the main machine is in place, adjust the main machine rolling.
⑨ Tighten the outer Kelly I and outer Kelly II to about 27 MPa.
⑩ Retract all shield supports.
⑪ Pull out the lower support of the shield to outer Kelly I, increase the pressure of the gripper by 0.2 ~ 0.4 MPa, and pull out the other shield to close to the tunnel wall.
⑫ Clamping cylinder.
⑬ Drag the back-up, and move the rear support forward.
⑭ Start the belt conveyor and start the main motor in turn while dragging the back-up.
⑮ Start the cutter head to rotate. At this time, the pressure of the support cylinder under the shield is about 12 MPa, which can be increased to 14 MPa.
⑯ Start tunneling.

In addition, when tunneling in soft rock or broken geological conditions, the thrust is small, the torque is large, and the vibration is large. During tunneling, the top shield and the wing shield can be supported to about 5 MPa to maintain floating support with the tunnel wall, which can greatly reduce the vibration, reduce the torque peak, and benefit the tool stress, the service life of motor and main engine.

8. Application of ventilation dust removal and other supporting auxiliary equipment

(1) Ventilation system

Related contents refer to 5.2.2.

(2) Dedusting System

Dedusting system is used during TBM tunneling. Because TBM will produce a large amount of extremely fine dust when breaking rocks, it will be fatal to operators and other equipment without dust removal. The working mechanism of dedusting system is dust removal with suction swirling water. During the TBM tunneling, the suction air duct sucks the dust in the front of the cutter head and in the front of the main engine into the dust collector, and discharges the dust-removed air through filtration and swirling water spraying, and the filtered water containing rock dust is recycled after sedimentation. The system is mainly composed of air suction pipe, industrial exhaust fan, noise-suppression equipment, water deduster, settling water tank, circulating water pump, etc.

(3) Installation system of bolt driller and ring beam

These two systems are mainly used to anchor and support the rock face when the surrounding rock is relatively broken during TBM tunneling. There are two bolt drillers in front of and behind the main engine to anchor the rock surface in the upper and lower working areas respectively. The ring beam erector is located at the rear of the cutter head, and is supported by the arch when the upper part of the self-shield of the broken rock surface is just exposed.

(4) Injection system

The system is located in the 20 m area behind the connection bridge, and mainly completes the secondary shotcrete reinforcement of the broken rock surface after anchoring and supporting. The concrete slurry is pumped by the transfer pump and input to the shotcrete manipulator for vertical shotcrete on the rock surface.

(5) Crane

In order to fulfill the lifting requirements of all parts of TBM, there are also several cranes. For example, the invert crane mainly completes the laying of the invert block at the lower part of the connection bridge at the rear of the main engine; The lower material crane is mainly used to lift other auxiliary materials' such as rails, arches, meshes, anchors, other than invert blocks transported from outside the tunnel; The upper material crane is mainly used to carry out the secondary lifting and placing of the materials transported from the lifting platform to the upper part of the connection bridge. Lifting platform is mainly used for lifting and transporting the materials in the upper and lower part at the connection bridge. Materials used in the upper part of the main engine can be lifted by this equipment to the upper platform of the connection bridge.

(6) Air compressor

It is equipped with $20m^3$/mm electric screw Ingersoll Rand air compressor, which mainly produces compressed air with a rated pressure of 0.8 MPa. There are three main types of equipment on TBM that use compressed air: First is the main bearing, the K-grease lubrication system, the second is the injection system, and the third is the use of air tools.

(7) Standby generator

The standby generator is an emergency standby power supply. It is only used for power supply of lighting, ventilation, dust removal and other systems, and is used for standby in case of sudden power failure of the grid, which is automatic switch.

The above is the functions of ventilation, dust removal and other auxiliary equipment on TBM. During tunneling, when used in hard rock and soft rock, the above equipment is used and maintained differently.

In the case of hard rock, because the rock is hard, TBM needs to consume more energy to break rock. Naturally, the part of energy converted into heat will be more, and the temperature in the tunnel will increase, so it is necessary to increase the power of the refrigerator of the ventilation system to reduce the temperature in the tunnel. With the increase of the temperature in the tunnel, it is possible to increase the temperature of the external air inhaled by the electric air compressor, resulting in the increase of the screw temperature and automatic downtime. The negative influence of the temperature increase in the tunnel caused by hard rock only affects the air compressor for ventilation, dust removal and other auxiliary equipment. In addition, due to hard rock tunneling, the rock surface is well formed, and the temperature increase has no special influence on the bolt driller, steel arch installation system, spraying machine, crane and standby generator. For supporting and spraying equipment, its workload and frequency of use will be relatively reduced.

In the case of soft rock, because the rock is soft or broken, after the cutter head breaks the rock, it is easy to form many small rocks. These small rocks are likely to cause crack damage of blades of industrial exhaust fan due to impact and reduce dust removal capacity after they are inhaled into the industrial exhaust fan. Because of the reduced dust removal capacity, the air in the tunnel will be worse than that in hard rock. Poor air in the tunnel requires increased ventilation volume. At this time, the ventilator outside the tunnel can be increased to increase the air volume so that the air with high dust content in the tunnel can be discharged outside the tunnel in time. The deterioration of the air in the tunnel also has an impact on the air compressor. Due to the high dust content in the inhaled air, friction heat is easily generated during screw operation, resulting in automatic downtime of the air compressor due to high temperature. When tunneling in soft rock, the rock surface is poorly formed and the workload of support and grouting is large. At this time, the workload of bolt driller, arch installation system and grouting system will increase, and the workload and frequency of each crane used for transporting materials will also increase.

5.3 Quality and Safety Management

5.3.1 Quality Management

Due to the change of construction methods, the content of construction quality management has also changed, which includes advance direction control, support quality, geological forecast, treatment of adverse ground condition, etc.

According to the characteristics of construction, TBM construction quality management must be controlled throughout the whole process, especially for the concealed project of invert block laying and grouting. Because it runs through the whole construction process, it is more necessary to implement quality control, and make quality records for each block to achieve quality traceability. Compared with the drill and blast, TBM construction quality management adopts "breaking up the whole into parts", and more manpower is required than the drill and blast.

The foundation of TBM construction quality management is good equipment performance and high-quality personnel. As a mechanized tunnel excavation, the equipment in good condition directly affects the quality control. If the dust removal effect of industrial exhaust fan is poor, it will lead to ZED laser refraction and deviation of advance direction; The performance of the injection machine affects the injection effect. The quality of construction personnel has a greater influence on quality management than the drill and blast.

As quality control runs through the whole process of construction, the quality requirements for construction personnel are improved accordingly. Construction personnel include related mechanical repair personnel, operators, technicians, managers, etc.

5.3.2 Safety Management

The main contents of TBM construction safety management include: Transportation safety management, equipment safety management and personnel safety management.

In addition to the fast operating speed of vehicles and insufficient space in the tunnel, another reason for transportation safety management is that the transport vehicles are connected in series with multiple mining vehicles (material vehicles), which are prone to decoupling and sliding due to various factors. It directly threatens the safety of equipment and personnel.

Equipment safety management: TBM occupies most of the space in the tunnel, and all construction personnel, vehicles and materials are gathered on TBM, so the probability of equipment damage is high. On the one hand, falling rocks may damage the equipment, on the other hand, equipment damage may occur due to improper operation of personnel.

Scientific technical support, strict organization and management, high-quality maintenance personnel, strict reward and punishment measures for construction discipline and advanced monitoring means make TBM maintenance work an organic whole, which greatly improves the maintenance quality and time utilization ratio of TBM and ensures that TBM can play its role efficiently.

In terms of safety and environmental protection management for TBM construction, the following work shall be focused on: TBM safety technical operating procedures shall be developed; Applicable ventilation methods, ventilation equipment and temperature control measures in the tunnel shall be selected to meet the requirements of national industrial hygiene standards; Lighting and firefighting facilities shall be provided; Adequate drainage equipment shall be prepared and emergency plan measures shall be provided; Safe passageway must be reserved at the operation location and site and smooth of the passageway shall be ensured; Corresponding preventive measures shall be adopted to reduce construction noise, vibration, water and soil pollution, control surface subsidence and reduce the possible impact of construction on the surrounding environment; In case of flammable or harmful gas in the tunnel, relevant detection instruments shall be provided and ventilation shall be strengthened to control the concentration of flammable or harmful gas within the safe allowable value; Wastewater treatment facilities shall be provided, and the wastewater discharged during construction shall be discharged after being treated to meet the standards.

In terms of disaster prevention management for TBM construction, the following work shall be focused on: Relevant emergency rescue construction plan shall be prepared; A special organization shall be established and necessary alarm, rescue and escape facilities shall be provided. Before commencement of construction, the emergency rescue construction plan training and rehearsal shall be carried out; Sufficient disaster prevention materials and equipment such as sand bags and water pumps must be prepared; Communication inside and outside the tunnel shall be kept unblocked; Sufficient portable dry fire extinguishers must be provided on the TBM, and special firefighting facilities must be provided at key parts; The back-up system is equipped with disaster prevention devices such as advance forecasting, warning devices, dust prevention, exhaust gas prevention, explosion-proof, noise prevention, fire prevention.

Questions

1. What principles should be followed in the preparation of TBM operative construction organization design?

2. What should be the main contents of TBM operative construction organization design?
3. What are the requirements for plane design of TBM construction site layout?
4. What aspects are included in the layout of temporary works at the TBM construction site?
5. What requirements should the assembly site outside TBM tunnel portal meet?
6. What requirements should the assembly site inside TBM tunnel portal meet?
7. Briefly describe the basic sequence of main machine assembly of open type TBM.
8. Briefly describe the basic sequence of main machine assembly of double shield TBM.
9. What requirements should the assembly of connection bridge meet?
10. What requirements should the assembly of the back-up gantry meet?
11. What requirements should the assembly of continuous belt conveyor meet?
12. Briefly describe the main commissioning contents of TBM.
13. What are the main contents of open type TBM commissioning?
14. What are the main contents of double shield TBM commissioning?
15. What requirements must be met for the acceptance of TBM main machine?
16. What requirements must be met for the acceptance of shield TBM segment installation machine?
17. What are the general working modes of TBM, and briefly describe their scope of application respectively?
18. What should we focus on in the safety and environmental protection management of TBM construction?
19. What should we focus on in disaster prevention management of TBM construction?
20. What are the main construction operation systems of TBM?
21. What are the main temporary facilities in TBM construction?
22. Briefly describe the contents of TBM construction preparation.
23. What are the preparations for TBM assembly?
24. Briefly describe the operation points of TBM boring.

Chapter 6
TBM Assembling and Tunneling

Key points of this chapter: This chapter introduces the transportation process of TBMs, then details the assembly and commissioning process of TBMs from pre-assembly preparation, site assembly and commissioning, and finally introduces the tunneling process of TBMs from the aspects of TBM tunneling procedures, operation, mode selection and parameter selection.

6.1 Transportation of TBM

TBM is a kind of large-scale construction machinery. There are many large pieces after the disassembly of the whole machine, such as cutter heads, drive components, etc. Due to its large mass and volume, there are corresponding requirements for the mode of transportation and the carrying capacity of roads, bridges and tunnels. Generally, the projects constructed with TBM are often located in mountainous areas, making the transportation problems more prominent.

The mobilization and transportation of the TBM is divided into two parts, one is the transportation of container and some bare parts. Containers and some bare parts are transported in batches by special container flat trailer and other gantries. Because there is no over-width, over-height, overweight and other situation, the transportation is relatively simple, and there is no need to carry out road blocking, bridge reinforcement, road obstacle removal and other treatments. Only one safety car needs to be arranged to divert traffic and transmit information when necessary. The other is the transportation of large pieces of the TBM, but the high mountains and dangerous roads, steep slopes and sharp curves are the key points and difficulties in the transportation of the TBM. For TBM transportation, a transportation plan shall be prepared in advance and safety measures shall be formulated to ensure the safety of TBM transportation.

K. Chen et al., *TBM Design and Construction*,
https://doi.org/10.1007/978-981-99-0059-6_6

1. Mobilization and transportation
(1) Prepare technical scheme for the transportation of TBM components in advance. The content shall include: advance investigation of the carrying capacity of the roads to be passed through; selection and configuration of equipment for large pieces transportation; safety measures in large pieces transportation and emergency plan for safety of large pieces transportation.
(2) Large pieces transportation shall be undertaken by qualified companies, and the carrier shall be determined by bidding.
(3) Safety measures shall be taken around the TBM lifting site, and all lifting equipment entering the site shall be carefully inspected and accepted. At the same time, safety inspection shall be carried out on all slings and various certificates of special operators shall be checked for qualification.
(4) During equipment transportation, the traffic must be ensured, and special personnel shall be assigned for safety and monitoring.
2. Organization structure
(1) Command structure

One commander-in-chief is assigned for TBM large pieces transportation, who is fully responsible for all matters related to the transportation; one executive commander is assigned to be specifically responsible for the organization and implementation of the transportation scheme, and is responsible for all specific work during transportation to ensure the progress and safety of transportation.

(2) Staff organization

For TBM large pieces transportation, 21 drivers of various types are responsible for parts transportation; 3 safety and quality supervisors are responsible for fleet guidance, clear road, order maintenance, and safety and quality inspection; 2 technicians are responsible for technical guidance, and arrangements for vehicle maintenance and repair; 2 liaison officers are responsible for internal and external liaison, including coordination and liaison in traffic, transportation management and traffic control along the way; 2 lifting workers are responsible for loading and unloading operations; 6 repairmen are responsible for the repair and maintenance of various vehicles and equipment; 2 security guards are responsible for the safety and security of vehicles and goods when parking; 1 doctor is responsible for the health care of all personnel; 2 logistics personnel are responsible for food and accommodation.

Reasonable, perfect and efficient organization of command and transportation agencies is important for the good transportation of TBM large pieces.

3. Transportation safety guarantee measures
(1) Survey in advance, and formulate reasonable, feasible and perfect transportation plans.
(2) Be fully prepared and troubleshoot in advance to ensure the implementation of the transportation plan.
(3) Fully understand the road conditions along the way, and entrust the local highway or transportation department to reinforce the road or remove obstacles in key sections in case of over-sized, over-wide and overweight objects.

(4) When loading, the TBM dismantling components and containers shall be placed stably on the flat trailer, and the components shall be fixed with the vehicle body to prevent slipping during transportation.
(5) During transportation, the accompanying personnel shall always observe the driving state of the vehicle and road conditions, and take corresponding measures in time.
(6) The driver of the transport vehicle shall strictly control the safety performance of the vehicle and carry out routine inspection before, during and after driving to ensure good performance of the vehicle.
(7) When lifting and unloading TBM components, the steel wire rope shall be hung first and then the chain shall be loosened.
(8) Reasonable security work shall be arranged to prevent loss and damage of goods.

6.2 Assembling and Commissioning of TBM

TBM integrates machinery, electronics, hydraulics and technology, with complex technology and huge structure. It is a factory-based tunnel production line integrating excavation, support, mucking, ventilation and drainage. The quality and accuracy of the assembly process is quite important for the future service performance, service life and even maintenance cycle of TBM. During assembly, the requirements of assembly process procedures shall be strictly met. The TBM sets were delivered to the site in bare parts and in containers. The main machines are delivered in bare parts, most of the accessories are delivered in boxes, most of the back-up systems are delivered in bare parts, and the key hydraulic and electrical components are delivered in boxes. These different forms and types of components are assembled with special machines and tools according to the accuracy requirements of TBM design documents, and the main machine, connection bridge, back-up and accessories are assembled respectively and connected into a whole with relevant components.

Based on the overall planning of the assembly site and the construction schedule and other conditions, the overall process of TBM assembly was determined. Carry out management, technical organization and arrangement of TBM assembly in different systems, parts and periods. Take safety guarantee measures to ensure the safe and smooth completion of TBM assembly.

The site assembly of TBM is divided into three main phases, namely the preparation phase, the assembly phase and the site commissioning phase.

6.2.1 Preparation Before Assembly

1. Basic requirements

Break down the assembly tasks, and determine professionals to be responsible for assembly and commissioning of each component of the TBM; Study the assembly drawings and technical requirements, and understand the assembly structure, characteristics and adjustment methods; Develop assembly process procedures and select assembly methods; Ensure the order of the assembly site, and determine the assembly site layout and reasonable assembly sequence; Prepare suitable assembly tools; Visually inspect the assembly, remove burrs and chamfer, and clean and lubricate it.

2. Data preparation before TBM assembly
(1) Site conditions and climatic conditions for tunnel site assembly.
(2) Technical data such as size, mass and position of TBM components.
(3) For the arrival of TBM components, the entering sequence of large pieces shall be determined according to the assembly sequence. During unloading, the components shall be hoisted and placed according to the ground diagram.
(4) Attached technical data and successful TBM assembly solutions and experiences from other previous projects.
(5) Prepare the assembly construction organization design according to attached data, determine the assembly personnel, assembly equipment, machines and tools, materials, etc., and check the preparation of various resources.
(6) Operative construction organization design of tunnel.
(7) Safety construction-related regulations of the state, the Ministry of Transport and the enterprise.

3. Technical preparation before TBM assembly
(1) Based on the size of the TBM diameter, a detailed and feasible TBM transportation, assembly and deployment scheme was developed. For the assembly of each large piece, we shall prepare the corresponding safety assembly plan, formulate the assembly requirements, personnel division and their responsibilities, safety guarantee measures and accident handling plan for each large piece in detail, and submit the assembly plan to the superior competent department for approval.
(2) The size of the assembly site shall be determined according to the TBM diameter, and the assembly site shall be rolled and leveled to meet the design requirements. Concrete with a thickness of not less than 20 cm shall be laid. The traveling track of large gantry crane shall be made of reinforced concrete, and the "three links" of water, electricity and roads shall be ensured in the site.
(3) Carry out technical training and technical disclosure, and technical testing and assessment for special types of work.
(4) Develop a reasonable plan for the amount of assembly materials, machines and tools needed.
(5) The assembly parts shall be clearly marked and neatly stacked, and cleaned well.
(6) A full-time safety officer shall be assigned to monitor the whole process of TBM assembly.
(7) A comprehensive inspection and maintenance shall be carried out on the assembly site, large lifting equipment, auxiliary assembly equipment, etc., and a written inspection and maintenance report shall be provided. Large gantry cranes for assembly must pass the inspection of the local safety supervision

department before use, and the use/start operation certificate shall be issued after it passed the inspection.

(8) The assembly site shall be enclosed and guarded by special personnel.
(9) Large assembly equipment and lifting slings shall be subject to simulated-quality trial lifting before assembly of large pieces, and the trial lifting scheme shall be prepared. The assembly can only be carried out after the requirements are met.
(10) Set up a leading group of assembly and divide it into a number of professional groups, so that the assembly work can be carried out in an orderly manner.
(11) Assembly site hardening to meet the requirements of equipment assembly, and construction of temporary storage site, preparatory tunnel, stepping tunnel, laying of large sling traveling track, etc.

4. Preparation of TBM assembly personnel and tools

(1) Staffing

According to the construction organization design and the characteristics of the TBM structure, we usually implement a three-shift operation and divide the work according to specialties. For each shift, more than 30 people, including electricians and crane drivers, are led by the shift leader and professional engineers to carry out assembly to ensure the assembly progress. Before each operation, the technical staff shall make technical disclosure to ensure that each staff understands the contents and dangers of the work to ensure that the work is carried out under safe conditions.

(2) Tool preparation

The assembly of equipment includes various steel structural components, oil pipes, wire connections, etc. To complete these tasks, relevant tools and materials must be used. The main tools include: various types of wrenches, pliers, chain blocks and electric tools, etc. Tools, materials, hydraulic pipe joints, special tools, etc. shall be placed nearby, and the following containers shall be placed in order facing the opening: maintenance room, storage room of screws and nuts, hydraulic room, electrical room and containers for storing various types of hydraulic and water pipe joints.

Lifting tools shall enter the construction site in advance and be placed as required. Large equipment such as gantry cranes shall be installed on site by companies with relevant qualifications and safety inspection shall be carried out. Hoisting equipment and lifting slings are shown in Table 6.1.

Due to the need for air, water and electricity for hydraulic tensioning, lighting, welding and other special tools during assembly, the TBM assembly site shall be prepared in advance, and a special construction scheme shall be prepared in advance for electricity, and the electrical engineer shall be responsible for standardizing the electricity system, strengthening organization and technical support to ensure electricity safety (Air: Prepare inflating pump, jackhammer and air impact wrench, and use them for quick fastening of connecting bolts during equipment assembly.; Water:

Table 6.1 Hoisting equipment and lifting slings

S. No.	Name	Qty	Specification
1	Gantry crane	2 sets	20t, 150t
2	Truck crane	2 sets	8t, 16t, 75t
3	Chain slings	1 set	32t × 4
4	Steel wire rope	16 for each	5t, 10t, 20t, 40t
5	Sling	12 sets	1–6t
6	Special lifting slings for drive assembly	1 set	
7	Cutter head lifting sling	1 set	
8	All kinds of lifting slings	4 sets	
9	Various types of lifting lugs	2 sets	M12, M16

Use as required; Electricity: Provide power for small electric machines, such as: inflating pump, electric drill, hand grinder, etc.).

In order to carry out electric welding operations, oxygen, argon and acetylene are indispensable. Before commencement, corresponding types and quantities of firefighting equipment and facilities shall be provided according to the needs of construction organization design and fire prevention measures. The distance between welding and cutting operations and oxygen cylinders, calcium carbide barrels and acetylene generators and other hazardous materials shall all meet the specified safe distance requirements. The fire use approval system shall be strictly implemented, and special personnel shall be assigned to ensure the safety of fire use.

At the same time, assembly materials shall also be prepared, mainly including steel wire rope, shackle, profile steel, square timber, planks and some emergency materials.

6.2.2 TBM Site Assembly

TBM site assembly is an extremely important task in the TBM construction organization. The Contractor's project management office shall prepare the construction site assembly plan one month before the TBM arrives at the site and submit it to the Supervisor for approval before implementation. The main working part of the TBM can be divided into the main machine part, the back-up part and the continuous belt conveyor part. Because the working positions of these three parts are different, the main machine and the back-up part of TBM are assembled in the assembly tunnel, and the continuous belt conveyor is assembled at its installation position (i.e. between the assembly tunnel and the muck discharge outlet).

The assembly of the whole TBM must be carried out in accordance with the installation order of its components to avoid rework caused by disordered sequence, which will affect the construction progress and assembly quality. Analyze the assembly methods of large parts and key parts, analyze and compare the welding methods of the cutter heads (tiled erection welding and vertical erection welding), and their impact on subsequent processes and welding accuracy; analyze the assembly methods of back-up sections and formulate a complete set of assembly sequence.

According to the determined assembly sequence, a detailed schedule is made for each link of the assembly progress, including the schedule of a series of processes such as main machine assembly, back-up assembly, belt conveyor assembly, complete machine commissioning and launching.

According to the assembly instructions and technical requirements of each TBM component, the construction methods of each assembly operation are developed. Requirements for the technical measures taken in key links, and the supply of resources are cleared.

Establish quality control objectives, divide and implement them into each functional department and site operator, and carry out measures, check, evaluation and acceptance. Quality control at the assembly site mainly includes welding quality inspection, physical and chemical testing and non-destructive testing, materials, equipment and processes, etc., so as to ensure that no designed function will be changed and no performance index will be weakened during TBM assembly.

Assembly of large equipment involves multiple types of special operations. All personnel involved must receive safety education and training, and pass the examination before going to work. In accordance with the types of operations involved in the assembly process, equipment operation procedures, cold-proof and skid-proof requirements for outdoor operations in winter, etc., and in combination with the assembly sequence, process and progress requirements, safety training for all personnel in the assembly operation shall be carried out, and courses such as "safety and technical training for work at heights", "safety training for crane lifting operation", "safety training for welding operation", "safety and operation procedures training for special equipment", "safety training for tunnel operation", "safety system training for projects" and "safety and technical disclosure training for temporary electricity" shall be set up. Special operators must have special work qualification certificate, prepare corresponding safety technical disclosure according to different types of work for each on-site operators, and prepare corresponding emergency plans for work safety accidents to ensure that no safety accidents occur during the assembly.

The TBM is assembled from front to back, from large parts to small parts, and from structure to pipeline. First, install the TBM shield, main machine and cutter head, in which the cutter head is assembled in blocks. First, install the center block, and then assemble 4 side sections in turn. At the same time, the overall assembly and component assembly of the back-up gantry are carried out simultaneously. The assembly and workflow of TBM in the tunnel are shown in Figs. 6.1 and 6.2.

1. Main machine assembly

The assembly of TBM main machine is often carried out in the assembly tunnel. The assembly tunnel is generally narrow, while there are many parts of TBM main machine. In order to ensure successful site assembly and avoid repeated handling, all

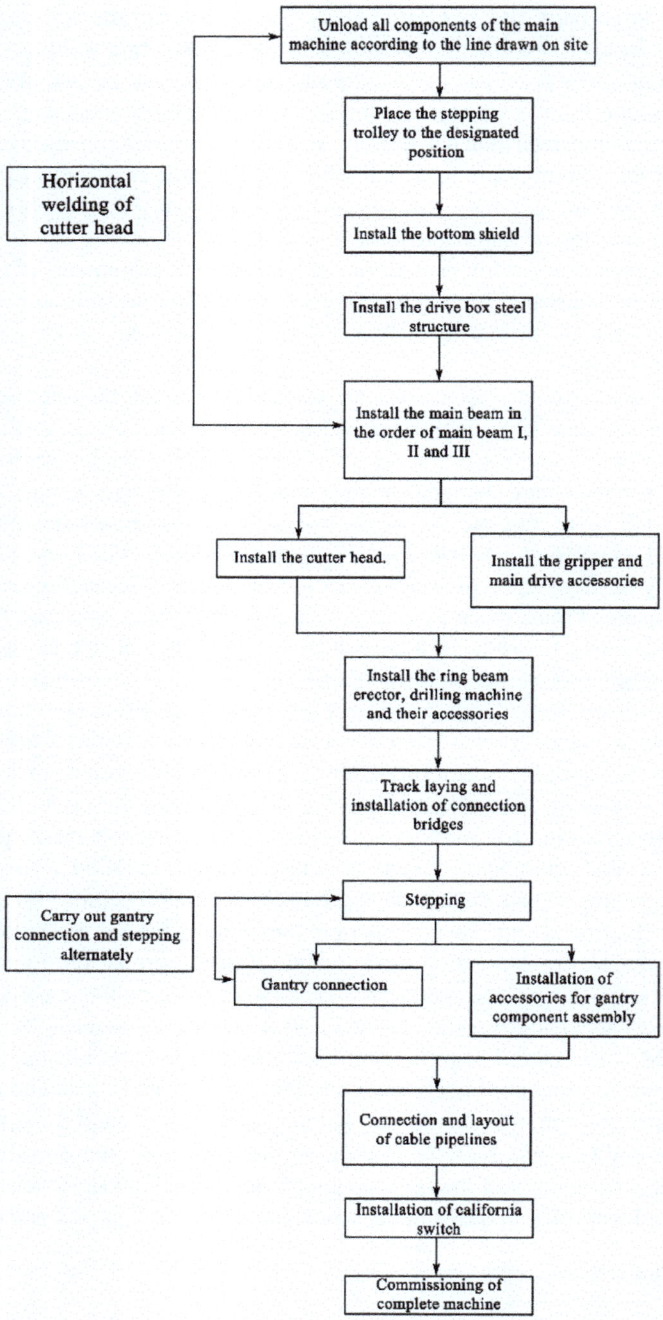

Fig. 6.1 Flow chart of TBM assembly sequence

Fig. 6.2 Assembly of TBM in tunnel

parts shall be arranged in advance according to the assembly position and sequence, and the site shall also be arranged. The distribution of the main machine parts is shown in Fig. 6.3.

2. Installation steps of TBM main machine

If installation steps are reasonable, the workload can be reduced, and repeated labor and a large number of reworks can be avoided. For the installation of TBM main machine, the assembly of large components of the main machine shall be assembled first, and then the assembly shall be carried out according to the installation sequence of main machine components before auxiliary engine components, main structural components before fragmentary components, parts before components, interior before exterior, and lower layer before upper layer. Finally, the layout of walkway, cover plate and steps, escalator, guardrail, sensor elements, control elements, signal lines, strong current lines, liquid, gas and water pressure pipelines shall be considered. For easy installation and efficiency of installation, the above sequence can also be carried out alternately without conflict.

The installation sequence of the main machine is shown in Fig. 6.4.

3. Installation methods and precautions

(1) Installation of stepping mechanism

The stepping mechanism is an important component for placing and fixing the main machine, which has high requirements for the flatness of the site. Before assembly, measurement shall be done in advance to ensure that the centerline of the stepping trolley coincides with the installation axis, and the flatness of the site shall be handled properly, otherwise the main machine may be unstable. At the same time, sufficient

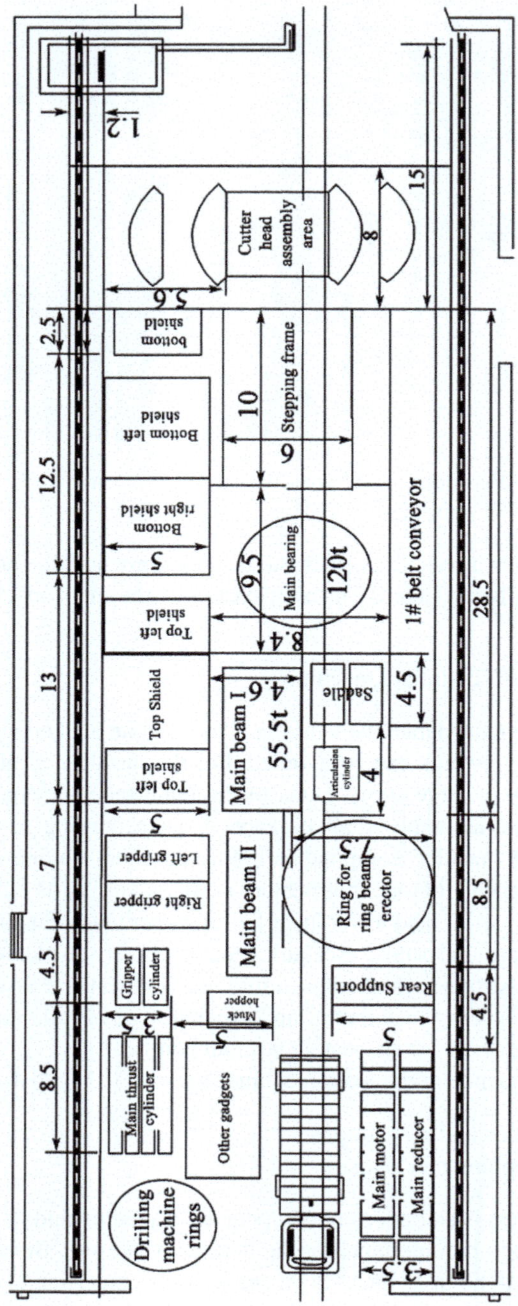

Fig. 6.3 Distribution of TBM main machine parts (unit: m)

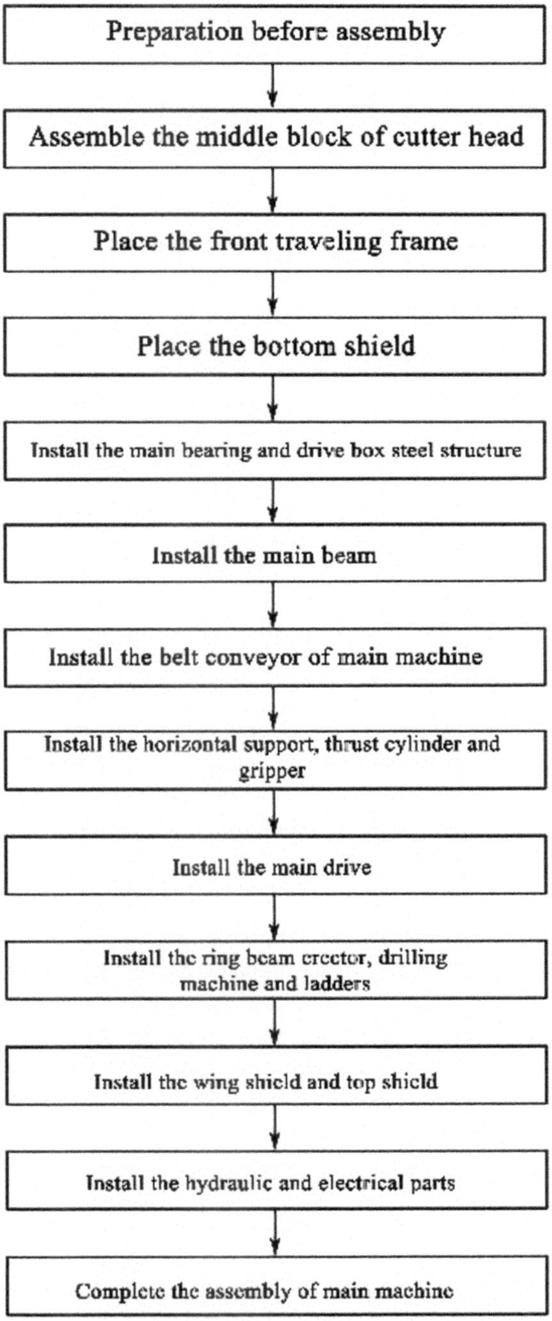

Fig. 6.4 Installation flow chart of TBM main machine

space shall be reserved at the placement position to avoid interference with the position of cutter head welding. After the bottom plate of the stepping mechanism is placed, the welding part of the inner surface needs to be polished to ensure that the friction between the bottom plate and the bottom shield is minimized during TBM stepping, and the joints are coated with lubrication grease.

(2) Installation of bottom shield

After the stepping mechanism is placed, the bottom shield can be placed (the position of the cutter head shall be reserved to ensure that the centralized gravity bearing points of the front main machine and the cutter head are located on the stepping frame after hoisting). After the placement, it shall be level and vertical, and all bolt holes and bottom shield pin holes shall be cleaned in time for anti-corrosion and anti-rust treatment on the contact surface with the main machine. The method is as follows:

① For installation, the bottom shield shall be placed on the stepping mechanism flatly. After the connection surface of the bottom shield is cleaned with cleaning agent by 2–3 persons, all machined surfaces and holes shall be thoroughly cleaned and the burrs shall be removed.
② Confirm that the mounting holes of lubricating oil return pipe are cleaned and intact. Slide the pipe into the hole and ensure that it fits well.
③ Install personnel access hole covers at the bottom of the bottom support.
④ Install the pins for the bottom shield. After necessary cleaning of the hole, install the connecting pin and tighten the bolts of the gland.
⑤ Install an anti-off key for the bottom shield pin and apply anti-self-locking agent. Level the bottom shield longitudinally and horizontally with a level ruler, and use steel plates to level any unevenness. After leveling, check the edge of the bottom shield to ensure that it will not roll when connecting with the drive box steel structure.

(3) Assembly method of cutter head rest

Place the cutter head rest horizontally on the timber pier, and clean the joint surface between the cutter head rest and the lower support before installation. Slowly lift the cutter head rest to a vertical state. After it is stable, slowly lift the cutter head rest to the right above the lower support, put it on seat and accurately align it with the bolt hole. Do not blindly tighten the bolt. Apply threadlocker to the bolts, and tighten the bolt to the specified value in a certain order with a hydraulic torque wrench according to the drawing instructions.

(4) Assembly method of the girder

Before assembly, the flange joint surface between the girder and the cutter head rest and the flange joint surface between the girders shall be cleaned, and the bolts and bolt holes shall also be thoroughly cleaned. In order to ensure that the girder is kept horizontal during lifting, the number of closed links of the lifting chain shall be repeatedly adjusted and trial lifting shall be carried out to find the ideal number of links. The height between the hook center and the upper surface of the girder shall be kept at least 3 m. After accurate alignment, install the bolts in time and tighten them according to the drawing instructions.

Use a jack to keep the girder level between the tail end wooden support and the lower bottom surface of the girder. After the support is stabilized, lift the middle section of the girder (the same as the lifting method of the girder). After tightening the bolts, use a jack to keep the girder level between the tail end support wooden frame and the lower bottom surface of the girder. After the support is stable, the tail section of the girder shall also be installed.

(5) Assembly method of belt bridge

Since the installation of the front of the belt conveyor and muck hopper does not conflict with the installation of the rear main machine, they can be installed simultaneously during the installation of the main machine, but must be installed before the cutter head is hung. The hoisting of the belt bridge is shown in the hoisting method of the girder. After installation, each section also needs to be kept level with a wooden support and a jack at its tail end. After the last section is installed, pull the No. 1 trolley to the lower part of the belt bridge tail, and seat the belt bridge. The muck hopper also needs to be installed smoothly by lifting the four corners with the steel wire rope and chain block. After lifting, level it with chain block, slowly move it to the installation position, adjust it with chain block to align it with the 4 pin holes, then apply anti-self-locking agent on the pin, thread it into the pin hole, install the cover plate and unload the crane.

(6) Installation sequence and tips of main drive assembly

10 main drive assemblies are located on both sides of the connection between the girder and the cutter head rest. Looking from the TBM into the tunnel, the left side is numbered from bottom to top as No. 5, No. 6, No. 7, No. 8 and No. 9, and the right side is also numbered as No. 4, No. 3, No. 2, No. 1 and No. 10. When lifting, it is required to bind firmly, with appropriate distance between the slings at both ends and stable center of gravity. The main drive is horizontally laid, and the flange directions of different main drives are different. No. 3, No. 4, No. 5 and No. 6 flanges are vertically upward, No. 1, No. 2, No. 7 and No. 8 flanges are vertically downward, No. 9 flange is horizontally left, and No. 10 flange is horizontally right. Due to the small space, the main drive shall be slowly lifted into place and then aligned with the connecting bolt holes one by one and screwed the bolts in one by one. Be very careful when hoisting, and protect the drive pinion from collision.

(7) Butting method of air duct

For square air ducts reinforced with thin iron sheets, sealing strips shall be pasted on one of the flange surfaces during butting. If it is rainy or the surface is wet, compressed air can be used to blow off the moisture before pasting. When the flange butt hole is slightly misaligned, first put on several screws, do not tighten them, and then tap them gently with a hammer. In case of serious deformation, tap the corresponding parts of the air duct with a hand jack inside the air duct, or turn over the air duct, and align it with the hole by its dead weight. After all screws are inserted, tighten them evenly and symmetrically.

For the round hard air duct inlaid with a flexible air duct, when it is slightly out of round, it can be rounded with a hammer. If the diameter of the hard air duct is slightly large and it is inconvenient for inlaying, ream several tapered openings with scissors in the circumferential direction of the round air duct, and then tap the opening with a hammer to inlay the flexible air duct. The metal collar of the hose bayonet lining can also be cut short by a small section, which will be more convenient for inlaying. After the flexible air duct is inlaid, install the clamp directly.

(8) Installation of ring beam

After the three sections of ring beam are assembled, a complete circular arch ring will be formed. Before the installation of the main bearing parts, the front part of the front inner frame shall be inlaid for temporary fixing (the front and rear parts shall be fixed with chain hoists and wires). When the shield assembly after the main bearing assembly is completed, the ring beam guide pulley and the ring beam rotary motor can be assembled. Before the ring beam is in place, only one ring beam mobile control platform can be assembled temporarily. Use a special crane sling to slowly hook the top of the ring beam and lift it, and move it forward to the ring beam guide pulley groove in the shield at the top of the cutter head, and then install the remaining ring beam mobile control platforms.

(9) Cutter head erection and hoisting

There are requirements for the environment of cutter head welding. It is necessary to build an insulation shed on the assembly site, and the sheds shall be structurally stable and easy to disassemble. The butting of the cutter head and the adapter ring has high requirements for both hoisting and assembly. When assembling the cutter head, the surface of the cutter head must be cleaned, the bolt holes need to be aligned precisely, and the connecting bolts need to be tightened to the specified torque after butting. The cutter head hoisting shall be carried out according to the special scheme for cutter head hoisting.

When connecting the cutter head flange to the adapter ring, fix the O-ring in advance. When assembling bolts, the connecting studs must be tightened twice with hydraulic tensioning tools, and 4 bolts shall be assembled symmetrically (1 on the upper, lower, left and right sides respectively). After tightening 4 bolts (the pre-tightening torque is 1/2 of the final torque), symmetrically assemble other bolts with the same torque, and finally assemble all bolts symmetrically in turn and tighten them with the final torque.

(10) Cutter installation

The installation of cutters is of great importance, and the installation quality will directly affect the service life and tunneling speed of cutters. When the cutter body is in place, higher accuracy and fit clearance for the mounting joint surface are required, otherwise the cutter body will be deformed. Therefore, collision is not allowed, and burrs, paint on the surface, dirt and rust on the inner hole of the screw must be removed without any blind spots. The surface shall not be cleaned with coarse abrasive cloth, but with metallographic abrasive paper and cleaning solution. In order to improve the cleaning efficiency, a high-speed rotating copper wire polishing wheel can be used to polish the joint surface. Besides, attention shall be paid to the orientation of the baffle ring of the cutter during installation. In order to make the force on the cutter ring more reasonable when cutting rocks, it is specially stipulated that the baffle rings of the center cutter shall face outward, while the baffle rings of other face disc cutters and side cutters shall face inward.

(11) Installation methods of dense hydraulic pipeline

Dense hydraulic pipeline flanges or round pipes are used for the bottom surface of each hydraulic console of the main machine, various hydraulic group valves, the connection between the connection bridge and the hose of the main machine, the splicing surface of the connection bridge and many other places. The pipelines are not only divided into upper and lower layers, but also have different thicknesses. During installation, attention shall be paid to the assembly sequence and the direction of the connecting pipe, and the pipelines shall be assembled flexibly without force. Assembly sequence shall be from bottom to top and from inside to outside. When there are enough space at both ends of the valve body, the pipe joint can be assembled from one end, otherwise it needs to be connected one by one from the center of the valve body interface to both sides. Multiple elbow joints on the same plane shall be arranged in a well-proportioned and beautiful manner as far as possible. When the wrench space is insufficient, pry up the pipeline to be tightened with a bar and then tighten it. For the crescent-shaped flange interface, install half of the flange first with a certain clearance, then clamp the round head of the oil pipe into the crescent notch. After alignment, install the other half of the flange, and then tighten the fastening screws evenly.

(12) Connection method of thick and long hose

As the pipe is long and thick, it is easy to be twisted and unnaturally stretched during connection and laying, and the threads are not easy to be fastened. Therefore, when screwing the joint, make sure the direction of the joint is correct, keep the axis of the pipe always perpendicular to the plane of the joint, and rotate the joint by hand as far as possible. Do not tighten the joints at both ends at the same time, but tighten one end first, then straighten the twisted pipeline, and then tighten the remaining end joints.

4. Main machine stepping

After the TBM main machine is completely installed, carry out the main machine stepping process. Step the main machine into the tunnel according to the following procedures:

(1) Confirm that the work before stepping is ready, and restore the main thrust cylinder, stepping jack, cylinder, connecting rod and rear support to the initial state.

(2) Start the stepping pump station and press the push button to start stepping. Other precautions:

The stepping commander shall always observe the sliding condition of the connection surface between the bottom shield and the stepping mechanism, and send command signals in time in case of problems. The stepping operator shall always keep in touch with the commander, and always note that whether the stroke of the main thrust cylinder is completed. After the stepping is completed for 1.8 m, press the stop button.

(3) Operate the stepping jack to lift the bottom shield about 20 mm away from the bottom plate of the stepping mechanism, and lock the jack; then lower the rear support using the group valve of the stepping pump station, so that the gripper at the bottom of the rear support is firmly supported on the stepping plane and locked after reaching the set pressure. The operator shall always pay attention to the pressure change of the rear support hydraulic system, and replenish the pressure in time until the temporary support of the saddle is lifted more than 20 mm above the ground.
(4) Operate the retraction handle of the main thrust cylinder to retract the main thrust cylinder for a stroke length. During the retraction of the main thrust cylinder, the saddle will be pulled forward together with the temporary support, so that the stepping thrust cylinder will be retracted by the force transmitted by the connecting rod to realize the stepping cycle reset.
(5) Release the pressure of the stepping jack, put the bottom shield down on the bottom plate of the stepping mechanism, and lift the jack to be 20 mm away from the ground; slowly retract the rear support until the bottom gripper is 20 mm away from the ground. Reset the stepping state, and extend the stepping rail (gauge 2000 mm) and locomotive running track (gauge 900 mm).
(6) Start the stepping operation of the next cycle according to the above process.
(7) After the cutter head is stepped to the cut-off line, stop the stepping.

5. Back-up assembly

According to different manufacturers and construction designs, the number of back-up gantries designed for each TBM is different. The back-up gantries are numbered in the order from the connection bridge to the tail, starting from No. 1, from small to large, where the shotcrete bridge is installed between No. 1 and No. 2. After the stepping of the main machine is completed, the installation and connection of No. 1 gantry and shotcrete bridge will be carried out. At the same time, the remaining gantry structural frame and accessories will be transported to the assembly area, and connected to the connection bridge from front to back in the order of No. 1 gantry, shotcrete bridge, No. 2 gantry and No. 3 gantry. When the assembly space is too small, a stepping operation can be performed every time a gantry is assembled until all gantries are connected. In the process of back-up assembly, all accessories shall be installed before the subsequent process.

6. Belt conveyor assembly
(1) Installation of main machine belt

After the installation and connection of all back-up jumbos are completed, the main machine, connection bridge and adapter belt shall be installed. First, install the main machine belt, adjust the position of each section of belt conveyor bracket and fix it, then install the belt, vulcanize and connect the belt, and then vulcanize and connect the connection bridge belt. Finally, vulcanize and connect the adapter belt.

Once the TBM belt conveyor is connected, the whole machine can be stepped to the launching chamber.

(2) Installation of continuous belt conveyor

The installation of the continuous belt conveyor can be carried out after the TBM back-up is fully stepped into the tunnel, and is divided into two working faces: the working face outside the tunnel and the working face inside the tunnel.

The parts to be installed inside the tunnel mainly include drive, support and adapter belt, which shall be installed according to the drawings. The parts to be installed outside the tunnel mainly include drive tower, adapter belt, unloading tower and corresponding belt conveyor support and drive, which shall be installed according to the site layout and drawings.

7. Pipeline connection of electrical and hydraulic systems

Once the TBM is stepped in place, the long and large cables of the electrical system can be laid. Each power supply system shall be laid in accordance with the requirements of designed cable channel and clear systems. The following principles shall be followed:

(1) High-voltage cables, power cables, lighting cables and control cables shall be laid separately according to the designed cable tray layout.
(2) The pipeline at the gantry connection must be reserved with radians to meet the requirements of curvature radius.
(3) Rubber shall be padded under the pipeline where it is easy to loosen and move to prevent long-term wear and failure of the pipeline.
(4) During the laying of electrical cables, attention shall be always paid to prevent abnormal cracking of cables, and waterproof covers shall be set up in each junction box in time.

Hydraulic oil filling of hydraulic system, lubricating oil filling of lubrication system, installation of grease system, gear oil filling of main drive reducer, etc. shall be added according to the designed oil filling amount. During operation, the fuel dispenser, pipeline and joint must be kept clean, and the environment shall be kept clean as far as possible to avoid filling when dust floats.

8. Principles to be followed during TBM assembly
(1) Installation of structural parts and components
① Site operators must comply with safety procedures.

The specified safety helmet must be worn, the safety belt must be fastened when working at heights, and the oil stains on the surface on which the personnel stand must be cleaned and dried in time to avoid injury caused by slipping, and fluke mind shall be avoided.

② During hoisting, special slings and riggings that are intact and match the tonnage of components shall be strictly used.

The large structural parts such as cutter head rest, lower support of cutter head, cutter head, left and right shields, top shield, front girder, rear girder, rear support, belt bridge, etc. are provided with fixed lifting lugs and holes for lifting. Such components shall not be tied with steel wire ropes, so as to avoid fracture after the wire ropes are sheared and damaged, thus causing malignant events during lifting.

When steel wire ropes are used for lifting, lifting lugs (rings) of corresponding lifting weight shall be used as far as possible. When the two ropes are very close to each other, try to separate them with sleepers to avoid instability caused by slippage during lifting.

It is strictly forbidden to use the obviously damaged wire ropes and cloth slings, and it is not allowed to use them again after knotting.

③ The hoisting commander shall strictly perform his duties.

In addition to the accurate, clear and concise gestures and commands, the commander shall carefully check whether the binding of components is reasonable and whether the suspension of lifting slings is stable before lifting. During lifting, the commander shall be observant and alert to find whether there are obstacles around the components, and especially remind irrelevant personnel to leave the site. After confirmation, lift the components slowly. During lifting, pay attention to the traveling route and lifting speed of components at any time, especially the safety of personnel and precision components. Once there are abnormal signs, handle them properly.

④ When lifting components, absolutely obey the commanders' orders.

Avoid multiple commands, which not only affects the efficiency of hoisting, but also is easy cause accident. Careful planning is required beforehand, especially for hoisting of large pieces, and the rationality shall be discussed, but no further arguments are allowed when hoisting.

⑤ The mounting surfaces of all components must be treated and cleaned before installation.

In the process of manufacturing, storage and transportation, machining tool marks, welding slag, rust and collision bulges of different degrees are inevitable on the surface of components. Especially for precisely matched surfaces, paint, cotton thread and oil stains are not allowed. Therefore, fine abrasive cloth or angle grinder is generally used to remove rust spots and paint on the surface of relatively rough joint surfaces. Welding slag and collision marks on the local bulges shall be polished and leveled. For precision surfaces, tooth surfaces and threaded holes, special cleaning fluid shall be used for repeated scrubbing to completely remove dirt and accumulated liquid. For the cleaning of the original thick oil film protective layer, in order to improve work efficiency, putty bucket lip or saw blade can be used until the smooth surface is exposed (paint spots shall be removed by polishing machine with copper brush), and finally we wipe the surface clean with cleaning fluid. Grease shall be applied to relative moving surfaces. It is better to clean and apply grease before installation to avoid further corrosion or dirt adhesion on the exposed mating surface for too long, thus increasing the amount of repeated cleaning operations.

⑥ Select the correct installation tool, and over-range use is strictly prohibited.

For thread fastening, spline end wrenches or socket wrenches are preferred, followed by open-end wrenches and adjustable wrenches. According to the arrangement of tightening torque range from small to large, the tools that can be used are torque wrench, adjustable small torque wrench, tube-in-tube adjustable torque wrench, proportional magnification wrench, impact electric wrench, impact pneumatic wrench, pneumatic oil hydraulic wrench and pneumatic oil hydraulic tensioning kit.

For the structural components with slightly misaligned bolt holes, after some of the bolts are set into the holes, the pins with corresponding taper can be gently driven into the misaligned bolt holes. After the other holes are aligned, the remaining bolts can be successfully inserted, and then the pins can be pulled out and the last bolts can be installed.

For structural components with serious misalignment, according to the components and stress, the hand chain hoist, hand jack or chain block of the corresponding grade can be used to tension or push in the appropriate direction for successful assembly.

When setting various pins, it is necessary to accurately align them with the holes before gently driving them in with a hammer. Try to use a small copper hammer, followed by an iron hammer and a sledge hammer. When knocking, apply moderate force and strike in the symmetrical direction of the components.

⑦ All bolts are tightened according to the specified method and torque and marked in time.

For many connecting bolts on the joint surface, a small amount of grease shall be applied before installation, and the bolts shall be tightened symmetrically and evenly from the center to both sides, and shall be tightened to the specified torque as many times as possible. For thread pairs coated with MoS_2, the tightening torque shall be reduced accordingly, and the specified data shall be checked. There shall be corresponding regulations for the bolts of many important parts, such as the bolts connecting the girder, the bolts connecting the cutter head rest to the girder, the bolts splicing the cutter head, the bolts connecting the cutter head and the bolts of connection bridge. These regulations shall be noticed during installation. To avoid omission, marks shall be made in time after bolts are tightened. For bolts tightened several times, the marks shall be different.

⑧ For rotary connectors of important components, note that whether there are marks between them and their fittings. For example, there shall be marks at the joint between the drive shaft and the main bearing inner ring and between the drive shaft and the transmission. They shall be carefully aligned to make the running components rotate smoothly as far as possible and achieve static and dynamic balance.

(2) Installation of hydraulic system

There are many components and pipelines of the hydraulic system, and correct installation and commissioning will directly affect its working performance. Improper installation of the hydraulic device and the connecting machinery will cause uneven wear, strain or breakage. The tightness, center coincidence, curvature radius, pipeline length, fixation and connection mode of the hydraulic line connection have an impact on the vibration, torsion, oil leakage and air intake of the pipeline, and pipes may also rupture under high pressure. Therefore, the following principles shall be observed during installation:

① The installation environment must be kept clean. It is strictly prohibited to use cotton cloth to scrub the pipe joints and exposed valve planes. Once fibers enter the system, the valve damping holes will often be blocked, which is risky.

② The control valve handle on the ground pump station shall be used by designated professionals. Without permission, it is strictly prohibited to be used by others to avoid accidents.

③ Ensure that all plugs, pipe seals, and plastic caps are gradually removed as the installation progresses.

④ Before the installation of the system, all components must be inspected and tested one by one, and untested components shall not be put into assembly. For hydraulic valves, pumps, various hydraulic rotary motors, rubber rollers of belt conveyors and other actuators used for the first time, an appropriate amount of hydraulic oil shall be injected before installation, especially for bearings, so as to avoid burning of components due to lack of oil.

⑤ Before the final assembly of the hydraulic system, all hydraulic components, auxiliary parts and pipelines must be cleaned and inspected, and special attention shall be paid to whether there are missing seals (rings) at each interface.

⑥ If there is obvious damage to the connecting threads of the important joints and plugs, or when the plug is loosened, the connecting threads do not move, but the joint rotates, it means that the seals of the thread pair have been damaged and the thread pair will definitely leak when used again. If it leaks, the joint shall be unscrewed and discarded and replaced with a new one. Therefore, in the assembly of hydraulic pipe joints, it is better to try to assemble it gently by hand. After it is almost screwed in place, tighten it with an open-end wrench. Note that the tightening torque shall not be too large. It will be OK as long as you feel that the joint is slightly tightened, and violent knock is not allowed.

⑦ As there are many hydraulic lines, the hose layout shall be orderly and hierarchical for future disassembly or fault finding. Separate the thick and thin oil pipes; separate the oil lines of left and right actuators; separate the main circuit and control oil lines; and separate the hydraulic and lubricating oil lines.

⑧ The assembly quality of hydraulic lines will directly affect the pressure loss, vibration, noise, oil seepage and air leakage of the system. Therefore, when the pipeline is too long, pipe clamps or straps shall be added and firmly fixed on the rigid support. The pipes shall be arranged in parallel and be kept as smooth as possible, without distortion, sharp turns, and additional stress on hydraulic components, so as to avoid vibration, noise and even looseness caused by uneven stress during operation. At the same time, it shall be noted that all kinds of pipes shall be far away from the sharp edges of the equipment, and the bending radius shall be sufficient when bending.

⑨ The measuring instruments in the installed system must be inspected and calibrated to ensure their accuracy and reliability.

⑩ After the system is installed, the hydraulic oil shall be filtered for many times before being filled into the system. Before the oil drum entering the oil tank of the ground pump station, it shall be filtered by the oil filter truck. Before the oil drum entering the large oil tank of the ground pump station, it shall also be filtered by the oil filter of the ground pump station. If conditions permit, the oil filter truck can be used to carry out self-operated circulation filtration on a circuit of the hydraulic system, and the oil injection of the lubricating oil circuit of the main bearing shall also be carried out according to this.

6.2.3 Commissioning

After the TBM is assembled, it enters the commissioning phase. The commissioning of the whole machine is divided into four stages: preparation before commissioning, no-load commissioning of TBM, joint commissioning of TBM and continuous belt conveyor and on-load commissioning.

1. Preparation before commissioning

When the TBM is stepped to the launching tunnel section, the following items shall be checked and confirmed before commissioning:

(1) Check the interference degree between the bottom shield, cutter head and the bottom plate of the stepping mechanism to ensure that there is no interference when the cutter head rotates.
(2) Confirm that the distance between the gripper and the launching tunnel wall meets the stroke requirements of the gripper cylinder.
(3) Confirm that the cable connection of the electrical system is normal, especially the state of switches of the high voltage system.
(4) Confirm whether the state of each gate valve of the hydraulic system is correct.
(5) Check and confirm whether the cooling water level, hydraulic oil level and gear oil level meet the TBM operation requirements.
(6) The personnel involved in commissioning and inspection must know the position of each emergency stop button and be equipped with hand-held interphones.

2. No-load commissioning of TBM

No-load commissioning shall follow the sequence of local commissioning before overall commissioning, and main power system before functional systems. Successively carry out the commissioning of electrical system, hydraulic system, joint commissioning and commissioning of various functional systems.

(1) Electrical system

The electrical system shall be commissioned in the sequence of high voltage, power and control. Check whether the live indication of each high voltage cabinet is normal in turn and confirm whether the power supply voltage is normal. Sequence of power transmission: switch on in sequence from top to bottom according to the voltage levels, and the operation of power transmission and power off must be performed by a licensed electrician. Commissioning steps:

① Check the main power supply line, recheck whether the connection is correct and whether there are risks. After confirmation, carry out 10 kV high-voltage trial power transmission to the high-voltage cable reel and switch cabinet, and then send it to the transformer and main switch cabinet after confirmation.
② After power transmission, commission the electrical control system, then comprehensively check the operation of the motor and pay attention to the operation direction of the motor.
③ Followed by the parameter setting and trial operation phase of each subsystem. This phase is the most critical and complex and may last for a long time. The most critical thing at this phase is the parameter setting and adjustment of various controllers (e.g.: frequency converter of foam system, flow meter, rotation angle controller of segment erector, etc.); calibration of display instrument; calibration of control circuit board; even partial change of PLC program, etc. The TBM subsystems generally include: segment installation, segment transportation, foam system, grease system, grouting system, cutter head drive system, thrust system, etc. Of course, except for the tube sheet transportation, other systems are interlinked. For example, if the commissioning of grease system has not been completed, the cutter head drive system cannot be commissioned. In addition, before commissioning, we shall note that whether the conditions for system commissioning are available. For example, in general, the cutter head drive system cannot be commissioned on the launching frame.
④ After the commissioning of each subsystem is completed, the whole machine can be put into trial operation.

(2) Commissioning of hydraulic power system

After the electrical system is confirmed to be correct, check the hydraulic lines and group valves to ensure that they are correct. Start each pump station in turn and observe the operation. Start accessories, and check the no-load operation. The commissioning steps of the hydraulic part are as follows:

① Check the oil level sensor of the tank and transparent hose of the liquid level display to see if the hydraulic oil is sufficient.

② Check whether all oil pump inlets are open.
③ Check whether all control valves are normal.
④ Check whether the hydraulic pump drive motor is turned correctly.
⑤ Check the hydraulic pumps one by one for proper operation.
⑥ Check whether the switch of each hydraulic valve group is sensitive and normal.
⑦ Check the interface of the hydraulic line for oil leakage.
⑧ Check whether the cooling system works normally.

(3) Joint commissioning

After the commissioning of electrical and hydraulic systems is completed, the joint commissioning of each system shall be carried out, and the interlocking and interlocking functions designed by the program shall be tested and inspected in turn to confirm that all interlocking and safety systems meet the design requirements.

(4) Functional commissioning

After the safety confirmation, the functional commissioning of each subsystem can be carried out. All moving parts such as ring beam erector, bolt driller, transportation trolley, main thrust cylinder, gripper, rear support and shotcrete system shall be subject to functional commissioning in turn, and shall be confirmed one by one to fully meet the design functional requirements and construction needs.

3. Joint commissioning of TBM and continuous belt conveyor

After confirming that the belt conveyor was installed according to the drawings and that it was safe, the communication control lines shall be first laid and tested. After successful communication, online commissioning can be carried out, and the commissioning steps are as follows:

① Power on all systems and check whether the power supply of each system is normal.
② Using the manual operation inching mode, start the tunnel opening drive system of the belt conveyor and continuous belt conveyor drive system at the tail of TBM in turn, and confirm that the rotation direction of the drive motor meets the drive requirements.
③ After the rotation direction of all drives is confirmed, the remote control mode can be used to start and stop the muck discharge system of the belt conveyor in the operator cabinet. During commissioning, check whether the functions of various emergency pull switches and anti-deviation switches are normal.
④ Gradually increase the running speed of the continuous belt machine and check whether the belt is running out of the drive pulley, and adjust them in time if any.
⑤ Adjust the tension of the tension device of the continuous belt conveyor until it is suitable for short-distance tunneling.

4. On-load commissioning

On-load commissioning can be carried out after no-load commissioning of TBM systems. The main purpose of on-load commissioning is to check the load capacity of various pipelines and seals, to make sure that working systems and auxiliary systems of TBM are in the working state meeting the normal production requirements. On-load commissioning of equipment is usually carried out during trial tunneling. Commissioning the use of each component under working pressure. Strict technical and management measures will be taken during on-load commissioning to ensure the safety, quality and accuracy of the Project.

6.3 TBM Tunneling

6.3.1 TBM Tunneling Procedures

Trial tunneling of TBM is required before normal tunneling of TBM. During the trial tunneling, the performance and coordination of TBM components shall be mainly checked, and the running-in of each equipment shall be carried out to make sure it is in the best state and is available for formal and rapid tunneling. Through the construction of trial tunneling, the construction operators can be basically familiar with the performance of the equipment, master the technical points of equipment operation and maintenance, and preliminarily summarize the selection of tunneling parameters and process control measures of the Project. Fully understand the whole construction organization to lay a foundation for smooth tunneling in the future.

After the trial tunneling, the machine shall be adjusted according to the condition, and then the official tunneling will be started. During TBM tunneling,

The horizontal gripper is tightened on the tunnel wall to provide counter force for the TBM tunneling. The cutter head is pushed forward under the thrust of the main thrust cylinder, and the back-up trolley is parked in the tunnel. As the cutter head bucket and scraper rotates, the rock muck cut off by the cutter head will be delivered from the bottom to the top, then be delivered to the belt conveyor in the middle of the cutter head along the muck chute, and finally be transported out of the tunnel by the TBM belt conveyor and the continuous belt conveyor. While the TBM is tunneling, the initial support and related supporting operations shall be carried out. When the cutter head tunnels 1.8 m forward, a tunneling cycle is completed. The process flow of the TBM tunneling is shown in Fig. 6.5.

6.3.2 TBM Tunneling Operations

The TBM control room is the control center for completing various tasks. 90% of the instructions on the equipment are completed in the operator cabinet, and the operation panel, display instrument, PLC system, steering display, etc. are installed inside. There are nearly 100 operating buttons and handles on the main operation panel to control the operation of equipment in different parts. Correct and reasonable operation is the key to give full play to the advantages of TBM construction, smooth

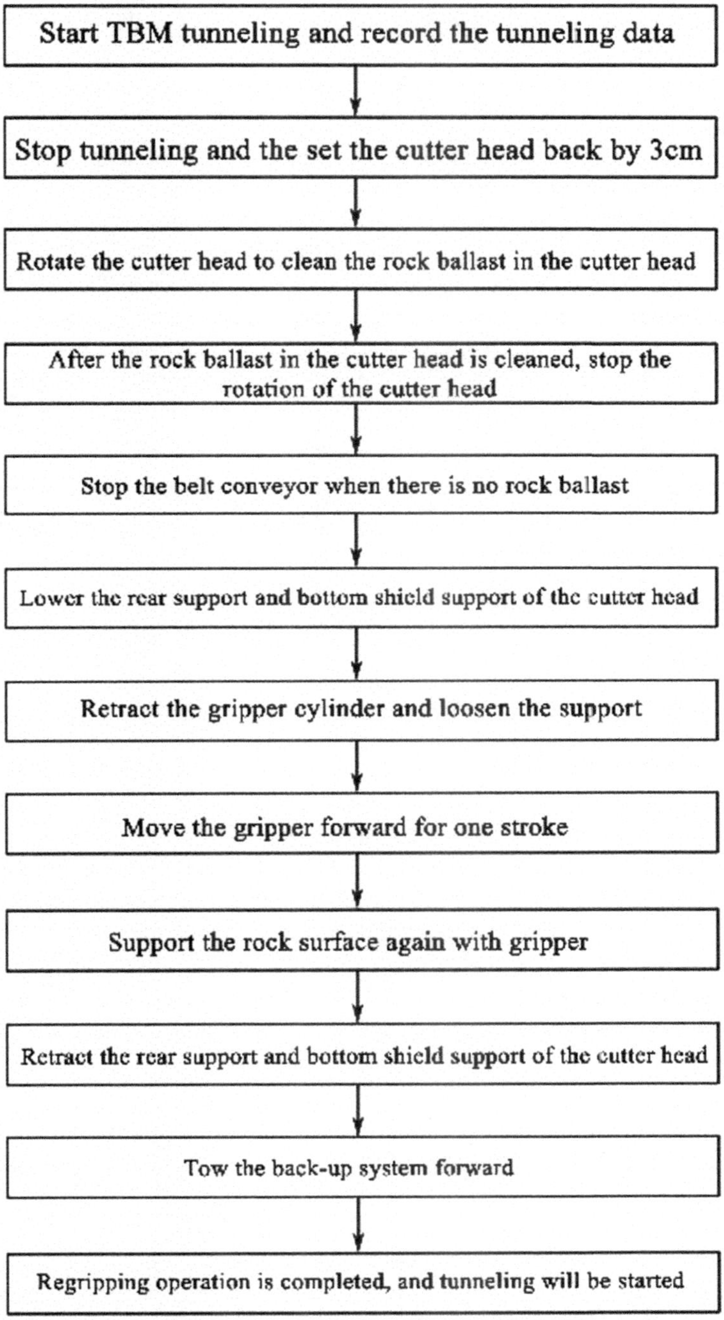

Fig. 6.5 Flow chart of TBM tunneling

operation and extended equipment service life. The features of the control room are as follows:

(1) Easy to operate and responsive. The centralized and reasonable arrangement of operating elements in the control room, numerous instrument arrangements and the information prompts on the operating state of the main equipment all provide the conditions for the operator to understand the state of the TBM in time and to take accurate measures for reasonable control.
(2) Perfect monitoring and safe operation. The main equipment of the TBM is started and stopped in the control room for easy monitoring. The sophisticated fault detection system timely and accurately displays the problems of each system on the monitoring screen, and the multi-angle video screen monitoring system reflects the working state of each main equipment and each operation team to the operator cabinet. The perfect PLC system detects every step of operation to avoid the adverse consequences caused by misoperation.

The operation of the TBM operator cabinet is mainly divided into 5 steps: preparation before startup, startup, tunneling, regripping and steering, and the detailed operation steps are as follows:

(1) Preparation before startup

In tunnel construction, electricity, water and wind are the three main factors, and this is also true for TBMs. After a trial run or a long shutdown, it is necessary to consider whether electricity, wind and water have been safely and correctly transmitted to the machine before startup. First, verify whether the medium-voltage power supply outside the tunnel is transmitted to the transformer of the machine and whether the primary side circuit breaker of the transformer has been connected.

After the power is turned on, it is also necessary to confirm whether the purified water outside the tunnel is available and is sent into the tunnel, and at the same time confirm whether the fresh air fan outside the tunnel is started and sends fresh air to the tail of the machine. After electricity, water and wind are available, observe and analyze the type of surrounding rock, choose reasonable tunneling parameters, understand the elevation deviation of the centerline and support in the last shift. When everything is clear, the preparation work is completed.

(2) Startup

After confirming that the control voltage is on, start the water purification pump (normal water pressure shall be about 0.7 MPa) and start the fan. Fans include relay fan, industrial exhaust fan and air cooling fan. When starting, fans can be started in groups by the start button or separately. Start the hydraulic power station after the fan has been started. In the same way as the fan, the hydraulic power pack can be started in groups or separately, and the air compressor shall be started on the operation panel at its power distribution cabinet. This is also need to be completed during this time period.

(3) Tunneling

Before starting tunneling, confirm the following work: The fan, pump station, motor and conveyor belt are started, the water system is normal, and oil and grease lubrication of the cutter head are normal (the above shall be completed at startup); The outer frame has been moved forward and supported tightly, the rear support has been retracted and moved forward, the shield clamping cylinder has been clamped tightly, the back-up system has been towed (the above shall be completed during regripping). The tunneling can be started after the above conditions are met.

(4) Regripping

After tunneling for one cycle, the PLC system will automatically stop advancing according to the signal from the sensor. Back the cutter head by 3–5 cm to make the cutter ring away from the rock surface. Then let the cutter rotate slowly to transport the residue out by belt conveyor. Then stop the cutter head from spraying water, stop the cutter head from rotating, stop the motor, and stop the conveyor after the muck on the conveyor belt is almost discharged. The corresponding buttons of the above controls correspond to the buttons at startup. At the same time, the towing of the back-up systems can be carried out.

(5) Steering

TBM direction control is an important part of construction. To ensure smooth construction of TBMs, laser guidance systems are usually configured to monitor the direction and position of the TBM at any time during tunneling. The laser guidance test is based on the laser emitted from the laser theodolite installed on the tunnel side wall of the tunnel. Through the interface unit, the laser data, distance and coordinates (x, y, h) of the theodolite are connected with the control system and are output to the control unit. When the laser beam hits the target unit installed on the TBM main machine, the target unit will measure the projection angle of the angle between the laser beam and the axis of the target unit on the horizontal plane and vertical plane, the position (x, y) of the laser beam on the target unit screen, the left and right lateral rolling angle of the target unit, and the distance between the theodolite and the target screen of the target unit measured by the prism in the target unit and the rangefinder. All these data will be transmitted to the main machine of the control unit through connecting cables. The current position of the TBM shall be calculated according to the tunnel design axis data and laser theodolite data input into the control unit, and the installation error between the target unit axis and the TBM main axis The x and y differences between the center of the TBM cutter head and the designed tunnel axis shall also be calculated. The operator controls the position of the TBM according to these data. As the TBM moves forward, the connecting cables between the control unit and the laser transmitter/theodolite/rangefinder are continuously released until the distance between them reaches the maximum value. At this time, the laser transmitter/theodolite/rangefinder and the interface unit shall be reinstalled.

The operator presets the position and guidance angle based on the data from the laser guidance system test, and adjusts the TBM tunneling direction by adjusting the guidance angle. The direction of the double-supported TBM is adjusted by two operating handles (front support and rear support control handles) in the operation room. The operation steps are as follows:

(1) Loosen the gripper to re-tighten the TBM. Before operating, operate the front support operating handle and raise the lower shield of the cutter head to adjust the direction.
(2) Based on the information provided by the guidance system (laser beam position of the target, horizontal position of the TBM, lateral rolling of the TBM), the tunneling direction of the TBM main machine is adjusted by two coordinated operating handles (front support and rear support).

Using the front support operating handle to lift or lower the lower shield of the cutter head, and using the integrated button on the operating handle to roll the lower shield of the cutter head laterally. When the handle is pulled to the left, the rear support of the machine swings to the left, that is, the machine swings to the right. When the handle is pulled to the right, the rear support of the machine swings to the right, that is, the machine swings to the left. The base holder of the cutter head shield is taken as the turning point. The changes of the steering are always visible on the display, e.g. current position, predicted position, machine guidance. When the displayed data is within the allowable range, the steering is completed.

Precautions for steering:

(1) The movement of the edge cutter shall not exceed 3 mm during each steering, otherwise the edge cutter will be damaged.
(2) The guidance angle of TBM shall be controlled within 8 mm/m, otherwise it will cause difficulties in laying invert blocks and passing of back-up equipment.
(3) The variation of the guidance angle of each steering shall be controlled within 2 mm/m. In extra-hard rock sections, the variation of the guidance angle shall be appropriately reduced to prevent damage to the edge cutter.
(4) The surface of the tunnel wall is regular (i.e., no traces or marks due to deflection are visible) in the correctly guided state. When the guidance is correct, there is no ridge-like surface on the tunnel wall. Step is a sign of excessive steering. If there is a step, the way of steering shall be corrected.

6.3.3 Selection of Tunneling Operation Mode

There are three working modes available in the TBM operator cabinet, i.e. automatic control thrust mode, automatic control torque mode and manual control mode. The working mode selected is determined by the operator according to the rock condition.

For homogeneous hard rocks, the automatic control advancing mode shall be selected. At this time, the machine will not overloaded and the advance speed will be the highest. Judgment basis for selecting this working mode: During tunneling, if the thrust reaches the maximum value first, but the torque does not reach the rated value when tunneling, it can be judged as hard rock and the automatic control advancing mode can be selected.

For homogeneous soft rock, generally, the thrust force is not too large, and the change of cutter head torque is the main factor, so the automatic control torque mode shall be selected. Judgment basis for selecting this working mode: During tunneling, if the torque reaches the rated value first, but the thrust force fails to reach the rated value or reaches the rated value at the same time, it can be judged as the soft rock state, and if the geology is relatively uniform, the automatic control torque mode can be selected.

If the state of rocks is uncertain, or the hardness of rocks varies unevenly, or there are many fracture zones, faults and fissures in rock joints, the manual control mode must be selected and the rock properties must be judged by operators.

In the operation process of manual control mode, if the rock is hard, the thrust reaches the rated value first, and the rock is relatively complete. At this time, the operation shall be based on the thrust operation mode, and the thrust pressure shall be limited to not exceed the rated value. If the rock joints are well developed, there are many cracks or fracture zones and faults, etc., at this time, the operation shall be based on the torque mode, and the tunneling parameters shall be selected mainly by torque changes combined with thrust parameters. The manual control mode is applicable regardless of the rock conditions.

6.3.4 Selection of Tunneling Parameters

Under different geological conditions, the tunneling parameters of TBM such as thrust, cutter head rotation speed and cutter head torque are different. Although TBMs are provided with automatic thrust and automatic torque operation modes, due to the relatively uneven rock, manual operation mode is usually used in TBM tunneling operations. The tunneling parameters of TBM shall be adjusted in time according to different geological conditions, so that TBM can safely and efficiently pass through sections of different geology.

When TBM enters weak and broken surrounding rock from hard rock, the corresponding main advancing parameters, muck quantity and particles of belt conveyor will change obviously. According to this change, the surrounding rock condition of the TBM cutter head working face can be roughly judged, and the manual adjustment operation mode can be used to adjust the tunneling parameters in time.

Advance rate (penetration): In the case of hard rock, the penetration is generally 9–12 mm/r. When entering the transition section of weak surrounding rocks, the penetration will slightly rise. Considering the mucking capacity of TBM belt conveyor, long penetration rise time is generally not allowed in on-site operation. At this time, the penetration decreases with the decrease of given advance rate speed. When completely entering the section of weak surrounding rocks, the penetration is relatively stable, generally 3–6 mm/r.

Thrust (advance pressure): In the case of hard rock, the advance speed is generally about 75% of the rated value, and the advancing pressure is in a corresponding proportion. When entering the transition section of soft and weak rocks, the advancing pressure decreases in an inverse parabola form, and the descending time is proportional to the length of the transition section. The advancing speed shall be reduced appropriately with the decrease of the advancing pressure. When the machine fully entered the section of soft and weak rocks, the pressure tends to be relatively stable. At the same time, the advance rate is generally maintained at about 40% of the rated value.

Torque: It is generally 50% of the rated value in the case of hard rock. When the machine is entering the transition section of weak surrounding rocks, the torque tends to slowly rise, and the rising time is proportional to the length of the transition section. When the machine fully entered the section of weak surrounding rocks, the torque will decrease with the decrease of the advancing speed and is generally about 80% of the rated value.

The cutter head rotation speed: It is generally about 6.0 r/min in the case of hard rock and shall be adjusted to 3–4 r/min when the TBM enters the end of the transition section to weak surrounding rock, and it is kept at about 2.0 r/min when the TBM completely enters the section of weak surrounding rocks.

Supporting force of gripper: In case of hard rock, it is generally rated value. When the gripper enters the transition section to weak surrounding rock, the gripper supporting force is generally adjusted to about 90% of the rated value. When the gripper enters the section of weak surrounding rocks (corresponding treatment is required on site), the gripper supporting force is generally adjusted to the minimum limit value. If necessary, the PLC program needs to be changed to set the limit value, and the advance rate is adjusted at any time according to the surrounding rock condition in front of the cutter head to ensure sufficient stability of the TBM.

6.3.5 Construction Labor Organization and Labor Allocation

The TBM construction team consists of 2 tunneling shifts, 1 servicing shift, 1 transportation shift (divided into 3 shifts) and 1 support shift (divided into 3 shifts). Each tunneling shift works for 8 h, mainly responsible for TBM tunneling in the tunnel and mucking operation of tunnel belt conveyor. Each servicing shift works for 6 h, mainly responsible for the repair and maintenance of TBM and belt conveyor. The transportation shift works simultaneously with the TBM tunneling and servicing shift, and is responsible for the transportation inside and outside the tunnel. The support shift works with the TBM tunneling and servicing shift, and is responsible for various support work outside the tunnel.

See Table 6.2 for the staffing of each shift.

Table 6.2 Staffing of tunneling shifts

Shifts	Type of work	Number of people	Work contents
Tunneling shifts	Foreman	1	Overall coordination of tunneling operations
	TBM operation	1	Be responsible for TBM operation
	Civil engineer	1	Be responsible for civil engineering in the tunnel
	Electrical engineer	1	Be responsible for electrical work of the machine
	Mechanical engineer	1	Be responsible for TBM mechanical work
	Temporary support shift	8	TBM upper support
	Track laying shift	4	Track laying
	Shotcrete shift	4	Shotcrete support operations
	Air, water and electricity shifts	7	Connection of air, water, electricity and muck removal
	Belt conveyor shifts	7	Tunnel continuous belt conveyor extension and belt conveyor operation inspection
Servicing shifts	Foreman	1	Overall coordination of servicing operations
	E&M engineer	4	Maintenance solutions and maintenance operations
	TBM operation	1	TBM operation
	Greaser	3	Grease filling
	Welder	1	Repair
	Oxygen welder	1	Repair
	Mechanic	4	Maintenance
	Maintenance of belt conveyor	10	Maintenance of TBM belt conveyor and tunnel continuous belt conveyor
	Cutter shift	9	Cutter inspection and replacement
	Air and water pipe shifts	2	Connection of air, water, electricity and muck removal
Transportation shifts	Foreman	1	Overall coordination of tunneling operations

(continued)

Table 6.2 (continued)

Shifts	Type of work	Number of people	Work contents
	Dispatching	2	Transportation dispatching of main and adits
	E&M engineer	1	Be responsible for overhaul and maintenance of electromechanical equipment
	Line engineer	1	Inspection and maintenance instructions for track lines
	Locomotive operation	5	Locomotive driving
	Switchman and signalman	5	Switch and signal control
	Track worker	8	Track inspection and maintenance
	Winch operation	1	Operation
	Porter	8	Material preparation and handling
	Craneman	1	Crane operation
Support shifts	Foreman	1	Overall coordination of support operations
	Mechanical engineer	1	Be responsible for maintenance of relevant equipment
	Electrical engineer	1	Be responsible for overhaul and maintenance of relevant electrical equipment
	Concrete mixing plant	5	Concrete production
	Material warehouse	3	Stock-out of materials
	Spare parts warehouse	1	Stock-out of spare parts
	Air, water and electricity supply	6	Operation and maintenance of fans, water pump stations and power supply facilities outside the tunnel
	Machine repair workshop	5	Processing of rock bolt, pad, wire mesh and emergency parts
	Mechanic	2	Equipment maintenance
	Car driving	10	Rock ballast transfer

6.3.6 Initial Support

The initial support of hard rock mainly consists of rock bolt construction, shotcrete support and local wire mesh support. The soft rock support mainly consists of wire mesh, shotcrete and rock bolt support, and steel support, and advance support such as advance ductule or advanced anchor bolts are used according to the surrounding rock conditions and design requirements. The materials used for construction support are processed outside the tunnel, and the materials for shotcrete are produced and supplied by large automatic mixing plants.

(1) Shotcrete process

The shotcrete support is completed by the shotcreting system of TBM, and the concrete is transported by rail tanker.

The shotcrete manipulator of the shotcrete system of TBM can move longitudinally and laterally in the shotcreting area, and the operator can complete the shotcrete operation required by the design by operating the control handle. The rotation frequency of the mechanical shotcrete manipulator is stable, and the distance between the nozzle and the sprayed surface is fixed. If the operators master the operation procedures, the rebound and flatness of the shotcrete will be excellent. In case of weak and broken surrounding rocks, it is necessary to quickly spray concrete to seal the surrounding rock after the shield leaves the rock surface. At this time, after the concrete pipelines in the shotcrete area on the TBM back-up system are extended to the shield, the initial shotcreting shall be carried out manually. After it enters the mechanical shotcrete area, the mechanical shotcrete method shall be used to spray again to the designed thickness.

(2) Rock bolt construction

The rock bolt construction is achieved by two rock drills equipped on the TBM main machine within the design range. Rock drill shall be used for drilling. Drilling requirements: The hole position shall be determined according to the design requirements, and the allowable deviation of the borehole position shall be 150 mm. If there are no special requirements, the drilling direction shall be perpendicular to the rock surface. The borehole depth shall not be less than the designed effective length of the rock bolt and shall not more than the effective length of the bolt body by 30 mm, and the borehole diameter shall be 42–45 mm. The accumulated water and rock powder in the hole shall be blown clean before filling the cartridge into the rock bolt holes completely. Finally, install the rock bolt. Before installation, check whether the specification, length and diameter of the raw material of the bolt body meet the design requirements. If they are qualified, manually or mechanically jack the bolt into the hole filled with cartridge. When inserting the bolt body, keep it in the middle, and the inserted length shall not be less than 95% of the design value. After the rock bolt is installed, it shall not be knocked at will.

(3) Wire mesh construction

The wire mesh shall be installed in the corresponding surrounding rock section according to the requirements of design support parameters. The reinforcement shall be straightened outside the site, cut into 2 and 3 m sections, fabricated manually on site and transported to the working face. Install it on site with the wire mesh erector provided with TBM, and fix it manually. The wire mesh shall be laid with the rock surface, with a distance of 3–5 cm from the rock surface. The wire mesh shall be firmly connected with rock bolts or other fixing devices, and shall not swing during shotcreting. The thickness of the shotcreting protective layer of wire mesh shall not be less than 2 cm.

(4) Ring beam construction

Ring beam production: The ring beam is made of I-steel required by the design and bent into a ring by a section steel bending machine. For easy transportation and installation, each ring of ring beam shall be fabricated in sections according to the design requirements, usually divided into 6 sections. The end of each section shall be welded with connecting steel plates, processed and formed by manual welding, and connected by bolts during installation. The ring beam shall be installed using the configured ring beam erector. The processed single ring beam shall be subject to trial assembly inspection according to the design requirements, and shall be stored in a centralized manner and clearly marked after passing the inspection.

Transportation and installation of ring beam: Generally, the ring beam is transported to the vicinity of the crane by the material transport vehicle, and then transported to the upper part of the beam erector by the crane. After the beam erector clamps the ring beams, rotate the erector with guide rail until the next section of beam can be fixed on the tail section of the previous section with bolts. Repeat this process until the whole ring is completed. When a ring is completed, the tensioning mechanism on the beam erector will expand the ring beam outward and wedge it with the rock surface.

Anchoring of ring beam: After the expanded ring beam is wedged with the rock surface, the rock bolt driller is used to drill holes for feet-lock bolts, and the rock bolt is locked according to the design requirements. Then it is longitudinally welded with the previous ring of ring beam through threaded reinforcement, and the circumferential spacing meets the design requirements.

(5) Advanced support

For the actual situation of gripper TBM tunneling in weak surrounding rocks, corresponding auxiliary construction measures shall be taken on the basis of advance geological forecast. The method of auxiliary construction measures is to carry out advance reinforcement before tunneling through, and the current method is to minimize the amount of collapse and spalling during tunneling and the amount of convergence of the surrounding rock out of the shield, and to effectively control the convergence and deformation of the surrounding rock by strengthening the initial support and other means. At present, we mostly use the advance rock bolt technology, which uses the TBM's self-configured probe drilling machine to drill holes. The advanced rock bolt is installed manually. It has the advantages of great initial support, easy for construction and high safety. It is applied to the surrounding rock that is predicted to collapse. The function of the advance rock bolt is mainly to limit the deformation of the surrounding rock and form a bearing ring in the surrounding rock. The rock bolt and the surrounding rock work together to bear the pressure of the surrounding rocks, thus maintaining the stability of the surrounding rock. This is actually the main guiding ideology of New Austrian Tunneling Method. It can be seen from the actual construction effect that the advance rock bolt does have the ability to prevent large deformation of surrounding rock. According to the analysis of the measurement results, the surrounding rock pressure on the sections with advance ductule construction is reduced by about 70% compared with that of the sections without advance ductule construction. In the construction, it is necessary to consider both the change of the surrounding rock with time and the load after the advancement of the tunnel face and the tilting of the strata. Due to the measures taken to strengthen the support to keep up with the tunnel face during construction, although the surrounding rock is difficult to achieve self-stabilization and the bedding is inclined to the inside of the tunnel, it is possible to increase the load of the surrounding rock. However, on the commencement date of tunnel measurement, the excavation and initial support of the tunnel face have been completed by about 20 m, equivalent to 8 times the diameter of the upper part of the tunnel. Since the stress relief has basically completed, the load on the temporary support has also been stabilized. The advance ductule is driven before the stress redistribution. Based on its constraint effect, the ductule provides a reaction force to the surrounding rock, thus maintaining and strengthening the self-stability of the surrounding rock. During excavation, due to the anchoring force of the advance ductule and the good tensile and shear resistance of the ductule body, the friction between the rock structural planes is increased, thus the stability of the surrounding rock is strengthened, the sliding between beddings is effectively controlled and the construction safety is ensured.

Questions

1. What are the two parts of TBM mobilization and transportation?
2. Briefly describe the measures to ensure the safety of TBM transportation.
3. What preparations shall be made before TBM assembly?
4. Briefly describe the site assembly sequence of TBM.
5. Briefly describe the characteristics of the TBM control room.

Chapter 7
TBM Construction in Special Stratum

Key points of this chapter: The tunnel construction technology with NATM as the core has gradually matured. TBM has been rapidly developed in various fields of engineering construction due to its advantages of rapidity, high quality, high efficiency, safety, environmental protection and automation. The main constraint of TBM application is the poor adaptability to adverse geological conditions, limiting its rapid development and promotion to a certain extent. In the projects with different purposes, different sections, long distance and deep overburden, such as Water Diversion Project from Gansu Datong River to Qinwangchuan District, Qinling Tunnel, Wanjiazhai Water Diversion Project and Dahuofang Water Tunnel built in China, and Yinhanjiwei Water Tunnel and Yinsong Water Tunnel under construction, many experience has been gained in handling adverse geological conditions including soft rock strata, very hard rock strata, water ingress and mud gushing strata, fault and broken zone strata, karst cave strata and others. This chapter describes the construction experience of TBM in different types of special geological tunnel sections through some construction measures and engineering examples.

7.1 TBM Construction in Soft Rock Strata

Soft rock is a complex rock mechanics medium with significant plastic deformation under a specific environment. It can be divided into two categories, namely geological soft rock and engineering soft rock. The so-called soft rock is the rock with uniaxial saturation compressive strength less than 30 MPa, also known as soft rock. It can be subdivided into softer rock (compressive strength 15–30 MPa), soft rock (compressive strength 5–15 MPa) and very soft rock (compressive strength less than 5 MPa).

K. Chen et al., *TBM Design and Construction*,
https://doi.org/10.1007/978-981-99-0059-6_7

Soft rock usually has low compressive and shear strength and is prone to small deformation. It has poor hydro-physical properties, easy to soften in water, and may produce disintegration and swelling. It also features obvious rheological effect, low long-term strength and others. In tunnel construction by TBM, soft rock is often encountered. Due to the structure and construction characteristics of TBM, it is more difficult to deal with soft rock deformation than those in drill and blast construction. This will not only cause delay in construction period, but also damage to expensive TBM equipment, and may even lead to TBM construction failure. In previous cases of TBM construction in soft rock strata, it was generally difficult to deal with the problem, and there were also painful lessons of switching to pilot tunnel. Therefore, the construction of TBM in soft rock strata is worth studying and drawing lessons from experience in order to ensure the construction period and save money.

7.1.1 Characteristic Analysis of Soft Rock

1. Physical properties of soft rock

   The physical properties of the rock are the main factors that determine the damage of the rock, and the clay content and structural surface in the soft rock control the engineering mechanical properties of the soft rock.

   (1) Softening properties
   Softening properties refer to the nature of the change in strength and stability of the rock after reacting with water, depending on the mineral composition, structure and structural characteristics of the rock. Rocks with high clay mineral content, large porosity and high water absorption rate are easy to soften and lose their strength and stability after reacting with water. The softening coefficient is used as an indicator of rock softening, and is numerically equal to the ratio of ultimate compressive strength in the saturated state to ultimate compressive strength in the air-dried state of the rock. The smaller the value, the more influence on the strength and stability of the rock by reacting with water.

   (2) Water absorption and swelling property
   The clay minerals such as montmorillonite, illite and kaolinite contained in the soft rock will expand when immersed in water, generating expansion strain and expansion pressure.

   (3) Disintegrability
   Soft rock will flake off and crack off under the action of physical, chemical and mechanical factors.

   (4) Prone to be disturbed
   Soft rock susceptibility to disturbance is mainly expressed as sensitivity to stress perturbation and environmental changes. Perturbation factors include the shape and size of the tunnel section, tunnel combination relationship, tunnel excavation method and excavation support procedures.

(5) Low rock strength
Soft rocks often show low rock strength. It mainly depends on the lithology, but also often is under the influence of joints and fissures and other tectonic factors.
(6) Plasticity
Soft rock has the property of producing irrecoverable deformation under the action of engineering forces.
(7) Rheological property
It means that the stress–strain curve of soft rock under long-term static loading varies with time. Especially at high stress levels, it is more likely to show rheological properties.

2. Mechanical properties of soft rock
The basic mechanical properties of soft rock include the softening critical load and the soft rock critical depth.

(1) Softening critical load
Softening critical load means the minimum load that causing significant deformation of the rock. Different types of rocks correspond to their objective softening critical loads. For rock engineering properties, the relationship between the level of load on the rock and the size of softening critical load can be used to distinguish engineering soft rock or engineering hard rock. The ratio of softening critical load to uniaxial compressive strength of soft rock: 0.3–0.5 for general expansive soft rock, 0.5–0.7 for high stress soft rock, and 0.4–0.7 for jointed soft rock.
(2) Softening critical depth
Corresponding to the softening critical load, there exists a softening critical depth. For example, the surrounding rock produces significant plastic deformation, large earth pressure and overburden with difficulty in support. The ground stress level at the softening critical depth is roughly equivalent to the softening critical load.

7.1.2 Mechanism of Deformation Evolution of Soft Rock in Tunnel

The large deformation of surrounding rock in the tunnel can be defined as the following: a type of plastic deformation damage with progressivity and obvious time effect of the tunnel and the surrounding rock of underground works. It differs from brittle failure caused by rock burst and from collapse, sliding and other damage limited to a certain structural surface control in broken rock zone of the surrounding rock.

The surrounding rock of soft rock tunnel always keep the primary structure of the rock mass, and under high stress the surrounding rock mass produces flow or plastic deformation, and with the effect of groundwater, the rock body softens and flows, and when the rock mass contains expansive minerals, it will undergo expansion and deformation. According to the development characteristics of the structural surface in soft rock, it is divided into homogeneous type, laminated type, interlaminated type, and soft rock type with swelling. The characteristics of deformation of soft surrounding rock are:

(1) The material condition of the surrounding rock is soft rock with low strength, and the structure of the rock has the characteristics of primary structure, and this type of surrounding rock also includes the swelling soft rock.
(2) The environmental conditions of the surrounding rocks have high ground stress problems of different degrees, and due to the low strength of the surrounding rocks, a high stress-to-intensity ratio is formed.
(3) The deformation of the surrounding rock is mainly the extrusion of the surrounding rock, while the expansion of the soft rock with expansibility is not significant.
(4) The softening effect of existing grounding water on soft rock plays an extremely important role in the deformation of the surrounding rock.
(5) The mode of deformation and damage of the surrounding rock is mainly plastic flow and bending deformation.

By analyzing the typical examples of large deformation of the surrounding rock, the evolution mechanism of soft rock deformation is generally summarized into the following nine aspects:

(1) Plastic flow of soft rock

The excavation of the tunnel leads to the adjustment of the stress in the surrounding rock, and the expansion caused makes the originally closed structural surface in the rock body open and slide and further fracture the surrounding rock mass, while changing the stress state and strength of the rock mass, the groundwater in the surrounding rock percolates along the open fissure and is softened, leading to plastic flow and large convergence displacement of the surrounding rock.

(2) Expansion and deformation of soft rock

The necessary conditions for rock expansion are expansive rock, water and stress. Expansive rocks contain large amounts of flaky expansive minerals such as montmorillonite, chlorite and kaolinite. For underground works in swelling rock area, excavation unloading leads to a significant reduction in radial stress, so that without support, tunnel wall σ_r(support pressure) = a(surrounding rock pressure) = 0, i.e., near the tunnel wall, the stress condition of surrounding rock expansion ($\sigma < \sigma_0$) is reached. In addition, the fissures generated by blasting, the original fissure opening and slipping, and even the new fissures generated by the stress adjustment process all provide the necessary channels for water to enter. Under the action of these two factors, the expansive surrounding rock then undergoes expansion and deformation, resulting in a slow increase in the load acting on the support.

(3) Bending deformation of plate and beam

For tunnels in laminated (especially thin-layered) rock, the deformation and damage mechanism can be explained by bending. Especially under the unloading conditions in the high ground stress area, the rock mass is prone to bending deformation. As the radial stress of the tunnel wall decreases and the tangential stress increases, the laminated rock deformation occurs in the form of plates under the action of transverse or longitudinal bending. Due to the gravity redistribution caused by the excavation of the tunnel, the laminated rock is bent, thus causing lateral deformation of the tunnel wall. In tunnel top and bottom plate, the vertical bending can occur under the horizontal stress, causing compression of the top and bottom heave.

(4) Plastic wedge

The plastic deformation of the jointed rock mass leads to the tracking of existing structural surfaces and the formation of glide planes, and such glide planes form a plastic wedge in the surrounding rock inside the tunnel and forms shear slip into the tunnel. With varying directions of main stress and lateral pressure coefficients, plastic wedges can appear in different parts around the tunnel, thus causing damage to the surrounding rock in these parts.

(5) Structural rheology

After loading, the rock is not immediately damaged, but can remain stable for a long time, indicating that the damage of the rock has diachronism. This is especially true for soft rocks. After excavation, the deformation characteristics of the surrounding rock and long-term monitoring data show that the deformation and stress distribution of the surrounding rock vary with time from the excavation of the chamber to several months or years, displaying obvious time-effect characteristics. This type of deformation is the characteristic of structural rheological surrounding rock.

(6) Progressive loosening expansion

For the rocks transformed by shallow epigenesis in the shallow epigenetic transformation zone, and the surrounding rocks are relatively broken, but there are ground stresses in the surrounding rocks. When the tunnel passes through this type of surrounding rock, the deformation of the surrounding rock features loose circle progressive expansion. If there is no timely support or proper support measures, large deformation will be caused.

(7) Differential loosening

In the tunnel underneath the mobile deformation rock mass transformed by shallow epigenesis near the surface, the surrounding rock stress has been released, and the surrounding rock mass is characterized by near loose structure, and the expansion process of deformation and damage of the surrounding rock is rapid and can be developed all the way to the surface.

(8) Tilt-settlement deformation

If there is underground goaf in the lower part of the tunnel project, and the rock layer is inclined, the equilibrium condition of the top rock layer is destroyed due to settlement deformation, resulting in sudden and continuous bending and collapse. And the collapse causes the destruction of the whole upper stratum and movement to the goaf, leading to the deformation and damage of the tunnel surrounding rock.

(9) Vertical Settlement Deformation

When the rock distribution of the underground goaf in the lower part of the tunnel project is nearly horizontal, the top plate of the goaf will bend and collapse, and the collapse will cause the upper strata to move towards the goaf, resulting in the bending damage of the surrounding rock and tunnel structure in the upper part of the tunnel.

7.1.3 TBM Construction Key Points in Soft Rock Strata

1. Establish a support system adapted to TBM construction
   When the geological conditions are poor, the intensity of support increases, the difficulty increases correspondingly, the consumption of materials increases, and the time affecting the excavation increases. Therefore, the geological forecast in the excavation is done according to the design documents, and the support plan is proposed in advance.

   In Class III or Class IV but with developed joints, TBM construction requires a large amount of support and a long duration of support time. For this reason, attention shall be paid to the following three points:

   (1) Improve the operational efficiency of rock bolts
   Due to the limitation of the cycle tunneling time, the installation speed should be accelerated, the consolidation effect should be improved, the anchoring requirements should be achieved quickly, the tunneling time should be shortened, and the tunnel excavation speed should be accelerated.
   (2) Proper temporary support
   In poor geological conditions, local rocks often prevent the TBM from tunneling normally, and if necessary, scaffolding support is adopted.
   (3) Timely shotcrete
   In general, manual shotcrete equipment near the cutter head should be used as less as possible, in order to reduce contamination of the main machine and reduce TBM downtime. However, when passing through the weak geology and sections with severe rock explosion, especially when the excavation surface is broken, due to the slowdown of the advance speed, the time for shotcrete construction of excavation face and No.2 platform manipulator is extended. Therefore, manual shotcrete has to be carried out, and the purpose is to close the excavation rock surface as soon as possible to reduce exposure time. Although the temporary shotcrete is only 20–30 mm thick, it forms

a tightly-bonded composite with the rock in a very short period of time, thus preventing the rock from loosening, releasing pressure and bending during the normal loosening process. The resulting effect is to transform the rock with lower strength into stable rock and achieve the goal of stopping the chemical changes inside the rock and the weathering of the rock. Due to the use of manual shotcrete, pumping distance increased, posing strict requirements for the wet spraying material, and corresponding preparatory work shall be done.

2. Reasonable selection of tunneling parameters
The geological conditions are the key to the advance speed. In addition to the compressive strength of the rock, the shear strength of the rock, quartz content and the degree of development of joints must also be considered. The cutting principle of TBM is that the rock is broken under the load of the cutter at the contact between the cutter and the tunnel face, and from this area, the rock is squeezed and cracked, and the rock is crushed into pieces between the cutters along these cracks.
During the tunneling, the rock mass is cut by TBM machine capability. If the rock has a low compressive strength, high penetration, and low and stable thrust, the relative advance speed will be high. If the rock has a high compressive strength, high thrust and low penetration, the corresponding tunneling speed will be low. In these homogeneous rocks, after selecting the appropriate tunneling parameters, it is possible to adopt an automatic operation system with a small thrust as a limit without the need to adjust tunneling parameters at any time. Cutters also show regular and normal wear.
For TBM tunneling in soft geology with developed joints, special attention shall be paid to the correct selection of tunneling parameters. The penetration increases when tunneling in fractured rock This is mainly due to the fact that some part of the cutters and the tunnel face are not in close contact and the cutter thrust increased (because the penetration increases with the increase of thrust). Generally, under normal circumstances, the cutter shall have a certain overload capacity. However, affected by the current cutter quality problems, the damage to the cutter increased a lot. In addition, sometimes the cutting surface of the cutter head is perpendicular to the weak face, and the weak face cutting resistance is small, the penetration increased under the initial thrust force. The cutter wear is small and the probability of damage is not high. When the cutting center of the cutter head is in the rock with fracture development or in the rock along the spalling surface with low shear strength, the block along the joints or spalling surface is cut. At this time, TBM has violent vibration and great torque fluctuations, and the cutter is subjected to high lateral load, and the bearings become tight, resulting in overheating and thus causing damage to the tool. For this reason, the thrust should be reduced now.
When the increase of penetration and torque is found in the excavation, it indicates the change of geological conditions, so the thrust should be reduced

and the penetration should be controlled as appropriate, so that the balanced productivity can be maintained and the cutter wear can be reduced.

7.1.4 TBM Construction in Soft Rock Strata

Gripper TBM is generally used for hard rock construction. In order to adapt to the geological changes in extra-long tunnels, gripper TBM is equipped with anchor machine, concrete sprayer, ring beam installer and probe drilling machine, to adapt to the construction in weak and fragile strata. When Gripper TBM is passing through weak and fragile strata, the method of anchoring after spraying or spraying after anchoring is adopted, and primary support of ring beam is erected (the workload of primary support is large). The following is an example of the construction technology of gripper TBM in soft and fractured strata in the Mogouling Railway Tunnel.

1. Project Profile

   Xi'an-Nanjing Mogouling Railway Tunnel is 6.114 km long. The surrounding rock is mainly marble, mica and schist, due to severe impact of tectonic compression and corrugation of regional F2 large fracture zone on the tunnel, the rock mass is broken, and there are hidden joints and developed joints. Extrusion structure surface and tunnel cross with each other by a small intersection angle, and there are long and dense joint belt and developed broken belt. The surrounding rock is weak and broken, with poor stability. Soft rock area accounts for 70.5% of the total length of the tunnel. The tunnel was constructed by China Railway Tunnel Group using a German Wirth ϕ8.8 m gripper TBM (previously used for the Qinling I line Tunnel). This project was started on March 10, 2000 and was completed on December 31, 2002. 4.653 km of this tunnel was dug with a TBM at a maximum monthly advance of 573.9 m. In this project, China Railway Tunnel Group mastered the tunneling technology of gripper TBM in soft ground.

   Mogouling Tunnel is located in the middle and low mountainous area of the Qinling Mountains, and crosses the watershed (Mogouling) from west to east between the Wuguan River and the Qingyou River. The entrance of the tunnel is at an earth slope on the left of National Highway 312 in Ximogou Village, Maoping Town, and the exit is on the north side of Qingyou Middle School in Qingyouhe Town.

   Mogouling Tunnel begins at IDK226 + 494 and ends at IDK232 + 606 on the north side of Qingyou Middle School in Qingyouhe Town. Except for 538 m entrance section located on the curve of $R = 1000$ m, the rest are located on a straight line There is a longitudinal gable slope composed of an upslope, with a length of 1936 m and a gradient of 3.5–10.5‰, and a downslope, with a length of 4176 m and a gradient of 4‰ down slope, basically inclining to the east. The maximum overburden of the tunnel is 313 m.

   Mogouling Tunnel is located in a terrain with large topographic relief and crossed gullies. The relative height difference is 100–300 m, and the project passes through Mogouling via a long tunnel. The maximum elevation is 824 m.

The water flow from gullies in the west of Mogouling flow into the Wukan River through surface runoff, and the water flow from gullies in the east of Mogouling flow into the Qingyou River through ditches.

2. General Construction Methods

   1 gripper TBM is advancing in one direction from the exit of the tunnel. Drill and blast method is used for the section (169 m) of IDK232 + 437–IDK232 + 606 at the exit, and this section serves as the preparatory chamber. Drill and blast method is used for the section (Class III surrounding rock, 10 m) of IDK232 + 427–IDK232 + 437, and this section serves as TBM starting point. Drill and blast method is used for the section of IDK226 + 494–IDK232 + 665 at the entrance, with a U-shaped cross section. Open cut is used for the section of IDK226 + 665–IDK232 + 725, and this section serves as the site for disassembling TBM. TBM is proposed for the section (5702 m) of IDK226 + 725–DK232 + 427. Considering strata conditions and terrain conditions at the tunnel portal, it is determined that TBM is disassembled at Taiyangpo Gully (IDK226 + 665–IDK232 + 725) after completing the construction.

3. Initial support construction

   (1) Initial spraying
   When the surrounding rock is broken and the self-stabilization ability is poor, the manual shotcrete system after the cutter head can be used, and the concrete is shot directly after the shield of the cutter head while digging, and the initial shotcrete is 3–5 cm thick to close the surrounding rock.

   (2) Ring beam and wire mesh installation
   The main machine of TBM is equipped with an annular ring beam mounter, which allows pre-assembly of 5 ring beams. The installation sequence: lift the first ring beam from the transport trolley with lifting device, put it on the top of the storage ring, use hydraulic cylinder to drive the lifting device to move horizontally and vertically, push the first ring beam to rest it against the storage ring and fix it with bolts, the other end is supported by a bracket, and then rotate the storage ring to leave space for the second ring beam, using the same process for the second, third and fourth ring beams, connect them with steel plate, and then rotate the storage ring again. Each ring beam is put on two brackets and one of them has a fixing clip. The fifth ring beam is connected with the fourth ring beam, and its free end slides into the end of the first ring beam, and fixed with G-shaped clip, and then rotate the storage ring by 180°, to make the ring beam overlap part located in the bottom of the tunnel. An external device is used to make the ring beam stand against the tunnel wall, and then a fastening plate is attached.

   Before the ring beam stand against the tunnel wall, wire mesh shall be installed behind steel extrados for the soft broken belt, and the mesh specification is 1.5 m × 0.9 m, grid 25 cm × 25 cm. Steel welding is used for longitudinal connection of ring beams, to ensure the quantity and quality, and prevent the surrounding rock instability.

(3) Rock bolt construction
The rock bolt can be driven into 4.0 m. The anchoring materials and rock bolt are installed manually, and the drilling rig assists in jacking them into place. The mesh is hung manually, and the hole is drilled with percussion drill, and the short reinforcement is fixed and connected to the rock bolt. The mesh is attached to the rock surface, and the debris is prevented from falling between the rock surface and the ring beam.

(4) Shotcrete
The wet shotcrete system is operated by electric remote controller, which can turn on or off the shotcrete and additives. The shotcrete is produced in the concrete plant, and transported to the tunnel by $6m^3$ concrete mixer. Except the manual spraying device after the cutter head shield, the manipulator spraying device set up on the platform of back-up gantry 2 can complete concrete spraying operation within the range of 300° above the invert and within the length of 5 m.

(5) Invert installation
The invert is a reinforced concrete structure. Its bottom is divided into two types, namely ordinary type and slot type. During laying, the invert is transported to the place under invert crane and rotated manually by 90°, and then transported to the invert placement area. Use invert crane to lift them and put them in place according to the designed center line and horizontal position, and then the water stop tape and mortar protection are used. When laying 4–5 inverts, use fine aggregate concrete to fill the 5 cm gap under the inverts. Concrete pouring is done by grouting machinery through grouting pipes. Before pouring concrete, clear the slag thoroughly and use water to clean, and seal the wall with sand bags to prevent the inflow of mud. During grouting, each grouting should make both sides of the invert stick out of the rock surface as appropriate. It is forbidden to let industrial exhaust fan waste water flow into the place just grouted during construction. In addition, the number of invert grouting shall be controlled. Half grouting of invert is not allowed.

7.2 TBM Construction in Hard Rock Strata

The tunnel passes through the hard and complete surrounding rock. The tunneling conditions are good, the lining is thin and even lining is not required. Therefore, the cost per unit length is relatively low. However, the high strength of the rock also brings difficulties in excavation and advance. For the tunnel constructed by using TBM, the surrounding rock is too hard and wear-resistant, causing great cutter wear.

This seriously affects the project progress, increases the cost of the project, and even limits the use of TBM.

According to some TBM manufacturers, TBM is applicable to the rocks with a strength of 300–350 MPa and can be used in almost all rock types. However, engineering practice shows that the most suitable range of rock strength for TBM is still 30–150 MPa. In the Wanjiazhai Water Diversion Project already completed, the unconfined compressive strength of the surrounding rock is generally within 100 MPa, and a high advance speed has been achieved. And according to the experience gained in a tunnel in Qinghai province and Qinling Railway Tunnel and other projects, when the compressive strength of the rock is greater than 100 MPa, but rock mass is also relatively intact, it will bring difficulties to TBM construction. The geology of the hard rock section of a tunnel in Qinghai is dominated by Proterozoic gneiss, granite gneiss, intrusive granite and intrusive quartzite, with a saturation compressive strength of 60–160 MPa and a solidity coefficient of 6–15. The stratum is mainly influenced by intrusive granite, and the surrounding rock is deeply metamorphosed, with high wear resistance. Due to the good integrity of the surrounding rock and undeveloped joints, for most of the time, TBM is tunneling in the intact surrounding rock with high strength, and the penetration was 2–3 mm/r, resulting in slow tunneling speed. Qinling Tunnel Line I passes through migmatitic gneiss and migmatitic granite. The rock compressive strengths of migmatitic gneiss and migmatitic granite are 79–325 MPa and 117–192 MPa, and the pure tunneling speed is 1–2 m/h. Therefore, same attention shall be paid to both intact hard rock with high strength and weak rock with poor stability where TBM passes through.

7.2.1 Criteria for the Classification of Hard Rocks

Standard for Engineering Classification of Rock Masses (GB 50,218–2014) stipulates that the rock with an unconfined saturated compressive strength not less than 60 MPa is hard rock; 30–60 MPa for harder rock; 15–30 MPa for softer rock; 5–15 MPa for soft rock; and less than 5 MPa for very soft rock. Different rock types can be classified according to the needs of different types of projects. In underground works, 60–120 MPa was classified as hard rock, and not less than 120 MPa as very hard rock, in order to adapt to the characteristics of tunnel construction. For full-section TBM construction, some sources classify rocks with a compressive strength (R_s) greater than 150 MPa as hard rock, 80–150 MPa as medium hard rock, and less than 80 MPa as soft rock.

The Specifications for Construction Planning of Water Resources and Hydropower Projects (SL 303–2017) lists the average weight, ultimate compressive strength and rock solidity coefficient of various types of rocks under natural humidity, and the compressive strength of common rocks is also listed in the book *TBM* for reference. However, in different regions, the tunnels are located in different strata, and the rock strength also varies significantly, so each tunnel project needs to conduct specific tests.

In addition to the compressive strength of the rock, the integrity of the surrounding rock, including the size of the structural face spacing, the shape of the rock, the properties of the joint face and the abrasion resistance of the rock have a great impact on TBM construction. For example, in the Wanjiazhai Water Diversion Project in Shanxi province, the Cambrian Fengshan Fm chert formation where No.4 south tunnel passes through has a high strength, with an unconfined saturation compressive strength of 100–140 MPa, but smooth tunneling is achieved. However, in a project in Qinghai province, although the unconfined compressive strength of the rock is 60–160 MPa, the rock mass is intact and the joints are not developed. But the advance speed is greatly affected.

7.2.2 Analysis of the Effect of Hard Rock on TBM Tunneling Efficiency and Its Cause

1. Reducing the efficiency of TBM digging

First, due to the large tunneling thrust force, small penetration, as well as cutter wear and serious abnormal damage, the tunneling efficiency is low. Second, if the overall strength and stiffness of the cutter head cannot meet the requirements, the cutter head face plate will be prone to cracking. Once the cutter head is cracked, it will be difficult to guarantee the repair welding quality due to restrictions such as conditions and environment. This will inevitably cause the insufficient performance of the thrust force, thus significantly reducing the tunneling efficiency.

2. Reasons affecting the tunneling efficiency

The first reason for the low advance speed of TBM in hard rock is the influence of geological conditions, such as the integrity, strength and wear resistance of the rock. The high quartz content in the rock causes great increase in cutter wear, and eventually limiting the use of TBM. The condition of the tunneling equipment is also very critical. Because in hard rock tunneling, the equipment is at full load or even overload, and accidents often occur in the main motor, frequency converter, reducer, hydraulic pump, hydraulic valve set, hydraulic cylinder, etc. The requirements for operators are also high. If there is a problem with any link, such as Lead Engineer, Operation Foreman, Mechanical Engineer or Electrical Engineer, it will increase the accident rate of the equipment and even cause damage to the main parts, and compounded by the frequent replacement of the cutters during the tunneling process, continuous operation of TBM is often interrupted.

7.2.3 *Key Points of TBM Construction in Hard Surrounding Rocks*

1. Key points

    (1) The physical parameters such as quartz content, abrasion resistance, compressive strength and integrity of the surrounding rock must be analyzed in detail by means of field sampling and testing, and a systematic database must be derived to determine the general mechanical index of the surrounding rock in the construction section.

    (2) For the cutters, the cutter ring shall be selected by combining the physical parameters of the surrounding rock for overall design. In the actual construction process, a variety of models of cutter rings should be selected for field boring tests under similar basic surrounding rock conditions, and finally selected by comparing the cutter consumption in terms of linear meters.

    (3) The collection of tunneling parameters should be carried out in accordance with each tunneling cycle to maintain the continuity of the data and form a database; and then screen and remove the values that deviate from the trend of the data to facilitate the analysis of the parameters.

2. Precautions for specific implementation

    (1) Systematic collection and analysis of physical and mechanical parameters of rock masses

    On-site sampling: on-site sampling is organized by the project laboratory and TBM tunneling team cooperates. The core drilling machine on TBM is used to take samples (Fig. 7.1), and core drilling machine is usually installed in the middle of the cutter head. The downtime of TBM is utilized to take sample ahead of schedule. However, on many occasions, because the core drilling machine is installed in the position of cutter head, on-site cutter replacement is affected. The core drilling machine can be adjusted to the platform of main machine after the shield. In this case, only radial coring can be carried out. In the hard rock section, the coring speed is relatively slow, and 50 cm coring length can be considered, and a group of samples is taken every 100 m or so. Within 100 m, the falling blocks can be collected, about a group every 10 m. The diameter of such blocks is about 30 cm, and the compressive strength is calculated by point load test (Fig. 7.2).

    Test and analysis: It is necessary to carry out test including compressive strength, quartz content, wear resistance value and integrity coefficient of the rock mass, etc. After the test results are obtained, make the test ledger, and then analyze the data curve of about 1000 m, and take the average value or trend line to make preliminary conclusion on the physical and mechanical index of the rock mass in front of the TBM boring.

    (2) Cutter Selection

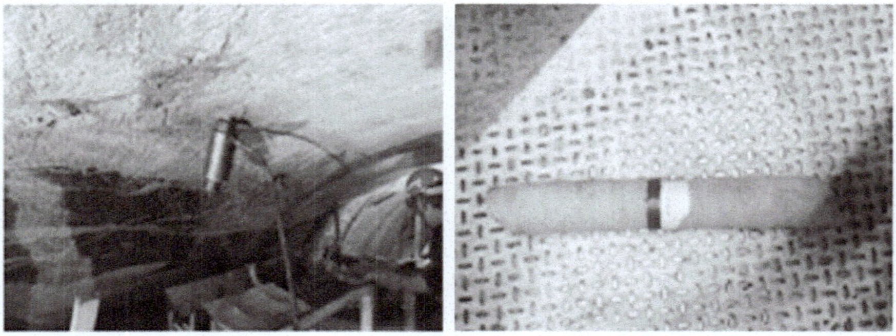

Fig. 7.1 Photos of on-site coring of TBM construction

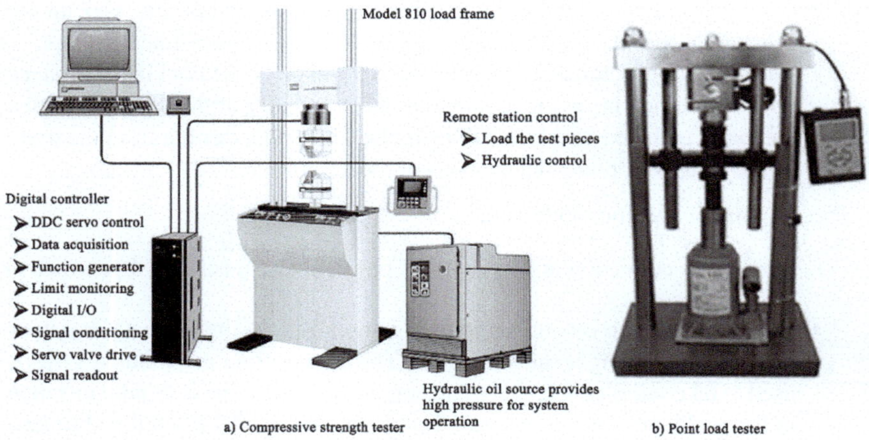

Fig. 7.2 Rock strength tester

There are many factors affecting the cutter consumption, and geological factors are the main reason. The structure of the cutter head, cutter spacing, number of cutters and control of tunneling parameters also affect the cutter consumption to a certain extent. In the process of breaking rock, the cutter ring directly cuts the rock. Under the joint action of huge load and high abrasive characteristics of the rock, wear is most likely to happen. The cutter ring material performance and configuration parameters determine the cutting performance of the cutter. Most of the cutter ring is made of hot abrasive steel containing Cr, Mo, V and Si and other alloy elements. It shall be made of good material, and have high hardness and uniformity, as well as good wear resistance and toughness. The diameter of the disc cutter is related to its rated thrust force. The increase in the diameter of the cutter can increase the rated thrust force of each cutter. every rotation of the cutter head will increase the cut depth of cutter, thus increasing the speed of

Fig. 7.3 TBM cutter head layout

equipment tunneling. In addition, the increased diameter of the cutter ring increases the allowable wear volume and extends the service life. Currently, 17in and 19in diameter disc cutters are the most common cutters used on TBM. According to the installation position of the disc cutters on cutter head, cutters can be divided into center cutter, face disc cutter and gauge disc cutter, as shown in Fig. 7.3.

(3) Selecting reasonable construction parameters during tunneling

TBM construction parameters should be reasonably selected and adjusted in a timely manner during the tunneling process. If the rock hardness encountered during the tunneling process is higher than expected, TBM thrust force should be increased in time within the cutter bearing capacity allowed, the rotation speed of the cutter head should be increased, and the torque should be reduced appropriately to avoid damage to cutter. Through the verification of the optimization adjustment of the tunneling parameters in different surrounding rock conditions at TBM construction site on the south side of the Yinhanjiwei Water Tunnel and a great deal of statistical analysis of the tunneling parameters, the following conclusion is reached. The selection of tunneling parameters for the section that has the surrounding rock strength above 150 MPa and no developed surrounding rock joints and fissures but good overall integrity: the thrust force is controlled at about 90% of the rated thrust force, the torque is controlled at 20% of the rated torque and the rotation speed is controlled at 5.4r/min. The selection of tunneling parameters for the section that has the surrounding rock strength in the range of 80–150 MPa and developed surrounding rock joints and fissures as well as adequate overall integrity: the thrust is controlled at 80% of the rated thrust, the torque is controlled at 30% of the rated torque and the rotation speed is controlled at 4.3 r/min.

7.2.4 Example of TBM Construction in Hard Surrounding Rock

1. Project Overview
The South Section of the Yinhanjiwei Water Tunnel is located in the territory of Simudi Town, Ningshan County, Shaanxi Province, starting from 1942 m downstream of the main tunnel extension from No. 3 adit. The total length of the bid section is 18.275 km, and the construction chainage is K28 + 085–K46 + 360. A ϕ8.02 m gripper TBM provided by Robbins is used for the construction, see Fig. 7.4 for the division of the bid section, the scope of the bid section includes:

(1) Construction of 18,275 m of the main tunnel (K28 + 085–K46 + 360). It consists of TBM jumbo installation shaft (100 m), TBM main machine installation shaft (80 m), TBM stepping shaft (220 m), TBM launch shaft (25 m), TBM maintenance shaft (60 m), 300 m drill and blast auxiliary section downstream of the maintenance shaft and TBM tunneling section (9860 m + 7630 m).
(2) No. 4 adit for auxiliary construction is constructed. The horizontal distance is 5784 m, the inclined length is 5820.21 m and the slope is − 11.96%. It is designed as a portal type, and the size is 6.7 m × 6.5 m. No.4 adit mainly solves the problems of long-distance construction ventilation, mucking and some construction material supply for the middle section, and the chainage of the main adit intersection is K38 + 400.
(3) Other works such as operation and maintenance of adits, blocking, protection of spoil area, prefabricated yard and material stockyard.

2. Construction

(1) According to on-site test and outsourcing test, the strength of surrounding rock in the TBM tunneling section is 80–216 MPa, and the average strength of surrounding rock is about 155 MPa. Due to the undeveloped joint fissures of rock mass, the overall integrity of surrounding rock is good. The penetration of cutter head is low, and the average penetration is only 3.4 mm/r.

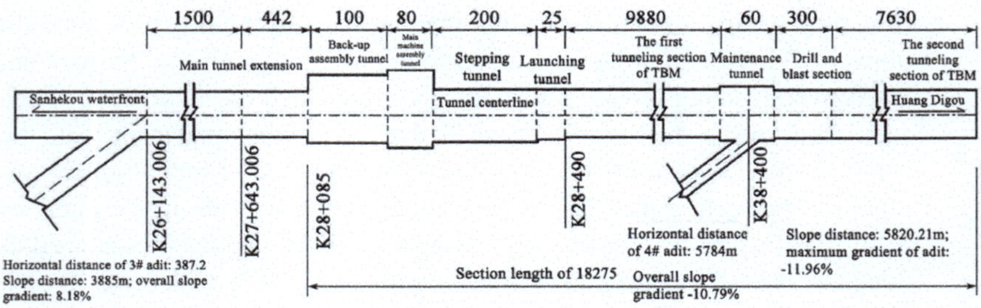

Fig. 7.4 Project layout (dimension unit in m)

According to the calculation of pure tunneling time, in the normal construction section, the tunneling speed is only 1.22 m/h in Class II surrounding rock section and 1.45 m/h in Class III surrounding rock section. Through the optimization and adjustment of tunneling parameters on site, the average penetration of Class II surrounding rock section is increased to 6.6 mm/r, the advance rate is increased to 1.8 m/h, and the daily advance rate is increased from 8.7 to 12.3 m. The average penetration of Class III surrounding rock section is increased to 9.2 mm/r, the advance rate is increased to 2.7 m/h, and the daily advance rate is increased from 11.6 to 16.5 m. Meanwhile, cutter consumption decreased by 30%.

(2) From May 17 to May 26, 2015, affected by rock burst, the tunnel face showed uneven pits, the cutter body was smashed, the cutter ring was broken and fractured and worn unevenly, and the cutter tooth were excessively worn, etc. The cutter consumption rose sharply, and the increase of cutter replacement led to a serious impact on the construction progress, and only 17.6 m of excavation was completed in 10 days. In the subsequent construction, for similar unfavorable sections, the tunneling parameters were adjusted (reducing cutter head thrust force and reducing cutter head rotation speed) and the cutter head inspection frequency was increased. In addition, the cutter head was inspected once every 2 cycles, and if necessary, once every cycle. The time for each cutter head inspection was controlled within 40 min, and the single cutter replacement time was controlled within 20 min. For this reason, the construction progress increased by 10% and the cutter consumption decreased by 15%.

(3) During the construction in the adverse geological section, TBM tunneling construction has been kept in a safe and controlled state by strengthening the on-site support, increasing the frequency of safety inspections, and arranging for safety personnel to supervise the key areas.

(4) TBM mandatory maintenance time of 4 h per day is ensured. The mandatory maintenance time is determined by Chief Mechanical Engineer, and followed by routine maintenance during TBM tunneling.

7.3 TBM Construction in the Strata with Water Inrush and Mud Gushing

Water inrush and mud gushing is a disaster often encountered in tunnel construction, and in practice, water inrush and mud gushing often cause serious economic losses to and delays in TBM construction. Therefore, familiarity with the characteristics of water inrush and mud gushing, as well as the analysis of actual construction cases, are the necessary conditions for TBM to smoothly pass through the strata with water inrush and mud gushing.

7.3.1 Characteristics of the Section with Water Inrush and Mud Gushing

Water inrush and mud gushing is one of geofluid disasters in tunnel construction. Compared with other disasters, it has the following characteristics:

(1) High frequency. Many tunnels are located in the groundwater-rich area and its soft ground. In the process of tunnel construction through the aquiclude or pressurized water, as long as the water channel exists, sudden water inrush and mud gushing is likely to happen.
(2) Serious consequences. Once massive water inrush and mud gushing in the tunnel occur during the construction process, the construction will be seriously hindered, and shallow groundwater and surface water depletion may happen, and even ground collapse and other environmental problems may occur.
(3) Difficult prediction. Due to the complex rock structure and significant non-homogeneous anisotropy of the surrounding rock, the calculation parameters obtained from partial borehole pumping and water pressure tests have certain limitations in spatial representativeness, and it is even more difficult for the sections with large overburden due to economic constraints. Therefore, it is exceptionally difficult to completely rely on the method of hydrogeological test to determine the seepage calculation parameters. Take a tunnel in Qinghai province for example, the maximum overburden is up to 1028 m, and the initial maximum water inrush in the section of the predicted chainage K20 + K768.00–K21 + 648.00 is 28392.03m^3/d, the regular water inrush is 9497.58m^3/d, but during the actual construction process, the water inrush is small. On the contrary, for some projects, due to the lack of advanced preparation, sudden water inrush and mud gushing during tunneling process caused great difficulties to the project. For example, the diversion tunnel of Zhangjiu River in Yunnan province has been forced to change the construction scheme due to multiple collapses, water inrush and mud gushing, as well as TBM jamming and shutdown, resulting in environmental disasters such as disappearance of surface spring water and drop of reservoir water level.

7.3.2 Danger of Water Inrush and Mud Gushing

(1) Hazardous to personnel and equipment safety. In the process of TBM construction or conventional construction, when the sudden water inrush and mud gushing exceed the drainage capacity of the tunnel and the mud clearing capacity of TBM, it poses a threat to the safety of the construction personnel and equipment inside the tunnel, especially in the long tunnels.
(2) Affect the stability of the surrounding rock of tunnel. In the tunnel section with high overburden, due to the high groundwater pressure, after tunnel excavation, the groundwater seeps into the tunnel in a concentrated manner, with large

hydraulic slope and strengthened outwash effect. In addition, the loose fillers in the fault fracture zone and intrusive rock contact zone are easy to deform or be taken away by high-pressure groundwater, forming mud gushing and inducing many collapses. Meanwhile, the connectivity of fissures is increased, thus increasing the seepage capacity and further increasing the water inflow. Furthermore, the long-term immersion in groundwater reduces the strength of the rock mass and softens the rock mass. In particular, the shale and clay rock swells when it meets water, and this swelling force acts on the surrounding rock or structure of the tunnel, causing damage to the surrounding rock or structure. In addition, water saturates the hydrophilic minerals on the structural surface of the surrounding rock, reduces the strength, plays a lubricating role, and causes deformation and collapse of soft rock in the tunnel.

7.3.3 Treatment Measures for Water Inrush and Mud Gushing Tunnel Section

According to the hydrogeological data in the early stage, before tunneling to the water-rich tunnel section of the fault zone, smear zone and intrusive rock contact zone, the water content of the rock stratum in front of the tunnel face shall be preliminarily judged by methods such as induced polarization, and then the water content within 30 m in front of the cutter head shall be determined by advanced geological drilling machine. When approaching the aquifer, the water volume and water pressure shall be detected by probe drill hole to judge the water gushing direction and the danger level of water inflow and mud gushing. The grouting water stop scheme shall be determined according to the water gushing and water pressure, and treatment measures such as grouting sealing and advanced reinforcement of the stratum shall be taken according to the actual situation on site.

(1) In general water-rich tunnel sections, when the water inflow of the tunnel is less than $100m^3/h$ or the water flow of a single hole is less than $10m^3/h$, TBM can directly pass through.
(2) In the water-rich tunnel section, when the water inflow of the tunnel is greater than or equal to $100m^3/h$, the treatment scheme shall be determined according to the principle of "blocking first, combining drainage and blocking, and comprehensive control".
(3) For the tunnel section with high external water, the scheme of pre-grouting and water plugging shall be adopted. The grouting around the excavation contour line shall be considered first, and the grouting range shall be extended when the water volume is large.

7.3.4 Construction Example of Water Inrush and Mud Gushing

For the 4th bid section of Yinsong Water Tunnel, on February 29, 2016, water gushing occurred in front of the tunnel face and mud gushing occurred at the tunnel face on March 24. The water level in TBM area rose rapidly to 95 cm, making it difficult for the support workers to install the ring beam at the bottom. The slag in front of the cutter head flowed out with the water, and the continuous belt system slipped, and TBM traveling and transportation track line was submerged and could not be extended. Meanwhile, the water level in the whole tunnel also rose rapidly, submerging some production and living areas at the exit, and TBM is forced to stop tunneling. Construction was difficult throughout March and April (Fig. 7.5).

In response to the above-mentioned dangers, the Project Management Office of the 4th bid section of Yinsong Water Tunnel has taken decisive and effective measures

a) Water ingress from the cutter chamber b) Water outflow from the left shield

c) Mud outflow from the cutter chamber d) Mud burst from the cutter hole

Fig. 7.5 Construction site of the section with water inflow and mud gushing

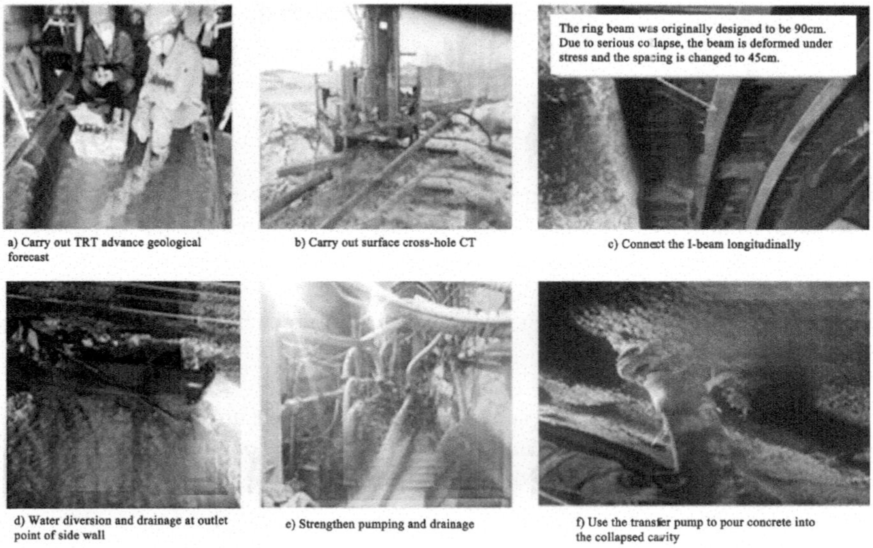

Fig. 7.6 Photos of response measures for the section of water Inflow and mud gushing

to minimize the downtime of TBM, thus achieving a major breakthrough. The main measures taken are mainly as follows (Fig. 7.6).

1. Before tunneling

   (1) Advance geological forecasting
   On the basis of daily advance geological forecasting of geophysical prospecting, joint forecast methods such as borehole forecast and surface cross-hole CT are continuously carried out on the tunnel face. Accurate forecasting of construction risks such as water inrush and collapse within a certain range in front of the tunnel face, and timely feedback, identification of construction risk sources and timely adjustment of tunneling parameters to guide the subsequent work.

   (2) Cutter head cleaning or solidification
   The manhole and muck scraping hole of cutter plate are partially blocked with welded steel plates to reduce muck discharge volume, and the cutter head is left idling before each tunneling to clean up the slurry of the cutter head. If the mud gushing amount in front of the cutter head increases, chemical grouting is adopted to consolidate mud gushing and crushed rock mass.

2. During tunneling
   The manual mode is adopted for tunneling, and the advance speed, cutter head rotation speed and tunneling thrust force shall be reduced at the same time to reduce the mucking volume and avoid belt blocking or emergency stop caused

by large mucking volume. In the process, the diversion, drainage, blocking and drainage methods are used for the leakage point to avoid water inrush and large water inflow.

3. After tunneling

   (1) Drainage
   The water gushing from the working face or the remaining water after grouting is drained away from the working face in time. The leakage of water on the sidewall is covered and drained to ensure the quality of shotcrete. After spraying concrete, due to the rise in water pressure, damage to primary support is likely to happen, and drainage measure or backfill grouting is used for sealing.

   (2) Strengthen the support
   Timely strengthening the support. Joint support measures, such as intensified ring beams (i.e. 45 cm/unit), steel slat, I-beam longitudinal connection, shotcrete closure, are used.

7.4 TBM Construction in Fault Fracture Zone

Because of its superior performance of high efficiency and safety, TBM has been increasingly applied in tunnel construction, but its adaptability to unfavorable geological conditions, such as fault fracture zones, is poor. These unfavorable geological conditions will have a great impact on TBM tunneling. This may not only reduce the utilization ratio of TBM tunneling operation time, but also even cause damage to TBM and difficulty in passing smoothly. Effective measures must be taken, otherwise improper handling will bring serious consequences to the construction. TBM cutter head jamming accidents have been encountered when TBM passes through the fault zone during construction of the headrace tunnel of Yunnan Tianshengqiao HPP, Shanxi Wanjiazhai Yellow River Diversion Project, Italy's Gossensa Railway Tunnel and Austria's Campolosso Railway Tunnel, leading to long downtime.

7.4.1 *Impact of Fault Fracture Zone on TBM Construction*

Fracture zones can have a very adverse impact on TBM construction, mainly including:

(1) Collapse and spalling of working face and arch crown after excavation. TBM cutter is buried and it is difficult to rotate the cutter head.
(2) The excavation face and side wall collapse, and the supporting force of the gripper is insufficient to provide sufficient supporting reaction force required for TBM tunneling.
(3) Bringing difficulties to shotcrete and rock bolt support.

(4) The surrounding rock is uneven in hardness, and the cutter head is prone to vibration when rotating, affecting the service life of the cutter and increasing the wear and consumption of the cutter.
(5) When the surrounding rock is uneven in hardness, it may intensify the swinging of TBM, bringing difficulties to the control of advance direction.
(6) If there is water gushing, the working face will collapse due to the action of water pressure, and the cutter head and the machine will be flooded by the water gushing, increasing the rotating torque of the cutter head. The side wall will collapse due to the action of water, the gripper cannot provide support, and the machine will be flooded by the water gushing.

7.4.2 Construction Measures for Fault Fracture Zone

When the gripper TBM passes through the fault fracture zone, the construction steps are shown in Fig. 7.7.

Before tunneling, advance geological forecasting shall be carried out to determine the edge, length, fracture degree and water content of the fracture zone. Different treatment measures shall be taken according to different conditions of the fracture zone. For slightly-fractured sections, if there is no impact on TBM, no treatment is required. For generally-fractured sections, the method of tunneling first and then treatment shall be adopted. For severely-crushed sections where TBM construction cannot be carried out, the tunneling shall be stopped, and the drilling and grouting equipment equipped with TBM shall be used for advanced reinforcement.

Then drill probe drill holes for inspection. If it is proved to be feasible, treatment shall be carried out before forward tunneling. If the crushed section is long and the crushing is severe, and TBM cannot be used for smooth passage after general reinforcement, the drill and blast method can be used to excavate an opening, and TBM can follow up to pass through.

TBM tunneling parameters reasonably selected during tunneling. Under different geological conditions, the tunneling parameters such as thrust force, advance speed, cutter head rotation speed, cutter head torque and gripper supporting force required for TBM are different. When TBM passes through the fault fracture zone, the tunneling parameters such as advance speed and cutter head rotation speed of TBM can be appropriately reduced. This can effectively reduce the disturbance to the surrounding rock, thus reducing or avoiding collapse. Furthermore, it is necessary to know well the change of surrounding rock ahead according to the relative

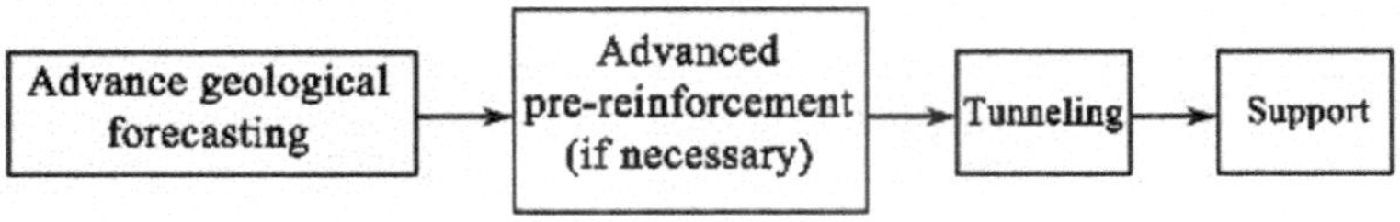

Fig. 7.7 Steps to pass through fault fracture zone

change of some tunneling parameters during tunneling. For example, the strength of surrounding rock can be calculated from the advancing pressure, and the integrity of surrounding rock can be calculated from the torque of cutter head, so as to adjust the tunneling parameters of TBM in time or take other measures to make TBM pass through quickly and safely.

Strengthen the support after tunneling. For the generally-crushed section, shotcrete or fiber concrete can be used, and rock bolts can be added locally for support. For severely-crushed sections, ring beams, wire mesh, shotcrete or fiber concrete can be used.

1. Advance Forecasting and Prediction
   Due to the limitations of geological exploration, TBM often encounters some unfavorable conditions not showed on geological maps during tunneling. In order to further ascertain the exact situation of the fault fracture zone in front of TBM, the advance forecast of engineering geology shall be actively carried out, so as to master the real situation of the fault fracture zone in detail to take reasonable measures. Common methods include:

   (1) Using the probe drilling machine equipped on TBM;
   (2) Utilization of the advance forecasting system;
   (3) Calculation from the geological conditions of parallel adit;
   (4) Using the outcrops, mucking conditions and abnormal conditions during tunneling for judgment.

   During the construction of Qinling Tunnel, the advance geological forecasting was carried out by the probe drilling machine (probe drilling of 20–30 m) and TSP advance geological detector of TBM, and the geological conditions ahead were comprehensively judged by the parallel adit of Line II, outcrops, mucking and abnormal conditions during tunneling, so as to better ascertain the situation of the fracture zone ahead and lay a foundation for the smooth passage of TBM.

2. Pre-reinforcement
   Determine whether it is suitable for TBM tunneling according to the geological conditions ahead. If it is suitable, continue tunneling forward; otherwise, reinforce the unfavorable geological sections ahead or adopt other methods to pass. If it is considered that the pre-reinforcement scheme is feasible after comparison with other schemes in terms of construction period and cost-effectiveness, TBM will be directly used to reinforce the unfavorable geological sections ahead, such as the fault fracture zone.

   The advance pre-grouting and advance pipe shed grouting reinforcement are carried out for the surrounding rock in the fault fracture zone in front of the tunnel by using the probe drilling machine and the grouting equipment equipped with TBM. Before drilling holes, in order to prevent surrounding rock collapse and slurry leakage on the tunnel face, a layer of concrete shall be sprayed by using the spraying system equipped with TBM after the excavation by the cutter head, but it is strictly prohibited for the cutter head to retreat to prevent the cutter head from falling and jamming. Before grouting, flush the borehole with water. During grouting, symmetrical grouting shall be carried out from the boreholes on both sides to the arch crown to prevent grout leakage. Grouting parameters are selected and determined according to the engineering geology and hydrogeology of surrounding rock (such as porosity, fissure rate, permeability coefficient, water inflow volume, water pressure, etc. of surrounding rock) and in combination with tests.

3. Treatment of collapse
   Different measures are taken according to the size of the collapse.

   (1) Small-scale collapse
   Several collapses or small-scale spalling occur on the top and face of the working face, but without expansion. Small stones fall between the cutter head shield and the rock wall, and small collapses occur in the arch or side wall, but there is no sign of further development or expansion. The tunneling is normal, the thrust force and torque change little, and the machinery (especially the main machine area) has no abnormal vibration or sound. The slag is evenly concentrated, and occasionally mixed with large rock chips. The following support measures are adopted: the part above the gripper shall be hung with wire mesh and provided with system anchor bolts, and ring beams shall be erected as appropriate.

   (2) Medium-scale collapse
   Severe spalling of working face, and severe collapse or local spalling of arch crown, but the cutter can still operate; severe collapse of gripper, difficulty in padding and regripping. A large number of stones falling between the shield and the rock wall; during tunneling, the mechanical vibration is large, there is abnormal noise, and the thrust force is weakening. The torque increases, with a tendency to go up and down. The large pieces on the belt conveyor increase, accompanied by a small amount of fine slag, and the slag piles increase or decrease now and then, and become uneven. The following support measures are adopted: the manual shotcrete system is used to spray concrete to the place of collapse to seal the surrounding rock in time and reduce the exposure time of rock. Install the circular ring beam, and hang the wire mesh above the gripper before installing the ring beam. The grouting anchor pipe is driven to improve the bearing capacity of the gripper.

   (3) Large-scale collapse
   Large-area collapses of arch crown and tunnel wall develop so rapidly that they cannot be controlled by conventional construction methods. The collapse of the arch crown was deep judging from the edge of the shield, and a large number of stones fell between the shield to the rock wall, and the collapse expanded to the rear area. The tunnel wall supported by the gripper collapses a lot, so that the reaction force cannot be obtained and the regripping cannot be achieved. The mechanical vibration is particularly large during tunneling, and the loud sound from the tunnel face can be heard from the operator cabinet. During advancing, the torque becomes large, and

the cutter is difficult or unable to rotate. A large amount of rock chips are generated, leading to blockage. In severe cases, the cutter head is stuck by stones and cannot rotate. The slag is mainly in big size, with few fine slag. The following support measures are taken: stop tunneling, take auxiliary countermeasures, drill holes from the rear of TBM, and grout at the collapsed part; remove the muck at the place of collapse, fill it with fine aggregate concrete and erect ring beam for support.

When Qinling Tunnel passes through F_{11} and F_{12} large faults, the following construction measures are taken: stop tunneling and use probe drilling machine for pre-grouting reinforcement of rock stratum; however, the drilling machine did not work, and there was no choice but to erect ring beam for support. Due to severe collapse, the ring beam is seriously deformed, and certain ring beam penetrated into the lining clearance by more than 5 cm. After the main machine passes through, the slag on the wire mesh shall be removed first, and the wire mesh shall be cut open, and I-steel support shall be added longitudinally at the place of collapse. The wire mesh shall be laid circumferentially, and ϕ22 reinforcement shall be inserted vertically and horizontally by using the rock gap, and welded with the wire mesh to form a reinforcement framework. Formwork erection and grouting are carried out on the original ring beam to form a closed reinforced concrete arch at the place of collapse. After concrete setting, the ring beam shall be cut off, the concrete invading the boundary shall be removed, and the ring beam shall be reinstalled.

In addition, the shotcrete in the fault fracture zone is very important. During manual shotcreting, relevant equipment must be protected to avoid contamination of main machine by concrete rebound materials. Shotcreting must start from filling holes and cracks on the rock surface. In the ring beam section, the gap between the steelwork and the surrounding rock must be filled with shotcrete.

4. Reinforcement technology of surrounding rock at gripper of TBM in weak and fractured ground
During TBM tunneling, the gripper supports the gravity of the equipment and transmits the reaction force of thrust force and cutter head torque to the rock wall of the side wall. When the strength of the rock wall of the side wall is sufficient to bear the pressure of the gripper, TBM can tunnel normally. Therefore, for a small range of side wall collapse, the supporting pressure on the surrounding rock is reduced by locking part of the gripper, and the thrust force and advance rate of TBM are correspondingly reduced for TBM to pass through the collapse section without stopping the TBM. If the side wall is relatively weak, sleepers are added at the gripper to increase the ground contact area for TBM to pass through. When the side wall of the tunnel collapses greatly or the strength of the surrounding rock of the side wall is insufficient to bear the pressure of the gripper, and the above measures cannot work, TBM shall be stopped first, and several grouting rock bolts shall be installed at the gripper to improve the bearing capacity of the

stratum after grouting. The tunneling can also be carried out after the support method combining shotcrete, wire mesh, steel arch and concrete pouring.

5. Subsidence treatment of weak zone
According to the geological conditions ahead, if it is judged that subsidence may occur and serious subsidence may still occur after simple treatment during construction, pre-grouting reinforcement treatment can be carried out, and tunneling can be carried out after a certain strength is reached. In case of TBM subsidence during tunneling, TBM shall move outside the fault weak area. Before moving, the excavation face and excavation section shall be provided with primary support to prevent collapse. Then install the sleepers, correct the attitude of the TBM with a jack, and then pour concrete for replacement. To make the concrete withstand the pressure of the gripper, the concrete must be poured up to the arch line.
6. Treatment of water inrush in fault fracture zone
During TBM tunneling, sudden water inrush can have a serious impact on construction, even endangering the safety of personnel and equipment. The principle to follow in the treatment plan is "Forecast First, Prevention Foremost, and Ensure Safety".
Before tunneling: probe drill hole can be carried out to detect the water flow and water pressure of the borehole in combination with the exploration holes in the fracture zone; determine the chainage of the water inrush point. Drill the advanced water drainage hole to discharge water. During the water discharge, always observe the change of water pressure and water volume. If the water pressure decreases, TBM will continue to tunnel under the condition of a good drainage system. If the water pressure and water volume of the drainage hole are not reduced, the working face and side wall will collapse or the drainage facilities will fall behind after excavation, grouting and water plugging measures must be taken.
After tunneling: The water inrush from the working face or the remaining water after grouting shall be discharged from the working face in time. Water leakage of the side wall shall be shielded and drained to ensure the quality of shotcrete. After spraying concrete, due to the increase in water pressure, there is a possibility of damage to the primary support, then the drainage method or backfill grouting method shall be adopted for blocking. When the water pressure is too high and the water volume is too large, the surrounding rock grouting is adopted to trap the water inside the surrounding rock.
For the exit section of Qinling Tunnel Line I, the following treatment measures are taken for the water inrush in the fault fracture zone:

(1) There is only slight water seepage or local dripping in the surrounding rock, and the treatment method of the section without water is generally adopted.
(2) In the sections with stranded and linear water inrush, permeable pipes shall be densified, and non-woven fabrics and waterproof boards shall be laid. The waterproof boards shall be pasted tightly without seams, and fasteners are not allowed to penetrate the waterproof boards. Waterproof concrete

with high impermeability shall be used to minimize construction joints, and waterproof connector shall be used for construction joints and settlement joints. Considering that the rock is soft and the subsidence of arch crown is found, full-round ring beams are locally erected with a spacing of 1.8 m, to prevent the surrounding rock from invading the lining clearance.

(3) For the section with large stranded water inrush, drainage holes with a depth of 1.5–2.0 m shall be drilled at a certain interval and rubber pipes shall be inserted. Water shall be led to the central ditch of the invert at the bottom of the tunnel through flexible permeable pipes for drainage, or water shall be intercepted, and the pilot tunnel shall be excavated, and the drainage is achieved through the parallel adit of Line II. The spacing between flexible permeable pipes shall be densified, and waterproof boards shall be laid on the top for waterproof protection and drainage. Waterproof concrete is also used. During shotcreting, the amount of cement shall be appropriately increased. Shotcreting operator shall master the amount of setting accelerator, and gradually spray concrete onto the section with water from far to near.

7. Treatment of regional faults
For large regional fault fracture zones, to make TBM pass through these sections quickly and safely and ensure the construction period, pretreatment can generally be carried out first, and then TBM tunneling or direct stepping can be carried out. The main technical measures of pretreatment are curtain grouting for the surrounding rock to reinforce the surrounding rock and block water.

7.4.3 Construction Example of Fault Fracture Zone

During the construction of No.7 Tunnel of Wanjiazhai Yellow River Diversion Project South Trunk Line in Shanxi Province, serious jamming occurred in Fault F_{65} and Fault F_{74}. For F_{74}, TBM stopped for 14 days and F65 for 95 days.

F_{65} fault obliquely crosses No.7 tunnel. The fault fracture zone is located at the section of K61 + 907.80–K61 + 930.80, with a length of 23 m. The fault affected zone is located at the sections of K61 + 785.60–K61 + 907.80 and K61 + 930.80–K61 + 972.60, with a width of 164 m. The total length of the fracture zone and affected zone is 187 m. Fault F_{65} has a strike of N10°E, a dip direction of SE and a dip angle of about 45°, and is a regional reverse fault. The rock mass of the fault footwall is the leopard limestone of the second member of the Lower Majiagou Formation of Ordovician, and the hanging wall is the dolomitic limestone of the first member of the Lower Majiagou Formation of Ordovician. The surrounding rocks of the fault fracture zone and the affected zone are mainly of Class V, and locally of Class IV in the tunnel section. And joints are developed. Affected by the fault, there are some irregular fissures, most of which are open and some are slightly open, and some are filled with red clay or yellow sub-clay. The rock mass in the fault affected zone is of stratified cataclastic structure, and the block diameter is generally less than 30 cm. Weathering is developed along the fissures, and the rock strength is medium. The attitude of rock stratum on the footwall is N85°W·NE∠24°. There are two groups of joints, with J_1 attitude of N55°E·SE∠65° and J_2 attitude of N30°W·SW∠87°. The surrounding rocks of the fault fracture zone are broken rocks in loose structure, composed of subangular, subrounded and partially rounded limestone breccia. Red clay and yellow mild clay are filled in the breccia gaps. The grain size of angular gravel varies from 0.05 to 2 m, generally larger than 30 cm.

On May 19, 1999, when TBM arrived at the chainage of K61 + 685.90, the surrounding rock changed from Class III to Class V, and then Class IV and V alternated. On May 28, 1999, the surrounding rocks at the chainage of K61 + 913.70 changed obviously. The construction slag materials were all massive rocks (limestone accounting for 60–70%) and a small amount of sand loam, rock debris and red clay. The diameter of the rock blocks was generally 10–40 cm, individually up to 50 cm, sub-rounded, and the surface was all weathered surfaces (occasionally fresh surfaces). A small amount of red clay was adhered to the surface of the rock blocks, and calcite veins were found individually. It was speculated that this section was in the middle of the fault fracture zone. Due to tunneling difficulties, TBM retreated by 60 cm, causing collapse in front of the machine head. After that, TBM continued to tunnel forward. When it reached the chainage of 61 + 917.172, the cutter head was stuck by the collapsed block and could not rotate, resulting in TBM shutdown.

On the afternoon of May 31, a total of 6 segments already installed were removed, and the machine head was moved backward at the same time, but the loose rock mass also collapsed. The space just emptied before the cutter head was filled, and the cutter head still could not rotate. It can be seen from TBM observation window that large stones (1.5–2 m in diameter) accumulated in front of TBM head, and small stones and a small amount of sand loam filled the space in between. The stones are in the shape of prismatic or sub-rounded. Continue to remove the installed segment, make the machine head move back, and start the cutter head at times, but all failed. Due to the backward movement of the machine head, the installed segments bulged about 6 cm towards the center of the tunnel, resulting in different damage to the 5 segments. The removed segments are seriously damaged and all are declared worthless. TBM head is moved backward by 2.56 m in total.

On June 3, the parties concerned discussed and studied the scheme of dealing with the jamming accident. In order to find out the geological conditions, probe drilling was carried out. Due to jamming and other reasons, the expected depth was not achieved in the 4 boreholes. Meanwhile, the collapsed rocks in the inner chamber of TBM cutter head are cleaned to open all disc cutters. And the rock conditions are directly observed. The rocks collapsed on the cutter head are cleaned through the manhole, mucking hopper and disc cutter mouth inside the cutter head to reduce the torque required by the cutter head when starting TBM. During this period, 72 holes for chemical grouting are completed intermittently, and 65 holes were inserted with PVC pipes with an inner diameter of 10 mm and a length of 2.5 m.

On June 16, the final treatment scheme below (Fig. 7.8) is determined according to the work carried out and the site geology:

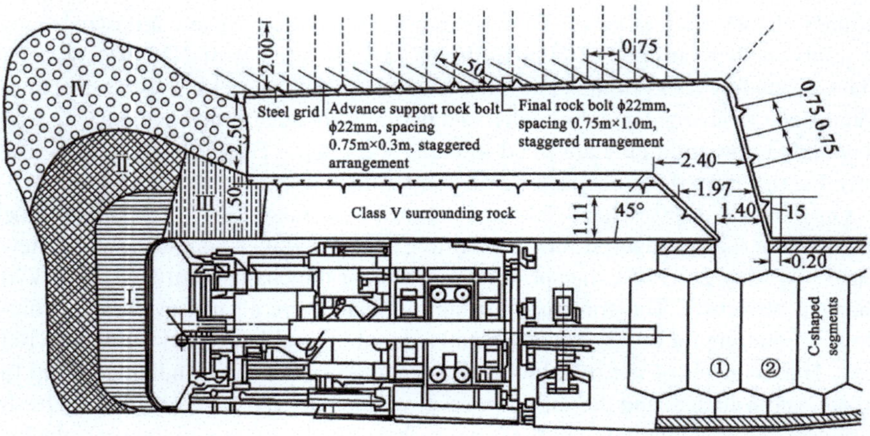

Fig. 7.8 Schematic diagram of F_{65} treatment of motianling major fault at the entrance of No. 7 tunnel of Wanjiazhai Yellow river diversion project South Trunk line (Unit: m). I—Epoxy resin fissure plugging layer; II—sand mortar; III—2 times injection of epoxy resin fissure plugging layer; IV—concrete backfilling

(1) PVC pipes of different lengths are inserted into the collapsed rock in front of TBM cutter head and chemically grouting is carried out with resin inside the pipes to form a 2 m thick consolidation layer in front of and around the cutter head. Then, 5–6 m PVC pipes are inserted into the upper part of the resin grouting layer, and the mortar is penetrated into the collapsed rock mass through the pipes to consolidate the collapsed rock mass. Meanwhile, a side tunnel is excavated from the place above the rear shield tail of TBM, so as to take further measures such as conventional construction when TBM still cannot be started after grouting.
(2) When the cutter head starts to rotate and the rock begins to collapse again and the resin grouting in front of the cutter head cannot control rock collapse, a grouting steel pipe with a length of about 15 m is driven from the top of TBM shield to reduce the pressure of the collapsed rock on TBM and prevent the rock from collapsing again.
(3) If TBM does not have enough torque to restart the cutter head, the side tunnel shall be extended to the front of TBM cutter head. And use conventional method to excavate the rock mass that has been consolidated by grouting before TBM until the cutter head can rotate freely.

According to the above scheme, centralized grouting was carried out twice from June 16 to July 8, with 82 barrels of polyurethane and 2.75m^3 of cement slurry injected for the first time. Due to serious grout leakage, grouting can only go on intermittently, indicating that resin grouting fails to block the flow of cement grout into the machine head.

The second resin grouting was carried out on June 28, with a total of 16 barrels/720 kg. Subsequently, 7-day cement grouting with 2–3% water glass was carried out in two stages, with a total grouting of 1.77 m^3.

The excavation preparation of the upper side tunnel started on June 21, and tunneling began on June 25. It is located at 1.5 m above the jacking segment and along the tunnel axis, with a width of 3.4 m and a height of 2.4 m. The planned excavation is 14 m. By July 10, when a total of 12.1 m was excavated, a cavity with a diameter of about 30 cm and a visible depth of about 2.5 m appeared in the lower right part of the tunnel face at the chainage of K61 + 912.10. The lithology at the tunnel face of the side tunnel is medium-bedded and thick-bedded limestone of Ordovician Majiagou Fm. ($O2s^2$). The original rock formation of the whole tunnel face is disordered. The rock mass is broken, with loose structure. The rock blocks are of different sizes and distinct edges and corners. Clay and some mylonite of 0.5–1.0 cm are interbedded between the rock blocks. Dissolution can be seen on the surface of some rock blocks, which are covered with calcite crystals, showing slight metamorphism and compression. Judging from the site conditions, the cavities are formed due to the geological tectonic action, and the rock blocks in the lower right corner of the tunnel face are hollow, indicating that this area is located in the Motianling Great Fault (F_{65}) area.

On July 11, the tunnel face was closed with shotcrete, and 57 boreholes (3–10 m deep) were drilled with hydraulic drills (Atlas drill, 50 mm in diameter) and 49 boreholes (3–6 m deep) were drilled with air leg drills (38 mm in diameter). Cement slurry of 126.3m^3 and resins of 126 barrels (3150 kg in total) are poured.

From August 4 to August 7, drilling and blasting was carried out in the side tunnel. On the evening of August 7, a large area of collapse occurred in front of the side tunnel when the tunneling reached about 0.5 m ahead of the cutter head, forming a cavity with a length of about 8 m, a width of about 7.5 m and a height of about 6.5 m. Subsequently, the collapsed rock mass in front of the tunnel face was closed, and four 2-inch (1 inch$\approx$3.33 cm) steel pipes with different lengths were embedded in the closed rock mass. After the closure was completed, the cavity formed by collapse is backfilled with grouting by four steel pipes. The grout seeped out of the pores at the bottom of the tunnel and flowed around TBM. Then, the rocks and consolidated grout around TBM cutter head were cleaned up, and TBM head was exposed on August 19, 1999.

On August 20, the first attempt to start TBM failed, and the excavation around TBM head continued. On August 21, the second attempt to start the cutter head was made, but TBM was pressed by large rocks on both sides of the cutter head, it stopped after 0.2 m of tunneling. Excavation around the cutter head was carried out again. On August 22, cutter head can rotate freely. However, as TBM shield tail could not move forward, by August 31, TBM only advanced for 0.8 m. Meanwhile, due to the strong thrust force of TBM, the connection between the shield partition and the shield tail behind TBM is damaged. After repair, on September 6, TBM fixture was ready for use and TBM advanced normally. By September 9, TBM advanced for 2.942 m in total. After that, it entered normal operation. It lasted for 95 days before and after this shutdown.

7.5 TBM Construction in Tunnel Section with Karst Development

The tunnel section with karst development is one of the most complex engineering geological problems in TBM construction. This unfavorable geological condition has a great impact on TBM tunneling. This not only reduces the utilization ratio of TBM operation time, but also may lead to TBM head subsidence and difficulty in smooth passing. Effective measures must be taken, otherwise improper handling will bring serious consequences to the construction. Therefore, it is especially important to know well the formation mechanism and classification of the tunnel sections with developed karst and analyze the previous engineering cases.

7.5.1 Mechanism of Karst Development

(1) General mechanism of karst development

The three necessary conditions for underground karst development include the nature of soluble rocks, geological structure characteristics and groundwater dynamic conditions, and soluble rocks are the material basis for karst development. The stronger the solubility of rock, the more conducive to karst development under the same conditions. As the tunnel is a linear project, the solubility of the soluble rock layer passing through each section of the tunnel controls the distribution characteristics of karst development along the tunnel. The significance of geological structure characteristics on karst development is that it not only controls the distribution and deformation characteristics of soluble rock strata, more importantly, creates basic conditions for selective corrosion of soluble rock by water flow. Therefore, the geological structure traces are the key objects to study the specific karst development and distribution.

(2) The relationship between karst hydrodynamic vertical zoning and tunnel karst development

The topography and landform of the tunnel area are often watersheds or inter-river land. The relationship between the elevation of the tunnel and the hydrodynamic zoning of karst controls the shape and water inflow of karst development in the tunnel. Han Xingrui et al. proposed the zoning mode by six groups based on years of research: ① surface karst zone, which is located on the intact soluble rock surface with relatively weak corrosion, and mainly developed with solution cracks, ponor, karst funnel and solution pore, etc.; ② vertical infiltration zone, which is located between the surface karst zone and the groundwater level in the region of the wet season. Since atmospheric precipitation and sediment carried by surface water enter the underground through this zone, vertical solution cracks and karst caves are largely developed, and some of them are filled with clay and gravel; ③ seasonal alternating zone, which is located between the groundwater levels in the region of the wet season and the region of the dry season. During the rise in the groundwater level, the groundwater flows horizontally, while during the drop in the groundwater level, the groundwater moves vertically. Because of the favorable hydrodynamic conditions, the karst effect is the strongest, and stratified horizontal and vertical karst caves and underground river systems are often developed; ④ horizontal pipeline circulation zone, which is located between the water level in dry season and the affected zone of underground river outlet. This zone is located in the upper part of karst aquifer, where karst is strongly developed, and some horizontal caves and underground river main channels are often developed; ⑤ pressure saturated zone is an aquifer located between the drainage surface of the underground river mouth and the affected zone of the drainage datum plane of the main local rivers. The distribution of karst caves is controlled by fault zones and various structural parts; ⑥ deep slow flow zone, with weak karst development, and karst cave or corrosion fault zone are controlled by large tectonic fault zone.

(3) Relationship between karst hydrodynamic horizontal zoning and tunnel karst development

According to the difference of karst groundwater dynamic conditions in the horizontal direction, the inter-river watershed and river valley are divided into recharge area, recharge runoff area and discharge area. Through combining the vertical zoning of karst hydrodynamic forces, the characteristics of karst development in each area can be grasped. The groundwater level in the recharge area is high and the seasonal variation zone is thick, but the karst development in the saturated zone is relatively weak and the development depth is shallow. The overburden of groundwater in the recharge runoff area increases, the channel development in the shallow saturated zone is strong, and the karst development depth is shallow. The vadose zone in the discharge area is thick, and the horizontal channel in the saturated zone is developed.

(4) Distribution law of karst morphology on tunnel route

Large karst caves, underground rivers and small karst channels are developed in key karst development areas, and small caves, karst fractures, karst grooves, joints and fissures of various sizes are developed around large caves. Outside key karst development areas, karst development forms include joints and fissures.

Therefore, on the basis of mastering the above-mentioned karst development mechanism and in combination with the hydrogeological conditions of the tunnel, the key sections of karst advance forecast and the corresponding karst development (or karst form) are divided, and appropriate geophysical prospecting methods are selected accordingly to achieve the purpose of improving the accuracy and economy of karst advance forecast.

7.5.2 *Classification of Tunnel Sections with Karst Development*

Due to the difference in the development conditions of karst caves, the types of karst caves are different. According to the filling conditions of caves, they can be divided into filled karst caves and unfilled karst caves. Among them, unfilled karst caves can be divided into small, medium and large types according to the size of karst cave space, and can be further divided into five types: karst channel, rock shelter, large rock shelter, rock hall and huge rock hall. According to the filling characteristics of karst caves, filled karst caves can be classified into four types: chemical sedimentary lithofacies, transport lithofacies, collapse lithofacies and seepage filling lithofacies (Table 7.1). Different filling types have different impacts on TBM construction, and different treatment measures are required.

Table 7.1 Classification of karst cave filling types

Level I	Level II	Level III	Classification basis
Unfilled caves	Small caves	Karst channel	Space size
	Medium-sized caves	Rock shelter	
		Large rock shelter	
	Large Cave	Rock hall	
		Huge rock hall	
Filled caves	Chemical sedimentary lithofacies	Drip-like chemical sedimentary lithofacies	Filling characteristics
		Flowstone type chemical sedimentary lithofacies	
	Transport lithofacies	Transport conglomerate lithofacies	
		Transport sandstone lithofacies	
	Collapse lithofacies	Fine clastic conglomerate lithofacies	
		Coarse clastic conglomerate lithofacies	
		Cracked enclosing rock lithofacies	
		Highly-cracked enclosing rock lithofacies	
	Seepage filling lithofacies		

7.5.3 Treatment Method for Tunnel Section with Karst Development

The karst cave can be permanently treated during tunneling, or only temporarily treated during tunneling and permanently treated in the later stage. Since the aspects, such as quantity, size, occurrence, position, hydrogeological conditions, whether there are fillers and filler characteristics of the karst caves that may be encountered during construction, are uncertain, probe drilling should be strengthened during tunneling to understand the characteristics of the karst caves to be encountered as soon as possible and formulate reasonable treatment measures. According to the construction characteristics of TBM, the principles of karst cave treatment are described below.

1. Shutdown treatment

   (1) For the small karst caves where the tunnel is advancing and development is stopped, TBM shall be stopped to retreat to the edge of the karst cave after 3–5 m advance, and part or all of the collapsed fillers and karst cave fillers originally supported by the shield shall be removed (if necessary, side tunnel can be drilled to assist the construction). If all of them are removed, materials with good groutability, such as masonry, pea gravel or concrete, shall be used for blocking and backfilling, and grouting shall be used for reinforcement. If some of them are removed, grid arch frame combined with shotcrete or other temporary supports shall be used, and then blocking and backfilling shall be carried out, and grouting shall be used for reinforcement. After TBM passes through, high-pressure consolidation grouting is carried out for the karst cave section through the segment backfilling hole, and rock bolts are installed for reinforcement.
   (2) For partially-filled karst caves below the tunnel bottom plate, if the distance between its boundary and the tunnel bottom plate is less than one time the tunnel diameter according to the probe drilling results, the fillers can be removed completely (small karst caves) or partially (medium karst caves), and the roof of the karst cave can be reinforced with pea gravel or concrete support, or grouting reinforcement can be carried out first. After TBM passes through, high-pressure consolidation grouting is carried out for the karst cave section through the segment backfilling hole, and rock bolts are installed for reinforcement. If the karst cave is beyond 1 time the tunnel diameter but within twice the tunnel diameter, the later high-pressure consolidation grouting reinforcement shall be carried out after TBM passes through.
   (3) For large karst caves with filler, pre-grouting treatment can be carried out for the whole cave according to the probe drilling equipment on TBM to prevent the filler from collapsing or subsiding when TBM passes through. After TBM passes through, high-pressure consolidation grouting is carried

out for the karst cave section through the segment backfilling hole, and rock bolts are installed for reinforcement.

(4) For large karst caves with empty or a small amount of fillers, the filling or reinforcement works are relatively large. If there are foundation construction conditions, aqueduct and box culvert measures for crossing can be taken by using beam, arch or bridge foundation. If the geological conditions are complex and other methods are difficult to deal with the karst caves, the karst caves can be bypassed by local rerouting.

2. Temporary treatment shall be carried out first during tunneling, and permanent reinforcement shall be carried out after passing through
   To reduce local subsidence and avoid other construction defects, special-type segments with bolt holes can be prefabricated. During installation, "banana-type" bolts can be used for connection at joints, or profile steel or other steel structures can be used for longitudinal integral connection at longitudinal joints in the whole karst cave section. In the later stage, high-pressure consolidation grouting is carried out for the karst cave section through the segment backfilling hole, and rock bolts are installed for reinforcement.

3. Karst Cave Water Treatment
   The main focus is drainage, supplemented by plugging, and combined with drainage and plugging. Drainage ditches, culverts or drainage tunnels can be adopted for drainage structures; advanced pre-grouting measures can be adopted for water plugging, and consolidation grouting can be adopted for water plugging after passing through karst caves.

7.5.4 Construction Example of Karst Development Tunnel Section

In the process of TBM tunneling in Lot IV of Songhua River Diversion Water Supply Project, it is necessary to tunnel the limestone tunnel section of 7 km. According to the geological exploration results and the actual construction process, it is judged that some tunnel sections are karst areas, and 12 karst caves have been identified, of which 3 are suspected karst caves. The height of the largest karst cave is 38.52 m, 5.7 m from the top of the karst cave; the height of the smallest karst cave is 6.3 m, 32.5 m from the top of the karst cave. During TBM tunneling in this section, it is easy to cause construction risks such as machine deviation, cutter head dropping, cutter head jamming, shield structure burying by mud ingress, electrical failure caused by water ingress, and overall machine jamming caused by convergence deformity. Therefore, it is very important to take strong TBM passing measures. The specific location of the karst cave is shown in Fig. 7.9.

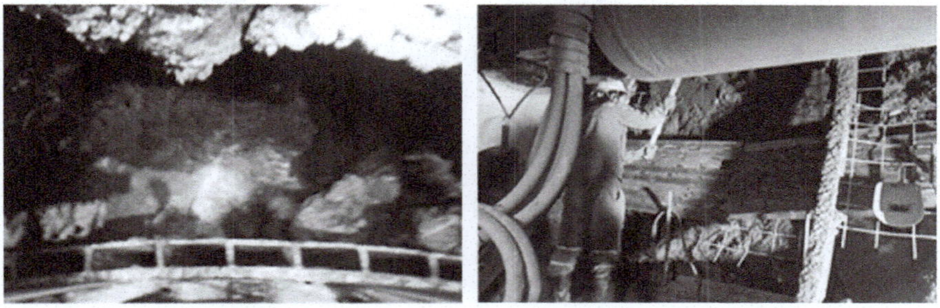

a) Karst caves and cavities at the top of tunnel   b) Karst caves and cavities at gripper position

Fig. 7.9 Construction difficulties of TBM passing through karst cave group development section

For the karst cave group development section, the project management office has held symposiums and expert meetings for many times to study the countermeasures, and the main technical countermeasures are as follows (Fig. 7.10):

1. Advance geological forecasting
   For this section, advance geological forecast shall be made in advance to accurately detect the spatial position, size and filling of the karst cave in front of the

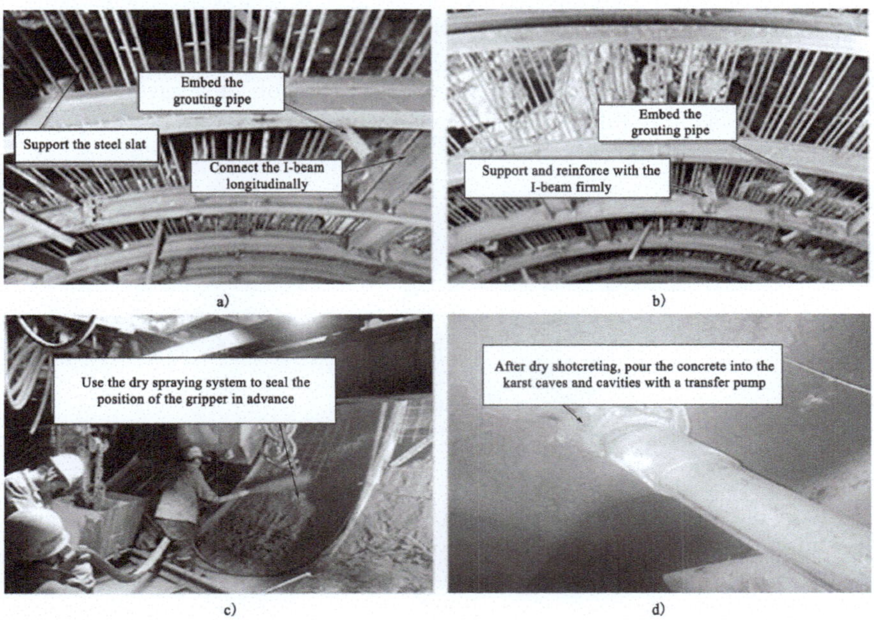

Fig. 7.10 Construction drawing of treatment measures for sections where TBM passes through karst cave group

tunnel face, and countermeasures and emergency material reserves shall be made in advance according to the detection results of advance geological forecast.

2. Karst Cave Treatment

(1) The karst caves beyond 3 m of the excavation contour line are not treated. If the excavation contour line is within 3 m, it is proved that the scale of the karst caves is not large, and TBM directly tunnels through and strengthens the primary support.
Arch crown position: At the moment when the karst cave comes out of the shield, joint support measures such as steel slat, ring beam, I-steel longitudinal connection, and additional welding support at the back of the arch shall be taken in time. If necessary, the dry shotcrete system shall be used to close the section. If there is a collapse cavity or karst cavity, concrete shall be backfilled, and finally backfill grouting shall be carried out to ensure compactness.
Position of gripper: combined support such as ring beam, hanging net dry shotcrete or pouring concrete shall be adopted for treatment in advance.
Tunnel bottom position: The karst cave at the tunnel bottom shall be controlled at 5 m outside the excavation contour line, and those greater than 5 m shall not be treated, and those less than 5 m shall be backfilled with mortar in advance for consolidation or replaced with concrete.
(2) Within 3 m of the excavation contour line, if a large-scale karst cave is identified, the drill and blast method shall be used in advance, and then TBM shall pass by stepping.
(3) According to the development characteristics of karst caves, the semi-filled and fully-filled karst caves are treated by static pressure grouting and injection of pure cement slurry.

Through the implementation of the above measures, TBM passes through the karst area with the lowest risk, which not only saves expenses, but also shortens the construction period.

Questions

1. Briefly describe the physical characteristics of soft rock.
2. Summarize the evolution mechanism of soft rock deformity.
3. Briefly describe the criteria for the classification of hard rocks.
4. Briefly describe the treatment measures for the tunnel section with mud and water ingress.
5. Briefly describe the impact of fault fracture zones on TBM construction.
6. Briefly describe the treatment method of TBM passing through the tunnel section with karst development.

Chapter 8
Cutter Management in TBM Construction

Highlights: This chapter firstly introduces the failure mode of TBM disc cutter from two aspects: normal wear and abnormal damage of disc cutter, then analyzes the influencing factors of disc cutter wear from geological, mechanical, geometrical and material aspects, introduces the theoretical prediction model and indoor test method of TBM disc cutter wear again, and introduces the TBM cutter management process from two aspects: inspection and replacement of TBM cutter in the tunnel, and inspection and maintenance of cutter workshop. Finally, the field application of disc cutter is analyzed for the Han River-to-Wei River Water Diversion Project.

8.1 Failure Mode of TBM Disc Cutter

In TBM construction, the wear condition is serious because the disc cutter breaks hard rock, and the wear condition is also complicated because of the complex working environment of the disc cutter. According to the analysis of the damage of the replaced disc cutter, the failure form of the disc cutter can be divided into normal wear and abnormal damage according to the wear form and wear degree.

8.1.1 Normal Wear of Disc Cutter

The normal wear of disc cutter is also called uniform wear of disc cutter, which means that the radial wear degree of all parts around the disc cutter ring is basically the same, that is, the radial wear volume of the cutter ring is basically the same, and the normal wear of disc cutter mainly occurs in the face disc cutter area with relatively single geological conditions. The rock breaking efficiency of the disc cutter is related to the width of the disc cutter edge. With the increase of the wear volume of the disc cutter, the width of the disc cutter edge gradually increases. When the

K. Chen et al., *TBM Design and Construction*,
https://doi.org/10.1007/978-981-99-0059-6_8

width of the disc cutter edge reaches a certain range, the rock breaking efficiency of the disc cutter and the advance speed of TBM will be affected. When the disc cutter is worn to the limit state, the disc cutter no longer has the rock breaking ability, and the disc cutter needs to be replaced at this time. The normal wear failure state of the disc cutter refers to that the edge width of the disc cutter exceeds the specified value when the disc cutter is evenly worn. The allowable maximum wear volume of disc cutters with different diameters is different, and the allowable wear of disc cutters in different areas of the cutter head is also different. This failure mode is the main form of disc cutter wear, accounting for 80–90% of the total.

When tunneling in hard rock with high strength and strong abrasiveness, the cutter ring is easy to cause large spot wear, the side wear of the cutter ring is very small, and the blade is easy to be ground flat. In this kind of high-strength rock, the cutter penetration is small, so the wear of the cutter ring is mainly the abrasion effect of fresh rock surface and cracks on the cutting edge. The rock strength is high, the pressure acting on the disc cutter is correspondingly large, and the cutter ring material is prone to plastic deformity; the depth of hard minerals pressed into the material is prone to exceed the critical depth to form indentation fracture, thus forming large point wear. In this kind of rock, the fracture area is shallow, and the rock powder has little abrasion on both sides of the cutting edge, so the cutting edge is ground flat. This wear form is the fastest form of cutter wear.

When tunneling in rocks with low strength and abrasiveness, most of the powder flows through the side of the cutting edge, resulting in large side wear and small point wear. The rock powder on both sides of the cutting edge is the main abrasive material. Since there are few abrasive minerals in the rock powder, the point wear is small. This kind of rock has low abrasiveness to cutters. The wear of cutters takes a long time, and both sides of the cutting edge are often worn to form sharp corners. The sharp-angled wear form is the slowest wear form of the cutter.

From the perspective of wear mechanism of cutter ring material, the main form of normal wear of disc cutter is abrasive wear, including abrasive wear based on plastic removal mechanism and abrasive wear based on fracture removal mechanism. The direct wear of disc cutter can be divided into two-body wear and three-body wear according to its contact conditions. Two-body wear refers to the wear caused by the movement of abrasive particles relative to a solid surface. Three-body wear refers to the movement of external abrasive particles between two friction surfaces, which is similar to the grinding effect. When the disc cutter breaks the rock, a broken body is first generated under the cutting edge and a crushing area is formed. The abrasive particles in the crushing area move between the disc cutter edge and the excavation face and transmit energy for rock breaking.

From the perspective of rock characteristics, when TBM tunnels in sections with low rock hardness, the impact force of the cutter ring on the rock is relatively small, and the wear of the disc cutter is mainly abrasive particle wear based on the plastic removal mechanism. The wear is mainly composed of primary chips formed by micro-cutting and secondary chips formed by furrow uplift and falling off, and the abrasive particle wear based on the fracture removal mechanism accounts for a small proportion. Therefore, for the disc cutter used in tunneling this stratum, the material selection should focus on the hardness index of the cutter ring material, and the high hardness of the cutter ring should be used to improve the efficiency of the cutter ring rolling and rock breaking. When TBM tunnels in sections with high rock hardness, the cutter ring will be strongly impacted by the rock, and the wear of abrasive particles based on the fracture removal mechanism will be intensified. Some microscopic furrows and cutting will be removed by a large fatigue spalling mechanism, which accelerates the wear of the cutter ring and reduces the wear resistance of the material. Therefore, for the disc cutter used for tunneling this stratum, the material selection shall focus on the comprehensive mechanical properties of the material, and a certain impact toughness shall be maintained while ensuring the high hardness of the material.

From the perspective of disc cutter ring materials, for disc cutter rings with low hardness and high toughness, when abrasive wear occurs, their plasticity is good and plastic deformity is easy to occur, so they are easy to be directly cut by abrasive or ploughed to both sides to form furrows, so the plastic removal mechanism is the main mechanism; However, for disc cutter rings with high hardness and low toughness, when abrasive wear occurs, deformity is not easy to occur due to low plasticity, but cracks are easy to occur due to low toughness, so the proportion of fracture removal mechanism increases. Therefore, when studying the prediction of disc cutter wear, it is necessary to consider both the abrasive wear of plastic removal mechanism caused by furrow deformity and micro-cutting, and the abrasive wear of micro-fracture removal mechanism caused by crack extension.

From the perspective of rock crushing effect, the wear of disc cutter can be divided into primary wear and secondary wear. The process in which the rock on the excavation face is crushed by the disc cutter and the rock is stripped from the rock mass to form rock debris is called direct rock breaking, and the wear caused by this process to the disc cutter is primary wear. The rock debris cut directly by rock breaking is not discharged in time, and some rock debris is accumulated between the cutter head face plate and the excavation face. Under the action of TBM thrust force, the disc cutter repeatedly breaks the rock. The wear caused to the disc cutter in this process is called secondary wear.

8.1.2 Abnormal Damage of Disc Cutter

Abnormal damage of disc cutter includes uneven wear of cutter ring, cracking of cutter ring, curling cutter edge of cutter ring, sliding of cutter ring, fracture of retainer ring, oil leakage of seal, wear of cutter body, damage of bearing, etc.

1. Uneven Wear of Cutter Ring

Uneven wear of cutter ring refers to the inconsistent radial wear degree of various parts around the disc cutter ring. Uneven wear of cutter ring is often encountered in cutter replacement/change, and the wear conditions are various. According to the uneven wear form of disc cutter, the uneven wear of disc cutter can be divided into chord uneven wear and edge uneven wear.

The disc cutter chord uneven wear can be divided into single-side uneven wear, multi-side uneven wear and bearing wear. The unilateral uneven wear of disc cutter refers to that when the disc cutter installed on the cutter head rotates, the disc cutter does not rotate with the cutter head at the same time; or at the initial stage of tunneling after the disc cutter is installed, the disc cutter rotates temporarily, and then the disc cutter stops rotating all the time, resulting in the unilateral chord wear of the disc cutter ring. Polygonal uneven wear of disc cutter refers to the uneven wear of the second side of the disc cutter caused by the instantaneous rotation of the wear surface by an angle due to the stress concentration at the uneven wear angle after the uneven wear of the single side chord of the disc cutter occurs. Such a cycle finally forms polygonal wear.

Uneven wear of disc cutter: From the perspective of wear shape, the wear on both sides of the disc cutter edge angle is uneven, the two sides of the cutter edge form different angles with the normal direction of the cutter ring, and the tip of the cutter edge shows an irregular curve. When the disc cutter is subject to uneven wear, uneven wear of string and edge may occur at the same time.

2. Cracking of Cutter Ring

Cracking of disc cutter ring refers to the radial cracking or blade collapse of disc cutter ring, which makes the disc cutter lose the rock breaking ability. If the force is large, the cutter ring will fall off. On the one hand, the fracture or blade collapse of the disc cutter ring is caused by excessive impact load on the disc cutter, and on the other hand, it is caused by improper heat treatment, insufficient toughness or excessive interference of the cutter ring body. If the cutter ring of the disc cutter has been worn, especially when uneven wear has occurred, it is more likely to crack the disc cutter ring.

3. Curling Cutter Edge of Cutter Ring

On the one hand, the curling cutter edge of the disc cutter ring is caused by the hard rock and the large tunneling thrust force of TBM; on the other hand, it is caused by improper heat treatment and insufficient hardness of the cutter ring.

4. Other Abnormal Damage

The cutter ring slides due to insufficient interference between the cutter ring and the cutter body, which generally occurs in the cutter body with multiple cutter rings replaced; The fracture of the retaining ring is caused by insufficient strength at the weld of the retaining ring, or due to insufficient groove depth at the retaining ring; Seal oil leakage due to damage of metal floating seal or failure of oil feeding screw plug; The cutter body wear volume is caused by the cutter ring exceeding the wear limit and not replaced. When the cutter ring is unevenly worn, it is often easy to cause the cutter body wear. Bearing damage is caused by silt entering the cutter body due to seal failure or bearing failure due to overload of the disc cutter.

8.2 Factors Affecting Wearing of TBM Disc Cutter

The factors affecting disc cutter wear can be summarized as follows: geological factors, mechanical factors, geometric factors and material factors, etc. Geological factors mainly include quartz content, compressive strength, integrity and geostress of TBM excavated rock. Mechanical factors mainly include the load of disc cutter, penetration of cutter edge, etc. Geometric factors mainly include installation radius of disc cutter, disc cutter diameter, cutter edge spacing and cutter edge width, etc. The material factors mainly include the hardness and impact toughness of the cutter ring material, and other factors mainly include the height difference and shape of the cutter edge.

8.2.1 Geological Influencing Factors of Disc Cutter Wear

Rock abrasiveness is the main factor affecting the wear of disc cutters, and the factors affecting rock abrasiveness mainly include rock strength, mineral composition, rock integrity and rock mass stress.

The rock strength is reflected by rock UCS (Unconfined Compressive Strength). The higher the unconfined compressive strength of rock is, the stronger the abrasiveness of rock is.

The rock mineral composition is reflected by the rock EQC (Equivalent Quartz Content), which is mainly determined by the Rosival mineral grinding hardness, the conversion factor of quartz is set to 1, and the conversion factor of other minerals is equal to the ratio of the Rosival hardness of the mineral to the Rosival hardness of quartz. The equivalent quartz content is calculated by multiplying the percentage content of multiple minerals in the rock by their conversion factors and then adding them together to obtain the equivalent quartz content of the rock. Rock EQC (Equivalent Quartz Content) reflects both the type of rock mineral composition and the converted content of each mineral composition. The abrasiveness of rock is also related to the size of mineral composition particles, and there is no clear quantitative index for comparison. The higher the rock EQC (Equivalent Quartz Content) is, the stronger the abrasiveness of the rock is.

The integrity of rock mass refers to the development degree of various geological interfaces dominated by fissures in rock mass, which is a comprehensive reflection of rock mass structure. The better the integrity of the rock mass is, the greater the bonding force inside the rock mass is, the greater the removal stress is required to remove part of the rock on the surface of the rock mass, and the more severe the cutter wears. The indexes used to measure the integrity of rock mass at home and abroad mainly include three categories: rock mass integrity coefficient based on elastic wave test, rock mass quality index RQD based on core drilling, and rock mass volume joint number based on structural plane statistics.

When TBM is tunneling in high geostress rock stratum, the stress release of rock mass causes the rock block to suddenly fly out of the tunnel face or tunnel periphery in the form of burst, forming a rockburst. The rockburst of the tunnel face has a great impact on the wear of the disc cutter of the TBM. After the rockburst of the tunnel face occurs, the middle part of the tunnel face collapses, and large rocks fall and smash against the disc cutter. In light cases, the disc cutter breaks, and in severe cases, the cutter ring breaks. After the center of the tunnel face collapses, only the cutter at the edge of the cutter head contacts the rock mass and is stressed, and the disc cutter in the middle of the cutter head does not contact the collapsed tunnel face. At this time, the TBM thrust force is not reduced, which is easy to cause overload of the gauge disc cutter; In addition, the falling large rocks smashed against the disc cutter will cause the blade to collapse and cannot be discharged from the muck discharge outlet. The gauge disc cutter needs to be squeezed and broken many times, which is easy to cause abnormal wear of the gauge disc cutter, especially uneven wear. The disc cutter wear caused by rockburst is mostly abnormal wear, and once it occurs, it often leads to the damage of the cutter body and the bearing together, resulting in serious economic losses.

8.2.2 Mechanical Influencing Factor of Disc Cutter Wear

The mechanical factors affecting the wear of disc cutter mainly include the load on disc cutter and penetration. The relationship between disc cutter load and penetration for different compressive strengths as given by Robbins, a US TBM manufacturer, is shown in Fig. 8.1. The shaded area in the figure is the economic area of TBM construction. It can be seen from the figure that when the unconfined compressive strength of rock is constant, the disc cutter load first increases rapidly with the increase of disc cutter penetration, and then slowly increases after reaching a certain penetration. It can be seen that under a certain compressive strength, the load of the disc cutter has a certain correlation with the penetration, and the penetration can be used as a mechanical influencing factor to measure the wear of the disc cutter.

In the process of rock breaking by rolling, the cutter edge of the disc cutter is mainly loaded in three directions of the rock, including the vertical force F_V, which is overcome by TBM thrust force; Rolling force F_R, overcome by cutter head torque; The lateral force points to the center of the cutter head and is generated by the lateral extrusion of the rock on the disc cutter edge and the centrifugal force of the cutter head rotation. The cutter head rotates at a low speed, so the centrifugal force is not considered, and the extrusion forces on both sides of the cutter edge can often be balanced with each other, so the lateral force is generally not considered. It has been shown that normal wear of disc cutter is dominated by the abrasive wear of the material. Professor Ernest Rabinowicz of MIT has established an abrasive wear model based on the micro-cutting hypothesis, where the abrasive particles are viewed as cones of hard particles sliding on soft materials and plowing a furrow, and the wear volume V of the material is:

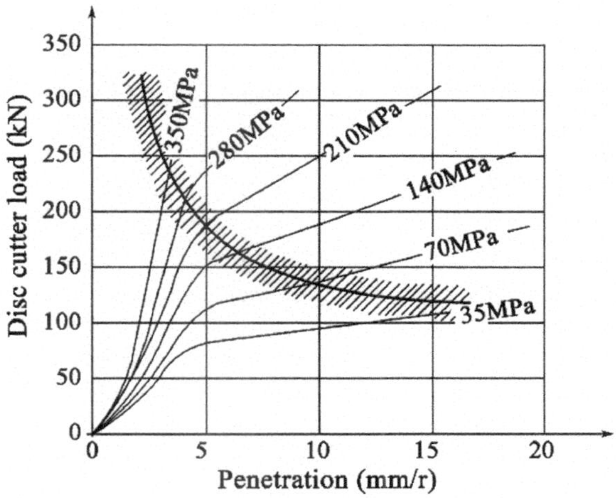

Fig. 8.1 Relationship between disc cutter load and penetration under different compressive strength

$$V = K \cdot \frac{P}{H} \cdot L \tag{8.1}$$

where:

- K abrasive wear coefficient;
- P normal load;
- H material indentation hardness;
- L sliding distance.

For the abrasive particle wear of the disc cutter, the normal load P in Eq. (8.1) is the vertical force F_V of the disc cutter. The model gives that the wear volume of the material is proportional to the normal load, so the slope of the curve in the figure is proportional to the wear rate of the disc cutter. Within the allowable load range of the disc cutter, the penetration of the disc cutter should be increased as much as possible, which is conducive to reducing the wear rate of the disc cutter.

8.2.3 Geometrical Influencing Factors of Disc Cutter Wear

The geometric factors affecting the wear of disc cutters mainly include the installation radius of disc cutters, the radius of cutter rings, the cutter edge spacing between and the cutter edge width. When the disc cutter edge spacing is the same, the larger the installation radius of the disc cutter installed on the cutter head is, the longer the rolling distance of the disc cutter is, the greater the rock breaking volume of the disc cutter is, and the faster the wear of the disc cutter is. The cumulative wear volume of the front disc cutter is counted, and the cumulative wear volume of the disc cutter at different cutter positions is approximately linearly increased with the installation radius of the disc cutter.

The larger the radius of the cutter ring is, the greater the allowable wear of the disc cutter is. For example, the maximum allowable wear volume of the 17 in face disc cutter is 25 mm, while the maximum allowable wear volume of the 20 in disc cutter can reach 35–40 mm; On the other hand, the larger the radius of the disc cutter is, the larger the allowable wear volume of the cutter ring material is. The wear process of the cutter ring is essentially the process of spalling off the cutter ring material from the surface of the cutter ring by hard rock particles. The more the allowable spalling volume is, the longer the service life of the cutter ring will be.

The smaller the distance between disc cutter edges is, the easier the rock between adjacent disc cutters is to spall off. For a single disc cutter, the wear volume is smaller. In low-strength stratum, the wider the cutter edge is, the more wear volume is allowed by the cutter ring, and the longer the service life of the cutter ring is. However, for high-strength hard rock stratum, the cutter edge is not easy to penetrate into the rock mass, and even causes curling cutter edge of cutter ring or bearing overload, instead it is not conducive to the wear resistance of the cutter.

8.2.4 Material Influencing Factors of Disc Cutter Wear

The indexes of disc cutter ring material mainly include surface hardness and impact toughness. The high hardness of cutter ring material is usually achieved by increasing the content of carbon and alloy elements. On the one hand, the addition of alloying elements will increase the production cost of the cutter ring and increase the resource consumption rate; on the other hand, the total content of alloying elements will be blindly increased, which improves the strength and hardness of the material, inevitably reduces the toughness of the material, and is not suitable for construction application in hard rock sections. For hard rock stratum, the content of alloying elements in the steel should be reduced to achieve a good match between hardness and impact toughness.

The medium carbon medium alloy steel is based on the tempering state of martensite. The carbides such as Mo_2C, VC and $M_{23}C_6$ precipitated in the structure have the characteristics of high hardness, high melting point and strong stability. These carbides are embedded in the tempering martensite matrix to improve the hardness of the cutter ring, but the impact absorption energy is low. In the process of TBM hard rock tunneling, in the early stage of tunneling, due to the relatively high hardness of the cutter ring matrix and carbide, the groove generated by the furrowing action is shallow and narrow, accounting for a small proportion. The hard particles in the rock mainly cut the cutter ring microscopically, and the hard carbide is broken and spalled off due to cutting, leaving pit marks on the material surface. At the same time, under the strong impact of the rock surface, the cutter ring is easy to initiate microcracks around the carbide. As the microcracks expand to the material surface, the material is worn by brittle spalling. When TBM tunnels in hard rock sections, although the medium carbon and medium alloy cutter ring maintains high hardness, the cutter ring material is removed by micro-cutting into a large fatigue spalling mechanism under actual working conditions, which accelerates the wear of the cutter ring and reduces the wear resistance of the material.

A large amount of fine carbides in the medium-carbon low-alloy cutter ring are dispersedly distributed in the lath martensitic matrix. Although the hardness value of the cutter ring edge is higher than that of the medium-carbon medium-alloy cutter ring, the impact absorption energy is far better than that of the medium-carbon medium-alloy cutter ring, and the strength and toughness of the cutter ring are well matched. Wear-resistant steels with good toughness fit have a reduced probability of direct cracking of the subsurface layer, and the material exhibits good wear resistance. Due to the relatively small amount of carbide formed by Cr and Mo in the structure of the medium-carbon low-alloy cutter ring, the distribution density of carbide in the matrix is relatively low, reducing the probability of microcracks initiation around the carbide and the probability of removing the medium-carbon low-alloy cutter ring with large fatigue spalling. During TBM tunneling, the cutter ring is under the micro-cutting effect of hard minerals under the strong impact of rock surface and the severe friction of hard particles. At the same time, the medium-carbon low-alloy cutter ring has good toughness, and the cutter ring is partially squeezed by abrasive to produce micro-uplift, which disappears and forms continuously, leaving grooves and deformity accumulation on the surface of the cutter ring. The high toughness of the medium-carbon low-alloy cutter ring matches with a certain hardness, which determines that the cutter ring is worn by cutting furrows and strain fatigue mechanism, and ensures the good wear resistance of the medium-carbon low-alloy cutter ring.

When TBM is constructed in sections with low rock hardness, the impact force of the cutter ring on the rock is relatively small, and the phenomenon of large fatigue spalling off of the cutter ring is reduced. Under this working condition, the medium carbon and medium alloy cutter ring characterized by high hardness can be constructed smoothly, ensuring the construction progress and reducing the construction cost. Therefore, the hardness index of the cutter ring material should be focused on when selecting materials, and the high hardness of the cutter ring should be used to improve the efficiency of cutting the cutter ring during rolling and rock breaking.

When TBM is constructed in a section with high rock hardness, the cutter ring will be strongly impacted by the rock. In order to reduce the probability of fatigue cracks in the tissue, it is necessary to improve the impact toughness of the material and avoid the wear of the cutter ring in the way of large fatigue spalling. When selecting the cutter ring, the comprehensive mechanical properties of the material should be emphasized, and a certain impact toughness should be maintained while ensuring the high hardness of the material. It is recommended to select a medium carbon low alloy cutter ring.

8.2.5 *Other Influencing Factors of Disc Cutter Wear*

In addition to the above-mentioned factors affecting the wear of disc cutter, other factors will also affect the wear of disc cutter, such as the edge shape of disc cutter and the height difference of disc cutter.

At present, there are mainly two types of similar uniform-section disc cutters: flat blade and curved shape blade. The transition fillet of the flat blade is small, and the contact surface between the blade and the rock is large when breaking rock. The transition fillet of the curved shape blade is large, and the contact surface between the blade and the rock is small during rock breaking. In addition, the width of the cutter edge is also different. The cutter edge width of the standard cutter ring is 3/4 in; the cutter edge width of the reinforced cutter ring can reach 1 in. For soft rock stratum, the wear life of flat blade and reinforced cutter ring is longer; for hard rock stratum, the service life of curved shape edge and standard cutter ring is longer. For different stratum, improper selection of edge shape of disc cutter will increase the wear of disc cutter.

The disc cutter rock breaking is a result of the cooperative operation of multiple disc cutters, after a certain disc cutter is excessively worn, a new disc cutter or a disc cutter that has not yet reached the wear limit is required for replacement. If the newly replaced cutter ring is much higher than the cutter edge of the nearby cutter ring with a certain wear volume, the newly replaced disc cutter will be overloaded and worn prematurely when TBM is advanced.

See Table 8.1 for the summary of factors affecting disc cutter wear.

8.3 Prediction Model and Indoor Test of TBM Disc Cutter Wearing Theory

8.3.1 *Prediction Model of Disc Cutter Wear Theory Based on Microscopic Wear Mechanism*

1. Basic Theory of Abrasive Wear

For abrasive wear, Professor Ernest Rabinowicz of MIT developed a model of abrasive wear based on the microcutting hypothesis, which views abrasive particles as cones of hard particles sliding on soft materials and plowing a furrow (Fig. 8.2). The cone half angle is θ, the cone bottom diameter is $2r$ (i.e., the width of the plowed groove), the normal load is F_V, the pressed-in depth is x, and the sliding distance is l.

The yield strength of the worn material is σ_s. The projection area in the vertical direction is πr^2, and only half of the conical surface (the conical surface in the forward direction) bears the load when the abrasive particles slide. Assuming that there are N micro-convex bodies on the abrasive particles on the contact surface, the normal

Table 8.1 Summary of factors affecting disc cutter wear

Classification of influencing factors	Specific influencing factors	Qualitative analysis of influencing factors
Geological factors	EQC (equivalent quartz content)	The higher the equivalent quartz content of the rock is, the stronger the abrasiveness of the stratum is, and the more serious the wear to the disc cutter is
	UCS (unconfined compressive strength)	The higher the rock compressive strength is, the harder the rock is to be damaged. The greater the load on the disc cutter during rock breaking is, the easier the disc cutter is to be worn
	Rock integrity K_V	The better the integrity of the rock is, the greater the bonding force inside the rock mass is, the greater the removal stress is required for rock breaking, and the more serious the wear of the cutter is
	Geostress	High geostress leads to rockburst and large rocks fall off from the tunnel face, which is easy to cause abnormal wear of disc cutters
Mechanical factors	Vertical force of disc cutter F_V	The greater the vertical force of the disc cutter is, the closer the contact between the rock and the cutter ring is, and the easier the hard particles in the rock will spall off the material on the surface of the cutter ring
	Penetration h	When the disc cutter load allows, the deeper the penetration is, the shorter the wear contact time between the disc cutter and the rock mass is, and the less likely it is to wear
Geometric factor	Cutter edge spacing S	The smaller the cutter spacing of disc cutter is, the easier the rock between adjacent disc cutters is to be spalled off, and the less the disc cutter wear is

(continued)

Table 8.1 (continued)

Classification of influencing factors	Specific influencing factors	Qualitative analysis of influencing factors
	Cutter edge width T	The wider the cutter edge is, the less likely the disc cutter is to be worn, and the less wear volume the disc cutter has; However, in high-strength hard rock, the cutter edge is wide and easy to be rolled, instead accelerating the wear of disc cutter
	Disc cutter radius R_0	The larger the radius of the disc cutter is, the more wear volume is allowed, the greater the load the bearing can withstand, and the longer the service life of the disc cutter is
	Installation radius R_i	The larger the installation radius of disc cutter is, the longer the linear distance of disc cutter movement during the cutter head rotation, and the greater the wear volume of disc cutter is
Material factors	Material hardness H_V	The higher the hardness of the disc cutter edge is, the better the wear resistance is. However, it is easy to break the edge when subjected to impact load, instead accelerating the wear of the disc cutter
	Material toughness J	The higher the toughness of the disc cutter edge is, the lower the hardness is, the stronger the impact resistance is and the weaker the wear resistance is. Therefore, both should be considered
Other factors	Edge shape of disc cutter	The sharper the disc cutter edge is, the easier it is to cut into the rock, but the less material the cutter edge allows to wear, the easier it is to wear

(continued)

Table 8.1 (continued)

Classification of influencing factors	Specific influencing factors	Qualitative analysis of influencing factors
	Height difference of disc cutter	Rock breaking of disc cutter is the result of cooperative operation of multiple disc cutters, and the higher the cutter edge of other disc cutters is, the easier it is to wear

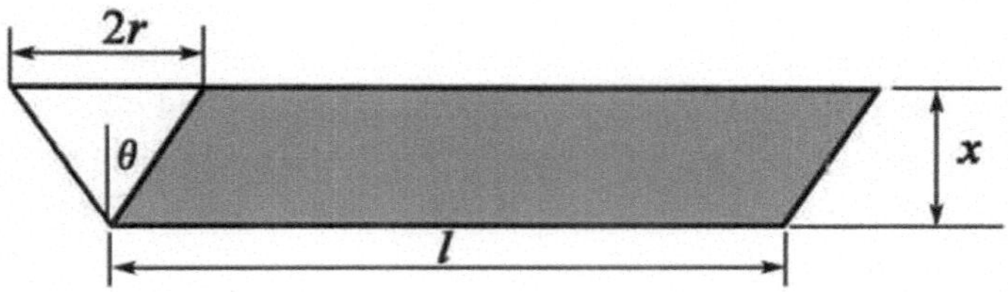

Fig. 8.2 Abrasive particle wear model

load of N micro-convex bodies is:

$$F_V = N \cdot \frac{\pi r^2}{2} \sigma_s \tag{8.2}$$

According to formula (8.1):

$$N = \frac{2F_V}{\pi r^2 \sigma_s} \tag{8.3}$$

The plowed volume is the wear volume, the projected area is a triangle, and the wear volume per unit sliding distance is:

$$Q_1 = N \cdot rx = \frac{2F_V}{\pi r^2 \sigma_s} \cdot r \cdot \frac{r}{\tan\theta} = \frac{2F_V}{\pi \sigma_s \tan\theta} \tag{8.4}$$

Considering the probability number K_1 of abrasive particles produced by the interaction of the micro-convex bodies, the wear volume per unit sliding distance is:

$$Q_1 = K_1 \cdot \frac{2}{\tan\theta} \cdot \frac{F_V}{\pi \sigma_s} = K_S \frac{F_V}{\pi \sigma_s} \tag{8.5}$$

where:

K_S abrasive wear coefficient, which is the product of geometric factor $2/\tan\theta$ and probability number K_1, K_S is related to factors such as abrasive hardness, shape and size, as well as the number of abrasive grains that play a cutting role.

Table 8.2 Abrasive wear coefficient

Year	Researchers	Type of wear	Abrasive particle size (μm)	Material	K_S ($\times 10^{-3}$)
1956	Saier	Two-body	70	Steel	16
1958	Khrushchev	Two-body	80	Multiple	24
1958	Toropov	Three-body	150	Steel	6
1961	Rabinowicz	Three-body	80	Steel	4
1961	Rabinowicz	Three-body	40	Multiple	2

The abrasive particle wear coefficient has a great influence on the accuracy of disc cutter wear prediction. The value of this coefficient is related to the type of abrasive particle wear, abrasive particle size, material characteristics and other factors. The abrasive particle wear coefficient obtained by some scholars through tests is shown in Table 8.2.

2. Derivation of Normal Load of Disc Cutter

The disc cutter interacts with the rock in three directions during rock breaking, including: Normal load, pointing to excavation face, provided by TBM thrust force; Rolling force, pointing to the tangential direction of the disc cutter, provided by the cutter head torque; The lateral force points to the center of the cutter head and is generated by the extrusion force of the disc cutter on the rock and the centrifugal force of the cutter head rotation. Its value is small and is generally not considered.

Among many rock breaking load prediction models of disc cutters, the accuracy of rock-breaking force prediction of disc cutters has been verified in many projects at home and abroad for the CSM model of rock-breaking force of approximate constant-section disc cutters widely used in shield and TBM in recent years. The normal load of CSM disc cutter rock breaking force model is:

$$F_V = C \cdot \frac{\varphi \cdot R_0 T}{1+\psi}\left(\frac{S\sigma_c^2\sigma_t}{\varphi\sqrt{R_0 T}}\right)^{1/3}\cos(\varphi/2) \tag{8.6}$$

where:

R_0 disc cutter radius (mm);
ψ pressure distribution coefficient of cutter ring top edge, generally − 0.2 to 0.2, taken as 0.1;
φ contact angle of disc cutter (rad), $\varphi = \cos^{-1}[(R0 - h)/R0]$;
h penetration of disc cutter (advancing distance per revolution of cutter head) (mm/r);
S cutter spacing (mm);
T cutter edge width (mm);
C dimensionless coefficient, taken as 2.12;
σ_c rock UCS (Unconfined Compressive Strength) (MPa);
σ_t shear strength of rock (MPa).

Since the penetration h differs greatly from the size of the disc cutter radius R_0, the angle of φ is generally small; according to the Taylor expansion of the trigonometric function sinφ:

$$\sin\varphi = \varphi - \frac{\varphi^3}{3} + \frac{\varphi^5}{5} + \cdots + (-1)^n \frac{\varphi^{2n+1}}{2n+1} \tag{8.7}$$

When φ is small, sinφ»φ can be obtained, then:

$$\varphi \approx \sin\varphi = \frac{\sqrt{R_0^2 - (R_0 - h)^2}}{R_0} = \frac{\sqrt{2R_0 h + h^2}}{R_0} \approx \frac{\sqrt{2R_0 h}}{R_0} \tag{8.8}$$

According to the test results of rock mechanics test, the value of shear strength σ_t of general rock is between 1/8 and 1/12 of compressive strength σ_c, and σ_t»$0.1\sigma_c$ is taken.

Due to:

$$\begin{cases} \cos\varphi = (R_0 - h)/R_0 \\ R_0\varphi \approx \sqrt{2R_0 h} \\ \sin(\varphi/2) = \sqrt{\frac{1-\cos\varphi}{2}} \\ \cos(\varphi/2) = \sqrt{\frac{1+\cos\varphi}{2}} \end{cases}$$

Therefore:

$$\begin{aligned} lF_V &= \frac{0.1^{\frac{1}{3}} C T^{\frac{5}{6}} S^{\frac{1}{3}} \cdot 2^{-\frac{1}{6}}}{1+\psi} \sqrt{2R_0 - h} \cdot \sigma_c h^{\frac{1}{3}} \\ &= 0.8 \cdot T^{\frac{5}{6}} S^{\frac{1}{3}} \sqrt{2R_0 - h} \cdot \sigma_c h^{\frac{1}{3}} \end{aligned} \tag{8.9}$$

Since the difference between the penetration h and the disc cutter diameter $2R_0$ is large, so $2R_0 - h$»D_0, then:

$$F_V = 0.8 \cdot T^{\frac{5}{6}} S^{\frac{1}{3}} \cdot D_0^{\frac{1}{2}} \cdot \sigma_c \cdot h^{\frac{1}{3}} \tag{8.10}$$

3. Derivation of Disc Cutter Wear Volume Prediction Model

Wear volume per revolution of disc cutter V_1:

$$V_1 = Q_0 l_0 = Q_0 R_0 \varphi \approx Q_0 \sqrt{D_0 h} \tag{8.11}$$

TBM tunneling distance L, radial wear volume of disc cutter X_1:

$$X_1 = \frac{V_1}{2\pi R_0 T_0} \cdot \frac{L}{h} \cdot \frac{R_i}{R_0} = \frac{K_S F_V L R_i}{\pi^2 \sigma_s T_0 R_0 \sqrt{2R_0 h}} \tag{8.12}$$

4. Derivation of Prediction Model for Cutting Depth Wear Rate of Disc Cutter

When the disc cutter rotates for one circle, the radial wear volume X_0 of the cutter ring is:

$$X_0 = \frac{Q_0 l_0}{\pi D_0 T} = K_S \cdot \frac{F_V l_0}{\pi^2 D_0 \sigma_s T} \tag{8.13}$$

where:

D_0 diameter of disc cutter (mm);
T cutter edge width of disc cutter (mm);
l_0 the sliding distance of the cutter edge on the rock after the disc cutter rotates for one circle (mm).

Since the wear of the disc cutter only occurs on the half arc where the disc cutter contacts the rock in the forward direction of the disc cutter, the sliding distance is:

$$l_0 = R_0 \varphi \approx \sqrt{D_0 h} \tag{8.14}$$

$$X_0 = 0.08 K_S \cdot \frac{S^{\frac{1}{3}} \sigma_c h^{\frac{1}{3}}}{T^{\frac{1}{6}} \sigma_s} \tag{8.15}$$

The wear rate of disc cutter cutting depth, i.e. the radial wear volume of disc cutter ring per unit tunneling distance of TBM:

$$\omega = \frac{X_1}{L} = \frac{X_0 R_i / R_0}{h} = 0.16 K_S \frac{S^{\frac{1}{3}} \sigma_c R_i}{T^{\frac{1}{6}} D_0 \sigma_s h^{\frac{1}{6}}} \tag{8.16}$$

where:

X_1 radial wear volume of disc cutter (mm);
L tunneling distance (m);
R_i installation radius of disc cutter on cutter head (m).

The disc cutter cutting depth wear rate prediction model formula (8.15) intuitively reflects that the disc cutter wear rate is proportional to the 1/3 power of the cutter spacing S, the rock compressive strength σ_c and the disc cutter installation radius R_i, and is inversely proportional to the 1/6 power of the cutter edge width T, the disc cutter diameter D_0, the cutter edge yield strength σ_s and the 1/6 power of the penetration h.

5. Derivation of Prediction Model for Groove Wear Rate of Disc Cutter

In order to compare the abrasion resistance of disc cutters when TBM is tunneling in different stratum, the concept of wear rate of disc cutter grooving is put forward.

The wear rate of disc cutter grooving refers to the radial wear volume of disc cutter from the rock breaking point.

During TBM tunneling, the movement track of the rock breaking point of the disc cutter is a cylindrical spiral with the installation radius R_i of the disc cutter as the radius and the penetration h as the lead. The movement distance l of rock breaking point of disc cutter is:

$$l = \sqrt{L^2 + 4\pi^2 n^2 R_i^2} \tag{8.17}$$

where:

L tunneling distance of TBM (m);
N number of revolutions of cutter head rotation.

$$l = \sqrt{L^2 + 4\pi^2 n^2 R_i^2} = L\sqrt{1 + 4\pi^2 R_i^2 / h^2} \approx \frac{2\pi R_i L}{h} \tag{8.18}$$

The groove wear rate of the disc cutter is:

$$\varepsilon = \frac{X_L}{l} = \frac{X_L h}{2\pi R_i L} \tag{8.19}$$

where:

X_L Cumulative wear volume of disc cutter for TBM tunneling L (mm).

$$\varepsilon = \frac{X_1}{l} = \frac{\omega L}{l} = 0.16 K_S \frac{S^{\frac{1}{3}} \sigma_c R_i}{T^{\frac{1}{6}} D_0 \sigma_s h^{\frac{1}{6}}} \frac{L}{l} = 0.025 K_S \cdot \frac{S^{\frac{1}{3}} \sigma_c h^{\frac{5}{6}}}{T^{\frac{1}{6}} D_0 \sigma_s} \tag{8.20}$$

The disc cutter groove wear rate prediction model formula (8.20) intuitively reflects that the disc cutter line wear rate is proportional to the 1/3 power of the cutter spacing S, the rock compressive strength σ_c and the 5/6 power of the penetration h, and is inversely proportional to the 1/6 power of the cutter edge width T, the disc cutter diameter D_0 and the cutter edge yield strength σ_s.

8.3.2 *Prediction Model of Disc Cutter Wear Theory Based on Macroscopic Wear Mechanism*

1. Basic Theory of Energy Wear

Rolling and rock breaking of disc cutter is realized by cutting into the rock to a certain depth under the action of vertical force, forming a groove on the tunnel face under the action of rolling force, and spalling the rock between two adjacent grooves. If the rock breaking of disc cutter is pure rolling movement, the wear volume of disc cutter during rock breaking is quite small; However, in the actual rock breaking process of disc cutter, very serious wear occurs. It can be seen that the cutting edge slides with the rock during the rock breaking process of disc cutter. Since the sliding friction is far greater than the rolling friction, the sliding friction is the main cause of disc cutter wear.

Fleisher put forward the theory of energy wear, which believes that wear occurs due to energy consumption in the friction process. When the accumulated energy of a certain volume of material reaches a certain critical value, it will spall off from the surface in the form of abrasive dust, so the wear of material is the process of energy conversion and consumption. The energy consumed by material wear is provided by the work done by friction force. Since pure rolling friction does not do work, the energy consumed by disc cutter wear is provided by the sliding friction work between disc cutter and rock. By establishing the relationship between friction work and wear volume, the wear of disc cutter can be predicted by sliding friction work during rock breaking.

2. Determination of Rock Breaking Slip Rate of Disc Cutter

In the process of completing a deep rock breaking, the sliding arc length of the rock breaking point of the cutter edge is a three-dimensional curve. Slip ratio refers to the ratio of slip distance to total contact distance. The analytical solution of the sliding distance S_i of each rock breaking point on the cutting edge for each revolution of the face disc cutter is:

$$S_i = \int_0^h \sqrt{\frac{R_0^2}{R_i^2} + \frac{2R_0}{2R_0 - h}} \mathrm{d}h$$

$$= \frac{2}{R_i}\left\{\begin{array}{l} \sqrt{\frac{R_0}{2}}\sqrt{2R_0R_i^2 + 2R_0^3} + R_i^2 \ln\left(R_0\sqrt{2R_0} + \sqrt{2R_0R_i^2 + 2R_0^3}\right) \\ -\frac{\sqrt{2R_0-h}}{2}\sqrt{2R_0R_i^2 + 2R_0^2(2R_0 - h)} \\ -R_i^2 \ln\left[R_0\sqrt{2R_0 - h} + \sqrt{2R_0R_i^2 + 2R_0^2(2R_0 - h)}\right] \end{array}\right\} \tag{8.21}$$

For each rotation of the disc cutter, the total contact distance between the rock breaking point of the cutter edge and the rock is half of the arc in the forward direction of the disc cutter, so the total contact distance l_0 is:

$$l_0 = R_0\varphi \approx \sqrt{D_0 h} \tag{8.22}$$

The ratio of slip distance to total contact distance is the slip ratio ξ of rock normally rolled by disc cutter:

$$\xi = \frac{S_i}{l_0} \tag{8.23}$$

When the disc cutter is worn normally, the slip ratio is small. When the disc cutter is eccentrically worn, the slip ratio is 1, and the wear of the disc cutter will increase greatly at this time. Therefore, it is necessary to avoid and reduce the abnormal wear of the disc cutter caused by eccentric wear as much as possible.

3. Derivation of Disc Cutter Wear Volume Prediction Model

Work done by sliding friction force per revolution of disc cutter W_0:

$$W_0 = FS_0 = 2\mu F_V \xi \pi R_0 \tag{8.24}$$

where:

S_0 slip distance of disc cutter per revolution;
μ sliding friction coefficient between disc cutter and rock.

Wear volume per revolution of disc cutter V_2:

$$V_2 = IW_0 \tag{8.25}$$

where:

I energy wear rate, i.e. the wear volume of disc cutter material caused by unit friction work.

TBM tunneling distance L, radial wear volume of disc cutter X_2:

$$X_2 = \frac{V_2}{2\pi R_0 T_0} \cdot \frac{L}{h} \cdot \frac{R_i}{R_0} = \mu I \xi \frac{F_V R_i}{T_0 R_0} \cdot \frac{L}{h} \tag{8.26}$$

where:

T_0 cutter edge width of disc cutter.

8.3.3 *Laboratory Test and Prediction Method of Disc Cutter Wear*

1. Rock Abrasiveness Test
   The abrasiveness of rock is used to describe the wear degree of rock cutter, and it is directly related to the wear degree of TBM cutter. The higher the strength of the rock is, the higher the percentage of hard minerals on the surface is, the denser the rock is, the higher the abrasiveness of the rock is, and the shorter the service life of the disc cutter is.

   At present, the tests commonly used to measure rock abrasiveness include: rock abrasiveness CAI index test, rock UCS (Unconfined Compressive Strength) test, rock mineral composition test, etc.

(1) Rock Abrasiveness CAI Index Test

CAI value (Cerchar Abrasiveness Index) is a parameter proposed by France Cerchar Institute in 1986 to measure the abrasiveness of rock, required to be measured by Cerchar scratch test. The test principle is to apply a load of 70 N to a steel needle with a 90° cone angle, and the hardness of the steel needle reaches 54–56HRC, so that it generates pressure on the tested rock sample, and then the steel needle moves 10 mm and scratches the rock surface, and the top of the steel needle cone will be worn. Measure the diameter of the top of the steel needle cone, and the CAI value of the rock sample is obtained by taking 1/10 mm as the unit.

According to the CAI value, the classification standard of rock abrasiveness is shown in Table 8.3.

In 1984, Büchi collected the test results of CAI values of different types of rocks, and obtained the CAI average values and value ranges of different types of rocks, as shown in Table 8.4.

Rock abrasiveness CAI test device consists of abrasiveness test device part and measurement and recording part, and abrasion test device part consists of frame, vise, lower fixture, steel needle, gearbox, servo motor and transmission system; measurement and recording part consists of vertical grating displacement sensor, horizontal grating displacement sensor, force sensor, EDC controller, computer, steel needle abrasiveness width measurement device (HD digital microscope). The test device is shown in Fig. 8.3, and the test process is shown in Fig. 8.4.

(2) Rock Compressive Strength Test

Unconfined compressive strength (UCS) of rock refers to the ultimate failure strength obtained when the sample is compressed in only one direction, that

Table 8.3 Classification standard of rock abrasiveness

CAI value	0.1–0.4	0.5–0.9	1.0–1.9	2.0–2.9	3.0–3.9	4.0–4.9	≥ 5
Abrasiveness	Extremely low	Very low	Low	General	Relatively high	Very high	Extremely High

Table 8.4 CAI values of different types of rocks

Rock type	CAI average	CAI value range	Rock type	CAI average	CAI value range
Limestone	1.2	0.5–1.8	Amphibolite	3.7	3.3–4.2
Sandstone (clay cementation)	2.5	1.5–3.5	Crystalline shist	3.2	2.2–4.2
Sandstone (SiO_2 cementation)	3.4	2.8–4.2	Gneiss	4.4	3.5–5.2
Basalt	2.7	2.0–3.4	Diorite/syenite	4.5	4.1–5.0
Andesite	3.0	2.8–3.2	Granite	4.9	4.5–5.3

Fig. 8.3 Rock abrasiveness servo tester of State Key Laboratory of Shield and Tunneling Technology

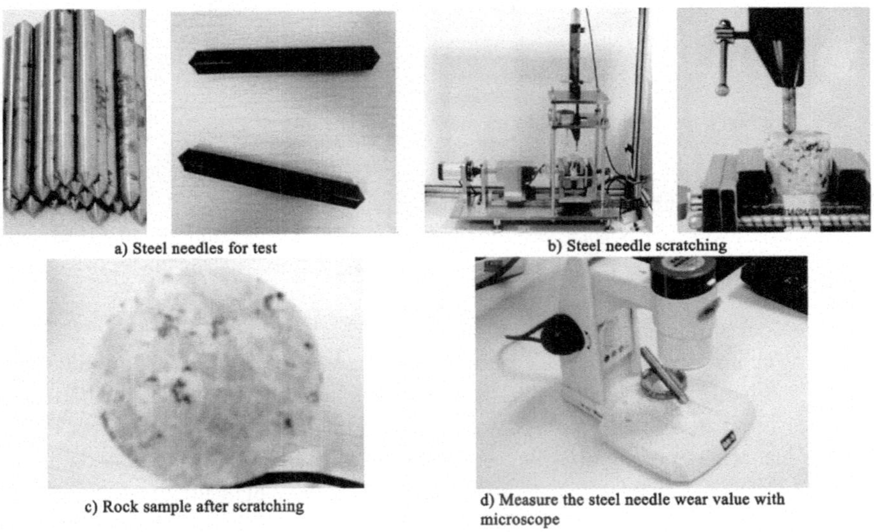

Fig. 8.4 CAI test process

is, the pressure strength value measured when the rock sample is placed between the upper and lower pressure plates for being pressurized until the sample is crushed. The determination is generally carried out with an UCS (Unconfined Compressive Strength) instrument, and a triaxial instrument can also be used as a uniaxial instrument.

Figure 8.5 shows the MTS rock mechanical property test device, which can be used to measure UCS (Unconfined Compressive Strength) and triaxial compressive strength. The MTS test system is accurate, the simulation field environment is high in accuracy, and the requirements for rock samples are high. Cylindrical rock samples with a diameter of 50 mm and a length of

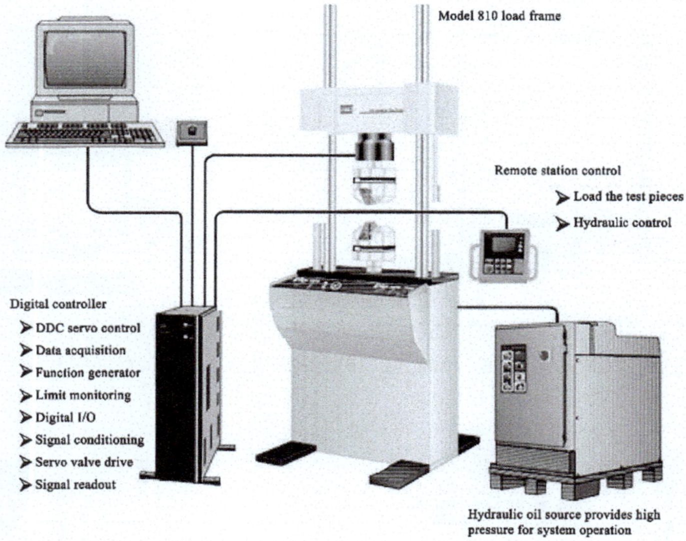

Fig. 8.5 MTS rock mechanical property test device

100 mm and a certain flatness are required. The laboratory can also use a press to process rock samples into cubes with a side length of 50 mm or 100 mm, and convert the compressive strength by measuring the breaking force.

(3) Rock Equivalent Quartz Content Test
The abrasiveness of rock is the result of a variety of mineral combinations with different properties in rock, including the mineral content and hardness of rock, the structure of rock, the size and shape of particles inside rock, and the connection mode between crystals inside rock. Quartz is a very common abrasion-related mineral found in the early stage, but the quartz content does not fully characterize the abrasiveness of rocks, so the concept of Equivalent Quartz Content (EQC) is introduced. The rock equivalent quartz content is obtained by multiplying the content of each mineral in the rock by Mohs scale of mineral hardness of this mineral relative to quartz. See Table 8.5 for the Mohs scale of mineral hardness values of common minerals.

The mineral content of rock can be obtained by rock powder diffraction test or by rock thin section identification tests. The samples are ground with professional grinding equipment to make the particle size meet the requirements of diffraction instruments (3–5 μm), and then loaded onto the sample bench for analysis. The diffractogram of rock sample composition can be derived from the analytical scan of diffractometer, and the rock powder diffractometer is shown in Fig. 8.6a; the rock thin section identification test can be carried out by using Leitz-ORTHOPLAN microscope, as shown in Fig. 8.6b.

Table 8.5 Mohs scale of mineral hardness of common minerals

Type of mineral	Diamond	Topaz	Corundum	Quartz	Orthoclase	Apatite	Fluorite	Calcite	Gypsum	Talc
Mohs scale of mineral hardness	10	9	8	7	6	5	4	3	2	1

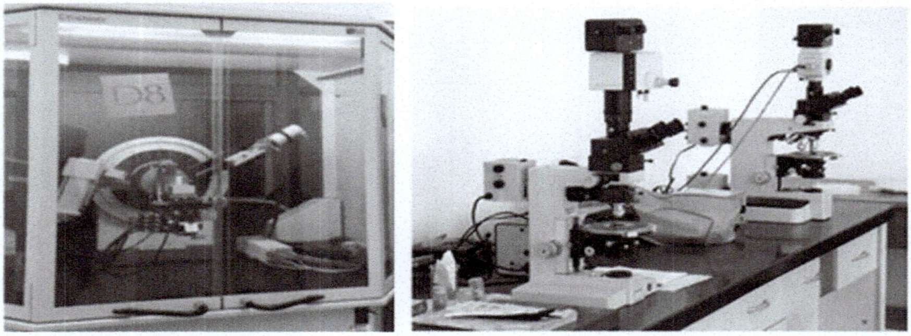

a) Rock powder diffraction test device b) Rock thin section identification test device

Fig. 8.6 Test device for mineral component content of rock

(4) Analysis of rock abrasiveness test results
By taking a total of 27 sets of samples from the construction sites of a water diversion tunnel in Qinling (11 rock samples in group A), a railroad tunnel in Guangzhou (11 rock samples in group B) and a water diversion tunnel in Northeast China (5 rock samples in group C), rock abrasiveness tests were carried out, and the results of testing each parameter are shown in Table 8.6, and the results of fitting CAI and UCS × EQC are shown in Fig. 8.7.

It can be seen from Fig. 8.7 that the rock abrasiveness index CAI has a positive logarithmic relationship with the rock UCS (Unconfined Compressive Strength) × EQC (Equivalent Quartz Content), and is highly correlated (correlation coefficient 0.956).

2. Wear Resistance Test of Disc Cutter
It is difficult to reduce the on-site working condition of the disc cutter indoors. In addition, wear is the effect of time accumulation, and the prototype test requires a huge amount of rock. Therefore, the wear resistance test device of small-size disc cutter with a scale of 1:10 has been developed, as shown in Fig. 8.8.

The wear resistance test device of small-size disc cutter consists of horizontal loading measurement and control subsystem, vertical loading measurement and control subsystem, rock sample and model disc cutter abrasiveness action mechanism and model disc cutter measurement part, as shown in Fig. 8.9.

Horizontal loading measurement and control subsystem consists of loading frame, servo pressure cylinder, load sensor, horizontal displacement sensor, servo valve, servo controller and servo oil source. The power supply servo oil source is used for horizontal pressurization, the gear pump is used for power, the guyed grating optical fiber sensor or LVDT displacement sensor is used for horizontal displacement measurement, the load sensor is used for horizontal force measurement, and the actuator is a low friction rigid actuator to accurately control the horizontal force. The EDC full digital servo controller PID closed-loop servo is applied to control horizontal displacement or horizontal force for closed-loop servo controller.

Table 8.6 Test results of rock abrasiveness test

No.	Lithology	UCS (MPa)	EQC (%)	CAI (0.1 mm)
A1	Granite	103	44.2	3.35
A2	Diorite	180	51.9	3.68
A3	Granite	210	45.0	3.87
A4	Granite	123	44.2	3.51
A5	Granite	125	48.4	3.53
A6	Granite	140	45.0	3.59
A7	Granite	193	46.5	3.54
A8	Granite	185	45.0	3.73
A9	Granite	160	49.2	3.42
A10	Shist	192	36.9	3.63
A11	Granite	173	50.0	3.76
B1	Argillaceous siltstone	16	49.0	1.00
B2	Argillaceous siltstone	23	48.9	1.40
B3	Argillaceous siltstone	24	33.9	1.11
B4	Argillaceous siltstone	26	41.8	1.32
B5	Sandstone	34	63.0	2.82
B6	Sandstone	45	66.0	3.04
B7	Sandstone	40	89.6	3.08
B8	Sandstone	44	58.0	2.98
B9	Sandy mudstone	8	24.1	0.34
B10	Sandy mudstone	11	29.9	0.59
B11	Sandy mudstone	13	32.1	0.78
C1	Diorite	59	30.2	2.39
C2	Granite	37	40.3	2.11
C3	Limestone	69	26.1	2.10
C4	Limestone	52	28.0	1.81
C5	Limestone	89	15.8	1.77

The vertical loading measurement and control subsystem consists of the bearing column, guide rail bearing platform, lead screw, AC servo motor, AC servo motor driver and vertical displacement sensor. AC servo motor, driver and guide rail lead screw are used to servo control the accurate linear motion of model disc cutter. AC servo motor adopts high-frequency reverberation AC servo motor; servo control is performed by closed-loop servo control of AC servo motor and driver through vertical displacement feedback.

The abrasiveness mechanism of rock sample and model disc cutter consists of rock sample, model disc cutter, rock sample fixture, rock debris box and rock debris scattering box. The rock sample is 320 mm long, 70 mm wide and 70 mm

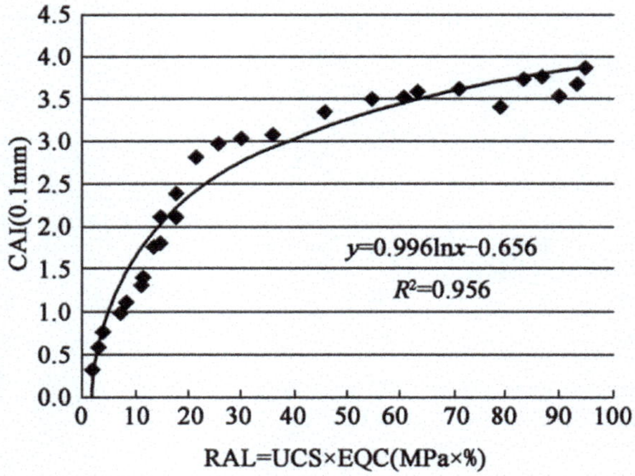

Fig. 8.7 Fitting relationship between rock abrasiveness parameters

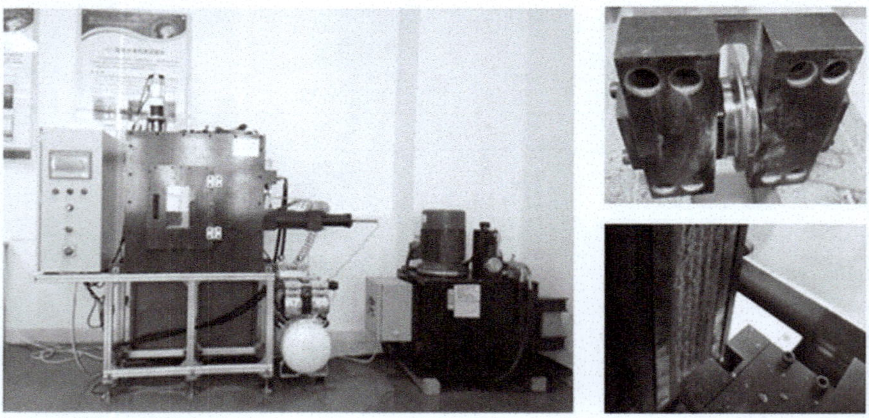

Fig. 8.8 Wear resistance test device of small-size disc cutter in State Key Laboratory of Shield and Tunneling Technology

thick. The rock sample is vertically fixed by the rock sample fixture. The model disc cutter is connected with the cutting box. The depth of the lower part of the model disc cutter in the cutting is controlled by adjusting the height of the cutting in the cutting box; Apply thrust force to the model disc cutter through the horizontal movement of the horizontal actuator on the rear side of the rock sample. The model disc cutter is connected to the servo motor drive through a lead screw for vertical movement.

The mass measuring device consists of an electronic micrometer scale, an electronic scale, a vernier caliper and a tape measure, which are used to measure the size and mass of the disc cutter and rock sample before and after the test.

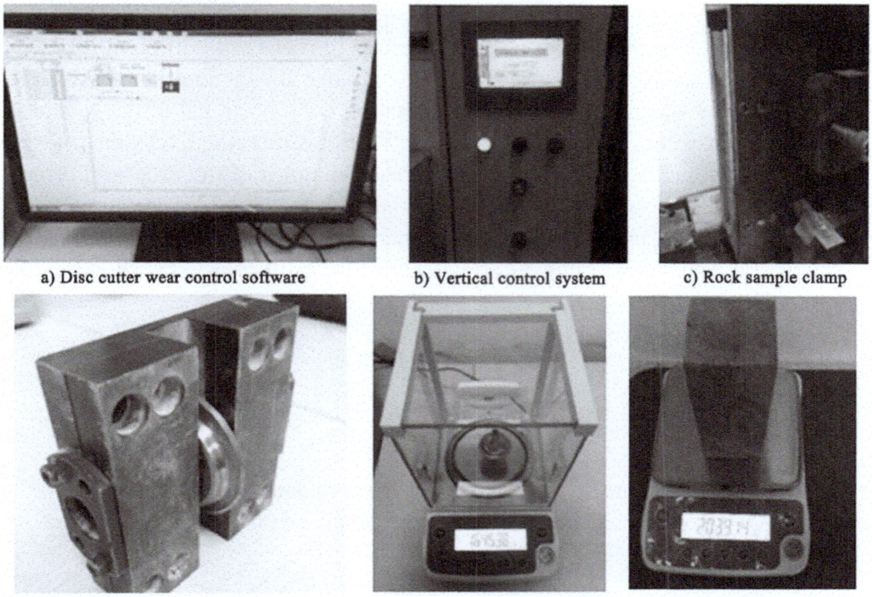

a) Disc cutter wear control software b) Vertical control system c) Rock sample clamp

d) Disc cutter fixture model e) Electronic micrometer scale f) Electronic percentage scale

Fig. 8.9 Main components of wear resistance test device of small-size disc cutter

3. Prediction Method of Disc Cutter Wear

   (1) Prediction Method Based on Wear Quality of Small Size Disc Cutter
   According to the laboratory test results of wear resistance of small-sized disc cutter, the mass loss of disc cutter is proportional to the rolling distance of disc cutter without abnormal wear. With this principle, the wear resistance test of small-size disc cutters can be carried out by collecting rock samples of the proposed project and those of the built project, and the wear quality of each 100 m can be measured. Then, the cutter consumption of the proposed project can be predicted by analogy according to the actual use of cutters on the site of the built project, as shown in the following formula:

$$\frac{W_1}{W_2} = \frac{\omega_2}{\omega_1} \tag{8.27}$$

where:

W_1 the service life of a single disc cutter in the built project, which can be measured by the rock breaking amount of each disc cutter;

W_2 Service life of single disc cutter of the proposed project;

ω_1 100 m wear mass of small-sized disc cutter in built project;

ω_2 wear mass per hectometer of the small-sized disc cutter of the proposed project.

(2) Auxiliary Prediction Method Based on Rock Abrasiveness
According to the abrasiveness test results of rocks with different lithology and the wear resistance test results of small-size disc cutters, the abrasiveness mass per 100 m of small-size disc cutters is proportional to the square of the abrasiveness index CAI of rocks. With this rule, the CAI value of rock samples can be tested by collecting rock samples of the proposed project and the rock samples of the built project; then, the cutter consumption of the proposed project can be preliminarily predicted by analogy according to the actual use of cutters on the site of the built project, as shown in the following formula:

$$\frac{W_1}{W_2} = \frac{\mathrm{CAI}_2^2}{\mathrm{CAI}_1^2} \tag{8.28}$$

where:
CAI_1 abrasiveness index of rock in the completed project;
CAI_2 Rock abrasiveness index of the proposed project.

8.4 TBM Cutter Management Process

8.4.1 Inspection and Replacement of TBM Cutter in Tunnel

When the TBM is in downtime state, the inspector shall first clean the cutter head with high-pressure water after entering the cutter head, in order to clearly check all components and reduce the temperature of the cutter head. Check whether the cutter fasteners are loose or fall off in turn, and re-tighten or replace them once found. Rotate the disc cutter (at least half a turn) to judge whether the torque of the disc cutter is normal, whether the cutter ring has blade collapse, uneven wear and fracture, and whether the disc cutter is hot or leaks oil. If there is any of the above abnormal damage, replace it in time. Then measure the wear volume of each disc cutter with a specific measuring scale, record the relevant data, and replace it according to the replacement standards in different areas. In addition, it is necessary to check whether the water spraying of the cutter head is normal, whether the cutter head is worn or damaged, and whether there are abnormal phenomena such as open welding at the welding position. After the inspection, clean up and count the tools, and do not leave them in the cutter head to avoid damaging the cutter head and related components during tunneling. Finally, the cutter head inspector takes out the cutter head and the cutter head inspection is completed.

The replacement standard of disc cutter is determined according to the size of disc cutter ring and the installation position of disc cutter. Figure 8.10 shows the replacement and installation standard of disc cutter at different positions of a TBM cutter head.

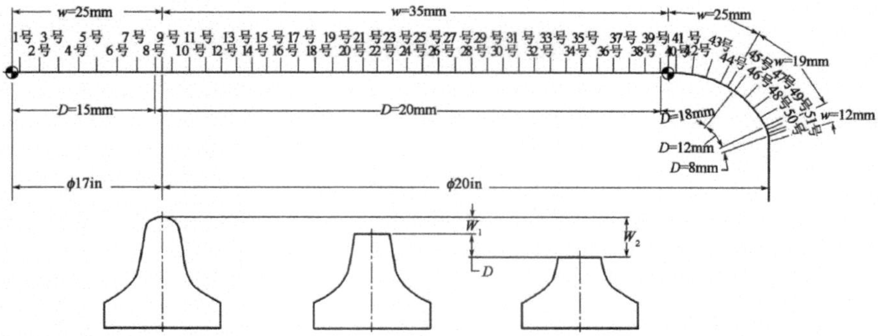

Fig. 8.10 Replacement standard of TBM cutter

In Fig. 8.10, *W* represents the allowable wear limit value of the regional cutter ring, and *D* represents the allowable limit value of the height difference of adjacent disc cutters. The center disc cutters of the TBM are four 17 in double disc cutters. The allowable wear limit value in this area is 25 mm, and the limit value of cutter height difference is 15 mm. The face disc cutter and gauge disc cutter of the TBM are 20 in single disc cutters. The allowable wear limit value in the front disc cutter area of No. 9–40 is 35 mm, and the limit value of cutter height difference is 20 mm. In the edge area, the allowable wear limit value of No. 41–44 cutter positions is 25 mm and the cutter height difference limit value is 18 mm; the allowable wear limit value of No. 45–49 cutter positions is 19 mm and the cutter height difference limit value is 12 mm; the allowable wear limit value of No. 50 and 51 cutter positions is 12 mm and the cutter height difference limit value is 8 mm.

After replacement, the disc cutter in the area with small allowable wear limit in the edge area can be installed in the face disc cutter area with larger allowable wear limit value after maintenance, and the cutter ring will be scrapped when the wear limit value of the cutter ring is reached. The scrapping limits of cutter rings of different sizes are different. The scrapping limits of radial wear volume of 19 and 20 in cutter rings are 30–35 mm, while the scrapping limits of radial wear of 17 and 18 in cutter rings are 20–25 mm.

After the radial wear of the cutter edge, the cutter edge of the cutter ring becomes wider and its ability to penetrate into the rock is significantly reduced. The torque and thrust force required for penetrating the same depth will increase significantly, increasing the load of TBM drive. Moreover, if the disc cutter is not replaced in time after wear, the muck scraper and cutter head face plate will contact the rock mass, resulting in more serious damage to TBM equipment. However, if it is replaced frequently when it is less than the change standard, it will not give full play to the due efficiency of each cutter ring and affect the machine-hour rate of TBM. The cutters shall be replaced in batches as much as possible to control the height difference between adjacent disc cutters. If the height difference is too large, it is easy to overload the disc cutter with small wear volume, which is not conducive to the service life of the disc cutter.

In addition to checking and replacing the disc cutter, TBM also needs to replace the surrounding muck scraper. There is a gap between the muck scraper and the rock wall. The gap between the rock wall formed by the rock breaking of the new cutter ring disc cutter and the muck scraper is large. When the cutter ring is worn, the gap between the muck scraper and the rock wall becomes smaller. When the cutter head swings too much, the muck scraper is easy to rub against the rock wall. In addition, the muck scraper will be worn. Therefore, the gauge disc cutter needs to be replaced in time after the wear and tear, in order to protect the effect of scraping plate.

Attention shall be paid to the following problems in the daily use of TBM cutters:

(1) The TBM cutter head must be cleaned every day. On the one hand, the rock and soil attached to the cutter edge and bucket edge shall be removed, and on the other hand, the wear degree of the cutter head, cutter and bucket scraper can be checked. Under special circumstances such as short maintenance time and lack of personnel, the cutter edge and bucket edge must be cleaned and excavated symmetrically to keep the operation balance of the cutter head.
(2) Check the fixing bolts of all cutters every day, tighten the loose ones in time, and replace the bolts with damaged threads in time. The bearings of each cutter must also be checked to see if they can rotate, and if they do not rotate well, the tool assembly must be replaced. When checking the bearing, use a special tool to rotate or test its temperature by hand. When the bearing cannot rotate or the temperature is higher than that of other cutters, the bearing must be damaged and the cutter must be replaced.
(3) The edge cutters must be checked and measured every day. When their wear reaches 10–15 mm, they need to be replaced in time. Otherwise, the diameter of the tunnel excavated by the TBM will become smaller, which will jam the TBM body or make it difficult to adjust the direction, affecting the normal tunneling of the TBM and causing serious wear of the bucket scraper.
(4) Once the replaced cutter is found, it must be replaced immediately. Otherwise, the ability of the cutter to cut and squeeze the surrounding rock will be reduced, which will inevitably increase the burden of cutting and squeezing the surrounding rock by adjacent cutters around it, and these adjacent cutters will also be damaged over time. If the vicious cycle continues, many cutters will be damaged at one time at the same time, which will cause great economic losses to the construction.
(5) When a new cutter is replaced, the TBM operator must be informed to slow down the advance rate during the first tunneling so that the cutter on the cutter head surface slowly contacts the rock. Because the edges of the cutters are not on the same plane at this time, the newly replaced cutters are protruding and do not match the old marks on the surrounding rock of the tunnel face, and the newly installed cutters will be easily damaged when the advance rate is fast.
(6) When the bucket scraper is seriously worn, it should be replaced in time, which is conducive to TBM mucking and can prevent the cutter head from being worn. If the reinstallation is not carried out in time, the base holder of the bucket scraper will be deformed due to serious wear, which will lead to difficulties in

reinstallation, and the TBM mucking will not be smooth, and the load of the drive motor will also increase, which will directly affect the normal tunneling of the TBM.

(7) During TBM tunneling, if the rock is found to be soft and the current of the TBM motor is large and unstable, it may be that the muck inlet in the cutter head or the lower hopper of the conveyor belt has been gradually blocked by the rock chips. At this time, it is necessary to stop the machine for inspection and cleaning. When the water content of rock is not large, manual pneumatic pick can be used for cleaning; if the water content of rock is large, high-pressure water gun can be used for cleaning. In either case, it is necessary to transfer the cleaned rock chips out of the cutter head in time to prevent the rock chips from depositing again and blocking the cutter head.

(8) When the surrounding rock is hard and TBM tunneling is difficult, the TBM operator must turn on the cutter water spray device to remove dust and cool the cutter head and cutter, thus prolonging the service life of the cutter. In addition, depending on the hardness of the surrounding rock, the TBM operator must stop the machine for every 5–10 strokes of tunneling, let the cutter worker enter the head to check the cutter and replace the damaged cutter in time.

8.4.2 Inspection and Maintenance of Disc Cutter Workshop

1. Non-disassembly Inspection of TBM Cutter

The removed disc cutter needs to be cleaned first and then inspected (Fig. 8.11a). First, the appearance inspection shall be carried out by flushing. If the disc cutter body is obviously damaged, it needs to be disassembled for inspection; if there is no obvious damage, it needs to be inspected without disassembly. The non-disassembly inspection is mainly to check the wear volume, start torque, lubricating oil and grease of the disc cutter and the tightness of the cutter body.

Check the radial wear volume of the cutter ring with a special inspection ruler for disc cutter ring (Fig. 8.11b). If the scrap limit of the cutter ring is reached, it is necessary to cut off the cutter ring with a cutting machine. If the scrap limit is not reached, check other items.

Check the start torque of the disc cutter with a torque wrench (Fig. 8.11c). If the measured torque value is within the specified range (such as 20–60 N m for Robbins 20 in disc cutter and 30–80 N m for Wirth 17 in disc cutter), it can be judged that the disc cutter assembly or internal parts are damaged.

Check the tightness of the cutter body through pressure test (Fig. 8.11d). Replace a pipe plug on the end cover with a seal test pressure gauge, and pressurize it to the specified pressure value with compressed air (for example, 35–70 kPa for Robbins 20 in disc cutter). If the pressure cannot be maintained for 5 min, it indicates the cutter body seal failure.

Check the grease state through the oil filling hole (Fig. 8.11e), and judge whether the grease needs to be replaced according to the color of the grease

a) Clean the disc cutter
b) Check the wear of cutter ring
c) Check the start-up torque
d) Check the airtightness of cutter body
e) Check the state of grease
f) Check the color of grease

Fig. 8.11 Inspection process of disc cutter

(Fig. 8.11f); judge whether the cutter body seal is damaged according to whether there is sand in the grease.

Finally, the above inspection results will be filled in the cutter inspection form and the disc cutter will be classified.

2. Removal of TBM Cutter

   (1) Removal of Cutter Ring
   If the wear volume of the cutter ring reaches the scrap value, the retainer ring shall be planed off and the cutter ring shall be cut off with a cutting machine

(Fig. 8.12a, b), but attention shall be paid to avoid excessive damage to the cutter body. Grind and derust the mating surface of the cutter body with a hand stone or grinder (Fig. 8.12c).

(2) Disassembly of Cutter Body
Remove the lock nut and end cover: apply oil to the thread interface area, then remove the lock nut with a nut removal tool, and remove the upper retainer with a lifting tool; remove the O-ring of the cutter shaft and discard it (Fig. 8.12d, e).

Remove the cutter shaft: Place the cutter on the cutter shaft sleeve and pressure platform, and apply pressure to the cutter shaft to make the bearing cone on the cutter shaft fall off (Fig. 8.12f, g); Place the small tool cup between the inner seal ring of the lower retainer and the bearing inner ring, and press the tool cup with a press. Turn the cutter body to remove the other seal ring, and remove the metal floating seal and discard it (Fig. 8.12h).

Remove bearing: If the bearing needs to be replaced, place the cutter body on the pressure platform, put the outer ring removal tool into the inner cavity of the cutter body (the pin faces downward and is between the outer ring removal tools), and press the bearing outer ring out with a press machine; Turn over the cutter body and press out the other outer ring. If the bearing can be used continuously, use the outer ring together with the roller and inner ring (8.12i, j).

3. Installation of TBM Cutter

(1) Installation of Cutter Ring
For the cutter body that has not been disassembled, the cutter ring shall be directly installed; for the cutter body that has been disassembled for maintenance, the bearing, seal, cutter shaft and retainer can also be installed after the cutter ring is installed.

Lay the cutter body flat with the groove of the opening ring facing upward; use a hand stone to make the contact surface between the cutter body and the cutter ring smooth, and avoid grinding the surface with electric tools if possible; use high-performance dissolving agent to remove the oil deposited inside the cutter ring and on the cutter ring supporting surface of the cutter body.

Use an induction heater or furnace to heat the cutter ring to 175 °C (maximum) (Fig. 8.13a). Note that the time when the cutter ring temperature is 150–175 °C cannot exceed 5 min, otherwise the hardness of the cutter ring will be reduced, thus shortening the service life of the cutter. Slide the cutter ring into the outer diameter of the cutter body, with the polished side facing downward; If the cutter ring is stuck, squeeze the shoulder of the cutter body with a special sleeve and cover plate.

Put the opening ring into the cutter body and place it in the slot with the welding slot upward (Fig. 8.13b); squeeze the opening ring with the clamp and align it with the welding slot; clamp the welding beater clamp on the

Fig. 8.12 Disassembly process of disc cutter

Fig. 8.13 Installation process of disc cutter

opening ring clamp; weld the separated area and forbid welding the opening ring on the surface of the cutter ring.

(2) Installation of Cutter Body

After the heat of the cutter ring diffuses to the whole cutter body, apply anti-seize agent to the inner diameter of the cutter body bearing and the outer ring of the bearing (or squeeze the bearing outer ring into the cutter body with the bearing outer ring installation tool on the mating surface between the cutter body and the bearing outer ring (Fig. 8.13c), and pay attention

i) Press in the rear end seal

j) Inlay the cutter body assembly into the cutter shaft assembly

k) Install the front bearing

l) Install the front end seal

m) Install the front end cover

n) Install the lock nut

o) Test starting torque

p) Check the sealing of the cutter body

q) Inject lubricating grease

r) Complete the installation of disc cutter

Fig. 8.13 (continued)

to ensuring that the bearing outer ring is stressed evenly. If the outer ring and the inner diameter are seriously misaligned, the outer ring and the inner diameter of the cutter body must be removed to check for damage.

Install the O-ring on the O-ring groove of the bottom flange of the shaft, place the shaft vertically with the threaded end on the top, wipe off all dirty oil on the shaft and the seal retainer hole, and apply anti-seize agent along the axial direction (Fig. 8.13d); Heat the lower retainer, and the maximum temperature shall not exceed 120 °C; otherwise, the O-ring will be damaged and shall be replaced. Slide the seal retainer onto the flange face of the shaft, and the maximum extrusion force of the seal retainer shall not exceed 100 kN (Fig. 8.13e).

Place the seal installation tool on the half of the seal face so that the sealed surface faces inward and the toric surface faces outward. It must be ensured that the suture of the compound curve is in the same direction as the sealing periphery; Install the seal half ring on the gland and push it evenly in the opposite direction of the seal installation tool until it is in place; After correct installation, the sealing surface will extend 8–10 mm beyond the lip surface of the seal cover plate (Fig. 8.13f); Install the O-ring to the inner seal ring.

Heat the bearing inner ring with an induction heater or oven to a temperature not exceeding 120 °C (Fig. 8.13g), and place the bearing inner ring on the bearing outer ring in the cutter body (Fig. 8.13h); Apply anti-seize agent 654-025 to the periphery of the seal ring and cutter body mounting seal, and press the inner seal into the cutter body with a special tool (Fig. 8.13i).

Place the shaft assembly vertically, with the threaded end on the top, and then apply anti-seize agent along the axial direction; lower the cutter body assembly to the cutter shaft, with the retainer ring upward; if it falls off from the seal cover plate area of the cutter body during the process of lowering the seal to the cutter shaft, stick the sealed toric surface to the seal cover plate area with a small amount of high-temperature silicon seal.
Place spacer sleeves on the shaft and bottom bearing cones, and put the heated bearing cones into the cutter body; install the bearing on the shaft and bearing cup, and apply a force of not more than 100 kN to the bearing cones (Fig. 8.13k).

Upper retainer and lock nut: Install the metal floating seal on the top inner seal ring, install the vent valve and vent plug to the upper retainer, install the metal floating seal on the upper retainer, clean the seal ring, and sleeve the upper retainer into the cutter shaft (Fig. 8.13l, m); Apply anti-seize compound to the threads of the nut and the cutter shaft, install the nut on the cutter shaft with a nut tightening tool, place the torque ring on one end of the opening ring of the cutter hub, place the hexagon bolt upward, and tighten the handle on the cutter body until it is tightened to the tail of the thread (Fig. 8.13n).

Install the torque wrench and test the start torque while tightening the nut until the start torque reaches the required range (Fig. 8.13o); Then conduct a seal test on the assembled tool. After the requirements are met (Fig. 8.13p), inject new lubrication grease (Fig. 8.13q). Finally, block the oil filling hole, and spray different color marks on the disc cutter according to the number of times the bearing is used to complete the assembly of the disc cutter (Fig. 8.13r).

8.5 Analysis on Field Use of Disc Cutter

8.5.1 Overview of Hanjiang-to-Weihe River Water Diversion Project and Field Use of Cutters

1. Overview of Hanjiang-to-Weihe River Water Diversion Project
   The Qinling Tunnel of Hanjiang-to-Weihe River Water Diversion Project is 81.8 km long in total, and the Lingnan TBM section is 18.275 km long in total. It consists of two tunneling construction sections: TBM back-up installation tunnel, main machine installation tunnel, stepping tunnel, launch tunnel, maintenance tunnel and TBM. The Project is located in the high-medium mountainous area of Qinling Ridge and the medium–low mountainous area of Lingnan, with topographic relief and a tunnel overburden of 500–2000 m. The lithology is mainly granite, with uniaxial dry compressive strength of 96.7–242 MPa, tensile strength of 2.1–3.3 MPa, internal friction angle of 51–68.5°, cohesion of 1.44–12.53 MPa and integrity coefficient of 0.59–0.89. The surrounding rocks are mainly of Class II, accounting for 72.9%. In addition, Class I accounts for 16.7%, Class III 6.5%, Class IV 2.9% and Class V 1%. In addition, the surface water is relatively developed. The groundwater is bedrock fissure water, which is rich in water and replenished by atmospheric precipitation. The maximum water ingress that may occur in the whole tunnel is 41,000 m^3/d.
   The Lingnan TBM section adopts gripper TBM tunneling with a diameter of ϕ8.02 m, and the excavation diameter is ϕ8.05 m when installing new cutters. Four 17 in center double disc cutters are arranged on the cutter head, with the blade numbers of 1–8 and the blade spacing of 101.6 mm. 43 20 in single disc cutters are arranged on the cutter head, of which 31 disc cutters with No. 9–39 are face disc cutters with a cutter spacing of 82.4 mm. The 12 disc cutters with cutter numbers of 40–51 are gauge disc cutters, and the cutter spacing decreases in turn, as shown in Fig. 8.14. W_{max} in the figure indicates the maximum allowable wear volume of disc cutters in this area.

2. Usage of Disc Cutter for Hanjiang-to-Weihe River Water Diversion Project
   The first 2000 m trial tunneling section of Lingnan TBM construction section of Hanjiang-to-Weihe River Water Diversion Project is divided into four stages, and the cutter consumption is counted every 500 m.

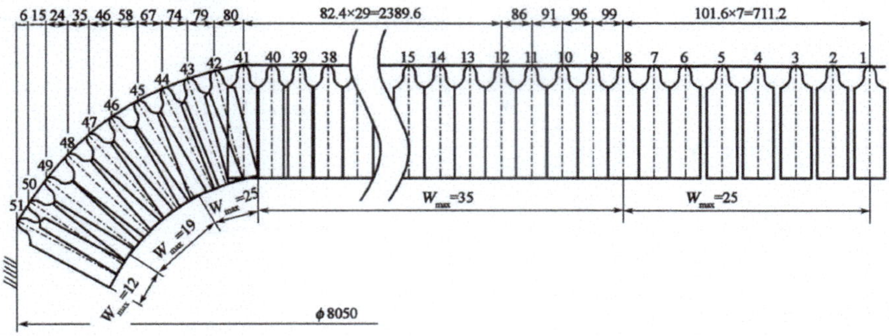

Fig. 8.14 Schematic diagram of arrangement of disc cutter wear of Hanjiang-to-Weihe River TBM (dimensions in mm)

(1) Statistics on Use of 0–500 m Cutters in Stage 1

Stage 1 See Table 8.7 for the replacement of 0–500 m disc cutters.

According to the statistical results, the disc cutter is replaced 401 times in total for the tunneling distance of 500 m from 0 to 500 m, of which 208 times are replaced with new disc cutters and the remaining 193 times are transition disc cutters, that is, the disc cutter has a certain wear volume but can still be used, and the replacement rate of new disc cutters is 51.9%; In the 401 replacements, the normal wear was 309 times and the abnormal wear was 92 times. The abnormal wear was mainly caused by uneven wear of the cutter edge, cutter edge collapse of the cutter ring, and fracture of the cutter ring and falling off of the retaining ring accounting for a small proportion. Abnormal wear accounts for 22.9% of the total number of cutter replacement/changes. The number of new cutters consumed per meter is 0.416; the tunneling distance of each new cutter is 2.40 m.

(2) Statistics on Use of 500–1000 m Cutters in Stage 2

Stage 2 See Table 8.8 for the replacement of 500–1000 m disc cutters.

Table 8.7 Replacement of 0–500 m disc cutters in Stage 1 of Hanjiang-to-Weihe river water diversion project

Tunneling time	March	April	May	June 1– June 22	Total
Tunneling distance (m)	13	103.5	160.5	223	500
Total cutter replacement/change times	15	81	187	118	401
Number of new cutter replacements	6	43	91	68	208
Uneven wear of cutter edge	0	2	16	12	30
Number of cutter ring blades	0	4	13	21	38
Number of broken cutter rings	6	3	4	0	13
Number of fallen retainer rings	0	3	4	4	11

Table 8.8 Replacement of 500–1000 m disc cutters in Stage 2 of Hanjiang-to-Weihe river water diversion project

Tunneling time	June 23–End of July	August	Early September–September 9	Total
Tunneling distance (m)	167.5	240	92.5	500
Total cutter replacement/change times	128	225	110	463
Number of new cutter replacements	91	85	20	196
Uneven wear of cutter edge	3	17	7	27
Number of cutter ring blades	4	4	9	17
Number of broken cutter rings	1	0	0	1
Number of fallen retainer rings	1	0	0	1

According to the statistical results, the disc cutter is replaced 463 times in total for the tunneling distance of 500–1000 m, of which 196 times are new cutters and the remaining 267 times are transition cutters, that is, the cutter edge has a certain wear volume, but it can still be used, and the replacement rate of new cutters is 42.3%. Among the 463 replacements, there were 463 normal wears and 46 abnormal wears. The abnormal wears were mainly uneven wear of the cutter edge, cutter edge breaking of the cutter ring, fracture of the cutter ring and falling off of the retaining ring, accounting for a small proportion, and the abnormal wears accounted for 9.9% of the total number of cutter replacement/changes. The number of new cutters consumed per meter is 0.392; the tunneling distance of each new cutter is 2.55 m.

(3) Statistics on Use of 1000–1500 m Cutters in Stage 3

Stage 3 See Table 8.9 for the replacement of 1000–1500 m disc cutters.

According to the statistical results, 363 disc cutters are replaced for a total tunneling distance of 500 m from 1000 to 1500 m, of which 185 are new cutters and the remaining 178 are transition cutters, that is, the cutter edge has a certain wear volume, but can still be used, and the replacement rate of new cutters is 50.1%. Among the 363 replacements, there were 347 normal wears and 16 abnormal wears. The abnormal wears were mainly uneven wear of the cutter edge, cutter edge breaking of the cutter ring, fracture of the cutter ring and falling off of the retaining ring, accounting for a small proportion, and the abnormal wears accounted for 4.4% of the total number of cutter replacement/changes. The number of new cutters consumed per meter is 0.370; the tunneling distance of each new cutter is 2.70 m.

(4) Statistics on Use of 1500–2000 m Cutters in Stage 4.

Stage 4 See Table 8.10 for the replacement of 1500–2000 m disc cutters.

Table 8.9 Replacement of 1000–1500 m disc cutter in Stage 3 of Hanjiang-to-Weihe river water diversion project

Tunneling time	September 9–End of September	October	Early November–November 9	Total
Tunneling distance (m)	167	257	76	500
Total cutter replacement/change times	153	168	42	363
Number of new cutter replacements	80	85	20	185
Uneven wear of cutter edge	0	5	0	5
Number of cutter ring blades	2	0	4	6
Number of broken cutter rings	0	0	0	0
Number of fallen retainer rings	0	5	0	5

Table 8.10 Replacement of 1500–2000 m disc cutter in Stage 4 of Hanjiang-to-Weihe river water diversion project

Tunneling time	November 9–End of December	October	Early November–November 9	Total
Tunneling distance (m)	129.8	262.7	107.5	500
Total cutter replacement/change times	156	252	58	466
Number of new cutter replacements	66	125	34	225
Uneven wear of cutter edge	7	7	9	23
Number of cutter ring blades	9	22	2	33
Number of broken cutter rings	0	0	0	0
Number of fallen retainer rings	0	0	0	0

According to the statistical results, the disc cutter is replaced 466 times in total for a tunneling distance of 500 m from 1500 to 2000 m, including 225 times of replacement with new disc cutters and the remaining 241 times of replacement with transition disc cutters, that is, the disc cutter has a certain wear volume but can still be used, and the replacement rate of new disc cutters is 48.2%. In the 466 replacements, the normal wear is 410 times and the abnormal wear is 56 times. The abnormal wear is mainly caused by uneven wear of the cutter edge, cutter edge collapse of the cutter ring, and the fracture of the cutter ring and the falling off of the retaining ring account for a small proportion. Abnormal wear accounts for 12.0% of the total number of cutter replacement/changes. The number of new cutters consumed per meter is 0.450; the tunneling distance of each new cutter is 2.22 m.

See Table 8.11 for the use of disc cutters in the above four stages.

8.5.2 Analysis of Wear Rule of Disc Cutter for Hanjiang-to-Weihe River Water Diversion Project

1. Analysis of Accumulative Wear Volume of Disc Cutter in Trial Tunneling Section of Hanjiang-to-Weihe River

The accumulated wear volume of each numbered disc cutter after multiple replacements in the four stages of TBM is shown in Figs. 8.15, 8.16, 8.17, 8.18 and 8.19.

It can be obtained from Figs. 8.15, 8.16, 8.17, 8.18 and 8.19: The cumulative wear volume of the front disc cutter increases approximately linearly with the increase of the installation radius of the disc cutter; The cumulative wear volume of No. 43 cutter in the transition area of the gauge disc cutter is the largest, up to 1996 mm, and the average wear volume per linear meter is 1 mm. The wear volume on both sides of the cutter position gradually decreases. The maximum allowable wear volume of No. 43 cutter position is 25 mm. According to this calculation, the cutter needs to be replaced every 25 m of tunneling at this cutter position. However, due to batch cutter replacement, abnormal wear and other reasons, the cutter at this cutter position is actually replaced 101 times in total, that is, the cutter is replaced every 20 m on average.

There are two views on increasing the edge cutter replacement/change distance. One view is that 1 disc cutter is added near the No. 43 cutter position to share the rock breaking task of the disc cutter in the edge area, reduce the wear rate of No. 43 cutter, improve the cutter replacement/change distance in the whole edge area, save the cutter replacement/change time and improve the time rate of TBM. In the design of the cutter head, a mounting hole for an unwelded cutter seat is reserved near the cutter position track, which meets the conditions for refurbishment. However, the stability of the cutter head after the cutter is increased should be ensured by calculating the overturning moment of the whole cutter head. Another view is that the secondary

Table 8.11 Summary of replacement of 2000 m disc cutters in four stages of diversion from Hanjiang river to Weihe river

Tunneling stage	Tunneling distance (m)	Cutter replacement/change times Time(s)	New cutter ring consumption Piece(s)	Replacement rate of new cutter (%)	Abnormal wear volume Time(s)	Abnormal wear rate (%)	Cutter consumption per meter (Piece(s)/m)	Tunneling distance per cutter (m/piece(s))
Stage 1	500	401	208	51.9	92	22.9	0.42	2.40
Stage 2	500	463	196	42.3	46	9.9	0.39	2.55
Stage 3	500	466	225	48.2	56	12.0	0.45	2.22
Stage 4	500	363	185	50.1	16	4.4	0.37	2.70
Total value (or average)	2000	1693	814	48.1	210	12.4	0.41	2.46

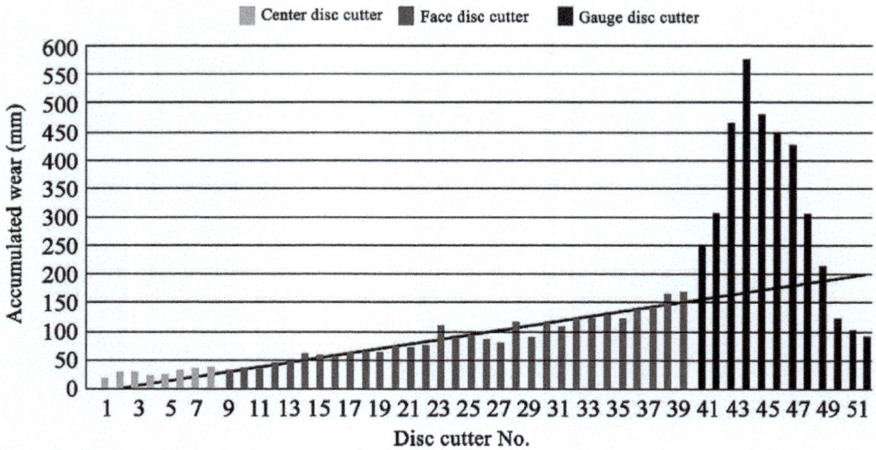

Fig. 8.15 Histogram of accumulated wear volume of disc cutter in Stage 1

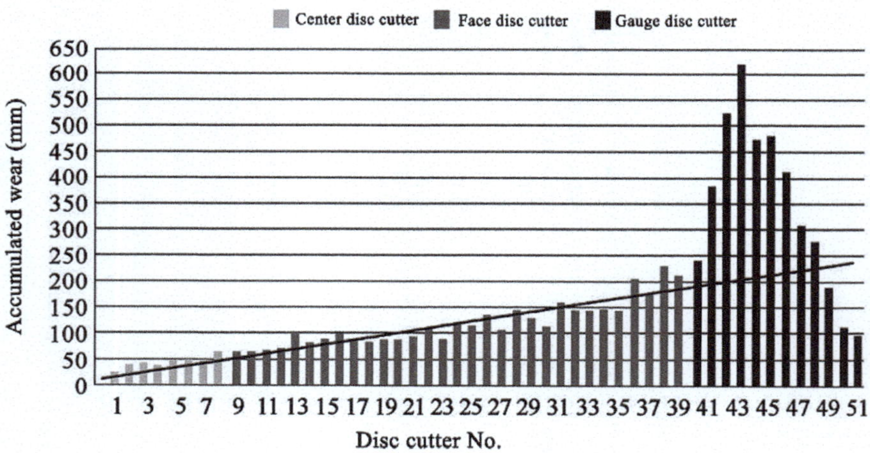

Fig. 8.16 Histogram of accumulated wear volume of disc cutter in Stage 2

wear of the edge area disc cutter is serious, increasing the cutter will further increase the consumption of secondary wear, and the effect of extending the gauge disc cutter change distance is not obvious, so it is not recommended to increase the cutter.

2. Analysis of Wear Rule of Disc Cutter in Trial Tunneling Section of Hanjiang to Weihe River Water Diversion Project

Some scholars use the radial wear volume of disc cutter per unit tunneling distance to measure the wear speed of disc cutter, and some experts use the radial wear volume of disc cutter per unit rock breaking volume to measure the wear speed of disc cutter. The rolling distance wear rate and rock breaking volume wear rate of each tool position are

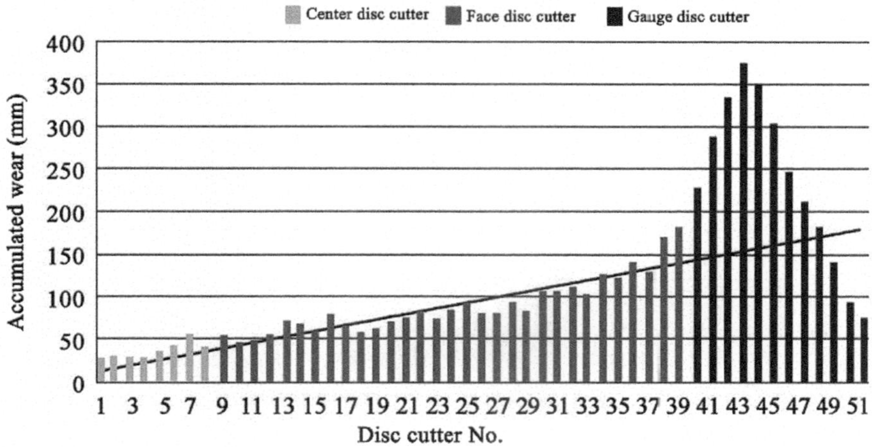

Fig. 8.17 Histogram of accumulated wear volume of disc cutter in Stage 3

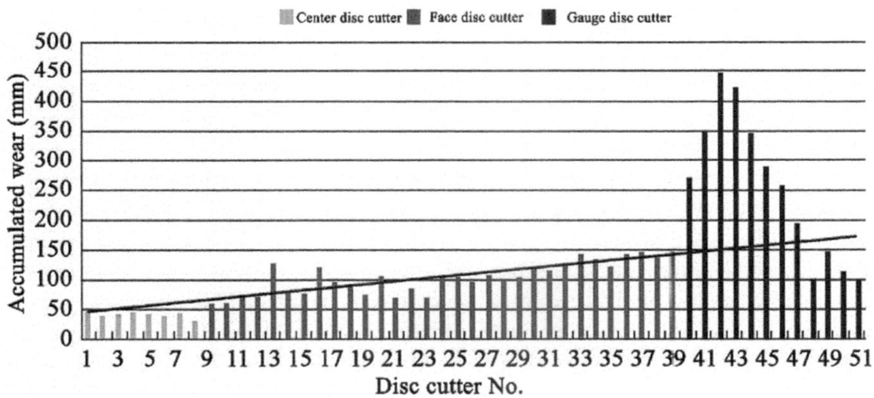

Fig. 8.18 Histogram of accumulated wear volume of disc cutter in Stage 4

calculated according to the accumulated wear volume of the first 2000 m. The rolling distance wear rate is the accumulated wear volume divided by the accumulated rolling distance of the cutter position. The calculation of the accumulated rolling distance is shown in Formula (8.29). The wear rate of rock breaking volume is the cumulative wear volume divided by the cumulative rock breaking volume of the cutter position. The calculation of the cumulative rock breaking volume is shown in Formula (8.30). Since the No. 1 cutter position is obviously abnormal when calculating the wear rate of rolling distance, and the No. 51 cutter position is obviously abnormal when calculating the wear rate of rock breaking volume, the above two cutter positions are eliminated during analysis.

$$L_i = 2\pi R_i L / h \tag{8.29}$$

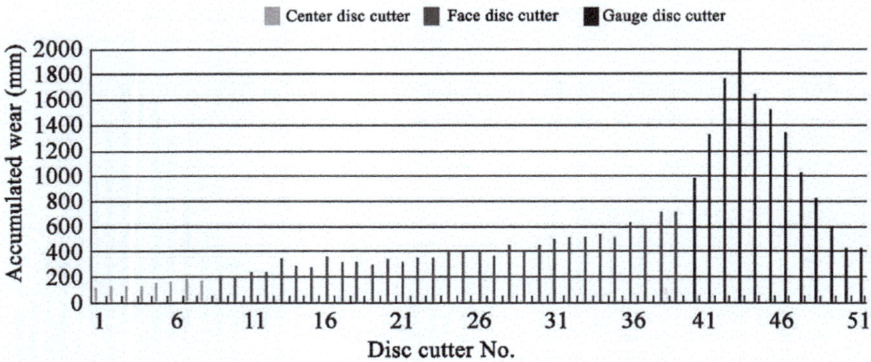

Fig. 8.19 Histogram of cumulative wear volume of 2000 m disc cutter before diversion from Hanjiang to Weihe river

$$V_i = 2\pi R_i S_i L \tag{8.30}$$

where: h—average penetration of disc cutter in the tunneling section. R_i and S_i are determined according to the design parameters of TBM cutter head and cutter, L and h are determined according to the actual tunneling situation, the tunneling distance L is 2000 m, and the average penetration is 6.25 mm/r after statistics.

It can be obtained from the rolling distance wear rate diagram (Fig. 8.20): The rolling distance wear rate values of face disc cutters at different cutter positions are relatively close, with an average value of 0.096 mm/km; However, the rolling distance wear rate of the gauge disc cutter has a similar distribution law to the accumulated wear volume, and the No.43 disc cutter position is the largest and gradually decreases to both sides; The wear rate of the rolling distance of the center disc cutter decreases gradually with the increase of the installation radius. It can be seen that the wear rate of disc cutter rolling distance cannot eliminate the wear rate influence of disc cutter arrangement on the cutter head. The wear of disc cutter is not only related to the rolling distance of cutting edge, but also related to the cutter edge spacing between adjacent disc cutters. The cutter edge spacing of face disc cutter is basically the same, so the wear rate of rolling distance can reflect the wear rate of disc cutter. However, although the rolling distance of disc cutter in the edge area is increasing, the cutter edge spacing is decreasing. At this time, the wear rate of rolling distance is not enough as an indicator to measure the wear rate of disc cutter.

According to the rock-breaking volume wear rate diagram (Fig. 8.21), the rock breaking volume wear rates of face disc cutters at different cutter positions are relatively close, with an average value of 0.188 mm/m^3. The rock breaking volume wear rate of No. 40–42 transition areas gradually increases, and the rock-breaking volume wear rate of No. 42–50 edge areas is relatively stable, with an average value of 0.496 mm/m^3. Compared with the rolling distance wear rate, the volume wear rate of rock breaking not only considers the rolling distance of disc cutter, but also considers the cutter edge spacing. In addition, although the volume wear rate of

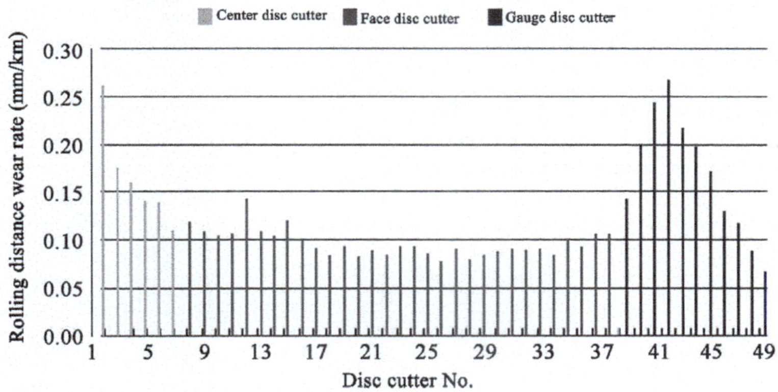

Fig. 8.20 Statistics of distance wear rate of disc cutter

front disc cutter and gauge disc cutter has a stable value, there are great differences between areas, indicating that there are other factors affecting the wear rate of disc cutter. The working conditions of the front disc cutter area and the gauge disc cutter area are analyzed. In addition to the wear of the rock mass on the excavation face, the broken rock chips are accumulated at the bottom in front of the TBM cutter head face plate. During TBM tunneling, the gauge disc cutter repeatedly breaks the broken rock block, resulting in secondary wear, while the disc cutter in the front area of the cutter head only has primary wear caused by direct rock breaking. Therefore, the volume wear rate in the edge area is higher, and the wear of the disc cutter in this area is the result of the superposition of the primary wear caused by direct rock breaking and the secondary wear caused by repeated rock breaking. Numerically, the average volume wear rate of rock breaking in the edge area is 2.6 times that in the front area, and it can be seen that the secondary wear rate is 1.6 times that in the primary wear rate, and the wear consumption of disc cutter is serious. For the central center disc cutter in the central area, the main reason is that the running radius of the disc cutter in the central area is small, the disc cutter edge has serious lateral sliding, and the disc cutter has serious sliding wear. The larger the installation radius is, the smaller the side sliding volume and the smaller the disc cutter wear volume is. It can be seen that if 1 disc cutter is added to the reserved hole near the track of No. 43 cutter, it can play a certain role in extending the cutter replacement/change distance in the edge area, but the secondary wear weakens more rock breaking capacity, so that the problem is still fundamentally solved.

Questions

1. Briefly describe the failure mode of TBM disc cutter.

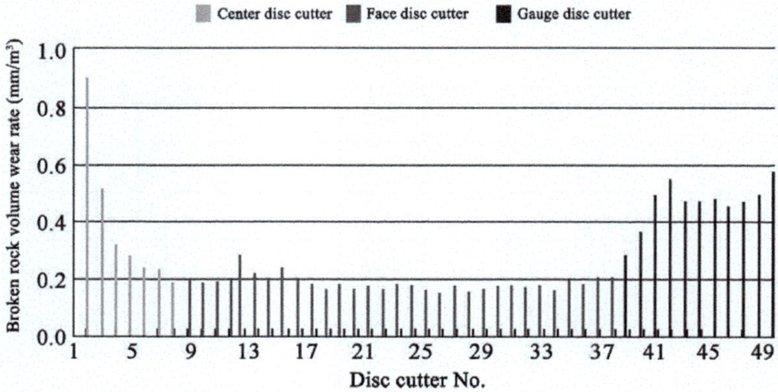

Fig. 8.21 Statistics of rock breaking volume wear rate

2. What does the abnormal wear of disc cutters include?
3. Briefly describe the influencing factors of TBM disc cutter wear.
4. What are the theoretical prediction models of TBM disc cutter wear?
5. Briefly describe the removal and installation process of TBM cutter.

Chapter 9
TBM Maintenance and Disassembly

The focus of this chapter: During TBM operation, a fault of any of its systems will lead to downtime, thus affecting the tunneling efficiency. Therefore, standardized operation, scientific management and normalized inspection and maintenance are indispensable to effective TBM tunneling. In addition, a scientific disassembly process after TBM tunneling is of great significance in extending the service life of TBM. In this chapter, TBM maintenance and disassembly procedures are explained in detail from the above two aspects, covering TBM maintenance, back-up maintenance of main machine, disassembly methods in the tunnel, and storage during disassembly to improve the scientific use of TBM and increase its service efficiency and service life.

9.1 Key Points of Standardized Operation of TBM Maintenance

9.1.1 Total Involvement

The standardized implementation of TBM repair, maintenance and condition monitoring involves not only the equipment maintenance department, but also the departments related to TBM construction. Only when the decision makers and executors at all levels understand the significance of repair, maintenance and condition monitoring, and the responsibilities, obligations and technical requirements for the posts concerning equipment maintenance and repair, can they achieve mutual trust and support, actively participate in TBM repair, maintenance and condition monitoring and thoroughly standardize relevant procedures to achieve normal operation of TBM.

K. Chen et al., *TBM Design and Construction*,
https://doi.org/10.1007/978-981-99-0059-6_9

9.1.2 Strengthening Maintenance

In terms of technical measures for maintenance, the concept that the earlier the mistake occurs, the deeper the harm will be should be kept in mind. Moreover, the key problems affecting the service life of the equipment should be addressed. Focus should be laid on oil–water management. Ultrafine filtration of oil should be applied constantly. Experience should be learned from enterprises with advanced equipment management. The principle of "cleaning for a clean and cleaner state" and "maintenance for effective and constant maintenance" should be strengthened; The selection of maintenance methods is based on the purpose of reducing maintenance time and cost.

9.1.3 Intensive Training and Assessment

In order to standardize the maintenance work, the training and assessment of maintenance personnel shall be strengthened to avoid human errors.

(1) Formulate standardized TBM operation, repair and maintenance procedures and processes, strictly implement post responsibilities and regulations, and continuously implement a level-by-level irregular inspection and assessment system.
(2) Relevant personnel shall be trained before taking up their posts, and training shall be carried out at different levels. Through continuous training, learning and enough practice, maintenance personnel cannot only understand the performance of TBM as soon as possible, but also know how to conduct maintenance and fault detection. This is an effective means to make up for the advancement of equipment technology ahead of the ability of maintenance personnel, and is also the premise for safe and normal operation of TBM. The contents and technical requirements of training can vary with different trainees.
(3) Special personnel shall be assigned to be responsible for the task in combination with group participation to avoid human errors. Special personnel shall be assigned to be responsible for the implementation while group participation shall also be carried out to facilitate mutual support and supervision. The prevention of human errors needs to be strengthened and improved from many aspects such as people's sense of responsibility, training level, technical ability, physical quality, psychological state, rules and regulations and working environment; Maintenance operations shall be carried out in strict accordance with operating procedures, and corresponding preventive measures shall be taken according to specific cases of human errors prone to occur in parts.

9.1.4 Compliance with Regulations

The standardization of maintenance work shall comply with the following regulations:

(1) Selection of Personnel. Select qualified construction site personnel, including shift leaders, mechanical engineers, hydraulic engineers, electrical engineers, operators of each station and assigned migrant workers, and specify the scope of responsibilities for equipment under their charge.
(2) Point selection. Determine the key points of maintenance on the equipment, and select the parts to be cleaned, inspected, lubricated and maintained.
(3) Item Selection. According to the characteristics of the selected parts, combined with the actual situation of the equipment, select several warranty items that must be carried out.
(4) Time Selection. Schedule the execution cycle and time of warranty items.
(5) Standard Selection. For each point of each operation content, formulate the corresponding implementation standards.
(6) Method Selection. Each operation shall have defined tools, methods and auxiliary materials.
(7) Shift Selection. According to the characteristics of the construction operation, the tunneling shift is dominated by tunneling, and mechanical engineers follow the shift to carry out emergency repair; Items that are difficult to handle or have a large amount of repair shall be left to the maintenance shift for repair; Daily maintenance, planned special maintenance and temporary maintenance are mainly carried out during the maintenance shift; The cutter shift is specially responsible for the daily inspection and replacement of cutters, and the repair of cutters and cutter heads. Since maintenance requires coordination of all parties, it is generally convenient to arrange in the daytime.
(8) Circuit Selection. According to the structural characteristics of the equipment, special inspection and repair shall be organized according to the main machine structure, back-up mechanical structure, hydraulic system, electrical equipment, water supply pipeline, high-pressure air circuit, cutters and auxiliary equipment and other independent circuits to avoid disorder and omission.
(9) Record. Make special records of equipment installation, oil change, testing, oil and material consumption as required; all important records shall be filed.

9.1.5 Reasonable Maintenance Division and Optimized Maintenance Method

(1) Major equipment faults shall be dispatched and commanded by the Chief Mechanical Engineer on site, and the technical room, tunneling shift, servicing shift and other relevant personnel shall be convened to discuss solutions and

formulate maintenance plans. When downtime is necessary, it shall be reported to the superior leader for approval before downtime for maintenance.

(2) The planned maintenance project adopts the form of centralized management, which is in the charge of the professional engineer of the technical office. The servicing team leader organizes personnel to carry out maintenance according to the plan arrangement and maintenance requirements of the technical office.

(3) Temporary repair of equipment fault during tunneling that affects tunneling shall be handled by the mechanical, electrical and hydraulic technical engineers under the leadership of the tunneling foreman. General minor faults shall be solved independently by the person in charge of each operation station in the tunnel as far as possible. When it is difficult to handle, the replacement and repair method shall be preferred to resume tunneling as soon as possible.

(4) Maintenance strategy the selection of TBM maintenance mode shall be determined according to the characteristics of the maintenance object. The selection of different maintenance modes depends on the maintenance cost and total benefit of the enterprise. Sufficient statistical data shall be accumulated, and calculation and selection shall be carried out according to the principle of minimum maintenance cost to determine the best maintenance method to be adopted. There are 7 maintenance modes as follows, which can be used alone or in combination of several modes.

① Condition-based Maintenance. For key components and parts that have a significant impact on tunneling construction and have guaranteed monitoring conditions and diagnostic accuracy, their deterioration trend can be accurately diagnosed through condition monitoring. The strategy of predicting condition-based maintenance is preferred, and maintenance is implemented after the equipment condition is obviously deteriorated, and the condition deterioration is reflected by the change of condition parameters. This strategy requires the monitoring of various parameters of the equipment to reflect the actual status of the equipment at any time. The monitored parameters can be alerted within sufficient lead time to allow appropriate maintenance measures to be taken. Generally, there is no fixed interval for this kind of maintenance operation, and the maintenance personnel will make a judgment according to the change trend of the test data and then determine the equipment maintenance plan.

Each equipment shall be provided with a set of monitoring or status inspection methods. Inspections can be periodic or continuous, and there are a variety of inspection methods. Schedule repairs only when data indicate that repairs must be performed. Since the fault state can be predicted, the maintenance plan can be carefully formulated and the preparation can be fully made, to greatly improve the maintenance efficiency and reduce the maintenance downtime. This method shall be promoted for key monitoring objects of TBM except for sudden faults.

② Regular Scheduled Inspection and Maintenance. The purpose of regular inspection is to understand the current state of the equipment, find out the existing defects and hidden dangers, and accordingly arrange the repair plan in a targeted manner to ensure a relatively high utilization ratio of the equipment. The key to regular inspection and maintenance lies in inspection. The type, frequency, position and technical requirements of the equipment to be inspected shall be properly arranged to form an overall and preventive regular inspection schedule. After inspection, the recent maintenance plan must be arranged and implemented. For example, the friction plate of the clutch of the cutter head drive gearbox is replaced by the servicing shift before reaching the maximum wear volume according to the regularly measured wear amount.

③ Regular Replacement and Repair. This is a preventive scheduled maintenance, which can prevent and reduce the occurrence of emergency faults. In actual implementation, the maintenance interval shall be determined according to the wear law. The conditions of the same type of equipment (such as transmission and clutch) are different, the repair contents and repair amount are different, and the repair interval shall not be the same. Such as the chain of TBM traction device, chain wheel, as the chain pitch becomes longer and the free gap becomes larger, the wear volume of the chain wheel increases, and when the predetermined wear limit is reached, it should be replaced regularly, which is an effective preventive measure.

④ Mandatory Maintenance. For key parts and components, their damage will cause huge downtime losses, or the position of the parts and components is difficult to disassemble and assemble (disassembly and assembly can only be carried out after other parts are removed). If such parts and components are replaced separately, it will also cause huge shutdown losses, so mandatory maintenance should be adopted. These parts are forced to be repaired or replaced by the servicing shift of tunneling construction, the downtime gap in normal maintenance operations (cutter replacement/change) or the planned long maintenance period to prevent unplanned downtime losses. The cost of such maintenance, although slightly increased, is a small percentage compared to the loss of downtime. This method is more effective for key equipment on TB880E TBM, especially some equipment with high utilization ratio, long maintenance time, tunneling shift that is not convenient for downtime maintenance but needs to ensure its long-term normal operation, such as clutch bearing, brake shaft seal, cutter head muck chute, cutter head bulkhead, etc.

⑤ Switching Repairs Using Spare Equipment or Parts. In the event of a fault of an equipment (component) or the discovery of signs of fault, etc., the spare equipment is started, and the faulty equipment (component) is stopped for maintenance, and then put in a standby state. There are 3 maintenance methods for in-service equipment of TB880E TBM (such as in-service gearbox, clutch assembly, motor, grease distribution valve, cutter, etc.): Regular downtime maintenance, downtime maintenance when symptoms of faults or defects are detected, and downtime maintenance when faults occur during operation all have little impact on tunneling and will not affect tunneling due to emergency stop. There is only a big difference in content, amount and cost of repair.

⑥ Active (improved) Maintenance. For key and important equipment and parts that affect tunneling, the mode of strengthening monitoring and active (improvement) maintenance shall be adopted, that is, on the premise of restoring the original technical performance status of the equipment, give on-site maintenance technicians certain flexibility, and integrate the idea of active maintenance and improvement maintenance into the standardized maintenance system. For example, the curved shape sliding block between the inner and outer frames of the main machine and the water spray pipeline of the cutter head are refurbished, and the small belt conveyor for cutter head muck removal is changed from motor drive to hydraulic drive.

⑦ Breakdown Maintenance. For some non-important production equipment or parts and components that have little impact on tunneling (such as belt carrier rollers on belt conveyors, small chiseling pits on conveying belts, low-value minor spare parts, etc.), they can be repaired and replaced by the servicing shift when the tunneling is finished.

9.2 Daily Maintenance Content of Main Machine

9.2.1 System Maintenance of Main Machine

(1) The inspection work of the servicing shift in the inspection tunnel shall be started before the maintenance of each shift and run through the maintenance process.

(2) Cleaning.

① The operator is responsible for cleaning the appearance and environment of the operating handle, instrument, window and other parts of the equipment in charge (such as the operation area of bolt driller, motor and gearbox housing, motor platform, steelwork mobile platform, upper platform of connection bridge, etc.).

② It is strictly prohibited to flush the accumulated muck with high-pressure water directly above the main bearing, so as to avoid siltation and jamming in the gaps on the left and right sides of the main bearing, which will cause wedge effect and out-of-round deformity of the main bearing, seriously affecting the service life of the main bearing.

③ In order to prevent oil seal damage and precision surface damage, the exposed surface of cylinder piston rod, guide rail surface, instrument surface, display window surface, surface of parts to be disassembled and oil nozzle of TBM shall be kept clean without rust spots, oil stains and muddy water, and a small amount of grease shall be applied after cleaning.

④ Purge the dust between the clutch pressure plate and the friction plate with high-pressure air every month, and properly cover the clutch cover with rubber to avoid falling into oil stains or dirt, causing the clutch plate to slip.
⑤ When necessary, dust-proof, waterproof, rust-proof and anti-smashing measures shall be taken. In particular, all hydraulic distribution valves, control valves and signal cables on the main machine shall be protected by welding iron plates or protective sleeves.
⑥ In order to prevent squeezing, jamming, deformity (wedge effect) or severe wear of structural components and relative moving parts due to rockfall and siltation, gravel and silt in each dead corner and slide rail surface shall be removed in time. For example, the guide sleeves of the vertical cylinder and horizontal cylinder of the steelwork moving platform shall be cleaned and oiled, and the bottom surface of the upper thrust cylinder and the lowest end of the steelwork jacking cylinder shall be kept clean.
⑦ The maintenance shift shall flush the steel wire filter screen and fiber filter screen in the deduster in time, and flush the dust in the air duct of the industrial exhaust fan; regularly open the small cover plate of the hard air duct above the rear of the cutter head, check the thickness of the dust in the air duct, and clean it in time.

(3) Lubrication.

① Each shift shall observe the oil level indicating devices of all components (gearbox, main bearing, main hydraulic oil tank, bolt driller oil tank, invert crane, material crane, etc.). If the oil level is insufficient, the cause shall be found out in time. If there is leakage, it can be supplemented by oil filter.
② The servicing team shall timely remove the dirt on the sliding surface of the guide sliding plate of the steelwork moving platform. All grease shall be applied at the same time.
③ Each shift shall check the working conditions of the outer frame and the main bearing grease pump (normal working pressure is about 16 MPa, and the number of pulses is 9 times). If it is too high, it is necessary to further check whether the grease inlet and outlet of the grease distribution valve are blocked. If they are blocked, they shall be cleaned and reinstalled. The grease shall be replaced in time when it is used up. For example, LZP2 lubrication grease is selected for the outer frame, and light yellow LZW25 lubrication grease is selected for the main bearing.
④ Enter the cutter head from the inner frame every week to check whether there is grease squeezed out at the inner and outer labyrinth seals of the main bearing. Pay attention to observation during downtime maintenance. If there is no overflow, turn on the grease pump to continuously pump grease, and then turn off the machine when the effect is satisfactory.
⑤ Clean the dust accumulated on the front and rear outer frame grippers and rear support guide pillars every quarter, and apply 5000 mm^2/s high viscosity gear oil.

⑥ Fill the main machine with lubrication grease. Add grease to the grease nozzles of each part of the main machine according to the requirements of Table 9.1. When filling grease into the grease nozzle, do not dip grease with hands or other unclean tools, and do not wipe with cotton cloth. The correct method is to open the stop valve to fill oil from the special grease nozzle of the grease pump of the outer frame. Never open the grease barrel cover to take oil, otherwise there will be infinite danger after falling into dirt.

When the TBM is stopped for more than one week due to failure, a small amount of grease shall be applied to the exposed surface of the oil cylinder.

(4) During the fastening and tunneling, all parts of the main machine vibrate greatly, and the bolts of structural components are easy to fall off and break off. The servicing team must check the parts with severe vibration at the front of the main machine, such as the joints between the upper and lower outer frames, the fixing sleeves at both ends of the transmission shaft, the steelwork rotary motor, the main motor, the clutch, the coupling, the thrust cylinder ball head, all steel structures and treads connected by bolts. If looseness is found, it shall be tightened in time according to the specified bolt tightening torque, and shall not be perfunctory, omitted or fastened only by feeling. Check the hydraulic tension bolts of all main machines with the hydraulic tension device according to the specified torque every six months.
(5) Adjustment: In case of belt deviation, excessive wear of clutch friction plate, inconsistency between the oil pressure of each circuit of the hydraulic system and the set working condition, adjustment shall be made in time.

9.2.2 Maintenance of Key Components and Equipment of Main Machine

1) Maintenance of Clutch

(1) Before maintenance, make sure that the hydraulic chamber is free of oil pressure, and disassemble the clutch after the tunneling is completed and the temperature drops. When cleaning the equipment, strictly prevent oil, grease or corrosive detergent from entering the clutch housing to avoid slipping when the clutch is engaged.
(2) Carefully check the hydraulic line for leakage and whether the rotating parts are well protected.
(3) Check the clutch engagement oil pressure to ensure that it is at least 2.8 MPa, the maximum allowable value is 30 MPa, and the maximum allowable time for clutch slipping is 2.5 s.
(4) Check the wear degree of the clutch plate. The maximum allowable wear volume of the 3 friction plates (double-sided) is 15 mm. Measure the clearance between the tappet seat plate and the outer friction plate. If it reaches 1810.5 mm, the inner friction plate must be replaced.

Table 9.1 Lubrication Grease Filling of Main Machine

Item		Number of parts × number of nozzles	Maintenance time	Remarks
Cutter Head Shield	Top shield cylinder	2 × 12	Weekly	Articulation joints at both ends
	Middle shield connection pin	2 × 4	Weekly	Articulation joints at both ends
	Wing shield cylinder	2 × 4	Weekly	Articulation joints at both ends
	Upper clamping cylinder	6 × 6	Weekly	Articulation joints at both ends
	Lower clamping cylinder	8 × 8	Weekly	Articulation joints at both ends
	Lower shield cylinder	4 × 8	Weekly	Articulation joints at both ends
	Steelwork guide wheel	4 × 4	Weekly	
	Material cleaning belt conveyor	1 × 2	Weekly	
Front outer frame	Top of front outer frame	1 × 8	Daily	Sufficient
	Steelwork moving sliding plate	2 × 12	Daily	Sufficient
	Bolt Driller	8 × 16	Daily	Sufficient
	Front outer frame gripper cylinder	16 × 32	Weekly	Articulation joints at both ends
	Guide pillar of gripper cylinder	4 × 8	Weekly	Both ends
	Steelwork mobile platform cylinder	4 × 4	Weekly	
	Steelwork transport trolley chain wheel	2 × 6	Weekly	
	Steelwork moving sliding plate cylinder	2 × 2	Weekly	Both ends
	Thrust cylinder	8 × 16	Weekly	Articulation joints at both ends

(continued)

Table 9.1 (continued)

Item		Number of parts × number of nozzles	Maintenance time	Remarks
Rear rack	Top of rear outer frame	1 × 8	Daily	Sufficient
	Bolt Driller	8 × 16	Daily	Sufficient
	Top of rear support	1 × 4	Daily	Sufficient
	Rear outer frame gripper cylinder	16 × 32	Weekly	Articulation joints at both ends
	Guide pillar of gripper cylinder	4 × 8	Weekly	Both ends
	Torsion cylinder (rear leg)	2 × 4	Weekly	Articulation joints at both ends
	Swing cylinder (rear leg)	2 × 4	Weekly	Articulation joints at both ends
	Connection bridge lifting cylinder	2 × 4	Weekly	Articulation joints at both ends
Rear rack	Rear support gripper cylinder	4 × 8	Weekly	Articulation joints at both ends
	Rear support lifting cylinder	2 × 2	Weekly	Articulation joints at both ends
	Rear support cylinder guide pillar	2 × 2	Weekly	Both ends
	Connection bridge and main machine connection cylinder	2 × 4	Weekly	Articulation joints at both ends
Connection Bridge Nearby	Upper material crane	1 × 4	Weekly	
	Material lifting platform cylinder	1 × 2	Weekly	
	Towing cylinder	2 × 4	Weekly	
	Connection bridge wheels	8 × 8	Weekly	
	Steelwork transport frame	1 × 11	Weekly	
	Lower material crane	1 × 5	Weekly	
	Invert block crane	1 × 11	Weekly	

(5) Check the appearance and free travel of clutch.
(6) When disassembling the clutch assembly again or every 800 h, check whether the lubricating oil in the coupling cavity is sufficient. If possible, replace the oil and seal ring.
(7) When disassembling and assembling the clutch assembly, the bolts should be tightened according to the specified torque and locked reliably.

2) Maintenance of Bolt Driller

(1) Perform functional checks and visual inspection of equipment defects every shift.
(2) Check the flushing box, hose and joint for leakage every shift.
(3) Add lubrication grease 2–3 times per shift.
(4) Every shift, check the tightness of the connection between the drilling machine and the sliding plate and the traveling chain, and check whether the chain mechanism is stuck. If it is stuck, loosen the front and rear fastening bolts, and then re-tighten them in the tightening sequence.
(5) Check the tightness of the important bolts on the curved shape track and rock drill every shift, and tighten them according to the specified torque.
(6) Clean up the gravels falling from the bolt driller every shift, and lubricate the drilling machine walking chain with waste gear oil at the same time; fill up the lubrication oil in the oil–gas lubrication glass oil cup at the right time.
(7) When cleaning the equipment, properly protect the parts (motor, oil pipe, power distribution box, etc.) that may bring danger and cause damage due to improper cleaning; restore the protective sleeve (cover) to the original position in time after maintenance.
(8) Before drilling: Drain the water from the water separator of the air pipe; Tighten all loose nuts and screws of the chain feed mechanism and check the tightness of the chain; Check the condition of the centering device on the chain feed mechanism; Lubricate the injector on the rotary boom and chain feed mechanism; Check the operating conditions of hose protective layer, drill rod lubricator and working light; Check the operation of emergency stop button; Check the drill bit.
(9) During drilling: check the return filter, and replace the filter element when the working temperature warning light is on or at least every 6 months; check the function of the impact timer, the sealing of the flushing head, the operating pressure and the oil leakage; check the operation of the stop and loop automatic devices.
(10) Check the inflation pressure of the accumulator of the rock drill every quarter (0.4 MPa for low pressure and 5 MPa for high pressure).
(11) See Table 9.2 for the pressure adjustment parameters of bolt driller.

3) Maintenance of Invert Crane and Material Crane

(1) Carry out crane function test every shift, and timely arrange repair when the action fails, does not act or the noise is too loud.

Table 9.2 Pressure adjustment parameters of bolt driller

Position	Pressure (MPa)	Position	Pressure (MPa)
Variable displacement pump	22.5	Full-stroke oil line pressure	17.5
Set value of pilot control oil pressure	3.5	Maximum rotary motion oil pressure	18.0
Rotation, impact, flushing and fast forward pilot control oil pressure	3.5	Maximum oil pressure for fast forward	18.0
Feed pilot control oil pressure	1.9–2.1	Anti-seize action pressure	1.3
Automatic return pilot control oil pressure	1.9	Maximum forward feed speed	0.3 m/s
Half-stroke oil line pressure	9.0	Maximum backward feed speed	0.5 m/s

(2) Clean the top of the crane from muck accumulation every shift. The top cover shall be tight to prevent muck falling, jamming or abrasion of the hanging chain; gear oil shall be often applied to the chain; muck accumulation at the idler wheel in the crane running area shall be cleaned.
(3) Observe whether the hydraulic joint, rotary joint or oil pipe leaks every day, check the cleanliness of the filter element, and add hydraulic oil in time when the oil level in the oil tank is low; check the No. 1, No. 2 and No. 3 hook gearboxes every week, and replenish oil and grease in due time.
(4) Check the wheel set wear, assembly wear and chain deformity of the traveling mechanism.

4) Maintenance of Industrial Exhaust Fan

(1) Clean the filter screen of the bulkhead, the inside of the filter chamber and the impeller every day, replace the filter screen as appropriate, and correct the deformity of the bulkhead; flush the dust accumulated in the air duct (including the hard air duct directly above the main machine).
(2) Check whether the nozzle is blocked. If it is blocked, clean it thoroughly and carry out water test for inspection.
(3) Clean the sedimentation tank weekly to remove silt.
(4) Check whether the flexible air pipe joint at the air duct connection along the way falls off or is damaged.
(5) Check whether the fixing bolts of the industrial exhaust fan motor are loose and whether the motor runs smoothly.
(6) Check the insulation value between lines and phases of the motor junction box every quarter; observe whether the rotating seal between the motor rotating shaft and the fan blade is reliable.
(7) Check the lubrication of the fan bearing every week and add grease every month.
(8) Clean the water pump filter element as appropriate.
(9) Check the universal joint every month.

(10) Check and clean the impeller half a year, and check the deformity of the impeller.

5) Maintenance of Steelwork Erector
(1) Every shift, lubricate moving parts and tracks and tighten bolts (mounting bolts of rotary motor are often loose).
(2) Check the working pressure of hydraulic system. Check whether the working pressure is normal: steelwork conveying trolley motor pressure 4.0 MPa, towing cylinder safety pressure 4.0 MPa in large cavity and 10 MPa in small cavity; lifting mechanism safety pressure 15 MPa; drive motor pressure 15.0 MPa.
(3) Whether the hydraulic joints and oil pipes of the hydraulic console are loose or leaking.
(4) Steelwork moving platform is flexible, there is no stagnation, if necessary, clean up the slide surface and lubrication.
(5) Check the bottom of the steelwork support pulley, whether the operation is flexible, whether there is the case of falling off.
6) Maintenance of Grease Pump
(1) Observe whether the working parameters of grease pump are normal values: maximum air pressure 0.9 MPa, minimum 0.3 MPa. Grease quantity is 47 cm^3/min for main bearing grease and 17 cm^3/min for outer frame grease.
(2) The top of the grease pump shall be properly covered to prevent water leakage, sand and other debris.
(3) It is strictly prohibited to bump the grease barrel during handling to prevent it from being dented or pitted, so as not to damage the wiper blade and pressure plate seal and affect the operation of the grease pump.
(4) When other lubricating grease is injected, it is specified that only the diverter valve of the grease pump of the outer frame can be used and the grease can be injected with a grease gun. It is strictly prohibited to directly use dirty hands or unclean tools to extend into the grease barrel.
7) Maintenance of No. 1 Belt Conveyor
(1) Observe the deviation of the belt every shift, and adjust the tension on both sides through the tension cylinders on both sides of the belt bridge at the tail of the belt conveyor. Take moderate tightness and no deviation during the operation of the belt as the standard.
(2) Observe whether the various models of belt rollers under the belt rotate flexibly every shift, whether the noise and radial runout of the drive pulley or driven pulley bearing are obvious, and whether the mounting bracket is fastened through the belt operation; check the wear condition of the roller and scraper, and replace them in time if they are excessively worn.
(3) Pull out the drive pulley drive motor every six months, check the wear degree of the gearbox gear, and replace with new gear oil.
(4) Clean up the dust, deposited concrete muck and silted gravel on both sides or under the belt of the belt bridge every week, especially around the belt carrier roller (the primary factor of rapid wear of carrier roller and belt); focus on checking the accumulation degree of rock ballast under the No. 1 belt of the

inner rack. If there is a large amount of accumulation, clean up it in time and find out the cause of accumulation.

(5) Check and adjust the deformity and damage of the triangular muck scraper of the lower belt and the muck scraper at the belt pulley and its tightness to the lower belt; replace the muck scraper if necessary.

(6) Check the damage of the belt surface, and select the renewal belt vulcanization method, local patching method, drilling suture patching method or filling patching method according to the damage degree.

8) Maintenance of TBM Electrical System

(1) Maintenance Process. Check according to the fault maintenance process.

(2) Key Points of Operation. Determine whether it is an electrical fault or a mechanical hydraulic fault according to the data recording system and the fault monitoring system. If it is a mechanical hydraulic fault, the mechanical hydraulic engineer shall check it. If it is an electrical fault, the following steps shall be followed.

① The fault alarm display is divided into power line fault, insulation fault and control system fault.

② Insulation fault is caused by the reduction of insulation value of a certain line on the power grid. When the insulation value in the power grid is less than 200 Ω, the alarm gives an alarm; when it is less than 50 Ω, the PLC system automatically shuts down the 400 V power grid, causing downtime. Such faults are generally diagnosed by troubleshooting one by one. Each line under the power grid shall be energized one by one, and the insulation value change of the insulation tester shall be monitored at the same time. When a certain line is powered on, the insulation test value of the insulation tester drops and an alarm is given, indicating that the cable or electrical equipment of the line has been damaged. Isolate the line for inspection, and the fault can be eliminated.

③ For the inspection of power line faults, the voltage and current in the circuit shall be measured, calculated and compared to determine whether they meet the reference range. The change of voltage and current is generally caused by the burnout and dampness of electrical equipment or the damage of electrical components, sometimes also caused by the damage and squeezing of cables.

④ During checking the control system fault, use the programmer to connect with the PLC program system to check the subroutine corresponding to the subroutine name displayed by the fault alarm. From the running state of the ladder diagram of the program, it can be seen whether it is an input or output fault.

(a) At the same time, judge whether there is signal input and output from the on and off of the LED of the input and output modules. If there is no input signal, the corresponding temperature, pressure, liquid level, stroke and other sensors or their connecting lines have been damaged.

(b) If there is an output signal, the output line and control valve must be checked when the actuator is not working. By measuring the voltage and current of the control valve and the manual operation of the control valve, it can be determined whether the control valve is damaged or the actuator is faulty.

(c) When an input signal is found but no output signal is found during the inspection, it must be the PLC program or module that is faulty. Sometimes, the program cannot be executed due to crash caused by voltage and current spikes of PLC power supply. In this case, the program in the computer can be restarted; If the program in the PLC is executed without output, it must be caused by the damage of the output module, and the output module must be replaced; If there is a signal input at the input port of the input module, but no signal is running in the program, it means that the input module is faulty and must be replaced.

(d) Loose connectors will also be the cause of such faults during the above circuit inspection.

⑤ After the fault inspection, the Fault Registration Form of TBM Electrical System shall be filled in as data for filing.

⑥ When the equipment is powered on for use, there is a small amount of electric energy loss in the electrical and magnetic circuit systems, which is converted into heat energy, increasing the temperature of the equipment itself and heating the surrounding air or medium. In case of short circuit, the circuit system will generate high heat (the temperature can reach 2000–3000 °C) through huge short circuit current, which will burn the equipment and even cause fire.

⑦ In case of poor contact, the contact resistance of the electrical connector will be increased, which will increase the calorific value. In severe cases, the connector will be melted and the equipment will be burnt.

(3) Maintenance of Main Motor.

① Motor cleaning: Regularly check the dirt and dust of the motor cooling system (dust will hinder heat dissipation, block the inlet and outlet of cooling air, and raise the temperature), and do not damage the motor protective paint during cleaning.

② Maintenance of rolling bearing: detect the bearing temperature and bearing running noise during operation. If the temperature and noise exceed the standard, stop the machine immediately and disassemble and check the bearing. During disassembly and inspection, if the bearing surface is found to be black, rough or worn, the bearing shall be replaced. See Table 9.3 for detection limit values of water-cooled motors.

Table 9.3 Detection limit value of water-cooled motor

Measuring position	Warning value		Shutdown value	
	Heating level			
	B	F	B	F
Stator coil temperature: heating allowed 80 K	130 °C	140 °C	135 °C	150 °C
Stator coil temperature: heating allowed 70 K	120 °C	130 °C	125 °C	140 °C
Rolling bearing temperature	105 °C		115 °C	
Housing vibration speed (rigid base)	5.5 mm/s		11.5 mm/s	

The warning and shutdown values given in the table above are the maximum values determined at an ambient temperature of 40 °C and an altitude of 1000 m. The maximum value may also be exceeded under certain insulation conditions, in which case the manufacturer's consent must be obtained to set a low hysteresis temperature.

In order to protect the motor, it is recommended that the warning temperature and shutdown temperature be 10–20 °C higher than the stagnation temperature respectively.

The standard for housing vibration management is based on the German standard. If the vibration value is between the warning value and the shutdown value, the motor can continue to operate under monitoring.

(4) Identification of Circuit Fault Characteristics.

① Motor: The commutator shall be checked for DC motor. If the commutator has serious carbon deposition, blue-purple spots, solder discoloration, and even burn marks such as solder splashing, metal melting, commutator tilting, and brush breaking, it is caused by overload and short circuit.

Determine whether the motor is deformed. Disassemble and check the corresponding surface paint layer and metal surface of the stator and rotor of the motor. If there are obvious scratches, it indicates deformity. If there are indentations on the stator surface or the rotor surface, but there are no scratches, it indicates no deformity.

In addition, the iron core of the relay may absorb iron chips and cause failure.

② Conductor: It refers to the conductor through which current flows when it is damaged, and the broken end of the metal wire is burnt and fused into a ball or several metal wires are fused together.

(a) Overload conductor: the current that the conductor can pass continuously without overheating, which is called the safe cross-sectional flow of the conductor or the safe current of the conductor. If the current flowing through the conductor exceeds the safe current value, the conductor temperature will exceed the maximum allowable temperature, which is called conductor overload. The damage of the conductor under overload is generally from inside to outside, and the insulation layer of the conductor is thermally discolored. As the temperature rises, the thermal color changes from milky white to yellow, brown and black. Conductor overload burns are caused by the heating of metal wires. The metal wires in the conductors are burned outward. The whole conductor will be burned from the power supply to the electrical equipment or at the short circuit, and the burn degree is basically the same. When the temperature exceeds 230 °C, the tin coating on the surface of the wire will diffuse into the copper, and the surface will become rough and even disappear. This is the main difference from the conductor that is burned by fire. The wire core changes from silvery white to the natural color of copper wire core. In case of short-time or small overload, the appearance of the conductor may be intact, but the insulation layer close to the metal wire becomes black or burnt, the insulation layer separates from the metal wire, and the core is exposed; or there is a phenomenon that the insulation material breaks through the cotton cloth and bulges on the surface of the conductor, and the broken end has the characteristics of power failure. The broken ends of overloaded conductors are fused together, which is more serious than that of normal energized conductors, and the core surface will be oxidized.

(b) The wire burned by fire is burned from outside to inside, and the whole wire has only intermittent burns. For the conductor with slight burn, only the surface of the insulation layer is thermally discolored, the insulation layer is tightly combined with the metal wire, and the metal wire is in the natural color of metal, which belongs to local burn. For the bundled conductors that are burned by fire, each conductor is usually burned within a certain range, but the strands covered inside will be less burned.

③ Electrical System.

(a) Observe the switch board fuse in detail. If the fuse is installed correctly and the capacity meets the requirements, it indicates that the circuit has no fault (unless the conductor is very thin, the circuit is long and the short circuit current is small).

(b) In the case of a short circuit fault, the switch housing was intact, but the interior was smoked and the bottom bakelite seat was burnt. The melt is broken, leaving splash marks of melted metal particles in the fuse. In case of overload, the melt temperature gradually rises and melts slowly, with a long residual part and no splash marks of metal particles.

(c) For example, the fuse link is installed in the fuse, and the narrow part between the fuse links is blown, which is caused by short circuit, because the narrow part has small cross-sectional area, large relative resistance and much heating, and is blown first. There are melting and peeling marks on the wide part of the melting piece, which is caused by serious overload. Because the heat dissipation of the wide part is more difficult than that of the narrow part, the temperature of the melting piece rises slowly.

(d) Pay attention to the color of each electrical connector for signs of change due to overheating.

(e) If there are marks of different combustion degrees on the upper, lower, both sides or both ends of the motor coil, and there is no short circuit mark, this is a phenomenon of external fire spread.

(f) If the combustible near the electrical circuit catches fire due to earthing and short circuit, the fault trace of the residual parts can be generally found, which is manifested as multiple fire points or strip fire points, and the strike is the same as that of the conductor. A magnifying glass can be used to observe the wire core and its cross section. If the cross section changes from thick to thin, the stranded wire has metal melting, adhesion and deformity, which is caused by the strength reduction and sag increase of the conductor after being burned by external fire, stretching first and then breaking. If it is torn off or cut off, the cross section of the conductor is characterized by cutting off the fracture, which is different from the short circuit arc burn mark. Pay attention to whether there is any trace of arc discharge at the place where the conductor is close to or in contact with the metal facility.

(g) Observe the wiring of metal hose or steel pipe. If the inner layer is seriously burned, it is caused by electrical fault; if the outer layer is seriously burned, it is caused by external fire spreading. It is also helpful to judge the direction of fire spreading from the combustion situation of inner and outer layers.

9) Hydraulic System Maintenance of Main Machine

The main process of hydraulic system maintenance in the downtime state of TB880E TBM is shown in Fig. 9.1.

The maintenance of the hydraulic system must be carried out by trained professionals, who must be familiar with the schematic diagram of the hydraulic system of the repaired part before maintenance; The maintenance and replacement of components shall be in strict accordance with the requirements of technical documents; The oil contamination degree shall strictly comply with ISO4406 standard.; When feeding oil, the oil feeding filter equipped with filter elements of corresponding filtering accuracy and grade must be used; Pay attention to pressure relief when overhauling the

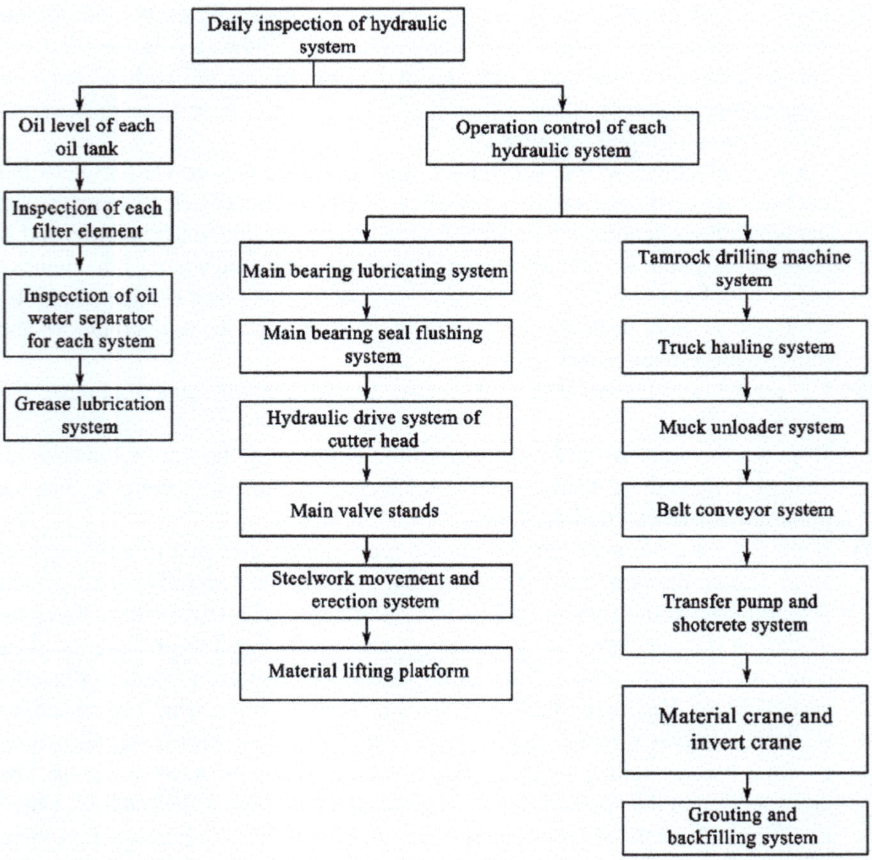

Fig. 9.1 Flow chart of daily maintenance of hydraulic system

hydraulic system to prevent high pressure from hurting people; Oil discharge and recovery shall be in strict accordance with national environmental protection regulations; When feeding oil, try to feed a barrel of oil at one time, and properly seal the remaining oil drum cover to prevent contamination; The oil shall be drained and replaced before cooling; Do not fill new oil into seriously aged or contamination oil; Necessary cleanliness shall be ensured during maintenance of the hydraulic system. Fiberless products or special paper towels shall be used as far as possible to wipe the bare parts of the hydraulic system components. Cotton cloth is prohibited.

(1) Observe the parameters of each pressure gauge during construction and downtime conditions. In case of any abnormal change, the nature shall be judged according to the situation, corresponding measures shall be taken and records shall be made to facilitate comprehensive analysis and treatment of the fault phenomenon.
(2) Corresponding protective measures shall be taken for oil pipes at vibration and easily worn parts, and regular inspection shall be carried out to replace worn oil pipes in time.
(3) Some hydraulic components (such as oil pumps) have a certain service life. The hydraulic components that reach the service life shall be replaced in time. The replaced hydraulic components can only be used after being sent to a professional maintenance organization for inspection or maintenance.
(4) Replace the failed filter element of each hydraulic system in time. The hydraulic filter element of TB880E TBM has the following three detection methods:
① Sensor Monitoring. Most filter elements of TB880E TBM are monitored in this way. When the machine displays the filter element blockage alarm information, the filter element at the alarm position shall be replaced in time.
② Differential Pressure Visual Indicator. This indicator is located on the filter. If the red button of the indicator pops up during cold start, it can be pressed to reset when the operating temperature is reached. If it pops up again during reset, its filter element should be replaced.
③ Filter Testing and Replacement. When the differential pressure alarm indication of the filter is found during operation, or the red button in the upper center of the filter jumps after being pressed several times, it indicates that the filter element has serious contamination and block, and must be replaced in time. If the false alarm is caused by other reasons of the sensor, it is not necessary to take out the filter element. However, if the filter element taken out has been contaminated (such as contaminated with sediment and placed anywhere), it cannot be reused regardless of whether the filter element is new or old.
(5) Pay attention to oil temperature changes during hydraulic system operation. If the oil temperature of the hydraulic system is too high for alarm or even downtime protection, the following inspections shall be carried out:
① Whether the hydraulic system operates under overload for a long time.
② Check whether the oil–water heat exchanger is dusty or whether the water circuit and oil circuit are unblocked.

③ The increase of working temperature indicates the increase of system friction and leakage. Check whether some hydraulic components work abnormally. If any, repair or replace them in time.

(6) Maintenance of Hydraulic Control and Regulating Device.

① Hydraulic cylinder.

(a) Hydraulic cylinders basically do not need maintenance, but must be lubricated for bearings, such as articulation holes or crossheads.

(b) Check for oil leakage during use.

(c) Moving seals are easily worn parts. If the internal or external leakage exceeds the allowable value, it is recommended to replace the seal and also to check the guide ring.

② Hydraulic Control and Regulating Device.

(a) After equipment is started, the operation and leakage of the valve shall be frequently checked according to certain rules.

(b) The oil level and oil temperature must be checked frequently. It is normal for the oil temperature to be between 60 and 70 °C.

(c) The built-in filter shall be cleaned once every 100 h of operation at the beginning, once a month at the same time, and also once every oil change.

(d) The initial oil change time is determined by the operating conditions and the condition of the oil.

(e) If the hydraulic equipment can be kept clean, there is no need to check the control components regularly.

③ See Table 9.4 for start-up and maintenance of hydraulic components after repair.

After the hydraulic components are repaired, the working temperature shall be measured regularly. If the working temperature increases, it indicates that the system friction and oil leakage increase. Inspection and maintenance shall also be recorded at all times.

(a) Precautions include: The oil shall be drained and replaced before cooling; Do not fill new oil into seriously aged or contamination oil; A filter (with a mesh width of 0.06 mm) or other filtering equipment should be used for oil feeding, preferably through the filtering system; The oil is sampled, its particle type, size and quantity are monitored and recorded.

Table 9.4 Startup and maintenance contents after repair

Maintenance content	Period
Check the oil level continuously, check the filter every 2–3 h, and clean it if necessary	At startup
Check the oil level, check the filter and clean it if necessary	Every day of the first week after start-up
Check the oil level, check the filter, and clean it if necessary; check the pressure of the main oil circuit and the control oil circuit intermittently every week	

(b) Maintenance of oil suction filter: Be careful when using the oil suction filter. Check in time after initial operation and clean at least once a week if necessary.
(c) Record all pressure correction values in the maintenance manual. Frequent correction indicates wear of the safety valve.
④ See Table 9.5 for filter maintenance.

(a) The filter is monitored by a differential pressure indicator, which starts to operate when the filter element needs to be replaced or the viscosity of oil is very high under cold start conditions.
(b) If the differential pressure relay is installed and generates a signal (red light) at cold start, check whether the red light goes out when the operating temperature is reached (if not, replace the filter element), which can be used as an indication of temperature rise.
(c) It is recommended to use the positive indication (green light) indicating that the circuit is normal and the negative indication (red light) indicating that the filter element is dirty for monitor and loop at the same time. If the differential pressure indicator acts, the continuous operation of the filter will cause contamination downstream of the filter, so such situation must be prevented.
(d) Unclean hydraulic system will quickly block a new filter element, so it is necessary to replace the filter element several times to stabilize its service life.
⑤ Maintenance of Heat Exchanger.
(a) The heat exchanger is prone to dirt accumulation and shall be cleaned regularly according to specific conditions. If there is slight fouling at either end of the tube, the heat exchange efficiency will be greatly reduced; in addition, if both sides of the pipeline are clogging, the tube will be broken due to the difference in thermal expansion caused by overcooling or overheating.
(b) If obvious pressure drop or temperature rise exceeds the normal value during operation, it indicates that the heat exchanger needs to be cleaned, and the maintenance cycle shall be determined according to this in daily maintenance. Since the difficulty of cleaning increases rapidly with the increase of scale thickness, the cleaning shall not exceed the maintenance cycle.
(c) If the heat exchanger is equipped with an anode protection device, its consumption must be checked regularly and replaced if necessary. It must be noted that the channel cover, shell cover, floating head or hood can be removed only after

Table 9.5 Filter maintenance contents

Every day	6 months since last filter element replacement
Check the dirt monitoring device of the filter element and replace the filter element if necessary; If there is external leakage, replace the O-ring at the leakage position; if there is more than one leakage, check the sealing surface and replace the failed component	Replace the filter element

Table 9.6 Maintenance contents of oil pump

Maintenance content	Period	Maintenance content	Period
Check the oil filter	Every day of the first week of initial start	Oil change	After 500 h of first operation
Check the oil filter	Every week of the first month of initial start	Oil change	Every 2000 h or every year
Check the oil filter	Every month		

the pressure in the heat exchanger is relieved and the tube and shell side fluids are drained during replacement.

⑥ See Table 9.6 for maintenance of oil pump and variable displacement pump.

Depending on the level of contamination and temperature of the fluid, maintenance may require shorter intervals. When changing oil, if the oil used cannot cross the temperature range of winter and summer, in order to maintain its low viscosity at the working temperature, it must be considered whether to use it in winter or summer.

9.3 Maintenance and Repair of Back-Up System

9.3.1 Adding Lubrication Grease

See Table 9.7 for the grease injection table of the back-up system.

9.3.2 Maintenance Process of Back-Up System

The maintenance process of the back-up system is shown in Fig. 9.2.

9.3.3 Maintenance of No. 2 and No. 3 Belt Conveyors

(1) According to the installation instructions of the belt scraper, the scraper is installed on the rubber plate and fixed on the bracket by bolts. When the bolts are screwed, the belt should just contact the full length of the scraper. If the scraper is installed on the torsion fastening device, turn 3 or 4 adjusting nuts to equalize the pressure on each side to ensure full contact between the scraper and the belt. The torsion fastening device is installed on the bracket and can also maintain equal contact pressure when the scraper is worn. In the initial running-in stage, the bolts shall be further adjusted (1 or 2 turns). After 3–4 weeks of installation,

Table 9.7 Grease injection of back-up system

Position		Period	Remarks
Back-up system	Muck unloader	Daily	
	Shotcrete bolt system	Every two days	
	Gantry idler wheels	Weekly	
	Cooling fan	Weekly	
	Concrete tanker crane	Weekly	
	Water pump	Every two weeks	Oil cup
	Air compressor	Weekly	
	Relay fan	Every two weeks	
Belt conveyor	Driven wheel (No. 1)	Daily	
	Guide wheel (No. 1)	Daily	
	Tension cylinder (No. 1)	Weekly	
	Driven wheel (No. 2)	Weekly	
	Tension cylinder (No. 2)	Weekly	Articulation joints at both ends
	Lift cylinder (No. 2)	Weekly	
	Driving wheel (No. 2)	Weekly	
	Driven wheel (No. 3)	Weekly	
	Driving wheel (No. 3)	Weekly	
	Red bearing pedestal	Weekly	

the scraper can be in the ideal position. When the scraper is worn and close to the plastic lateral plate, remove the tubular scraper bracket, loosen the scraper fastening bolts and replace the worn scraper. If a large amount of muck rock is accumulated at the scraper, the scraper shall be cleaned to obtain the maximum belt cleaning capacity. The contact pressure between the belt and the scraper should be checked monthly. If the contact pressure between the belt and the scraper is too low, the cleaning efficiency of the scraper will be affected. On the contrary, if the scraper is installed too tightly, it will lead to belt damage and

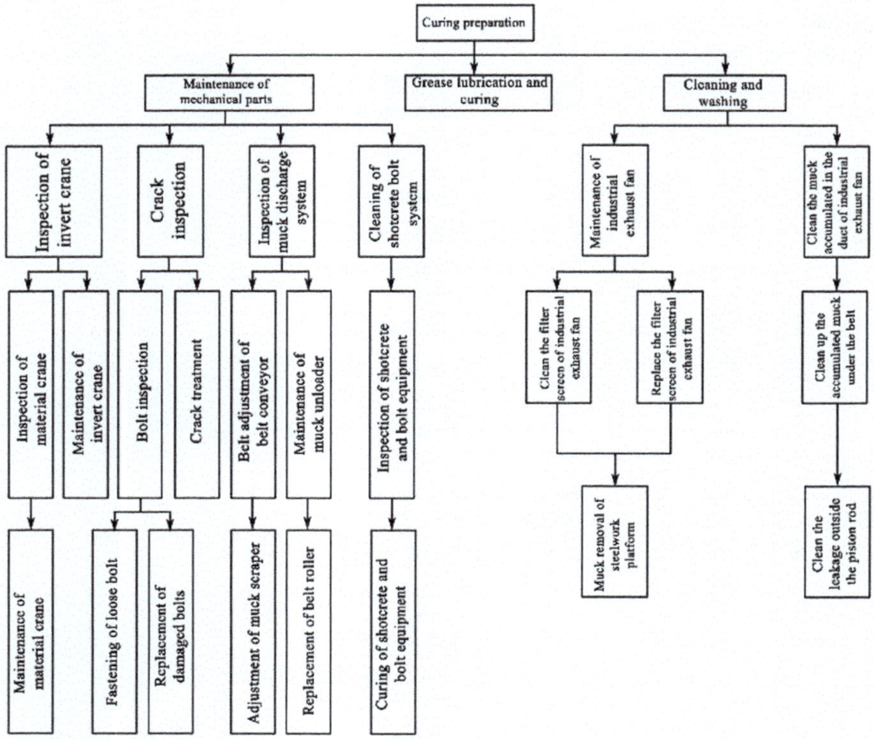

Fig. 9.2 Maintenance flow chart of back-up system

excessive scraper wear. The contact pressure can be determined by observing the surface condition of the belt. When the surface of the returned tape is basically clean, it indicates that the contact pressure is appropriate, otherwise the nut should be adjusted. When the contact pressure is properly adjusted, the lock nut should be tightened.

(2) Belt Damage Inspection and Repair. Possible causes of belt damage are as follows:

① As the slag is mixed between the belt and the pulley, the belt is bulged, causing the belt to be sheared, scratched and penetrated at the slag scraper, resulting in breakthrough damage and breakage. Focus on checking whether the protective covers on both sides of the muck hopper above the driven pulley at the front of No.1 belt are tightly covered and firm, otherwise, replace them in time.

② The breast plates on both sides of the belt bridge become sharp due to wear, and the belt is easy to be cut after deviation, and even cut through transversely, resulting in premature scrapping.

③ When TBM enters the zone with weak surrounding rocks and developed joints and fissures, the belt will form pits or even be chiseled through due to the impact of falling stones.

④ When the belt muck scraper is adjusted too loose, the wedged stones will squeeze or cut the belt; when the belt muck scraper is adjusted too tightly, the wear-resistant alloy sheet of the muck scraper will also cut the belt.

According to the degree of belt damage, the renewal tape vulcanization method, local patching method, drilling suture patching method or filling patching method are selected.

The damage to the belt is mostly in the form of scratches and glue, especially breakthrough scratches. For chiseling and gluing small wounds, each time after cleaning the wound and blowing dry, apply chlorine with adhesive and dry with hot air, repeatedly until the pit is filled. For breakthrough defects, draw a wide strip of curved shape at both ends in the surrounding area of the breakthrough part (the longitudinal length exceeds the scar by 6 cm, and the transverse width exceeds the scar by 8–15 cm depending on the damage), cut off the rubber above the fiber layer of the belt with a paper cutter, and polish it flat with a polishing machine; If the length of the scar is long, the scar shall be stitched and filled with glue first, and then a piece of rubber with the same shape, the same thickness and an arc edge shall be cut for adhesion (where there is repair space, the back of the belt shall be treated as far as possible). The key to belt repair is that the wound surface should be cleaned up without residual oil, foreign substance and moisture; the timing of adhesive application and drying and belt bonding should be appropriate. After bonding the belt, use a glue hammer or hammer to flatten it and fill the seam with glue for drying. If the surface rubber of the belt is thinned due to long-term use, a piece of reinforced air duct cloth of the same size can be cut and directly pasted.

If the belt vulcanization process is used to repair the belt damage, a complete set of tools, adhesive materials (raw glue, thick belt, acetone or xylene, chemical fiber towel, etc.), small tools (including paper cutter, sharp drag bit, sharp cutter oilstone, scissors, steel plate ruler and needle-nose pliers, combination pliers, polishing machine, hair dryer, small flashlight, etc.) shall be prepared for vulcanizing the belt.

Select a suitable position to place the vulcanizing machine, and there should be a gap of 3–4 m on both sides of the vulcanizing machine along the running direction of the belt; The height shall at least ensure that there is a clearance of 30 cm on the top surface of the belt fastening binder plate, so as to avoid interference or collision when installing the fastening bolts of the binder plate; Remove the belt breast plate and carrier roller on the belt bridge where the vulcanizer is placed; Thoroughly clean the surrounding environment without floating dust and oil stains to ensure that the surface of the belt is clean.

Fabrication of belt joint: The total length of overlapping and buckling at both ends of the belt joint shall be $0.8B$ (B is the belt width). The front and back sides of the tape must be distinguished, and the belt shall be peeled into corresponding steps along the cord layer of the lining. The length of each step shall be evenly divided according to the number of layers of the cord layer of the belt lining. The width of each cutting of rubber should be 50 mm, so that the rubber can be torn off from the cord layer in strips. The slope of the cutting at the starting position of the belt shall be consistent with the slope of the vulcanizer. In order to ensure the buckling quality, the rubber on the surface of the lining cord layer shall be completely removed during peeling. The surface adhesive can be carefully polished with a copper wire polishing wheel, and then the surface shall be carefully cleaned with acetone or xylene. After the liquid volatilizes or is dried by a blower, the adhesive tape shall be spread flat on the vulcanizer plate. Evenly brush a layer of green glue black slurry on the surface of the peeled step to make it fully absorbed before brushing again; After slightly drying, lay a layer of raw rubber closely and carefully buckle and butt joint to ensure that the joints are aligned and tightly closed. There shall be no bubbles in the bonding layer. Carefully cut the raw rubber at the bulge with a blade, and then compact it tightly without leaving any gap.

Fasten the binder plates of the vulcanizing machine in turn, tighten the binder plate bolts evenly, correctly connect the temperature sensor and the heating device, press with a hydraulic press until the specified instrument reading is reached, turn on the heating power supply at the same time, and carry out the automatic vulcanization and curing procedure; when the specified vulcanization temperature is reached, start timing and heat preservation for about 40 min. During this process, pay attention to observe the water pressure at any time. If the pressure is insufficient, pressurize at any time to keep the binder plate closely attached. After the holding time is reached, the cooling stage of the vulcanizing machine begins, and the cooling should usually be slow. When the temperature is lower than 60 °C, the cooling process ends. The vulcanizing machine can be removed, and the joint quality of the vulcanizing belt can be checked. After confirmation, the original state of the belt conveyor can be restored, and the belt tension can be adjusted to resume normal operation.

(3) Belt Pulley Repair. The failure parts of the belt pulley are mostly uneven wear of the bearing holes at both ends and serious wear of the rotating shaft journal. The repair method is as follows: Carry out weld overlay on the inner surface of the bearing hole, and re-grind to the required size after alignment; Alternatively, apply metal repair adhesive to the inner hole, location the bearing with the fixing effect of adhesive, and directly reassemble the bearing and shaft; Repair the worn parts of the rotary shaft by weld overlay and turning; The bearings and shaft seals at the shaft ends should be replaced at the same time as each pulley repair.

(4) Drive Hydraulic Motor or Motor Output Spline Shaft Wear Repair. In order to strive for tunneling time, such wear shall be repaired by directly installing spline sleeves or couplings and applying adhesive to the gap between the shaft and the sleeve. The key to the bonding process is to thoroughly clean the contact face of the dual parts, i.e. first remove rust with abrasive cloth or wire brush to show metallic luster, then repeatedly scrub with acetone or carbon tetrachloride, apply strong metal wear-resistant adhesive after drying, and ensure that it is fixed in the correct engagement position without skew. Carefully remove excess glue when the binder is initially set. It can be put into use immediately after completely reaching the bonding strength, and the curing time of the adhesive depends on the type of adhesive.

9.3.4 Material Truck Towing System Maintenance

(1) Check the function of each shift (whether the action is normal, whether the towing is weak, etc.).
(2) Clean up the chain rail and chain groove muck.
(3) Check whether the drive slewing motor sprocket and guide small sprocket are worn out, whether the towing chain is unhooked, deformed or stretched; check the chain connecting block and chain groove wear degree. Tension the chain with a sprocket tensioning mechanism as appropriate, or cut, weld and replace worn parts.
(4) Check the operating pressure of the towing pump station and the heating of the pump body every shift. If the temperature is too high, it shall be stopped immediately for cooling and the operator shall be informed. After use, it shall be stopped in time.

Most of the common faults of the towing system are chain stretching and deformity, sprocket and chain groove wear and misalignment. For the fault of chain deformity and elongation, measures can be taken to remove the obviously deformed chains by gas cutting, then weld the broken chains and adjust the tension of the towing device at the same time; Observe the worn parts of the chain slot every day, and trim the worn and deformed parts at the bottom and top of the slot by electric welding if necessary.; When the sprocket and bracket are seriously worn, only new methods can be taken.

9.3.5 Maintenance of Muck Unloader

(1) Perform functional checks per shift.
(2) Add lubrication grease.
(3) Flush the accumulated stones at the muck bucket, muck bucket door and muck bucket breast plate gap, pulley, etc. with water.
(4) Observe the surface wear and jamming of the flap, and repair welding in time in serious cases.
(5) Check the fitness between the muck scraper and the belt and adjust it in time.
(6) Check whether the connecting bolts of the traveling motor are loose or broken; whether the traveling idler wheel slips. If slipping occurs, the idler wheel tread shall be intermittently weld overlaid with convex edges along the circumferential direction to increase the traveling adhesion.
(7) Clean up the dust on the movable cable trough rail, and repair the wear and deformity of the cable trough guardrail and idler wheel track in time.
(8) Damage inspection and repair of water pipes and movable chains.
(9) After the back-up equipment enters the curve section, each shift shall check whether the running cable slides flexibly in the chute and whether it is stuck. If not, it shall be adjusted and corrected in time.

9.3.6 Maintenance of Air Compressor

(1) Check the coolant liquid level in each shift, and replenish it in time if it is insufficient.
(2) Check whether the safety valve is normal every week.
(3) Replace the coolant filter after the air compressor runs for 150 h for the first time, replace the coolant filter every 2000 h thereafter, and replace the air filter every year or when the indicator light flashes.
(4) Check the connecting bolts of each part.
(5) Check the differential pressure when the separator element indicator light flashes. When the compressor is running with load, press the pipeline/oil tank selection button. If the differential pressure is equal to zero or greater than 100 kPa, the components need to be replaced. When replacing the components, remove the ventilation pipe, loosen the connection between the ventilation pipe and the separator tank cover, and then remove the ventilation pipe; Remove all pipes from the separator tank cover; Remove the positioning screw fixing the box cover and remove the box cover; Remove the used old components and put them into a seal bag for proper disposal; Clean the sealing surfaces of the box and the box cover; Install replacement parts. Reinstall the new components in reverse order; start the compressor and check for leaks.

In this process, it should be noted that sealant should not be applied to the surface of the separator box or box cover; at the same time, do not remove the staple from the anti-static sealing ring of the separator element, because it plays a earthing role and is used to remove any possible static electricity.

(6) Purge the cooler and its surroundings with high-pressure air regularly every month.
(7) Quarterly check all hoses for breakage and aging, and replace all hoses after 2 years of use.
(8) Calibrate the pressure sensor every half a year and the safety valve every year; replace the separator element and coolant filter at the same time of replacing the coolant every 2 years. The coolant replacement method is as follows:

① Coolant liquid level detection. Frequently observe through the glass observation port installed on the side of the separator box, the liquid level should usually be in the middle position when the machine is running with load.

② Adding coolant. After running for at least 40 s before the compressor is shut down, add coolant to make the liquid level reach the middle position of the glass observation port.

③ Replacing the coolant filter. Loosen the filter with a suitable tool; Remove the filter from the housing and put it into the sealing bag for proper disposal.; Clean the assembly surface of the housing; Remove the new filter from the protective package; Install the used filter; Apply a small amount of lubricant to the filter gasket; Screw in the filter until the sealing gasket contacts the housing, and then screw in by hand for half a turn; Start the compressor and check for leaks.

④ Coolant replacement. Immediately after the compressor stops running, discharge the coolant, because at this time the liquid is the easiest to discharge, and the contaminants are in suspension. Put a suitable container under the discharge plug; filter, separator element; install the plug and refill the coolant. Remove the plug and drain the coolant completely; replace the coolant and start the compressor to check for leaks; check the coolant liquid level and replenish it if needed. If the compressor operates under poor conditions, the coolant discharge interval should be shortened. It should be noted during the above operations that oil cannot be mixed with coolant.

(9) Inspection steps of high-pressure air temperature switch.

Routine inspection of temperature switches (1TAS and 2TAS) is as follows:

① Air cooler: Turn off the circuit breaker of the fan motor to stop the cooling fan.

② Water cooler:Cut off the cooling water. The machine should be shut down at 109 °C (standard application) or 100 °C (TUV application). It should be noted that the discharge temperature cannot be out of range under any conditions. If the machine cannot be stopped when the fault occurs, please check immediately.

The temperature should be recorded when the machine is stopped for future comparison at the end.

(10) High speed bearing temperature switch. When the air end bearing temperature reaches 109 °C (standard application) or 100 °C (TUV application), it shall be shut down.

(11) When adding grease to the motor bearing, clean the area around the air inlet and outlet, remove the cock, and use a grease nozzle to add grease to the two grease fittings marked as port B and port L respectively. Add 40 g to port B and port L of the drive end bearing, 20 g to port B and 40 g to port L of the non-drive end bearing. Fill once every 4000 h at port B and once every 2000 h at port L. After filling, first screw on the air inlet plug, and then screw on the exhaust port plug after the equipment has been running for 10 min.

9.3.7 *Maintenance of Shotcrete Bolt Equipment*

(1) See Table 9.8 for the maintenance contents of the transfer pump hydraulic pump station.

(2) See Table 9.9 for the maintenance contents of the transfer pump mechanism.

(3) Maintenance of shotcrete manipulator.

① Add lubrication grease to the automatic grouting machine, longitudinal mobile gantry, steelwork, recovery plate, longitudinal mobile guardrail, longitudinal drive device and sprocket, bearing, longitudinal mobile guide wheel and guide rail and laterally move related devices every day.

Table 9.8 Maintenance contents of transfer pump station

Maintenance Content	Period		Other intervals	Performed by
Check for safety devices and contamination	Daily			
Check the fuel level of fuel tank	Daily			
Check the level of contamination of the vacuum gauge filter	Daily			
Check the seal between hydraulic joints	Daily			
Check the tightness of hydraulic joints and bolts	Daily			
Check the high pressure of pressure gauge	Daily			
Record the conversion pressure between the accumulator and the oil pump	Daily			
Observe whether the cable is damaged	Daily			
Wipe off the condensate on the surface of the hydraulic oil tank	Initial operation for 50 h	Every 500 h of operation	Observed	
Check the oil water cooler dirty condition		Every 500 h of operation		
Observe dirt in all screens, sieves and orifice covers, clean or replace if necessary		Every 2000 h of operation	Q1	
Check the oil pump star filter		Every 2000 h of operation	Half a year	
Check whether the float switch in the oil tank is normal		Every 2000 h of operation	Half a year	

(continued)

Table 9.8 (continued)

Maintenance Content	Period		Other intervals	Performed by
Measure the oil pump parameters and compare them with the theoretical values		Every 2000 h of operation	Half a year	
Oil change			Annually	
Empty the oil sump of hydraulic oil and clean it				
Check the function of overload valve			Annually	
Check the function of overload valve accumulator			Annually	
Check the function of starting valve			Annually	
Check the ventilation equipment of motor				
Lubricate motor lubrication points				
Check the bearing and sealing device of the motor				

② Inspection and maintenance of oil level, filter and oil pipe leakage of manipulator pump station.

③ The manipulator operator shall:

(a) Operate in strict accordance with the operating procedures to prevent pipe blockage: when starting, turn on the air first, then feed the material, and supply the accelerator after the concrete is sprayed from the nozzle; when stopping, turn off the accelerator metering pump first, then stop feeding, and stop the air supply after the small amount of concrete and accelerator remaining in the nozzle is completely blown away.

(b) The operator of the concrete transfer pump shall control the pumping frequency and pressure. Generally, it is set to 12 r/min, the pressure is kept between 6 and 10 MPa (cylinder inlet pressure), and the pumping speed is 6 m^3/h, so that the work is relatively stable and does not affect the construction progress. When TBM passes through unfavorable geological sections and requires advanced shotcrete and pressure grouting, the pumping pressure shall be increased to 10 –12 MPa due to the long conveying pipeline.

Table 9.9 Maintenance contents of transfer pump mechanism

Maintenance content	Period		Other intervals	Performed by
Lubricate all lubrication points	Daily		Every 4 h	
Check for safety devices and contamination	Daily			
Check the water volume of the water tank and add it if necessary	Daily			
Check whether the cable is damaged	Daily			
Check the seal of conversion cylinder	Daily			
Check the lubrication of conversion cylinder bearings	Daily			
Check the leakage of the connecting flange between the oil pump and the conversion cylinder	Daily			
Check the oil pan and chip pan of the hydraulic system	Daily			
Check the seal of hydraulic cylinder			Every month	
Check the bearing seal of hydraulic cylinder			Every month	
Check the centralized lubricating machine	Initial operation for 50 h	Every 1500 h of operation	Quarterly	
Check the main control board	Initial operation for 50 h	Every 1500 h of operation	Quarterly	
Check screw unit and line protection on conveying piston		Every 250 h of operation		
Check the position and function of the electronic contact switch		Every 500 h of operation		

(continued)

Table 9.9 (continued)

Maintenance content	Period		Other intervals	Performed by
Check the torque of S-type conveying pipe bolts		Every 1000 h of operation	Quarterly	
Check the mounting bolts of the hydraulic cylinder				
Checking the wear of double orifice plates and wear rings		Every 1000 h of operation		
Check the welds of conveying pipe and double orifice plate		Every 2000 h of operation	Quarterly	
Check cycle time		Every 2000 h of operation	Quarterly	
Check the conveying piston for wear		Every 2000 h of operation	Quarterly	
Check the main control board		Every 2000 h of operation	Quarterly	
Check the pull rod installation		Every 2000 h of operation	Quarterly	
Check the connecting bolts of the conveying pipeline			Quarterly	
Check the wall thickness of S-shaped conveying pipe		Every 2000 h of operation	Half a year	
Check the conveying cylinder for wear		Every 2000 h of operation	Half a year	
Check the wear of conveying cylinder bearing		Every 4000 h of operation	Annually	
Check the bearing at the pressure transfer port		Every 4000 h of operation	Annually	
Check S-tube bearing		Every 4000 h of operation	Annually	
Check the conveying pipeline for wear			Annually	

(continued)

Table 9.9 (continued)

Maintenance content	Period		Other intervals	Performed by
Check the sealing of conveying pipeline (water pressure detection)			Annually	

(c) In order to prevent concrete blockage, the quality of concrete shall be strictly controlled, and unqualified concrete shall not be used.; Before concrete transportation, the pipeline shall be blown clean with high-pressure air to ensure that there is no residual gravel and water in the pipe. If there is water residue, the concrete will be diluted by water during transportation, so that gravel, sand and cement are separated from each other, and free gravel is easy to accumulate, resulting in blockage; After the shotcreting and anchoring work is completed, the pipeline shall be carefully cleaned. First, the pipeline shall be flushing repeatedly with water, and then flushing with water driven by high-pressure air to ensure that there is no residue in the pipeline.

9.3.8 *Maintenance of Emergency Generator*

(1) Before each startup, check the oil and fuel levels and the coolant liquid level, and replenish them in time if insufficient.
(2) Check the electrolyte position and voltage of the storage battery to ensure that it is fully charged at any time.
(3) Check whether the air filter is blocked.
(4) Regularly clean the dust on the surface of the engine (including the intercooler).

9.3.9 *Maintenance of Air Conditioner System*

(1) During normal operation, observe the operating parameters of the air conditioner each shift (working pressure and temperature, and the cooling water flow shall be 20 m^3/h per unit).
(2) Check all pipelines for leakage (refrigerant, compressor oil, etc.), and open the evaporator cover plate to check whether the copper pipe is contaminated.
(3) Check screws and circuit connectors.
(4) Replace the compressor oil and oil filter every 500 h.
(5) Before the equipment is sealed, start the air conditioner equipment, take back the Freon refrigerant into the system, disconnect the condenser pipeline, and immediately seal the exit channel of the air conditioner cooling system tightly and properly.

(6) Generally, the compressor oil shall be replaced every 5000 h. If it has been disassembled halfway, it shall be replaced every 3000 h. The oil replacement method is as follows: After shutdown, close the inlet valve and pressure valve of the compressor, depressurize the compressor through the air stop valve, open the oil filling (draining) valve of the compressor for oil drain, remove the mounting cover of the crankcase, remove and clean the oil filter and clean the crankcase groove. The filter block in the filter drier must be replaced at the same time as each oil change. The compressor of TB880E TBM air conditioner system uses KES100 oil with a total capacity of 17 L. After the compressor is shut down, connect the vacuum pump and the oil filling pipe to the air release valve and the oil filling valve respectively, and start the vacuum pump to pump out the air in the oil chamber and suck in the new compressor oil.

9.3.10 Maintenance of Dumper

(1) Perform weekly functional checks. Including the inspection of mechanical mechanisms such as muck car fixing and positioning devices, rotary drives and chains, the inspection of indicator lights and limit switches of electrical control parts, the inspection of hydraulic system pipeline leakage, pump station pressure, oil level, filter, etc.
(2) Clean the accumulated rocks of the dump mechanism and the muck car positioning mechanism at any time every day to avoid interference and wear of these accumulated rocks.
(3) Add grease to all parts to be lubricated once a week, or add grease to some lubricating points in sequence every day.
(4) Check and adjust the position and sensitivity of all sensors of the dumper when necessary.
(5) Check the bolt connection and the wear condition of chain and pulley sprocket.

9.4 Disassembling Preparation and Personnel Allocation

9.4.1 Disassembling Preparation

According to the actual situation of the construction site, the TBM will be disassembled in the tunnel. Due to the complex structure of the TBM and the large tonnage of the main machine parts, the following preparations must be made during disassembly:

(1) Prepare the TBM removal site (construction tunnel);
(2) Organize the disassembly personnel;
(3) Adjust and implement the TBM disassembly scheme according to the actual situation on site;

(4) Make a reasonable plan for disassembly and transportation of materials and accessories;
(5) Make TBM parts identification and relevant records;
(6) Crane mobilization.

9.4.2 Personnel Allocation

TBM is dismantled in the tunnel, with complex works, high technical requirements and long cycle. Two-shift operation is adopted. See Table 9.10 for personnel arrangement during dismantling.

Table 9.10 Disassembly personnel schedule

Teams	Assembly personnel allocation	Remarks
Mechanical team	Technical personnel: 4	
	Technical workers: 20	
Hydraulic team	Technical personnel: 4	
	Technical workers: 8	
Electrical Team	Technical personnel: 4	
	Technical workers: 6	
Supply team	Technical personnel: 2	Including transport truck driver
	Technical workers: 8	
Dispatching	4	
Safety officer	4	
Personnel consolidation	64	

9.5 Construction of Disassembly Tunnel

9.5.1 Function of Disassembly Tunnel

TBM tunneling can only move forward but not backward. For tunnels with short one-way construction length, after TBM tunneling is completed, TBM can be pulling or stepping in and out of the tunnel for disassembly outside the tunnel; For long tunnels, when two TBMs are used for construction in opposite directions, dismantling tunnels must be set up in sections with good surrounding rock conditions near the breakthrough surface, hoisting and transportation equipment must be installed and arranged in the tunnel, and TBMs must be disassembled and transported out of the tunnel in batches. The bridge crane and its running rails are installed in the disassembly tunnel. The crane trolley is used to move left and right and the crane is used to move forward and backward. The disassembled main machine and back-up large pieces are lifted, displaced and loaded on the transport flat car for outward transportation. Therefore, the disassembly tunnel is the place for disassembly, hoisting and outward transportation after TBM tunneling, and it is an indispensable part of the construction of two TBMs facing each other. The disassembly tunnel must have sufficient space and structural strength to install the bridge crane. The crane running rail foundation must have reliable bearing capacity to ensure the safe and stable hoisting of TBM large pieces.

9.5.2 Structure and Section Size of Dismantling Tunnel

The length and section size of the disassembly tunnel can be determined according to the TBM model, the technical requirements for TBM disassembly and the technical parameters of the crane used. The length and section of the disassembly tunnel shall be determined in accordance with the principle of not only meeting the space size requirements that must be occupied by large parts such as cutter head, main bearing and inner and outer racks of TBM during lifting, disassembly and assembly operations, but also minimizing the construction cost of TBM disassembly tunnel.

9.5.3 Construction of Disassembly Tunnel

The section of the disassembly tunnel is large, and bridge cranes and other equipment shall be installed in the tunnel. The construction is complicated, and the construction period is generally about 5 months, which must be completed before the arrival of TBM. The construction method of TBM disassembly tunnel depends on specific conditions. If Qinling Tunnel is 2 parallel single-track tunnels, the parallel adit of Line II has been constructed in advance. Therefore, the parallel adit of Line II enters the dismantling tunnel through the transverse passage for construction in advance; TBM is adopted for the entrance section of Taohuapu No. 1 Tunnel, drill and blast is adopted for the exit section, and the TBM dismantling tunnel is constructed by the exit construction team and completed before the arrival of TBM.

The section of the dismantling tunnel is generally mushroom-shaped and large. The lifting point for crane installation shall be embedded in the arch, and the joist of the crane running rail shall be constructed on the arch foot platform. The requirements for the size of the excavation contour line are very strict, especially the surrounding rock at the arch foot platform shall not be damaged, so as not to affect the bearing capacity of the joist foundation. Therefore, in drilling and blasting, the spacing and charging density of peripheral holes shall be strictly controlled according to the design to reduce over excavation and under excavation. For arch foot platform, the platform forms a right angle against the vertical wall, making the blasting and forming difficult. Therefore, the technique of reserved smooth blasting layer for secondary blasting and excavation expansion and forming can be used to minimize the disturbance and damage to the joist foundation parts.

To ensure sufficient structural strength and stability of the disassembling cavern, the disassembling cavern should be reinforced by composite lining using rock bolts and shotcreting. Additional arched steelwork and embedded parts are added to the initial support within a certain range of the lifting point of the crane. During construction, the embedded parts and steelwork are welded into a whole part to ensure the safety and reliability of crane during installation and operation. To improve the stability of the joist area, the rock bolt spacing of the side wall within 2 m above the joist is determined as 1.0 m $\times$ 1.0 m, and the length is increased to 4.0 m; the rock bolt spacing of the side wall within 3 m below the joist is increased to 0.5 m $\times$ 0.5 m, and the length is also increased to 4.0 m; two additional rock bolts are installed on the side wall of the joist itself to connect with the side wall.

To enhance the stability of the side wall of the disassembling cavern, the rock bolts in the straight side wall area are exposed 35–40 cm so as to extend into the second lining of the side wall and connect the side wall with the surrounding rocks. The second lining, the arch and the side wall of the disassembling cavern are constructed separately. For the construction of arch lining, scaffolding and combined steel formwork are used to form the arch first by pouring on the arch foot platform; the side wall lining is poured symmetrically in sections along the longitudinal direction after the bottom plate is prepared well. In addition, a 35 cm thick plug wall is poured with C20 concrete at the end face of the intersection between the disassembling cavern and the TBM circular section.

9.5.4 Construction of Traveling Track Joist of Crane

The joist is the foundation of the crane traveling track, so it is of significance to ensure the joist construction quality. During construction, the joist is required to be straight, with flat top surface, so that the traveling track is evenly stressed, thus ensuring a smooth traveling of crane.

Before formwork erection and pouring, the rocks surrounding the arch foot platform should be checked first to see whether they are intact and stable, and if any obvious joints and broken rock blocks are found in the platform against the straight side wall, all of them should be removed. The joist rebar should be placed in strict accordance with the design size and position, and the concrete should be tamped and compacted when poured. The two rows of rebars on the joist extend at least 1.5 m into the side wall lining to ensure the connection between the joist and the side wall lining. After the joist is constructed, the gap between the joist and the side wall should be backfilled with C20 concrete.

9.6 Management of Technical Documents About Disassembly

9.6.1 Identification and Photographing of Mechanical, Electrical and Hydraulic Systems Before Disassembly

Identifications, photographs and registration information of mechanical, hydraulic and electrical systems are the most valuable reference information for reassembly, and they must be given great attention.

(1) The mechanical parts and components of the TBM differ greatly from each other, so different identification schemes are adopted according to their different characteristics. However, they share one thing that two ways of identification are adopted, namely identification with painted words and hanging nameplate except for the back-up frame structural components.
(2) Based on the technical documents of the TBM, the original identifications are fully reserved, and the connecting cables and electrical equipment are coded and registered according to the actual situation and the customary identification method, which are simple to understand and easy to search.
(3) TBM has numerous and complex hydraulic system pipelines, which are distributed in various parts of the machine. In some positions, oil pipes of the same specification are in parallel; pump station is mostly located in the back-up No. 1 gantry, while some are arranged in the appropriate location according to the actual needs; actuators are naturally distributed in the corresponding positions. Therefore, before the disassembly of the TBM, the identification scheme for the pump station, oil pipelines, valves and actuators should be developed according to the pipe laying diagram and system diagram for the hydraulic system in the attached technical documents. For systems without pipe laying diagram in the technical documents, the pipe laying diagram should be drawn with reference to the hydraulic system diagram and the actual pipeline layout. To

simplify the identification code, the diagram number of the pipe laying diagram is replaced by English letters, and the identification work should be carried out by personnel familiar with the TBM hydraulic system, who will fill in the identification registration form carefully and conscientiously. The identification registration form for pump station, valve, cylinder and motor should include the names of the parts, specifications/models, identification codes, and installation positions; identification registration form for oil pipeline should include the diagram name, diagram number, diagram code, oil pipeline identification code, pipe diameter, model, length, starting and ending positions and route.

(4) To facilitate view and reference during assembly, before TBM disassembly, the positions of mechanical, electrical, hydraulic systems, their interrelationships and pipeline layout should be photographed at appropriate angles to accurately show their positional relationship. If necessary, 3–4 photos can be taken for a part. The photo process should be well documented. Photo registration form should include the photo number, the name of the photographing part or component, identification code or nameplate, and position.

9.6.2 Archiving of Technical Data About Disassembly

The technical data about disassembly include the following four parts:

(1) Disassembly records made by each shift.
(2) Transportation and loading records.
(3) Packing records.
(4) Memorandum of special meeting during disassembly.

9.6.3 Statistical Data About Disassembly

Statistical data about disassembly should include the following parts:

(1) TBM post-disassembly maintenance plan.
(2) Statistics of defective parts during TBM disassembly and repair.
(3) Statistics of various installed bolts, inventory statistics and order list during TBM disassembly.
(4) Summary of information about TBM hydraulic system (including identification registration form for hydraulic system, photo registration form, photo information, pipe laying diagram and packing registration form for TBM oil pipeline).
(5) Summary of information about TBM electrical system (including identification records for electrical system, photo catalogue, disassembly registration form and packing registration form).
(6) Packing statistics of TBM spare parts.
(7) Identification and sorting of TBM bare parts.

(8) TBM outsourced maintenance and machining plan.
(9) Packing statistics of TBM spare parts by part.
(10) Pending issues about TBM.

9.7 Disassembly in the Tunnel

For Qinling tunnel, two TBMs were used to tunnel from the entrance and exit in opposite directions. Since the middle section of the tunnel was constructed by drill and blast, the two TBMs could only be disassembled in their respective disassembly caverns after breakthrough. In this section, we will take the exit of Qinling Line I Tunnel and Taohuapu Tunnel #1 as examples to specify the mechanical equipment, schemes and precautions to be taken during the disassembly of TBM in the caverns.

According to the preliminary TBM disassembly scheme provided by Wirth, and the TBM disassembly scheme proposed by the Construction Contractor based on the TBM assembly experience, the TBM main machine was disassembled inside the cavern, and its back-up was towed out of the cavern by locomotives in several times, and then disassembled outside the cavern using the gantry crane used during TBM assembly. Length of the disassembling cavern was determined to be 55 m based on the length of the main machine, the parking length of the main machine components, the traveling length of the crane used for disassembly and the safety distance. The contour surface of the disassembling cavern is determined to be a mushroom-shaped cross section according to the size of the crane and the space required for the disassembly of the TBM, with a maximum net width of 15.04 m and a maximum net height of 17.66 m.

9.7.1 Lifting and Transportation Equipment

1) Bridge crane inside the cavern

Technical performance parameters of 150t bridge crane are shown in Table 9.11.

2) Lifting and transportation equipment outside the cavern
(1) Means of transportation

Table 9.11 Technical performance parameters of 150t bridge crane

Loading mass (t)	2 × 75	Operation mode of crane	Self-propelled
Span (m)	12.8	Trolley type	Mobile, with a movement distance of 8 m
Lifting height (m)	12	Operation mode of bridge crane	Ground cord control
Structure	Convex		

Table 9.12 Allocations of lifting and transportation equipment outside the cavern

S/N	Name	Qty	Specification (t)	Remarks
1	Gantry crane	1 set	150	2 × 75t + 15t
2	Truck crane	1 set	20	For assembling the gantry crane
3	Transportation vehicle	2	5	
4	Locomotive	2	45	
5	Platform gantry	1		For modification of back-up platform gantry

Rail transportation is used as the means of transportation for the construction of Qinling Line I Tunnel, and the main steel structural components disassembled from the TBM main machine are transported by Line I. Trackless transportation is used as the means of transportation for the construction of Qinling Line B Tunnel. The small structural components and related accessories from the main machine and various connecting parts are transported out of the cavern by Line II trucks without affecting the construction of Line B.

The main machine parts of TBM used for Taohuapu Tunnel #1 are loaded by the back-up platform vehicles, and are towed out of the cavern by the locomotives in turn for disassembly. According to the site conditions of Taohuapu, parts are disassembled and transported simultaneously, and containers are rearranged during the implementation, and the transportation sequence is arranged according to the disassembly sequence.

(2) Transportation equipment

In the case of rail transportation for the Qinling Line I Tunnel, the TBM back-up platform gantries are used as flatbed trucks for transporting large parts and components, which are additionally installed with double traction devices, namely two 45t locomotives for traction. For the trackless transportation used for the Qinling Line II Tunnel, two ZQ140 transportation vehicles are used, as large transportation vehicles cannot move due to limited conditions of the construction site.

3) Allocations of lifting and transportation equipment outside the cavern

Allocations of lifting and transportation equipment outside the cavern are shown in Table 9.12.

9.7.2 Disassembly Scheme

1) Preparations

(1) Material preparation. The materials to be prepared include the back-up hooks for transportation, the wheels used on the connection bridge, the

lower shield, the frame, the rear support stepping mechanism bolts, various lifting lugs and hooks, etc.

(2) Identification preparation. The identifications to be prepared include hydraulic and electrical identifications.
(3) Personnel arrangement (crane operators working in three shifts).
(4) Crane arrangement (truck crane).
(5) Site and container arrangement.
(6) Track arrangement outside the cavern.
(7) Placement of water pipes and cables inside the cavern.
(8) Information preparation.
(9) Arrangement of electricity and air supply inside the disassembling cavern.
(10) Disassembling tools and machines. The disassembling tools and machines include the following parts:
    ① Special lifting sling (ring gears of cutter head, bearing and ring beam), various lifting chains (according to the number of links).
    ② Pneumatic and hydraulic socket wrenches.
    ③ Pneumatic and hydraulic tension device.
    ④ Auxiliary hydraulic pump station.
    ⑤ DC welder for carbon arc gouging, air compressor, gouging carbon rod.
    ⑥ Chains: 2 for each of 5t, 3t and 1.5t.
    ⑦ Crowbar: 4.
    ⑧ Open-end wrench: 10–36 mm (3 wrenches), 41–75 mm (2 wrenches); box wrench: 10–36 mm (3 wrenches), 16–18 mm (3 wrenches).
    ⑨ Steel wire rope (for lifting).
    ⑩ 8 slings (for 1t and 5t each).
    ⑪ Sleepers, for support of the inner frame and placement of large parts and components.
    ⑫ Plugs for oil pipes, plugs for hydraulic joints (Table 9.13).
    ⑬ Sleeve, 3 sets (including ID 14–36 mm small and medium square sleeves, ID 36 mm above medium and large square sleeves).
    ⑭ Palm ropes.
    ⑮ 3–4 M16, M20, M24, M30 lifting lugs each.

2) Method of disassembling the main machine inside the cavern

(1) The main machine enters the disassembling cavern.

The steps for the main machine to enter the disassembling cavern are as follows:

① Put the lower shield of the cutter head into the entrance of the disassembling cavern, move the main machine forward to the lower shield bracket position, and install the lower shield bracket.

② Move the main machine forward under the help of the lower shield of the cutter head, and install the stepping mechanisms of the front and rear frames in turn, and move the main machine forward to the predetermined position under the help of the stepping mechanisms.

Table 9.13 Plugs for hydraulic joints

S. No.	Description	Specification/model	Planned quantity (set)	Inventory quantity (set)	Actual quantity (set)
1	Plug	DN08 hose plug, sw19	492	174	646
2	Plug	DN12 hose plug, sw30	320	118	418
3	Plug	DN16 hose plug, sw36	184	152	316
4	Plug	DN20 hose plug, sw46	63	71	114
5	Plug	DN25 hose plug, sw50	54	106	140
6	Plug	DN32 hose plug, sw60	20	59	60
7	Plug	DN50 hose plug, sw75	55	23	58

(2) The main machine is disconnected from the connection bridge.

The steps for the main machine to disconnect from the connection bridge are as follows:

① Retract all the grippers of the front and rear frames and try to lock the gripper in place.

② Rotate the cutter head to the hoisting angle.

③ Disconnect the cylinder between the main machine and the connection bridge; disconnect the lubricating oil line, hydraulic oil line, gas line and water line between the main machine and the connection bridge.

④ Remove all the lubrication pipeline, hydraulic pipeline, water pipeline, corresponding control desks and numerous valve blocks, coolers, oil filters and other control components (the re-action of the hydraulic actuators will rely on the ground pump station).

⑤ Remove the circuits, various sensors and signal lines of the complete machine.

3) Disassembly of main machine peripherals

The steps for disassembling the main machine peripherals are as follows:

(1) Remove the cutter from the cutter head.
(2) Disassemble the belt conveyor, the invert crane, the slag remover and the corresponding tracks.
(3) Disassemble the probe drilling machine.
(4) Disassemble the steelwork mobile trolley.
(5) Remove the safety fence on the top shield.
(6) Disassemble the railings, traveling treads, top moving platform and the top duct of the main machine.

(7) Disassemble 4 bolt drillers and standing platforms at the front and rear sides.

4) Disassemble the main parts of the main machine

The steps for disassembling the main parts of the main machine are as follows:

(1) Lift and hold the cutter head, remove the bolts connecting the cutter head to the main bearing, lift the cutter head and place it horizontally in the front end of the disassembling cavern.
(2) Disassemble the connecting parts between the inner frame and the connection bridge (including the connection bridge lifting cylinder and other components).
(3) Tow the No.1 Belt Conveyor from the inner frame and transport to the rear part of the disassembling cavern.
(4) Disassemble the steelwork mounting mechanism assembly and the steelwork moving sliding plate assembly.
(5) Disassemble 8 gearboxes, couplings and 8 motors.
(6) Disassemble top shield, left and right shields, central shield and corresponding cylinders and transport them to the rear part of the disassembling cavern in time.
(7) Remove the muck chute in the front of the main bearing.
(8) Disassemble the top 6 clamping cylinder assemblies.
(9) Lift and hold the main bearing, remove the connecting bolt between the main bearing and the inner frame, disconnect the transmission shaft from the main bearing, lower the main bearing and place it horizontally on the upper plane of the cutter head (the upper plane of the cutter head is padded with the skid of appropriate thickness).
(10) Disassemble the upper and lower clamping blocks of the main bearing.
(11) Disconnect the rear support from the rear outer frame steering cylinder, lift away the rear support assembly (including the rear support stepping mechanism), and transport as a whole out of the cavern before disassembly.
(12) Disassemble the 4 front and rear thrust cylinders from the top of the main machine.
(13) Disassemble the 8 sliding block assemblies between the inner frame and the front and rear frames.
(14) Remove the rear upper outer frame.
(15) Disconnect the transmission shaft from the inner frame, and lock the No. 3, No. 4, No. 5, No. 6 transmission shafts to the upper part of the front outer frame.
(16) Disassemble the upper part of the front outer frame.
(17) Disconnect the 4 thrust cylinders in the lower part of the front and rear frames from the inner frame.
(18) Lift the whole inner frame to the special conveying platform trolley, and then disassemble the connecting bolts of the front and rear inner frames and transport to the outside of the cavern respectively.
(19) Remove the rear thrust cylinder.
(20) Remove No. 2 and No. 7 transmission shafts.

(21) Remove the front thrust cylinder and No. 1 and No. 8 transmission shafts.
(22) Disconnect the stepping mechanism from the front and rear lower outer frames respectively.
(23) Load and transport the stepping mechanism and lower shield out.
(24) Transport the cutter head and bearing out.
(25) Divide the cutter head into several portions and transport out.

5) Back-up disassembly method

The method for disassembling back-up system is as follows:

(1) Remove the rear ramp and tail carriage first and place them on No. 17 and No. 18 platform gantries, and drain the residual water in the two water tanks after the TBM is stepped into place.
(2) Disconnect the connection between platform gantries, flexible connection of air ducts, high-pressure air ducts, high-pressure water pipes, cable connection of each section, disassemble the belt and place it on the truck and tow out.
(3) Remove equipment, railings, cable troughs and other objects on the gantry that may interfere with the steelwork and place them reliably.
(4) Transport 2–3 sections of back-up at a time, connect with a special towing hook and flatbed truck, and tow out by locomotive.
(5) Disassemble the back-up outside the cavern.
   ① Disassemble the special equipment as a whole to avoid damage and bruising; and place its fixing bolts together with the equipment.
   ② Cut away the U-bolt belt bridge and disassemble the belt bridge as a whole.
   ③ Split the welds and bolts of the platform gantry, and disassemble in the order of cross beam first and then side frame.
(6) Disassemble the muck unloader.
   ① Lower the muck unloader as a whole.
   ② Disassemble the belt conveyor rollers and belt bridges at all parts.
   ③ Disassemble all parts of the muck unloader (the pant-type funnel should be in one piece).
(7) Disassemble the shotcrete platform truck.
   ① Disassemble the pump station, oil tank, motor, manipulator, conveying pipeline and other equipment.
   ② Lower the upper fanshaped track.
   ③ Disassemble the lower cross beam.
(8) Disassemble the connection bridge.
   ① Disassemble the control cabin of the main machine, disconnect the cable oil pipe from the control cabin, disassemble the lower bolts, and lower the control cabin of the main machine.
   ② Disassemble the railings and corresponding equipment on the upper layer of the connection bridge.
   ③ Disassemble the steel structural components of the upper tread and the left ventilation duct.

④ Disassemble the equipment on the upper layer and disconnect the cables and oil pipes of each part (the fixing bolts of the disassembled equipment should be properly stored).

⑤ Disassemble the cables passing through the connection bridge (complete before towing out of the cavern).

⑥ Disassemble the connecting cylinders between the main machine and the connection bridge, and between the connection bridge and the back-up.

⑦ Disassemble the No. 2 belt conveyor and other devices and disassemble the sections of the connection bridge after towing out the back-up.

6) Disassembly and transportation sequence for large parts and components

The disassembly and transportation sequence for large parts and components is as follows:

(1) Disassembly and tow-out sequence of connection bridge.

① Disassemble and place the TBM control cabin upright that may interfere with part of the steelwork.

② Remove the equipment, railings and frames on the left and right sides of the upper layer to facilitate towing out.

③ Install the lower bracket and crawler tug.

(2) Disassembly and transportation sequence of main machine.

① Disassemble the upper auxiliary parts, including the steelwork platform, various components of the bolt driller, group valve, oil pipeline, cable, cylinder, various small devices, motor, gearbox, motor platform and air duct of industrial exhaust fan.

② Take precautions when disassembling large parts and components of the main machine:

Weld the lifting points of each part first when disassembling the top shield, wing shield and cutter head parts. General sequence of disassembly: cutter head → shield (wing shield first and then top shield) → main bearing → rear support → two inner frames in outer frame bucket (including two sections, front and rear sections; rear section first and then front section) → two lower outer frames (independent from stepping mechanism).

(3) Main machine parts.

① 8 thrust cylinders (loaded into 8 small containers A-24).

② 3 main group valves, 1 small group valve, 8 hydraulic motors, 1 group valve for auxiliary drive unit, 1 shield cylinder, 1 rear support towing cylinder, and 1 steering cylinder.

③ 2 sets of structural components of bolt driller and 1 set of steelwork transportation part.

④ 1 set of each of connection bridge, industrial exhaust fan, auxiliary hoist and pump.

⑤ 1 set of each of invert crane and lower material crane.

⑥ 2 grease pumps, 5 connection bridge towing cylinders, 2 drilling machine consoles (including maneuvering valve and control valve).
⑦ 8 main motors and 8 clutches.
⑧ 8 gearboxes (after disassembly of ring gear).
⑨ 8 sets of transmission shafts and sleeves.
⑩ 1 set of cylinder each for gripper sliding block, sliding plate, moving beam of drilling machine and steelwork (system), and 3 water pumps.
⑪ 1 set of lifting platform and auxiliary parts of connection bridge each.
⑫ 1 set of No. 1, No. 2 and No. 3 belt conveyor rollers each.
⑬ 1 set of No. 1, No. 2 and No. 3 belt conveyor structural components each.

(4) Other matters.
① Accessory storehouses are loaded into 4 containers.
② Air ducts are loaded into 3 containers.
③ TBM air and water pipes are loaded into 2 containers.
④ 3 fans are loaded into 1 container.
⑤ Cutter workshop and 4 cutter storehouses are loaded into 5 containers.
⑥ Packing parts and components (railings, etc.) are loaded into 4 containers.
⑦ Disassembly and packing of special tools, disassembling tools, lifting slings, etc.

9.7.3 Precautions for Disassembly

1) General precautions

(1) Preparations before disassembly

Preparations before disassembly include the following contents:
① During equipment assembly disassembly, clean, decontaminate (including oil, mud, dirt), dry and pave with cushions, and cover the site to be disassembled and the surrounding environment to prevent personnel from slipping, prevent dirt from mixing and prevent heavy objects from falling.
② Before disassembly, power off the equipment. Release the air pressure, oil pressure and other pre-compression forces of the elastic components in the enclosed oil tank formed during the operation of the TBM, and drain the oil if necessary. Protect the electrical components and parts prone to oxidation and rusting.
③ Assign a special personnel to be responsible for making identifications and markings to the parts to be disassembled (including system category, name, assembly drawing number, original installation position, interface symbols, etc.), and filling in the disassembly registration form, in order to understand the disassembly sequence, know the progress, facilitate

the handover, and also lay a foundation for the subsequent installation and facilitate the search when encountering the installation problems.

④ Don't allow disassembly personnel to carry out cross-discipline, cross-system disassembly; assign a special personnel to be responsible for the same system thoroughly in order to maintain the continuity of work.

⑤ For structural components and large parts at high altitude that are easy to fall and out of human control, take appropriate support and lifting measures, tie securely, hang reliably, before disassembly and lifting.

(2) General sequence of disassembly

The general sequence of disassembly is as follows:

① Disassemble the electrical, signal and communication systems first; then the hydraulic and pipeline systems, and finally the mechanical parts.

② Strong current part before weak current part.

③ From top to bottom, from outside to inside.

④ Main part before branch part.

⑤ Main structure before frame structure.

⑥ Equipment first, then the assembly, then the components, and finally the parts.

(3) Disassembling tools and equipment

High-pressure disassembling equipment (such as disassembling device for hydraulic pre-tightening bolts and electric and pneumatic torque wrenches, etc.) cannot be put into use until they pass the operation test.

For thread fastening, spline end wrenches or socket wrenches are preferred, followed by open-end wrenches and adjustable wrenches. Tools that can be used according to the arrangement of tightening torque range from small to large are torque wrench, adjustable small torque wrench, adjustable sleeve torque wrench, wrench with (planetary gear) torque multiplier, electric impact wrench, pneumatic impact wrench, pneumatic oil hydraulic wrench and pneumatic oil hydraulic tensioning kit.

(4) General principles of disassembly

① From a practical point of view, the parts that may not be disassembled shall not be disassembled as far as possible. In order to reduce the disassembly workload and avoid damaging the fitting performance, the parts that can still ensure the operational performance may not be disassembled, but necessary tests or judgment shall be carried out to ensure that there are no latent defects.

For the connectors that cannot be disassembled or whose accuracy will be reduced after disassembly, attention shall be paid to protecting them during disassembly.

If the internal technical condition is uncertain, the parts must be disassembled and inspected to ensure the maintenance quality.

For some equipment assemblies, hydraulic pump stations and other connecting lines and pipelines, they should be not disassembled but maintained as they were as long as the hoisting and transportation are not affected.

② Attention should be paid to the cleaning of the hydraulic system. During disassembly, special care, caution, attention should be paid to the appearance of components and environmental cleanliness, especially the pipe joints whose plugs and protective caps should be installed immediately after disassembly at any time to avoid man-made contamination.

③ For the shaft-hole-type assembly parts, the principle of using the same force for disassembly and assembly should be followed. During the disassembly of the shaft-hole-type assembly parts, the force used for assembly should be used for disassembly in general. In case of any abnormality, find out the cause to prevent the parts from being bruised, roughened or even damaged during disassembly. Shrinkage fitting parts should be disassembled with heat. In general, destructive disassembly is not allowed.

④ The conditions for assembly should be created for disassembly. If the technical data is not complete, the disassembly process must be recorded, so that the principle of "disassembly first and then reassembly" can be followed during reassembly. For the disassembly of precision or complex components, assembly sketch should be drawn or necessary marks should be made during disassembly to avoid incorrect assembly.

⑤ The disassembled parts should be stored by category, so as to find easily and to prevent damage, loss or mistake. The disassembled parts and components should be stored separately by assembly, component, part, by electrical, hydraulic, mechanical parts and components, by large and small parts and components, and by coarse and precision parts and components.

The components subject to temporary repair can also be placed sequentially on a clean workbench in the sequence shown in the assembly drawing.

According to the structural characteristics of the parts, slender parts should be hung to prevent bending deformity; for non-interchangeable parts or high-speed rotating disc parts, adverse effects of changes in operating performance (such as decentration, mass imbalance, static and dynamic imbalance, etc.) should be avoided and they should be stored in groups or marked with a marking pen.

2) Special precautions for TBM disassembly

Special precautions for TBM disassembly are as follows:

(1) Before the TBM breakthrough, a comprehensive and careful recheck is required to complete the identification of mechanical, electrical and hydraulic parts.
(2) In addition to the equipment, tools and machines used in the assembly, a special coupling device for traction of gantry dedicated for TBM disassembly and connection bridge support wheel frame should be prepared completely.

(3) The insufficient quantity and specification of various pipe joints and plugs should be checked, supplemented and machined.
(4) Before TBM breakthrough, on-load performance test should be carried out on the main machine, back-up and auxiliary equipment to fully identify the performance of each mechanism and equipment and provide a basis for timely maintenance and repair after disassembly as well as the formulation of spare parts plan.
(5) Parts and components of TBM main machine are loaded on the back-up platform gantry and towed by locomotive. Attention should be paid to the loading mass and tunnel limit dimensions.
(6) TBM back-up gantry and connection bridge should be towed out of the cavern at a speed ≤ 5 km/h.
(7) Packaging and storage of TBM parts and components.
   ① Mechanical parts and components.
      (a) Clean the dirt and rusty spots on the contact face and fitting pair, coat with anti-rust oil or grease, perform dust treatment on important parts and components, and then pack with wooden boards for protection.
      (b) Clean and paint the surface of parts.
      (c) Wrap with plastic cloth and pack in wooden cases before sealing.
      (d) Clean disassembled bolts, nuts and washers, soak in oil and then store them.
      (e) Package the bolts, nuts, washers and other connecting parts of separate equipment separately and mark them.
   ② Electrical components.
      (a) Dedust and decontaminate the surface and interior of the components.
      (b) Protect electrical components well from dampness, dust and freezing (water-cooled motor).
      (c) Clean up the surface of the cable, wrap its terminals with plastic cloth and seal with tape.
      (d) Protect the distribution cabinets, distribution boxes and other important electrical control facilities from water, dampness, dust, package them in fully enclosed plastic bags with desiccant (bagged) inside.
      (e) Pack in wooden cases for protection and then put into enclosed containers for storage.
   ③ Hydraulic components.
      (a) Dedust and decontaminate the surface of the hydraulic components.
      (b) Fill the oil pumps and oil motors with at least 3/4 of the original grade of mineral oil. If the storage period is more than 9 months and special oil in reserve must be filled with, the oil port should be plugged with a steel plug (joint).
      (c) Fill the hydraulic steel pipe with hydraulic oil and seal with a joint.

(d) Seal the hydraulic hose with a joint and roll up for storage.
(e) Fill the hydraulic valve with oil and seal each oil port with a joint.
(f) Ensure that the inner surface of the hydraulic oil tank is covered by oil film, and seal each oil port.

9.8 TBM Repair and Storage

9.8.1 Repair Items and Contents

Repair items and contents are shown in Table 9.14.

9.8.2 Typical Part Trimming Process

1) Disassembly and assembly processes of main bearing and drive assembly

(1) Preparations

① Tool preparation. Including tool skid, several iron blocks, 8 special lifting hooks, large lifting chain of main bearing, 5t flat sling, hydraulic tension device (including VG10 special hydraulic oil, oil seal of tension cylinder) and corresponding molds, inflating pumps (0.6–0.7 MPa) and air pipes, plastic films or colored strips for covering, plastic package bags, cleaning fluid, chemical fiber towel, heavy-duty sleeve, torque wrench and other common tools and dial gauge, steel tape, level ruler, outside micrometer (to measure the drive shaft diameter) and other measuring tools.

② The main bearing is padded with skid at its bottom near the outer labyrinth at intervals of 120° and placed stably (with the inverted taper end facing downward and the connecting end of the drive shaft facing upward for main bearing with inner and outer labyrinths); the height of the skid at the labyrinth bottom is 1 m, while the gap between the inner labyrinth band and the ground is padded with wooden board at intervals of 120°. Please pay attention to avoiding the position of labyrinth mounting screw when padding.

③ The 4 support cylinders of the lower shield are removed and all the clamping blocks should be removed when conditions permit, so that the bearings can be lifted horizontally when disassembled.

④ External dirt must be cleaned (including the cleaning of lifting chain). In particular, the dirt on the upper surface of the bearing and in the drive shaft hole must be thoroughly cleaned.

(2) Disassembly steps

Table 9.14 Repair items and contents

Repair items	Repair contents
Cutter head	Welding the muck chute, perform flaw detection on cutter hole, welding cutting teeth, repairing water spray system, welding the cutter head crack, inspecting and maintaining the scraper seat, welding and machining the cutter seat hole, perform weld overlay on worn parts of cutter head
Main bearing and drive assembly	Disassembling and inspecting, replacing seals, cleaning labyrinth, and changing the positions of wear ring
Transmission shaft	Correcting No. 6 transmission shaft, handling the deformity of each sleeve, checking worn parts of drum-shaped gear (ferrules)
Clutch	Uniformly replacing friction plate, spring shaft, mandrel, coupling top pillar and lock nut
Gearbox	Removing mud (water), disassembling and inspecting one by one, replacing defective parts, seals and bolts, enlarging borehole of sleeve at the fixed position
Main motor	Removing mud, cleaning water jacket, disassembling and inspecting, drying windings, inspecting or replacing tail slewing bearing
Inner frame	Derusting, trimming, finishing the slide rail, curved shape sliding block, steelwork platform slide, tapping valve block at the fixing position, treating No.1 belt conveyor support pulley seats in the inner frame
Outer frame	Derusting and checking the gripper cylinder leakage, sorting the guide pillar, fixing the position of the curved shape sliding block, and replacing the upper and lower connecting bolts
Thrust cylinder	Cleaning the piston rod, replacing the valve, connector, oil pipe seal, greasing the ball head position, and replacing the oil inlet and outlet pipes of the front lower thrust cylinder that is deflated due to excessive steering adjustment during TBM tunneling
Rear support	Same as the outer frame; checking the fixed position of the cylinder of the oil pipe
Bolt driller	Checking mobile cylinder (slipper), checking and replacing curved shape track and chain; replacing defective parts, correcting feed beam, replacing chain and fixing bolt, correcting console, repairing rock drill complete machine, replacing all grease nozzles and all oil pipes
Hydraulic system	Replacing P3, P4, P6 pump stations, changing oil in main oil tank completely, cleaning and descaling cooler thoroughly, replacing DN50 pipe joint, replacing DN40 pipe joint seal, replacing each leakage valve seal, checking and commissioning hydraulic equipment and system in each independent pump station and changing oil

(continued)

Table 9.14 (continued)

Repair items	Repair contents
Connection bridge	Handling the damaged parts, water leakage, air leakage of industrial exhaust fan; disassembling and checking fan motor; refurbishing the electro-hydraulic system of invert crane; handling rotary joint of material crane; sealing pump system; reasonably planning air and water pipelines; replacing and repairing defective parts of grease pump
Back-up	Refurbishing towing system, checking towing motor performance, welding and repairing severely worn chain groove, improving chain wheel system, thoroughly checking air compressor, cleaning deposit dust, replacing air filter element and worn parts, checking and correcting fan blades, welding and repairing all the worn parts of the belt conveyor and correcting deformity; replacing retaining wall (plate) and damaged roller of belt conveyor, vulcanizing belt pulley again, welding and repairing all the worn parts of muck receiving hopper, re-welding the flap of muck unloader, replacing bearing, checking and replacing carrier roller, checking and repairing the drive part of No. 3 belt conveyor (rear drive pulley, motor, transmission mechanism), repairing california switch (air cylinder, air pipeline), stopping leakage of water tank, cleaning silt in the tank, checking, welding and repairing each section of the gantry wheel pair of the back-up, replacing wheel pair bearing, changing No. 9 pump station coupling assembly into a new assembly
Air duct check, deformity correction	Correcting the deflated air duct from the inside out with a rubber hammer and restoring to original state
Steelwork installation system	Repairing and maintaining rotary motor installation base and lower support seat; shaping, welding and repairing 2 mobile platforms; checking, repairing, sealing and replacing each jacking cylinder

① Remove all connecting oil pipes (lubricating oil pipes, grease pipes) and valve bodies from the main bearing, and seal and wrap them properly with plugs or cover plates, and record the disassembly positions of the oil pipes.

② Remove the connecting rib plates and guard plate of the muck chute from the hollow part of the bearing.

③ Remove the 8 drive gear shafts.

First, disconnect each lubricating oil pipe of drive gear shaft bearing, remove the gland screw of drum-shaped gear sleeve and pull out the drum-shaped gear sleeve, and remove the 6 screws of the gear shaft flange gland. Next, install the lifting lugs in the center hole of the drive gear shaft, slowly lift the drive gear shaft until the bottom bearing breaks away from the seat hole, and then remove the remaining drive gear shaft in turn. Finally, disassemble the parts on the drive gear shaft, including removing the snap spring on big-end bearing retainer ring with a circlip plier; pressing out the big-end bearing retainer ring, big-end bearing and small-end bearing successively by drawing method or padding method.

④ Remove the 2 eddy current testing sockets; disconnect wires completely from the eddy current testing sockets during removal.

⑤ Remove the hexagon socket mounting screws (part No. 29) from the circumferential position of the bottom inner labyrinth to facilitate smooth lifting of the drive components without traction and smooth release of the inner labyrinth ring from the inner labyrinth.

⑥ Disassemble the 72 hydraulic tension bolts from the outer ring of the main bearing (adjust the working pressure of tensioning to about 1.7×10^8 Pa); remove the 6 socket head cap screws on the circumference of the tension bolts (these screws connect the drive components and the double-row thrust roller path).

⑦ Use 75t gantry crane and lifting chain of main bearing to slowly lift the main bearing drive assembly until it is completely separated from the main bearing, place 3 iron blocks arranged at 120° in the empty space, pad with thin wooden boards, and place the drive assembly on it stably.

⑧ Install the lifting sling and connection pin that match the main bearing lifting lug support.

⑨ Lift the main bearing vertically with a large hook and use the small hook of the crane to turn over twice in the air to make the labyrinth surface of the main bearing face upward and place it on the skid stably.

⑩ Remove the inner labyrinth seal ring, three inner lip seals, four spacer rings and housing ring assembly in sequence.

⑪ Remove the outer labyrinth pressure ring screw, the outer labyrinth seal ring and the ring gear connecting screw, remove the outer labyrinth pressure ring and outer labyrinth ring, and take out the outer lip seal and the four spacer rings.

⑫ Assemble the outer labyrinth seal ring and the pressure ring again as they were after the outer seal ring is inspected completely.

⑬ Initially clean up the old bearings, properly pack the old bearings with the spare main bearing wrapping film, vacuum them while quickly hot press and seal them, and put them into the bearing packing wooden cases for sealing and wait for repair.

(3) Check

① Check the wear conditions of the inner lip seal and the contact surface between the inner labyrinth sliding seal ring and housing ring assembly, and consider repairing, renewing or replacing the four spacer rings for the third time, as appropriate, after measurement.

② Check the wear condition of the inner labyrinth coupling and decide whether to replace or machine a new one.

③ Check the wear conditions of the outer labyrinth seal ring and housing ring assembly, and keep records for archiving.

④ Check and measure the wear conditions and fit dimensions of the big-end and small-end bearings, spacer rings, oil seal on the flange gland, gear shaft spline teeth and drum-shaped gear sleeve on the drive gear shaft assembly, and decide whether to renew or repair.
⑤ Check the wear and aging conditions of all the lip seals and O-rings in the list of spare parts, and decide whether to renew.

(4) Installation preparation
① Prepare 20 L of lubricating oil, half barrel (about 100 L) of main bearing lubricating grease, whole roll of plastic film, special tapered air nozzle, 2300 N·m torque wrench, half barrel (about 100 L) of clean diesel, and imported cleaning agent.
② Clean all the parts to be installed with cleaning solution (especially for the drive assembly, open all the round cover plates to clean thoroughly), clean all the mounting surfaces of the spare bearing with anti-rust oil, purge all the inner bores, screw holes or counterbores with compressed air clean, wipe the mounting surface with chemical fiber towel without leaving any residual dirt and gravel; clean and pack parts not installed temporarily with plastic films and plastic bags properly for protection, in principle, clean, install and cover for protection simultaneously at any time; dress clean for work at heights to avoid secondary pollution.
③ Perform performance (function) test on the bridge crane to ensure that the lifting of the large and small hooks, traveling of trolley, movement of crane are error-free, accurate and reliable during the hoisting of large parts and components.

(5) Installation
① Assembly of drive shaft assembly. Before assembly, soak the whole assembly in diesel and clean it, then lift it out and dry with a towel, and purge the threaded holes with high-pressure gas; perform shrinkage fitting on big-end and small-end bearings, i.e., bake with a blowtorch at a uniform speed along the circumferential direction, or heat to a suitable temperature in an adjustable thermostat, then quickly install them onto the journal bearing, while pouring new lubricating oil onto the bearing rollers, rotate the bearings while pouring, so that they are evenly lubricated; install the retainer rings of inner rings and snap springs of the bearings onto it in sequence.
② After the drive assembly is thoroughly cleaned, connect the distribution valve and each grease pipe to each grease port of the drive assembly, and pump the grease continuously with the grease pump until clean grease can be squeezed out of all the grease ports of the drive assembly.
③ Put the lifting ring into the center hole of the spline shaft of the drive shaft, lift with a crane, hold and align the center hole of the drive assembly by one person at the top surface and lower down slowly while slightly rotating the gear shaft, align the upper gear of the shaft with the engagement position of ring gear of main bearing while lowering down slowly, observe the precise alignment of small-end bearing with

the bearing seat hole by another person from the facade of the drive assembly until they are put in place; align the flange pressure plate of drive shaft according to the original marks made during disassembly or oil port connection positions, tighten the flange pressure plate according to the specified torque, and connect the lubricating oil pipes well; align the drum-shaped gear sleeve with spline gear sleeve and set in place, apply thread locker to the threads of the plate bolts, install the binder plate, tighten the bolt according to the specified torque; adopt the same method to install the remaining drive shaft assembly.

④ Install the top round cover plate and tighten the bolts symmetrically and evenly; disconnect the drive assembly from all external connecting oil pipes.

⑤ Lift the drive assembly with the large hook of the bridge crane, use the small hook of the crane to turn over the drive assembly twice in the air (with the mounting surface of cutter head facing upward); increase the height of the bottom surface by 1 m by padding to facilitate the assembly of hydraulic pre-tightening bolts; clean the mounting surface of the drive assembly again.

⑥ Clean the spare main bearing (remove the set screws of the outer labyrinth lip seal pressure ring, and confirm whether there are lip seal and spacer ring inside), lift the spare bearing (with inverted taper end facing upward) with special lifting lugs, align the mounting position of drive assembly (mark well in advance), and lower down the large hook to the bottom completely and slowly and accurately align (try to align the 3 hydraulic tension bolts from bottom to top along the circumferential position at 120°, and align accurately if possible).

⑦ Install 4 pre-tightening bolts at intervals of 90°, and use the hydraulic pre-tightening device to pre-tighten symmetrically to 1×10^8 Pa, then install the remaining bolts in the same method; pre-tighten symmetrically to 1.4×10^8 Pa for the second time, and finally pre-tighten to 1.7×10^8 Pa.

⑧ Install 12RD ± 0.5NBR-65 elastic pillar (O-ring) in the housing ring assembly, or with the help of 6 persons, cut vertically when the inner circumference of the O-ring groove of the housing ring assembly is tight a bit (hold the O-ring evenly by 6 persons and do not make the O-ring twist before cutting), use glue to accurately align and bond and then put into the groove, feel the right tightness, and then apply a little hydraulic oil on the surface of the O-ring.

⑨ Hoist the housing ring assembly into place according to the mounting marks and tighten the set screws (part No. 19) symmetrically and evenly according to the specified torque.

⑩ Decide whether to replace the position of the inner labyrinth spacer ring according to the check results of the inner labyrinth ring. The O-ring installation method is as follows:

Hold any one new lip seal by more than one person (more than 8 persons) evenly with its opening facing downward, gently put one part of the seal ring into the labyrinth ring groove firstly with the depth of 5 cm, find the symmetry points of the seal ring in a circle of 180° and then gently plug into the position of 5 cm, then the two points at vertical direction into the ring groove by this method, do not twist the lip seal, and then find the four points 45° apart from this four points, and plug gently like this, plug the remaining points evenly until the seal ring is attached flatly and closely; push the lip seal with a clean rubber hammer slowly and gently by 8–10 persons along the circumferential direction until it is in place. Subsequently, push the spacer ring into the ring groove horizontally with the ring opening facing downward, place the second lip seal horizontally with the its opening facing upward, put it into the ring groove in the same way as above, install the second spacer ring, the third lip seal and the third spacer ring into the ring groove in sequence, and finally install the fourth spacer ring onto the O-ring and then into the ring groove.

⑪ Install the newly machined inner part of the labyrinth ring and tighten the socket head cap screws.

⑫ Install or re-bond a new O-ring into the O-ring groove of the inner ring connected with the muck chute in the same manner as above.

⑬ When installing the outer part of the inner labyrinth ring, apply a little hydraulic oil on the exterior of the O-ring during alignment, keep the ring horizontal when hoisting the inner labyrinth ring, do not skew it, gently and slowly lower it until the O-ring is smoothly squeezed in and the ring body is placed in place.

⑭ Install the lubricating oil pipes, grease pipes, coarse oil return pipes and their respective valve bodies in sequence according to the lubrication pipeline arrangement route.

⑮ Hold the eddy current socket lead and sponge sleeve properly in advance in the process of installing the spare bearing onto the driving assembly, lead (guide) the bearing into the eddy current monitoring hole gradually until they are in place during the bearing dropping process, and then connect and fix the eddy current socket and connecting wires as they were to ensure that they are not damaged during the hoisting process.

2) Disassembly and assembly processes of gearbox

(1) Preparations

Prepare a clean and dust-free operating environment with enough space; machine a solid support frame to place the gearbox smoothly on it, which should be easy to move, preferably in the form of a trolley, and prepare appropriate hoisting equipment. To ensure that the gearbox can be lifted horizontally, three points (or two points) will be used for lifting and each lifting sling should have a lifting capacity of not less than 1000 kg. In addition, to ensure sufficient clearance, the distance between the hook and the upper surface of the gearbox support frame should be not less than 2.0 m when the hook reaches the highest point. Clean the surface of the gearbox and drain the lubricating oil inside the gearbox; remove the 4 positioning pins on the upper and lower joint surfaces of the brake and gearbox and part of the end cover fastening bolts; put the gearbox on the support frame and move it into place so that the upper support surface of the support frame is in a horizontal position (adjust with a level ruler).

(2) Disassembly

① Remove the remaining fastening bolts on the end cover.

② Lift the gearbox slightly, observe whether the gearbox is lifted stably, and adjust the position of the lifting point if necessary; if two points are used for lifting, trial lifting is required before the bolts are removed to ensure that the gearbox can be lifted horizontally.

③ Tighten the 4 top bolts at the lower end cover to separate the end cover and the one-stage planetary gear reduction mechanism from the gearbox body; continue to lift the gearbox body horizontally until the one-stage reduction mechanism can be moved horizontally.

④ Remove the support frame and the one-stage planetary gear reduction mechanism, clean the oil on the surfaces of the gear and end cover, and remove the sealant at the joint of end cover with a scraper; clean with a cleaning solution that does not contain aromatic hydrocarbon materials to avoid damage to the seal.

⑤ Remove the planetary carrier and planetary gear as a whole.

(3) Repair or replacement

The general defects of a gearbox mainly include deformity of gearbox body, cracks, worn bearing hole and gear damage.

① If the warpage of the upper plane of the gearbox body is small, the gearbox body should be placed upside down on the grinding platform for grinding and leveling. If the warpage is large, grinding or milling should be used for leveling. At this time, leveling should be carried out with the hole as the reference to ensure the verticality of the machined plane against the axis.

② When the parallelism error between the hole center distances exceeds the maximum permissible error, the counter boring and bush fitting method can be used for repair so as to restore the relative position accuracy among shaft holes.

③ When the gearbox body has cracks, it should be welded, but the deformity of gearbox body and the resulting white cast irons should be minimized.

④ Worn bearing holes of gearbox body can be repaired by standard-size repair method and bushing method or machining method in combination with electroplating, brush plating or spraying method.

⑤ The remaining parts can be replaced or repaired depending on the damage conditions.

(4) Assembly

① Install the one-stage reduction planetary carrier and planetary gear.
② Move the support frame and the one-stage reduction mechanism back to their original positions and place them horizontally (the bottom of the support frame is padded and leveled with a level ruler) so that the edge of the end cover is aligned with the edge of the gearbox body vertically.
③ Apply an even layer of sealant to the joint surface of the end cover.
④ Level the gearbox body with a level ruler, slowly and horizontally lower the gearbox body, and slightly rotate the auxiliary drive input shaft to make it engage properly when the two-stage reduction sun gear is about to engage.
⑤ Continue to lower the gearbox body to make it engage with the end cover, install the 4 positioning pins, and tighten the fastening bolts symmetrically to the specified torque.
⑥ Turn over the gearbox and install the brake.
⑦ Add 40 L lubricating oil

(5) Run after installation

① Strengthen the monitoring of the lubricating oil and timely know the internal state of the gearbox through oil sample analysis.
② Regularly clean the oil and dust on the surface of gearbox, improve the heat dissipation capability, and ensure the good performance of the lubricating oil.

3) Clutch repair

Damages to the clutch mostly occur in the friction plate wear or uneven wear, damage to bearings at both ends, obvious abrasion marks on the spring shaft and mandrel, weakened spring stiffness, scratches of pressure plate, internal gear wear of coupling, O-ring breakage and deformity. The damaged parts must be replaced during repair to avoid incomplete separation or uneven wear and vibration when the clutch is working. The key to repair is to ensure the accuracy of assembly, the height difference among 6 springs in the same layer should not exceed 0.5 mm; the overall thickness of clutch is controlled within the range of (210 ± 0.5) mm; it is not allowed to use hammer but the rubber hammer or top pressing method can be adopted during bearing assembly; the torque of set screws should be appropriate and uniform and the locking plate must be locked well; key screws should be applied with thread locker.

4) Trimming of steel structural components

Metal structural components in the role of dynamic load are prone to having weld cracks, structural deformity and fastener damage, loosening and other problems.

The steel structural components are trimmed mainly based on the inspection results of damages.

① Deformity correction: For example, for rod (shaft) bending and twisting, if the bending amount is less than 8/1000 of the length, the cold correction method can be used for correction; for shafts having higher requirements, needing accurate correction or having large bending amount, the hot correction method can be used for correction. For seriously deformed parts, the deformed parts can be completely removed with a flame cutting gun and then re-welded or spliced into a new structure (the same treatment method can also be applied to the plates).

② Wear: For the treatment for wear, the surface is cleaned, followed by wear-resistant weld overlay and forming by grinding.

③ Rust: For the treatment for rust, grinding and sandblasting are carried out.

④ Cracks and fractures: For the repair of cracks, the original weld metal should be removed, and then the welding repair is carried out on the base metal, the welding rod should be selected according to the different materials of and stresses on the components. If the crack is large or there are many cracks, a plate can be welded onto the entire area of the crack for reinforcement; for minor cracks (length of radial cracks does not exceed 10% of the diameter of the shaft), the groove welding can be conducted at a crack before tempering, or the groove can be made at a crack, then bonded for repair and machined after curing.

⑤ Bolt hole damage: The damaged bolt holes can be drilled large and paired with larger bolts. If the structural strength is affected, all the steel structural components can be redrilled after welding repair and their surfaces are subject to grinding and derusting, and the surfaces are coated with antirust paint or the primer in a color as consistent as possible with that of the original parts according to the requirements for anticorrosion.

5) Repair of transportation vehicles

Debris left on the mining vehicle and concrete tanker should be cleaned immediately after the breakthrough, the wheel pairs, brakes and air pipelines should be checked carefully and grease should be injected for maintenance, the vehicle surfaces should be derusted and painted and then sealed and stored. Locomotives, man cars and flatbed trucks should be thoroughly cleaned and maintained after they are disassembled completely, and engine repair and surface spray painting work should be carried out to eliminate existing faults and then sealed and stored.

6) Keyway damage

When the keyway has small dents, burrs and minor wear, it should be trimmed with a fine file, oilstone or scraper; otherwise, groove should be opened or enlarged in the original position or 90°–180° rotational position and re-keyed with a new key.

7) Hydraulic components

(1) Hydraulic cylinder

The surface of inner hole of hydraulic cylinder body has very shallow linear friction scratches or dotted scars locally. If its usage is not affected, very fine sand strip, oilstone or polishing and grinding head can be used to eliminate them; it can also be repaired by cold welding method in combination with adhesive.

(2) Valves

Hydraulic valves and other components can be repaired well due to their standardization, generalization, and higher degrees of serialization. Although parts may be subject to wear and fatigue or seal may be subject to aging after being used for a period of time, disabling their technical performances to meet the design requirements, as long as they are not completely scrapped, they should be repaired early and used continuously after repair.

① Repair of valve elements and valve holes of slide-valve-type components. When the fit clearance is 20–25% larger than the specified value for new parts, the valve element must be repaired by increasing its size and then ground.

② Repair of valve elements and valve holes of cone-valve-type components. When the seal contact surface of conical seat valve is poor, grinding is needed only for the repair, as the gap can be compensated by the conical valve automatically under the action of the spring.

③ Valve components should be repaired if they are stuck or have burrs or grooves.

④ Seal repair.

9.8.3 Storage of Parts and Components After Disassembly and Repair

After the TBM main machine and its back-up parts and components are disassembled and repaired, the parts and components can be packaged and warehoused, while being registered and archived in the TBM technical office. During the storage of parts and components, the storage period, the terrain of the storage location, transportation conditions, environmental and climatic conditions, IP grade and safety precautions, etc. should be taken into account. Otherwise, unnecessary injury or damage may occur due to improper storage. The disassembled and repaired parts and components should be stored in such a manner as to ensure that the mechanical and technical conditions of the parts and components are intact. Since TBM is generally stored for more than 3 months, most parts and components should be stored in accordance with the requirements for seal and storage of mechanical equipment.

(1) Bare large steel structural components should be stored in relatively dry positions with flat terrain and relatively high terrain which are conducive to drainage, and the centers of gravity of the structural components should be as low as possible to avoid instability (movement or tilt) of the stationary components caused by accidental interference. The bottom surfaces of the structural components should be padded with skids or green bricks; water is easy to be accumulated at the bottom of the structural components, which should be properly covered or drained to avoid rusting. Special attention should be given to the rust protection and greasing on the precision contact faces. Precision contact faces cannot be exposed but wrapped tightly and stored in a dry room after rust protection, application of rust preventive oil, grease or paint.

(2) Valuable parts and components (various cylinders and other precision parts and components) must be stored the containers. If no overweight occurs, their centers of gravity should be maintained in the middle (in left and right, front and rear directions) as much as possible and they should be reasonably placed and fully filled to avoid collision among each other, and fixed reliably (fixing the anchors with square timbers, binding the upper and lower parts with steel wires and fixing the relative positions of components with jack timbers) before being locked and stored.
(3) In addition to the provisions of Article (2), main bearing and drive assembly should be tightly wrapped in plastic film at their peripheries, and their storage should be regularly checked. When spare bearings are assembled, the surfaces of precision parts and components including rollers, inner and outer roller paths and wear rings should be painted with antirust paint respectively and then wrapped with high-strength coating after the assembly is completed. The air inside should be pumped out by vacuum pump, then spare bearings are sealed tightly and then lifted into a solid wooden packaging box which is provided with some silica-gel desiccants and then reliably bound.
(4) Hydraulic valves, electrical components, sensors, precision parts, seals, communication devices and imported parts for the warehouse should be classified and stored separately, by their size. A batch of wooden packaging boxes should be fabricated and numbered for their storage, while special attention should be paid to moisture protection.
(5) When the air duct cloth is folded, in addition to the flat placement, special attention should be paid to the flexible air duct to see whether the steel hook is left on it, which should be taken off or taken out timely if found, so as to avoid inadvertently scratching the duct, resulting in accidental losses.
(6) Rubber hoses (hydraulic oil hoses, high-pressure water pipes) should be put into boxes and warehoused in to prevent exposure to the sun, because the UV rays in the sun's rays will accelerate the aging of rubber, making it hard and brittle. All kinds of oil products (oil drums) should also be warehoused in.
(7) Special personnel should be assigned to guard and prevent theft and fire. The parking place should be guarded by special personnel to prevent theft. The parking place should be provided with fire safety equipment to prevent fire.

Questions

1. Please briefly describe the main process for the maintenance of hydraulic system when TBM is shut down.
2. What preparations should be made before disassembly?
3. What are the disassembly technologies, disassembly statistics and archived data?
4. Please briefly describe the disengagement steps between the main machine and the connection bridge.

Chapter 10
Risk Control in TBM Construction

This chapter focuses on the introduction of classification of TBM construction risks and construction risk analysis, with special attention to the adverse ground condition for TBM construction and its countermeasures and innovative design.

10.1 Classification of TBM Construction Risks

Risks in TBM construction always look for or utilize the vulnerabilities such as geological complexity, TBM inadaptability, human perception limit and irrationality of proposal and measures as the breakthrough to cause engineering accidents. Therefore, the risks of TBM construction are divided into three main categories: geological risk, equipment risk and man-made risk.

The main risk factors affecting TBM construction and their proportion are:

(1) Geological risk—40%. As an essential prerequisite, detailed and reliable hydrogeological data directly determine the success of TBM project. Hydrogeological data determine whether it is feasible to adopt TBM, the type selection of TBM, the main parameters of TBM, the selection of auxiliary construction equipment and the formulation of emergency plan.
(2) Equipment risk—30%. TBM endowed with advanced technology and reliable quality and experienced TBM manufacturers with professional services are the key to the success of the TBM project. For TBM, professional manufacturing and professional services are required. Professional manufacturing includes advanced technology and reliable quality. Only advanced technology can make the construction safer and more efficient, which is one of the key factors to ensuring the construction period; professional services include experienced and professional services, because rich experience is required in dealing with the risk of tunnel project. Therefore, TBM manufacturer is required to have rich experience; professional services include timely technical support and timely supply of spare parts.

K. Chen et al., *TBM Design and Construction*,
https://doi.org/10.1007/978-981-99-0059-6_10

(3) Man-made risk—30%. An experienced, scientific, professional and efficient construction team is the contributing factor to the success of TBM project. For the risks of underground works, rich experience is required, so the construction team is required to have rich experience. Moreover, scientific management is needed to ensure the efficiency of TBM, save costs and create benefits due to the tight construction period of TBM project. Therefore, the construction team is required to adopt scientific management method. The TBM construction process is arranged compactly, and efficient professional operators are required for efficient and advanced TBM, which is one of the key factors to ensure safety, quality and construction period. Therefore, the construction team is required to be professional and efficient.

10.2 Analysis of TBM Construction Risks

Considering the rapid construction and safety protection, both open-type TBM and shield TBM have geological risks that cannot be effectively overcome during the construction of weak and broken surrounding rocks.

10.2.1 Construction Risks of Open-Type TBM

Open-type TBM is more effective in hard rocks and has a faster construction speed. However, it is less effective in weak surrounding rocks, resulting in collapse, slow excavation speed, short advancement, vertical ring beam and large amount of backfill concrete required, long muck removal time, gripper failure, ring beam failure, easy to cause equipment damages and personnel injuries, etc. (Fig. 10.1).

When the open-type TBM tunnels the soft ground, the weak surrounding rocks cannot provide sufficient supporting force for the gripper, which causes the gripper to damage the support results, thus causing secondary collapse. This is specifically reflected as:

(1) The gripper interferes with the already erected ring beam;
(2) After the gripper passes, the ring beam is inoperative, resulting in aggravated collapse;
(3) A large amount of backfill is required when the gripper collapses or becomes unstable, which seriously affects the construction period.

10.2.2 Construction Risks of Shield TBM

The double shield TBM has a cylindrical shield protective structure, which allows the installation of segments while advancing and has a faster advance speed under

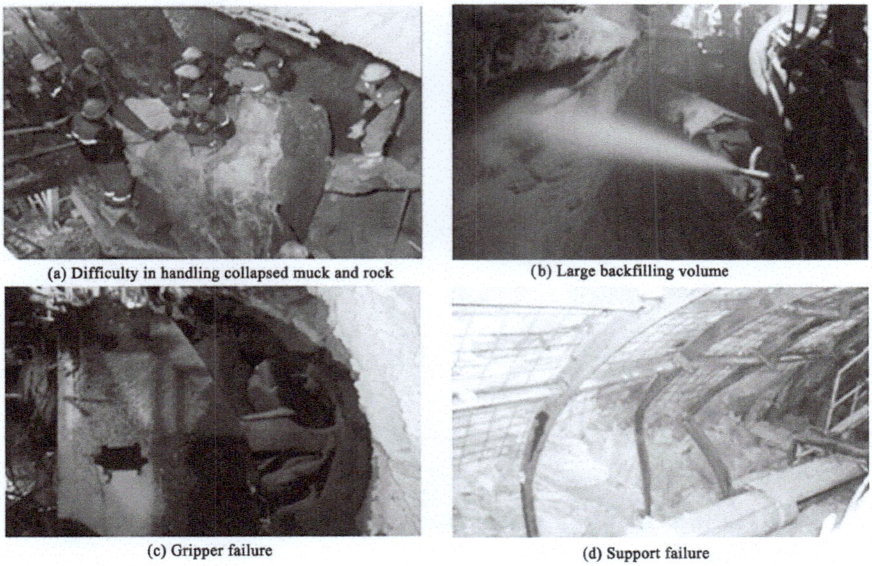

Fig. 10.1 Common faults of open-type TBM

relatively stable, moderate rock compressive strength and less abundant groundwater conditions. However, if it cannot pass quickly and promptly when passing through the broken and blocky surrounding rocks with large changes in ground stress, the cutter and shield may be faced with the risk of being stuck; once the TBM jamming occurs, it is very difficult to deal with it, which often leads to a long-time shutdown. During high-pressure water ingress, it is easy to cause segment breakage, thus leading to personnel and equipment safety accidents.

The main construction risks of double shield TBM include:

(1) Cutter head jamming: When the surrounding rocks of the tunnel face are broken and unstable by itself, resulting in collapse;
(2) Front shield jamming: When the surrounding rocks are greatly deformed and the friction generated by the surrounding rock stress acting on the shield is greater than the maximum thrust of the main thrust cylinder;
(3) Supporting shield jamming: When most of the surrounding rocks are deformed and the supporting shield is squeezed, the friction generated by the surrounding rocks acting on the supporting shield cannot be effectively overcome by the pressure of the auxiliary thrust cylinder during regripping;
(4) Tail shield jamming: The surrounding rocks continue to converge and deform under the action of ground stress, squeezing the tail shield to contract inward, resulting in a too small gap between the tail shield and the segment;
(5) Segment breakage: during water ingress under high water pressure.

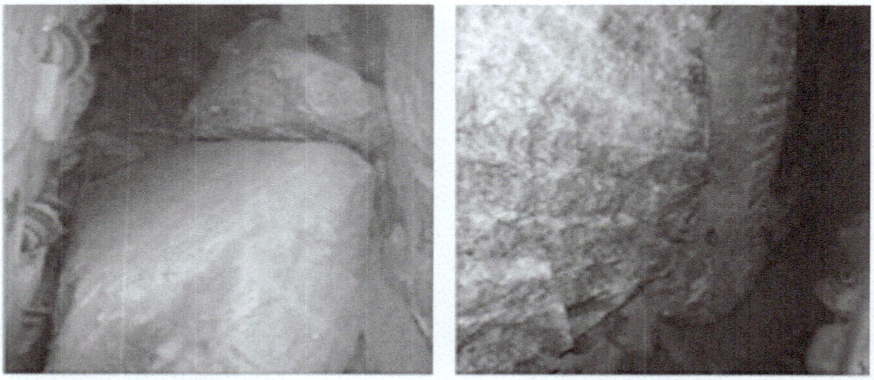

(a) Gotthard Tunnel (b) Lötschberg Railway Tunnel

Fig. 10.2 Blocky rocks encountered at tunnel faces

10.3 Adverse Ground Condition and Countermeasures in TBM Construction

10.3.1 Blocky Rock Problems and Countermeasures

The problems may be encountered during advancement, including uneven or unstable tunnel face and the gripper's inability to support effectively. The blocky rocks encountered at the tunnel faces of the Gotthard Tunnel and the Lötschberg Railway Tunnel are shown in Fig. 10.2.

1. Influence of blocky rocks on TBM tunneling

The main influences of blocky rocks on TBM tunneling include:

(1) High frequency and large vibration to the cutter;
(2) Fierce collision among the cutter, the cutter head and the blocked blocky rocks;
(3) Abnormal vibration to the cutter head;
(4) Cutter and bearing damages;
(5) Abnormal wear of mucking transportation system.

2. Countermeasures for blocky rocks

The commonly used countermeasures for blocky rocks are shown in Table 10.1.

10.3.2 Squeezing Ground Problems and Countermeasures

1. Problem analysis

Table 10.1 Countermeasures for blocky rocks

Item	Measures
Cutter profile	Flat
Cutter spacing	The ratio of cutter spacing to penetration is 10–20
Number of cutters (10 m in diameter)	60–65
Disc cutter diameter	17–19 in.
Cutter distribution	Double spiral, impact protection
Cutter head shield	Shield as required (keep 12.5 cm distance from the tunnel face)
Scraper	Periphery
Rotation speed (10 m in diameter)	155–190 m/min
Speed control	Variable
Operation control	Reducing tunneling and advance speeds when encountering blocky rock

In the short term, the high pressure applied to the cutter head/shield can cause TBM blocking. In the long term, the high pressure on the lining support can lead to deformation of the tunnel structure and failure of the tunnel function. If the squeezing degree is too large, the tunnel needs to be redesigned.

2. Countermeasures to prevent TBM from being blocked by the squeezing ground

   (1) The annular clearance between the shield and the rock can be expanded to 15–25 cm by expanding mill.
   (2) Stepped shield with large diameter of the front part of the shield can be used to prevent the rear part of the shield structure from being squeezed.
   (3) Lubricant (bentonite) can be used to reduce friction between the shield and the rock.
   (4) Powers of the primary thrust system and auxiliary thrust system can be increased to overcome the friction caused by the squeezing ground.

10.3.3 Problems of Ground with High Water Pressure and Countermeasures

Close mode is more suitable for tunneling strata with high water pressure (e.g., jointed rock mass). The strata with high water pressure were successfully tunneled with a dual mode TBM used in Holland (slurry/TBM) (water pressure: 13 bar).

10.4 Innovative Design in Respond to TBM Construction Risks

To improve the adaptability of TBM in variable and extreme geological environments, the design principles of multifunctional and highly adaptable TBM are as follows:

(1) The safety of personnel and TBM must be ensured.
(2) Various difficulties encountered in construction can be effectively solved.
(3) Various geological risks can be dealt with, such as the risk of rock burst, collapse and large deformity of surrounding rocks under high ground stress conditions with large overburden, the risk of collapse in fracture zones and weak surrounding rocks, the risk of water seepage and water ingress.
(4) TBM is highly adaptable with partial specialization.

10.4.1 General Compact TBM

General compact TBM is a more adaptable type than open-type TBM, which is suitable for rocky grounds, faults and squeezing grounds with unstable excavation surfaces. It is suitable for rock tunnels supported by New Austrian tunneling method (NATM) (Fig. 10.3). General compact TBM is usually used for tunnel excavation under good surrounding rock conditions without the need for segment lining installation; with the installation of NATM support systems made of prefabricated ring beams allowed, the TBM can be supported and operated and thus has a strong tunneling capability in unstable surrounding rocks. It can replace open-type TBM. As the compact TBM is designed to integrate the entire tunneling system, from the external equipment to the back-up and the TBM itself, the transportation, installation and operation costs at the TBM site can be reduced while the efficiency is increased.

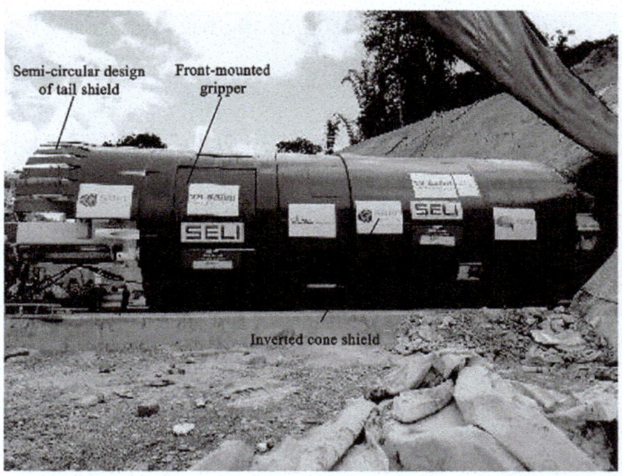

Fig. 10.3 General compact TBM

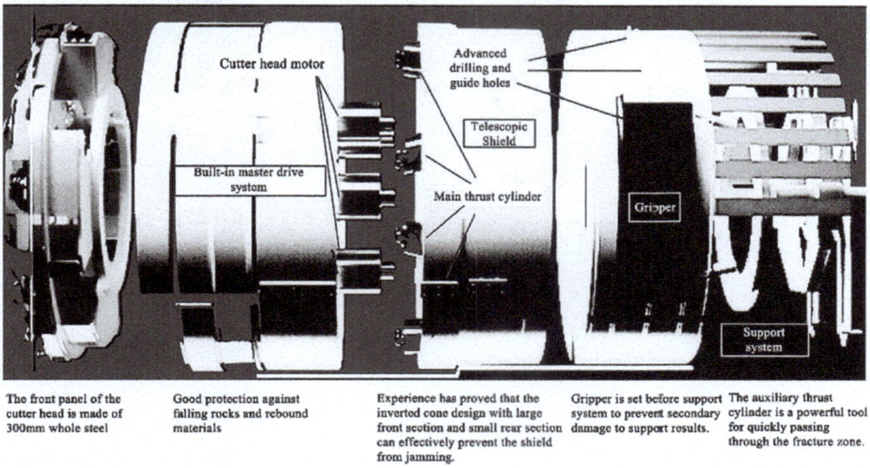

Fig. 10.4 Structure of general compact TBM

In terms of structure, the main thrust cylinder and the gripper are inside the shield structure. The gripper is arranged in front of the ring beam erector and equipped with an auxiliary thrust cylinder system. The inverted cone shield structure is designed to effectively prevent the TBM from jamming, and the auxiliary thrust cylinder can effectively ensure the fast passage through the fracture zones. In addition, it is equipped with a very compact back-up system and tunneling system, which allows for quick assembly, operation and disassembly as well as easy transportation (Fig. 10.4).

1. Function design

   (1) The main thrust cylinder and gripper are inside the shield structure.
   (2) The gripper is in front of the support system such as the ring beam erector.
   (3) It is equipped with an auxiliary thrust cylinder system; when the gripper cannot effectively support the tunnel wall under adverse ground conditions, the fast passage is achieved by the auxiliary thrust cylinder system.

2. Front-mounted gripper

The front-mounted gripper located at the head section can effectively avoid the interference between the gripper and the support of open-type TBM. It can replace open-type TBM and also has the advantage of construction safety of shield TBM, as shown in Fig. 10.5.

3. Anti-jamming shield design

The compact and optimized design is carried out to minimize the length of the shield. The inverted cone design is adopted for the main machine. The tail shield is designed as a semi-circular finger shield to effectively prevent the tail shield from being jammed and steel tile support is adopted for sections with severe geological

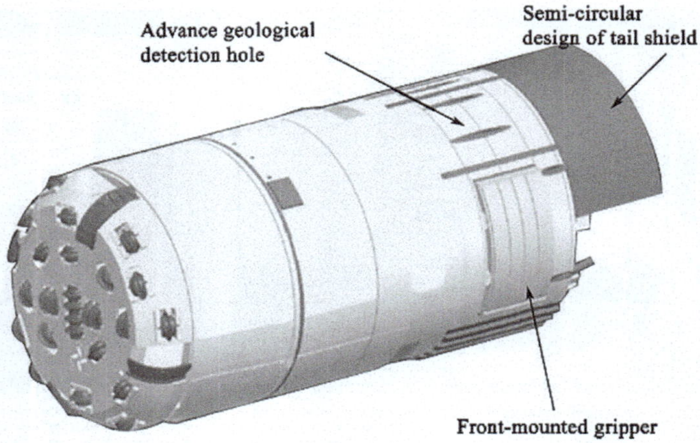

Fig. 10.5 Schematic diagram of front-mounted gripper

conditions (Fig. 10.6). Steel tiles with certain minor deformities allowed can play a better eggshell effect, which can effectively reduce and prevent the occurrence of shield jamming.

4. Advanced geological exploration

The General Compact TBM is equipped with a 360° dual probe drilling machine system (Fig. 10.7) near the location of rock outcrop to perform geological coring, construction of drainage holes and pressure relief holes, and other construction operations, making advanced geological exploration extremely easy.

5. Wet shotcrete system

The TBM is designed with a double wet shotcrete system close to the tail shield. When necessary, the shotcrete support is established close to the outcropped surrounding

Fig. 10.6 Steel tile support

Fig. 10.7 Dual probe drilling machine system

rocks in time when the surrounding rocks outcrop, so as to avoid the situation that core components in the working area should be protected during the wet shotcrete conducted by the open-type TBM and carry out operations quickly. Moreover, this system is closer to the tunnel face than the open-type TBM wet shotcrete system, which is conducive to rapid and safe construction.

10.4.2 Double-Shield Multifunctional TBM

Double-shield multifunctional TBM has a compact short shield, conical shield structure, telescopic hinge, main/auxiliary thrust system, strong combination performance, adjustable overcut volume and other structural features, which enable it to be applicable to a variety of complex rock tunnels, such as soft rock strata, hard rock strata, through faults, strata with high-pressure water ingress and other strata. However, it has a shortcoming of not being able to balance the tunnel face pressure. In addition, it has a highly adaptable stratum treatment system, including treatment of the full face and stratum treatment equipment installed at the shield structure and cutter head, and the rock support system can adapt to a wide range of geological conditions, enabling the TBM to tunnel on severely squeezing ground and rock burst area. Under the geological conditions with weak and broken rock strata, the double-shield multifunctional TBM tunnels in double-shield mode and converts to single-shield mode when the rock stratum is stable itself, while in the faults, it can also handle the rocks in front of the shield. With full consideration of the needs of stratum treatment functions, double-shield multi-functional TBM is able to reinforce the tunnel face with its cutter head, fill the empty cavity and stabilize the tunnel top with its telescopic shield structure, and grout through tail shield grouting and advanced pipe-shed grouting. It is particularly suitable for water ingress fissure rock strata and faults, which require rock reinforcement treatment.

Fig. 10.8 A Double-shield multifunctional TBM

The double-shield multifunctional TBM (Fig. 10.8) is suitable for tunneling under soft rock, hard rock, through faults, high-pressure water ingress and other conditions, but cannot effectively balance the tunnel face pressure.

1. Design features:

    (1) Compact short shield;
    (2) Conical shield structure (smaller tail shield than front shield);
    (3) Multi-module design, strong combination performance;
    (4) Telescopic hinge;
    (5) Equipped with main and auxiliary thrust systems;
    (6) Adjustable overcut volume;
    (7) Highly adaptable stratum treatment system, including treatment of full face and stratum treatment equipment installed at the shield structure and cutter head;
    (8) Rock support system, adaptable to a wide range of geological conditions;
    (9) Capable of tunneling on severely squeezing ground and rock burst areas;
    (10) Double-shield mode for tunneling on weak and broken rock strata;
    (11) Single-shield mode in extreme geological conditions;
    (12) Treatment of rocks in front of the shield in the faults.

2. Stratum treatment design

    (1) By cutter head (reinforce the tunnel face);
    (2) By telescopic shield structure (fill the empty cavity and stabilize the tunnel top);
    (3) By tail shield (grouting/advanced pipe-shed grouting).

3. Common classification

(1) Stratum treatment type

Fig. 10.9 Stratum treatment type double-shield multifunctional TBM

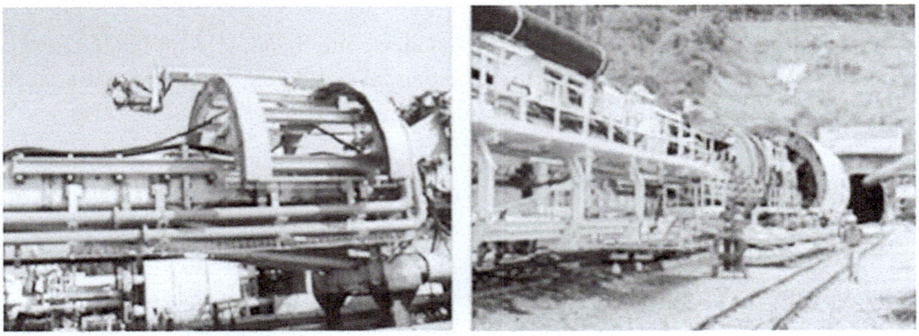

Fig. 10.10 Rock support type double-shield multifunctional TBM

Additional stratum treatment equipment is provided using the openings at the shield and cutter head. This type is applicable to water ingress fissure rock strata and faults, and it can solve the problem of water ingress, as shown in Fig. 10.9.

(2) Rock support type

It is equipped with a set of rock reinforcement equipment for rock reinforcement. The rock support type double-shield multifunctional TBM is shown in Fig. 10.10.

10.4.3 Dual Mode TBM

The most important feature of the dual mode TBM lies in that the TBM itself has the two or more tunneling modes, which can be switched in time when adverse ground conditions are encountered. Dual mode TBM can be divided into two categories: earth pressure/TBM and slurry/TBM, which can be switched between earth pressure balance and TBM mode, slurry balance and TBM mode, respectively.

Dual mode TBM (earth pressure/TBM) selects the mode according to the geological conditions under which the TBM passes to achieve good rock integrity and a self-stabilized excavation surface. This type mainly adopts the TBM mode, which is supplemented by earth pressure balance mode for soft grounds with local soft soil, very soft rock and other conditions. It solves the shortcomings of traditional earth pressure balance shield such as usage limitation in complex strata and inapplicability of TBM to soft soil strata, and has two tunneling modes on hard rock strata and soft soil strata to integrate shield technologies and TBM technologies. It is a new hybrid TBM integrating hard rock TBM tunneling technology and soft soil shield technology, which expands the geological adaptation range of the equipment and effectively reduces construction risks. It is simple to operate and has good economic benefits. In addition to the common earth pressure balance technology, the TBM has its own characteristics in design.

1. Dual mode TBM (earth pressure TBM)

Equipped with a composite earth pressure balance cutter head, it is applicable to soft rock and soft soil strata and the tunnel face can be stabilized. It can be operated in two modes: TBM mode and earth pressure balance mode.

(1) EPB double-shield TBM.
(2) EPB single-shield TBM: The single-shield TBM mode is used for tunneling on rock strata, and the EPB shield mode is used for tunneling on soft soil strata, as shown in Fig. 10.11. The schematic diagram of main machine in shield mode is shown in Fig. 10.12.

Dual mode TBM tunneling is characterized by single shield mode, which can significantly improve the tunneling efficiency. The muck in the excavation chamber is basically idle, which can greatly reduce the wear of cutter head, cutter and screw conveyor components and reduce the drive torque. The soil conditioner is not used basically, and only high-pressure water is sprayed in the excavation chamber for

Fig. 10.11 Robbins earth pressure balance single-shield TBM

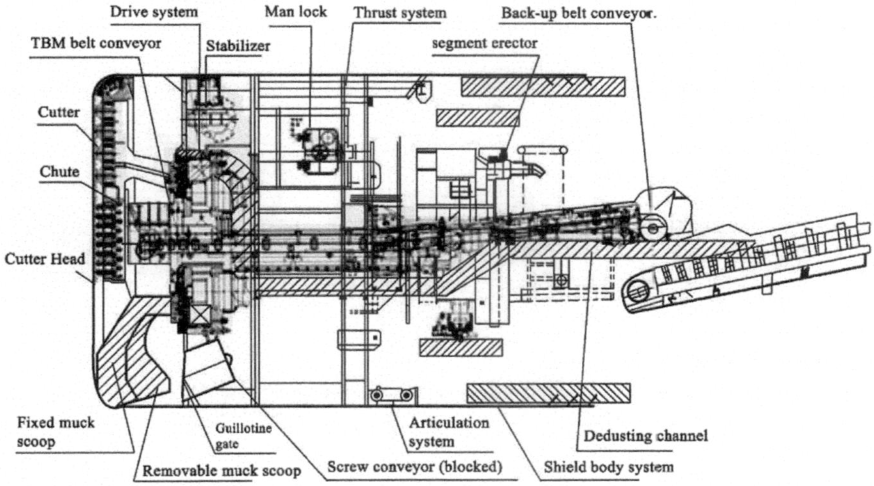

Fig. 10.12 Schematic diagram of main machine of dual mode TBM in single-shield mode

cooling and dust removal. The excavation chamber works under atmospheric conditions, with reduced grease consumption by main bearing and significantly reduced thrust force load on the cutter head. Therefore, the single shield mode has relatively lower tunneling efficiency, drive torque, total drive thrust, and tunneling cost. In the earth pressure balance boring mode, the excavation chamber is a closed system, and the balance between the excavation chamber pressure and the water and earth pressure at the tunnel face is established during the tunneling process to achieve tunneling in earth pressure balance mode. When tunneling mode switching is required, mode switching-related work is required.

Dual mode TBM (earth pressure/TBM) can realize such mucking functions as mucking through the screw conveyor in the earth pressure balance boring mode, soil conditioning through center rotary joint channel and mucking by TBM belt conveyor in the single-shield tunneling mode. Moreover, it has the man lock air pressure regulation function in earth pressure balance mode and dust removal function in open mode. The upper compartment of the front shield is designed with a dismountable mixing bar, a high-pressure water spray nozzle and a dedusting channel. The drive center can be installed with chutes, center rotary joints and TBM belt conveyors, and the lower part can be installed with a screw conveyor. Among them, the man lock, mixing bar, center rotary joint and screw conveyor are used in earth pressure balance mode and removed or blocked in open mode; the dedusting channel, stabilizer, chute, dismountable muck scoop and TBM belt conveyor are used in single-shield mode and removed, left idle or blocked in earth pressure balance mode; the mixing bar, chute, pressure-bearing compartment of center rotary joint and dedusting channel are designed to be dismountable or separable for easy removal and transportation during mode switching (Fig. 10.13).

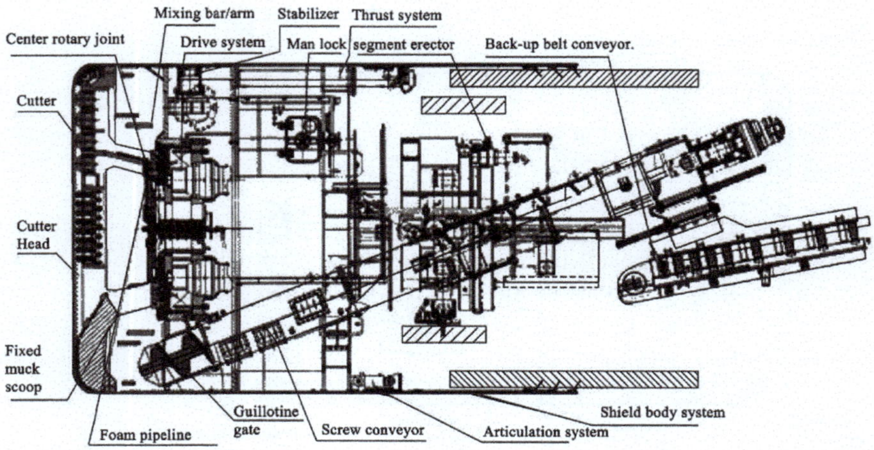

Fig. 10.13 Schematic diagram of main machine of dual mode TBM in earth pressure balance mode

The cutter head is designed to be a cutterhead for mixed ground, which is suitable for tunneling in weathered rock and soft soil strata (Fig. 10.14). The soil conditioning channel is arranged on the cutter head for earth pressure mode. Moreover, the cutter head structure, muck scraping cutter and muck scoop are symmetrically arranged. The cutter head has the functions of two-way rotation and mucking, which can effectively avoid the impact of the main machine and segment rolling on the tunneling construction during the tunneling process of the conventional single shield.

2. Dual mode TBM (slurry/TBM)

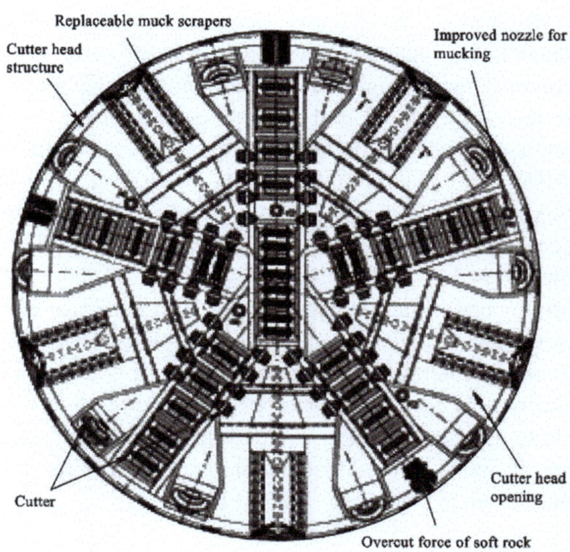

Fig. 10.14 Cutter head of dual mode TBM (earth pressure/TBM)

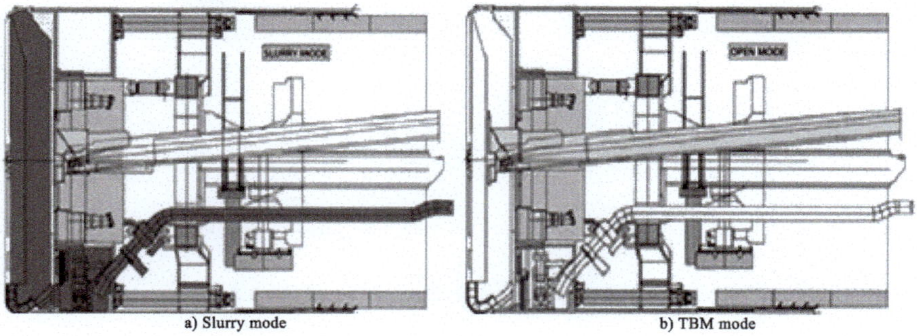

Fig. 10.15 Schematic diagram of dual mode TBM used in Holland

Slurry balance single-shield TBM: The slurry mode is used for tunneling high-water pressure fissure rock strata, and the TBM mode is used for tunneling rock strata without water or with little water. The dual mode TBM used in Holland is shown in Fig. 10.15.

The dual mode TBM (slurry/TBM) has a slurry mode used for tunneling of high-water pressure fissure rock strata, and its TBM mode is used for tunneling of strata without water or with little water. When the stratum conditions change, it can flexibly realize the switching between slurry and TBM modes to meet the needs of safe and rapid construction.

Detailed and reliable hydrogeological data are fundamental to the success of TBM project; TBM equipment with advanced technology and reliable quality and experienced TBM manufacturer with professional services are the keys to the success of TBM project; an experienced, scientific, professional and efficient construction team is the contributing factor to the success of TBM project. The highly adaptable multifunctional general compact TBM, dual-shield multifunctional TBM and dual mode TBM have advantages that conventional TBMs do not have. When they are used under suitable geological conditions, geological risks in TBM construction can be effectively avoided.

Questions

1. Please briefly describe the classification of TBM construction risks.
2. What are the construction risks of open TBM?
3. What are the construction risks of shield TBM?
4. What are the innovative designs used in response to TBM construction risks?

Chapter 11
TBM Design and Construction Examples

Key points of this chapter: focus on the Tianshengqiao Hydropower Station water diversion tunnel, Qinling Line I railway tunnel, Mogouling railroad tunnel, West Qinling railroad tunnel, Zhongtianshan railroad tunnel, Chongqing Rail Transit Line 6, Switzerland Vereina railroad tunnel, St. Gotthard Base Line railroad tunnel, water supply diversion tunnel from Songhua river to the central cities of Jilin province, Gaoligong Mountain railroad tunnel and other projects using gripper TBM engineering examples; focus on the water diversion tunnel of Datonghe to Qinwangchuan, water diversion tunnel of Yellow River to Shanxi province, Qinghai water diversion tunnel of Datong River To Huangshui River, water diversion tunnel of Hongyan River to Shitou River, Taiwan Snow Mountain Highway Tunnel, South Africa Lesotho Plateau diversion tunnel and other projects using double shield TBM engineering examples; focus on the Gansu Yintao water supply diversion tunnel, Chongqing Rail Transit Line 5 and other projects using single shield TBM engineering examples.

11.1 Gripper TBM Project Examples

11.1.1 Tianshengqiao Hydropower Station Water Diversion Tunnel

1. Project Overview

In 1985 and 1988, 2 American Robbins ϕ10.8 m gripper TBMs were successively introduced for the water diversion tunnel project of Guangxi Longlin Tianshengqiao II Hydropower Station, which was the first tunnel constructed with large section TBMs in China and the machine was the largest full-face hard rock TBM in the world at that time. Due to geological reasons, high equipment fault rate and low

K. Chen et al., *TBM Design and Construction*,
https://doi.org/10.1007/978-981-99-0059-6_11

progress, the average advance per month was 150 m at the beginning of the construction, and the highest advance per month was 242 m, but later due to numerous karst caves, fault zones and rockbursts, the disc cutter and cutter seat were damaged or smashed off, the bucket lip was seriously worn, the TBM working condition was deteriorated, the maintenance time was prolonged, coupled with the lack of management, and the late average advance per month significantly reduced to only about 60 m. Although the superiority of the TBM was not fully exploited in the construction of the Tianshengqiao Hydropower Station tunnel, the advantages of the machine itself and its relevance to the geology were proven.

2. Reasons for TBM Solution Selection

The Tianshengqiao II hydropower station is a water diversion power station with a design head of 176 m and an installed capacity of 1320 MW. There are 3 water diversion tunnels, each with an average length of 9555 m; the inner diameter ϕ8.7–9.8 m. The 8105 m tunnel section from the inlet near the sub-tunnel passed through the limestone stratum, which was mostly II and III type of surrounding rocks, and the overburden was 300–760 m; the 1450 m tunnel section below the sub-tunnel to the surge shaft passed through the sand shale stratum, which was III–IV type of surrounding rocks, and the overburden was 150–300 m. The axes of 3 tunnels are parallel to each other, with a spacing of 40 m for the limestone section and 50 m for the sand shale section, and an average slope drop of 3.31%.

During the preliminary design phase, the tunnel excavation scheme was always a major factor affecting the tunnel design, determining the selection of tunnel axis and the final determination of the tunnel diameter.

In the preliminary design report in 1976, the tunnel was excavated with the selection of drill and blast. The tunnel axis is arranged as a big broken line scheme along the river bend, with the inner diameter of 9 m and each tunnel length of 11.2 km. 6 support tunnels was set up to facilitate the "short tunneling of long tunnels". After the project resumed in 1982, TBM was introduced as the comparison scheme of tunnel excavation. Through consultation by the domestic experts and repeated comparison of designs for nearly 2 years, the excavation scheme with TBM excavation as the main method and drill and blast excavation as the auxiliary method was finally determined, and 3 construction support tunnels were arranged. The No. 1 and No. 2 main tunnels were tunneled 6400 m upstream by 1 TBM each into the No. 2 branch tunnel near the sub-tunnel; No. 1 and No. 3 construction branch tunnels were tunneled the remaining 3155 m tunnel section length with drill and blast.

At that time, two advantages of the selection of TBM scheme were mainly considered:

(1) Fast Advance Speed

This advantage can give full play to the characteristics of the tunnel line arranged by cutting and straightening the river bend topography of the project, each tunnel length can be shortened by 1645 m, so that the tunnel length is reduced to 9555 m. Although it cannot open as much tunnel working surfaces as the drill and blast for short tunneling of long tunnels, it shoulders the construction of a longer blind advancing tunnel section, but its fast advance speed can still guarantee the schedule, and save the amount of temporary construction work. The cost of TBM excavation is approximately double that of drill and blast, but this disadvantage is offset by its fast advance speed.

(2) Low Disturbance to the Surrounding Rock and High Excavation Quality

On the one hand, the amount of overcutting (overcutting within 5 cm) and the corresponding amount of concrete overfilling are reduced; on the other hand, the lining structure can be optimized. The preliminary design scheme is considered as 1/3 concrete lining and 2/3 shotcrete bolt lining. The 1984 cost estimate calculated that the TBM straight line solution saved about CNY 145 million compared to the drill and blast method broken line solution.

3. TBM Construction Process

The project introduced the ϕ10.8 m gripper TBM made by Robbins Company of the United States, and it was the largest full-section hard rock TBM in the world at that time. 2 TBMs with factory numbers 353–196 and 353–197 TBM respectively were used in the Chicago waste water treatment project.

The TBM construction process is as follow:

The TBM main machine is 16.6 m long, after the 11 m connection bridge behind the main machine, there are 16 platform gantries and 4 ramp gantries together to form a track-type back-up gantry group (No. 1 back-up gantry is 110 m long, No. 2 back-up gantry is 145 m long), the lower wheels of the back-up gantry are placed on the tunnel excavation track and follow the TBM main machine forward. The gantry is set on the platform of the back-up gantry to divide it into upper and lower layers. The lower platform is used for loading muck and shunting, with double tracks laid and track gauge 1066 mm. The upper track of the ramp at the tail of the platform is connected to the track laid in the tunnel, and the ramp gradient is not more than 5%, so that the mucking car and transport vehicle can get on and off the platform. The upper platform is arranged with muck transfer belt conveyor, variable pressure equipment and air ducts, etc. It is also the main passage for staff to enter various parts of the machine.

The TBM cutter head is equipped with 69 disc cutters, and the cut rock block is dumped by 12 buckets on the cutter head into the trough and transferred to the 42 in wide belt conveyor, then transferred to the hopper at the end of the transfer belt on the platform frame of the back-up gantry and unloaded into the muck car on the lower platform of the back-up gantry. With a volume of 19.6 m^3 per muck car, each train, consisting of 6–8 mucking muck cars, are pulled out of the tunnel by a 35 t diesel locomotive made by Fuji Heavy Industries of Japan. There are muck dumpers at the entrance of the branch tunnel, and the muck cars enter the muck dumper (which can unload 2 at the same time) and turn over 180° to unload the stone muck into the muck pit, and then the stone muck is loaded into a 20 t dump truck for transportation by loaders.

TBM thrust stroke is 1.8 m, thrust force is 13800 kN, the horizontal support force acting on the side wall of the tunnel is 32560 kN per side, the cutter head drive power is 1790 kW, the total mass of the machine is 734 t, the maximum weight (inner cutter head support) is 88.5 t, the voltage level is 480 V, and the frequency is 60 Hz.

4. TBM Assembly

The design assembly plan was to assemble on the special Robbins trolley, then pulled into the traction section in the tunnel. The trolley was removed from the delivery trough, and the cutter head was installed to start the tunneling for 3 months.

Later, 16 Hilman rollers were purchased from the United States along with the main machine making the assembly method different. The Hilman roller is an extremely handy tool for installing major machinery and equipment and moving over short distances and it is suitable for TBM assembly. No. 1 TBM was assembled outside the tunnel, while No. 2 TBM was assembled inside the tunnel.

(1) No. 1 TBM Assembly

No. 1 TBM assembly was carried out outside the entrance of No. 2 branch tunnel. Before assembly, the No. 2 branch tunnel was excavated 24 m long by drill and blast method, and the branch tunnel section was lined with a diameter 10 cm larger than that of the TBM excavation diameter, with a full circle within 4 m from the working surface, followed by a 0.6 m long, 0.6 m deep, 5 m wide column removal groove on the bottom plate, then later, the tunnel section gradually changed into a city portal type, and the tunnel bottom plate then became a flat bottom with a width of 8.6–10.9 m, with a 3 m wide, 0.5 m thick bottom plate concrete poured in the center and extended to 30 m outside the tunnel, and 2 No. 32 I-beams were buried longitudinally in the concrete bottom plate, with a spacing of 2.28 m, as a track for the Hilman roller group. In addition, in the longitudinal axis of the concrete bottom plate at a distance of 1.5 m, a 30 cm deep, $\phi 10$ cm steel pipe was pre-buried as a fixed hole for the thrust cylinder, and the TBM was installed on the Hilman roller group on this bottom plate, lifted by a 150 t crawler crane of KH-700 type and a 75 t tyre crane from Hitachi Japan, and then lifted by a thrust cylinder which was fixed in the fixed hole of the bottom plate with a thrust force of 220 t on the TBM, and then removed from the Hilman roller group at the column removal groove. The assembly time was 56d and the tunneling of No. 2 branch tunnel was started on March 18, 1985.

(2) No. 2 TBM Assembly

During the construction of No. 2 branch tunnel, No. 1 TBM encounters rock burst, karst cave and fracture zone for many times. After temporary support, the diameter of some tunnels was less than the standard diameter of 10.8 m, and the diameter of individual sections had been reduced by as much as 50 cm from a single point, so that the No. 2 TBM main machine can no longer pass through as a whole.

The final method of assembly in the tunnel was adopted through the program demonstration. Using the natural space formed by the intersection of the branch tunnel and the No. 2 main tunnel, a 12 m long, 25.5 m wide and 16.3 m high installation chamber was first excavated, and a 187 m TBM preparation section with a diameter of ϕ11.2 m was excavated upstream along the No. 2 main tunnel using drill and blast. Then, the in-situ steering device and TBM thrust device were embedded with 20# I-beam and 20 mm thick steel plate in the installation chamber and positioning section. After dismantling the track in the branch tunnel and cleaning the tunnel, the part of the main machine which had been assembled outside the tunnel (there were still 10 big parts not yet installed) was made to enter the installation chamber along the branch tunnel by itself in 7d time, and the head and tail of the machine were jacked up respectively with 6 electric hydraulic jacks of 1000 kN, and a total of 10 Hilman rollers were installed on the front and rear supports and moved to the in-situ steering device. Due to the poor geological conditions of the installation chamber, the 40 t crawler crane of the advanced tunnel was used to install the left and right supports, top shield pedestals, double-side and middle shields and 4 external cutter heads and 10 large pieces, with a maximum component mass of 29 t. After the main machine was installed, it was transferred to the embedded track of the positioning section of No. 2 main tunnel through the in-situ steering device, and No. 2 TBM was pushed to the working face of No. 2 main tunnel in the advancing mode of No. 1 TBM at the tunnel portal.

Due to the limitation of geological conditions, the turning radius of the turnout track with a width of 25.5 m is only 65 m, which is far less than the requirement of 500 m turning radius required by the whole mobile mucking platform. Therefore, the platform vehicle is towed by a winch section by section to the positioning section of No. 2 main tunnel, and then integrally connected with the TBM. No. 2 TBM was put into operation on July 3, 1988, after 85d of work from the tunnel entrance to the start of operation.

5. TBM Dismantling

There are three types of TBM dismantling methods. One way is to remove at the exit of the mountain from the tunnel through TBM; the other way is to remove the whole machine from the exit of the tunnel from the original tunneling road and remove it at the entrance; the third way is to remove it inside the tunnel. The tunnel entrance of the Project to No. 2 branch tunnel is advanced by TBM and drill and blast in opposite directions, and the tunnel is broken through in the mountain. Due to the shrinkage of the tunnel section caused by the support of some tunnels advanced by TBM, and the tunnel lining concrete behind the TBM has been followed to block the passage, so the removal scheme in the tunnel is selected.

The demolition site is located in the drill and blast excavation section near the end of TBM tunneling. The construction unit compared the 150I crawler crane solution and the bridge crane solution, but the crawler crane solution had a large excavation section (23 m high and 17 m wide) and was not convenient to operate. The bridge crane scheme was adopted for construction, and the size of the demolition site was $L \times W \times H = 31.5\ m \times 12.2\ m \times 19.4\ m$. The heaviest piece to be lifted was the cutter head support (88.5 t), which was loaded into a platform gantry after removal, and the No. 7 connecting tunnel near the upstream of the No. 1 TBM removal site (section size: $W \times H = 8\ m \times 6.5\ m$, turning radius of 9–17 m) entered No. 2 main tunnel, and then was transported upstream from No. 2 branch tunnel portal, lasting 22d, and completed on February 23, 1992. The No. 2 TBM, on the other hand, was transported directly from the No. 2 main tunnel upstream to the No. 1 branch tunnel after being dismantled.

6. TBM Construction Implementation Effect

1. The No. 1 TBM, including the No. 2 branch tunnel, tunneled a total of 4.6 km, including the No. 1 main tunnel 3.3 km in 77 months, with an average advance per month of 60 m; the No. 2 TBM tunneled the No. 2 main tunnel 3.1 km in 48 months, with an average advance per month of 65 m. The average advance per month of the two TBMs was only 60–65 m, which was relatively large difference from the design average advance per month of 200–250 m in the preliminary TBM scheme. During construction, only about half of the tunneling length specified in the design had been completed. The superiority of TBM has not been fully exploited due to the harsh geology.

2 sets of full-face large-diameter hard rock TBMs were introduced into the Project, and a total of 7.7 km long tunnel was driven. This was the largest TBM construction in China at that time, accumulating rich construction experience for the promotion of TBMs in the future. It was a beneficial exploration and attempt.

Construction practice had proven that higher advance speed and minimal disturbance to the surrounding rock are two of the main advantages of using TBMs. In general, the advance speed of TBM in this project was relatively low, but the maximum advance per month reached 242 m (February 1986), so a high advance speed could be achieved. In addition, the quality of the tunnel was quite high, with over-excavation within 5 cm and high finish. However, it also proved that TBMs were particularly sensitive to geology and difficult to cope with harsh geology. For example, on July 23, 1985, at K0 + 619 of No. 2 branch tunnel, a karst cave with a span of 7.4 m that was basically orthogonal to the tunnel line appeared. The karst cave extended about 200 m to the upper right at an included angle of about 30° with the horizontal line, and was filled with clay. The sliding soil mass was about 10,000 m^3, and it had been treated for 174 days before passing. The reason for the difficulty in handling was that the TBM blocked the channel between the karst cave and the outside world, and only bypass channel could be used for handling at the karst cave.

7. Analysis on the Reasons for Failure to Give Full Play to the Superiority of TBM

(1) Geological factors

During the construction of water diversion tunnel of Tianshengqiao II Hydropower Station, unforeseen and abnormally developed karst caves and rock burst were encountered, which often led to difficulties in TBM excavation and failed to give full play to the mechanical efficiency of TBM.

From July 23, 1985 to January 11, 1986, the No. 2 branch tunnel K0 + 619 karst cave dark river channel was handled, No. 1 machine was shut down for 174 days, nearly 6 months, 4 pipe pillar piles were completed, 7.3–18.5 m long, 150 m^3 of concrete was poured, 5450 m^3 of yellow mud was cleared and 50 t of steel support was erected.

From February 21, 1986 to April 9, 1986, rockburst occurred in K0 + 935–K0 + 980 section of No. 2 branch tunnel, 421 m^3 of muck was cleared, 35.1 t of steel support was erected, 161 m^3 of concrete was sprayed and 624 rock bolts were laid, causing No. 1 machine to be shut down for 48 days.

From April 12, 1986 to July 10, 1986, in the large karst cave of K0 + 984.3–K1 + 002 section of No. 2 branch tunnel, 2503 m^3 of yellow mud was removed, 49 t of steel support was erected, 512 m^3 of shotcrete was placed, and 406 rock bolts were installed, resulting in 90 days of downtime of No. 1 machine.

The No. 2 branch tunnel constructed by No. 1 TBM, the II and III type of surrounding rock accounted for 89.3%, and the V type of surrounding rock only accounted for 1.3%, the adverse geology tunnel section mainly had karst caves, fracture zones, rock bursts and water leakage. In TBM tunneling section, except that it was difficult to deal with the creep mass filled in the karst cave K0 + 619 of No. 2 branch tunnel, the rest of the karst caves encountered are basically filled with dense clay, which affected the construction period. Tunnel rock burst frequently, due to the random nature of its occurrence cannot be prevented, TBM excavation on the surrounding rock disturbance small, the perimeter of the surrounding rock stress cannot be fully released with the excavation, so the rock burst on the construction of a greater impact than the drill and blast section, according to statistics from March 1985–September 1989, only rock burst treatment accounted for 24% of the construction period. In addition, the water ingress also had a negative impact on the construction schedule. According to statistics, during the 77-month construction of No. 1 machine, the time of direct shutdown due to geological influences such as karst caves and rock burst accounted for 28.85%. Especially in 3 years from 1985 to 1987, TBM encountered serious unfavorable geological conditions, with an average annual shutdown time of 44.58% and a high of 56.83% in 1985. Geological factors are undoubtedly an important factor restricting the excavation speed of TBM.

(2) Matching Conditions of TBM

In general, although the direct shutdown time in the last 3 years is less than that in the previous 3 years due to geology reasons, there are more rockburst sections, and rockburst rockfall damages the disc cutter or causes the dislocation of the cutter seat, resulting in the deterioration of the working condition of the disc cutter, frequent cutter changes and increased time.

As the cutter is not suitable for the geology, the cutter consumption is large, with an average cutter consumption of 0.32 pieces/m. Due to the high quality of cutters required by TBM, but the quality of domestic cutters does not meet the requirements, it is necessary to import, and sometimes the supply is not timely and the construction period is affected. After 1988, the normal downtime increased significantly, which seriously affected the operational efficiency of the TBM. According to statistics, the normal downtime rate of No. 1 TBM was 10.49% in 1985, 11.08% in 1986, 19.34% in 1987, 43.36% in 1988, 54.11% in 1989, and 33.86% in 1990.

The power supply frequency is not matched with the power supply frequency (the machine is American cycle wave 60 Hz), so the power cannot be fully played.

Due to the introduction of second-hand machines and obsolete equipment, electrical faults account for a large proportion every year, mechanical faults fluctuate greatly, and hydraulic faults and lubrication faults showed a rising trend in construction. Especially in the last 3 years, the fault rate was relatively high. In 1988, the fault rate of the main machine alone reached 27.11%.

(3) Man-hours Consumed for TBM Assembly and Removal

The assembly outside the No. 1 TBM tunnel lasted 56 days, and it took 6.6 months from the excavation of the dismantled chamber to the removal of the machine outside the tunnel. The construction period for the assembly chamber in the No. 2 TBM tunnel was 8 months, and it took 85 days for the TBM to be put into operation from the tunnel entrance, which affected the construction of No. 1 TBM by 70 days.

(4) Weak Management Level and Technical Level

The high degree of mechanization of the TBM requires a high level of management and technical skills of the construction unit. However, due to the first use of TBM by the Construction Contractor, this aspect is still weak.

As geology disasters cause great difficulties to tunnel construction, geology survey shall be further strengthened, disaster prevention and control shall be strengthened, and geology forecast shall be carried out. Auxiliary construction methods corresponding to geology conditions shall be selected, surrounding rocks shall be reinforced in time, and ring beam support, probe rock bolt, probe canopy tube support and other work shall be carried out.

11.1.2 Qinling Line I Railway Tunnel

1. Project Overview

Qinling Super-long Tunnel is located on the Xi (Xi'an) Kang (Ankang) Railway, with a total length of 18.456 km. The elevation difference between the entrance and exit of Qinling Tunnel is about 155 m. It crosses the east–west tectonic zone of the Qinling Mountains and has undergone multiple phases of tectonic movement, metamorphism, magmatic activity and mixed lithification, and its geology structure and stratum lithology are very complex. The lithology is dominated by hard rocks such as mixed granite and mixed gneiss, with a dry compressive strength of 78–325 MPa. After demonstration and comparison of various construction schemes, it is decided to adopt TBM method and drill and blast for construction of the tunnel of Qinling Line I. 1 gripper TBM tunneling, shotcrete bolt support and composite lining are adopted at the entrance and exit respectively, and the construction scheme of secondary mold pouring concrete lining is carried out by full-circle traveling formwork jumbo. The tunnel is designed as a circular section, with an excavation diameter of $\phi 8.8$ m and a finished tunnel diameter of $\phi 7.7$ m, see Fig. 11.1.

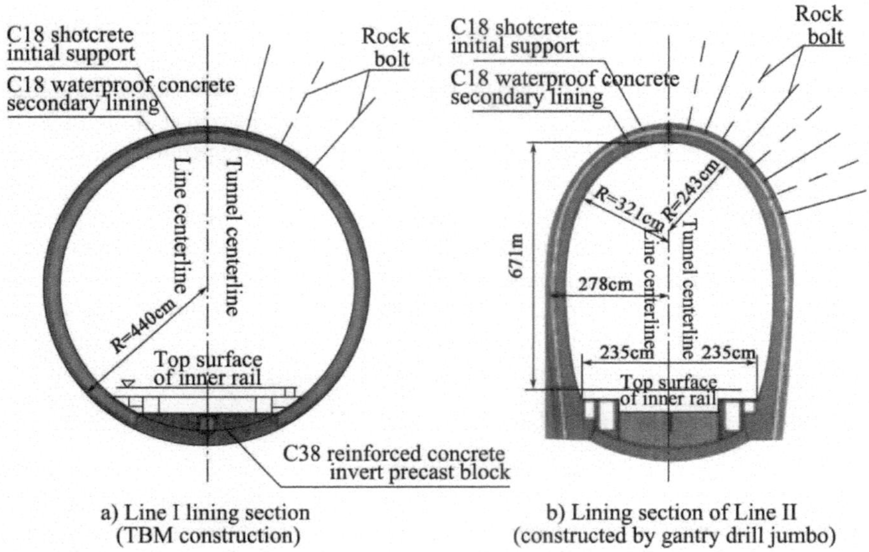

Fig. 11.1 Cross-sectional view of Qinling Tunnel

The engineering geological conditions of Qinling Tunnel from the entrance to the exit can be basically divided into 4 sections (Fig. 11.2). There are mainly two types of lithology: one lithology is mainly mixed gneiss, interspersed with gneiss and shist residue, compressive strength of 78–325 MPa, most of the rock body is not greatly affected by the structure, good lithological integrity, few joints, fissures; the other lithology is mainly mixed granite, compressive strength of 117–192 MPa. The cumulative width of the tunnel across the fault is nearly 1000 m, and the joints are developed.

According to Railway Tunnel Design Specification (TBJ 3–1984), 90% of the TBM construction section is above Class VI, and 10% of the section is accompanied by many extruded weak structural surfaces and jointed dense zones. In addition, rock burst, water ingress, geothermy energy and instability of surrounding rock, etc. may occur during construction. V and IV type surrounding rock joints are more

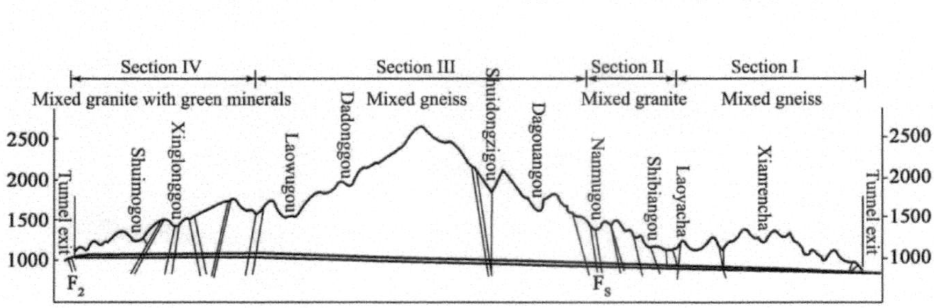

Fig. 11.2 Schematic diagram of geological longitudinal section of Qinling Tunnel

developed–not developed, the rock body is large block masonry structure–giant block state structure, very poor drillability, 1.8 m per cycle, the pure digging time is 3–5 h; fault unfavorable structural surface, rock explosion and other adverse geology caused by the destabilization of the surrounding rock after excavation, in order to ensure the safety of the machine and personnel, need to stop frequently to strengthen the support, so the TBM utilization ratio is low.

The northern entrance of Qinling was constructed by China Railway Tunnel Group, with a TBM tunneling length of 5.244 km, it started at 8:00 am on January 19, 1998 and was connected to the southern entrance at 7:32 am on August 29, 1999, with an average monthly progress of 252.36 m in 1998, 290.64 m in 1999, and the highest monthly progress of 528.48 m; the southern entrance of Qinling was constructed by China Railway 18th Bureau Group, with a TBM tunneling length of 5.621 km.

2. Technical Parameters and Characteristics of TBM

2 TB880E gripper TBMs made by Wirth Germany were introduced to the Qinling I tunnel, as shown in Fig. 11.3, and their main technical parameters are shown in Table 11.1. TB880E TBM is composed of 3 parts: main machine, connection bridge and back-up system, integrating functions such as tunneling, supporting, mucking, ventilation, drainage, cooling and lighting. TBM construction is very different from drill and blast construction in terms of the sequence of operations. TBM construction is operated as an industrialized construction system. It has 3 basic characteristics: coordination, continuity and intensity, determining the principles of its construction organization.

(1) Coordination

As an industrialized construction system, TBM is closely related to each working unit and is very orderly. Any uncoordinated work process will have a significant impact on the TBM construction efficiency. Quantification of work unit capability

Fig. 11.3 TB880E Gripper TBM made by Wirth, Germany

Table 11.1 Main technical parameters of TB880ETBM

System name	Parameter name	Technical parameters
TBM main machine	Tunneling diameter	ϕ8.8 m
	Overall dimension	22 m × 8.8 m × 8.8 m
	Advance speed	(Saturated compressive strength 260 MPa)1.0 m/h
	Cutter head power	3440 kW
	Cutter head rotation speed	2.7/5.4 r/min
	Maximum thrust	21,000 kN
	Torque	5500 kN m
	Machine stroke	1800 mm
	Maximum supporting force of supporting system	60,000 kN
	Support earthing pressure	1.4–2.8 MPa
	Number of cutters	6 single-edged center disc cutters, 58 single-edged disc cutters
	Cutter Spacing	About 75 mm
	Cutter bearing capacity	25 t/pcs
	Transformer capacity	4000 kV·A
	Conveyor transfer capability	780 m^3/h
Back-up system	Back-up gantry gauge	2980 mm
	Transport train gauge	900 mm
	Back-up length	210 m
	Minimum curve radius	500 m
	Belt transfer capability	770 m^3/h
	Muck loading chute moving distance	80 m
TBM auxiliary equipment	Lifting capacity of invert block crane	13 t
	Probe drilling machine working range angle	Top range 80°–90°
	Probe drilling machine elevation angle	0–10°
	Probe drilling machine borehole diameter	ϕ64 mm, depth 30 m
	Bolt driller borehole diameter	ϕ38 mm, depth 3.5 m
	Operating range	Top of TBM front end 0–150°; below arch of TBM rear end
	Number of bolt drillers	TBM front-end 2 sets; back-end 2 sets

(continued)

Table 11.1 (continued)

System name	Parameter name	Technical parameters
TBM back-up equipment	Continuous agitator	6 m^3/h
	Agitator	160 L
	Grouting pump	160 L/min
	Grouting pressure	1–50 bar
	Grouting pump piston diameter	ϕ120 mm
	Turbo Mixer	5.5 m^3/h
	Borehole diameter of grouting rod	54–120 mm
	Grouting pipe	ϕ50 mm
	Wet shotcreting machine	3–14 m^3/h
	Shotcrete system air volume consumption	10 m^3/h
	Shotcrete system air pressure	6 bar
	Number of manipulator systems	2 pcs
	Horizontal movement of manipulator system	5 m
	Manipulator system arc movement	± 150°
	Horizontal telescoping of manipulator system	2 m
	Emergency generator power	200 kVA/160 kW
	Output power	380 V/3 phase/50 Hz
	Air volume of air compressor	20 m^3/min
	Air pressure of air compressor	8.5 bar
	Air compressor power	132 kW

matching and work organization, as well as the orderly arrangement of work time for each process, are the main elements of coordination requirements.

(2) Continuity

TBM construction is a continuous parallel cycle of individual operations alternately, unlike the drill and blast method of construction, each individual operation in the process is an intermittent cycle. This is determined by the mechanism of continuous rock breaking by TBM. In TBM construction, the fault of any one process or single operation can lead to a production stoppage of the entire mechanical construction system. Therefore, the time requirement of continuous operation between processes is the main content of the continuity requirement of construction organization.

(3) Intensity

As the rock breaking method for continuous tunneling construction of the industrialized system, the TBM production equipment is required to concentrate all the functions of tunnel construction according to the 3 basic work of tunneling, supporting and transportation. Therefore, the intensity of single operation of TBM construction is one of its important characteristics.

It can be seen from the above 3 characteristics that the production system of TBM not only has strict requirements on hardware configuration and mode, but also has higher requirements on software configuration in construction organization management. The construction unit has done a lot of research in hardware on the outbound transportation of excavated muck, support and energy assurance; in software, it has carried out various management on each link and achieved great effect and achievement.

3. Analysis of Operation Decision System for TBM Construction Organization

The fundamental objective of tunnel construction is to obtain the best progress, construction period and construction quality at an appropriate cost. To achieve this, each construction process of TBM and the relationship between them must be studied first.

The advancing process of TBM equipment is also the tunneling process of construction. The two processes of "tunneling → drainage → transportation" and "transportation → support → tunneling" are the basic operations of gripper TBM construction. The guarantee system of tunneling construction operation includes direct and indirect parts, consisting of material organization and transportation and TBM maintenance.

TBM construction is an integrated process compared to the drill and blast, based on a high degree of integration of the process by the TBM and guaranteed by the performance and quality of the TBM product. People deal more directly with machines and indirectly with rocks. In terms of approach, TBM construction has a completely different concept compared to the drill and blast due to the high automatization of the TBM. The organization and management of working procedures are more strict and compact in time sequence. Due to different rock breaking mechanisms, the TBM tunneling and muck transportation are synchronized. For gripper TBM only in the time of support, people will deal with the rock like drill and blast.

The fundamental process and relationship of tunnel construction are illustrated in Fig. 11.4. As can be seen from the figure, tunnel construction is based on 3 tasks, such as tunneling, support and drainage and transportation, and supported by the power supply, air and water systems. The environmental protection system is the basic environment required to meet the premise necessary for construction; the monitoring system is used to ensure automated construction.

(1) Tunneling and Muck Dumping Operation System

TBM tunneling is the process of cutter head advancing → rotating → cutting. The cut rock enters the cavity of the cutter head through the bucket and scraper on the cutter head, and then is transported out through the belt conveyor. This working process is the core operation process of the TBM, which determines the tunneling speed of the

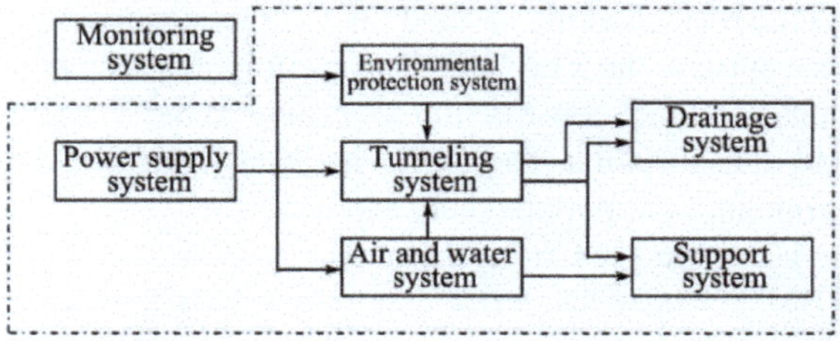

Fig. 11.4 TBM construction relationship and process

project construction and the connection with the corresponding supporting processes. Therefore, the pure tunneling speed of TBM system and the tunneling time of single tunneling stroke are put forward, and the basic requirements of material transportation organization are put forward.

(2) Tunnel Support and Construction Safety System

During tunnel construction, various problems of stratum stability will inevitably be encountered to be solved. The stability of surrounding rocks will not only endanger the safety of personnel and equipment, but also affect the success of the project. At the same time, it also includes the impact of the use of equipment on the human body and life, the impact of the construction environment on the safe use of equipment, etc.

The support mode is determined according to the hydrogeological conditions of the project. Its supporting structure, parameters and measures are also determined according to the geological conditions. In addition to meeting the engineering quality requirements and corresponding equipment and manual operation requirements, the support shall also be planned and designed according to the advance speed and time sequence. This is very important for avoiding and reducing work interference and waiting for support time, and improving the ratio of pure tunneling time and project speed, as shown in Fig. 11.5.

(3) Transportation System

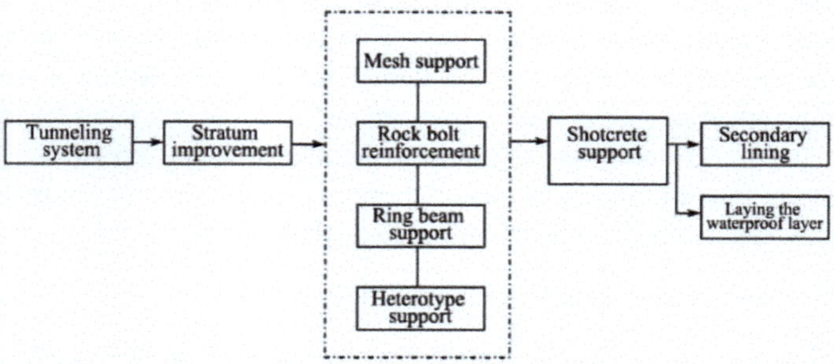

Fig. 11.5 Tunneling and Supporting Operation Process. *Note* Special-shaped support includes curved shape channel steel and ϕ22 steel processing single layer simple grid and inserted grouting backfill reinforcement

It mainly consists of the mucking transfer and the transportation of materials such as invert blocks. Materials mainly include engineering materials and equipment maintenance materials. Transportation system is an important system of tunnel construction, a link between the tunneling process and the work outside the tunnel, and an important link that cannot be ignored.

(4) Energy Supply and Labor Environment System

Electricity is an important energy source of TBM. Therefore, the TBM system operation is subject to the quality of the power supply and power distribution system. The power distribution must be set according to the power distribution standard of electric equipment and the operating condition requirements of the tunneling system.

Wind, water and electricity, etc. are the fundamental energy that must be consumed in both tunnel construction and TBM construction. Therefore, the supply of wind, water and electricity is the fundamental condition for normal and effective construction.

In the process of tunnel construction, there will be dust, harmful gas, noise and humidity, which are harmful to human body, equipment and the surrounding environment. Therefore, the protection of the labor environment is necessary to prevent equipment failure and personal injury, as well as to comply with environmental health and labor regulations.

TBM industrialized construction management is reflected in coordination and organization of the operation sequence of the macro system and control of the time and quality of a single micro-operation. System operation and process management include monitoring of hardware system, construction organization of production system, operation organization of guarantee system, troubleshooting for abnormalities, etc. The efficiency of traditional drill and blast is subject to the cycle time of the whole process, while that of TBM construction is subject to the operating time of the machine.

4. Main Technical Indicators of TBM Tunneling at North Portal of Qinling Mountains

(1) Construction Period and Completion Chainage

① Maximum Tunneling Record
② Breakthrough date: 7:32 a.m. on August 29, 1999;
③ The total tunneling length completed was 5244 m (design boundary length, including 71 m of TBM commissioning tunneling from December 18, 1997 to January 18, 1998). Among them, the TBM passed 709.4 m of soft and weak surrounding rocks, 1819 m of V and IV hard surrounding rocks, and 24 faults.

(2) Maximum Tunneling Record

① Maximum Tunneling Record
② Monthly record 528.48 m (July 26-August 25, 1999), chainage DK69 + 669.2–DK70 + 197.68.

(3) Average Tunneling Indicators
① Monthly Indicators.

According to the calendar statistics, the total tunneling time (calculated from the trial tunneling started on January 19, 1998) was 11.4 months in 1998, and 7.9 months in 1999, totaling 19.3 months. In 1998, the tunneling depth was 2878.92 m, with an average monthly footage of 252.36 m; in 1999, the tunneling depth was 2296.08 m, with an average monthly footage of 290.64 m.

If the downtime of 2023 h caused by design and manufacturing quality of the Wirth Company is deducted, for 84.3 days in total, equivalent to 2.81 months (one month is calculated as 30 days), the downtime of 0.94 months in 1998, the tunneling time after deduction is 10.46 months, and the average advance per month is 275 m; in 1999, the downtime of 1.87 months, the tunneling time after deduction is 6.03 months, and the average advance per month is 380.78 m. The total average monthly advance dimension is 313.71 m.

② Average Daily Advance Rate.

According to the calendar construction days, the TBM construction days were 347 in 1998 and 240 in 1999, totaling 587 days. The average daily advance rate in 1998 was 8.291 m; the average daily advance rate in 1999 was 9.57 m; and the average daily advance rate during the TBM construction period was 8.81 m.

After passing through the F_4 fault (DK69 + 010–DK69 + 390), the section from the through point to DK70 + 245 is 855 m long (from July 9, 1999 to August 29, 1999) in 50 days, with a daily average of 17.1 m.

If we deduct the downtime caused by the design and manufacturing quality of Wirth, the average daily advance rate in 1998 was 9.02 m; in 1999, the average daily advance rate was 12.5 m; and the average daily advance rate during the construction period of TBM was 10.29 m.

③ Machine Time Utilization Ratio.

The daily maximum was 75%; the monthly maximum was 58.31%; and the total average was 38.7% (after deducting mechanical failures and other factors).

④ Equipment Availability Ratio.

The average was 78.82% in 1998; 53.77% in 1999; and 65.74% in the overall average.

⑤ Advance Speed.

In 1998, it was 1.2 m/h; in 1999, it was 1.38 m/h. The total average for both years is 1.3 m/h.

(4) Cutter Consumption

See Table 11.2 for cutter consumption. The data in the table has included 71 first dress of cutting tools.

(5) See Table 11.3 for the statistics of accessories, cutters and power consumption during TBM tunneling.

Table 11.2 Statistics of TBM Cutter Consumption at Northern Mouth of Qinling

Name		Unit	Qty	Total
Center disc cutter	Assembly	Pcs	49	98
	F.R.L	Pcs	12	
	Cutter ring	Pcs	37	
Face disc cutter	Assembly	Pcs	430	1852
	F.R.L	Pcs	36	
	Cutter ring	Pcs	1386	
Edge cutter	Assembly	Pcs	60	214
	F.R.L	Pcs	10	
	Cutter ring	Pcs	144	

Table 11.3 Statistics of TBM accessories, cutters and power consumption at north port of Qinling Mountains

Item	Qty	Average unit length tunneling consumption	Remarks
Random imported cutter	JPY 364,234,039	JPY 69,457.29/m	Unit price is FOB Germany, JPY
Random imported accessories	JPY 262,783,814	JPY 50,111.33/m	Unit price is FOB Germany, JPY
Domestic procured accessories	CNY 9,726,634	CNY 1854.84/m	Calculated in CNY
Total power consumption	12239200 kW h	2338 kW·h/m	Including invert production, electricity consumption for ventilation outside the tunnel, etc
Total tunneling	Length 5244 m		

5. Mobilization and transportation of TBM

TBM is a large-scale tunnel construction equipment with complex structure. After assembly, its total mass reaches more than 3000 t and the total length is 256 m. Considering the carrying capacity of domestic railway transportation and highway transportation, the arrival port from Germany is Shanghai Port and Lianyungang Port.

(1) Introduction to Mobilization and Transportation

The TBM used in the Qinling Tunnel was produced by Wirth, Germany, but some of the back-up equipment and back-up auxiliary equipment was produced by the former Baoji Engineering Machinery Factory of the Ministry of Railways. Therefore, after the processing and manufacturing of this part of equipment are completed, the equipment will be directly transported to the construction site by the Baoji Construction Machinery Plant by vehicles. There are 12 platform vehicles (#7–#18) (including upper components), 26 sets of 20 m^3 mucking cars, and the rest of the equipment/components will be transported from Germany to domestic ports. If the equipment/components can be transported by railway, the goods will be converted from Shanghai/Lianyungang Port to Xi'an by railway after arrival, and then transported to the construction site by vehicles.

All goods transported by railway that are out of limit (over-height, over-width, over-weight and over-length) shall be transported to the construction site by truck after arriving at Lianyungang. In order to improve the transportation of large parts of the TBM, to meet the domestic railroad and highway transport conditions, at the beginning of the TBM purchase contract, the TBM design contact group formed by the relevant units of the former Ministry of Railways for the overall dimension and quality of the TBM components had repeatedly negotiated with Wirth to change the design to facilitate the transport of large parts, such as the original design of the equipment bridge for a total length of 26 m steel truss structure, later changed to a combination of 3 sections, the longest section is only 10.3 m, which greatly facilitates the transport through capacity, and finally determined that 11 pieces arrived in Lianyungang directly to the construction site (Table 11.4). Other equipment/components are transported by rail from Shanghai Port or Lianyungang Port to Xi'an Station and then to the construction site by truck.

(2) Transportation Organization of TBM Mobilization Equipment

Table 11.4 Large equipment directly arrived at site by TBM at northern mouth of Qinling Mountains from Lianyungang Port, Jiangsu Province

Serial number	Name of large components	Quantity (piece)	Dimension (L × W × H, mm)	Mass (t/piece)
1	Cutter Head	2	8550 × 4400 × 2300	65
2	Main bearing box	1	5700 × 5400 × 2200	86.6
3	Spare main bearing	1	5600 × 5350 × 1400	36.6
4	Front inner Kelly	1	8600 × 3110 × 3250	64.5
5	Rear inner Kelly	1	10,070 × 3970 × 3630	65
6	Outer Kelly (lower)	2	8000 × 2400 × 3450	60
7	Outer Kelly (upper)	2	8220 × 2400 × 3350	62
8	Cutter Head Shield	1	7100 × 4100 × 2200	32.2
9	Total	11		658.8

To ensure safe transportation of TBM mobilization equipment to the destination, in early 1997, the former Xikang Railway General Command of the Ministry of Railways set up a TBM transportation coordination group to be responsible for/coordinate the entire TBM transportation work, and convened several meetings to lay out the work in stages. At the same time, the work of arrival/transit, transshipment and customs clearance was entrusted to the Ministry of Railways Materials General Corporation, reducing the transshipment link.

The Xikang Railway Command of China Railway Tunnel Group also designated special personnel to be responsible for this work, and to carry out this work under the guidance of the coordination group of the former Xikang General Command of the Ministry of Railways. During the period, it participated in the preparation of the tender for the transportation of large items and the signing of the contract for the transportation of large items.

According to the content of the bidding document for TBM heavy-cargo transportation issued by the General Command Office of Xikang and the comprehensive strength of the tender, the reputation and qualification of the enterprise and the degree of response to the bidding document, etc., it was finally determined that Beijing Heavy-cargo Transportation Company would be the successful bidder for TBM heavy-cargo transportation, undertaking the automobile transportation task from Lianyungang to the construction site. TBM heavy equipment was shipped from Lianyungang Port on August 17, 1997 and arrived at the construction site safely at 2:00 pm on September 1, 1997. During the whole transportation process, the carrier achieved safety, high quality, and intactness of the carrier equipment. After the TBM transport was started from Lianyungang, close contact was maintained with the construction site every day. When the transport fleet arrived in Xi'an, site staff and vehicles were sent to meet them.

In order to earnestly carry out the receiving work of TBM, a TBM commodity inspection team was formed on the construction site to be responsible for the commodity inspection of TBM equipment, spare parts, accessories, etc. arriving at the construction site. Members of the team cooperated with the representatives of the foreign construction site and the personnel of the Commodity Inspection Bureau to carry out unpacking inspection one by one in a serious and responsible manner, and recorded the problems found in the commodity inspection, such as shortage, damage and specification discrepancy one by one, and asked the representatives of the foreign construction site to confirm.

(3) Transportation Safety Matters of TBM Mobilization Equipment

In order to ensure that the TBM is intact and delivered to the construction site safely with high quality, Xikang General Command Center and the TBM heavy-cargo transportation coordination team have repeatedly mentioned in the invitation for bid for TBM heavy-cargo transportation that the work quality requirements are: Large TBM components shall be transported to the construction site at the entrance and exit of Qinling Tunnel on schedule without damage and with quality and quantity guaranteed. It was also confirmed in the contract of carriage, and the carrier performed this purpose during the entire transport operation. In order to properly implement the purpose of TBM heavy-cargo transportation safety, the carrier shall survey the transportation route for many times before tender and after winning the bid. If there are obstacles to the safe transportation of bridges, culverts, road conditions, obstacle clearing, etc. along the route, the carrier shall negotiate with the provincial, municipal highway, traffic police and other departments along the route, and strive to seek the support and help of the traffic police departments along the route to formulate multiple disposal plans in advance. The bridge and culvert reinforcement and roadway barrier removal were prepared in advance so that the large TBM pieces received timely help and support when passing through provincial and municipal areas along the way. The local traffic police actively cooperated to dredge the roads for the passing sections and escort them. Since the transportation safety purpose was implemented and the preventive measures were proper, the shipper and the carrier cooperated closely to jointly implement the purpose of delivering the goods to the construction site in good condition and on schedule (Fig. 11.6).

After the large pieces arrived at the construction site, according to the requirements of the TBM assembly sequence, the unloading sequence and storage placement were reasonably arranged. The best technical operator of the 75 t × 2 + 15 t gantry crane manufacturer was selected to undertake the unloading of the large pieces of TBM. Under the guidance of the General Director of Xikang, the whole unloading was carried out safely from 5:00 p.m. on September 1 to 8:30 a.m. on September 2, so that the whole transportation and loading/unloading of large pieces of TBM were completed successfully.

6. Preliminary Supporting Projects

(1) Assembly Site, Temporary Storage Site, Preparation Chamber and Stepping Chamber

The length and size of the stepping tunnel depends on the diameter of the TBM cutter head and the distance from the front of the cutter head to the front of the TBM outer Kelly. The length of stepping tunnel of TB880E TBM is ≥ 10 m.

The temporary storage site should be built as close as possible to the assembly site, and should be able to store at least 50–70% of the TBM parts.

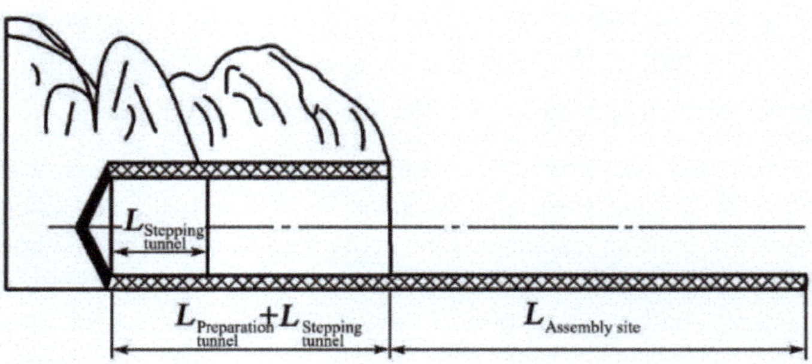

Fig. 11.6 Relationship between Assembly Site and Preparatory Tunnel and Stepping Tunnel

The size of the preparation tunnel depends mainly on the overall dimension of the TBM and the natural topography of the tunnel entrance. Its length is:(see Fig. 11-6 Relationship between Assembly Site and Preparatory Tunnel and Stepping Tunnel)

$$L_{\text{preparatory tunnel}} = 0 \sim (\text{TBM}_{\text{total length}} - L_{\text{assembly site}} - L_{\text{stepping tunnel}}).$$

The length of the assembly site is at least equal to the length of the TBM main machine plus the length of the connecting bridge plus the length of the first trailer plus the length of the towing cylinder plus the storage length of the main bearing plus the storage length of the cutter head plus the motorized length, and the width can at least meet the requirements of TBM assembly and access of large pieces. The assembly site at the north entrance of Qinling I Line is about 90 m long. The relationship between the preparatory tunnel, stepping tunnel and assembly site is as follows:

$$\text{TBM}_{\text{total length}} = L_{\text{assembly site}} + L_{\text{Preparation tunnel}} + L_{\text{Stepping Chamber}} \tag{11.1}$$

$$\begin{aligned} L_{\text{assembly site}} \geq\ & \text{TBM}_{\text{TBM length}} + L_{\text{Connection bridge}} + L_{\text{Towing cylinder}} \\ & + L_{\text{First trailer}} + L_{\text{Main bearing}} + L_{\text{Cutter head}} + L_{\text{Maneuver}} \end{aligned} \tag{11.2}$$

L_{maneuver} is generally 10 m.

(2) Invert Block Prefabrication Plant, and Concrete Mixing Plant

The invert block prefabrication plant is built at the place where the transportation track must pass, and is connected with the main transportation track by branch lines. The demoulding site of the prefabrication plant is set according to the needs of TBM construction, and the construction needs of the TBM are measured according to the strength hardening period of the invert block to meet the fastest construction speed of the TBM.

The production capacity of the concrete mixing plant depends on the size of the precast plant and the construction speed of the TBM, while taking into account the needs of the later lining construction of the TBM. The concrete mixing station has sufficient storage site for aggregate. The concrete mixing plant cannot only meet the needs of the invert block prefabricated plant, but also meet the needs of the initial support of the TBM construction and the needs of the grouting of invert block.

(3) Cutter Storage Warehouse, and Material and Accessory Storage

The cutter is a necessary consumable for TBM construction, requiring a large amount and frequent maintenance. The storage warehouse and maintenance site for cutters are determined according to the type, quantity and turnover of cutters on the TBM cutter head. The turnover of cutters is 2–3 times the quantity of cutters on the cutter head.

The TBM construction is fast, requires a large amount of materials, and has many types and quantities of accessories. There must be a suitable site to meet the construction needs. The material and accessories warehouse must store enough materials and accessories and be close to the transportation track. The storage capacity of materials shall not be less than the usage in 6 months.

(4) Installation of Transportation Track and Muck Dumping Machine

The length of the transport track is determined by the minimum needs of TBM construction and the natural topography of the tunnel entrance:

$$L_{\text{transport track}} \geq L_{\text{assembly site}} + L_{\text{slag disposal machine}} + L_{\text{1 locomotive}} + L_{\text{10 muck cars}}$$

The muck dumping machine is the muck unloading equipment for the last process of TBM tunneling construction. The installation site shall be as close to the opening as possible to shorten the muck unloading time. Figure 11.7 shows the ROWA muck dumping used at the north port of Qinling.

During the operation of the muck dumping machine, 2 muck cars with full load of 20 m^3 rock ballast respectively are clamped at the same time and turned over by 180° to achieve the purpose of muck unloading. In the process of muck dumping, the base foundation of the muck dumping machine shall bear the impact force and shall meet the following requirements:

① Static load $P_1 \geq 1260$ kN.
② Dynamic load $P_2 \geq 1600$ kN.
③ Lateral pressure $F \geq 120$ kN.
④ After installation, the flatness of the base plane of the muck dumping machine is 0–1 mm.

Fig. 11.7 Muck dumping machine

⑤ The diagonal length between the bases of the muck dumping machine after installation is (14.856 ± 2) mm.

The installation procedure of muck dumping machine (Fig. 11.8) is as follows:

① 4 bottom plates were placed, 2 seats were installed, the seat position was measured with theodolite and steel ruler, and adjusted with adjusting cushion blocks until the requirements are met.
② For anchor bolt anchorage, when fast coagulation is required, quick coagulation agent can be added to the anchored concrete.
③ The idler wheel support ② of the rolling body shall be installed (total 2 sets of 4 pieces).
④ The connecting beam ③ shall be installed.
⑤ The turning body shall be assembled on ground.
⑥ The crane shall be used to hoist the turning body to the idler wheel support; the turning drive chain shall be installed; the rails through which muck cars get in and out of the muck dumping machine shall be installed, with rail joint clearance of not more than 5–10 mm; the control room of the muck dumping machine shall be installed. ⑪

7. TBM Assembly

Due to the large volume and mass of TBM main machine components (the heaviest single piece is about 86 t, and the assembled cutter head is 130 t) and the high requirements for the flatness of the site, the TBM main machine assembly site is flat at the portal of Line I, and flat at the north portal (the site area at the inlet end is 20 m × 110 m). Among them, the 75 m long portal is used as the unloading storage

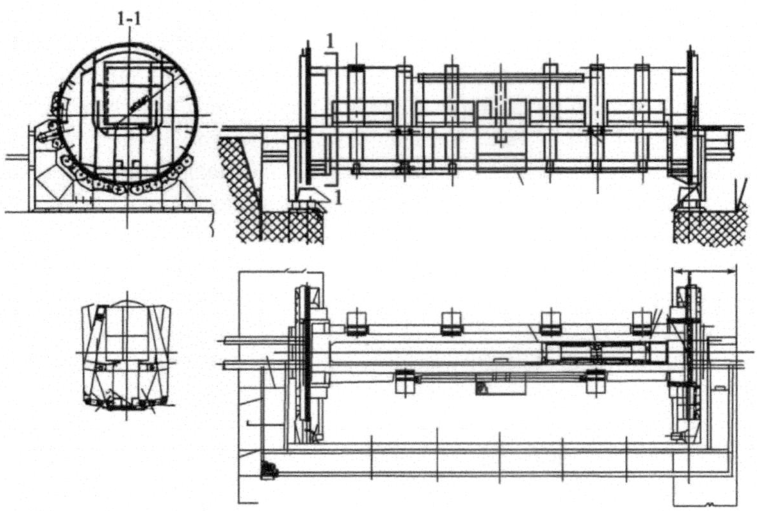

Fig. 11.8 Installation Diagram of Muck Dumping machine

and assembly site for large parts of the main machine, with a concrete hardening thickness of 30 cm. A 2 × 75 t + 15 t gantry crane with a span of 20 m is installed in the site for unloading and assembly. A 50 t truck position is reserved near the bridgehead on the left side of the line for assembly in cooperation with the gantry crane.

The assembly of the back-up system is assembled on the Shibiyu bridge by using 50 t truck cranes. For some large parts that cannot be hoisted by truck cranes, the gantry cranes at the tunnel entrance are assembled during the stepping after the TBM main machine is assembled, and the track of the back-up platform gantry is laid within the length range of the whole back-up assembly site.

According to the TBM assembly and commissioning requirements, a 250 m preparatory tunnel and a 10 m departure tunnel were set up in the inlet section of the tunnel constructed in advance by the drill and blast. The mass of a single TBM piece reaches 86 t. In order to ensure the bearing capacity of the assembly site, 300# concrete is used for bottoming, with a thickness of 30 cm and a flat slope on the top. Flatness is required to be 1 mm.

The assembly equipment mainly includes: one 75 t × 2 + 15 t gantry crane, one 6 t forklift, one 5 t and 40 t platform gantry respectively, and one 25 t and 40 t truck crane respectively.

The basic assembly sequence of TBM main engine is as follows: Positioning of stepping mechanism → connection of thrust cylinder with front and rear outer frames → connection of lower part of front and rear outer frames (with thrust cylinder) with stepping mechanism → installation of front and rear couplings for transmission shaft, and installation of front and rear positioning protective sleeve → installation of transmission shaft with protective cover at the lower part of front and outer frames, and preliminary fixation on the outer frame → installation of front and rear inner frames on the front and rear outer frames, and connection of front and rear inner frames → installation of bolt driller tracks on both sides of front inner frame → installation of front and rear outer frames (with thrust cylinder) → installation of transmission shaft with protective cover at the upper part of front and outer frames, and preliminary fixation on the outer frame → Placement of ring beam erector → cutter head lower shield in place → Installation of the rear support with stepping mechanism on the rear inner frame → Installation of the inner rack tail → hoisting of main bearing → installation of upper shield support → Installation of the collecting hopper in the drive assembly → hoisting cutter head → Installation of the wing shield and top of the cutter head → Installation of the drive shaft and thrust cylinder in the correct position → Installation of drive mechanism → installation of ring beam erector and drive device → installation of TBM belt conveyor → installation of main machine working platform and other platform → installation other auxiliary equipment and accessories.

The assembly procedure of the equipment bridge is as follows: Ground connection of equipment bridge → installation of auxiliary frames on both sides and the upper part on the equipment bridge → installation of rear support of equipment bridge on the extended track → connection of hoisting equipment bridge with the main machine → installation of auxiliary equipment on the equipment bridge → installation of working platform on the equipment bridge → installation of operator cabinet → connection of each system with the main machine → installation of belt conveyor on the equipment bridge.

The assembly procedure of the back-up gantry is as follows: assembly of each section of the gantry and installation of the corresponding auxiliary equipment on the gantry → connection of each gantry → installation of the material truck towing system → installation of the belt → installation of the electrical system, and hydraulic system → connection with the main machine for commissioning.

8. Construction Management

There are 2 tunneling teams, 1 maintenance team and 1 transport team. The division of work is clear. The maintenance team is divided into main machine and back-up team; the tunneling team is divided into supporting team, and invert laying, grouting, track extension team, etc. The TBM tunneling cycle mainly consists of excavation, mucking, transportation, support, installation of invert blocks, TBM regripping, laser positioning, ventilation and dedusting, air, water, electricity pipeline and track extension. A unified dispatching and command system shall be established during construction to coordinate the relationship between each process inside and outside the tunnel and ensure the continuity of construction.

(1) Excavation

The average cycle time of TBM tunneling is 80–90 min with 1.8 m advance per cycle. Affected by the category of surrounding rock and the development degree of joints, the cycle time is 35 min at least and 5 h at most. As an important part of construction management, the core of excavation management is the selection of tunneling parameters and the control of advance direction. The main driver is required to select reasonable thrust force and torque according to the change of geological conditions to improve the tunneling efficiency.

(2) Mucking Transport

The tunnel is transported by rail. The transportation track is of four-track double line, with a track gauge of 900 mm. 43 kg/m standard rail is directly laid on the precast invert block. California switch is provided at the back-up tail for passing vehicles. According to the circulating muck volume (about 180 m^3), the number of muck car groups is 10, each with a capacity of 20 m^3, and the 35 t imported diesel locomotive is used for traction and transportation.

The mucking process is as follows: rock chips through the bucket → TBM belt conveyor (length: 22.5 m, width: 1.2 m) → equipment bridge belt conveyor (length: 26 m, width: 1.2 m) → back-up system belt conveyor (length: 160 m, width:1.2 m) → two-way muck chute (with moving distance 80 m, muck car does not need to move when loading muck) → muck car → muck dumping of muck dumping machine outside the tunnel → trucks dumping to the spoil area.

After the allocation of sufficient transport vehicles, the focus of transport management is on the maintenance of transport vehicles, muck dumping machines and the maintenance of transport tracks.

(3) Support

Supporting operations are generally carried out simultaneously. TBM is equipped with sufficient support equipment, but when passing through special geology sections such as faulted and broken zones and soft water ingress sections, the lack of sufficient experience in the beginning of construction results in long latency time for downtime, which will extend the cycle time and restrict the advance speed of TBM. There was essentially no impact during the subsequent construction. The key to support operations is to develop reasonable technical measures according to different geological conditions, and gradually establish a support system suitable for TBM construction.

(4) Elevated Invert Block Installation and Track Extension

The invert blocks are transported from outside the tunnel to the invert crane under the TBM equipment bridge and lifted by the crane, with laser positioning, taking about 7–8 min per block. After the installation of the invert block, the gap between the invert block and the surrounding rock is injected with fine stone concrete under 0.3–0.4 MPa pressure through the grouting hole on the invert block every 5–7 blocks to ensure the compactness of the substrate. The transport track needs to be extended once for every 7 elevated invert blocks installed. During the construction, the contact with the traffic dispatching outside the tunnel must be strengthened to ensure the timely supply of the invert block and grouting material.

(5) Construction Survey

The TBM advance direction is controlled by the ZED-260 laser guidance system. The guidance system is moved forward once every 100 m, and the invert laser is moved forward once every 80 m. This work is performed during downtime for maintenance and does not take up cycle time. The TBM main machine position and the invert laser are linked and calibrated to reduce survey tolerances and control tunneling deviations and invert block installation deviations.

(6) Ventilation and Dedusting

The TBM ventilation supplied by the main air duct is sent to each working area of the TBM through the secondary ventilator at the back-up tail via relay fan. The dedusting is achieved by water spraying from the cutter head and dedusting device, and the purified air is blown back to the tunnel as clean air.

During the tunneling period, the ventilation and dedusting system is always in working condition. During construction, whether there is air leakage in the air duct shall be checked, and the filter screen cover of the industrial exhaust fan shall be checked and replaced to create a good and comfortable working environment.

(7) Wind, Water and Electricity Pipeline Extension

The air duct storage cylinder contains 100 m air duct, 50 m ϕ100 mm water pipe and 400 m 10 kV HV cable, all of which are coiled on the reel. The construction is followed by the forward movement of the back-up system, and the wind, water and electricity pipelines are automatically extended forward.

9. Construction Technology of Fault Fracture Zone

See Sect. 7.4.2 for the construction facilities of gripper TBM when passing through fault fracture zone.

10. Tunneling Operation Procedures

(1) An alarm is sounded before starting the equipment to draw the attention of personnel in each work area.
(2) High pressure water pump and cutter head spray pump station is started.
(3) Hydraulic pump stations No. 6, No. 7, No. 3, No. 4, No. 1, No. 5, No. 8 and No. 2 is started in sequence.
(4) All ventilation, dedusting and cooling systems are started.
(5) High/low speed of the cutter head is select according to the geological condition, soft start of the motor is selected, and the start interval of the motor is controlled by the soft starter.
(6) 8 main motors are started with a start interval of 1 min. When 2 main motors are still not started, belt conveyors No. 3, No. 2 and No. 1 are started the same time.
(7) After the motor is running normally, the cutter head rotating is started.
(8) Start tunneling: manual control is selected, the advance speed is adjusted to 20%, and after the tunneling parameters are relatively stable and the actual tunneling is 5 cm, the advance speed starts to be increased in 10% increments; when the thrust oil pressure reaches 25 MPa and is basically stable, the advance speed is increased in 5% increments until the thrust oil pressure reaches the rated value of 26.5 MPa.
(9) After tunneling a cycle of 1.8 m, the cutter head is backed off 2–3 cm to stop the cutter head rotation after the cutter head idling for 10–20 s, and the motor and belt conveyor are stopped in sequence to carry out regripping operation.

11. Regripping Procedure.

(1) Set the cutter head back 2–3 cm.
(2) Idle the cutter head for 10–20 s, stop the cutter head rotation, power off the motor and stop the belt conveyor.
(3) Pull out the bottom shield (also known as lower support of the shield) to increase the cylinder pressure to 18 MPa.
(4) Release the shield clamping oil cylinder. At this time, the bottom shield pressure drops and should continue to rise to 14–16 MPa.
(5) Pull out the rear support and raise the pressure of the rear support to about 15 MPa.
(6) Relax the front outer Kelly and rear outer Kelly.
(7) Move the outer Kelly forward and make sure it is in place.
(8) Adjust the deviation of the upper, lower, left and right sides of the TBM according to the guidance system. During adjustment, as the laser detection is slow, the operation can be carried out after the data display is stable. Each adjustment shall not exceed 3 cm.

(9) Tighten the front outer Kelly and rear outer Kelly to about 27 MPa.
(10) Retract all shield supports.
(11) Pull out the bottom shield to front outer Kelly, increase the pressure of the gripper by 0.2–0.4 MPa, and pull out the other shield to close to the tunnel wall.
(12) Clamp the clamping cylinder.
(13) Drag the back-up, and move the rear support forward.
(14) Start the belt conveyor and start the main motor in turn while dragging the back-up.
(15) Start the cutter head to rotate. At this time, the bottom shield oil cylinder pressure is about 12 MPa, which can be increased to about 14 MPa.
(16) Start tunneling.

12. Reaming Operation Procedure.

The reaming operation is carried out when the edge cutter needs to be changed. Only one reaming cutter is required each time. The reaming diameter of TB880E TBM is 8.9 m. The reaming procedure is as follows:

(1) Set the cutter head back 1 m.
(2) Check and confirm that the reaming operation valve is operating properly.
(3) Check and confirm that the reaming cylinder and reaming knife retract freely.
(4) Select the motor to rotate the cutter head at high speed and continuously extend the reaming cutter to make the reaming cutter contact the tunnel wall slowly.
(5) The reaming speed shall be controlled at 10% at the beginning. After 5 cm of tunneling, the speed can be adjusted to 10%–20%, with a maximum of 25%.
(6) The reaming length is generally 60–80 cm, and during the reaming operation, oil supply should be carried out to the reaming cylinder continuously to keep the extension pressure at 20 MPa.
(7) After the reaming is completed, the reaming oil cylinder shall be recovered, the machine shall be stopped, and the cutter head shall be backed by 20 cm for replacement of the edge cutter or scraper.

13. Selection of Excavation Mode.

The gripper TBM has 3 tunneling modes, namely, automatic control thrust force mode, automatic control torque mode and manual control mode. The tunneling mode shall be selected according to the rock condition during operation.

(1) Automatic control thrust force mode is selected for tunneling in homogeneous hard rock conditions. The selection of this mode of excavation is based on the judgment that the surrounding rock can be judged as hard rock if the thrust force reaches the maximum value first and the torque does not reach the rated value during excavation. During operation, the equipment will not be overloaded and can ensure a high tunneling speed.

(2) When tunneling under surrounding rocks with developed joints, the equipment thrust will not be too large, but the cutter head torque is very high, so the automatic control torque mode shall be selected. The selection of this mode of excavation is based on the following criteria: when excavating, the torque reaches the rated value first and the thrust force does not reach the rated value or reaches the rated value at the same time, there is no large muck output on the belt conveyor, and the surrounding rock can be judged as homogeneous soft rock.

(3) If the state of surrounding rock cannot be judged, or the hardness of surrounding rock varies unevenly, or there are fracture zones and faults in surrounding rock joints, the manual control mode must be selected.

14. Selection of Tunneling Parameters under Different Geological Conditions.

(1) Operation under hard rock geological conditions with developed joints

The compressive strength of surrounding rock itself is large, and it is not easy to be broken. If the advance speed is too low, it will cause a large amount of wear of the cutter and cutter ring. If the advance speed is too high, it will cause overload of the cutter, resulting in oil leakage or string wear.

① Select the motor high speed (5.4 r/min).
② At the beginning of tunneling, the advance speed shall be 15%, and the speed can be increased only after 5 cm of tunneling.
③ Under normal circumstances, the advance speed is generally $\leq 35\%$.
④ When the torque is about 30%, the advance speed is $\leq 25\%$ and the corresponding penetration is 1.7 mm.
⑤ When the torque is about 35%, the advance speed is $\leq 30\%$ and the corresponding penetration is 2.2 mm.
⑥ When the torque is about 40%, the advance speed is $\leq 35\%$, and the corresponding penetration is 2.9 mm.

(2) Operation under Geological Conditions of Soft Rock with Developed Joints.

The thrust force is small during tunneling, so the automatic control torque mode shall be selected to closely observe the torque change and adjust the optimal tunneling parameters.

① The 8 main motors are all in use, and the advance speed can be adjusted to about 80%, but the torque value must be ensured to be $\leq 80\%$ and the variation range $\leq 10\%$, and the corresponding penetration is 10 mm.
② The main motors are not all put into use, and the torque of at least one motor shall be reduced by 5%. For example, when 6 motors are driven, the torque shall be $\leq 70\%$.

(3) Operation under Geological Conditions with Developed Joints and Large Hardness Changes.

Select the manual control mode and closely observe the change of thrust force and torque.

① Selection of operating parameters: thrust force $\leq$ 17000 kN, torque $\leq$ 55%, torque variation range $\leq$ 10%, corresponding penetration of about 6 mm.
② For tunneling under such surrounding rocks, the thrust and torque are constantly changing, and fixed parameters cannot be selected. Therefore, close observation shall be carried out to adjust the tunneling parameters at any time. If the vibration suddenly intensifies, the torque changes greatly, and there are irregular polyprisms in the muck, the cutter head rotation speed can be changed to low speed (2.7 r/min). After the vibration is reduced and returns to normal, the cutter head can be changed to high speed tunneling.
③ When the torque and thrust force change greatly, the advance speed shall be reduced as much as possible and controlled at about 30%.

(4) Operation under Geological Conditions such as Relatively Developed Joints, Many Cracks, or Fracture Zones and Faults.

For such surrounding rock tunneling, the torque control mode shall be the main mode to select and adjust the tunneling parameters, but the torque change, current change, thrust force value and rock condition shall be closely observed.

① Selection of operating parameters: when the motor is at high speed, the advance rate is less than 50% and the torque variation range is less than 10%; when the motor is at low speed, the advance rate starts to be 20%. After the rock change is stable, the advance rate can be increased by more than 45% and the torque variation range is less than 10%.
② During the tunneling process, closely observe the mucking of the belt conveyor. When polygonal prismatic rock blocks with a diameter of about 30 cm appear on the belt conveyor, and the proportion of polygonal prismatic rock blocks is 20–30%, the advance speed shall be reduced and the penetration shall not exceed 7 mm. When a large number of polygonal prisms appear on the belt conveyor and continuously output to the outside, the pushing shall be stopped, the motor rotation speed shall be replaced to 2.7 r/min, the low-speed pushing shall be carried out, and the penetration shall not exceed 10 mm; When the rock condition changes greatly, the cutter may be locally subject to axial load during tunneling, which will affect the service life of the cutter. Therefore, the torque change range must be strictly controlled not to exceed 10%, and the fault shall be steadily transited at a low advance rate. In general, the advance rate $\not>$ 55% and the penetration degree does not exceed 7 mm.

15. TBM Disassembly.

The construction of Qinling Tunnel is carried out by 2 TBMs from both ends of the tunnel to the middle, and finally the TBM is disassembled in the tunnel. According to the space dimensions that must be occupied when dismantling some large pieces of TBM, such as cutter head, main bearing, inner and outer casings, etc. in the tunnel, and at the same time to meet the space/safety dimensions during the installation and operation of the dismantling crane, the construction cost of the TBM dismantling tunnel is minimized, and the working dimensions of the dismantling tunnel are finally determined to be 55 m long, 12.8 m wide and 18.16 m high.

(1) Technical Parameters of Bridge Crane

The bridge crane used for TBM removal is manufactured by Hangzhou Dali Machinery Technology Engineering Co., Ltd. of the Ministry of Water Resources. See Table 11.5 for main technical parameters.

The TBM was disassembled on September 12, 1999, the 130 t cutter head was safely removed and leveled on September 27, and all auxiliary equipment and accessories of TBM were transported out of the tunnel on October 13.

(2) Key Points for Installation of Hoisting Equipment in the Tunnel

The bridge crane for dismantling TBM is transported from the excavated Qinling Tunnel Line II into the dismantling tunnel. In order to facilitate the installation and transportation of the bridge crane, the length of the component shall not be greater than 6 m and the mass of a single piece shall not be greater than 20 t. The actual lifting trolley of the bridge crane is 13 t, with an overall dimension of 4.5 m × 3.2

Table 11.5 Technical parameters of bridge crane

Item	Parameter
Model	Q75 t + 75 t/12.8 m bridge crane
Overall dimensions (L × W × H)	13500 mm × 6400 mm × 2100 mm
Whole machine form	Double beam, double trolley, bridge type
Motion mode	Both the crane and trolley lifting mechanisms can act independently. When necessary, the double trolleys can operate synchronously, and the double lifting machinery can lift and lower synchronously
Minimum hook spacing less than	2 m
Lifting height	1.2 m
Gauge	12.8 m
Working speed of lifting mechanism	4 m/min, eddy current speed regulation 0.8 m/min
Trolley operating speed	13 m/min; variable frequency stepless speed regulation
Crane operating speed	12 m/min
Power supply	Three-phase AC 380 V, 50 Hz
Whole machine power supply mode	Suspension Cable
Operation mode	Ground cord control
Maximum wheel load	32 t
Mass of whole machine	74 t
Maximum single piece length and mass	Not more than 6 m and not more than 20 t

m × 2 m and 2 pieces in total. It is difficult to transport the trolley into the tunnel by rail transit vehicles, and the trolley must be transported into the dismantling tunnel by trucks in two times.

In order to facilitate the installation of the dismantling crane and the unloading of components in the tunnel, it is preferred to install anchor points and ground anchors on the top and ground of the dismantling chamber. The rock bolt is of ϕ22 threaded steel. The rock bolt is required to be perpendicular to the rock as far as possible, and the mutual tolerance of each group of rock bolts shall not be greater than 2 mm. After the construction of each rock bolt is completed, a pull-out test shall be carried out, and the pull-out resistance of each rock bolt shall not be less than 50 kN. Pre-machined 15/5 t lifting lug plate is installed.

(3) Selection of Transport Equipment

The disassembly of the TBM main machine part is completely carried out by a 75 t × 2 bridge crane. The transportation equipment is mainly the supporting flat jumbo and traction locomotive on the TBM. When pulling the flat jumbo outside the tunnel during the disassembly, 4–5 flat jumbos are taken as a group, and the accessories on the flat jumbo that hinder access are removed in advance,

Or removed and put on the flat jumbo. Other auxiliary equipment, main machine accessories and equipment on the TBM are also removed in advance and placed on the flat jumbo and transported out of the tunnel with the vehicle. When the flat jumbo is transported out of the tunnel, it is usually pulled by double locomotives. During the transportation equipment bridge, since one end of the equipment has a row of traveling wheels and the other end has no supporting points (connected with the main machine) under the working state, 2 groups of transportation frames are processed on site as supporting points, and 2 groups of 20 m^3 wheelsets of muck cars are installed under the transportation frames. The transportation frames are made of 25 mm × 250 mm H steel and 30 mm thick steel plates by tailor welding.

11.1.3 Mogouling Tunnel

Mogouling Tunnel of Xi'an Nanjing Railway (Fig. 11.9) has a total length of 6.114 km. The surrounding rocks are mainly marble, mica and schist. As the tunnel is seriously affected by the structural extrusion and crumpling of the regional F2 large fault zone, the rock mass is broken, the recessive joints and joints are developed, the extruded structural planes intersect with the tunnel at a small angle, the long and large joint concentrated zones and the extruded fracture zones are developed, the surrounding rocks are weak and broken, and the stability is poor. The soft rocks account for 70.5% of the total length of the tunnel. The tunnel was constructed by China Railway Tunnel Group Co., Ltd. with ϕ8.8 m gripper TBM (previously used for Qinling Line I tunnel) of Wirth, Germany, and commenced on March 10, 2000 and completed on December 31, 2002. 4.653 km of this tunnel was dug with a TBM

Fig. 11.9 Mogouling Tunnel constructed by gripper TBM

at a maximum monthly advance of 573.9 m. In this project, China Railway Tunnel Group mastered the tunneling technology of gripper TBM in soft ground.

1. Geological Conditions

(1) Geographical Locations

Mogouling Tunnel is located in the middle and low mountainous area of the Qinling Mountains, and crosses the watershed (Mogouling) from west to east between the Wuguan River and the Qingyou River. The entrance of the tunnel is at an earth slope on the left of National Highway 312 in Ximogou Village, Maoping Town, and the exit is on the north side of Qingyou Middle School in Qingyouhe Town.

(2) Scope of Works

Mogouling Tunnel begins at IDK226 + 494 and ends at IDK232 + 606 on the north side of Qingyou Middle School in Qingyouhe Town. Except that 538 m of the inlet section is located in a curve with $R = 1000$ m, the rest is located in a straight line. The longitudinal slope of the tunnel body is 3.5–10.5% of the upslope with a length of 1936 m and 4% of the downslope with a length of 4176 m, basically inclining to the east. The maximum overburden of the tunnel is 313 m.

(3) Topography

Mogouling Tunnel is located in a place with large topographic relief and alternating between gullies and beams, with a relative height difference of 100–300 m. The line passes through Mogouling in a long tunnel. The maximum elevation is 824 m. The water flow from gullies in the west of Mogouling flow into the Wukan River through surface runoff, and the water flow from gullies in the east of Mogouling flow into the Qingyou River through ditches.

(4) Climatic Conditions

The average temperature of the coldest month in the area where the tunnel is located is 16 °C, and the maximum freezing depth of the soil is 0.5 m.

(5) Stratum Lithology

The bedrock at the tunnel is mostly exposed, consisting of Middle Devonian quartz shist and marble intercalated with mica quartz shist. There are Quaternary alluvium and diluvium on the gully bed and slope at the inlet and outlet. The stratum and characteristics are as follows:

Quaternary Holocene diluvial sandy clay (Q_4^{dl2}): It is mainly distributed at the exit end, with a thickness of 5–20 m, brownish red, of uniform soil texture, mainly composed of clay particles, with good viscosity, containing a small amount of crushed stones, hard plastic, Grade II ordinary soil, Class II surrounding rock, $\sigma_0 = 120$ kPa.

Quaternary Holocene diluvial sandy clay (Q_4^{dl2}): It is mainly distributed on the first terrace of Qingyou River at the exit end, 2–5 m thick, brownish yellow, relatively uniform, mainly clay particles, high sand content, mixed with boulder, round gravel, hard plastic, Grade II ordinary soil, $\sigma_0 = 150$ kPa.

Quaternary Holocene diluvial sandy clay (Q_4^{dl7}): It is distributed at the exit end, 5–8 m thick, light yellow, mainly composed of marble and shist, angular, with granular composition of ϕ20–60 mm accounting for about 45%, ϕ60–120 mm accounting for about 10%, $\phi > 120$ mm accounting for about 10%, filled with soil, loose–medium dense, wet, Grade II ordinary soil, Class II surrounding rock, $\sigma_0 = 500$ kPa.

Quaternary Holocene diluvial boulder soil (Q_4^{dl8}): It is distributed in the acorn ditch at the exit end, with a thickness of about 7 m, gray–grayish yellow. The boulder is mainly composed of marble and quartz shist, in round edge–perfectly round shape. The grain size composition is about 20% for ϕ20–200 mm, 60% for $\phi > 200$ mm, and the maximum grain size is about 2500 mm. It is filled with sandy soil, medium dense, wet–saturated, $\sigma_0 = 700$ kPa.

Quaternary Upper Pleistocene Alluvial Pebble Soil (Q_3^{dl7}): It is distributed in the lower part of the sandy clay at the exit end, with a thickness of 3–10 m, gray, with good rounding of pebbles, mainly composed of marble and quartz shist, with the grain size composition of about 20% of ϕ20–40 mm, about 15% of ϕ40–60 mm, about 15% of ϕ60–80 mm, about 10% of ϕ80–100 mm, about 5% of $\phi > 100$ mm, filled with sandy soil, wet, medium dense, Class II ordinary soil, $\sigma_0 = 600$ kPa.

Middle Devonian quartz shist (Q_{2c}^{dSc7}): It is distributed in DK226 + 494–DK231 + 410 of the tunnel body, grayish yellow and brownish yellow, of crystalloblastic texture and schistose structure, with mineral texture dominated by quartz and mica, with quartz content of 55–69%, feldspar content of less than 5%, mica content of 22–35%, with schistosity and joints relatively developed–developed, and weathered seriously–heavily. The weathered layer is 2–5 m thick, with the maximum thickness of about 10 m, Grade IV soft rock, $\sigma_0 = 500$ kPa, intact rock resistivity $\rho = 800$–1500 Ω m, seismic wave velocity $v_p = 4380$ m/s, Grade V secondary hard rock, Class IV–V surrounding rock, $\sigma_0 = 1000$ kPa.

Middle Devonian marble mixed with quartz shist (Q_{2c}^{dSc7}): It is distributed at DK231 + 410–DK232 + 602 of the tunnel barrel. Marble: grayish white, white, of crystalloblastic texture and massive structure, mainly composed of calcite, with a small quantity of mica and quartz. The calcite content is 84–89%, the quartz content is 2–7%, and the mica content is 2–4%. The joints and bedding of the rock mass are relatively developed. The rock mass is seriously–heavily weathered. The weathered layer is 1–2 m thick, with a maximum thickness of about 10 m. Grade IV soft rock, $\sigma_0 = 500$ kPa, and the intact rock is Grade V secondary hard rock and Class IV surrounding rock, $\sigma_0 = 1000$ kPa. The marble is mostly located at the exit end, locally intercalated with quartz shist, of crystalloblastic texture and schistose structure, with a layer thickness of 1–2 m. The rock mass is relatively broken under the influence of F2 fault on the left side of the line, with a resistivity of $\rho = 200$–900 Ω·m and a seismic wave velocity of $v_p = 3820$ m/s. In the surface investigation, there are local dissolution marks in the marble, and there are karst fissures or small karst caves along the bedding, joints or contact zone with quartz shist.

Cataclastic rock (Tγ): mainly exposed in Mogouling fault and F_2 fault zone, mainly composed of cataclastic marble, gray, with obvious cataclastic structure, broken rock, locally mixed with mylonite. Rock extrusion fissures are developed, the bonding strength of fissure layers is low, and the fissures are easily broken by hand. It is Grade IV soft rock, with $\sigma_0 = 600$ kPa, and Class III surrounding rock. According to the geophysical prospecting data, the seismic wave velocity $v_? = 3000$–4000 m/s and the resistivity $\rho_s = 200$–300 Ω m.

(6) Geological Structure

Folds.: According to the regional data, the tunnel construction site is located in the Liquan-Zhashui Variscan fold belt, and is the north wing of a compound syncline. According to the field investigation and analysis, there is no large fold structure in the tunnel, the schistosity stratum is developed, the schistosity attitude is relatively disordered and changes greatly, and only a relatively obvious anticline shape is shown at the entrance section of the tunnel. The attitude of the Xi'an side of the anticline is NW inclined to the north, and the core is located near the south vertical slope platform (DK228 + 000). The rock mass here is seriously weathered and the attitude is unstable, and the stratum is middle Devonian quartz shist.

Fault (F_2): The Danfeng-Shangnan Fault (F_2) is distributed on the left side of the tunnel. It is a regional fault and is the main fault divided into two tectonic units in the Qinling Tectonic Zone, namely the North Qinling and the South Qinling Tectonic Units. To the north of the fault is the stratum of the Qinling Group in the Middle Proterozoic Erathem, and to the south is the stratum of the Chigou Formation in the Middle Devonian Erathem. This fault shows obvious linear image on the satellite photograph, and the structural features are very obvious. On the aerial photograph, the F_2 fault zone is a low-lying trough landform with a width of 200–400 m and locally up to 800 m. The north side of the construction site is a fault valley. The fault attitude is N60°–80°W/60°–69°N in this section. The fault material is mainly cataclastic quartz schist, cataclastic marble and mylonite, intercalated with fault gouge and breccia. The main fault carries water. Since the Caledonian, the F_2 fault has experienced many times of tensioning and compression, and has multi-period activity. Since the Cenozoic era, the F_2 fault displays the nature of tensile normal fault, and its activity is characterized by dislocation, fracture and refurbishment of the consolidated rocks and mylonite in the zone. Although the tunnel avoids the F_2 fault fracture zone, the 1.5 km long section at the exit end is about 130 m away from the fault boundary, and the wave velocity of rock mass near the geophysical prospecting test is low, about 3820 m/s. Therefore, the F_2 fault still has a certain impact on the tunnel, especially the barrel at the exit end. In addition, some NE and NW small faults matching with F_2 may be encountered at the barrel, but these small faults are not obvious on the surface due to limited exposure width and the influence of vegetation coverage.

Bedding and schistosity characteristics: Since F_2 fault passes through the left side of the line, under the influence of the structure, the bedding and schistosity attitudes in the rock mass are unstable, especially the kneading phenomenon in the shist stratum is very developed. The whole tunnel:

The attitude of IDK226 + 498–DK231 + 250 section is mainly N50°–75°W/60°–80°S.

The attitude of IDK231 + 250–DK232 + 606 section is mainly N60°–80°W/65°–85°S, with mica quartz shist and developed joints. The marble is generally more than 50 cm thick, mixed with thin-bedded mica quartz shist.

According to the statistics of each observation point of the tunnel barrel, it is found that the joints in shist mainly include:

J_1: N28°–80°W/51°–86°S, $d = 0.05$–0.6 m, $l > 0.4$ m, slightly open;
J_2: N10°–73°E/39°–80°S, $d = 0.05$–0.6 m, $l > 0.1$ m, slightly open;
J_3: N40°–84°W/30°–66°N, $d = 0.1$–0.5 m, $l = 0.05$–3 m, slightly open.
Joints in marble mixed with quartz schist mainly include:
J_1: N45°–65°W/35°–90°S, $d = 0.1$–0.5 m, $l > 0.4$ m, slightly open;
J_2: N50°–80°W/42°–90°S, $d = 0.05$–0.5 m, $l > 0.1$ m, slightly open;
J_3: N20°–60°E/50°–90°S, $d = 0.1$–0.5 m, $l = 0.1$–1 m, slightly open.

According to the statistics of joint density of the tunnel barrel, the joint JV (average value) of the barrel is = 8/m^3, and the bedrock is relatively intact.

(7) Hydrogeology

According to the ground survey and hydrogeological calculation results, the tunnel can be divided into two hydrogeological zones:

① Bedrock Fissure Weak Water-rich Area.

IDK226 + 494–DK231 + 400 section, with a length of 4.906 km long and overburden of 40–310 m, is mainly composed of mica quartz shist of single lithology. It is predicted that the unit water ingress of the tunnel is 200.8 m^3/d and the maximum water ingress is 1988.9 m^3/d.

② Bedrock Fissure Medium Water-rich Area.

IDK231 + 400–DK232 + 606 section, with a length of 1.206 km and overburden of 40–120 m, is mainly composed of marble mixed with mica quartz shist and cataclastic marble. It is predicted that the unit water ingress of the tunnel is 495.2 m^3/d and the maximum water ingress is 1190.4 m^3/d.

The maximum total water ingress of the tunnel is 3179.3 m^3/d. Groundwater is non-erosive.

2. Design Overview

(1) Preparatory Tunnel (IDK232 + 437–DK232 + 606,169 m)

According to the requirements of TBM assembly and commissioning, it is necessary to set up a preparatory tunnel and use the drill and blast method to apply it in advance. In combination with the geological conditions of the portal, the 169 m exit end is the preparatory tunnel. The section of the preparatory tunnel is of a circular inclined wall section. The primary support is mainly of shotcrete bolt mesh. The arch wall is provided with rock bolts and wire meshes. The rock bolts are 3 m@1.0 m × 1.0 m long and the wire meshes are 8@25 cm × 25 cm, supplemented by 3 pieces/2 m profile steelworks. The secondary lining is of molded concrete. After the section excavation, the initial support was carried out, and the secondary lining was applied in time after the TBM passed. According to the national defense requirements, the secondary lining of IDK232 + 535–DK232 + 606 section adopts 50 cm thick fully enclosed reinforced concrete structure, and the secondary lining of IDK232 + 437–DK232 + 535 section adopts 50 cm thick plain concrete structure.

(2) Departure Tunnel (IDK232 + 427–DK232 + 437,10 m)

A 10 m departure tunnel is set at the exit of the tunnel to meet the requirements of TBM tunneling. The departure tunnel is constructed by drill and blast method. The section after lining is a circular section with an inner contour of 8.9 m. The lining is a molded concrete lining with a thickness of 45 cm. The construction support is shotcrete bolt support. Rock bolts are set above the maximum span, 1–2.5 m@1.2 × 1.2 m. In order to ensure that the TBM enters the departure tunnel from the preparatory tunnel, the bottom of the departure tunnel shall be 15 cm higher than the bottom plate of the preparatory tunnel.

(3) TBM Construction Tunnel (IDK226 + 725–DK232 + 427,5702 m)

According to the layout of elastic monolithic track bed, central ditch, various pipe trenches and equipment, the circular section with a basic inner contour diameter of 700 cm in the straight section of the tunnel is determined. Since the superelevation of the outer rail in the curve section makes the basic inner contour of the circular lining straight line cannot meet the requirements of normal tunnel limit, according to the engineering geological and hydrogeological conditions in the curve section, the shotcrete thickness is up to 8 cm, and the shotcrete bolt mesh and steelwork support are strengthened. The contact mesh (horizontal suspension) adopts the method that the suspension points must be densified to reduce the sag and meet the requirements of curve construction clearance.

The circular section of the TBM construction tunnel is 8.8 m, and the composite lining is adopted. The primary support is mainly of shotcrete bolt mesh. In the section design of the TBM construction tunnel, a construction tolerance of 10 cm is considered, including a construction and survey tolerance of 5 cm for TBM and a construction tolerance of 5 cm for molded concrete lining.

3. General Construction Methods

TBM construction is carried out by using 1 gripper TBM single-port tunneling at the exit of the tunnel, of which the 169 m section of IDK232 + 437–DK232 + 606 at the exit end constructed by drill and blast is the preparatory chamber; For section IDK232 + 427–DK232 + 437, Class III surrounding rock of 10 m is constructed by drill and blast and is a TBM departure tunnel; Inlet IDK226 + 494–DK226 + 665 section is constructed by drill and blast, with horseshoe section; IDK226 + 665–DK226 + 725 section is constructed by open cut, which is the TBM dismantling site; For section IDK226 + 725–DK232 + 427, it is proposed to adopt TBM for construction at 5702 m. According to the stratum conditions and the topographic conditions at the portal, it is determined that the TBM shall be removed in the solar slope ditch (IDK226 + 665–DK226 + 725) after construction.

4. Construction Preparation

The main construction preparations are as follows: Site layout outside the tunnel, construction of invert prefabrication plant, design and construction of temporary power supply, design and construction of turnover platform, mobilization and transportation of TBM, parking and assembly site layout of TBM parts (structures), assembly and commissioning of TBM, construction of TBM preparatory tunnel and departure tunnel, and configuration of TBM assembly, hoisting and transportation equipment.

5. Portal Drill and Blast Construction Section

The TBM preparatory tunnel and departure tunnel are constructed by drill and blast. The preparatory tunnel has a single-center horseshoe-shaped section. The design section of the departure tunnel is circular with a diameter of 8.9 m, and its supporting reaction force is required to be greater than 2.4 MPa. The length of the preparatory tunnel and the departure tunnel is determined according to the site conditions outside the tunnel and the actual construction geological conditions. Since the total length of the TBM is 256 m, in order to form the production capacity, it must meet the requirements of TBM erection, tunneling material feeding and mucking. Therefore, there must be a corresponding safety distance between the feeding siding and the TBM back-up ramp rail; At the same time, as TB880E TBM is gripper hard rock TBM, the requirements for geological conditions are very strict to achieve the tunneling capacity. The actual surrounding rock conditions play a crucial role in determining the length of the preparatory tunnel and the departure tunnel, and it is required to avoid the soft surrounding rock section and fault section at the portal. Therefore, the determination of TBM preparatory tunnel and departure tunnel is mainly restricted and affected by the above two factors.

Limited by the portal site, the chainage intersection point between the main track of the outer rail line of the Mogouling Tunnel and the loading sidings in the production area from DK232 + 750 to the chainage to the portal DK232 + 608 is about 142 m. Theoretically, the total length of the preparatory tunnel and the departure tunnel is up to (256–142 + safety distance 5) 119 m, which can be met. However, since the TBM will be assembled synchronously during the construction of the preparatory tunnel, the portal should also reserve the necessary distance range (at least 20 m) for construction to reduce conflicts. Therefore, the total length shall be 139 m. The site layout and design of the preparatory tunnel are 71 m long and the departure tunnel is 10 m long, 81 m in total. There is a great conflict. In addition, when the actual construction reaches the cut-off chainage of the designed departure tunnel (DK232 + 525), the geological conditions show that it is a fault fracture zone, with developed joints and fissure water, and is not suitable for TBM tunneling. Therefore, according to the actual geological and hydrological conditions, the cut-off chainage of the preparatory tunnel is gradually extended to DK232 + 437 in the section with good geological conditions during construction, and the departure tunnel reaches DK232 + 427.

According to the geological conditions, the construction of the preparatory tunnel and departure tunnel of Mogouling Tunnel can be roughly divided into the following five stages:

(1) Sandy Clay Section (DK232 + 570–DK232 + 608)

The method of three-step micro-bench circumferential excavation with reserved core soil is adopted, with short advance rate, weak blasting and strong support. At the same time, it is reinforced with probe rock bolts, with a circumferential spacing of 0.3 m and a length of 3 m. For every 0.5 m of short advance rate cycle, I_{16} ring beam shall be erected and concrete shall be poured into the tunnel immediately.

When the upper bench enters the tunnel for 8 m, it is found that the shotcrete of the upward slope of the portal is seriously cracked, the surface subsidence of the vault is large, the convergence speed of the tunnel wall is fast, and the situation is very dangerous. The excavation is stopped immediately, and the following reinforcement measures are taken: Grouting and shotcreting of roof; 10 cm thickening layer of shotcrete for full section in the tunnel; Increased measurement frequency; Increase the excavation radius from 5.25 m to 5.35 m, and reserve enough deformity; The originally planned molded concrete is changed into shotcrete with the same thickness.

After reinforcement, the excavation continued. It was found that the convergence was still aggravated. When excavating the middle and lower benches, it was found that the geological conditions were different from the design, and the sand clay mountain on the top of the tunnel had an overall downward trend. The situation was critical, so the excavation was stopped immediately and the portal section was reinforced timely and concentratedly. The measures to be taken are as follows: additional cross bracing and vertical support to close the tunnel face; connection of 6 m of open cut tunnel as retaining wall of vaulted mountain; grouting treatment of arch and side wall for overall reinforcement; timely construction of reinforced concrete invert to form a ring-shaped closed stress structure.

After the reinforcement of portal, the three-step benched excavation shall be carried out forward according to the design. The construction procedure of three-step benched excavation method is as follows:

① Construction Process of Upper Bench.

Surveying and setting-out → setting-out of advanced pipe shed or probe rock bolt (select as required) → manual excavation of upper bench → initial shotcreting of 5 cm → erection of ring beam → installation of system rock bolt → welding of longitudinal connecting bar and wire mesh → concealed inspection → repeated shotcreting to the design thickness → surveying and setting-out.

② Construction Process of Middle and Lower Bench.

Surveying and setting out → mechanical excavation of core soil → manual staggered excavation of soil on both sides of the bench → initial shotcreting of 5 cm → extension of ring beam → installation of system rock bolt → welding of longitudinal connecting bar and wire mesh → concealed inspection → repeated shotcreting → surveying and setting out.

③ Invert Construction Process.

Surveying and setting out → mechanical excavation of soil mass at the bottom → manual excavation of invert to the design position → installation of invert grid ring beam → welding of connecting reinforcement → concealed inspection → pouring of invert concrete → curing and formwork removal → surveying and setting out.

(2) Earth-rock Transition Section (DK232 + 557–DK232 + 570).

When the upper bench was excavated to DK232 + 570, cataclastic marble and fault mylonite already appeared in front, with developed joints, but it was difficult to carry out manual excavation. In order to speed up the construction progress, the method of reserving surrounding layers by weak blasting is adopted for construction of upper bench. The specific procedures are as follows: surveying → setting up pre-grouting pipe → welding with erected ring beam → installing system rock bolt → welding connecting bar and wire mesh → shotcreting → drilling → blasting → initial shotcreting → surveying.

The pre-grouting pipe is ϕ40 mm, 3.5 m long and 30 cm spacing, arranged on both sides of the steelwork, with an external angle of 5°–10°. The length of the two overlapping is not less than 1 m. The cement-sodium silicate binary slurry is adopted for grouting, and the mix proportion is cement: water: water glass = 1:1:0.5, and the final grouting pressure is 0.6 MPa.

Blasting design: The thickness of the surrounding layer is reserved to be 40–50 cm, and short advance rate and weak blasting are adopted. The fire detonator is blasted in several times, and the explosive is ϕ32 mm and 20 cm long latex explosive.

(3) Section Change (DK232 + 481–DK232 + 557).

The lithology of this section is mainly composed of cataclastic marble, intercalated with quartz mica shist, with developed joint fissures, severely weathered, joint plane intercalated with fault gouge, and fissure water ingress in some sections. The measured water ingress is 0.2–0.3 m^3/h, and is extremely easy to collapse. Therefore, the excavation radius is still 518 cm, and the three-step micro-bench method is adopted for construction. The principle is still "short footage, weak blasting, strong support, frequent measurement and early closure", and the method is the same as above, and staggered excavation is adopted. The invert construction is of C20 concrete structure, closely followed by the lower bench excavation, with a distance of about 20 m. The half width is excavated and poured with concrete, with a length of 10 m, and the other half width is used as a construction access road. After the concrete strength of the invert reaches 70%, it can be used as a transportation access road, and then the concrete of the other half width is constructed.

(4) Section Change (DK232 + 437–DK232 + 481).

The excavation radius of section DK232 + 440–DK232 + 481 changes to 505 cm. The section DK232 + 437–DK232 + 440 gradually changes to 475 cm. The excavation method is the same as above.

(5) Departure Tunnel Construction (DK232 + 427–DK232 + 437).

The designed excavation section is round, with an excavation radius of 4.9 m, a molded concrete lining thickness of 45 cm and an inner contour radius of 445 cm. Considering the technical requirements of the departure tunnel, as long as the bearing capacity and section smoothness meet the requirements, and in combination with the actual construction situation, it is decided to adopt shotcrete lining: the excavation radius is 475 cm and the shotcrete thickness is 450 cm. The specific steps are as follows:

① Excavation: The upper and lower micro-bench method is adopted for excavation, and smooth blasting is adopted. After excavation, 5 cm concrete is sprayed first, and then the radial system rock bolt of 0.8 mX0.8 m is installed. It is arranged in quincunx, with mesh hanging and ϕ22 mm full-round reinforcement ring installed. The excavation continues to DK232 + 427, with the same support.

② Shotcrete lining: hang another layer of wire mesh with a spacing of 20 cm with the first layer, and weld it with short reinforcement; add longitudinal connecting reinforcement, with a circumferential spacing of 1.0 m, and weld it firmly; install inspection rings, accurately positioned and welded, with a spacing of 1.0 m; spray concrete in layers to the design radius of 450 cm.

The surrounding rock of the departure tunnel is still composed of cataclastic marble intercalated with mica-quartz shist, with relatively developed joints. After excavation, no large volume block falling occurs, the rock surface has water seepage, and the integrity of the surrounding rock is relatively good.

After the construction of the departure tunnel is completed, it is estimated that the shotcrete strength can reach more than 70%, that is, the strength can reach about 15

MPa, while the ground pressure of the gripper to the rock wall during TBM tunneling is only required to be not less than 2.4 MPa. The shotcrete strength has fully met the requirements, and the flatness after shotcreting also meets the requirements of TBM tunneling.

6. TBM Stepping Construction

(1) Filling of Road Muck and Laying of Track in the Portal Section

In order to meet the needs of TBM stepping, a tunnel waste muck subgrade with a width of 5.0 m and a thickness of 0.8 m shall be laid on the erection yard in the portal erection yard section, and a layer of crushed stone ballast with a thickness of 0.3 m shall be laid on it. A total of 200 m^3 of waste muck, 70 m^3 of ballast, 176 sleepers with a length of 1.5 m, 352 iron pads, 1060 spikes, 16 rails, 16 sets of rail connecting cleats and their supporting bolts are required.

Laying roadbed and road muck, extending the track while stepping, all materials are supplied from both sides. In order to save time during construction, waste muck, sleepers and rails can be stored on both sides of the machine in advance. The dumping and paving were carried out by loader, and the road muck was backed up by dump truck, paved by loader, and cooperated by manpower.

Track laying shall be carried out according to the design gradient, and shall be smooth and straight, with a gauge tolerance of $\pm$ 4 mm.

The subgrade shall be filled densely without large settlement. The waster muck shall have good water permeability and the content of fine-grained soil shall not exceed 15%.

Elevation control method: Piling and leveling on both sides of the line, and pulling ropes on both sides to determine the elevation of the rail surface.

(2) Laying of Arch Block and Construction of Lower Filling Concrete

A centerline stake shall be set every 100 m, and the sideline shall be set out. The rail surface line or the top surface line of the invert block shall be drawn at the elevation first, and the centerline of the invert shall be aligned with the set point. The elevation shall be controlled by a line rope or level gauge. The centerline and elevation deviation shall be $\pm$ 5 mm, the staggering shall be less than 5 mm, and the gap shall be less than 10 mm. M20 mortar shall be used and water-stop shall be plastered. Concrete pouring shall be carried out by pumping C20 concrete, supplemented by shuttle groove. Steel formwork shall be used for formwork erection. The plug and side formwork shall be firm without formwork shifting, and the concrete shall be poured densely without holes. The bottom of the invert block shall be cleaned thoroughly before laying, and the invert block shall be placed correctly without sinking. If there is slab staggering after the invert block is placed, it shall be re-laid in time.

7. TBM Construction

The TB880E ϕ8.8 m gripper TBM of Wirth, Germany is used in the Project, and it was once used for the construction of the north portal of Qinling Line I tunnel.

(1) TBM Tunneling Principle

TBM consists of two parts: main machine and back-up system. The main machine is used for rock crushing and muck collection and transfer; the back-up system is used for muck loading and support.

The tunneling system is mainly composed of cutter head with cutter, cutter head drive and thrust system. The TBM main machine uses its two outer Kelly supports, rear supports, and 4 fulcrums of the cutter head shield for support and regripping operations during the tunneling process. The inner Kelly support is of box section, with the front connected to the cutter head, the middle connected to the drive assembly, and the rear connected to the rear support. There is a large space inside inner Kelly support to conveniently arrange the conveyor belt (No. 1 belt conveyor). The quenched guide rail on inner Kelly support has an outer bearing seat with X-shaped gripper to slide longitudinally on it, and the inner and outer Kelly supports are connected by a push cylinder. The inner Kelly support transmits thrust force to the cutter head guidance and transmits machine torque to the outer Kelly support. There are 2 groups of outer Kelly supports and grippers. The grippers are X-shaped. The X-shaped grippers arranged horizontally are conducive to improving the stability of the whole machine.

During tunneling, the gripper firmly locks the outer Kelly support on the excavated tunnel wall to bear the pressure of cutter head torque and thrust force. The thrust cylinder takes the outer Kelly support as the fulcrum, and the thrust force is applied to the inner Kelly support and the cutter head to push the cutter head for rock breaking and tunneling. Under the action of thrust force, the disc cutter installed on the cutter head tightly presses the rock surface. With the rotation of the cutter head, the disc cutter rotates around the center axis of the cutter head and rotates around its own axis. Under the strong thrust and torque of the cutter head, the single-blade disc cutter rolls on the concentric slit of the tunnel face. When the thrust force exceeds the strength of the rock, the rock under the disc cutter is directly broken, the disc cutter penetrates into the rock, and the tunnel face is squeezed and broken by the disc cutter to form a concentric groove of the tunnel. With the increase of groove depth, the cracks on the rock mass surface deepen and expand. When the shear and tensile strength of rock is exceeded, the rock between adjacent concentric grooves spalls in pieces. The rock chips collapsed at the bottom of the tunnel is lifted by 8 side shovels evenly distributed on the cutter head and 4 central buckets and scrapers and transported to the top of the cutter head with rotation, and falls onto the inner Kelly support belt conveyor along the muck groove of the cutter head, and is discharged to the mucking train by the mobile muck chute on No. 3 belt conveyor.

The thrust cylinder extends for one stroke, and the cutter head and inner Kelly support (all main machine parts except outer Kelly support) move forward for one stroke accordingly, that is, 1.8 m of a tunneling stroke is completed.

After the tunneling process is completed, carry out the regripping operation, and complete the adjustment of the advance direction of the next cycle during the regripping process. First read and record the position of the machine, operate the cutter head of the main machine to retreat about 30 mm from the tunnel face, and the cutter head and belt conveyor stop running successively. Then lower the rear support frame and the cutter head shield support, retract the gripper cylinder, loosen the support, and then retract the thrust cylinder, so that the two outer Kelly supports move forward along the inner Kelly support slide rail for one stroke. According to the position of the machine at the end of the previous cycle, take the cutter head shield as the fulcrum, operate the telescopic cylinder at the rear support, so that the tail of the machine swings up, down, left and right and twists along the machine axis to reach the predetermined position, thus completing the adjustment of the advance direction. Then operate the machine to make the 2 outer Kelly supports tighten the tunnel rock wall again, retract the rear support and the cutter head shield, and at the same time pull the back-up system forward for one tunneling process (1.8 m), start the machine, to carry out the next cycle of tunneling operations.

(2) TBM Construction Process

TBM is a construction machinery integrating tunnel excavation, support and muck loading. During tunneling, it completes mucking, primary support, installation of invert blocks, air, water and electricity, track laying and other work. According to the equipment configuration of TB880E, its tunneling construction can be divided into 4 parts: excavation, mucking, support and auxiliary operation.

① Excavation Operation.

The two X-shaped supports of the TBM main machine on the outer Kelly supports are used to tightly press the tunnel rock wall and make the main machine fixed. Then, under the action of the thrust cylinder and cutter head drive device, the disc cutter is pressed against the rock surface and crushed with the cutter head rotation; the bucket on the cutter head rotates continuously to shovel the rock chips and unload it into the belt conveyor inside the inner Kelly support for transportation. The thrust cylinder extends for one stroke, that is completing a tunneling cycle of 1.8 m. When the tunneling length completes 1 cycle of 1.8 m, the TBM main machine uses its 2 outer Kelly support and rear support, and the cutter head shield support 4 fulcrums to complete the forward "crawling" of the main machine by 1.8 m, and then drags the back-up system forward by 1.8 m to start the next cycle.

② Mucking Operation.

At the same time of tunneling, the rock chips on the tunnel face is unloaded into the No. 1 belt conveyor of inner Kelly support through the bucket, and then unloaded to the belt bridge conveyor (No. 2) behind the main machine by the belt conveyor, and then sent to the No. 3 belt conveyor on the back-up gantry by the belt bridge conveyor, and unloaded to the transport train by the bidirectional chute on the No. 3 belt conveyor. After the muck is unloaded by the muck transport train at the dumping platform outside the tunnel, it is transported to the spoil area by truck.

The moving distance of the two-way muck chute is 80 m, and it can meet the muck loading requirements of stationary trains.

③ Support Operation.

While the main machine is tunneling, the main machine and the back-up auxiliary equipment complete the primary support operation and the installation of the invert block of a tunneling cycle.

The ring beam is pre-assembled and installed by the annular installer and jacking device behind the cutter head. 4 rock drills complete the construction of rock bolt holes. 2 of them are located behind the cutter head shield and are responsible for rock bolt boreholes within the range of 2 × 75° on the top of the tunnel. The mesh is processed and formed outside the tunnel. The mesh in the tunnel is welded and fixed at the tail of the rock bolt, and the mesh is welded with the ring beam. Shotcreting is completed by 1 back-up shotcreting manipulator, with a spraying range of 300° at the upper part of the tunnel and a longitudinal movement distance of 5 m.

For the installation of invert block, the invert block crane under the belt bridge lifts the invert block from the invert block vehicle and transports it forward to the required installation position. The gap under the invert block shall be filled with cement mortar through the grouting pump.

④ Auxiliary Operation.

Track extension: Every 7–10 pieces of invert blocks are installed to extend the track once, using the material crane in place to fix manually.

Water supply pipeline extension: The maximum water consumption of TBM is 60 m^3/h. The TBM is equipped with a 50 m soft water pipe, and the main water supply pipe is extended once every 50 m of tunneling.

Power supply cable extension: TBM is equipped with 400 m flexible cable, and the main power supply cable is extended every 400 m of tunneling.

Construction ventilation: TBM equipped with 1 complete ventilation system is connected with the tunnel ventilation system through an air duct box. The air duct box stores 100 m air ducts, and the air duct box is replaced once every 100 m of tunneling.

Construction drainage: The construction wastewater of TBM is pumped into the central ditch of the invert block by a pump under the belt bridge, and discharged outside the tunnel by the central ditch.

(3) Operation Content and Equipment Allocation.

See Table 11.6 for the operation mode and equipment allocation of each process of tunnel construction.

(4) Rock Chips Loading and Material Transportation.

Rock chips loading and material supply in the tunnel are completed by the equipment in the back-up system, and mucking and material transportation in the tunnel are completed by rolling stocks. The transportation system adopts double-track track transportation, with the left track as the mucking line and the right track as the feeding line.

The transport track is made of 43 kg/m steel rail with 900 mm gauge. A pair of single california switch shall be added every 3 km in the tunnel to facilitate vehicles to meet under abnormal conditions or other transportation faults in emergency.

The mucking train transports the rock ballast from the working face to the dumping platform, and then the dumper unloads it to the spoil area, and then the dump truck is used to transport it to the spoil area for the second time.

Table 11.6 Operation Mode and Equipment Allocation for Each Tunnel Construction Process

Serial number	Construction content	Equipment allocation and operation mode
1	Tunnel Excavation	TBM mainframe and cutter head rotary extrusion cutting rock breaking
2	Mucking	Cutter head scraper, bucket muck collection, three-stage belt conveyor muck unloading, 10×20 m^3 muck car for mucking, traction by 35 t diesel locomotive, and transportation by truck outside the tunnel
3	Support	The ring beam erector is used to install the ring beam; there are 4 bolt drillers, and the system bolt drillers are completed at the rear part. The shotcrete system is divided into cutter head shield, and then the shotcrete is directly sprayed manually with a nozzle and electrically remotely sprayed with a manipulator on the back-up platform. The wire mesh is manually laid
4	Adverse ground condition pre-treatment	1 probe geological drilling machine can drill ϕ64 mm and 30 m drilling holes or ϕ102 mm and 20 m deep pre-grouting holes within 80° of the tunnel top, and is equipped with a double mud grouting pump for pre-grouting
5	Grouting operation	TBM is equipped with a set of dry material feeding system and grouting system. It can be used for filling under the invert block and pre-grouting in adverse ground condition
6	Installation of invert precast blocks	1 set of 13 t monorail invert crane for manual installation
7	Material supply	Materials are transported by the material vehicle in the rolling stock. After the material vehicle arrives at the TBM unloading area, the materials are transported to the place of use by the 7 lifting equipment equipped
8	Track extension	The TBM rail storage rack can store 12 rails, and the rail laying is completed manually with a crane every 12 m
9	Air duct extension	The TBM is equipped with an air duct storage cylinder with a memory air duct of 100 m. The manually buried hook air duct automatically extends with the TBM advance, and the air duct box is replaced every 100 m

(continued)

Table 11.6 (continued)

Serial number	Construction content	Equipment allocation and operation mode
10	Water pipe extension	The TBM is equipped with a water hose reel with ϕ100 mm, 50 m hose, and the main water supply pipe is extended every 50 m
11	Cable extension	The TBM cable reel is equipped with 400 m cables, and the tunnel cable is extended every 400 m
12	Auxiliary chamber	It is equipped with 1 mobile electric air compressor, manual excavation by drill and blast with hand-held pneumatic drill, and manual muck loading with belt conveyor

Tunneling muck loading: The muck is loaded into the rolling stock through the belt conveyor and the mobile two-way muck car. The back-up gantry 5–14 are the muck loading area, and the track is a double-track, and it can accommodate 2 trains parked in parallel. The belt conveyor is used to load the muck of 1 tunneling stroke into 1 muck car and transport it out, and the other one is waiting for the mucking of the next stroke.

Cleaning of the rock chips at the bottom of the invert block: it is manually cleaned into the muck bucket, unloaded to the belt conveyor behind the cutter head with the monorail bucket, and sent to the front of the cutter head with the belt conveyor.

Cleaning of rebound material: The rebound material is manually collected into the muck hopper and lifted to No. 3 belt conveyor from the lower part by the muck hopper system.

Composition of the rolling stock:1 invert block vehicle + 1 ring beam/rail vehicle + 1 shotcrete tank truck + 10 muck cars + 1 man car + 1 35 t locomotive; 1 invert block vehicle + 1 concrete tank truck + 10 muck cars + 1 35 t locomotive + 1 man car.

The train formation can be changed according to the needs of maintenance, grouting, rail supply, etc. The train formation structure remains unchanged during tunneling. Materials such as rails, arches, support materials and air duct cassettes can be supplied during maintenance.

The transportation of rock chips and materials is carried out by 2 locomotives in combination with 3 sets of rolling stocks. 1 locomotive and 1 set of material trucks are used for loading outside the tunnel. During normal tunneling, 1 set of train is always waiting for muck loading, 1 set of train is running, and 1 set of train is unloading muck. The operation mode of the train is considered as 25 km/h inside the tunnel, 5 km/h on the trestle and 10 m/h outside the tunnel.

Shunting operation outside the tunnel: In order to reduce the time of shunting and loading operation outside the tunnel, the train formation is divided into 3 locomotives, 3 groups of muck cars and 4 groups of material vehicles (plus 1 group of spare material vehicles) for operation and formation, and the material vehicles left by the first train before each train enters the tunnel are used for shunting and loading in advance. After the material car is uncoupled from the muck car, it is pulled to the charging line by the locomotive outside the tunnel, and the material is loaded along the line according to the needs of material supply in the tunnel. The specific loading procedure is as follows:

① The material vehicle is first pushed into the material shed by the locomotive, and the wire mesh, sodium silicate, rock bolt and other materials are installed;
② Pull to the storage yard of prefabricated blocks and finished materials, and install materials such as invert prefabricated blocks, rails, air duct boxes, water pipes, ring beams and cables;
③ Enter the concrete production area along the loading line and load concrete, dry mix and other materials as required;
④ Through the crossover, the material vehicle is pushed by the locomotive to the parking place of the material vehicle of the feed line, and the unloaded muck is pulled into the tunnel by the muck car.

Shunting operation in tunnel: When the muck car stops in front of the crossover of the back-up inclined platform of TBM (No. 16 and No. 17 gantries), the crane system unloads the air duct box and cables to the corresponding positions, and then the locomotive pushes the ore vehicle to the muck loading area to wait for muck loading. The skip car is unhooked, and the chain traction system tows the skip car to the loading and unloading place of the mixing tank (No. 4 gantry). The third material vehicle is unhooked from the first 2 skip cars, and the sprayed materials are unloaded here. The first two material cars will continue to be towed to the unloading area under the equipment bridge for unloading by the material crane. After unloading, the material vehicle is towed back by the chain traction system and coupled with the muck car to leave the tunnel.

Loading, transportation and storage of materials in the tunnel: completed by 6 cranes, 2 lifting platforms and 2 transportation tools of back-up equipment.

(5) Initial support construction.

① Initial Spraying.

When the surrounding rock is broken and the self-stability is poor, the manual shotcrete system behind the cutter head can be used. At the same time of tunneling, the shotcrete is directly behind the cutter head shield, and the initial shotcrete is 3–5 cm to seal the surrounding rock.

② Installation of Ring Beam and Wire Mesh.

The main machine of TBM is equipped with an annular ring beam mounter, which allows pre-assembly of 5 ring beams. The installation sequence is: Lift the first section of steel arch from the conveying jumbo and place it on the top of the storage ring. The hydraulic cylinder drives the lifting device to move horizontally and vertically, pushes the first section of steel arch and leans against the storage device, and then fixes it with bolts. The other end is supported on a support, and then the storage ring rotates to leave space for the second section of steel arch. Then the 2–4 sections of steel arch are lifted from the conveying jumbo and placed on the storage ring, connected with steel plates, and then the storage ring rotates again. Each steel arch is placed on 2 supports, with 1 fixing clamp. The fifth section is connected with the fourth section, and its free end slides into the free end of the first section and is fixed with G-shaped clamps. Then the storage ring rotates 180° to make the overlapping part of the steel arch at the bottom of the tunnel. Use an external line device to make the steel arch tight to the tunnel wall, and finally bolt the fastening plate.

Before the ring beam stand against the tunnel wall, wire mesh shall be installed behind steel extrados for the soft broken belt, and the mesh specification is 1.5 m × 0.9 m, grid 25 cm × 25 cm. Steel welding is used for longitudinal connection of ring beams, to ensure the quantity and quality, and prevent the surrounding rock instability.

③ Rock Bolt Construction.

The installation of rock bolt can be drilled for 4.0 m, the anchoring cartridge and rock bolt are manually installed, the drilling machine assists in jacking in place, the net is manually hung, the impact drill is used for drilling, the short reinforcement is fixed and connected with the rock bolt, and the mesh frame is attached to the rock surface to prevent gravel from falling between the rock surface and the arch frame.

④ Shotcrete.

The wet shotcrete system operated by an electric remote controller can turn on or off the shotcrete and additives. The shotcrete is transported to the tunnel by a 6 m^3 mixing tank truck after production in the concrete factory. In addition to the manual spraying device behind the cutter head shield, the manipulator spraying device set on the platform of back-up No. 2 gantry bridge can complete the shotcrete operation within 300° and 5 m above the invert.

⑤ Installation of Invert Blocks.

The invert is a reinforced concrete structure. Its bottom is divided into two types, namely ordinary type and slot type. During laying, the invert is transported to the place under invert crane and rotated manually by 90°, and then transported to the invert placement area. Use invert crane to lift them and put them in place according to the designed center line and horizontal position, and then the water stop tape and mortar protection are used. When laying 4–5 inverts, use fine aggregate concrete to fill the 5 cm gap under the inverts. Concrete pouring is done by grouting machinery through grouting pipes. Before concrete grouting, the muck shall be removed thoroughly, the wall shall be flushed clean, and sandbags shall be used to seal the wall to prevent slurry from flowing in. During grouting, it is advisable to make both sides of the invert block arch out of the rock surface. During construction, waste water from the industrial exhaust fan shall be prevented from flowing into the place where grouting is just carried out. In addition, the number of invert grouting shall be controlled. Half grouting of invert is not allowed.

8. Key Points of Construction of Soft and Weak Surrounding Rocks

(1) Establish a support system adapted to TBM construction

When the geological conditions are poor, the intensity of support increases, the difficulty increases correspondingly, the consumption of materials increases, and the time affecting the excavation increases. Therefore, the geological forecast in the excavation is done according to the design documents, and the support plan is proposed in advance.

In Class III or Class IV but with developed joints, TBM construction requires a large amount of support and a long duration of support time. For this reason, attention shall be paid to the following three points:

① Improve the operational efficiency of rock bolts

Due to the limitation of the cycle tunneling time, the installation speed should be accelerated, the consolidation effect should be improved, the anchoring requirements should be achieved quickly, the tunneling time should be shortened, and the tunnel excavation speed should be accelerated.

② Proper temporary support

In poor sections, local dangerous rocks often make it impossible for TBM to tunnel normally, and if necessary, shed support is adopted.

③ Timely shotcrete

In general, manual shotcrete equipment near the cutter head should be used as less as possible, in order to reduce contamination of the main machine and reduce TBM downtime. However, when passing through the weak geology and sections with severe rock explosion, especially when the excavation surface is broken, due to the slowdown of the advance speed, the time for shotcrete construction of excavation face and No. 2 platform manipulator is extended. Therefore, manual shotcrete has to be carried out, and the purpose is to close the excavation rock surface as soon as possible to reduce exposure time. Although the temporary shotcrete is only 20–30 mm thick, it forms a tightly-bonded composite with the rock in a very short period of time, thus preventing the rock from loosening, releasing pressure and bending during the normal loosening process. The resulting effect is to transform the rock with lower strength into stable rock and achieve the goal of stopping the chemical changes inside the rock and the weathering of the rock. Due to the use of manual shotcrete, pumping distance increased, posing strict requirements for the wet spraying material, and corresponding preparatory work shall be done.

(2) Reasonable selection of tunneling parameters

The geological conditions are the key to the advance speed. In addition to the compressive strength of the rock, the shear strength of the rock, quartz content and the degree of development of joints must also be considered. The cutting principle of TBM is that the rock is broken under the load of the cutter at the contact between the cutter and the tunnel face, and from this area, the rock is squeezed and cracked, and the rock is crushed into pieces between the cutters along these cracks.

During the tunneling, the rock mass is cut by TBM machine capability. If the rock has a low compressive strength, high penetration, and low and stable thrust, the relative advance speed will be high. If the rock has a high compressive strength, high thrust and low penetration, the corresponding tunneling speed will be low. In these homogeneous rocks, after selecting the appropriate tunneling parameters, it is possible to adopt an automatic operation system with a small thrust as a limit without the need to adjust tunneling parameters at any time. Cutters also show regular and normal wear.

For TBM tunneling in soft geology with developed joints, special attention shall be paid to the correct selection of tunneling parameters. The penetration increases when tunneling in fractured rock This is mainly due to the fact that some part of the cutters and the tunnel face are not in close contact and the cutter thrust increased (because the penetration increases with the increase of thrust). Generally, the cutter should have a certain overload capacity under normal circumstances, but due to the influence of current cutter quality problems, the damage of the cutter has increased a lot; In addition, sometimes the cutting face of the cutter head is tunneled perpendicular to the weak surface, and the cutting resistance of the weak surface is small, thus increasing the penetration under the original thrust. In addition, the cutter wear is small and the probability of damage is not large. Another reason is that when the cutting center of the cutter head is located in the rock mass with well-developed fissures or the rock mass with low shear strength along the spalling surface, the cutting block is cut along the joint or spalling surface. At this time, the TBM has violent vibration and large torque fluctuation, making the cutter bear high lateral load and the bearing tightens, resulting in overheating and thus causing cutter damage. At this time, the thrust force should be reduced.

When the increase of penetration and torque is found in the excavation, it indicates the change of geological conditions, so the thrust should be reduced and the penetration should be controlled as appropriate, so that the balanced productivity can be maintained and the cutter wear can be reduced.

11.1.4 West Qinling Railway Tunnel

1. Project Overview

The total length of the West Qinling Tunnel of the new Lanzhou-Chongqing Railway is 28.236 km. The ϕ10.2 m TBM construction is adopted for the DIK421 + 239–DIK415 + 588 section (the first tunneling stage) and the DIK411 + 388–DIK403 + 590 section (the second tunneling stage) at the tunnel exit, with a tunnel section area of 82 m^2. The 2113 m preparatory tunnel at the portal section and the 4715 m preparatory tunnel between the two tunneling sections are constructed by drill and blast. The engineering plan is shown in Fig. 11.10.

2. TBM Synchronous Lining

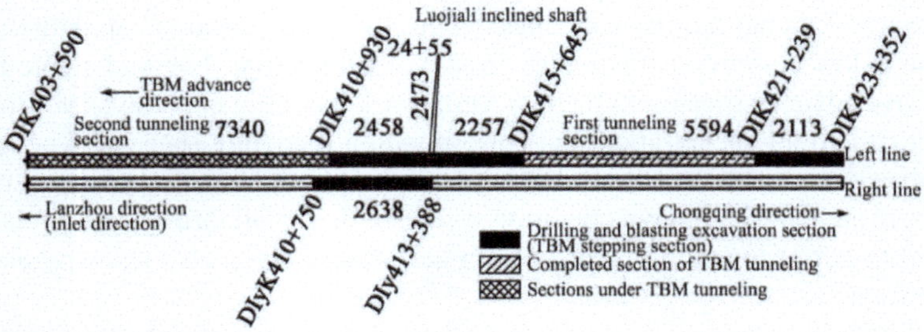

Fig. 11.10 Schematic Diagram of the Engineering Plane (dimension unit: m)

(1) Difficulties in Synchronous Lining

① The lining jumbo will pass through the continuous belt conveyor, air duct, construction cable and communication cable under construction during the traveling operation, and the removal of these facilities will affect the tunneling of the TBM ahead;
② Prefabricated invert blocks, steel supports, shotcrete materials required for front tunneling, and how transport vehicles pass through the lining jumbo, front waterproof board operation platform and reinforcement operation platform;
③ How to adopt an effective management system in the construction to minimize the interference between the jumbo running and TBM tunneling construction.

(2) Overall Construction Scheme

Since the mucking of the second tunneling section carried out in Luojiali inclined shaft has little impact on the lining of the first tunneling section, synchronous lining will be carried out in the later stage of the first tunneling section (about 3 km). Considering the length of TBM and back-up facilities, the placement distance of air supply machinery for initial shotcreting of tunnel face and back-up repeated shotcreting, transverse passage construction and other factors, the synchronous lining jumbo is about 2 km behind the TBM tunneling face, and the synchronous lining jumbo is 16.5 m long.

Since the advance speed is faster than the synchronous lining speed of concrete of a lining jumbo equipment, a lining jumbo equipment shall be added after the lining is delayed for a certain distance.

In order to ensure that the lining will not affect the double-track transportation at the same time, the walking tracks of the lining jumbo and the rack are fixed on the low side walls on both sides, and the air duct slides through the jumbo and the rack when passing through.; Removal of communication and power lines to the ground inside the lining jumbo; The belt conveyor underpinned by bearing slides on the jumbo and bench does not affect the operation of the belt conveyor. The waterproof board system and reinforcement binding operation are carried out on the bench 25 m in front.; The 16.5 m integral hydraulic steel formwork jumbo is produced in a centralized manner at the 90 mixing station outside the concrete tunnel, transported by rail-mounted concrete trucks, pumped into the formwork by concrete transfer pumps, tamped and compacted by immersion vibrator, and removed for curing after the concrete reaches the strength.

(3) Secondary Lining Construction Process

The secondary lining adopts an integral self-propelled formwork jumbo with a length of 16.5 m, and 1 rack is arranged in front of and behind the jumbo respectively. The front workbench is used for reinforcement binding, waterproof board hanging, non-woven fabric, air duct removal and belt conveyor bracket removal; the rear workbench is used for concrete surface repair, air duct hanging and belt conveyor bracket installation.

If reinforced concrete lining is required for Class V surrounding rock section, the front workbench shall be used to bind the reinforcement. See Fig. 11.11 for the specific construction process flow.

(4) Construction Method

① Construction of Low Side Wall.

Before lining construction, low side wall shall be constructed first. The low side wall is 2.2 m away from the tunnel centerline and 56 cm high. The top elevation of the low side wall is the bottom elevation of the operating ditch. See Fig. 11.12 for the structure of the low side wall.

The main processes are as follows: clearing the loose muck on both sides of the invert → laying waterproof board → connecting the longitudinal blind pipe with the drain hole of the invert block → filling both sides of the inverted arch block with concrete → erecting the formwork of the low side wall → pouring the concrete of the low side wall.

The virtual muck on both sides of the invert shall be cleaned manually. After cleaning, the waterproof board of the low side wall and the longitudinal and transverse blind pipes shall be installed. The overlapping length shall be reserved for the waterproof board and blind pipes, and they shall be welded with the waterproof board of the arch during lining. After the waterproof board and blind pipe are completed, concrete shall be poured and filled, and the concrete surface shall be flush with the invert surface. After the filling concrete reaches the strength, the survey team shall determine the accurate position of the low side wall, and then erect the formwork and pour the concrete. The low side wall formwork shall be set-shaped formwork and poured at one time. Back-sticking plastic water-stop and mid-buried steel-edge water-stop shall be embedded in the concrete joint.

② Track Laying.

After the concrete of the low side wall reaches the strength, the jumbo walking track is laid on the foundation of the low side wall, with a gauge of 5 m. The track laying of the work bench, formwork jumbo and repair bench is all set on the foundation of the low side wall to ensure double-track transportation in the tunnel. When pouring concrete, the concrete transfer pump is parked on the track on one side, and the track

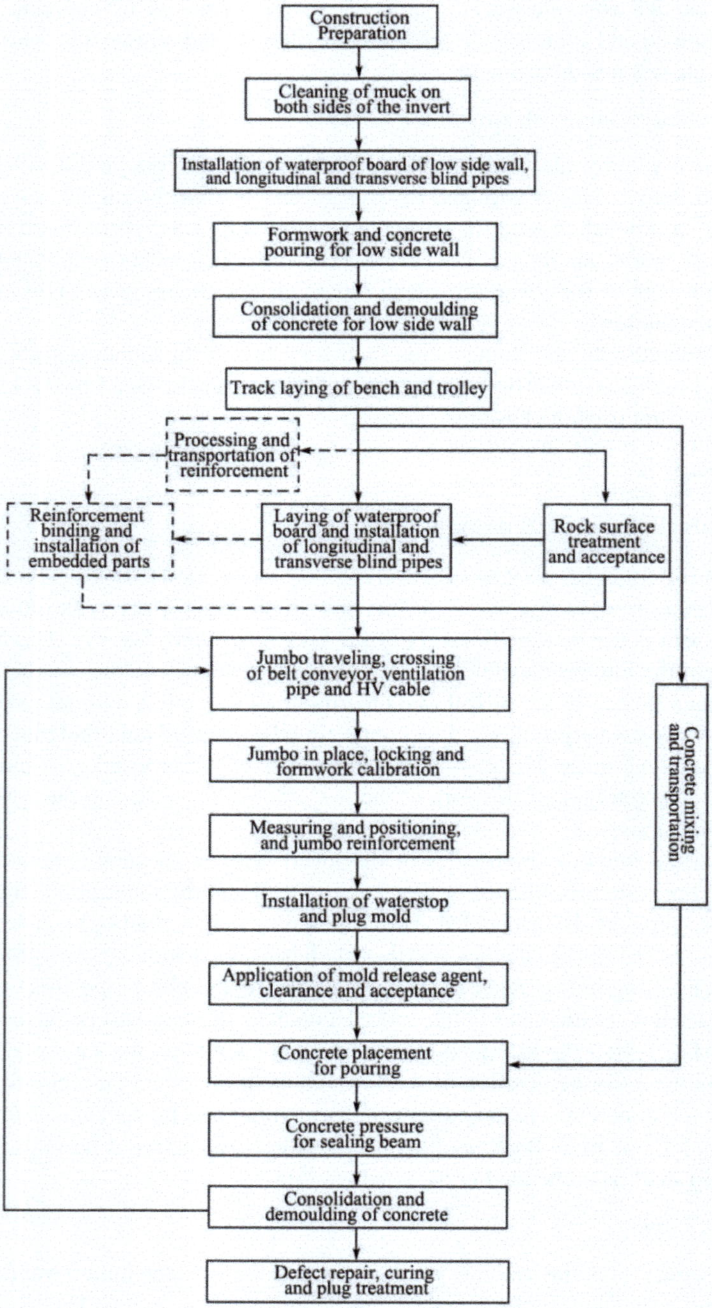

Fig. 11.11 Construction flow chart

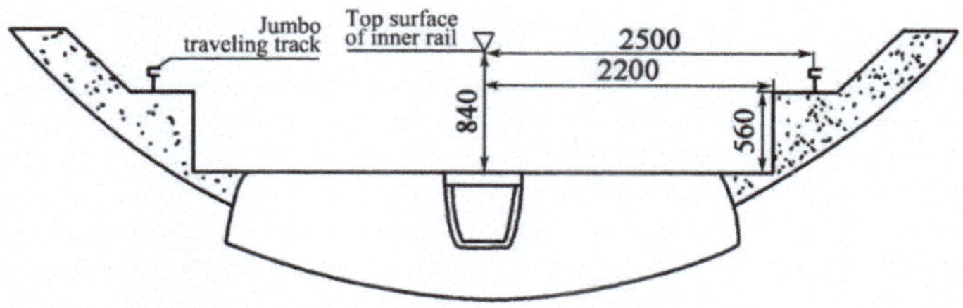

Fig. 11.12 Schematic diagram of low side wall structure (dimensions in mm)

on the other side is used for transportation in the tunnel. Movable california switches are set at 100 m in front of and behind the jumbo to ensure that the transportation train in the tunnel passes.

③ The operation bench shall be moved through.

A workbench with a length of 25 m is arranged in front of and behind each jumbo. A telescopic support is installed between the rack and the jumbo. The length of the telescopic support is 25 m, and the length of the whole jumbo and bracket is about 120 m.

After the jumbo is positioned, the front workbench is hung with waterproof boards, the air duct is removed, and a group of formworks is installed. The front workbench travels forward the length of a group of formworks, and the belt conveyor bracket on the side wall is removed while traveling, and the traveling telescopic bracket of the front workbench is extended. After the formwork jumbo is demoulded, walk forward the length of a group of formwork. The front telescopic support is shortened and the rear telescopic support is extended. After the formwork jumbo is positioned, the rear workbench moves forward, and the belt conveyor support and air duct are installed while walking. The removal and installation of the belt conveyor support will not affect the normal operation of the formwork jumbo through the expansion and contraction of the expansion support.

A bearing slide is arranged on the working frame to underpin the support frame of the primary side wall of the belt conveyor, so that the support frame of the primary side wall of the belt conveyor removed when the working frame moves forward can continue to be used without being affected. The removed original support frame shall be transported to the rear and installed on the side wall of the lined section.

④ Air Duct Picking and Shifting.

When the front workbench moves forward, remove the air duct fixed on the vault, slide the air duct through the jumbo and the bench, and re-hang the air pipe through the bench behind the jumbo.

⑤ Removal of Communication and Power Lines.

The communication and power lines on the original side wall shall be manually removed to the ground inside the jumbo and bench during TBM tunneling and maintenance, and wrapped with waterproof and anti-electric insulation pipes.

⑥ Construction of Waterproofing System.

Waterproof sheets and horizontal and longitudinal blind pipes are hung through the front bench.

⑦ Fabrication of Reinforcement.

According to the design requirements, C35 cast-in-place reinforced concrete lining is adopted for Class V surrounding rock, and the reinforcement binding is carried out by the front operation bench after the tunnel waterproof sheet is laid.

⑧ The jumbo is moved in place.

a. After the formwork removal of the jumbo meets the traveling conditions, the traveling track of the jumbo shall be calibrated first to see whether it is straight. If the deviation of the traveling track of the jumbo is too large, the continuous belt will deviate seriously in the jumbo section. See Fig. 11.13 for the photo of the lining jumbo.
b. Before the jumbo travels, move slowly, hold the bearing slide rail underpinning device at the front end of the jumbo to hold the running belt conveyor, observe whether there is a gap, make fine-tunning, then remove the triangular support and carrier roller of the nearby main tunnel belt conveyor, and transport them to the tail of the jumbo to prepare for the installation of the belt conveyor support in the lining section.

Fig. 11.13 Lining Jumbo

c. Run the jumbo, and observe the deviation of the belt conveyor on the jumbo support roller. If the deviation is serious, slightly adjust the left and right positions of the belt pulley for deviation adjustment. At the same time, the installation of the belt conveyor support in the lined section shall be carried out. The installation of the belt conveyor support and the carrier roller shall be consistent with the installation method during TBM tunneling. The borehole position fixed by the support shall be set out while the jumbo is measured and positioned, and marked on the lining wall of the tunnel.
d. Concrete production and pouring, formwork removal and curing.

Concrete is produced centrally by 90 mixing station outside the tunnel, concrete is poured by delivery pump inside the tunnel, and watered for curing after formwork removal.

⑨ Clearance Dimension Requirements of Jumbo.

When the formwork jumbo is arranged with double tracks, considering the size of the locomotive, there shall be enough space to cross the formwork jumbo, and the clearance dimension of the formwork jumbo can be designed as: The height is 3.08 m, the width is 4.4 m, and the maximum dimension of locomotives (including invert block transport vehicles, pedestrian vehicles, concrete tankers, etc.) is 2.75 m (including the height of rails), the width is 1.8 m, the distance between the central ditches is 0.4 m, and the side track gauge of two locomotives is 2.9 m, to ensure that two locomotives can pass through the formwork jumbo at the same time.

(5) Quality Assurance Measures

① Experienced technicians, quality inspectors, management personnel and operators shall be provided; sufficient construction machinery and equipment shall be provided, and the specifications, quantity and performance of construction equipment shall meet the needs of construction and production.
② Operators of special processes and special types of work working with certificates is ensured.
③ Establish a quality evaluation system, regularly evaluate the construction quality, timely feedback the project quality information, and take the evaluation results as one of the basis for formulating the project construction plan.
④ Establish a quality reward and punishment system, clarify the reward and punishment standards, achieve clear reward and punishment, and prevent quality accidents.
⑤ Strictly implement the project supervision system, carry out self-inspection by the operation team, re-inspection by the Project Management Department, and timely notify the Supervisor to inspect and sign for confirmation after passing the inspection. Concealed works must be signed for confirmation by the Supervisor before concealment.
⑥ Strictly manage the construction discipline and control the quality of the working procedure. If the working procedure is unqualified, the construction of the next working procedure cannot be carried out.

⑦ Adhere to the three-level review system of survey to ensure that the centerline, level and structure size and position are correct. Each measuring pile point shall be carefully protected.
⑧ All incoming materials must have factory certificates and quality inspection reports, and can only be used after passing the inspection.

(6) Safety Assurance Measures

① Establish and improve the work safety responsibility system, formulate various work safety rules and regulations and operating procedures, improve work safety conditions, strengthen work safety supervision and management, eliminate work safety accidents, and ensure the safety of life and property.
② Strengthen the construction safety education for employees, compile and print the safety protection manual and distribute it to all employees. Workers should be tested and assessed for safety operation before starting work, and only those who are qualified are allowed to start work.
③ Receive, transport, store and keep blasting equipment.
④ Adequate safe passageway and working space shall be reserved around electrical equipment and electrical lines.

3. Mucking Technology of TBM Supporting Continuous Belt Conveyor Outside the Tunnel

(1) TBM Construction Mucking Method

A continuous belt muck discharge system is adopted in the tunnel to match TBM tunneling, with a belt width of 914 mm and a design transfer capability of 600 m^2/h. Compared with the rail muck car transportation system, the continuous belt transportation in the tunnel has the following obvious advantages: The continuous belt is a continuous transportation without time interval, and the mucking efficiency is high; The continuous belt energy uses electric energy and has no contamination to the environment and a good construction environment.; Continuous belts rarely break down in daily life, with low maintenance cost and low difficulty in daily management.

During the construction of the first tunneling section of TBM of West Qinling Tunnel TBM Project of Lanzhou-Chongqing Railway, the average truck haul distance from the temporary spoil yard outside the tunnel to the spoil yard is 5.5 km long. The muck is transported to the portal by TBM continuous belt, transported to the muck separator of the temporary spoil yard by the horizontal cross belt conveyor at the portal, and then transported to the spoil area by dump truck.

In the construction of the second tunneling section of TBM, the tunneling working face is 10.5 km away from the tunnel exit. In order to reduce the transportation distance of waste slag, the Luojiali inclined shaft excavated in advance between the first and second tunneling sections is used as the tunnel mucking channel. In the tunnel, TBM continuous belt is used to horizontally transport the muck to the bottom of the inclined shaft and the three forks of the main tunnel, and then the muck is transferred to the entrance of the inclined shaft through the transfer belt in the inclined shaft. The slope from the entrance of the inclined shaft to the muck yard on the mountain is relatively large, the average truck haul distance of the extension line is 3.5 km, and the haul distance is relatively long. A large-slope uphill transfer belt is set at the entrance of the inclined shaft to directly transfer the slag to the slag yard on the mountain. Loaders are used in the slag yard for short-distance transportation and pushing.

(2) Mucking Method Outside the Tunnel for TBM Tunneling in the First Stage

The tunneling length of TBM in the first stage is 5594 m, and the continuous belt conveying length in the tunnel is 2113–7707 m. Due to the narrow site at the portal, the temporary spoil yard cannot be arranged, so a 130 m horizontal cross belt conveyor is set at the portal to transfer to the temporary spoil yard, and a muck separator is set at the end of the belt to realize automatic muck loading of transport vehicles. The cross belt conveyor at the tunnel portal has a belt width of 1000 mm, a drive power of 90 kW and a transfer capability of 1100 t/h. The muck separator is provided with 3 loading ports and 1 relief port. In case of no truck at the loading ports or excess muck, the relief port is used to temporarily discharge the muck which then will be loaded into trucks by loaders or excavators and transported outside. The layout of the muck separator is shown in Fig. 11.14.

The road between the muck separator and the spoil area is clay-bound macadam pavement with an average haul distance of 5.5 km and a round trip time of about 60 min for a dump truck. The equipment configuration is shown in Table 11.7.

Fig. 11.14 Layout of Muck Separator outside Tunnel Portal

Table 11.7 Mucking equipment configuration outside the tunnel portal during first-stage tunneling of TBM

Serial number	Equipment name	Specification	Qty	Remarks
1	Muck separator		1	Muck loading at temporary spoil yard
2	Dump truck	18 m^3	21	Outward transfer of muck
3	Loader	ZLC50	1	Muck loading at temporary spoil yard
4	Excavator		1	Muck loading at temporary spoil yard

The first stage of TBM tunneling started on June 25, 2010 and ended on May 28, 2011, lasting 11 months and 3 days, with a monthly average advance rate of 508 m, a maximum monthly advance rate of 841.8 m, a maximum daily advance rate of 42.69 m, a muck loosening factor of 1.4, a maximum daily muck disposal volume of 4900 m^3, and a daily average muck disposal volume of 1944 m^3. 8:00 am to 12:00 am is the time for mandatory equipment maintenance, during which the TBM stops tunnelling and 12:00 am–8:00 am of the next day is the normal tunnelling time. Due to the influence of geological conditions and equipment status, the advance speed generally fluctuates between 0 and 4 m/h, leading to uneven unloading of the muck separator, and significant fluctuation of mucking volume between 0 and 460 m/h. Through analysis, the following problems exist:

① The significant change in advance speed has led to a dramatic change in the demand for muck trucks. During maintenance or low tunneling of TBM, a large number of muck trucks are idle, while during fast tunneling, even if all the trucks are put into operation, they cannot meet the discharging demand of the muck separator.

② It is very difficult to meet the requirements for mucking at the peak speed of TBM tunneling. When TBM tunneling reaches the peak speed, the unloading capacity of the muck separator is lower than the transfer capability of the belt conveyor, resulting that some of the muck cannot be automatically unloaded into the dump trucks in time through the muck separator, and can only be unloaded at the temporary spoil yard through the relief port of the muck separator, but the temporary spoil yard has a limited capacity so it cannot hold a large amount of muck.

According to the field practice with a muck separator arranged, about 70% of the muck can be automatically loaded into trucks through the muck separator, and the remaining 30% at the relief port can be loaded into trucks by loaders. According to the statistics, the mucking cost indicators are shown in Table 11.8.

In the early stage of the project, because the temporary spoil yard was not formed, no muck separator was set, and all the muck was loaded by loaders continuously

Table 11.8 Mucking cost indicators with a muck separator

Item	Unit price	Quantity (m^3)	Amount (CNY)	Remarks
Muck loading cost	CNY 7/m^3	137,612	963,286.8	
Truck transport cost	CNY 14.5/m^3	458,708	6,651,266	Average haul distance 5.5 km
Muck separator manufacturing and installation	CNY 900,000/set	1	900,000	
Total		458,708	8,514,552.8	
Unit cost for muck transfer outside the tunnel			CNY 18.6/m^3	

Table 11.9 Mucking cost indicators without a muck separator

Item	Unit price	Quantity (m^3)	Amount (CNY)	Remarks
Muck loading cost	CNY 7/m^3	458,708	3,210,956	
Truck transport cost	CNY 11.5/m^3	458,708	5,275,142	Average haul distance 5.5 km
Total			8,486,098	
Unit cost for muck transfer outside the tunnel			CNY 18.5/m^3	

throughout the day. On the basis of practice, it is found that if the temporary spoil yard is relatively large, the number of trucks required to transport the muck outside during TBM tunneling is reduced by 4–5, and the unit price of truck transport cost can be reduced to CNY 11.5/m^3. The mucking cost indicators are shown in Table 11.9.

In summary, by setting a muck separator, most of the muck can be automatically loaded by the muck separator so that the muck loading workload of the mobile machinery is reduced. By setting a relief port, temporary mucking can be carried out when the unloading capacity of the muck separator is insufficient, to meet the needs of the overall mucking capacity.

The setting of muck separator belongs to fixed cost, and the muck loading and truck transport costs are variable costs. The setting of muck separator depends on the break-even point of cost and mucking volume. The analysis of the break-even point of cost and mucking volume is shown in Fig. 11.15.

According to the break-even analysis, in the first stage of TBM tunneling, the mucking volume of 473,684 m^3 is the break-even point of cost and mucking volume. When the mucking volume is greater than 473,684 m^3, the muck separator can bring obvious economic advantages; when it is less than 473,684 m^3, the muck separator cannot bring such advantages. If the temporary spoil yard outside the tunnel is small due to terrain restriction, the installation of muck separator can effectively solve the

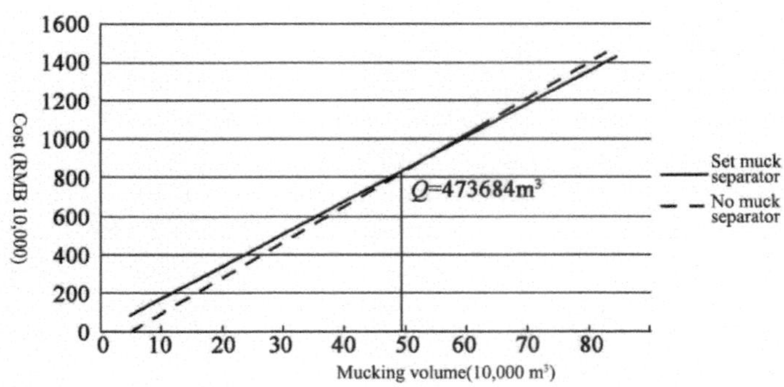

Fig. 11.15 Analysis of break-even point of cost and mucking volume

problem that the normal tunneling is restricted by the small capacity of the temporary spoil yard. The main purpose of setting a muck separator in West Qinling Project of Lanzhou-Chongqing Railway is to solve the problem of small site of the temporary spoil yard, realize fast muck loading and reduce muck accumulation, so as to meet the need of mucking for rapid TBM tunneling.

(3) Muck transfer outside the tunnel in the second stage of TBM tunneling

The length of the second stage of TBM tunneling is 7420 m, and the length of conveying belt inside the tunnel is 2478–9898 m. The slag is transferred from Luojiali inclined shaft bottom to shaft mouth via 2455 m conveying belt. The site at Luojiali inclined shaft mouth is narrow, surrounded by high mountains and deep valleys. The slag yard is at the gully above the inclined shaft. The straight-line distance between the gully mouth and the inclined shaft is about 300 m, and a 60 m high ridge stand between the gully mouth and the inclined shaft. Therefore, it is very difficult to build an access road, as a length of about 3.5 km is required to meet truck transport demands, and slope gradients at many positions has to be around 7%. With all things considered, a loading conveying belt is directly installed at the inclined shaft mouth, with a maximum elevation angle of 21°, a horizontal distance of 277.5 m, a height difference of 59.15 m, a belt speed of 2.5 m/s, a belt width of 1000 mm, a 200 kW variable frequency head driving, and a mucking capacity of 500 t/h. The slag is directly transferred to the slag yard on the hill through the transfer belt, and an extension belt is set up in the slag yard, equipped with 1 loader, 1 excavator and 3–4 dump trucks. The layout of the transfer belt is shown in Fig. 11.16.

The maximum round trip of dump truck in the slag yard is not more than 1.8 km, and the overall slope gradient is 6–8%. The equipment configuration is shown in Table 11.10.

The second-stage tunneling started on August 28, 2011, and ended on May 31, 2012, lasting 9 months and 3 days, with an accumulated tunnelling of 4000 m, an average monthly advance rate of 444 m, a maximum monthly advance rate of 569 m, and a maximum daily advance rate of 33.48 m. The mandatory maintenance system

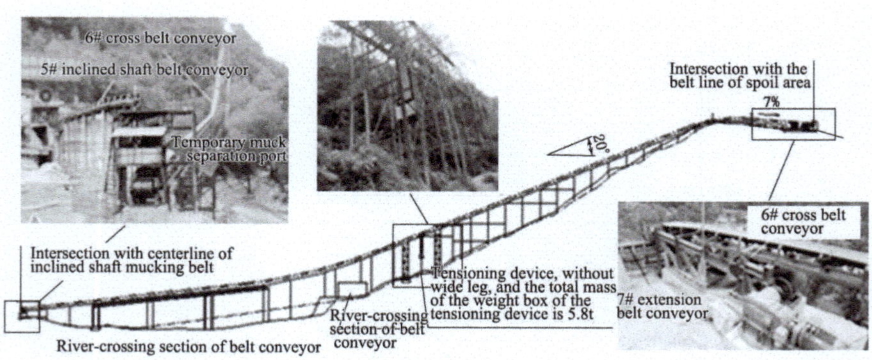

Fig. 11.16 Schematic diagram of the layout of transfer belt outside the tunnel

Table 11.10 Configuration table of slag discharge equipment outside the tunnel second stage

Serial number	Equipment name	Specification	Qty	Remarks
1	Transfer belt	200 kW	1	Transfer slag from the tunnel portal to the slag yard
2	Extension belt	90 kW	1	Extended slag conveying in the slag yard
3	Dump truck	18 m^3	4	Slag dumping
4	Loader	ZLC50	1	Loading at slag yard
5	Excavator		1	Loading at slag yard

for the first-stage TBM equipment still applies, and the mucking rate has changed significantly.

Through the engineering practice, by setting up transfer belt and extension belt, the distance is reduced from the average haul distance 3.5 m from the shaft mouth to the spoil area to the average round trip of about 0.9 km in the slag yard, which greatly reduces the number of vehicles required for slag transport.

By setting up the transfer belt, the slag is dumped with a short distance in the slag yard by mobile machinery, and the mucking cost indicators is shown in Table 11.11.

If the muck separator is set up in the spoil area in the second stage, it will also significantly increase the demand for the number of transport trucks, and the extension belt of the slag yard cannot be properly moved forward with the accumulation of slag in the process of mucking due to the limitation of the muck separator, which is not technically feasible. Therefore, muck separator is not provided in this scheme.

If we do not use the transfer belt, and directly transport the slag to the slag yard by truck at the shaft mouth, the mucking cost indicators are shown in Table 11.12.

Table 11.11 Slag discharge cost indicators for setting up transfer belt

Item	Unit price	Qty	Amount (CNY)	Remarks
Muck loading cost	CNY 7/m^3	608440 m^3	4,259,080	
Truck round trip cost	RMB4.3/m^3	608440 m^3	2,616,292	4 trucks provided
Belt installation cost	RMB3206000	1	3,206,000	Consider 30% of residual value
Electricity bill	0.75 RMB/kWh	1879200 kWh	1,409,400	Operation for 15 h per day, 18 months for construction period
Total			11,490,772	
Unit cost of slag discharge			RMB18.89/m^3	

Table 11.12 Slag discharge cost indicators without setting up transfer belt

Item	Unit price	Quantity (m^3)	Amount (CNY)	Remarks
Muck loading cost	7	608,440	4,259,080	
Truck transport cost	14.5	608,440	8,822,380	Large slope gradient
Total			13,081,460	
Unit cost for muck transfer outside the tunnel			RMB21.5/m^3	RMB/m^3

Here is the comparison between Transfer Belt and Direct Truck Transport. There is a fixed input of belt conveyor for setting up transfer belt, and there are expenses of building a long access road for direct truck transport. If not considering the cost of building an access road, the analysis chart of the cost break-even point of mucking volume is shown in Fig. 11.17.

Through the break-even analysis, for the second stage of TBM tunnelling, the mucking volume of 314313 m^3 is the cost break-even point of mucking volume. When the mucking volume is more than 314313 m^3, it has obvious economic advantage to set up the transfer belt; when it is less than 314313 m^3, it is not suitable to set up the transfer belt. If the transfer belt cannot directly reach the slag yard, but also need to be transported 4–5 km, it is preferable to choose the scheme composing of transfer belt at the shaft mouth + temporary slag storage yard with muck separator + dump truck transport.

(4) Summary

Under the condition of good geological conditions as well as long, continuous and fast TBM tunneling, the use of continuous belt for mucking inside the tunnel can give full play to the efficacy of fast TBM tunneling for large amount of mucking. However, if the capacity of slag loading and transport system outside the tunnel is not inadequate, it will greatly limit the fast tunneling capacity of TBM, and different slag loading and transport solutions outside the tunnel will produce great differences in economic effects, so the actual construction must be based on the site traffic conditions, terrain

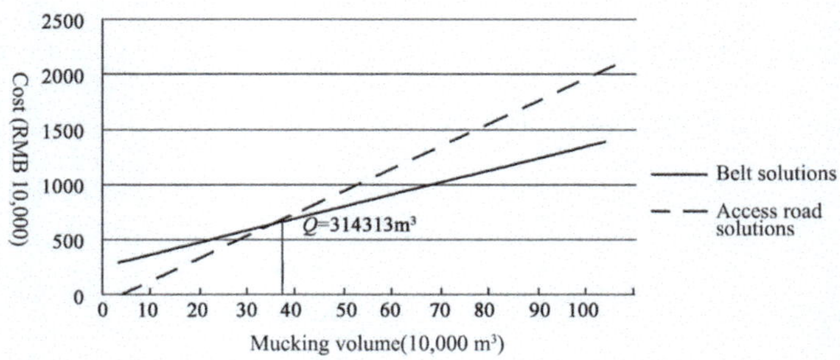

Fig. 11.17 Analysis chart of cost break-even point analysis of slag discharge volume

conditions, transport distance, etc. By considering the transport mode (belt or truck), capacity of temporary spoil yard, slag loading method (mechanical loading or muck separator), the amount of mucking and other aspects, and through the cost break-even analysis, select a feasible and applicable, reasonable and economic method of slag loading and transport outside the tunnel.

4. TBM main bearing seal damage fault and handling

(1) Problems and Descriptions

During the period from October 2011 to March 2012, the main drive lubrication system often broke down, as shown by:

① Wheel lubrication pressure, main bearing lubrication pressure, pinion front bearing lubrication pressure, pinion rear bearing lubrication pressure, lubrication oil return pressure and lubrication oil return filter pressure were often high, manifested as the corresponding filter displaying red (namely PLC alarm), sometimes even shortly after the replacement of new lubricant, the same thing happened, as shown in Fig. 11.18.

② The relief valve regulating the pinion lubrication pressure, main bearing lubrication pressure, pinion front bearing lubrication pressure and pinion rear bearing lubrication pressure were often out of order, and returned to normal after cleaning the relief valve spool.

③ In addition, sediment was found in the magnetic filter element, and also in the bull gear ring when checked from the drive pinion spare window, as shown in Fig. 11.19.

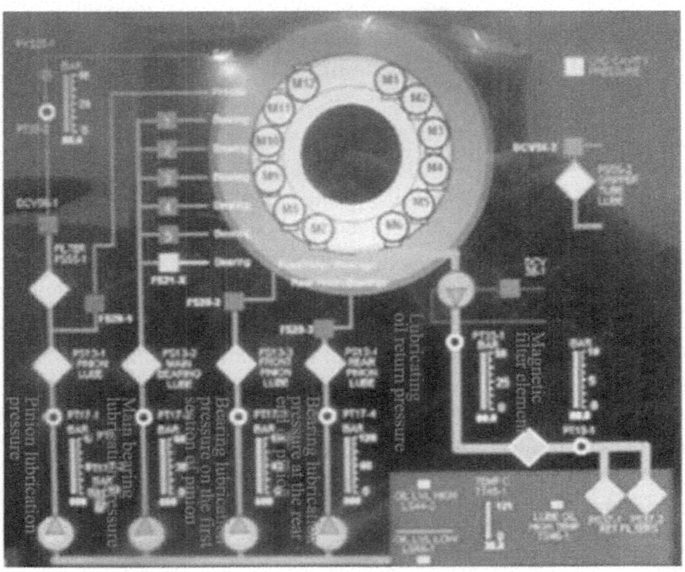

Fig. 11.18 Operation panel of operator cabinet

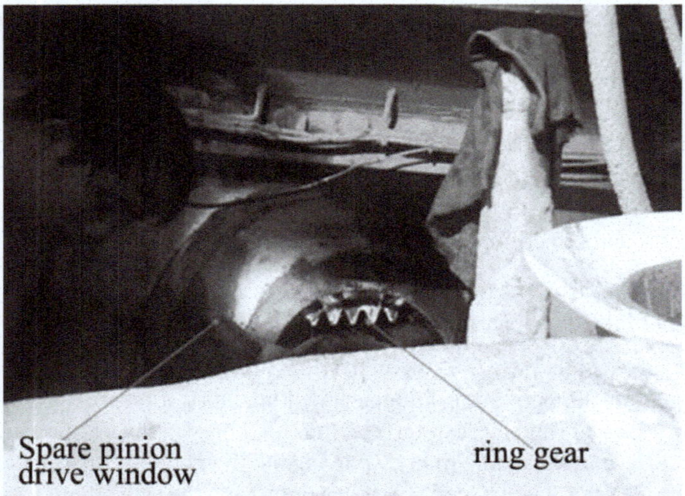

Fig. 11.19 Disassembly process

(2) Analysis of Causes

The main reason for frequent problems with the main drive lubrication system: it is likely that the inner seal is worn, and dust and debris from the tunnel face enters the bull gear chamber from the inner seal (Fig. 11.20).

After the dust and other dirty things enter the bull gear chamber, the lubricant flowing through the bull gear chamber is contaminated, and the contaminated lubricant will enter the lubricating system again after returning to the oil tank, and the dirty particles in the lubricant will block the valve spool and the filter element, so that the fault occurs.

On February 18, 2012, after dismantling the end cover of the inner seal and the end cover of the drive box steel structure, wear was found in the inner seal and there was sediment in the seal chamber, and the wear-resistant alloy steel belt in contact with the inner seal was also worn, which proved that the previous analysis was correct.

Because it takes about one month for the wear-resistant alloy steel belt to ship from USA, after cleaning the inner seal chamber with machine oil, we put the original inner seal on temporarily and continued digging, and then installed the new inner seal and wear-resistant alloy steel belt when the wear-resistant alloy steel belt came back.

From March 25 to March 27, 2012, we spent 3 days completing the replacement and installation of the new inner seal and wear-resistant alloy steel belt.

(3) Replacement preparations and steps for main inner bearing seal and wear-resistant alloy steel belt

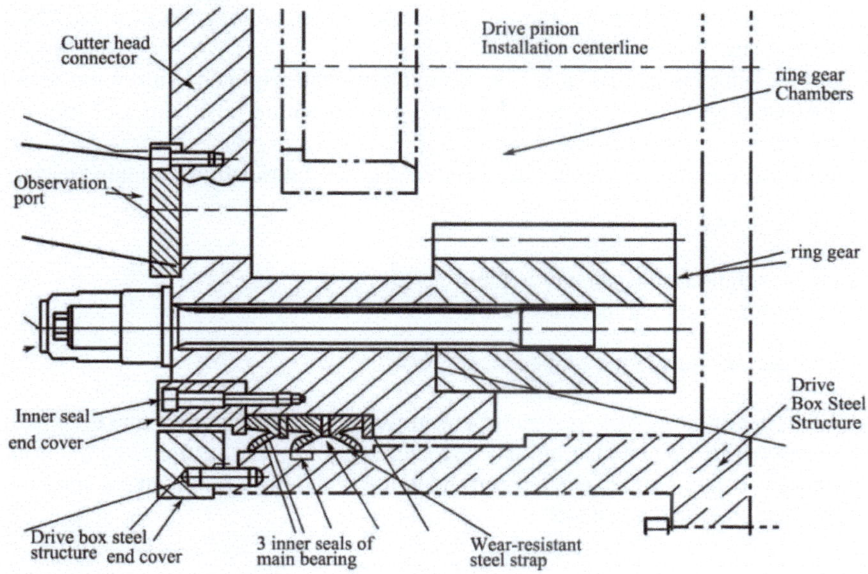

Fig. 11.20 Drawing of Main Bearing Seal

According to the site observation, the first section of the front of the slag bucket can be cut out to let the wear-resistant belt pass, the specific replacement scheme is as follows:

① Ensure that the machine is stopped in good geological conditions, push the cutter head onto the tunnel face to prevent it from rotating, and ensure that the cutter head brakes properly.
② Clean main machine belt, cutter head and slag bucket of the slag, clean them with water.
③ Weld 2 lifting lugs on the top of the slag bucket, use two 3 t hand chain hoists to lift the front of the slag bucket, pay attention to protection, and cut the slag bucket on both sides at the same time.
④ Remove the inner seal and wear-resistant alloy steel belt: weld wire rod around the perimeter and remove the original seal, split ring and wear-resistant belt. Clean the dirt and dust from the inner seal. Brush a small amount of lubricant over the periphery of the inner seal.
⑤ Weld $\phi 12$ mm × 110 mm wire rod on the inside of the main bearing, with an interval of 300 mm, hang the wear-resistant belt on it, start heating evenly up and down, left and right, when detecting that the heating temperature reaches 150–200 °C, start installing the new wear-resistant belt, all personnel must wear welding gloves before pushing, push hard all around at the same time, if part of it is not in place, use the wooden stick with hammer impact to fix it in place.
⑥ Install the new seal after the wear-resistant belt cools down, make sure the first direction on both sides of the inner seal both is opposite to the second and the

third direction, tighten the clamping ring bolts, and tightening torque of each bolt is 180 N·m.

⑦ Welding slag bucket, hoist slag bucket cut out to the original position, use Φ150 steel welding positioning brace, to ensure the accurate positioning of the two slag bucket during welding.

⑧ Retrieve all tools in the cutter head and replace the lubricant in the main bearing.

(4) Tools and Staffing

The tool configuration is shown in Table 11.13.

The staffing is shown in Table 11.14.

(5) Precautions during the replacement process

① When heating the wear-resistant belt, move the welding gun to make the belt heated evenly. When heating, pay special attention to the surrounding environment to prevent mutual interference and ignition of other equipment, and provide six fire extinguishers on site.

Table 11.13 Tool configuration

Serial number	Name	Specification/model	Unit	Qty	Remarks
1	Arc-welder's gloves		Pairs	8	
2	Slotted screwdriver	350 mm	Pcs	2	
3	Thread locker		Bottle	2	
4	Glass glue		Bottle	2	
5	Welding gun		Piece	4	
6	Chisel	250 mm	Piece	2	
7	Cotton cloth		Bag	1	
8	Batten	300 mm × 40 mm × 20 mm	Pcs	4	
9	Hexagon	ϕ12	Piece	2	
10	Torque wrench	350 mm or 500 mm	Piece	1	
11	Wire rod	ϕ12 mm × 110 mm	Pcs	40	
12	Electric welding machine		Set	2	
13	Oxygen		Bottle	6	
14	Acetylene		Bottle	6	
15	Welding gun		Piece	4	
16	Coal gas		Bottle	4	With 4 sets of tubes
17	Scrubbing brush		Piece	2	
18	Hand chain hoist	3 t	Pcs	2	

Table 11.14 Staffing

Serial number	Personnel	Number	Remarks
1	Welder	4	
2	Operator	8	
3	Technical officer on duty	2	
4	TBM operator	2	
5	Site supervisor	2	
Total		18	9 persons per shift

② To avoid the influence of welding current on the main bearing, it is required that the welding bond strap must be on the slag bucket. In order to ensure the quality of welding, it is required that the welding position should be heated before welding.

11.1.5 Zhongtianshan Railway Tunnel

1. Project Overview

Zhongtianshan Railway Tunnel is a key project of Tuku railway line 2 of Southern Xinjiang Railway, It is 22.467 km long, with a maximum overburden of 1728 m. The geological conditions are very complex, with high ground stress, mud gushing, rock storms and water ingress and other adverse ground conditions in the tunnel. Therefore, gripper TBM and drill and blast method is used for the tunnel.

The overall topography of the area where the tunnel passes through is high in the middle and low in the north east and south west, with an elevation of 1100–2950 m, a highest elevation of 2951.6. The terrain is cut obviously, with gullies and ravines, sparse vegetation, and a relative height difference of 800–1200 m. The maximum overburden of the tunnel is over 1700 m, and the design slope is 1.1% uphill. The mouth area is constructed with ϕ8.8 m open-type TB880E TBM imported from Wirth, which has already applied in the construction of Qinling Tunnel and Mogouling Tunnel, and has undergone construction for nearly 10 km in total before entering the site. A total of 12,476.7 m of tunnelling was done in the right line of Zhongtianshan Tunnel.

From the geology of the surrounding rocks after construction, three sets of stratigraphy (five sections) were revealed, namely Silurian metamorphic sandstone interspersed with schist (three sections), Silurian angular porphyry, and Variscan granite. The grade and length of each surrounding rock are shown in Table 11.15. See Table 11.16 for the main performance indexes of rock.

2. TBM main bearing bull gear repair technology

(1) Problems and Descriptions

Table 11.15 Grade and length of various types of surrounding rock of TBM construction

Surrounding rock grade	II	III	IV	V	Total length
Construction length (m)	1532	9550.7	1249	145	12,476.7

Table 11.16 Main Performance Indexes of Rock of TBM Construction

Rock/soil description	Density (g/cm^3)	Compressive strength (MPa)		Rock integrity index *KV*	Quartz Content (%)	Wear resistance value (1/10 mm)
		Drying	Saturated			
Sa metamorphic metasandstone	2.7–2.79	74.2–178	60.2–159	0.3–0.75	5–30	2.76–3.91
Eclogitic angular porphyry	2.69–2.79	52.9–75.4	37.4–63.6	0.4–0.75	20–40	3.73–4.74
Variscan Granite	2.6–2.72	87.3–90.9	70.2–73.8	0.30–0.75	20–45	4.13–5.48

During the construction of TB880E in Zhongtianshan Tunnel, oil leakage was found in the No. 6 pinion gear, and then there was an abnormal noise from the main bearing.

Through disassembly inspection after stopping the machine, it was found that front and rear bearings of No. 6 and No. 7 pinion gears have fallen apart, the pin roller has fallen off, with a small part crushed. Some part of the pin roller has fallen into the gap between the pinion gear and bull gear, resulting in abnormal damage to the pinion gear and bull gear, and some teeth were severely damaged. Below are photos of the damage:

① The teeth with severe damage is shown in Fig. 11.21.
② The less damaged part of the teeth is shown in Fig. 11.22.

(2) Material analysis of bull gear teeth and various technical parameters

① The material of the bull gear provided by Wirth Group is 42CrMo4.

It is found that the material has high strength, high hardenability, good toughness, small deformation during quenching, and high creep strength and lasting strength at high temperature. Used for manufacturing higher strength and tempered cross-section forgings, such as bull gears for locomotive traction, supercharger transmission gears, rear axles, extremely-loaded connecting rods and spring clips, etc., chemical composition:

Carbon (C): 0.38–0.45%;
Silicon (Si): 0.17–0.37%;
Manganese (Mn): 0.50–0.80%;
Sulfur (S): Allowable residual content $\leq$ 0.035%;
Phosphorus (P): Allowable residual content $\leq$ 0.035%;

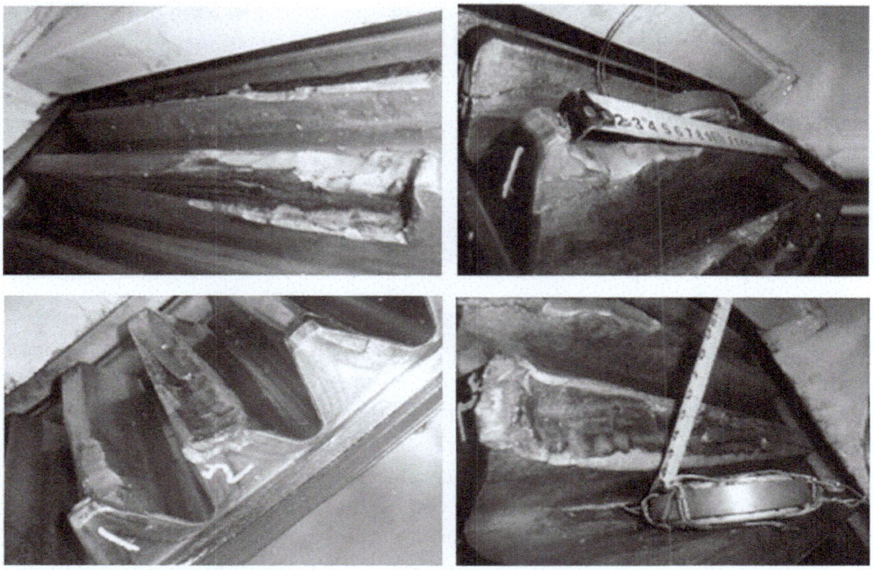

Fig. 11.21 Severely-damaged Gear

Fig. 11.22 Less damaged Gear

Chromium (Cr): 0.90–1.20%;
Nickel (Ni): Allowable residual content ≤ 0.030%;
Copper (Cu): Allowable residual content ≤ 0.030%;
Molybdenum (Mo): 0.15–0.25%.

② Both LYC Bearing Corporation and Yanshan University analyzed the material of TB880E main bearing, proving that the content of metal elements is matched with 42CrMo4.

③ The tooth surface hardness of the bull gear is HRC45–50, the surface hardness of the pinion teeth is HRC61–62, and the core hardness of the damaged teeth is HRC30–31.

④ The number of teeth of the bull gear is 174, the number of teeth of the pinion is 14, and the modulus of the teeth is 22 calculated from the dimensions of the bull gear.

(3) Difficulties in repairing bull gear

① Bull gear damage tooth repair process (overlay welding method, laser melting or inserted teeth, etc.). After analysis and demonstration and from the analysis of the repair conditions on site, only the overlay welding method is used.
② 42CrMo4 is a medium-carbon tempered high-strength steel with poor weldability, which can easily cause thermal cracking in the weld layer metal.
③ How to ensure the accuracy of the repaired bull gear? Limited by the site conditions, only manual grinding is used for ensuring the accuracy of the repaired bull gear, and the mold having the profile identical to bull gear profile shall be processed.
④ In the process of repairing the bull gear, the working space is limited, so try to work from the two round holes at the end of the drive shaft, if it is impossible to work, then the holes must be cut in the structural parts for operation.
⑤ Operators must lie on their fronts and put their heads into the hole, with the shoulders squeezed in the middle of the hole, the operation must be done by shifts, and each shift lasts shortly.
⑥ There are seals at both ends of the bull gear, the welding temperature must not be too high, and the seals must be protected.
⑦ The cracks in the damaged tooth are deep, and the grinding workload is large, and it is time-consuming to carry out flaw detection while grinding.
⑧ After welding a layer, it is necessary to carry out grinding and then flaw detection, and the workload is large.

(4) Analysis and test of the selection of process for repairing bull gear

① Zhengzhou Precision Work Surface Technology Co., Ltd. went to the site to try welding repair on broken part of No. 2 and No. 5 pinion gear, and two methods were used.

A. Overlaying weld repair by DC welding machine.

The mode of welding rod used is stacking 212, chromium-molybdenum type overlaying weld electrode. The broken parts of three teeth were welded respectively, the first tooth used continuous rapid welding, and grinding was carried out after welding, and welding crack was found; for the second tooth, slag cleaning was carried out for each welding, and cracks were found through stress relief by knocking; the third tooth was provided with stress relief welding, and the surface hardness reached HRC50.

B. Overlaying weld repair by imitation laser welding machine.

Fig. 11.23 WS Series Imitation Laser Welder

For imitation laser welding machine, welding surface heat generation is less, and there is no welding cracks, but the surface hardness is HRC28–29, almost equal to and the pinion gear core hardness HRC30–31.

WS series imitation laser welding machine (Fig. 11.23) function description:

Arc discharge: DC current accumulated in the capacitor is discharged with a period of 10^{-2}–10^{-1} s and melted with an ultra-short time of 10^{-4}–10^{-3} s.

Electrode heating and melting: at the position where the electrode material and the welding material meet, the temperature can reach 2800 °C. Metallurgical fusion; the welding material, after plasma, fuses with the workpiece diffusion layer via metallurgical fusion to diffuse in and penetrate into the lower layer of workpiece surface, to achieve high-strength adhesion.

Two major features:

Feature 1: Low heat input, unprecedented low heat input overlay layer is achieved by pulsed ultra-short time discharge.

Feature 2: High-strength adhesion, because a strong diffusion layer is formed and does not peel off after processing, a high-strength adhesion overlay layer is achieved.

The advantages of WS imitation laser welding machine are as follows:

a. Extremely-high welding accuracy.

Since the output current and time of the welding machine are precisely controlled, the current (2–200 A) and time (1–500 ms) can be freely adjusted, and the 2 A small current can be operated stably, so even 0.1 mm or 0.2 mm welding wire can be welded perfectly, reaching the welding accuracy of the laser welding machine.

b. Extremely-small welding repair impact.

Because of the application of patented technology, the arcing current and time of the welding machine are several times lower than those of general-purpose argon welding, and the impact of the welding process on the work can be overcome, so that the work can be repaired even if there is no machining margin.

c. Very small thermal impact.

Since the welding current and time are precisely controlled, the input energy can be precisely controlled to ensure that the input energy is only sufficient for the fusion between the welding wire and the workpiece. No excessive energy will be applied to the workpiece, achieving the ideal repair effect.

d. Extremely-high bonding.

When this machine is used for welding, metallurgical fusion are achieved for the welding material and the foundation. The bond after welding is extremely high. It can be applied to all kinds of processing methods, and the phenomena like unfirm bond or falling off caused by other types of cold welding machine will not happen.

e. Suitable for repair welding of different positions.

(a) Repair welding of dents, holes and openings in flat area.
(b) Fine seams, grooves.
(c) Edges and corners, ridges, and sharp points.
(d) Surrounding dents after weld porosity and normal argon arc welding.
(e) Mold after repair welding electrical discharge machining, nitriding and nitro-carburizing treatment.
(f) Scope of application: stainless steel manufacturing, mold industry, casting industry, various workpiece parts repair and various precision welding.

② Tangshan Xinchuan Machine Repairing Company conducts test welding repair for the pinion, the method used is DC overlay welding repair, the welding rod used are two self-developed welding rods. From the effect of test welding, the surface has certain defects such as the pattern in the shape of chicken claw, and the surface hardness is HRC32–35.

③ Zhengzhou Precision Work Surface Technology Co., Ltd. adopts the imitation laser welding method to achieve the effect of no cracks and high hardness material. The welding material is three kinds of welding wire (Thyssen brand) imported from Germany, the priming layer is TIG-SKD888, with good combination of strong tensile strength, the transition layer is TIG-SKD61, and the surface layer is TIG-SKD61, with high hardness, equivalent to the quenching layer. After test welding, there is no crack, and the surface hardness can reach about HRC50.

(5) Repair solution for bull gear

① Prepare materials for repair.

Such as argon gas, DC welding machine, imitation laser welding machine, angle grinder, sander, ventilator, cotton cloth, towel, probing agent, etc.

② Choose a professional manufacturer for gear repair.

According to the test welding repair of the damaged teeth of the pinion, Zhengzhou Precision Work Surface Technology Co., Ltd. was selected as the professional manufacturer for the main bearing repair. With three shifts, at least three laser welders, and professional grinding personnel and repairing personnel for the bearing seat.

③ Technical group, logistic support group and production coordination group were established in the work area, and the engineer on duty is arranged.

④ Pinion repair and installation of four pinions driving the cutter head.

A. Install No. 2 and No. 8 pinion, replace with new bearings and seals (No. 8 arc damaged, replacement required), and then install the corresponding couplings, transmission shafts, gearboxes and main motors.
B. Clean up the slag on the cutter head to prevent the cutter head from getting stuck.
C. First repair No. 4 and No. 5 pinion, install them inside, replace with new bearings and seals, and then install the corresponding transmission shaft, gearbox and main motor.
D. Then repair the teeth of No. 1, No. 2, No. 3, No. 7 and No. 8 pinion, and perform crack detection. Overlay welding for the shafts of No. 6 and No. 7 pinion, and then outsourcing processing.
E. Remove the cutter head fixings.
F. Try to turn the cutter head, if it still cannot turn, then investigate the cause.
G. Carefully clean the large iron pieces inside during the process of inching the cutter head.

⑤ Repair of bull gear.

A. Build work stands at the holes of No. 1 and No. 7 pinion as required by the gear repair manufacturer. No. 1 is the hole for tooth grinding and repair, and No. 7 is the hole for tooth welding repair.
B. Installed lighting for repair operations inside No. 1 and No. 7 pinion boxes. The repair manufacturer processes the mold according to the shape of the intact teeth.
C. Turn the severely-damaged tooth of the bull gear to No. 7 pinion hole.
D. Prepare a wet cotton cloth or towel, and place it under the work position to prevent iron chips from falling into the main bearing.
E. The damaged teeth will be polished, and the polished surface must be rounded and smooth, without right angles.
F. Use oxygen acetylene to remove surface oil.
G. Clean the repaired surface of the tooth with acetone.
H. Tooth repair welding is done by laser welding, three types of welding materials are used for the welding process, a special priming layer of laser welding rod is used for priming, then the core of the tooth is welded, and finally the hard surface layer is welded.
I. After welding, the teeth are shaped by grinding.
J. The repaired teeth will be clearly marked and recorded.

⑥ Installation and test run after repair.

A. Organize personnel to inspect the cleaning of iron debris inside the main bearing and conduct a thorough inspection of the repair quality.
B. Restore main drive of NO. 1, No. 3 and No. 6 pinion. Seal No. 7 pinion hole.
C. Add gear oil for cleaning.
D. Fill gear oil.
E. Perform test run and observe if there is any abnormal condition.
F. Resume digging status.

⑦ Measures to monitor the condition of the bull gear after repair.

A. Observe the oil leakage of the transmission shaft in time, if there is oil leakage, pull out the pinion gear in time to check the condition of pinion bearing.
B. Observe the clutch damage in time. If the clutch is replaced twice in a row within a week, the pinion bearing corresponding to this clutch should be checked.
C. Increase the frequency of vibration monitoring of the main motor and gearbox to reflect the situation and analyze it in time if there is any abnormality.
D. The gear oil and lubricating oil circuits are checked once or twice a week for smoothness.
E. For the bull gear teeth that are severely damaged, the flaw detection of these teeth is carried out regularly according to the marks made after repair.
F. Two persons are responsible for monitoring the condition of the bull gear of main bearing, one for the maintenance and one for equipment and materials, but a checklist is made for each item to be monitored and filled in by these two persons.
G. Replace pinion bearing once every four to five months.

(6) Implementation process of bull gear repair

① Repair is done for No. 3, No. 4 and No. 5 negative addendum pinion shaft with no spare parts. Overlay welding is done for the worn optical axis of No. 6 pinion, and repair is commissioned, as shown in Fig. 11.24.
② The welding work position is opened, and the repair is shown in Figs. 11.25 and 11.26.
③ Grinding and flaw detection are done for the severely-damaged teeth, see Fig. 11.27.
④ Overlay welding, grinding and flaw detection are done for the teeth being ground, see Fig. 11.28.
⑤ After rough grinding and overlay welding of the 7 teeth with serious damage, polish them.
⑥ There was also a small amount of damage to 162 teeth, and crack removal was performed. The tooth can function as usual once its engagement length reaches 80% of the original tooth length (280 mm), and no welding repair is needed.

(7) Acceptance and condition inspection after bull gear repair

Fig. 11.24 Gear repair

Fig. 11.25 Opening at bull gear welding work position

The teeth were checked for engagement, and the indented parts were polished so that most of the teeth could be engaged. The acceptance record of TB880E main bearing is shown in Table 11.17.

After the repair, flaw detection is conducted for the repaired teeth at the time of digging for 30 m, and no cracks are found, and the quality of repair is good, and they can function normally in the future.

(8) Lessons learned

Although there is some experience in the repair of gears in China, it is the first time that the repair of gears in the main parts of such large equipment as TBM is implemented, and the structure of the main drive is complex, and the repair is restricted by space and location, making maintenance extremely difficult, and after repair, it has to withstand working conditions of high torque and high thrust force,

Fig. 11.26 Schematic diagram of the opening above No. 71 pinion

Fig. 11.27 Flaw Detection of Gear after being Ground

making maintenance extremely risky. However, if the new main bearing is replaced with, the only way to replace it is by building a detour pilot tunnel to the front of the cutter head of tunnel face, which takes at least six months, leading great risk in construction period.

Fig. 11.28 Gear Repair

From the current situation of inspection during the tunneling process, the 7 teeth repaired by overlay welding have no cracks or deformation, the repair quality is good, and the equipment condition has been restored, laying a good foundation for the successful completion of the construction task of Zhongtianshan Tunnel.

3. TBM main bearing and cutter head repair

(1) Problems and Descriptions

Since November 2010, TBM main bearing and cutter head have faults one after another, making it difficult to continue digging without repair.

On November 30, 2010, as geological conditions permitted, the Project decided to shut down TBM and repair main bearing and cutter head, which was completed on December 28.

(2) Reasons for repair

① The cutter head and cutters are severely worn, and cracks appear in large numbers.

A. The face plate of the cutter head is badly worn, partly worn through, and the steel plate previously welded felt down and is deformed, and two holes of 400 mm × 400 mm are found. The other 150 mm × 150 mm small piece of steel plate welded last time has 1/3 worn off, see Fig. 11.29.
B. The total length of the crack in the cutter is about 6610 mm, and the distribution is shown in Fig. 11.30.

Table 11.17 TB880E main bearing acceptance record

Inspection content	Test result	Maintenance personnel
Bull gear repair condition; Pinion repair condition; Cleaning of foreign substance in the main bearing; Repair of No. 6 bearing seat	169 teeth of the bull gear are repaired, including 7 more serious broken teeth, the remaining 162 are small cracks in the end, after processing, all of them have been repaired. After flaw detection, there are no cracks, the test hardness is HRC49–52, meeting the expected standards, regarded as accepted Four upper pinions (No. 3, No. 4, No. 5 and No. 6), there are broken teeth, and all of them have been repaired. No cracks after flaw detection, regarded as accepted The main bearing was damaged due to the pinion bearing, which fell into the roller path and is broke, and after many times of cleaning and washing, the foreign substance such as iron chips were not seen in the compartment using the camera observation No. 6 bearing seat has a slight deformation, after the application of gluing, it works again	Zhengzhou Precision Work Surface Technology Co., Ltd. and our personnel

Fig. 11.29 Cutter head face plate worn through

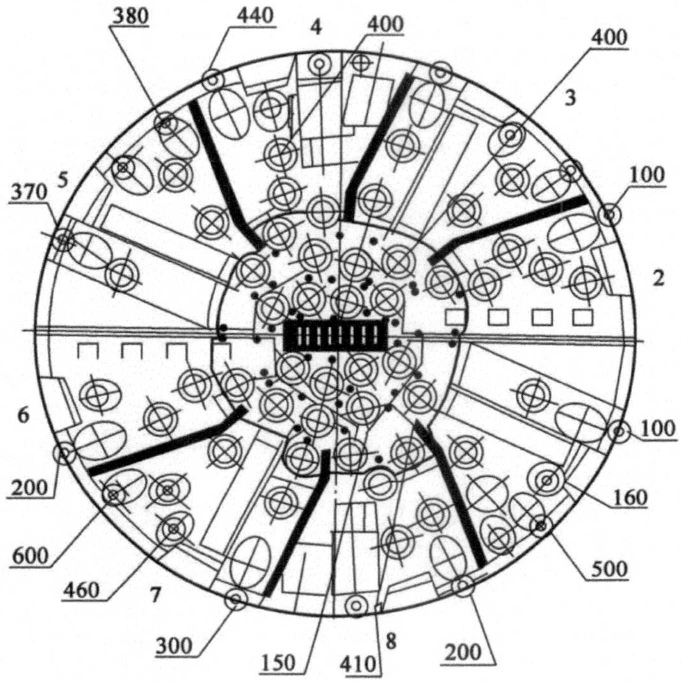

Fig. 11.30 Distribution of Cracks in the Cutter Head (Unit: mm)

The specific length of cracks: 100 mm for No. 1 transverse crack, 150 mm for No. 10 longitudinal crack, 180 mm for No. 12 transverse crack, 900 mm for No. 14 transverse crack, 350 mm for No. 15 transverse crack, 120 mm for No. 15 longitudinal crack, 140 mm for No. 16 longitudinal crack, 270 mm for No. 17 transverse crack, one 40 mm diameter hole and one 20 mm diameter hole for No. 25 cutter hole, 40 mm for No. 36 transverse crack, 60 mm for No. 37 transverse crack, 370 mm for No. 41 transverse crack, 70 mm for No. 42 longitudinal crack, 380 mm for No. 43 transverse crack, 60 mm for No. 49 transverse crack, 100 mm for No. 49 longitudinal crack, 460 mm for No. 50 transverse crack, 300 mm + 300 mm for No. 54 transverse crack, 500 mm for No. 56 transverse crack, 200 mm for No. 59 transverse crack, 200 mm for No. 61 transverse crack, 100 mm for No. 61 longitudinal crack,100 mm for No. 62 transverse crack, 110 mm + 290 mm + 40 mm for No. 63 transverse crack, 130 mm + 50 mm for No. 63 longitudinal crack, 100 mm for No. 64 transverse crack, 440 mm for No. 65 transverse crack and 320 mm for center cutter frame.

C. Most of the cracks in the compartment are new cracks that had not been welded before, and 70% were at the bottom of compartment. The total length is about 7160 mm.

No. 1 compartment bulkhead crack 700 mm, No. 2 scraper seat is severely damaged.

No. 2 compartment bulkhead crack 400 mm.

400 m crack for the bulkhead next to No. 57 cutter of No. 3 compartment, bulkhead bottom crack 700 mm, 600 m crack for bulkhead next to No. 60 cutter, bulkhead bottom crack 900 mm.

No. 4 compartment bulkhead crack 300 mm, bulkhead bottom crack 700 mm, 100 mm crack for No. 4 scrapper seat, and bulkhead top broken.

No. 5 compartment bulkhead bottom crack 800 mm.

280 mm crack for bulkhead next to No. 46 cutter of No. 6 compartment, 300 mm crack for bulkhead next to No. 63 cutter, bulkhead bottom crack 120 mm, and 200 mm crack for No. 4 scrapper seat.

500 mm crack for bulkhead next to No. 54 cutter of No. 7 compartment, 160 mm crack for bulkhead next to No. 61 cutter, and 70 mm crack for No. 4 scrapper seat.

No. 8 compartment bulkhead bottom crack 300 mm.

D. Eight (No. 1, No. 2 and No. 3) scraper seats are seriously worn, see Fig. 11.31, most of them cannot be used.
E. The left arc plate of the slag bucket was partially worn through, and a new steel plate needed to be welded.
F. Some of the face disc cutter seat are worn, and the serious wear is mainly concentrated in No. 50 to No. 65 cut seats.
G. The cutter drum of the high cutter position is severely worn, see Fig. 11.32.

② The main bearing is in extremely poor working condition.

Since repaired in June 2010, the effect of the main bearing is not satisfactory, the gear oil mechanical foreign substance and water often exceeds the standard, and after September, the gear oil has to be replaced once every week or so. In November, the

Fig. 11.31 Cutter head scraper wear

Fig. 11.32 Cutter drum wear of high cutter position of cutter head

gear oil leakage was also quite serious, sometimes a barrel of gear oil a day, but due to the geological conditions, no repair was carried out.

A. When replacing the oil, pull out some iron and copper chips from the bottom of the oil tank and the observation hole.

In the main bearing gear oil, there were iron pieces falling from the ball and copper chips falling from the holder, and the roller path also started to have larger iron pieces falling, the situation is very serious.

B. Grease injection.

The outer seal is in good condition, and the inner seal turns to F4 oil channel to fill HBW grease, but at present F3 oil channel cannot be filled with grease, and HBW grease has entered the main bearing gear oil, indicating that the wear-resistant steel belt on the original support ring has been displaced.

Most of the fixing bolts M16 × 35–10.9 on the inner side of the labyrinth binder plate of the inner seal are damaged, resulting in a large gap. The labyrinth ring is currently fixed by replacing some of the bolts and welding.

Through inspection, the gap between main bearing and cutter head joint surface inspection is large, when tightening the bolt of the cutter found, there is water and dust in the bolt hole, indicating that the dust ring may be loose or damaged.

③ The main problem of the main bearing.

A. The flange surface of the cutter head is not tightly combined with the main bearing, and there is a large deformation, leading to frequent loosening of the cutter head bolts, and the impact on the main bearing during the digging process will accelerate the damage of the main bearing.

B. Large gap of main bearing. During the lip seal replacement in January 2009, the technician from foreign side mapped the site and found that the pinion bearing ball had wear trace and the bull gear has settled for 2–3 mm.
C. The main bearing ball and roller path are damaged, the ball drops out, the soft belt of the roller path are certainly damaged, and the roller path damage will be more and more serious after repeated rolling.
D. The wear-resistant steel belt of the inner seal support ring is damaged, resulting in damage to the lip seal, which cannot play the role of sealing, and dust, HBW grease and water continuously enter the main bearing gear oil.
E. No. 24 bolt in Fig. 11.35 is loose and broken, resulting in a large gap between the labyrinth binder plate and the bulkhead, and dust and water can enter the system.
F. The large clearance between the flange of the cutter head and the main bearing joint surface may cause dust and water to enter the gear oil lubrication system through the dust ring position.

(3) Main bearing repair process.

① Cutter head fixation and separation from main machine.

Cutter head backs 1.5 m, double rows of #120 I-beam welded on 6 points in front of the cutter head supports the tunnel face, the lower part of the cutter head is supported by an additional steel plate, the upper part is fixed by the rock bolt, see Fig. 11.33.

Loosen the cutter head coupling bolts and slag bucket bolts, the slag bucket ix fixed on the cutter head, with 50 kN jack to assist in detaching main machine and cutter head, main machine backs 1.2 m.

② Main bearing disassembly and inspection.

The outer ring greasing is good, dustproof sealing effect is good, but the cutter head and the main bearing coupling surface is not tightly combined, the coupling surface has rust, the cutter head coupling bolt hole has water and dust, gear oil has obvious leakage. The condition of the inner seal is very poor, a lot of dust has entered the

Fig. 11.33 Photo of cutter head fixing

inner labyrinth, there is no grease accumulated inside, and the labyrinth binder plate is very badly worn.

The wear-resistant steel belt of the inner seal support ring has fallen off and broken into many sections, causing serious damage to the inner seal lip seal.

After removing the dust seal binder plate, the base surface of the inner seal support ring was badly worn, and there were deep dents after the wear-resistant steel belt was removed.

No. 4 main bearing pinion bearing is damaged.

③ Main bearing repair.

Seal the base surface of the support ring with the plasticizer, wait for the consolidation and grind, and then reapply the plasticizer after shrinkage.

After grinding, as the injection hole is in the inner ring of the support ring, the support ring is lifted out.

After cleaning, coated with adhesive, and use special clamps and two packers to fix wear-resistant steel belt, tighten the packers and use rubber hammer to make packers firm by knocking, wait for consolidation, weld the joint, inject glue, add plug, and install support ring.

After inspection of the outer seal disassembled, outer seal is in good condition, and the original lip seal is installed, and the inner seal lip seal is put in place.

Due to the serious damage of the labyrinth binder plate of the inner seal, the holes of the fixing bolts of the labyrinth binder plate of the inner seal are deformed, so some bolt holes are added, and the original M16 bolts are changed into M18 bolts.

The labyrinth binder plate of the inner seal was repaired, but because the deformation was not properly controlled, it could not be installed after the seal was installed, and then the correction was continued, and it took two more days than planned to install it.

④ During the repair of main bearing, the welding of the cutter head face plate was carried out.

The main thing is to add 40 mm thick steel plate on the face plate, because the cutter head is not connected with the main bearing, in order to avoid deformation of the cutter head coupling flange, no compartment welding repair is carried out. The repair of the disc cutter base was also not carried out due to the difficulty of smoke extraction for two working surfaces. After the repair of main bearing is completed, the reinforcement work of adding welded steel plate to the cutter head face plate is basically completed, see Fig. 11.34.

(4) Cutter head repair

① Dismantle the main bearing work stand, dock the cutter head, remove the device fixing the cutter head and the cutter head work stand, install the new side cutter, and dig forward for 1 m.
② Move cutter head backward, blast-hole drilling for three chambers, to facilitate the repair of cutter head scrapper seat and the cutter drum of the high cutter position. It will take a long time to drill blasthole and clean slag.

Fig. 11.34 Cutter head face plate welding work

Fig. 11.35 Repair of cutter drum retainer and disc cutter seat

③ Then the cutter head props up the tunnel face, using the rock wall cut by new edge cutter for positioning scraper seat.

④ After positioning, the cutter head is moved backward and the scraper seat, cutter drum, cutter drum retainer and disc cutter seat are repaired, see Fig. 11.35.

(5) Repair result achieved and lessons learned

① The main bearing seal repair is satisfactory.

The new inner seal lip seal was replaced with, and the support ring wear-resistant steel belt was restored to its initial condition when Zhongtianshan entered the site.

The labyrinth ring was repaired by welding, and the control gap was small, and the greasing effect was good. There is no looseness in dustproof seal binder plate, so the possibility of dust entering is relatively small.

② The condition of the main bearing main push roller path is not ideal, inspected from the observation port with an endoscope inspection, quenching layer spall is serious.
③ The cutter head repair took a long time, but it created conditions for smooth TBM tunneling.

A. The cutter head was repaired by welding, and the structural strength of the cutter head was restored.
B. The scraper seat of the cutter head was repaired, so that the frequency of damage to the high cutter position was greatly reduced.
C. The cutter drum, disc cutter seat and cutter drum retainer of the cutter head are restored so that the cutter will not be damaged abnormally.
D. The water spraying on the cutter head was restored so that the temperature at the front of the cutter head would not rise too quickly.
E. The cutter's compartment and slag bucket were repaired so that the strength of the cutter head was guaranteed so that the slag would not spill into No. 1 belt pulley to cause abnormal damage to No. 1 belt.

④ The difficulty of repairing the inner seal labyrinth was underestimated, resulting in a long delay.
⑤ The main bearing and the cutter head were repaired at the same time, and many processes interfered with each other, resulting in poor smoke exhaust, which makes the process arrangement failing to reach the ideal condition.
⑥ Insufficient understanding of the severity of the cutter head, such as the difficulty of repairing the disc cutter base and cutter barrel, also contributing to the long repair time.
⑦ All the cutters in the high cutter position will be replaced to be used as transition cutter, and the time to remove and install the cutter is longer, which is also one of the factors affecting the construction efficiency.

11.1.6 Chongqing Rail Transit Line 6

1. Project overview

The stratum where Chongqing Rail Transit Line 6 passes through is mainly sandstone, mudstone, sandstone with mudstone and tuff stratum, etc. The rock quality is hard, and TBM is used for the first time in the construction of urban railway tunnel in China.

The TBM test section of Chongqing Rail Transit Line 6 Phase I is an open section of Wulidian-Shangyanggou Reservoir, passing through seven stations, including Red Land, Huangnipang, Hongqihegou, Floral Park, Dalongshan, Ranjiaba and Optoelectronic Park, with a chainage of CK17 + 438–CK29 + 128 (Fig. 11.36), and a total length of 12.122 km. The test section tunnel is constructed by two gripper TBM, one for each tunnel, each with a cumulative length of 8.261 km.

The geology of this section is mainly IV grade surrounding rock, accounting for 99.4%. The tunnel mainly traverses sandstone and muddy sandstone, sandstone (4.6%), and sandstone mudstone and mudstone (88%). Sandstone mudstone and mudstone rock saturation compressive strength is 12.6–17.2 MPa, sandstone saturation compressive strength is 29.9–36.7 MPa. Its lithology is softened and mud is formed when meeting water, muddy sandstone lithology has a great impact on TBM advance speed.

2. Analysis on influencing factors of TBM fast drilling

(1) Large mud content on the tunnel face, mud formed on the cutter head

In the process of digging in mudstone, sandy mudstone and muddy sandstone section, gripper TBM is affected by water spraying from cutter head and underground water, which softens the mudstone, smudges the cutter head, blocks the cutter hole, and enter slag bucket and inner Kelly, causing the increase in digging torque and the difficulty in slagging. No. 1 (inner Kelly) belt conveyor cannot start normally, and the single machine current exceeds the limit. Shut down TBM for slag cleaning under the pressure of circumstances. The longest shutdown was 36 h to clean the cutter head (Figs. 11.37 and 11.38).

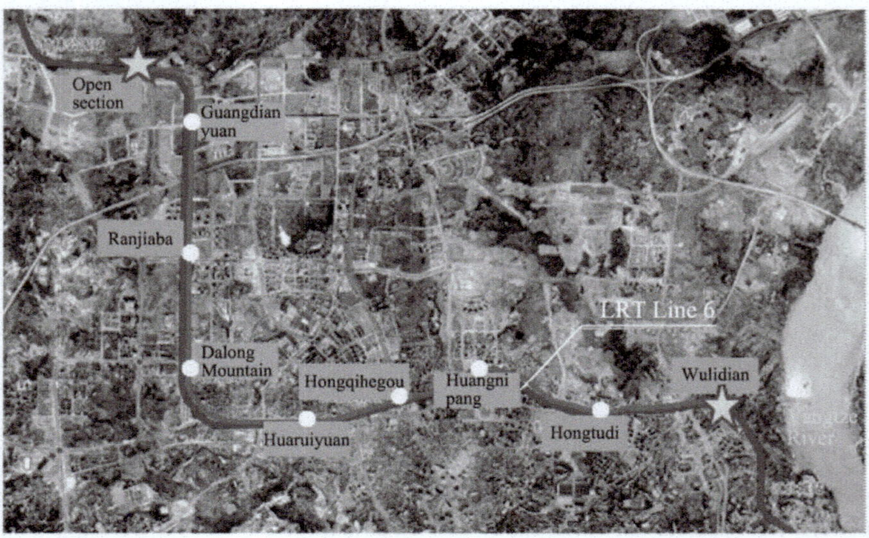

Fig. 11.36 Project location map

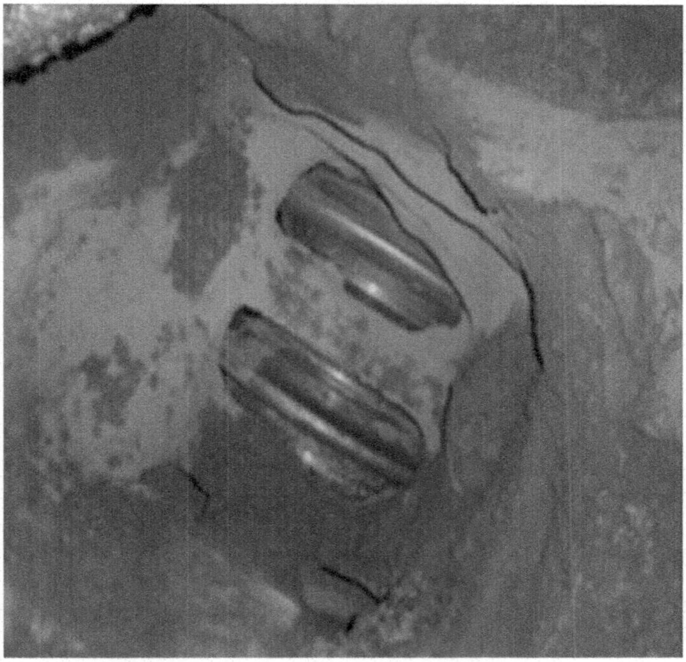

Fig. 11.37 Cutter hole severely blocked

(2) Poor integrity of the surrounding rock, easy to fall off

The mudstone and muddy sandstone sections are mostly broken, and the integrity is not good, easy to fall off. When TBM was digging in the yellow and red area, the cavity collapsed is up to 4 m deep, and TBM must stop working to support and clean the slag blocks when rocks fell off, which seriously affected the progress of digging (Figs. 11.39 and 11.40).

3. Supporting Measures for Gripper TBM Tunneling in Mudstone

During TBM tunneling in the mudstone section, due to the special nature of the surrounding rock, if controlled by the parameters of normal tunneling, it will bring difficulties to the tunneling. In the process of tunneling, main problems are prevention of mudstone softening and mud formation by water, tunneling parameters control and construction environment control in the tunnel. If solving these three problems, you can effectively ensure continuous and fast TBM tunnelling through the mudstone section, to achieve the purpose of efficient and rapid production.

(1) Preventing mudstone from softening and forming mud in contact with water

① Main problems.

Fig. 11.38 Clearing serious clogging of cutter hole

Fig. 11.39 Collapsed cavity with serious block falling

Fig. 11.40 Shutdown to strengthen the support

The composition of the surrounding rocks in this section is mainly mudstone, sandy mudstone and muddy sandstone, which are purple-red in color and have natural compressive strength of 20–26 MPa, and are soft rocks–relatively softer rocks, and the rocks are relatively complete. In this section, after the surrounding rock is cut by the cutter head, there are some small particles, which are easily muddied by water. Under the condition of normal water spraying on the cutter head, the cutter head is easily muddied by the caked slag, which makes mucking difficult and affects TBM tunneling (Figs. 11.41 and 11.42).

② Solutions.

In mudstone geological conditions, to ensure that TBM cutter head is not covered by mud and achieve dust reduction, water spraying of cutter head is stopped during tunnelling. Water spraying device is added at the place where No. 1 belt conveyor and No. 2 belt conveyor meet, which is opened during tunneling. The water can be sprayed into water mist, which plays an effective role in dust reduction (Fig. 11.43), thus greatly reducing the possibility of clogging of cutter and mucking outlet and increasing TBM tunnelling efficiency under the condition of mudstone. In addition, water spraying of cutter head is stopped to make the tunnel face dry as far as possible, thus markedly reducing the frequency of cleaning cutter head and buying more effective time for TBM normal tunnelling.

Controlling the amount of water spraying on the cutter head is the key to prevent cementation of mudstone. When the surrounding rock is dry and there is no influence of groundwater, the amount of water spraying on the cutter head is controlled according to the principle of ensuring the safety of the cutter head, and in practice, it is basically controlled at 1/2 of the amount of water spraying during normal

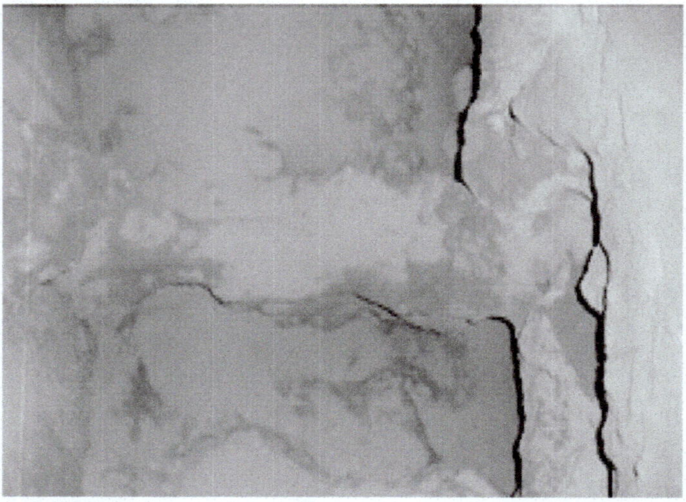

Fig. 11.41 Mud on Cutter Head

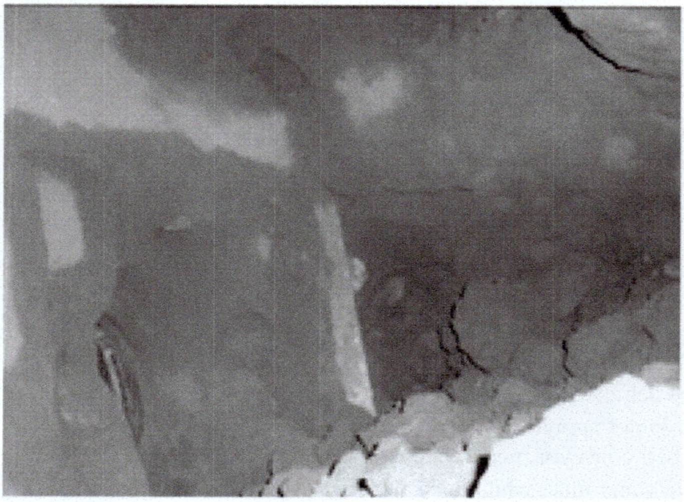

Fig. 11.42 Clogging of Cutter Hole by Mud

tunnelling. When there is groundwater influence on the surrounding rock, it is controlled according to the size of the groundwater and the wetness of the slag, and in practice, it is controlled at 1/4 of the amount of normal water spraying, and even water spraying is not required.

When the cutter head is sprayed with little or no water, the slag is basically granular (Fig. 11.44). And no caked mud is found when there is no influence of water, ensuring the normal work of the cutter head and belt conveyor. However, when the

Fig. 11.43 Water mist for dust reduction

local groundwater is too large and mud is inevitable, through communication with TBM manufacturer, a set of cutter head foam spraying system is added to achieve dust reduction and prevent the mud, displaying good effect.

(2) Control of tunnelling parameters

Fig. 11.44 Mucking effect of cutter head without water spraying

The selection of tunnelling parameters is mainly based on the mucking situation to determine the rotation speed of the cutter head. When there are no large pieces of slag and no collapse in the process of tunnelling, generally high-speed tunnelling is selected. If there are large pieces of slag and collapse of the surrounding rock, generally low speed tunnelling is selected. This, on the one hand, is to reduce the disturbance of the cutter head to the surrounding rock. On the other hand, this is to control the amount of mucking of the belt conveyor to avoid the fault of the belt conveyor due to too fast mucking and too big rocks.

The surrounding rocks of this project are relatively uniform, and the compressive strength values of various surrounding rocks are within the range of 15–40 MPa, and the strength of most of the sandstone, muddy sandstone, sandy mudstone and mudstone are within the range of 20–30 MPa, and the compressive strength of the rocks is small, easy to break and easy for tunnelling. TBM rotation speed can be divided into two types: high speed mode (5.44–11.97 r/min) and low speed mode (0–5.44 r/min).

① Stratigraphically stable soft rock (stable mudstone, sandy mudstone strata) adopts lower speed, small thrust force and large penetration.
② Uneven stratigraphic softness (sandstone, mudstone junction strata, collapse, falling, water seepage) adopts lower rotation speed, larger thrust force and large penetration.

The project adopts gripper TBM construction, and the selection of various tunnelling parameters is the key to fast and efficient production. During TBM advance, the tunnelling parameters shall be adjusted based on the results of the advance geological forecast, different geology and overburden to determine the stability of the surrounding rock, diggability.

During tunnelling in mudstone section, due to the soft rock, the surrounding rock cutting is relatively easy, so TBM can quickly advance. However, limited by the capacity of TBM mucking equipment, if TBM advance speed is too fast, beyond the capacity of the mucking system, the belt conveyor is clogged by slag, and cleaning the slag will cost a lot of manpower, material and time. If TBM advance speed is too slow, although the mucking system can operate normally, but TBM's boring capacity is not fully released, TBM's economy and advanced performance is greatly reduced. Therefore, during tunnelling in mudstone section, how to choose a reasonable tunnelling parameters becomes the key to achieve the value of TBM.

The main parameters of TBM tunneling in general hard rock section are as follows: cutter head thrust force 8000 kN, cutter head rotation speed 10 r/min, penetration 3 mm, each digging cycle takes 50–90 min (depending on the hardness of the surrounding rock), when advancing to the junction of mudstone section and hard rock section, the cutter head thrust force suddenly drops, penetration increases, followed by a large amount of mucking of the belt conveyor, blocking the belt conveyor outlet in a very short time, and No. 1 belt conveyor also stops due to excessive slag, and it takes 36 h to clean up the slag at the slag outlet and the slag stored on the belt conveyor, which greatly affects the normal progress of tunneling. The design of the maximum mucking capacity of TBM belt conveyor can reach 1.8 m^3/min, which can calculate the theoretical maximum advance speed of TBM:

1.8

In the mudstone section, It can be controlled at 90% of TBM's maximum tunneling capacity (considering a 10% safety factor):

$$0.057 \times 90\% = 0.051\ \text{m/min}$$

So in the mudstone section, the maximum advance speed of TBM is controlled at 0.05 m/min. At this time, the parameters of TBM tunnelling are as follows: thrust force 5000–6000 kN, cutter head rotation speed 7 r/min, penetration about 7 mm, a tunnelling cycle about 30 min, and the mucking system keeps normal operation.

During the routine cutter head inspection, after testing, TBM cutter wear is normal, and there are no problems such as spin wear and uneven wear caused by fast advance speed, so the above tunnelling parameters are reasonable and feasible in the construction. If we do not take into account the impact of the initial support process, TBM advance speed will be very considerable, fully meeting or even exceeding the requirements of the construction schedule, and the economy and advanced performance of TBM is leveraged.

To sum up, the selection principle of TBM tunnelling parameters mainly needs to consider the capacity of the mucking system and the cutter safety.

(3) Construction environment control in the tunnel

During the normal operation of TBM, the water spray from the cutter head can basically eliminate the dust generated by the cutter pressing against the surrounding rock. However, during the construction in mudstone section, water spray from cutter head cannot be controlled as usual, the amount of water spray is small, dust increases, air quality in the tunnel is poor and visibility is poor. All these brings adverse influence on the health of workers and various sophisticated electronics, and seriously affects the measuring accuracy of TBM tunnelling (TBM uses PPS automatic measuring system). Therefore, it is difficult to control excavation direction and quality accidents are very likely to happen. In response to these situations, the following methods are used to control them:

① Add dust reduction device at the outlet of No. 1 belt conveyor.

No. 1 belt conveyor is the connecting belt that conveys the slag from the cutter head to the mucking belt of TBM back-up system, and this position is the place where dust mostly is generated. Meanwhile, the outlet of No. 1 belt conveyor is near the measuring prism on the cutter head, and the dust directly affects the measuring accuracy. For this reason, this position is the key to dust reduction. Main measures are as follows: on the one hand, water mist device is added at the end of No. 1 belt conveyor for dust reduction. On the other hand, the outlet of No. 1 belt conveyor is covered with canvas to reduce dust. Good results were achieved (Figs. 11.45 and 11.46).

Fig. 11.45 Water spraying for dust reduction at the outlet of No. 1 belt conveyor

Fig. 11.46 Covering dust at the outlet of No. 1 belt conveyor

② Refurbishment of ventilation equipment to reduce dust.

When dust reduction device is added at the outlet of No. 1 belt conveyor, modification is made to TBM ventilation equipment at the same time. The air duct is extended by 5 m towards main machine, and the ventilation hole directly faces main machine operation platform, and secondary ventilation outlet is added. Secondary fan is added under the shotcrete bridge, and TBM ventilation equipment greatly improves the upper construction platform environment of main machine, reduces temperature, and greatly improves air quality and operation environment, which is good for better construction of workers (Figs. 11.47 and 11.48).

4. Summary

In the process of TBM tunnelling in mudstone section, main problems are prevention of mudstone softening and mud formation by water, tunneling parameters control and construction environment control in the tunnel. If solving these three problems, you can effectively ensure continuous and fast TBM tunnelling through the mudstone section, to achieve the purpose of efficient and rapid production. Through adding water spraying device for TBM, adjusting water spraying timing of cutter head and control water spraying amount, the difficulty that cutter head is smudged by

Fig. 11.47 Relocation of the outlet of TBM secondary ventilation outlet

Fig. 11.48 Addition of secondary fan under the Shotcrete Bridge

mudstone softening and mud formation by water is overcome, thus greatly reducing the frequency of cleaning cutter head and buying more effective time for TBM normal tunneling. TBM maximum advance speed in the mudstone section is achieved by optimizing TBM tunneling parameters. Through modifying ventilation equipment and adding secondary ventilation outlet, the environmental quality in the tunnel is greatly improved, the temperature is reduced, air quality is greatly improved and operation environment is markedly improved, good for better construction of workers and faster construction speed.

11.1.7 Vereina Railway Tunnel, Switzerland

1. Project overview

The Vereina Tunnel in Switzerland is a 19,062 m long railway tunnel passing through the Alps. The northern entrance is a 2.16 km long Zugwald tunnel. Due to geological conditions, route requirements and schedule and other considerations, the two tunnels were divided into six lots for tendering. The horizontal curtain grouting method was used for the construction of the overburden of the Zugwald tunnel, and this method is characterized by a series of horizontal holes drilled along the upper perimeter of the tunnel prior to excavation, followed by the injection of high-pressure cement slurry into the holes, under which the tunnel was excavated under a protective shelter. The middle bedrock section (sedimentary rock) was then constructed using TBM method. The Vereina tunnel was constructed by drill and blast method from the northern entrance and the southern exit at the same time, and the northern end was excavated by TBM method after the completion of the two lines. The southern end is still excavated by the drilling and blasting method.

The construction schedule of these two tunnels was very tight and reasonable. After the completion of the entrance section of Zugwald tunnel by horizontal grouting, TBM was used to break through the Zugwald tunnel, and after the breakthrough, TBM and its supporting equipment were moved to the northern end of the Vereina tunnel, 3 km away, and TBM tunnelling is conducted from north to south. Meanwhile, at the southern end of the Vereina tunnel, the drill-and-blast method was used to construct the tunnel in opposite direction.

Construction of such two tunnels began simultaneously in 1991, the Zugwald Tunnel was completed in 1995 and the digging of the Vereina Tunnel was completed on March 26, 1997, and the Vereina Tunnel project was completed in 1999, five months ahead of schedule. The Vereina Tunnel was excavated 2133 m from the north by drilling and blasting, and then dug 9451 m by TBM; 7478 m from the south by drilling and blasting. The total cost of the Vereina tunnel project is CHF 340 million. When the 2 tunnels were constructed by TBM method, TB770/850E gripper TBM made by Wirth was used.

The Zugwald tunnel was dug 1650 m by TBM, and the rock stratum is mostly sedimentary rocks of dolomite and mud shale. Here, there are also unfavorable terrains consisting of glacial alluvium such as cobbles and conglomerates. The maximum daily advance rate of TBM was 25 m during the construction of the Zugwald tunnel.

The rock formations encountered during the construction of the Vereina tunnel include both locally-distributed weak stratum and moderately-hard sedimentary rocks, as well as very hard igneous rocks consisting mainly of gneisses and amphibolites, and the Vereina tunnel also traverses several fault zones formed by Alpine orogenesis. When digging into igneous rock, TBM advance speed is 600 m per month, and the maximum daily advance is 42.7 m. The Vereina Tunnel was constructed with TBM on May 12, 1995 and the tunnelling was completed on March 26, 1997. The longitudinal sections and design construction progress of these 2 tunnels are shown in Fig. 11.49.

2. Construction organization and management

The Vereina Tunnel was designed and contracted by Amberg Engineering and has a very lean organization, as shown in Fig. 11.50.

The above-mentioned personnel are the management staff of the permanent organization, including the mechanical and electrical engineers and construction engineers on duty 24 h a day to ensure that TBM construction site situation is always in hand. The management procedure is as follows: for the mechanical and electrical

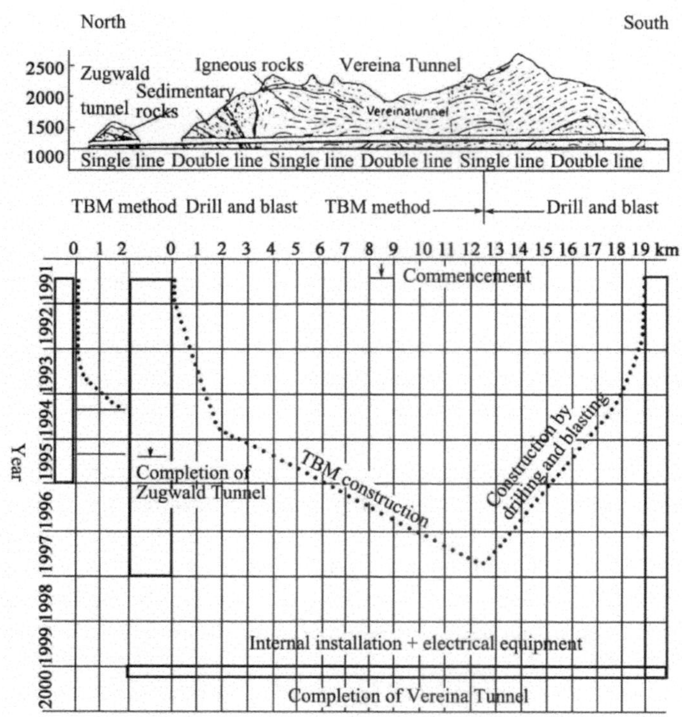

Fig. 11.49 Longitudinal Section of the Tunnel and Design Construction Progress Chart

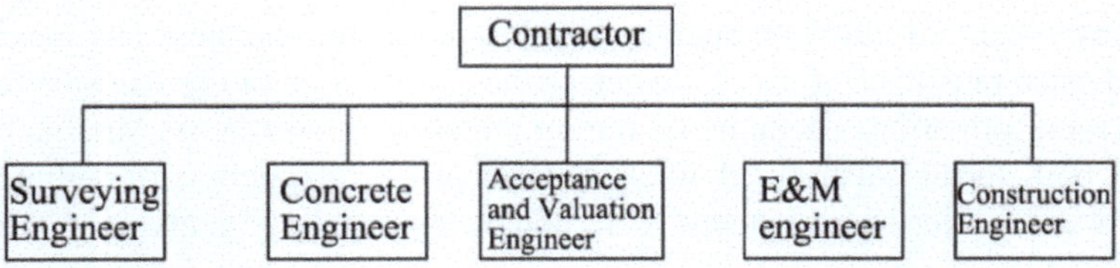

Fig. 11.50 Construction organization and management of Vereina Tunnel

problems arising during the maintenance shift, the construction site directly informs the mechanical and electrical engineer of these problems, and the mechanical and electrical engineer is responsible for solving these problems; for the problems arising from the tunnelling shift, the construction site informs the construction engineer of these problems, and the construction engineer is responsible for solving these problems. The managers of all permanent organizations are responsible to the Project Manager, who is responsible for the overall work and for the coordination of all departments.

(1) Workforce organization of the tunnelling shift

During the construction of the Vereina tunnel, two tunnelling shifts every day, each consisting of 21 people, with the following staffing:

1 shift leader: responsible for the work arrangement of the personnel; inspection and coordination of the work of the personnel, and timely reporting of geological conditions.

1 TBM operator: responsible for operating TBM, filling in the digging report, receiving information from each operation point at any time during the digging, dealing with some minor faults, and judging changes in geological conditions.

1 mechanic: responsible for welding work in TBM construction, water supply and drainage, air duct and water pipe extension work, and responsible for dealing with all kinds of mechanical faults, and responsible for inspection and inspection of all parts of TBM.

1 Electrician: responsible for lighting in all parts of the tunnel, maintenance of power supply equipment in the tunnel, extension of cables, and responsible for the maintenance of various meters and display instruments in the control room, and the maintenance of PLC faults, and quickly finding the electrical fault points.

7 operators for bolt-shotcrete support: 5 operators work on the upper part of main machine platform behind the cutter head, responsible for drilling anchor holes, anchoring anchor rods, hanging mesh, installing ring beam or initial spraying concrete when necessary, and reporting geological changes in a timely manner. 2 operators work on the lower part of main machine platform behind the cutter head, responsible for cleaning the slag at the bottom behind the cutter head and connecting the lower steelwork.

3 operators for invert installation and track laying: responsible for the installation and grouting work of the invert block, and dragging the expansion rail forward, carrying and laying the rail and lifting the invert block.

2 operators for shotcrete: responsible for loading and unloading concrete tanks, 1 operator responsible for operating shotcrete machine, and 1 operator responsible for operating concrete manipulator.

3 Internal combustion operators: each person is responsible for 1 mucking train, and is responsible for loading slag in the tunnelling, transporting switches along the track, and operating slag disposal machine for unloading slag.

1 person on standby: called at any time in the tunnel where needed.

1 person for material supply: responsible for supplying all kinds of materials needed inside and outside the tunnel, loading and unloading the invert blocks and stacking them neatly, loading the materials needed in the tunnel and transferring them to the predetermined rail line.

The above-mentioned posts are fixed so that they can master the performance of the equipment and become highly skilled operators of the operation to meet the needs of TBM construction. However, the scope of work is not fixed, nor is it a matter of each person doing his/her own job and ignoring anything else, but that the shift leader makes coordination according to the needs of the work, assists other posts to complete the work of the digging cycle to achieve the best use of personnel.

(2) Workforce organization for maintenance shift

One maintenance shift is arranged every day, the main task of the shift is to check and repair the main machine, back-up system and accessories, and conduct a comprehensive inspection of the machine, electricity and oil, and be responsible for the inspection and replacement of cutter.

6 inspectors for main machine: responsible for a comprehensive inspection of all parts of the main machine, and responsible for lubrication maintenance, as well as cutter inspection and replacement.

4 inspectors for back-up system: responsible for a comprehensive inspection, repair and maintenance of the accessories of the back-up system, inspection and maintenance of hydraulic system, lubrication of the moving parts, etc.

1 Electrician: responsible for the comprehensive inspection and maintenance of the electrical part, and participate in the inspection of the main machine and back-up system, and responsible for the extension of the cable and communication line.

1 internal combustion engine operator: responsible for driving and transporting the rails as well as transporting materials necessary for maintenance, as well as for the extension of the wind pipes and water pipelines.

The repair and maintenance shift has a very tight schedule and must complete all the inspection and maintenance tasks of TBM on time and as per scheduled quality and quantity, to ensure the normal digging work of the tunneling shift. The above-mentioned personnel are required to work closely together and are also responsible for filling in maintenance records. After the maintenance work, they have to go back to work in the workshop outside the tunnel to repair the supporting facilities outside the tunnel and make preparations for TBM.

(3) Workforce organization for cutter shift

The maintenance of the cutter is exclusive to 3 skilled workers. 1 cutter shift leader is responsible for the file of each cutter and good supply of spare parts, and 1 person is specially responsible for disassembly and cleaning, and 1 person is responsible for installation and testing, and 3 people should sign on the cutter maintenance record at the same time. Each cutter shall be filed, so that each cutter can be utilized to the fullest, and the cutter consumption costs are controlled from the management point of view.

(4) Supply of materials and preparation of spare parts during excavation

Before the tunnelling shift starts work, the maintenance shift must store the rails and the necessary support materials for the tunnelling shift. All kinds of materials needed during tunnelling shall be prepared outside the tunnel, and some wearing parts are required.

TBM should be stocked with all kinds of cutters, including a side cutter, a center cutter and four face disc cutters, and if a cutter is replaced during tunnelling, a new cutter must be brought in immediately from outside for replacement at any time.

(5) Construction surveying in the tunnel

For the tunnel with TBM construction, because of its fast advance speed, it is required to control the construction direction of TBM every day, so there should be enough time for surveying work every day, which is usually arranged during the maintenance time to reduce mutual interference, and avoid surveying errors. For this reason, two independent surveying systems are used. That is, automatic surveying systems and manual surveying systems. The two methods are used for crosschecking to ensure the accuracy of the surveying work.

(6) Slag disposal

The treatment method used in the Vereina Tunnel was to make full use of the slag in the tunnel to control costs and to reduce the number of spoil area for environmental protection. The slag transported from the tunnel is screened immediately after unloaded from the slag dumping machine, and the slag ($\geq$ 16 mm) is processed and used as tunnel slag or concrete aggregate, while the slag ($<$ 16 mm) is treated as slag and dumped at the spoil area.

(7) On-site communication system

The Vereina Tunnel has three communication systems: first, a public telephone network, connected to the offices; second, an internal telephone network to communicate with the sites outside the tunnel; and third, a wireless telephone system, mainly used inside the tunnel and for transportation. A relay station is set up every 2 km in the tunnel to ensure that the entire construction site is connected, so that all areas of the site can get in touch with each other. The internal combustion engine dispatching adopts wireless communication system and is commanded by the unified control.

3. Main Technical Characteristics of TBM

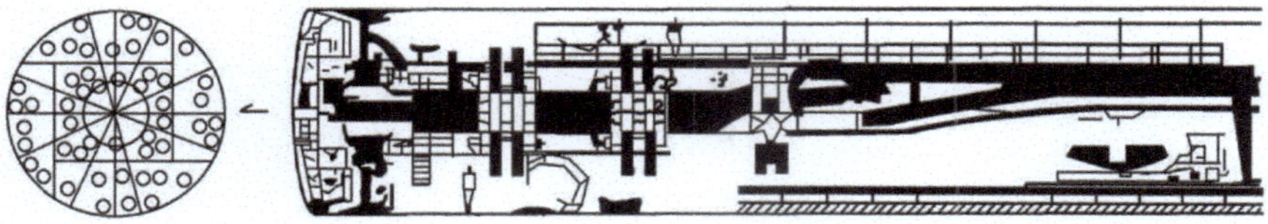

Fig. 11.51 Gripper TBM used in the Vereina Tunnel

The Vereina Tunnel was constructed using a TB770/850E ϕ7.64 m gripper TBM manufactured by Wirth. The gripper TBM consists of a main machine and a back-up system, see Fig. 11.51.

The maximum excavation diameter of TBM cutter head is ϕ7.89 m, the maximum torque is 5970 kN·m, the thrust force is 16,500 kN, the large cutter head is equipped with 57 disc cutters, the diameter of the disc cutter is 490 mm, and the mass of the whole machine is 750 t. The maximum length of one advance is 1800 mm. The power of the main machine of TBM is 3200 kW, and the power of the back-up system is 960 kW, the total power of the main machine plus the back-up system is 4160 kW. The back-up system is made by Rowa.

TBM construction ventilation adopts press-in hose ventilation, the diameter of the hose is ϕ2.5 m, and the main fan is set at the entrance of the tunnel portal, and the booster fan is set at 3 km inside the tunnel, the fan is Kofman fan made in Germany.

The main technical features of TBM are as follows:

(1) The axial thrust force provided by TBM is very high, with each disc cutter having a thrust force of about 30 t, giving it sufficient cutting power to break extremely hard rock.
(2) TBM is a gripper TBM and has a short cutter head and shield, which facilitates timely support. The contact area with the surrounding rock around the tunnel is small, which reduces the risk of getting stuck for cutter head.
(3) TBM is equipped with a pair of reaming disc cutter with a 5 cm extension around the cutter head. It is possible to obtain the required section size by expanding the section in case of excessive deformation of the surrounding rock, and to correct the deviation if there is direction or elevation deviation after one cycle of excavation.
(4) TBM provides a maximum torque of 5970 kN·m and a normal operating torque is 3980 kN-m, and automatic stepless speed adjustment (0–7.5 r/min) can be ensured under the condition of constant torque. If the cutter head is stuck by the surrounding rock, the maximum torque can make it escape.
(5) TBM is designed as a gripper TBM and is equipped with support equipment adapted to different geological conditions. These equipment include: one probe drilling machine and one pregrouting equipment; one ring beam installer; two rock bolt drillers; and one shotcrete pump and one automatic concrete spraying equipment. In addition, TBM is equipped with equipment for lifting, transporting and laying pre-cast concrete invert blocks.

(6) The distance between the front and rear supports can be adjusted, so that the spacing between the rows of ring beams can be changed according to the different geological conditions.
(7) The wet shotcrete system is equipped with high spraying efficiency, up to 12 m^3/h, and the strength of the shotcrete can reach 50 MPa. The equipped belt conveyor has a mucking capacity of up to 560 m^3/h.
(8) TBM's tunneling and support can operate in parallel, which minimizes downtime and improves machine utilization ratio and tunneling efficiency.

4. TBM construction technology

The Vereina Tunnel was excavated 2133 m from the north end by drill and blast, and 9451 m was bored by TBM. The key to the success of TBM construction in this tunnel is that the geological conditions, rock characteristics and measures to be taken during TBM construction were fully considered during the design phase. And the relevant measures are integrated into the selection and determination of the main technical parameters and design of TBM. The rock formations excavated in TBM construction range from locally-distributed soft strata and moderately-hard sedimentary rocks to very hard igneous rocks, which are mainly composed of gneiss, granite and amphibolite. TBM has to pass through several tectonic fracture zones formed by the Alpine orogeny; the overburden of these fracture zone sections are up to 1200 m.

The selected TBM is capable of cutting hard rock, but also has the adaptability to pass smoothly through the fracture zone and deal with the large deformation of the tunnel due to high ground pressure. The success or failure of TBM construction depends not only on the selection and design of the TBM itself, but also on the reasonable combination of TBM and the back-up system. As far as construction management is concerned, it depends on whether the construction managers have a good knowledge of the geological conditions, and whether they are skilled in TBM construction and maintenance techniques. Over the years, attention has been focused on developing and improving the performance of TBM, while the importance of the interplay between the main machine and back-up system has been underestimated, resulting in the efficiency of TBM often not being fully utilized. The recognition that TBM need to be equipped with efficient supply and delivery systems has only been attached great importance to in recent years. During TBM construction in the Vereina Tunnel, the geological factors, tunnel support mode, material and slag transportation methods, form of ventilation, measuring methods and construction safety requirements were fully taken into account. The details are summarized as follows:

(1) Geological factors

For the surrounding rock through which TBM passes, 86% is very hard igneous rock (average compressive strength of 100–250 MPa); 10% is moderately hard sedimentary rock and gneiss; 4% is very soft fracture zone and adverse geological zone; and the overburden of Vereina Tunnel is 1200 m, with the need of considering the factor of large tunnel deformation.

(2) Support Mode

The support mode used for TBM construction are rock bolts, ring beam and concreting, the amount of which is related to the rock type, as shown in Table 11.18.

(3) Means of transportation

The material and slag transportation in TBM construction is carried by double track with a gauge of 2900 mm, and the track is fixed to the precast concrete invert block laid immediately after TBM.

Table 11.18 Rock Type and Support Mode and their Positions relative to the Tunnel Face

Rock type	Percentage (%)	From TBM shield to the first support zone of TBM rear support point (at about 15 m)	From the starting point of the back-up system to the second support zone 45 m to the tunnel face	Third support zone
I–III	86	Only steel arches are erected in the arch, 2–4 glass fiber rock bolts are installed on the side, and precast concrete foundation is laid at the bottom	Spraying 6–10 cm concrete in the arch only	Reinforcement According to the rock and deformation
IV–V	10	Circle frame is equipped with flexible ring beam and wire mesh, with a steel arch spacing of 1.6 m. 4–6 glass fiber grouting rock bolts are installed for the circle frame, with an interval of 1.6 m, and the bottom is laid with precast concrete foundation	10–13 cm concrete is poured for the cycle frame	
VI–VII	4	Circle frame is equipped with flexible ring beam and wire mesh, with a steel arch spacing of 0.8 m. 3–4 glass fiber grouting rock bolts are installed for the circle frame, with an interval of 0.8 m, and the bottom is laid with precast concrete foundation	25–27 cm concrete is poured for the cycle frame	

(4) Form of Ventilation

The specifications of TBM construction ventilation and dust removal equipment must meet the construction requirements for air volume and meet the health standards for allowable dust content in the air. The selection of ventilation equipment must also consider the influence of rock temperature, diesel locomotive exhaust gas discharge, air duct layout location and other factors.

(5) Measuring Method

Laser guidance technology is used to control the advance direction of TBM, and the relative position of TBM tunnelling curve to the theoretical tunnel curve is displayed on a computer screen in time. TBM construction needs to take into account the measurement factors, and reserve the space required for guiding laser placement and measurement to meet the requirements of tracking the actual position and trajectory of TBM, as well as be able to display the relevant measurement data and deviation from the design position on the computer screen at any time.

(6) Construction Safety

TBM construction needs to consider safety measures, and must be equipped with gas detection and alarm, automatic power cutoff, fire extinguishing and rescue facilities.

5. Tunnel section and lining design

The tunnel cross-section of the TBM construction section of the Vereina Tunnel is shown in Fig. 11.52. The tunnel diameter of TBM construction section is ϕ7.64 m, which is based on a comprehensive consideration of the railroad vehicle limits. It is based on the maximum support thickness, the estimated tunnel deformation as well as the construction and measurement errors. The Vereina tunnel lining was designed according to the principle of single-layer shotconcrete support, with a maximum lining thickness of 35 cm. The sum of tunnel deformation and construction error allowed in difficult areas is 23 cm. Compared with composite lining, the advantage of single-layer shotconcrete support is the flexibility of use. Under the premise of meeting the lining quality requirements, the amount of tunnel excavation and the consumption of lining materials can be reduced, thus reducing construction costs. The key to the successful application of this support principle is the demand of using high quality lining materials, of which the mix ratio and quality of shotcrete and the selection of rock bolt support materials are particularly important. As the initial support, the shotcrete shall have a high compressive strength (more than 30 MPa), and has a good waterproof performance as well as frost and acid resistance; as the secondary support, the concrete, in addition to the high waterproof performance as well as frost and acid resistance, shall have a compressive strength more than 40 MPa. For this reason, wet shotconcrete technology was used in the Vereina tunnel project. Compared to dry concreting technology, wet shotconcrete technology also has the advantages of high efficiency (12 m^3/h), low rebound (less than 10%) and good labor conditions (low dust content).

The design of the tunnel lining according to the single-layer shotconcrete support principle also requires the use of rock bolt materials with permanent load-bearing

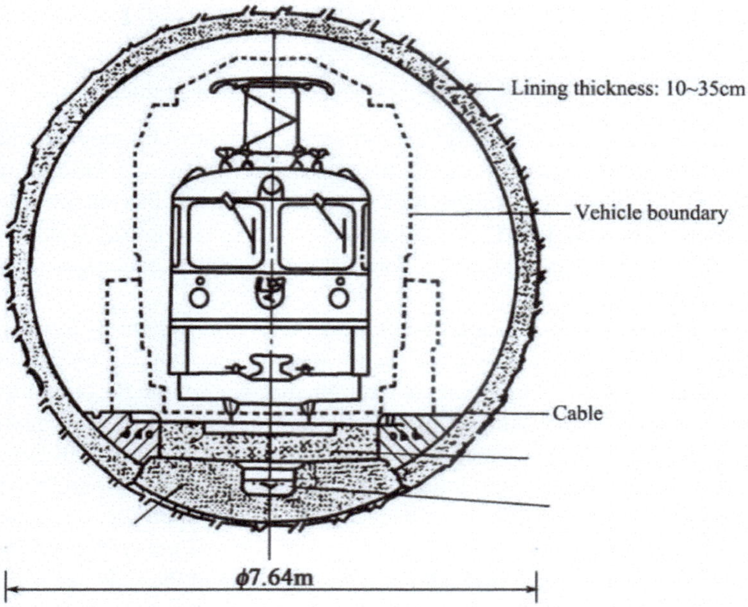

Fig. 11.52 Tunnel cross-section

capacity and corrosion resistance. For this reason, glass fiber reinforced resin-coated or grouted plastic rock bolts are commonly used in the Vereina tunnel. Compared with mortar rock bolt, this rock bolt is light in weight, easy to install, corrosion-resistant, and has good bending and deformation ability, and does not damage the cutter when cutting the rock bolt.

In order to improve the strength and stability of the tunnel foundation, a 1.6 m precast concrete element with a central drainage ditch was used as the tunnel ballast bed, and the joints between the precast elements were sealed by grouting. Compared with cast-in-place concrete ballast bed, precast concrete ballast bed is stronger and has a high laying efficiency.

The success of the application of single-layer shotconcrete support also depends on the ability to solve the tunnel's waterproof performance and drainage problems. The Vereina tunnel waterproof performance is mainly achieved through the improvement of the material waterproof performance (structural waterproofing) and other measures such as water drainage and plugging. For the cases like small water ingress and water dispersion, shotconcrete material waterproof performance is improved to prevent the tunnel arch and side wall seepage. For this reason, special requirements are imposed on the waterproofing properties of all structural materials (shotconcrete, grouted concrete, precast concrete). When the water seepage is concentrated locally and the amount of water is large, a flexible drainage hose is set up at the leakage point to drain it to the side ditch at the bottom of the tunnel. For fault and other fracture zone drainage, in addition to the above measures, the measure of grouting for water plugging is added.

6. Geological advance forecasting and project quality management

Due to the poor adaptability of TBM construction method to lithological changes, there is a risk that TBM will get stuck due to excessive deformation or falling when the initial stress in the rock is high and the rock is unstable. To avoid this phenomenon, the seismic measurement system TSP developed by Amgerg Group was systematically applied in the Vereina tunnel project to forecast the geological conditions in front of the tunnel face.

The principle of TSP advanced geological prediction system is based on the transmission and reception of seismic wave signals. The specific operation is to arrange a certain number of boreholes at a certain distance behind the tunneling face, and then conduct micro-blasting in the boreholes to generate acoustic signals. The blasting-induced seismic waves propagate as waves in the rock mass in all directions, with some of waves propagating towards the front of the tunnel. When the seismic wave meets the interface of lithology change in front of the tunnel, a part of the wave will be reflected back from the interface. By receiving and analyzing the received reflected wave signal, we can find out the spatial location of the reflected signal and its relative position to the tunnel, and then determine the distance of the reflected interface from the tunnel face and the angle of intersection with the tunnel axis.

Compared with the drilling method, this measurement method has the advantages of long forecasting distance (more than 150 m), does not affect tunneling, and can provide three-dimensional information on the lithological change interface and rock strength change (elastic modulus and Poisson's ratio).

The systematic application of TSP advanced geological prediction survey method in the Vereina Tunnel Project has already brought significant economic benefits for the safe and efficient construction of the project. Its economic benefits are mainly in the ability to greatly reduce unorganized construction, reduce the incidence of accidents and downtime, guide tunnel lining, and facilitate the deployment and management of materials.

The geo-advance prediction survey is only one element of the total quality management implemented in the Vereina tunnel project. To ensure the quality of the project, the comprehensive quality management and cost management mechanisms were systematically introduced in the Vereina project. Quality control includes: checking the quality of shotconcrete and other concrete components at regular intervals, such as the strength, water resistance, anti-freezing performance and acid resistance of the precast concrete ballast bed; checking the particle distribution of the concrete aggregates selected; checking the quality of tunnel excavation (over-excavation and under-excavation) using a laser profiler; applying geophysical methods to forecast the geological conditions in front of the tunnel face to guide the tunnel construction; arranging measurement points in a planned manner to record and track the tunnel deformation in a timely manner; regularly checking the acidity and alkalinity of the ground water; and automatically measuring the concentration of harmful gases in the tunnel. Cost management includes: comprehensive records of the actual amount of various support materials, changes in the amount of water ingress in the strata and its impact on construction, changes in rock types and over-excavation due to geological reasons, measurement of the footage of each cycle of TBM, and using them as a criterion for judging the drillability of the rock, etc. By adopting the above-mentioned comprehensive quality management and cost management mechanisms, the actual cost per unit length of tunnel and the quality of work achieved can be checked at all times.

7. Application of TBM in tunnel construction

In the section completed with TBM of the Vereina Tunnel, the average daily advance rate was 2.5 m in the soft glacial alluvium composed of glacial alluvium, and the average daily advance rate was 12 m in the sedimentary rocks such as dolomite, and a maximum daily advance rate was 25 m. From the above-mentioned average daily advance rate, TBM and its back-up system selected for the Vereina Tunnel are technically feasible. The technical performance of TBM and its back-up system can be evaluated from the following five aspects:

(1) Rock cutting ability and tunnelling efficiency

TBM selected for the Vereina Tunnel was designed primarily to cut hard rock. For each disc cutter, the axial thrust required to cut moderately hard sedimentary rocks is 100–120 kN, and the axial thrust required for all the disc cutters mounted on the cutter head is about 30% of the maximum thrust force that TBM can provide. The wear rate of the disc cutter in the tunneling is closely related to the geological condition, 1 disc cutter is replaced at the average tunneling of 550 m^3 (compact measure). The pure boring efficiency of a TBM is generally evaluated in terms of its advance per 1 cycle. The application effect of TBM passing through moderately hard sedimentary rocks has shown that its tunnelling efficiency is ideal. The experience gained is that TBM are less efficient in cutting loose rock formations, especially when the loose formations are interspersed with large amounts of cobble-like glacial or riverbed alluvium. The cutter head is often stuck by the pebbles during cutting, and the result is that these pebbles cannot be broken, but also the cutter and bucket are damaged by the impact. Removal of these pebbles often requires shutdown, which greatly affects the efficiency of TBM's excavation.

(2) Reliability of rock bolt and ring beam support device equipped for TBM

The most important difference between TBM construction method and the conventional drill and blast method is that TBM construction process of rock breaking, support and mucking can be operated in parallel. Due to the space constraints of TBM, the tunnel lining construction method is also different from the conventional drill and blast method. The support following the tunnelling surface is one ring beam and 2 rock bolts installed at a certain interval, to strengthen and stabilize the surrounding rock. This type of support can effectively prevent the roof plate from collapsing when TBM passes through moderately stable surrounding rock, ensuring the safety of personnel and equipment. And there is no impact on TBM's tunnelling efficiency. For unstable and soft strata, TBM's tunneling efficiency is deeply influenced by the support efficiency, and the adopted support forms mainly include: pre-grouting in the rock layer in front of the tunnel face, erecting round ring beam and system rock bolt right after TBM shield. In addition, in order to prevent the roof plate from destabilizing and falling, it is also necessary to apply rock bolt support and shotconcrete by manpower subsequently. TBM used in the Vereina Tunnel is equipped with the above-mentioned support devices to deal with adverse geological conditions, and the rock bolts and ring beam equipped can meet the construction needs of the tunnel, and the construction practice proves that this set of support devices is reliable. The excavation advance rate (average excavation advance 2.5 m/d) achieved when passing through soft ground is also better compared to other methods.

(3) Transfer of slag in TBM and its back-up system

A single belt conveyor is used to transport the slag out of the tunnel during TBM construction. The slag is first transferred by TBM's cutter head to the belt conveyor of the back-up system through the belt conveyor equipped with the main machine, and then further transferred to the subsequent belt conveyor and the silo at the entrance. Some tunnel projects do not use the subsequent belt conveyor and silo at the entrance of the tunnel, but the belt conveyor of the back-up system, and the hopper will transport the slag to the train consisting of large slag carts to transport slag. The current experience shows that the transport of slag in the main machine and back-up system of TBM is successful; the transport in the subsequent belt conveyor is still problematic due to the small radius of the tunnel curve, mainly manifested as the difficulties in lateral adjustment of the belt conveyor under different operating conditions (full load, wide load, uneven dryness of the slag).

(4) Control of TBM advance direction

To control TBM advance direction, laser guidance technology is used. The guidance system is able to display the relative position of TBM tunnelling axis to the theoretical axis of the tunnel on a computer screen in a timely manner, using a measurement technique that proved to be very effective. This is well illustrated by the small horizontal and vertical deviations when completing the tunnelling of the Vereina tunnel.

(5) Operation reliability of automatic shotcrete equipment

The equipment used for wet shotcrete support was a shotcrete pump and an automatic concrete spraying equipment manufactured by Cifa. During the first 200 m tunnelling, the shotconcrete quality is bad. After adjusting the efficiency of the pump, the diameter of the delivery pipe, and controlling the amount of additives, the concrete spraying quality was greatly improved and the operation of the equipment became normal.

The above is an evaluation of TBM and its back-up system used in the Vereina Tunnel from the perspective of rock breaking, support and transportation. Therefore, the design of TBM and its back-up system is a success. It is important to note that in the case of small tunnel curve radius and large tunnel length, there are many problems in using the subsequent belt conveyor and entrance silo after TBM for slag transportation, so these problems shall be avoided. From other super long tunnels using TBM construction, we know that instead of using the subsequent belt conveyor and entrance silo, the belt conveyor of the back-up system is used to funnel the slag into the large mine carts, and the slag train composed of 10 large mine carts with a volume of 15–16 m^3 will carry the slag out of the tunnel, and then the mine carts of the slag train will be turned over by the tippler at the spoil area or the slag is dumped by mine carts at the spoil area. This type of slag transport method has achieved satisfactory results in the construction of super long tunnels.

For the construction of the Vereina Tunnel, TBM method was used together with the drill-and-blast method. For TBM construction method, TBM has passed through very hard igneous rocks, moderately hard rock formations, and also through soft strata and tectonic fracture zones. Tunnels constructed with TBM are usually long or super long, and constructing long tunnels inevitably has to pass through rock layers with different geological conditions and different characteristics. Therefore, the development direction of TBM manufacturing and construction technology is to enable TBM to pass through rock strata with different geological conditions and characteristics. As can be seen from the above, TBM supplied for the Vereina Tunnel has taken into account the requirements of different geological conditions in a comprehensive manner. In terms of rock breaking, TBM can provide sufficient axial thrust force and torque, and can automatically adjust the speed steplessly. In terms of support, TBM is equipped with probe drilling and grouting equipment, rock bolt driller and ring beam installers, as well as wet shotcrete equipment. In addition, in order to ensure the full efficiency of TBM, the design of its back-up system shall be also attached great importance to. The design and construction of the main machine and the back-up system of TBM was successful, judging from the fact that TBM reached a maximum daily advance rate of 25 m through sedimentary rocks and soft strata.

To avoid sudden encounter with weak fault zones or other adverse geological zones during TBM construction, TSP advanced geological forecasting and measurement technology was systematically applied in the Vereina Tunnel, greatly reducing the unorganized construction, accident incidence and abnormal downtime of the construction, and reducing investment and construction time for the project.

11.1.8 Gotthard-Basis Tunnel

1. Project Overview

(1) Engineering Background

The Gotthard-Basis Tunnel, GBT for short, is a route (tunnel) passing through the Gotthard Pass, one of the most important channel passing through the Alps on the north–south axis of Europe. Since 1980, traffic volumes increased more than tenfold, and the existing road and rail tunnels have reached their limits of use. To provide a faster and flatter passage through the Alps, the voters in Switzerland decided to build this tunnel through the Gotthard mountain. This tunnel is almost at the horizon and 600 m (1969 ft) lower than the existing railroad tunnel.

It took 17 years and SwF 12 billion (about EUR 11 billion) to construct this tunnel. It is about 35 mile long (about 57 km) and passes through the bottom of the Alps, 8000 ft (about 2438.4 m) from the ground. It surpasses Japan's Seikan Tonneru (about 53.9 km long), making it the longest and deepest tunnel in the world (including railroad tunnels and road tunnels).

(2) Construction History

The history of the Gotthard-Basis Tunnel dates back to the eighteenth century and has evolved over the centuries to eventually become one of the longest tunnels of the century. Its construction history is shown in Table 11.19.

(3) Engineering Geological Conditions

Table 11.19 Construction history of Gotthard-Basis Tunnel

Time	Construction History
1707	Morettini, a master builder from Ticino, used cutting and blasting to dig a 64 m-long tunnel through the Chilchberg mountain, the famous Urner Loch
1872	Construction of the first railroad tunnel passing through the Gotthard Mountains began, directed by Engineer Louis Favre from Switzerland
1880	tunnelling completed, which was extremely accurate by the standards of the time
1882	The 15 km long railroad tunnel was put into operation, which was the longest railroad tunnel in the world at the time
1969	Construction of the first road tunnel passing through the Gotthard Mountain began
1980	The first road tunnel was put into operation, linking Göschenen and Airolo
1993	Exploration borehole started in the Piora Basin
1996	Initial preparation and exploration work for the Gotthard-Basis Tunnel began in Sedrun
1998	The Swiss Government approved funding for New Alpine Transversal (NEAT) Project
1999	With the first blast, excavation work began at Sedrun
2001	Herrenknecht was awarded the first equipment order
2003	4 Herrenknecht gripper TBM started digging
2006	Herrenknecht TBM tunnelling from the northern and southern ends of the tunnel reached their first receiving shafts, up to nine months ahead of schedule
2008	Successfully passing through the intimidating Piora Basin
2009	Successful completion of mechanized tunneling operations at the northern end
2010	Passing through the eastern tunnel between Sedrun and Faido
2011	Meeting the main line in the western tunnel between Sedrun and Faido
2016	Gotthard-Basis Tunnel, the core part of the New Alpine Transversal (NEAT) Project, was put into operation

The uniaxial compressive strength of the rock in the Gotthard-Basis Tunnel is up to 257 MPa, the point load strength is up to 17 MPa, and the abrasion resistance index is greater than 4.

2. Construction Scheme

Due to geological conditions, line requirements and schedule considerations, the Gotthard-Basis Tunnel was divided into four main sections for excavation, namely Amsteg-Sedrun Section, Erstfeld-Amsteg Section, Bodio-Fadio Section, and Fadio-Sedrun Section. Figure 11.53 is a schematic diagram of the Gotthard-Basis Tunnel, 57.1 km long. It is mainly constructed by TBM method, with part of the tunnel constructed by the drill-and-blast method.

The total length of the Amsteg-Sedrun/Erstfeld-Amsteg section is 35689 m, using two gripper TBM with a diameter of 9580 mm, a cutter head power of 3500 kW, a 17 in blade tool of cutter and a cutter head thrust force of 27500 kN, as shown in Fig. 11.54.

The total length of the Bodio-Faido/Faido-Sedrun section is 49729 m, and two gripper TBM are used. The diameter of TBM used in the Bodio-Faido section is 8830 mm, and the diameter of TBM used in the Faido-Sedrun section is 9430 mm,

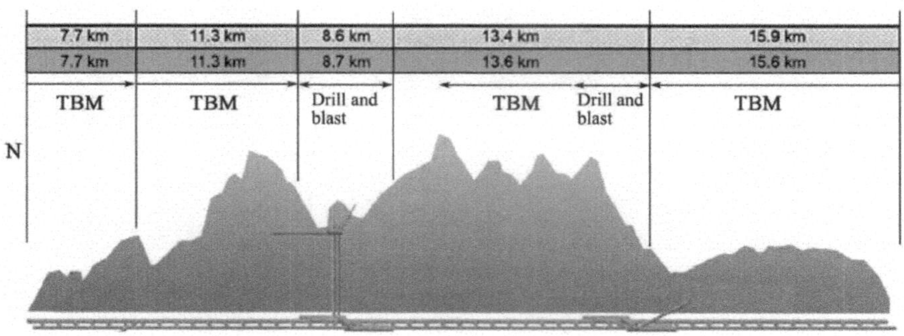

Fig. 11.53 Schematic diagram of the Gotthard-Basis Tunnel. *Note* The arrow represents the direction of excavation

Fig. 11.54 TBM with a diameter of 9580 MM

the power of the cutter head is 3500 kW, 17 in blade is used for cutter, and the thrust force of the cutter head is 27,500 kN. Figure 11.55 shows TBM used in these two sections. Table 11.20 shows the main technical parameters of TBM.

3. Construction difficulties

The Gotthard-Basis Tunnel project entails the construction of a total of 152 km of tunnels and shafts. Planners have divided 2 main tunnels and approximately 180 construction access roads into 5 bid sections. 5 bid sections can be constructed simultaneously, so the total construction period can be greatly reduced. However, this also requires a great deal of logistical support, the construction of entrance and supply tunnels, and the excavation of a large number of underground chambers for boring operations or drill-and-blast construction.

At the construction site of the Gotthard-Basis Tunnel, there are three entrance and supply tunnels, namely the Amsteg tunnel in the north, the Sedrun transition tunnel in the center and the Faido tunnel in the south. It is relatively easy to enter the tunnel

a) Diameter: 8830mm   b) Diameter: 9430mm

Fig. 11.55 Other two TBM

Table 11.20 List of main technical parameters of TBM

Length	450 m (including back-up system)	Length	450 m (including back-up system)
Weight	2700 t	Advancing stroke	2 m
Diameter	8.8 m (Bodio), 9.5 m (Faido)	Advance rate	Up to 40 m/d
Drive	3.5 MW	Price	Approx. CHF 30 million
Thrust force	Max. 2500 t	Manufacturer	Herrenknecht, Germany
Support force	7000 t	Shifts	17 people
Torque	> 6000 kN·m	Running cycle	2 × 9 h excavation + 6 h maintenance
Cutter head rotation speed	6 r/min	Time of service	320 days/year

from the tunnel portal near Bodio and Erstfeld in the north. At Amsteg, only a 2 km short tunnel is for entering the tunnel. The entrance tunnel at Faido is 2.7 km long and has a gradient of approximately 13%. The construction of the Sedrun transition tunnel was quite complex. First, it has to dig a horizontal entrance tunnel about 1 km long above the main tunnel, and then builds two shafts (about 800 m) running downwards to the tunnel. Finally, a 450 m long ventilation shaft is built to complete this spectacular project.

(1) Adverse ground condition

During the tunnelling of a tunnel as large as the Gotthard-Basis Tunnel, unexpected situations can occur at any time. Despite a lot of research and preparation, unexpected difficulties may arise from time to time. For example, in February 2003, when tunneling has just begun from the southern end of the Gotthard-Basis Tunnel, just 200 m advance, the loose rock strata blocked construction. The two TBM (TBM Sissi for the east and TBM Heidi for the west) that started work at Bodio encountered a zone of brecciated rock. The geology is too soft for a gripper TBM designed for hard rock conditions, and it is almost impossible to dig quickly. Every 1 m of tunnelling has to go through a complex set of procedures for safety support. It was not until August 2003 that the two TBM were able to complete about 400 m of tunnelling and were able to pass through these broken zones. As shown in Fig. 11.56, it is the adverse geological zone of the Gotthard-Basis Tunnel.

According to the construction plan, geologists estimated that the construction shall be stopped for 4 months to deal with Intschi zone. Fortunately, this zone is about 50% shorter than the estimated length. The construction team operated "Gabi1" and "Gabi2" TBM through this zone, albeit at a slower pace. Sometimes the cutter head can advance as much as 12 mm a week in one cycle and achieve the excellent advance of 560 m a month, while sometimes it only advances 140 m a month in difficult areas, which means that the cutter head can only advance 3 mm a week in one cycle.

In June 2005, "Gabi2" TBM encountered difficulties during the tunneling process. Without any warning, loose rocks mixed with water instantly entered the cutter head of "Gabi2", which was working in the west tunnel. At first, the tunneling crew tried to manually clean the powder and fines from the cutter head area and retract TBM by a few centimeters, but failed. Finally, the cement and bentonite mixture was injected

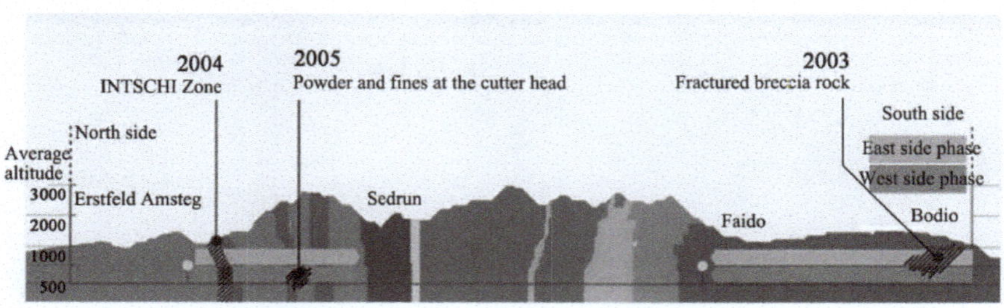

Fig. 11.56 Adverse geological zone of the Gotthard-Basis Tunnel

into the unstable area in front of TBM cutter head for reinforcement. At the same time, the tunnel builders dug a 50 m-long westward connecting tunnel from the front of the east tunnel already dug, then dug backward in the direction of "Gabi2" in the west tunnel, and finally unloaded the "Gabi2" cutter head from the front. After a five-month pause, normal excavation work was not resumed until November 2005.

(2) Piora Basin

Another complex area during the excavation is the Piora Basin (Fig. 11.57). Piora Basin is a funnel-shaped formation, filled with sugar-grain dolomite and water, reaching deep into the Sierra rock. This geological situation was detected long ago, and exploring here has been more intensive than those in any other area of the Gotthard Range. Since there was no way to know how deep the "funnel" went into the mountain range, it was decided to conduct test drilling. A horizontal tunnel approximately 5.5 km long was bored from the state highway into the disturbance zone. In 1996, the exploration tunnel reached the fault zone. The exploration revealed that the loose rock grains are under a huge pressure of about 150 bar. A cloudy column of water mixed with dolomite shot out of the mountain and flooded the road.

19 oblique boreholes were drilled from the exploration tunnels to the vicinity of the future baseline tunnels, which eventually gave the engineers a complete view of the subsurface. These oblique boreholes reach deep into the hard rock and do not encounter water pressure. By observing drilling core, temperature and seismic observations, geologists have concluded that the base of the Piora Basin is closed by gypsum. Then, on October 12, 2008, we received good news from "Sissi": "It Successfully Passing through the Piora Basin". "Heidi" soon also made it.

(3) Pilot tunnel

Excavating the pilot tunnel from Faido to Sedrun with conventional drill and blast method was also an extreme technical challenge. Here, the tunnel builders build the massive underground station (Fig. 11.58) through blasting the rock to serve as a dedicated parking port for trains in case of emergency. There is also a bifurcated tunnel for trains to go from one tunnel to another during operation. Through a complex set of cross passage and connecting passage, it is possible to ensure that the smoke is discharged and fresh air can enter the tunnel. The two multi-purpose stations

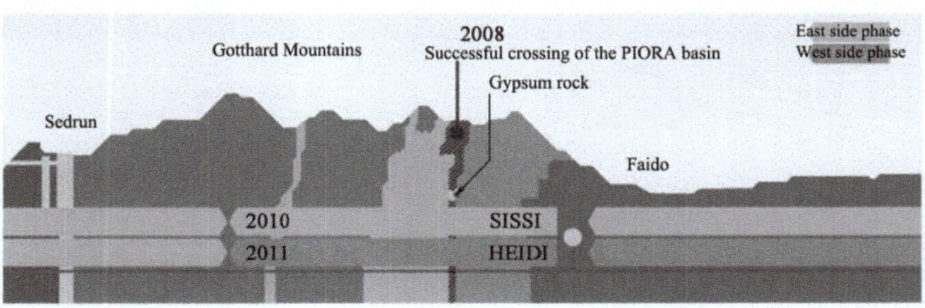

Fig. 11.57 Piora Basin

Fig. 11.58 Underground station

themselves are huge construction sites. Here, the tunnel builders have to face the special challenges posed by the drilling and blasting of the rock.

(4) High ground temperature

For TBM construction and instrument working environment, a certain temperature shall not be exceeded. However, in the deep overburden, the temperature inside the tunnel is expected to rise to 55 °C, and the actual measured temperature is over 40 °C. According to the Swiss safety regulations amended according to the Gotthard Tunnel, the temperature in the working environment should not exceed 28 °C. There are no other engineering experiences available at such deep overburden and high temperatures. Under such high temperature conditions, especially at the excavation face where a lot of heat is generated by mechanical rock crushing and some machines that generate a lot of heat, a lot of cooling devices are required. The ventilation device of the tunnel cannot work as a cooling device because the gas is heated during the flow inside the tunnel. Therefore, modern tunnel cooling units are needed, absorbing heat through cold water and discharging it through large steel pipes.

Four return pipes are required for the tunnel double tracks, which means a 100 km long pipeline is required. The power consumption of the power required to pump water every day costs $10,000 (about RMB 62,928). Outside the tunnels, sedimentation tanks for water treatment and cooling towers are required, and six large cooling towers were eventually installed to deal with the high temperatures.

(5) Rock burst

The overburden of the tunnel in front of Fadio is as high as 2300 m, and strong rock bursts occur under higher ground stress. Therefore, a 9.43 m diameter TBM was selected in this area to support the rock burst caused by the higher ground stress. The pressure on the mountain is so high that the conventional method of increasing the excavation section does not work. In general, this pressure deforms the mountain, and the greater the mountain deformation, the greater the released mountain pressure.

However, in the critical area of the Gotthard-Basis Tunnel, the required stability could not be achieved by simply letting the mountain deformation take its course. A force against the mountain is needed to inhibit the complete closure of the tunnel chamber. The solution is to use overlapping sliding brace rings. The principle of this method is that two steel half-rings with the function of overlapping sliding are connected to form a complete bracing ring in the tunnel section. The tunnel wall was over-excavated by about 70 cm, and under the pressure of the mountain, the rings slowly slid toward the center until the two half-rings are interlocked and support each other.

4. Construction process

The opening ceremony in 1993 at the Piora Basin elaborates on the exploration of its location and extent, as well as the engineering and hydrogeological conditions, to determine the most favorable route for the Gotthard-Basis Tunnel. Exploration revealed that the base of the Piora Basin is closed by gypsum rocks.

In 1996, the initial preparation and exploration work for the Gotthard-Basis Tunnel was started in Sedrun, and the AlpTransit Information Center was opened to the public to provide information on the progress of the project.

The first exploration hole passing through the Piora slope at the horizon of the Gotthard-Basis Tunnel encountered the rocks of the Gothenburg Massif to the north on January 30, 1997. This has led to the conclusion that the Piora slope is up to 250 m wide and consists mainly of water-saturated dolomite with high water pressure.

Exploratory work on the Piora Basin was completed in May 1998. At the horizon of the Gotthard-Basis Tunnel, the rock of the Piora slope is hard, intact, and has a low water content and is dry enough for TBM tunnelling, or drilling and blasting excavation.

Construction of 800 m deep shaft began on February 4, 1999, providing a construction passage to the tunnel section. On March 30, DETEC officially started the construction of the Gotthard-Basis Tunnel. Tendering process started in Amsteg, Sedrun, Faido and Bodio sections. The construction of a short tunnel of about 2 km at Amsteg on November 4, 1999 signaled the official start of construction of the northern part of the Gotthard-Basis Tunnel.

On March 1, 2000, the construction of the 800 m-deep shaft at the tunnel horizon was completed. On July 10, a 1.2 km long bypass tunnel was completed at Bodio.

The Amsteg-Sedrun East Line was constructed by TBM since May 22, 2003, and the tunnelling length was 10,722 m, and the tunnelling was completed on June 5, 2006; the West Line was constructed by TBM since August 11, 2003, and the tunnelling length was 10,702 m, and the tunnel was completed on October 9, 2006. The Erstfeld-Amsteg East Line was constructed by TBM since December 6, 2007, and the tunnelling length was 7148 m and the tunnelling was completed on June 16, 2009; the West Line was constructed by TBM since March 10, 2008, and the tunnelling length was 7116 m and the tunnelling was completed on September 16, 2009. Figure 11.59 shows the completed tunnelling of the east line and the west line of Erstfeld-Amsteg tunnel.

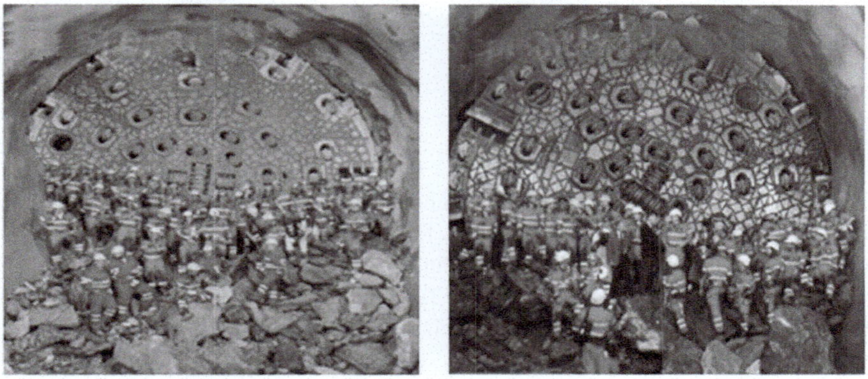

Fig. 11.59 Schematic diagram of completed tunnelling of the Erstfeld-Amsteg Tunnel

The Bodio-Fadio East Line was constructed by TBM since January 9, 2003, and the tunnelling length was 13,426 m, and the tunnelling was completed on September 6, 2006; the West Line was constructed by TBM since February 22, 2003, and the tunnelling length was 14,088 m, and the tunnelling was completed on October 26, 2006. The Fadio-Sedrun East Line was constructed by TBM since July 6, 2007, and the tunnelling length was 11,134 m, and the tunnelling was completed on October 15, 2010; the West Line was constructed by TBM since October 12, 2007, and the tunnelling length was 11,088 m, and the tunnelling was completed on March 23, 2011. Figure 11.60 shows the completed tunnelling of the east line and the west line of the Fadio-Sedrun tunnel.

The Gotthard-Basis Tunnel was put into operation in 2016 as a core part of New Alpine Transversal (NEAT) Project.

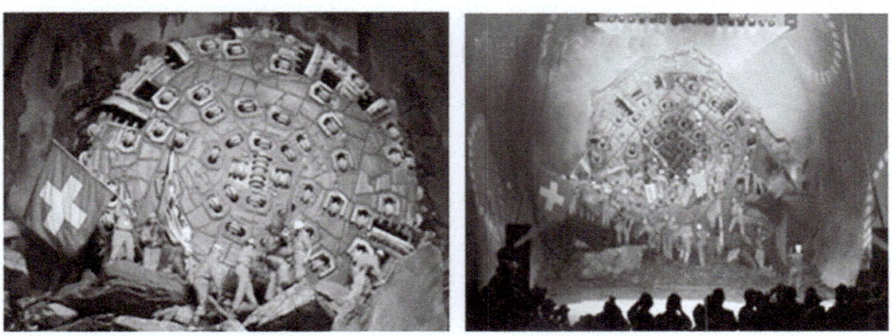

Fig. 11.60 Schematic diagram of completed tunnelling of Fadio-Sedrun Tunnel

11.1.9 Songhua River Water Diversion Tunnel

1. Project Overview

(1) Project Profile

The main construction contents of the Songhua River Water Diversion Tunnel include Fengman Reservoir water intake project, Fengjialing water diversion hub project, the main water transmission line, Changchun trunk line, Siping trunk line, Liaoyuan trunk line and ancillary works along the Tunnel. The total length of the project is 263.45 km, including 133.98 km of tunnels and 129.47 km of pipelines (PCCP, SP, cast-in-place culverts). The overall work quantities are 31,970,000 m^3, including 100,271 t of steel reinforcement. The design diversion flow of the intake is 38 m^3/s, and the design multi-year average diversion volume is 898 million m^3.

The fourth bid section of the trunk line of the Songhua River Water Diversion Tunnel of Jilin Province is located between Chalu River and Yinma River in Jilin City, Jilin Province. The chainage is K48 + 900–K71 + 900, with a total length of 23 km. The location of the project is shown in Fig. 11.61.

The fourth bid section of the Songhua River Water Diversion Tunnel is responsible for the 23 kmlong diversion tunnel. This tunnel is constructed by a gripper TBM with a diameter of ϕ7.93 m together with manual drilling and blasting method. The drilling and blasting method uses loaders and dump trucks for mucking, while TBM method uses a continuous belt conveyor for mucking, and the transportation uses a combination of trackless and tracked transportation system. The main works mainly includes:

① 424 m of main tunnel by drill and blast method at tunnel outlet, 58.298 m of surge shaft at Yinma River, 17.488 km by TBM, including 3125 m from outlet to Jiancaodianzi, 1564 m from Jiancaodianzi to Xiaoheyan, 3886 m from Xiaoheyan to No. 8 adit, 8913 m from No. 8 adit to No. 7 adit;

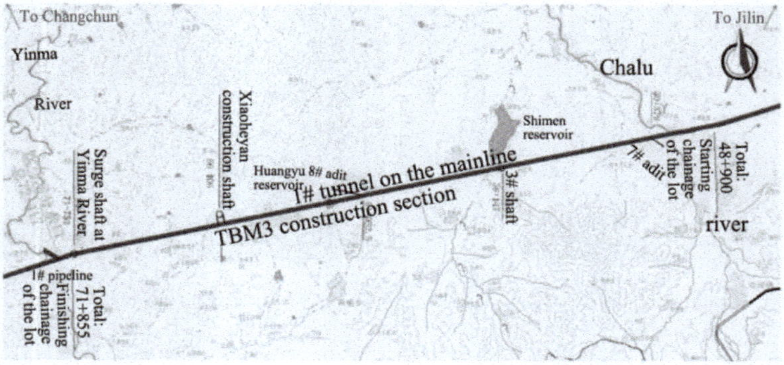

Fig. 11.61 Location of the fourth bid section of the Songhua River water diversion tunnel

② Jiancaodianzi shaft: 33.9 m shaft, 147 m adverse ground section with small burial depth handled at main tunnel;
③ Xiaoheyan shaft: 43.94 m shaft, 12.86 m horizontal passage, 284 m adverse ground section handled at main tunnel;
④ No. 8 adit: 1192.2 m, 475 m of maintenance room and service cavern of main tunnel; 1371 m between drill and blast section and TBM section;
⑤ No. 3 ventilation shaft: vertical shaft 52.7 m;
⑥ No. 7 adit: 518.01 m, 80 m disassembly room, and 1455 m of the drill and blast construction of Chalu river section. From ending chainage to TBM section: 1270 m in total.

(2) Engineering Geological Conditions

This bid section is the section with the most complicated geological condition, the most difficult construction and the most difficult operating environment of the trunk line of the Songhua River Water Diversion Tunnel. The stratigraphic lithology mainly includes tuff, conglomerate, carboniferous tuff, sodic porphyry, quartz amphibole and granite. The whole tunnel passes through several gullies and valleys, and the overburden of the tunnel is 38–150 m, and 7 km of tuff karst section is developed with dissolution cave group and carbonaceous slate. There are 49 faults in the whole section, with 31 in the tuff section. There are 12 proven caves in the Graystone section, with 3 suspected caves by physical exploration. The analysis concluded that Fw24-1, F24-2, F28, F38, F41 and other fracture faults are water-blocking or water-conducting fractures, and are located in gullies, with water catchment conditions, and may form links with surface water, and there are problems of collapse and mud gushing in the fracture zone.

Most of the strata where the tunnel passes have deep overburden, and the surrounding rock is basically slightly weathered, fresh rock, generally water-poor, with weak permeability, and has no contact with surface water bodies, or poor contact. The permeability and water-richness is not good, basically not likely to produce water ingress. However, there may be water ingress in the strata that have close contact with surface water bodies, and are located in the gullies and have the conditions for water catchment.

Along the tunnel, the following sections are prone to water ingress and mud gushing: ① shallow section of river valley, Chalu river section; ② tectonic development of the valley section, fault broken belt, etc.; ③ gully shallow section with tuff karst developed, Xiaoheyan (K66 + 498–K66 + 803), Jiancaodianzi (K67 + 914–K68 + 280), etc.; ④ crossing routes, section near the reservoir (Huangyu Reservoir).

The tuff stratum Xiaoheyan gully and Jiancaodianzi gully have revealed the water ingress during the construction process. The maximum water ingress is 1200 m^3/h when TBM passes through Xiaoheyan gully, and the maximum water ingress is 1500 m^3/h when TBM passes through Qijianfang gully. For No. 7 adit, the water ingress reaches 1100 m^3/h when the tunnel passes through Chalu river section.

(3) Project difficulties

① Large-span and large-scale project management.

The main tunnel of this project is 23 km long, including 2 inclined shafts and 4 vertical shafts, which are constructed simultaneously by TBM + drill & blast method. And there are nearly 600 people working in 5 work areas, the farthest distance from the project management office is more than 40 km. Building trades are complex, working environment varies from each other, work quantities are large and management points are diversified. Therefore, an effective project management is the top priority of this project.

② TBM passing through tuff karst area.

According to geological data, nearly 7 km from K63 + 964 to K70 + 823 in this bid section is a limestone cave section, and some part of limestone cave section is a karst area: chainage K63 + 884–K65 + 978, TBM tunnelling section has 4 karst caves, the maximum karst cave is 38.52 m high, 5.7 m from the top of the cave; the minimum cave is 6.3 m high, 32.5 m from the top of the cave; chainage K65 + 978–K67 + 913, Xiaoheyan drill and blast section has 3 karst caves, the maximum cave is 5.1 m high, 2 m from the bottom of the cave; chainage K67 + 913–K71 + 046, TBM tunnelling section has 2 karst caves, the maximum cave is 18.8 m high, 17 m from the top of the cave. During TBM tunneling in this section, it is easy to cause construction risks such as machine deviation, cutter head dropping, cutter head jamming, shield structure burying by mud ingress, electrical failure caused by water ingress, and overall machine jamming caused by convergence deformity. Therefore, taking effective measures to safely and smoothly pass through the karst area of the tuff section is the major difficulty of this project.

③ TBM passing through fault fracture zone.

According to the tendering documents and the design geological survey report, the geological structure of this section is complex. The whole tunnel passes through a total of 39 faults and low resistance anomaly belts (including 17 identified faults and 19 anomaly belts by geophysical prospecting and 3 faults by remote sensing interpretation), including 32 faults in the TBM excavation section. In the lithological contact zone, the joints and fissures are developed and the rock body is also broken, with strong water-richness. There are 5 faults with great influence on the project, namely Fw24-1, F24-2, F28, F38 and F41. F41 and F38 are located within the TBM tunnelling section. When TBM is advancing in the fault and broken influence zone, the surrounding rock deformation is large, easy to cause collapse. In addition, the direction of tunnelling is difficult to control, and the gripper of TBM cannot advance.

④ Construction in the section of sudden mud gushing and water ingress.

For this bid section, the locations that are prone to water ingress and mud gushing are: river valley, valley shallow section, Chalu River, Beigou, Xiaoheyan and Jiancaodianzi, valley section with tectonic development, F28, F38, F41 and other structures and gullies with low-resistance zone development, crossing routes, sections near the reservoir, Shimen Reservoir and Huangyu Reservoir. For the general rock body, in the process of excavation, the common phenomena are seepage or drip. It is preliminarily estimated that the water ingress is < 10 L/min for each 10 m of the excavation. Some part of the tunnel section passes through the highly-weathered or fully-weathered rock, and lithological contact zone. The overburden of the tunnel is shallow, and has hydraulic contact with surface water, so the water-richness is also strong, and It is preliminarily estimated that the water ingress is < 100 L/min for each 10 m of the excavation. The maximum water ingress Qo at the Chalu river is 1620–3500 m^3/d. Water ingress will have a major impact on TBM tunneling, and in severe cases will flood the equipment.

⑤ Multiple construction methods and application of new technology.

This project involves a variety of construction methods such as drill and blast method and TBM method in the tunnel construction, including adit, shaft and main tunnel, etc. The construction process is complicated, and the working faces cross with each other, and the key routes are closely connected. Therefore, through advance planning and application of new technology, production capacity is balanced. And benign process connection are the priorities and difficulties to achieve the project schedule.

2. TBM Targeted Design

(1) Adaptability design of cutter head

This tunnel passes through complex geological conditions, and the surrounding rock strength changes greatly. This requires that TBM cutter head design and cutter configuration can adapt to the maximum saturation compressive strength of 100 MPa for tunnelling in hard rock and the saturation compressive strength of 2 MPa for tunnelling in soft rock, and the cutter head and cutter should have high wear resistance, to reduce the frequency of cutter replacement and achieve continuous rapid tunneling. In the cutter head design and cutter configuration, the selected TBM is mainly targeted at the following:

① The enclosed cutter head can effectively support the tunnel face. This is helpful in preventing the collapse of the surrounding rock in the stratum with poor stability, and can provide safety guarantee for personnel to inspect and change tools in the cutter head.
② There are two manholes on the cutter head face plate, through which personnel can enter the tunnel face to remove obstacles and observe the geological situation in front of the tunnel face in time.
③ The cutter head is designed to rotate clockwise in one direction to excavate evenly. There are 8 muck discharge outlets arranged around the cutter head, and one side of the outlet is equipped with a number of buckets made of wear-resistant materials. The cutter head has a targeted wear-resistant design, and the muck inlet is welded with wear-resistant composite steel plates. In front of the cutter head, many wear-resistant composite steel plates are welded to protect the cutter head.
④ The structure of the cutter head is a heavy-duty welded steel structure to suit long-distance tunneling and hard rock geological conditions. The use of thick steel takes full account of the high dynamic load of the disc cutter when breaking rock.

The center section and side section are made of 270 mm forged thick plate, and the flange of cutter head is made of 340 mm forged thick plate (Fig. 11.62), and the welding process is strictly controlled by flaw detection (Fig. 11.63). Prior to machining, the design of the cutter head requires the use of finite element analysis to calculate the expected severe service loads. All flange surfaces are machined and connected with special bolts to ensure sufficient strength and reliability of the cutter head.

⑤ The cutter adopts non-linear arrangement to make the force on the cutter head more uniform to avoid larger stress concentration and prolong the service life of the cutter head. The vibration is relatively smaller under the condition of the same rock strength, which is conducive to reducing the cutter consumption (Fig. 11.64).

⑥ The design of small spacing between cutters (Fig. 11.65) is adopted. The spacing between cutters of the center disc cutter is 89 mm, face disc cutter 75 mm or 80 mm. Under such conditions, the cutter head has a strong capacity of breaking rocks. The small spacing between cutters is not only conducive to breaking rock, but also can effectively reduce the vibration of the cutter head and extend the service life of the cutter.

⑦ The dust is controlled by spraying water on the tunnel face through the nozzles on the cutter head, which can achieve the cooling of the cutter at the same time. The nozzle is mounted on a large nozzle holder, and a wear-resistant plate is welded to avoid blockage and damage by broken rock. Rubber hoses are used for the water entry to avoid cracking of the steel pipes under the vibration of

Fig. 11.62 Cutter head face plate flange forged with thick steel plate

Fig. 11.63 Welding process strictly controlled by flaw detection

Fig. 11.64 Picture of cutter head

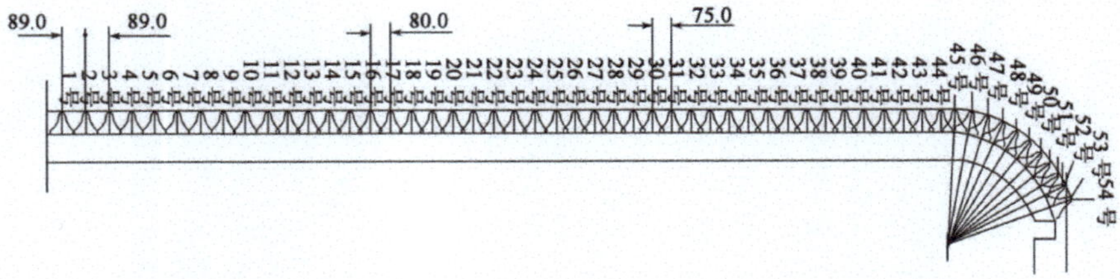

Fig. 11.65 Design of small cutter spacing (Unit: mm)

the cutter head, and all pipelines are protected by heavy-duty steel structures to avoid damage.

⑧ The cutter seat is machined directly on the 270 mm thick plate (Fig. 11.66), with no welding. This cannot prevent fatigue cracks, and achieve high strength and rigidity of the cutter head. The cutter seat is machined directly, there is no welding deformation, and the disc cutter is installed with higher precision, and more reliable.

(2) Optimization of auxiliary system and back-up system

① Relocation of the water hose reel.

Necessity of modification: as the water hose reel is configured in No. 15 gantry, the gantry is followed by the California switch, and the extension of the water pipe is blocked in many places, which delays the time for extension, thus affecting the efficiency of TBM tunneling.

Refurbishment measures: Move the water pipe reel to the ramp rail position and directly connect it with the rear water pipe, thus reducing the extension time of the water pipe (Fig. 11.67).

② Replacement and support of rear gantry platform pillar.

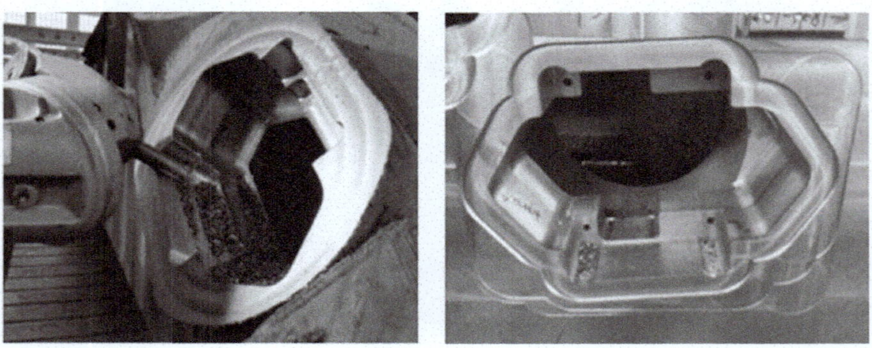

Fig. 11.66 Direct machining of disc cutter seat

Fig. 11.67 Relocation of water pipe to ramp rail

Necessity of reconstruction: As the HV cable is hung on the right side wall, when the cable is extended to the pillar of the rear gantry platform, the cable placed on the flat plate needs to be manually placed on the flat plate again by bypassing the pillar, thus delaying the time of cable extension and affecting the tunneling efficiency of TBM.

Refurbishment measures: cut off the affected part of the pillar, use a single small pillar as support, use a jack to support the upper pillar when extending the cable, take out the single small pillar for cable to pass through, and then place the small pillar in the original position (Fig. 11.68).

③ Refurbishment of shotcreting pipeline.

Necessity of refurbishment: The shotcreting pipeline of TBM is long and is often blocked on site, affecting the tunneling efficiency of TBM.

Refurbishment measures: shorten the shotcreting pipeline and weld hose support near the shotcreting nozzle to reduce the impact time of pipe blockage caused by hose bending during the shotcreting, thus improving the tunneling efficiency (Fig. 11.69).

(3) Optimized design of continuous belt conveyor for TBM turnaround

In the original design, the belt muck discharge system of the adit after turnaround is undertaken by T2, T3 and T4 (emergency belts). After T2 receives muck from muck transfer tower at the main tunnel intersection, the muck will be transported to the T3 belt of the muck divider at the tunnel opening, and then it will be transported to

Fig. 11.68 Replacement and support of rear gantry platform pillar

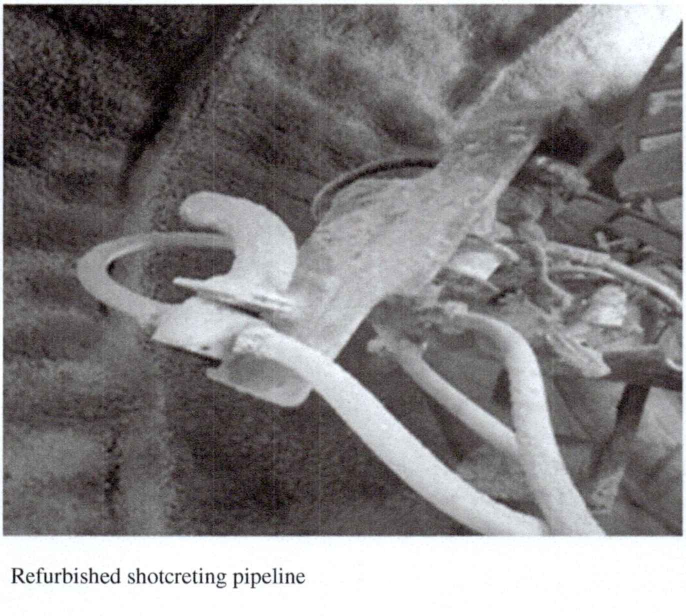

Fig. 11.69 Refurbished shotcreting pipeline

the spoil area by dump truck through the terminal diversion muck device of the belt conveyor.

Optimization measures: cancel the T3 belt and muck divider, extend the T2 belt, and directly extend the adit belt to the belt terminal diversion muck device, and transport the muck to the spoil area by dump truck.

Necessity of optimization: According to the statistical analysis of the construction process time and construction experience of the first tunneling section, the failure rate of continuous belt increases with the increase of tunneling distance, and often causes blockage of the blanking port, which results in a large workload and longer time for cleaning the blanking port, and has a greater impact on the construction efficiency. To ensure the normal operation of the muck discharge system, reduce the continuous belt failure rate and improve construction efficiency. Cancel T3 belt and muck divider according to actual situation.

3. TBM construction technology

(1) Treatment technology for shallow water-rich gully section of tuff karst

TBM passes through shallow water-rich gullies in the limestone section for 5 times. The Yongshengxing gully is the longest section with a length of 466 m, and the groundwater is extremely developed. The burial depth of the Xiaoheyan gully and the Jiancaodianzi gully is the shallowest. The minimum burial depth is 28 m, and there is a section of full-section soil that invades the main tunnel body respectively.

Technical ideas for dealing with the shallow water-rich gully section of tuff karst: when TBM tunnels to the vicinity of this section, long-distance advanced geological forecast (such as TSP, TRT, etc.) shall be arranged in advance, combining with medium and short-distance geological forecast for detailed surrounding rock conditions in front of the tunnel face. The medium and short-distance geological forecast shall be strengthened, continuously forecasted and used for guiding construction. Carefully study the emergency plan for TBMs passing through sections with adverse ground condition and do a good job of training, detail the possible risks and prepare targeted handling measures. Sort out and check the drainage pipelines, grouting pipelines and contingency reserve of equipment and materials in advance. When the TBMs arrived at the front of the section with adverse ground condition, we shall arrange the belt vulcanization, cable reel extension and cutter head maintenance in advance to avoid the long shutdown in the most unfavorable position of the gully. To pass through this section, the team, management personnel and technical personnel shall think in a unified way, strengthen the support and shall not be over-bold. The main countermeasures include:

① Advance geological forecast: Adequate advance geological forecasting, a variety of complementary means, a combination of geophysical prospecting and supplementary exploration of the surface to accurately detect the geological structures of this section and assess the TBM construction risk, and the preparation of the corresponding emergency supplies.
② Strengthen the initial support: according to the results of the advance geological forecast and the surrounding rocks of the exposed shield, take timely measures

to strengthen the support, using the ring beam + steel slat + connecting bars + shotcrete joint support. If necessary, use the dry shotcreting system to seal in advance.

③ Water ingress treatment: In case of water ingress, start the emergency plan, strengthen the pumping and drainage, and adopt water plugging scheme of radial grouting at the water ingress position.

④ Monitoring and measurement: After the initial support and front reinforcement are completed, the measuring points will be set up in time, and the data of deformation, convergence and rate change will be fed back to the technical staff of the engineering department. The technical staff will decide whether to take secondary reinforcement measures according to the feedback data.

(2) Treatment technology of cave group in tuff karst area.

The TBM has tunneled nearly 7 km of limestone cave section, and part of it is a karst area. 12 caves have been identified, of which 3 are suspected karst caves. The height of the largest karst cave is 38.52, 5.7 m from the top of the karst cave; the height of the smallest karst cave is 6.3, 32.5 m from the top of the karst cave. During TBM tunneling in this section, it is easy to cause construction risks such as machine deviation, cutter head dropping, cutter head jamming, shield structure burying by mud ingress, electrical failure caused by water ingress, and overall machine jamming caused by convergence deformity.

According to the dangers occurred in the 7 km tuff tunnel section during TBM tunneling and the analysis of unknown construction risks, the project department held several special discussion meetings and expert meetings, and the main technical ideas are as follows:

① Advance geological forecasting.

According to this geological condition, we shall carry out the advance geological forecast, and take a combination method of long and short distance (long distance TRT and short distance induced polarization advance geological forecast) to accurately detect whether the spatial position and size of the karst cave in front of the tunnel face are filled, etc., and make countermeasures and reserve emergency supplies in advance according to the results of the advance geological forecast.

② Dissolution and karst cave treatment measures.

a. Dissolution and karst caves at the vault.

The karst cave appears at the vault, and ϕ12 or ϕ16 steel slat and I16 ring beams (the spacing can be 45, 90, 180 cm) shall be installed for support before the shield is exposed. If the cave is filled and accompanied by falling blocks, in order to prevent the deformation of the steel slat and the convergence of the ring beam, the ϕ22 mm connecting reinforcement shall be changed to I16 I-beam for longitudinal connection with the ring beam, and the spacing of the ring beam shall be reduced if necessary. If the cave is not filled, after the shield is exposed, one end of the I16 I-beam support shall be welded to the ring beam and the other end shall be tightened to the rock

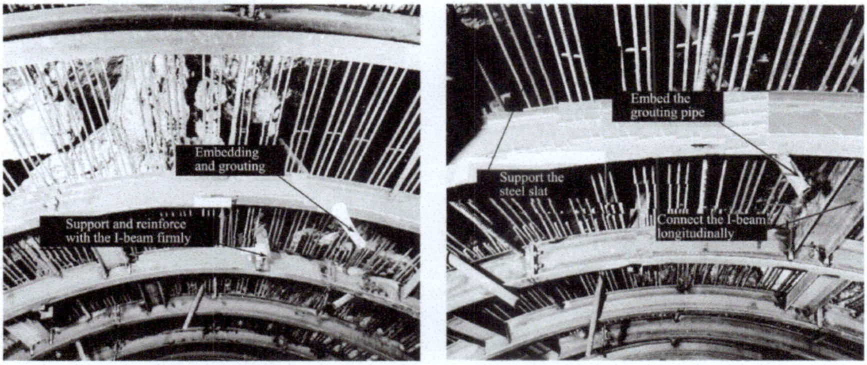

Fig. 11.70 Karst caves and cavities at the vault position

surface. After the I16 I-beam is firmly welded, the ϕ8 mm wire mesh shall be filled into the empty cavity, the cave shall be closed with iron sheet and the ϕ42 mm grouting pipe and exhaust pipe shall be installed. The karst caves and cavities shall be backfilled with C20 fine stone concrete, and the backfilling grouting shall be carried out at the backup section (Fig. 11.70).

b. Dissolution and caves are at the gripper position.

When the caves are located at the gripper position, they shall be treated in two cases:

If the cave is filled and is relatively broken, H150 steel or I16 I-beam shall be placed at the TBM gripper position to provide sufficient support for the TBM gripper, and the mesh or steel slat shall be installed in advance. Emergency shotcrete shall be used for shotcreting at the karst cave position, and the shotcrete thickness shall be flush with the inner arc of the ring beam. After the concrete strength meets the requirements, the TBM can tunnel through in a slow speed (Fig. 11.71).

If the cave is not filled, we shall place H150 steel or I16 I-beam on the back of the ring beam and weld it, or stuff the cave with folded ϕ8 mm wire mesh and ϕ22 mm reinforcement. Mesh shall be hung and shotcrete shall be used in this place, and the thickness of the shotcrete shall be flush with the ring beam. When the concrete strength meets the requirements, the TBM can tunnel through in a slow speed.

c. Dissolution and caves are at the bottom of the tunnel.

In the process of TBM tunneling, the prediction of the surrounding rocks of the tunnel face shall be strengthened, and the possibility and scale of the karst cave at the bottom of the tunnel shall be inferred in combination with the geophysical prospecting geological forecast. Then, the emergency pump station and pipeline shall be used to go deep in front of the cutter head and backfill the karst cave at the bottom of the tunnel, so as to realize the technical treatment measures of slowly advancing while backfilling. When the caves are at the bottom of the tunnel, the ring beam is longitudinally connected with I16 I-beam at the bottom to prevent uneven

Fig. 11.71 Karst caves and cavities at the gripper position

settlement of the ring beam and the track panel to ensure the safety of the support and locomotive operation.

③ Dissolution, cave cavity backfill and grouting construction.

a. Connect backfill and grouting pipelines.

The backfill and grouting pipes shall be embedded and installed in advance behind the shield using the bolt drillers, and the grouting pipes are connected with the small conduit by threading. The grouting casing is equipped with an air outlet pipe and a slurry inlet pipe, and the switch is controlled by a valve. Install 20 mm plastic pipe as air exhaust pipe, connect various pipelines such as grouting pipe, and use anchoring agent and cotton cloth to seal the pores of the shotcrete surface to prevent slurry leakage. After the pipeline is connected, a grouting test shall be performed to check and determine the effectiveness and reliability of the grouting equipment, pressure gauges and pipelines, and the backfill and grouting pipeline must be extended to the position of connection bridge and as far forward as possible.

b. Backfilling and grouting.

After the initial support surface is shotcreted and closed, backfilling and secondary grouting reinforcement shall be carried out at the connection bridge in time. We shall use the grouting pump to pour grouting into the hole from low to high. Before grouting, the drilled grouting holes shall be temporarily blocked to prevent grout spilling and ensure the grouting pressure. After the designed final pressure is reached and grouting continues for 10–15 min, stop grouting (Fig. 11.72). If the pressure rises

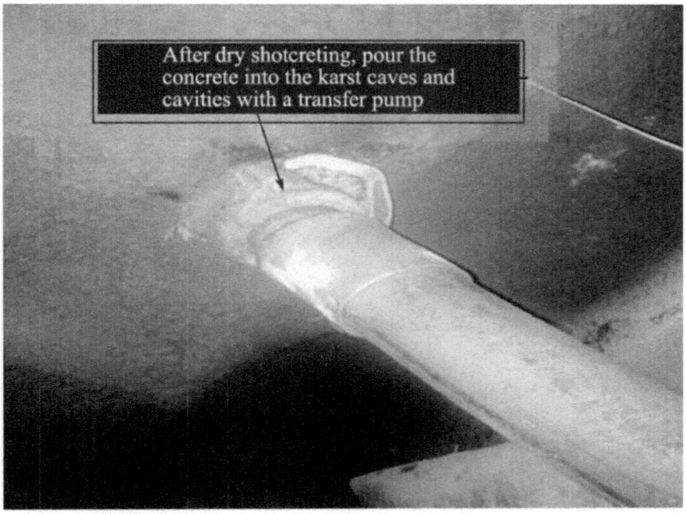

Fig. 11.72 Concrete backfilling at the vault

suddenly during grouting, the pipe may be blocked and the machine shall be stopped for inspection.

c. Hole sealing.

Seal the hole with a grout excluder valve, connect the grout excluder valve with the grouting pipe, close the valve after the single-hole grouting is completed, and remove the valve after the grout is solidified.

④ Monitoring and measurement.

a. Layout of measuring points

According to the surrounding rock grade, tunnel face excavation method and other factors, determine the number of measuring points and lines, and monitor and measure the peripheral displacement and vault sinkage. Monitoring and measuring points shall be densified in sections with large karst caves and large collapse caves on the vault. 1–3 vault sinkage measuring points shall be arranged on each section, and the measuring points shall be placed in or near the center of the vault. The layout of measuring points is shown in Fig. 11.73.

b. Buried.

The measuring piles at each point shall be buried in the cross section of the same mileage. Each measuring point shall be buried as close as possible to the working face. The initial reading shall be taken within 12 h after the surrounding rock is exposed, and shall not exceed 24 h at the latest. The initial deformation value reading must be completed before the next cycle of excavation.

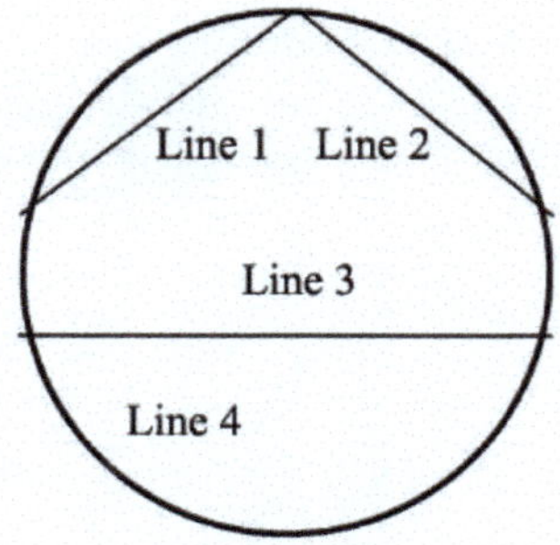

Fig. 11.73 Layout of arch monitoring and measuring points

Measurement of vault sinkage and clearance horizontal convergence shall be carried out in the same section, and the sinkage shall be measured by total station. When the geological conditions are complex, the sinkage or bias pressure is obvious, in addition to the measurement of the vault sinkage, the hance sinkage and basement uplift shall also be measured.

For the protection of measuring points, corresponding marks are required after the monitoring and measuring elements are buried to reduce damage during construction. For some monitoring and measuring points damaged in special sections (with weak surrounding rocks and fast displacement), the elements shall be buried again in time and the measuring frequency shall be increased.

(3) Construction technology of TBM passing through fault fracture zone

When TBM tunnels through the limestone sections, 32 fault fracture zones have been discovered, of which 2 are greatly affected, with the maximum range of 200 m, and there is another fault superimposed with carbonaceous slate. The top of this section collapses seriously, with the deepest collapse cavity of about 6.5 m. It is difficult to support on site and there are a lot of muck need to be cleared. Several symposiums have been held for the project in this section to discuss countermeasures, and the main countermeasures are as follows:

In the process of TBM tunneling, adverse ground conditions such as fault fracture zone are encountered many times. During the on-site construction, the "principles of preventing the development of free face, preferring strong support to weak support, short advance cycle, strong support and frequent measurement" are formulated in combination with the existing resources of the Yinsong and the opinions of all parties involved in the construction.

① Advance geological forecasting.

We will make advance geological forecast detection and increase the frequency of advance geological forecast during tunneling to determine the edge, length, crushing degree and water content of the fracture zone, and adjust the tunneling parameters, attitude and emergency material reserve in the caves in advance according to the results of the advance forecast.

② Tunneling parameters.

Table 11.21 For value range of tunneling parameters in different geological conditions

Geology	Advance speed (mm/min)	Cutter head torque (kN·m)	Cutter head rotation speed (r/min)	Penetration (mm/r)	Thrust pressure (bar)	Gripper pressure (bar)
Carbonaceous slate	10–26	400–3800	1.8–4.5	3–10	60–120	194–254
Fault fracture zone	52–70	1630–3360	4.8–5.9	9–13	112–165	255–294
Normal tunneling section	62–80	2200–3310	6–6.7	9–13	134–235	273–323

When the TBM passes through the fault fracture zone, the advance speed, cutter head rotation speed, advance thrust, gripper pressure and other advance parameters of the TBM shall be reduced appropriately (Table 11.21), so as to effectively reduce the disturbance to the surrounding rocks, thus reducing or avoiding collapse.

③ Strengthen the support.

a. Shield exit position: For the general sections with broken rocks, use the steel slat, ring beam, connecting bars, shotcrete and other joint supports. For the severely broken section, use the steel slat, densified ring beam or improved ring beam, and replace the connecting bars, welding support on the back of the ring beam and other joint supports (Figs. 11.74 and 11.75) with I-beams. If necessary, use the dry shotcreting system to seal this section in advance. For the position with a large range of cavity collapse, fill it by pouring or shotcreting, and carry out grouting to ensure compactness.
b. Position of gripper: If there are falling blocks or cavity collapse at the position of the gripper, and no reaction force of the gripper is provided, the following three methods can be adopted: Adding square timber at the position of the gripper; hanging mesh and using the dry shotcreting system at the position of the gripper for sealing; pouring concrete to fill it tightly (Figs. 11.76 and 11.77).

④ Monitoring and measurement.

After the reinforced support is completed, the measuring points will be set up in time, and the data of deformation, convergence and rate change will be fed back to the technical staff of the engineering department. The technical staff will decide whether to take reinforced support measures according to the feedback data.

(4) Construction treatment technology of TBM in case of mud burst and water ingress

According to the design geological data, nearly 7 km cave section of the four lot (K63 + 964–K70 + 823 m) of Yinsong water supply is a tuff section. The lithology

Fig. 11.74 Steel slat support

Fig. 11.75 Welding supports on the back of ring beam

Fig. 11.76 Adding square timber at the position of the gripper

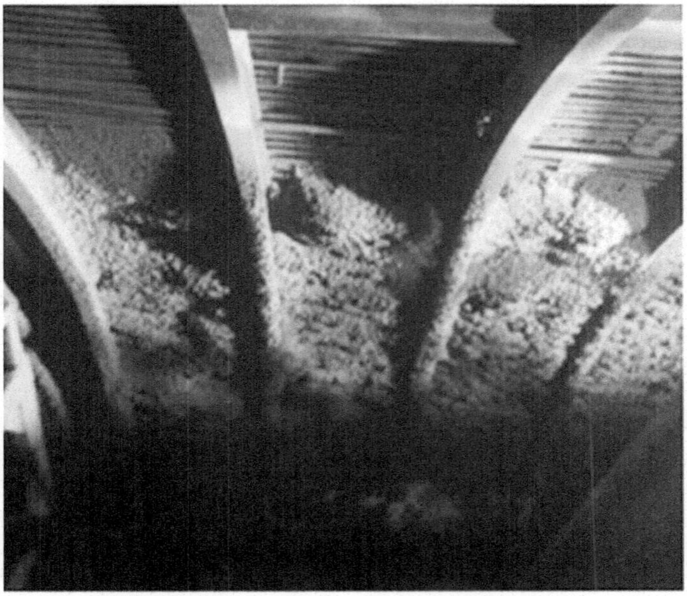

Fig. 11.77 Spraying treatment at the position of gripper

of the strata passing through is mainly devonian and carboniferous micrite limestone. There are many fault fracture zones, low resistance abnormal zones and karst caves in the limestone section, and some shallow gully sections are passed through, with the risk of mud burst and water ingress. On February 29, 2016, when tunneling through the 7 km limestone section K66 + 348, water burst in front of the tunnel face and mud burst on March 24. The amount of water ingress reached 950–1000 m^3/h, which made it difficult for support workers to install ring beams at the bottom. The rock ballast in front of the cutter head flows out with water (Fig. 11.78), the continuous belt system slips, and the TBM traveling and transportation tracks are flooded and cannot be extended, resulting in flooding at the exit of the tunnel opening (Fig. 11.79). The materials cannot be transported normally, and the TBM is forced to stop tunneling, resulting in difficult construction in March and April.

In response to the above-mentioned dangers, the Project Management Office of the 4th bid section of Yinsong Water Tunnel has taken decisive and effective measures to minimize the downtime of TBM, thus achieving a major breakthrough. The resulting rescue ideas and the main measures taken are shown in Fig. 11.80.

a. Continuously improve the emergency drainage system, increase the investment of emergency supplies to ensure that the rescue is carried out quickly.
b. Sort out shaft escape passages, improve safety and rescue measures (such as escape rope ladders, escape jackets, etc.).

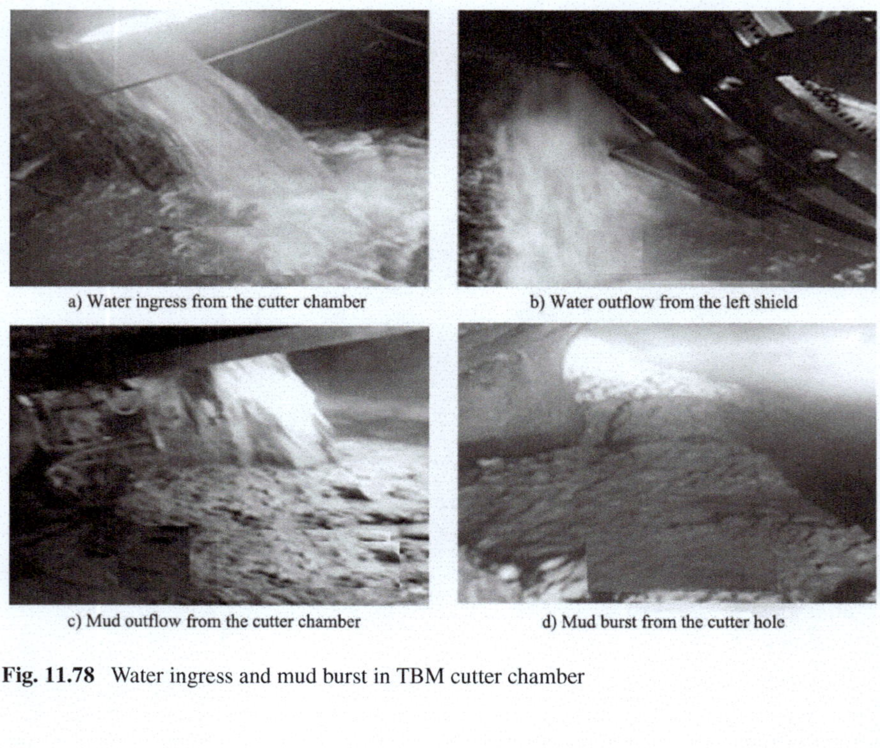

a) Water ingress from the cutter chamber
b) Water outflow from the left shield
c) Mud outflow from the cutter chamber
d) Mud burst from the cutter hole

Fig. 11.78 Water ingress and mud burst in TBM cutter chamber

Fig. 11.79 Flooding at tunnel exit

Fig. 11.80 Countermeasures for Construction in mud burst and water ingress section

c. Monitoring and control of safety, tunnel deformation, water ingress, stability of surrounding rock of tunnel face, surface sinkage and collapse shall be carried out throughout the process.
d. Find out the geological conditions of the surrounding rock ahead to prevent the further expansion of water burst.
e. Combined with the results of the geological forecast, the organization will hold an expert meeting to study the next step of the pre-control program and the treatment program of water ingress and blocking, and resume tunneling according to conditions.

Specific countermeasures mainly include measures before, during and after tunneling.

① Before tunneling.

a. Advance geological forecasting.

On the basis of the daily advance geological forecasting by geophysical prospecting, the joint forecasting methods such as borehole forecasts and surface cross-hole CT are continuously applied at the tunnel face. Accurately predict whether there are construction risks such as water burst and collapse within a certain range in front of the tunnel face, timely feedback information, identify construction risk sources and adjust tunneling parameters in time to guide follow-up work.

b. Cleaning or consolidation of cutter head.

Welded steel plates in the cutter head hole and the muck hole shall be partially blocked to reduce the muck discharge quantity, and the cutter head shall be idled before each tunneling to clean up the slurry on it. If the mud gushing amount in front of the cutter head increases, chemical grouting is adopted to consolidate mud gushing and crushed rock mass.

② During tunneling.

The manual mode is adopted for tunneling, and the advance speed, cutter head rotation speed and tunneling thrust shall be reduced at the same time to reduce the mucking amount and avoid belt blocking or emergency stop caused by large mucking amount. In the process, the diversion, drainage, blocking and drainage methods are used for the leakage point to avoid water burst and large water ingress.

③ After tunneling.

a. Drainage.

The water gushing from the working face or the remaining water after grouting is drained away from the working face in time. The leakage of water on the sidewall is covered and drained to ensure the quality of shotcrete. After spraying concrete, due to the rise in water pressure, damage to primary support is likely to happen, and drainage measure or backfill grouting is used for sealing.

b. Strengthen the support.

Strengthen the support in time, and take joint support measures such as densified ring beam (i.e. 45 cm/truss), steel slat, I-steel longitudinal connection, shotcrete sealing, etc.

4. Summary

For the limestone karst strata, we can determine the conditions of surrounding rocks in front of the tunnel face by advance geological forecasting, adjust and optimize the tunneling parameters in time during tunneling, adjust the support measures according to the actual situation of surrounding rocks after the shield is exposed, and carry out backfilling and grouting in time after the initial support is completed, thus reducing the construction risks such as low cutter head and jamming during the TBM tunneling, and reducing the safety risks such as initial support deformation and damage, and vault collapse, and making long-distance tunneling feasible when tunneling in the limestone stratum with open type TBM.

For the construction section with extraordinarily large water ingress and mud burst where the TBM of four lot of Yinsong Water Supply passes through limestone of 7 km, appropriate disposal measures were taken before, during and after TBM tunneling, so that the TBM successfully passed through this dangerous section. We minimized property loss and there is no casualty, and we also bought valuable time for timely completion. At the same time, the measures taken when open type TBM passing through the construction section with water ingress and mud burst, will also provide a reference for the construction of similar projects.

11.1.9.1 Targeted Design of TBM in Gaoligong Mountain Tunnel

1. Project Overview

(1) Tunnel Project Overview

The new Dali-Ruili railroad is located in the western region of Yunnan Province, east from the end of the Dali station of vast railroads, west to Ruili, and the total length of the line is about 330 km. Gaoligong Mountain Tunnel, a key project that restricts the completion of Dali-Ruili Railway, has a total length of 34.5 km. It is the first extra-long single-track railway tunnel under construction in China and the seventh longest tunnel in the world.

The terrain and geological conditions of Gaoligong Mountain Tunnel are extremely complex. The tunnel entrance section is connected to Nujiang Bridge, and the terrain is relatively steep. The exit section is close to the Longchuan River, and the terrain is relatively wide. In the tunnel, there are crisscross valleys on the surface, with great topographic relief, alternating mountains and rivers. The surface elevation is 640–2340 m, the relative height difference is about 1700 m, and the topographic relief is great.

The Gaoligong Mountain Tunnel has the geological characteristics of "three highs" (high geothermy, high ground stress, high seismic intensity) and "four active" (active neotectonic movement, active geothermal water environment, active external dynamic geological conditions, active shallow surface transformation process of the bank slope), which brings great challenges to the construction of this project.

The exit section of Gaoligong Mountain Tunnel is proposed to be constructed by large diameter TBM in the main tunnel + small diameter TBM in the pilot tunnel. The TBM excavation diameter of the main tunnel is 9.03 m, the total tunneling length is 12.37 km, the maximum slope is −9%, and the maximum buried depth is 1155 m. Among them, 2 sections with a total length of 300 m are constructed by drilling and blasting, and then passed by stepping. 2 sections with a total length of 140 m are expanded excavation sections, and the expanded excavation diameter is increased by 10 cm. The excavation diameter of TBM in pilot tunnel is 6.39 m, and the total tunneling length is 10.18 km, including 2 sections with a total length of 180 m, which are constructed by drilling and blasting, and then passed by stepping.

The purpose of designing the small diameter TBM for construction in pilot tunnel is firstly to use its probe to explore the geological conditions for the large diameter TBM in the main tunnel. Secondly, the two major faults of Laodongpo and Guanglinpo can be quickly reached and treated by drilling and blasting. When the large-diameter TBM arrives, it can pass the faults by stepping. The construction risks and the latency time of TBM shutdown are also reduced. Finally, the tunnel can be connected with the pilot tunnel in the exit direction of No. 2 shaft as soon as possible and the difficulty of shaft construction can be reduced. The construction layout of the Gaoligong Mountain Tunnel is shown in Fig. 11.81.

(2) Geological characteristics of the TBM section of the tunnel

In the TBM construction section at the exit of Gaoligong Mountain Tunnel, the surface is sporadically covered with Quaternary Holocene landslide deposit, diluvial-colluvial, alluvial-proluvial, diluvial-proluvial, diluvial and eluvial, and Upper Pleistocene alluvial-proluvial soft soil, silty clay, coarse sand, gravelly sand, fine round (angular) gravelly soil, coarse round (angular) gravelly soil, crushed stone, cobbly soil, boulder soil, block stone soil and other strata. Underlying Upper Tertiary, Yanshan Stage granite, mixed granite of unknown age, diabase veins and fault breccias, crushed rocks and alterated rocks in fracture and fault fracture zones of various periods. The geological longitudinal section is shown in Fig. 11.82.

The strata crossed during TBM construction are mainly Yanshan Stage granite (8.81 km, 73%), schist, slate, phyllite mixed with quartzite and metasandstone (1.44 km, 22%). The surrounding rocks are mainly of Class III, but surrounding rocks of Class IV and Class V account for up to 40%. It is predicted that the normal

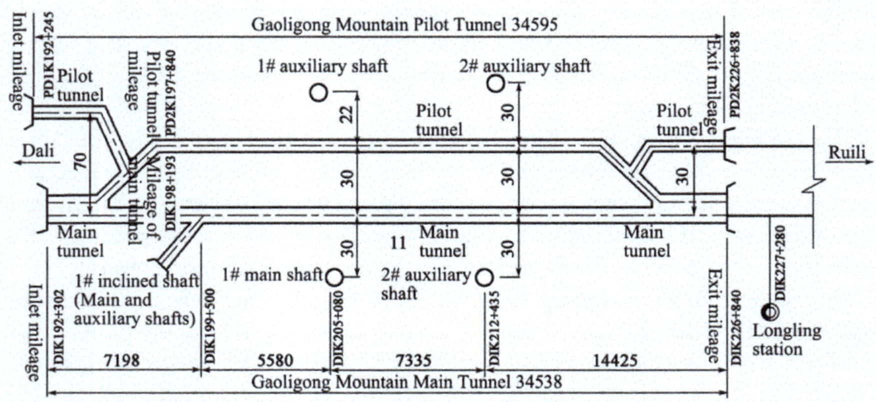

Fig. 11.81 Construction layout plan of Gaoligong Mountain Tunnel (Unit: m)

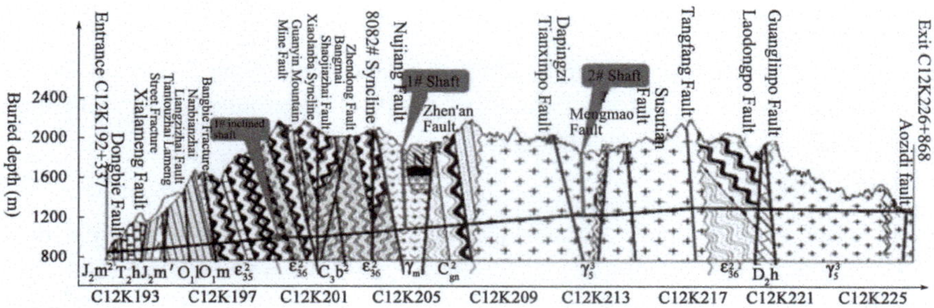

Fig. 11.82 Geological longitudinal section of Gaoligong Mountain Tunnel

water ingress of the main tunnel is 127,700 m^3/d and the maximum water ingress is 192,000 m^3/d. The lithology and main parameters of the surrounding rock in the TBM tunneling section are shown in Table 11.22.

(3) Analysis of key and difficult points based on TBM construction of the project

Table 11.22 Lithology and main parameters of surrounding rock in TBM Tunneling section

Serial number	Stratigraphic lithology and geological structure	Length (m)	Wear resistance (1/10 mm)	Uniaxial saturated compressive strength (MPa)	Intactness coefficient
1	Yanshan Stage granite	8810	3.34–3.59	46–65	0.25–0.85
2	Dolomite, limestone mixed with quartz sandstone	290	3.34	65	0.69
3	Class V geophysical prospecting anomaly zone	840	3.34–5.18	8–40	Contains 2 faults
4	Limestone, dolomite mixed with quartz sandstone	460	3.34	40	0.45–0.85
5	Schist, slate, phyllite mixed with quartzite, metamorphic sandstone	1440	3.34	20–40	0.45–0.85
6	Alterated rocks	140	3.34	4.6–15	i.e. expanded excavation section

① Adverse ground conditions of the tunnel.

The adverse ground conditions in the TBM section of Gaoligong Mountain Tunnel include: high ground stress, fragmentation of stratum, high ground temperature, high seismic intensity, lava and water burst, etc.

a. High ground stress.

The maximum buried depth of the tunnel is 1155 m and the regional stress field is relatively high. The relationship of three-way main stress values near the tunnel trunk in the measurement area is: vertical main stress (about 31 MPa) > maximum horizontal main stress (20–29 MPa) > minimum horizontal main stress (13–19 MPa). The angle between the dominant direction of the maximum horizontal main stress and the tunnel route is 12°–21°, and rock burst and large deformation of soft rock may occur in the tunnel.

According to the prediction, the total length of the rock burst section of the main tunnel of Gaoligong Mountain Tunnel is 2020 m, including 1250 m medium rock burst and 770 m slight rock burst. The total length of large deformation section of soft rock in the main tunnel is 3185 m, of which the length of slightly large deformation section is 1435 m. In the fault fracture zone, local fault clay particles are prone to medium and large deformation, with a length of 1750 m.

b. Stratum fragmentation.

There are 20 sections in the TBM construction section of the main tunnel of Gaoligong Mountain Tunnel, with a total length of 1,280 m, which belong to sections with broken–extremely broken rocks. Among them, 2 sections, with a total length of 300 m, are constructed by drilling and blasting and then passed by stepping. There are 2 expanded excavation sections with a total length of 140 m. There are 15 sections in the TBM construction section of the pilot tunnel, with a total length of 980 m, which belong to sections with broken–extremely broken rocks, among them, 2 sections with a total length of 180 m are constructed by drilling and blasting and then passed by stepping. The percentage of IV and V surrounding rocks is as high as 40%, and TBM tunneling is not suitable for the broken strata.

c. High ground temperature.

The cross-mountain section of Gaoligong Mountain Tunnel is located in the Mediterranean-South Asia geothermal anomaly zone, and is a regional high heat flow area. There are 123 hot spring groups exposed in the mountain-crossing section, including more than 20 thermal springs, hot springs and boiling springs closely related to the line.

12.44 km in the middle of the tunnel is the section with abnormal rock temperature, and its longitudinal temperature distribution is 28–50 °C. The temperature of the rock temperature anomaly section is high in the middle section and low at the two ends. Among them, the length of the middle section at 37–50 °C is 700 m, the length of the rock temperature anomaly section at the entrance section at 28–37 °C is 2.9 km, and the rock temperature anomaly section at the exit section (TBM section) at 28–37 °C is short.

d. High seismic intensity.

The tunnel area is located in the southwest of China, crossing the Xianshui River-East Yunnan Seismic Belt, Southwest Yunnan Seismic Belt and Himalayan Seismic Belt. Most sections of the tunnel are located in the Southwest Yunnan Seismic Belt, and a small part of sections are located in the Xianshui River-East Yunnan Seismic Belt. The regional earthquake is very active and frequent. The peak ground acceleration of the survey area is 0.20 g, and the period of the seismic response spectrum is 0.45 s.

e. Lava and water burst.

The lava strata that the tunnel passes through are mainly limestone, dolomite sandstone, dolomite limestones with quartz sandstone in the middle of the tunnel. The karst caves cannot be seen from the borehole at tunnel exit (TBM section), and only part of denudation pores of the core can be seen. The karst is weakly developed. According to the prediction of the water burst, TBM will cross the sections where water may suddenly burst.

② Tunnel construction risk analysis.

In view of the special geological conditions of "three high and four active" in Gaoligong Mountain, conventional TBMs cannot meet the construction requirements of this project. Through the risk analysis of TBM construction in sections with adverse geological conditions, we propose a highly adaptable TBM design solution.

The possible construction risks of TBM construction in Gaoligong Mountain Tunnel include: TBM jamming in tunnel section with weak and broken surrounding rock, rock burst in surrounding rock and palm face caused by high ground stress, support damage caused by convergence and extrusion deformation of surrounding rock, high-pressure water burst in TBM construction, and thermal damage to equipment and personnel caused by high-temperature thermal damage.

a. Risk of TBM jamming in sections with weak and broken surrounding rocks.

There are 20 sections in the TBM construction section of the main tunnel, belonging to sections with broken–extremely broken rocks, 1280 m in total. There are 15 sections in the TBM construction section of the level pilot tunnel, which belong to sections with broken–extremely broken rocks and are not suitable for TBM tunneling sections, 980 m in total. In the sections with alterated rocks and dense granite joints or fault breccia, the rock is extremely broken, and is hard to realize self-stabilization and easy to collapse. There is a risk of TBM getting stuck during construction.

In the sections with broken rocks, the strength of the rock is low and most of the groundwater is abundant, which makes the TBM easy to collapse after tunneling, thus forming a great resistance to the rotation of the cutter head. TBM is used for construction in the broken surrounding rock stratum, except for the Laodongpo fault and Guanglinpo fault. There are 150 m in each of these two sections that need to be treated in advance by drilling and blasting, and the rest of the tunnel sections are all dealt with tunneling. At present, during the domestic construction process, TBM is often stuck. Once the TBM is stuck, the construction progress will be affected, causing safety and quality hazards, and may even damaging the TBM equipment and causing significant economic losses.

b. There is a risk of rockburst in high ground stress surrounding rock and tunnel face.

Due to the high ground stress, there is a risk of rockburst during TBM construction in tunnel of the Gaoligong Mountain. In order to prevent and control rock burst in TBM construction, the rock burst encountered in TBM construction can be divided into four levels according to the strength of the rock burst in the cutter head area and the L1 area. This grading methods is different from the grading of rock burst in the "Code for Water Resources and Hydropower Engineering Geological Investigation (GB 50,487–2008)", as shown in Fig. 11.83.

Minor level of rock burst: Sound, falling small fragments; slight burst and impact inside the rock and crackle. The rock burst of this level has little impact on TBM construction, but will increase the amount of muck to be cleared at the bottom of the tunnel.

a) Minor rock burst b) Primary rock burst c) Medium rock burst d) Violent rock burst

Fig. 11.83 Grading of rock burst in TBM construction

Primary level of rock burst: Due to the pressure, the thickness of flakes fell from the rock were 5 cm, and there are dust clouds. The rock burst of this level is destructive to the initial support of the tunnel, resulting in deformation of the wire mesh and steel slat.

Medium level of rock burst: In the area without support, rocks in the tunnel soffit or tunnel face fall or crack due to stress. There are large irregular rocks fall, accompanied by a lot of noise. The rock burst of this level occurring in front of the cutter head may result in abnormal damage to the cutter; occurring in the shield position may result in damage to the machine and injury to personnel.

Severe level of rock burst: Loud noise accompanied by rock burst; rocks suddenly fall accompanied by loud noise in the tunnel soffit (both in support and unsupported areas); the rock burst of this level may result in damage to the support material (shotcrete cracks, broken top rock bolt, ring beam deformation, etc.).

c. Risks caused by convergence and extrusion deformation of surrounding rocks, TBM jamming and support damage.

During TBM construction, in tunneling sections of shist, slate, phyllite mixed with quartzite, and metamorphic sandstone, as well as the Lisu Tian fault, Tangfang fault section, the soft rocks may be extruded and deformed. The total length of soft rock deformation of the tunnel is expected to be 3185 m, including a slight deformation of 1435 m, and moderate deformation of clay particles in the local fault of fault fracture zone (1750 m). The maximum length of surrounding rock convergence deformation is predicted to be 50 mm.

In tunnel construction, the original equilibrium of the rock is changed after excavation and the rock is exposed to the air, so it will be relaxed and deformed. The self-supporting capacity of the soft rock section is not enough, so this section is more prone to large deformation. If the surrounding rock deforms too fast, there may be a risk of the TBM shield structure being stuck. If the support is not timely, support strength or rigidity is not enough, the initial deformation amount will be too large, exceeding the reserved deformation amount, resulting in the beam replacement before lining to ensure the thickness of the secondary lining. Under some high ground stress conditions, the soft rock still undergoes continuous and slow deformation for a long time even after the secondary lining is completed, resulting in cracking of the secondary lining and intrusion of the structure into the limits, and needs to be reworked for treatment.

d. Risks of high-pressure water burst during TBM construction.

According to the design, the predicted normal amount of water ingress in the tunnel is 127,700 m^3/d and the maximum amount of water ingress is 192,000 m^3/d. During TBM construction, high-pressure water burst mainly leads to TBM stop for drainage due to the large amount of water ingress, and the equipment will be flooded if the drainage is not timely.

e. Risk of high-temperature thermal damage to equipment and personnel.

The cross-mountain section of Gaoligong Mountain Tunnel is located in the Mediterranean-South Asia geothermal anomaly zone, and is a regional high heat flow area. There may be high temperature and high pressure hot water (steam) and high temperature rock and other thermal damage problems. Due to the high temperature, the efficiency of personnel and construction equipment is more likely to be reduced, and the quality of the engineering entities will be directly affected. Failure to take effective measures for prevention and control of the high temperature during construction may lead to personal injury or safety and quality accidents, thus affecting the construction period and cost.

2. TBM Targeted Design

(1) Design of the cutter head variable-section lifting excavation system

In view of the large deformation stratum of soft rock in Gaoligong Mountain, we have different requirements for the lining thickness of local sections, and have technical requirements for variable sections in the excavation section of the tunnel using TBM methods, which require the cutter head and cutting system of the TBM have the function of expandable excavation. Currently there are two main ways for gripper TBM in construction of variable sections.

① Replace the side section of cutter head.

Block form of 4 + 1 shall be used in the design of cutter head, and the space for transporting the new side sections to the main machine section shall be reserved in the backup design. When it is necessary to expand the variable sections, firstly, manually expand the cavity in the area of the cutter head, then remove the side sections and transport them outside the cavity in turn by lifting slings and lifting points, and then mount the new side sections to expand the variable section.

Advantages of this method: a large variable-diameter range and a large direct variable section. Disadvantages of this method: It is necessary to prepare side sections for variable sections, which is costly. Because it is impossible to prepare one set of each series of side sections of various diameters, the size of the variable diameter is limited. The replacement of side sections in the tunnel is complicated and the cost is high. The bottom shield needs to be adjusted for a wide range of expanded excavation, which is time-consuming and laborious.

② Add cushion blocks to the edge cutter.

By adding cushion blocks to the edge cutter of the cutter head and moving the edge cutter outward, the cutter head can be resized. During normal tunneling, upward tunneling of the main machine shall be kept so that the "bowing" of the main machine will not happen due to expanded excavation at the bottom.

Advantages of this method: The operation is relatively simple and flexible. Only the edge cutter needs to be reinstalled, and no manual extensive expansion of the tunnel is required. It is economic. Only a set of pads is required, and less manpower and time are required for expanded excavation. Disadvantages of this method: Due to the small range of variable sections limited by the structure of the pad and cutter housing, the tunnel can only be expanded by 50 mm (radius direction) at the maximum.

For the above two kinds of variable sections, there are limitations to using TBM to expand excavation: Either the operations are complex, time-consuming and laborious, or the amount of variable section expansion is small. Therefore, the TBM prototype in this study combines the advantages of the above two expanded excavation methods and the structural features of the 9.03 m cutter head, and is optimized on the basis of the second expanded excavation method. Minimize the complexity of the operation while increasing the expanded excavation amount of cutter head.

This TBM combines the reserved cutter housing + cutter cushion block + 20′ disc cutter + offset disc cutter, which can achieve expanded excavation of 50–100 mm in the radius direction. To prevent the machine from sinking after the bottom is expanded, a synchronous lifting mechanism is added between the bottom shield and the drive box to ensure that the bottom of the tunnel will not be over-excavated.

③ Expanded excavation design of TBM in Gaoligong Mountain.

a. Design scheme of cutter head diameter expansion.

When the radius of the cutter head is expanded by 50 mm, only No. 57–62, CW1 and CW2 gauge disc cutter need to be adjusted, and other disc cutters do not need to be adjusted, as shown in Fig. 11.84. For the cutter seat requiring additional gaskets, after the L-shaped block is removed, the surface of the cutter seat and the surface of the L-shaped block shall be cleaned before reassembly.

When the cutter head is expanded by 100 mm in the radius direction, the installation of No. 56–62, CW1 and CW2 gauge disc cutters needs to be adjusted, while other disc cutters do not need to be adjusted. No. 56, No. 57 and No. 58 disc cutters can be raised by adding gaskets, and after the L-shaped block is removed, the surface of the cutter seat and the surface of the L-shaped block shall be cleaned before reassembly. No. 59, 60, 61, 62, CW1 and CW2 disc cutters can be raised by replacing the L-shaped block. The 20in disc cutters used in Fig. 11.85 are all 19in cutter bodies with 20in cutter rings, of which CW2 uses 20in disc cutters with an eccentricity of 40 mm, as shown in Fig. 11.85.

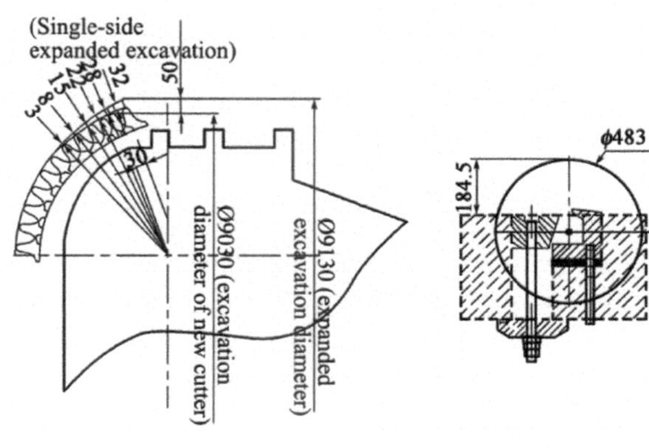

Fig. 11.84 Implementation plan of single-side expanded excavation of 50 mm (Unit: mm)

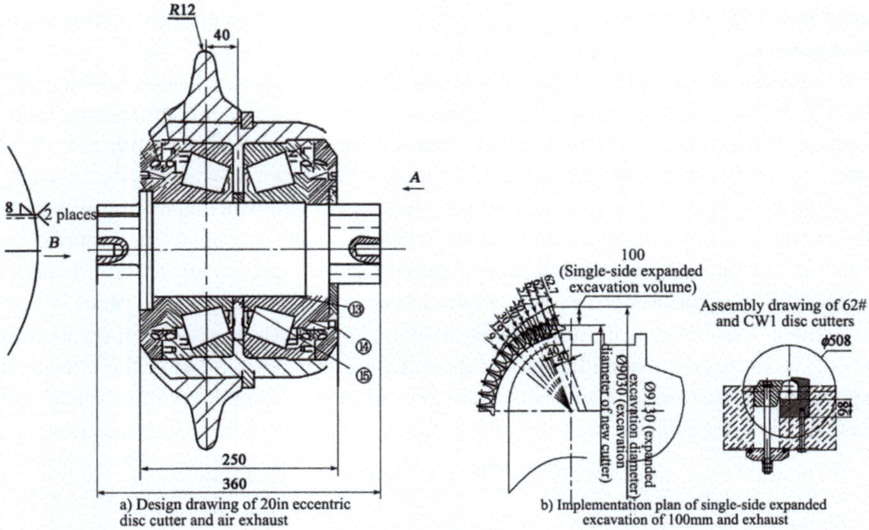

Fig. 11.85 Implementation plan of eccentric disc cutter and single-side expanded excavation (Unit: mm)

b. Design scheme of drive lifting.

To achieve continuous long-distance expanded excavation of the TBM, the cutter head diameter needs to be variable. At the same time, the drive unit shall be lifted to prevent the spacing between the bottom shield and the surrounding rock after the cutter head diameter becomes larger. This TBM bottom shield is rigidly connected to the drive box steel structure by bolts. If excavation with variable diameter is required, first remove the connecting bolts between the bottom shield and the drive box steel structure, lift the drive box steel structure to a certain height with the lifting cylinder, fill steel plates with corresponding thickness between the drive box steel structure and the bottom shield, and finally rigidly connect the drive box steel structure and the bottom shield again. The drive lifting device is shown in Fig. 11.86.

(2) Normalized front-mounted L1 zone shotcrete design

① Existing TBM construction plan for sections with soft and weak surrounding rocks and the risk of rockburst.

a. Improve the operational efficiency of rock bolts.

Due to the limitation of the cycle tunneling time, the installation speed should be accelerated, the consolidation effect should be improved, the anchoring requirements should be achieved quickly, the tunneling time should be shortened, and the tunnel excavation speed should be accelerated.

b. Proper temporary support.

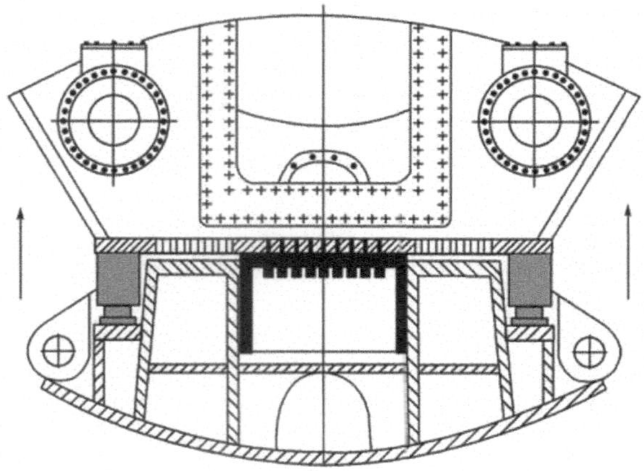

Fig. 11.86 Schematic diagram of drive lifting device

In poor sections, local dangerous rocks often make it impossible for TBM to tunnel normally, and if necessary, shed support is adopted.

c. Timely shotcrete.

In general, manual shotcrete equipment near the cutter head should be used as less as possible, in order to reduce contamination of the main machine and reduce TBM downtime. However, when passing through the weak geology and sections with severe rock explosion, especially when the excavation surface is broken, due to the slowdown of the advance speed, the time for shotcrete construction of excavation face and No. 2 platform manipulator is extended. Therefore, manual shotcrete has to be carried out, and the purpose is to close the excavation rock surface as soon as possible to reduce exposure time. Although the temporary shotcrete is only 20–30 mm thick, it forms a tightly-bonded composite with the rock in a very short period of time, thus preventing the rock from loosening, releasing pressure and bending during the normal loosening process. The resulting effect is to transform the rock with lower strength into stable rock and achieve the goal of stopping the chemical changes inside the rock and the weathering of the rock. Due to the use of manual shotcrete, pumping distance increased, posing strict requirements for the wet spraying material, and corresponding preparatory work shall be done.

Although certain results can be achieved with the traditional TBM construction method, there are still some shortcomings in the manual and timely shotcrete construction in the L1 zone, such as uneven concrete injection and limited injection range, which greatly affects the safety and timeliness of TBM construction. In view of this, based on the new TBM designed for the construction of the Gaoligong Mountain tunnel, the immediate shotcrete in the L1 zone was fully designed and studied.

② Shotcrete design for L1 zone of TBM in the Gaoligong Mountain.

When TBM passes through unfavorable stratum, the collapse and convergence of surrounding rock shall be reduced after it is out of the shield, and the wet shotcrete manipulator shall be designed in the main machine section to seal the surrounding rock in time. Through structural and spatial optimization, the arc rail and rack are added to the ring beam bracing mechanism. After gear-rack engagement of the traveling trolley, the wet shotcreting trolley is able to travel along the arc track. The wet shotcreting nozzle is installed on the wet shotcreting trolley, and the spacing between the tunnel arm and the nozzle can be adjusted.

The wet shotcreting material in L1 zone is transported to the two concrete transfer pumps on the right side of the connection bridge through the L2 zone mixed shotcreting pump, and then the material is pumped to the wet shotcreting nozzle in L1 zone through the transfer pump on the right side of the connection bridge. For wet shotcreting feeding in L1 zone, concrete is pumped by relay. The design scheme of immediate shotcreting device in L1 zone is shown in Fig. 11.87.

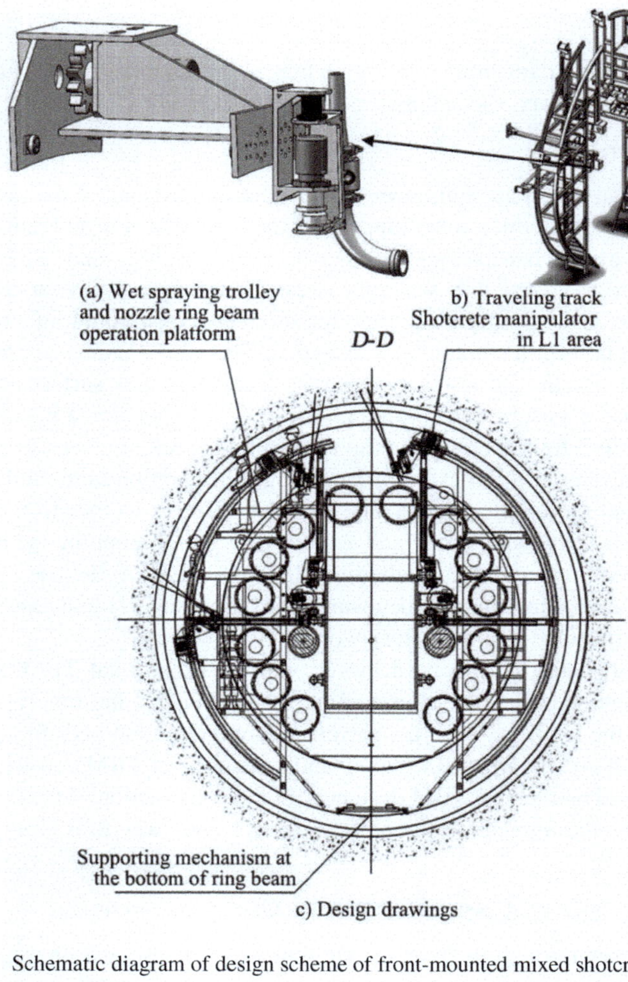

Fig. 11.87 Schematic diagram of design scheme of front-mounted mixed shotcreting device

(3) Design of reinforced rapid supporting system

TBM, featuring excellent performance such as high efficiency and safety, is increasingly applied in tunneling, but its adaptation to unfavorable geological conditions such as rockburst is poor. These unfavorable geological conditions will have a great impact on TBM tunneling. This may not only reduce the utilization ratio of TBM tunneling operation time, but also even cause damage to TBM and difficulty in passing smoothly. Effective measures must be taken, otherwise improper handling will bring serious consequences to the construction. Such accidents as TBM cutter head jamming have been encountered when TBM passes through the fault zone during construction of the headrace tunnel of Yunnan Tianshengqiao HPP, Shanxi Wanjiazhai Yellow River Diversion Project, Italy's Gossensa Railway Tunnel and Austria's Campolosso Railway Tunnel, leading to long downtime.

Therefore, as for the possible rockburst stratum during construction of the Gaoligong Mountain Tunnel, a new TBM with rapid supporting function is designed based on the traditional TBM. Rapid supporting is mainly concentrated at the TBM section, including systems such as steel slat supporting, wire mesh storage, ring beam erector, bolt driller, shotcret, etc. See Fig. 11.88 for the specific integrated supporting system.

① Steel slat supporting system

As for the steel slat supporting system of the new TBM, the space for steel slat storage is reversed within 250° of the shield structure, so that supporting measures can be taken timely once the rocks are beyond the range of the shield, to prevent stone falling (Fig. 11.89).

② Wire mesh storage system

See Fig. 11.90 for the wire mesh storage system.

The wire mesh storage system of the new TBM can achieve better supporting effect through effective combination with the steel slat supporting system. Specifically, the

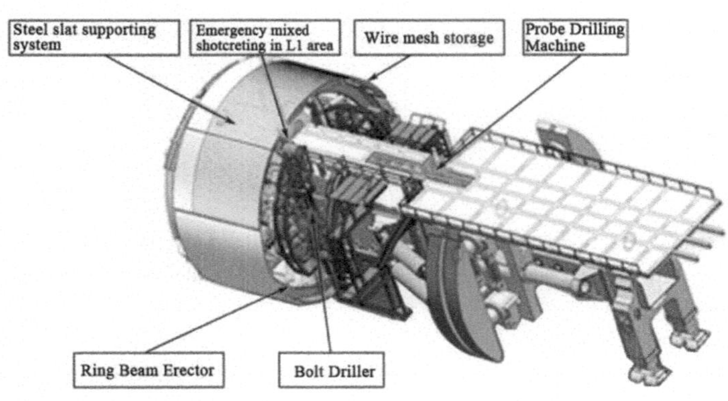

Fig. 11.88 Design scheme of integrated supporting system

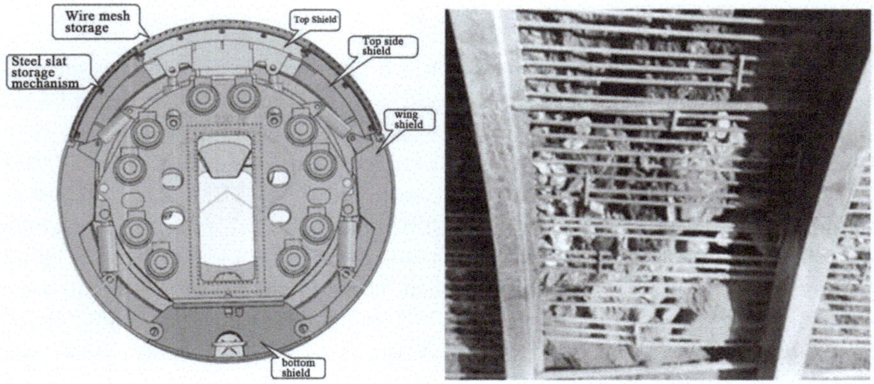

Fig. 11.89 Steel Slat Supporting Range and Effect

Fig. 11.90 Wire mesh storage system

space for wire mesh storage is reserved within 250°, so that supporting measures can be taken timely as appropriate once the rocks are beyond the range of the shield, buying time for subsequent procedures.

③ Ring beam erector

As for complicated geological conditions of the Gaoligong Mountain Tunnel Project, well-targeted improvement has been designed based on the ring beam used for the Songhuajiang River Diversion Water Supply Project according to the advantages of the ring beam erector.

The ring beam erector is installed behind the supporting shield, so that the ring beam can be erected timely through hydraulic control under the protection from the top shield. The ring beam erector can be used to install the ring beam divided in sections through lifting, rotary erection, in-place, longitudinal movement, top and lateral supporting and tensioning and sealing of bottom opening. The ring beam erector is mainly made up of the erection ring and supporting devices. Moreover, ring gears which can serve as sliding rails of the wet shotcrete manipulator in Zone L1 are designed on the ring beam. See Figs. 11.91 and 11.92 for the design drawings and specification parameters.

④ Bolt driller rapid supporting system

The traditional TBM bolt driller is distant from the shield during installation since affected by the ring beam erector, which will greatly affect the bolt-shotcrete supporting speed while encountering unfavorable stratum, increasing the risk to the construction personnel and equipment. The bolt driller can be moved forward as much as possible during design based on the optimized ring beam platform without interference from such platform. Bolt supporting can be made once the rocks are beyond the range of the shield for 1 m. Besides, two independent gear-ring track travelling systems are used for the bolt driller and the advance driller, which remarkably improves the working efficiency of the bolt driller, providing strong guarantee for rapid supporting during construction. See Fig. 11.93 for the design and installation drawings of the bolt driller.

(4) Design of concealed full-perimeter probe drilling machine system

Most tunnels for which TBM is used will encounter certain soft rocks as well as various extremely adverse geological conditions such as fracture zones, fault zones,

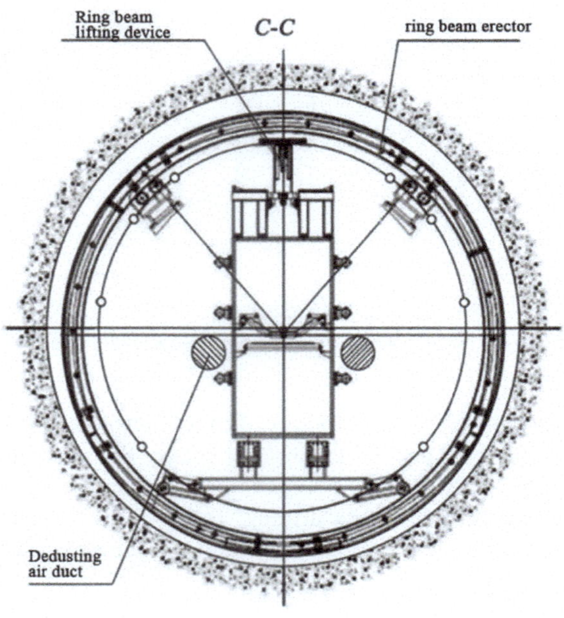

Fig. 11.91 Design drawings of erection rings of ring beam erector

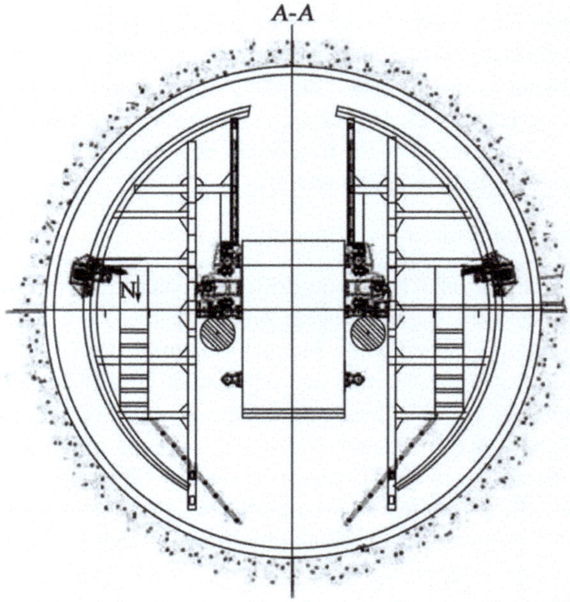

Fig. 11.92 Design drawings of supporting devices of ring beam erector

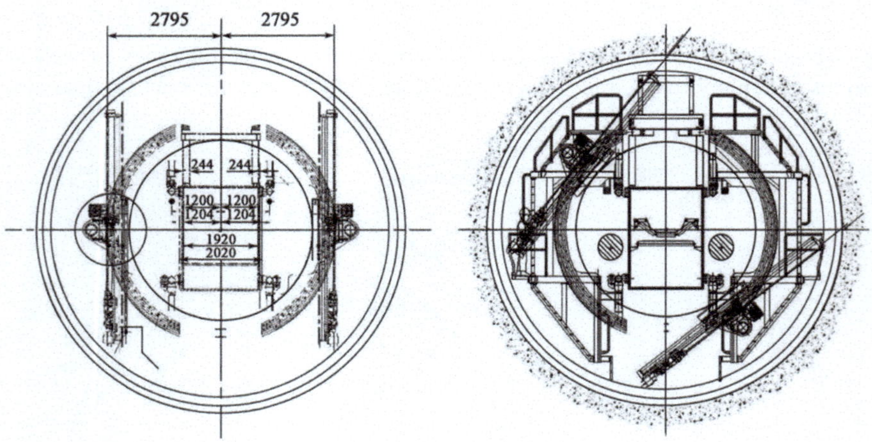

Fig. 11.93 Design and assembling drawings of bolt driller in zone L1 (Unit: mm)

pebble layers, etc. In this case, the rocks must be subjected to advance support before tunneling to prevent collapse of rocks. The current advance supporting methods mainly include probe bolts, canopy tubes and pre-grouting. The precondition of all these methods is probe drilling. To solve this problem, the conventional practice is to provide a probe drilling machine in Zone L1 of TBM.

① Previous cases of probe drilling machine

The Wirth TB880E open TBM is equipped with a TAMROCK HL500S probe drilling machine during construction of the Mogouling Tunnel of the Xi'an-Hefei Railway, achieving favorable effect.

a. Structure and installation of HL500S probe drilling machine

The hydraulic drilling machine is installed on the feed beam, and can move back and forth on it (Fig. 11.94). The feed beam is installed on an arc-shaped beam with guide rails, and the arc-shaped beam is fixed on the K1 platform. A hydraulic travelling motor is provided on the feed beam support, making the feed beam and the drilling machine move in a circular manner along the arc-shaped beam, so that the area within 83° at the upper half of the rock wall can be drilled. There is a pitch cylinder on the feed beam to regulate (0–8°) the pitch angle (the angle between the bore and the rock wall) of the drilling.

The hydraulic drilling machine can be installed with an eccentric bit for drilling with casing, or be installed with a ball-tooth bit for ordinary drilling.

b. Advantages of probe drilling machine

The probe drilling machine is characterized by a low fault rate, reliable performance and fast drilling since the adoption of advanced hydraulic drilling machines and hydraulic operation system. According to statistics on numerous drilling operations, the average speed of drilling with casing is 6.61 m/min, while that of ordinary drilling is 5.43 m/min, including the time for positioning, drilling and removal of drilling tools.

Since two drilling methods are available, especially drilling with casing, it can adapt to various adverse geological conditions. The tube used can be reserved in the hole upon completion of drilling with casing, which saves the time required to re-install the canopy tubes and avoids collapse of holes, thus remarkably improving the utilization rate of holes, achieving favorable supporting effect.

Only one driller and two personnel for installation and removal of drilling tools are required, which saves the labor forces and realizes favorable economic returns.

c. Disadvantages of probe drilling machine

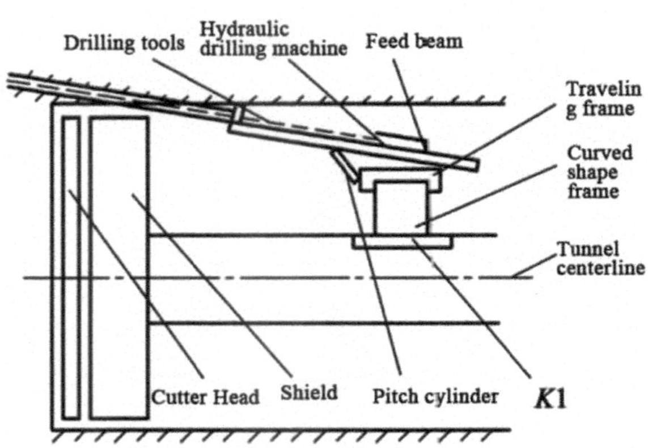

Fig. 11.94 Structure of probe drilling machine

Removal of drill rods: Such problem as failure in removing the drill rod occurs due to too tight engagement with the adapter upon completion of drilling during initial use of the probe drilling machine, which severely affects the operability of the machine.

Space squeeze: Installation of the bolt probe drilling machine in the limited working space of the TBM Zone L1 accounts for a large proportion of space, which remarkably narrows the working space of the bolt driller and the ring beam erector, bringing inconvenience for construction.

Limited working range of probe drilling machine: Since the feed beam and the drilling machine can only move along the arc-shaped beam in a circular manner in this design, drilling can only be made within 83° at the upper half of the rock wall, making construction beyond this range impossible, which remarkably narrows the use range of the probe drilling machine.

② Design of TBM probe drilling machine system for Gaoligong Mountain Tunnel

The design scheme of a new TBM concealed probe drilling machine is proposed based on previous problems in the TBM probe drilling machine during use according to the geological characteristics of the Gaoligong Mountain Tunnel.

See Figs. 11.95 and 11.96 for the design and structure of the concealed probe drilling machine. A probe drilling machine, a feed beam and a ring gear beam are designed below the main beam platform separately. The pump station of the probe drilling machine and the bolt driller system in Zone L1 are shared. The probe drilling machine is concealed below the main beam platform with one gear-ring track beam during normal tunneling, with two gear-ring track beams being fixed respectively on both sides. If the probe drilling machine is required, it will be lifted with the gear-ring track beam jointly for connection with the ring gear beam on both sides through the oil cylinder on both sides, forming an integral gear ring. The probe drilling machine may travel for one circle at an external angle of 7° along the ring gear beam for probe bolts, ductules or canopy tubes. It can also coordinate with the pre-grouting equipment in advance support of rocks in front of the cutter head. The probe drilling machine is linked to the base through bolts to facilitate removal.

(5) Design of high-ground-temperature forced refrigeration system.

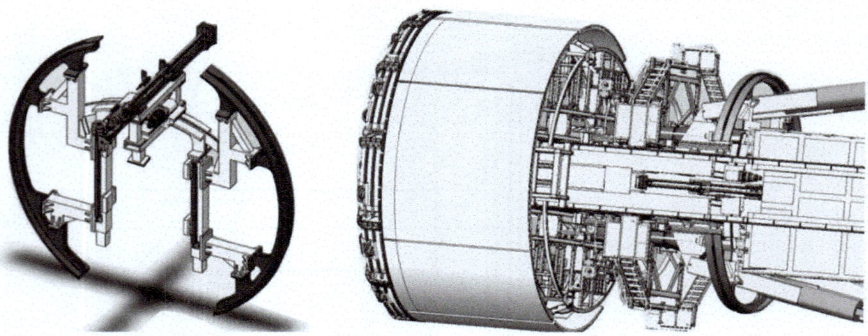

Fig. 11.95 Structure of concealed probe drilling machine

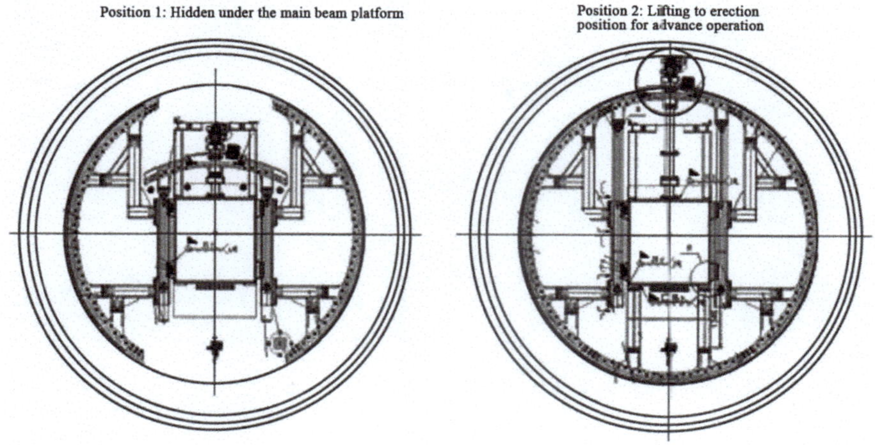

Fig. 11.96 Design of probe drilling machine

① Selection of refrigerating unit

Model: RTHE200 refrigerating unit (three sets);

Cooling capacity: 3 × (705–715 kW); performance factor: 4.8; input power (power consumption): 149.2 kW;

Refrigerant: R134a;

Size: 3601 mm × 1089 mm × 1905 mm;

Water inflow required for external circulation water (providing three specifications below);

a. 3 × 147.68 m^3/h; water inlet temperature: 30 °C; water outlet temperature: 35 °C;
b. 3 × 73.84 m^3/h; water inlet temperature: 25 °C; water outlet temperature: 35 °C;
c. 3 × 55 m^3/h; water inlet temperature: 25 °C; water outlet temperature: 38.5 °C;

Volume required for frozen circulation water: 3 × 120.77 m^3/h; water temperature: 7–12 °C.

② Air cooler

The processing capacity is 1824 m^3/min, and the processing capacity when the thermal power is 715 kW is required when the chilled water passing through the air cooler is 7 °C.

According to the tunneling length and relevant tunneling standards, the forced refrigeration equipment selected for the dedusting system, secondary ventilators, fresh and dirty air ducts are shown in Fig. 11.97.

3. Summary

As for the complicated geological conditions of the Gaoligong Mountain Tunnel, the TBM is designed in a well-targeted manner, realizing design of the cutter head

Fig. 11.97 Refrigerating unit and air cooler of new TBM

variable-section lifting excavation system for weak rocks, normal front shotcret in Zone L1, concealed full-perimeter probe drilling machine system, reinforced rapid supporting system and high-ground-temperature forced refrigeration system.

11.2 Double Shield TBM Project Cases

11.2.1 Headrace Tunnel of Datong River-Qinwangchuan Diversion Project

1. Overview of the project and the 30 A tunnel

The project is located in Yongdeng County, north of Lanzhou City, with the 312 National Highway and Lanzhou-Xinjiang Railway passing through the project area and Lanzhou Zhongchuan Airport in the irrigation area, which is convenient for the construction of the project.

The project is a large inter-basin irrigation project diverting water from the Datong River to Qinwangchuan in the north of Lanzhou. The total length of the main canal is 86.9 km, including 33 tunnels, with a total length of 75.11 km. Other buildings include 2 inverted siphons, 9 flumes, as well as canal buildings and open channels and other works. The design diversion discharge is 32 m^3/s, and the enlargement discharge is 36 m^3/s. The total irrigation area is 860,000 mu (1 mu $\approx$ 666.67 m^2, the same below). The #1 East Canal is 49.56 km long, with a design discharge of 14 m^3/s and an irrigation area of 310,000 mu, while the #2 East Canal is 54.12 km long, with a design discharge of 18 m^3/s, an enlargement discharge of 21.5 m^3/s and an irrigation area of 500,000 mu. The rest irrigation areas are located along the main canal. The main canal, the #1 East Canal and the #2 East Canal have a total length of 190.58 km, with 71 tunnels (110 km in total length), 45 branch canals (675 km in length), more than 200 buildings such as gates, head falls, bridges and culverts on the canal and branch canals as well as supporting works with an irrigation area of 57,300 ha. in the field. The total quantities of all earth-rock, concrete and masonry works are 27,400,000 m^3. The project is characterized by inter-basin water diversion, gravity irrigation, complicated geological conditions, great construction difficulties and high technical requirements. The #1 East Canal, the main canal and the #2 East Canal have been completed and put into service in 1993, October of 1994 and October of 1995 successively, and are now playing their roles. In total, 140,000 mu farmland has been irrigated up to 1995, and above 500,000 mu farmland can be irrigated at present. Among the many tunnel clusters, the drill and blast method and the TBM method are selected according to the geological characteristics, tunnel length and external construction conditions. Among them, the double shield TBM used in the 30 A Tunnel and the #38 Tunnel is the most effective. The 30 A Tunnel is located between Shuimogou and Dashagou in Yongdeng County, Gansu Province, also known as Shuimogou Tunnel, with a line length of 11.649 km and a design discharge of 32 m^3/s and an enlargement discharge of 36 m^3/s. Considering the complicated landform and staggered gullies on the surface and that even a single tunnel is longer than 10 km, problems regarding ventilation, transportation and construction period during construction are hard to be solved. Therefore, the detour scheme has been considered during preliminary design, and the excavation principle of "short tunnel with numerous working faces" is adopted. The canal detours from the outlet of the inverted siphon at Shuimogou to the inlet of the flume at Dashagou with a total length of 14.962 km. In total, 7 tunnels are included, i.e. #30–#36 tunnels 11,600 m long. The tunnels are linked with flumes and open channels (4 flumes 518 m long and 2844 m open channels). According to repetitive demonstrations by experts, the construction scheme of detour for the canal from Shuimogou to Dashagou is changed as linear crossing via TBM before commencement. The total length of the line is shortened for 12,012 m, including one tunnel (30 A) 11.649 km long, one 70 m flume and 293 m open channels. The length of all tunnel canals is 2950 m shorter than that in the preliminary design scheme (Fig. 11.98).

The 30 A and #38 Tunnels are built by the joint venture of Italy CMC and China Huashui Hydropower Development Co., Ltd. through international competitive tendering. The Robbins TBM-188–227 double shield TBM is used for excavation, and the prefabricated reinforced concrete segment lining method is used

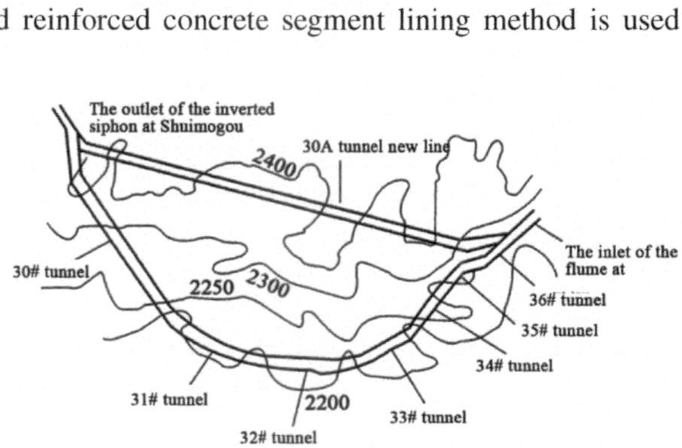

Fig. 11.98 Re-routing scheme of 30 A tunnel

for construction. The 30 A Tunnel was commenced in December 1990, and was completed in January 1992, only lasting for 13.5 months. The average monthly advance per working face is 860 m, and the maximum monthly advance is 1300 m, reaching the first-class construction speed at that time and realizing favorable construction benefits and safe and civilized production of no casualties during construction of long tunnels.

The #38 Tunnel is 5.4 km long, with medium-hard sandstone as rocks and little groundwater. The tunnel sections for which TBM is used are 4947.6 m long. Tunneling was commenced in early April 1992 and was completed in mid-August, only lasting for 4.5 months. The average monthly advance is 1100 m, the maximum monthly advance is 1400 m and the maximum daily advance is 75.2 m, reaching a record high in China and the world advanced record at that time.

2. Geological conditions and engineering characteristics of 30 A Tunnel

30 A Tunnel is a long tunnel with uneven soft and hard lithologies. One of the characteristics of the project is complicated geology. Presinian crystalline limestones and slate intercalated with phyllite are developed from the inlet to outlet of the stratum successively, and this section is 1680 m long, with a dry density of 2.59–2.84 g/cm^3, uniaxial compressive strength of rocks of 26–133.7 MPa and a softening coefficient of 0.77–0.97. The 9790 m long Tertiary system contains boulder, conglomerate, glutenite, argillaceous siltstones and sandstones, with a dry bulk density of 2.01–2.48 g/cm^3. Fault structures are not developed here, and the uniaxial compressive strength of rocks is 2.79–15.29 MPa (natural state). The rocks are soft and will be softened in case of water, with a softening coefficient of less than 0.35, a φ value of 35.2°–43.3° and a C value of 0.18–0.21 MPa and a disintegration time of 0.77–1.48 h. This stratum is the main stratum through which the 30 A Tunnel passes. The Quaternary loess about 150 m long is developed at the outlet and will be softened in case of water, with a dry density of 1.31–1.44 g/cm^3, a φ value of 22°–32° and a C value of 0.01–0.04 MPa. The Presinian stratum is greatly affected by the structure with broken rocks and 23 large fault fracture zones are developed here, with groundwater along this stratum. The fault fracture zone at the contact zone has the largest water burst, and the maximum water burst of about 150 L/s was identified during tunneling, and the rest is bedrock fissure water which is non-corrosive to ordinary cement concrete. Moreover, the tunnel walls also traverse 3 karst caves (77 m^3) unique to limestone layers. Second characteristics of the project: The 30 A Tunnel is 11,649 m long, and the problems regarding ventilation and transportation in the tunnel are hard to be solved during construction. Third characteristics of the project: Since high mountains and gullies are developed on the surface along the tunnel axes, the landform conditions for excavating inclined shafts or shafts for construction in sections are not met. Fourth characteristics of the project: Since the tunnel entrance is located on the steep hard-rock slope, it is hard to arrange the platform at the tunnel opening and construction facilities, with groundwater being developed here. Based on this, the construction deployment of one-way tunneling from outlet to inlet can only be adopted.

3. Engineering design of 30 A Tunnel

(1) Section structure selection

The 30 A Tunnel is of the horseshoe-shaped section originally, and the hard-rock sections are excavated through the drill and blast method. Except that 90 m at the entrance is reinforced with reinforced concrete, the shotcrete bolt lining method is adopted for the rest tunnel sections. Cast-in-place concrete is used for all bottom plates. Both the soft-rock section and the exit section pass through the Quaternary stratum and are excavated via a roadheader, with bolt-shotcrete supporting and cast-in-place concrete lining or reinforced concrete lining.

Meanwhile, the alternative of circular section with excavation through the drill and blast method is also proposed. Use of the alternative (circular section) through excavation via full face double shield TBM was proposed by the joint venture of Italy CMC and China Huashui Hydropower Development Co., Ltd. during tendering. The full face double shield TBM has already been applied successfully in other countries, but it has not been used in China. The double shield TBM featured by safe construction, fast tunneling, simultaneous excavation and lining and lower investment is selected finally through repetitive comparison, demonstration and analysis. The excavation diameter of TBM is ϕ5.53 m. Prefabricated reinforced concrete segment lining is adopted for the reinforced section, soft-rock sections and Quaternary stratum tunnels at entrance and exit. The segment has a thickness of 30 cm, a width of 160 cm and an arc length of 400 cm and each segment is about 5 t (Fig. 11.99). Four prefabricated segments are arranged in a quincuncial manner for each ring, i.e. stagger-jointed longitudinally and horizontally. The segments are of C25 reinforced concrete, and the 25MnSi reinforcement with a design strength of $\sigma = 3800$ kg/cm^2 is used during construction. Segments are classified into heavy and light ones based on steel content. The light segments, with a steel content of 67.5 kg/m^3, are used for grade-III/IV rock lining. The heavy segments, with a steel content of 77.2 kg/m^3, are used for grade-V rocks and the tunnel sections with groundwater. Asphalt PU sealing strips are pressed into the groove reserved at segment joints to guarantee the sealing state of grouting. The average gap between surroundings outside the segment and the tunnel face is 6.5 cm, and pea gravels with a diameter less than 10 mm shall be backfilled behind the TBM shield immediately. Afterwards, neat cement grout shall be used for LP grouting. For the tunnels at the hard-rock sections, except that the prefabricated reinforced concrete heavy segment lining above is adopted for the reinforced section 90 m long at the entrance and the contact zone of the limestone section with the Tertiary stratum with rich groundwater and karst caves, light segment lining is adopted for the rest limestone sections without water. In other words, prefabricated reinforced concrete segment lining is adopted for the 30 A Tunnel in total length.

(2) Preliminary design and review of section structure

① The prefabricated reinforced concrete segments used for the 30 A and #38 Tunnels are designed by Lombardi, a company in Switzerland, entrusted by the joint venture of Italy CMC and China Huashui Hydropower Development Co.,

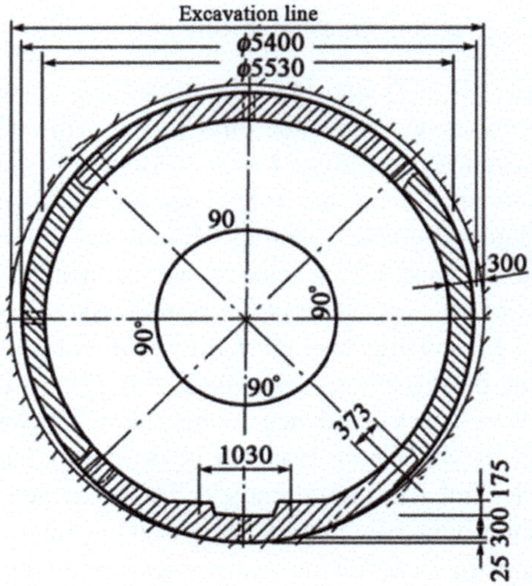

Fig. 11.99 Structure of segment lining of 30 A tunnel (Unit: mm)

Ltd. The *Technical Report for Stability Analysis and Proposed Lining Dimensions of International #2 Lot Tunnel of the Datong River-Qinwangchuan Diversion Project* was proposed in June 1988. According to the test data concerning rock specimen of the 30 A and #38 Tunnels in the *References*, a tendering document, crushing degree of rocks, and whether clay occurs as well as the experience of Lombardi in research on similar rocks, the geomechanical parameters (Table 11.23) of 3 types of rocks are selected in this report, to calculate the rock pressure (Table 11.24) of lining under the short-term and long-term load conditions of the tunnels featuring 3 such types of rocks under different overburden thicknesses through the characteristic curve method. According to the rock pressure of the lining structure, without considering the influence of groundwater, assuming that the horizontal rock pressure at the arch foot is 0.8 times of the vertical rock pressure at the vault, the horizontal rock pressure calculated is 42.4×10^{-2} MPa as per the rock pressure of 53×10^{-2} MPa at vault of the grade-2 rock, which is poorer among these two types of rocks. The lining segment thickness and the steel content of concrete proposed preliminarily are 30 cm and 70 kg/m^3 (design strength of reinforcement: $\sigma = 440$ MPa, which is the Swiss standard $FeB_{44}K$) through calculation of internal force.

② The *Technical Report for Stability Analysis and Proposed Lining Dimensions of International #2 Lot Tunnel of the Datong River-Qinwangchuan Diversion Project* proposed by Lombardi is reviewed by Gansu Water Conservancy and Hydropower Survey and Design Research Institute, the Design Institute of the project. It is believed that no complete theories and methods have been proposed for lining design of tunnels up to now, and any theory has its limitations. The lining dimensions and parameters are preliminarily determined through theoretical analysis and engineering analog generally, with timely adjustment of design parameters based on the geological conditions revealed with the progress in

Table 11.23 Geomechanical parameters of rocks

No	Grade of rocks	Elastoplastics	Unit weight γ (t/m^3)	Short-term load				Long-term load			
				Internal friction angle φ(°)	Cohesion C (MPa)	Modulus of deformation E (10^4 MPa)	ΔV (‰)	Internal friction angle φ(°)	Cohesion C (MPa)	Modulus of deformation E (10^4 MPa)	ΔV (‰)
1	Conglomerate/sandstone intercalated with claystone	Elastic	2.4	25	0.1	0.1	0.5	25	0.05	0.1	1
		Elastic	2.4	30	0.2	0.3	0.5	30	0.1	0.3	1
2	Crushed Conglomerate/sandstone	Elastic	2.4	30	0.1	0.3	0.5	30	0.05	0.3	1
		Elastic	2.4	35	0.2	0.5	0.5	35	0.1	0.5	1
3	Moderately crushed conglomerate/sandstone	Elastic	2.4	35	0.1	0.5	0.5	35	0.05	0.5	1
		Elastic	2.4	40	0.2	0.7	0.5	40	0.1	0.7	1

Table 11.24 Rock pressure of lining

Grade of rocks	Overburden thickness (m)	Steady pressure under short-term load (10^{-2} MPa)	Steady pressure under long-term load (10^{-2} MPa)
1	350	59.5	70.0
2	350	41.5	53.0
2	200	20.8	31.0
3	350	22.0	31.0

construction. The interaction force (the stable rock pressure acting on lining) between rocks and lining is analyzed through the characteristic curve method of elasticoplastic deformation in this report, with different geomechanical parameters of rocks under short-term and long-term load being considered. Besides, different deformation characteristics of the elastic and plastic zones of rocks and the influence of tunnel overburden thickness on rock pressure are considered in this report, which has its own characteristics. However, just as other calculation methods, this calculation method is also hard to accurately simulate the practical conditions of a complicated tunnel project. For instance, the claystones with intercalations shall be treated separately for local reinforcement, and shall not be treated as homogeneous elastoplastic bodies. Moreover, basic calculation parameters such as geomechanical parameters of rocks and initial deformation can be selected in a more arbitrary manner. In the paragraph when lining structure starts bearing load after excavation of rocks, the rock convergence is taken as 1 cm, and the distance with the tunnel face (at the TBM cutter head) is 15–20 m. If calculating as per the TBM tunneling speed, excavation only lasts for around 8 h. The convergence of 1 cm accounts for 9.7% of the total convergence of 10.3 cm after excavation of grade-1 rocks, 41.7% of the total convergence of 2.4 cm after excavation of grade-2 rocks and 100% of the total convergence of 1 cm after excavation of grade-3 rocks. Moreover, the 1 cm greatly affects the rock pressure acting on the lining. Therefore, the theoretical calculation results can also serve as reference for determining or adjusting design parameters of lining. In this report, rocks are classified into 3 grades, which is slightly rough if comparing with the practical engineering geological conditions of the 30 A and #38 Tunnels. The rock pressures of 3 such grades of rocks calculated under long-term load conditions are 0.7 MPa, 0.53 MPa and 0.31 MPa, i.e. 29 m, 22 m and 13 m respectively if converting into load (collapsing arch) height. According to the Protodyakonov's loose media theories, the collapsing arch height is $H = B/2F$, and the firmness coefficients F of 3 such grades of rocks are 0.09, 0.125 and 0.21 respectively through back calculation, which only equivalent to the firmness coefficient of silt or quicksand.

On the contrary, according to the Terzaghi's theory, the load arch height is $H = A(B + H)$, and the coefficients A of 3 such grades of rocks are 2.6, 2.0 and 1.2 respectively through back calculation, which equivalent to sand, soil or completely crushed rocks. Therefore, it can be believed that the rock pressure in this report is conservative. Three such grades of rocks in this report can cover all possible grades of rocks in the 30 A and #38 Tunnels. Limited by the materials, the influence of groundwater is not considered in analytical research. However, the geomechanical parameters of rocks selected in Table 11.23 are greatly impaired if comparing with the test data of rock blocks, and the circular lining structure adopted can withstand a huge external pressure, so it is available. Since limited by the excavation diameter of TBM and the segment prefabrication equipment, it is hard to change the segment thickness of 30 cm. The requirements of different strata can only be met through different steel contents of segments. The prefabricated segment lining with an average steel content of 70 kg/m^3 can be used for the tunnel sections with grade-2 rocks and an overburden thickness of 350 m through calculation, i.e. the tunnel sections with a rock pressure of 0.53 MPa. For the 30 A Tunnel, the tunnel sections with a buried depth of the Tertiary stratum above 200 m are about 3000 m long, only accounting for 30% of the prefabricated segment lining section. Therefore, the estimated average steel content of 70 kg/m^3 is proper.

(3) Design and review of construction drawings of section structure

① In May 1989, the *Technical Report for Prefabricated Segment Lining of International #2 Lot 30 A and #38 Tunnels of the Datong River-Qinwangchuan Diversion Project* was proposed by Lombardi. In this report, another method is adopted to calculate lining internal force, and the segments are classified into "light" and "heavy" ones based on the steel content. The following factors are considered in this calculation:

a. The longitudinal average compressive stress of the segment calculated based on the total thrust (2400 t) exerted by TBM to segments is 5 MPa. Considering the concentrated stress, when the direction of the TBM is adjusted, the maximum compressive stress can reach 8–10 MPa. Moreover, the axial force can only be delivered through the contact surface narrowed between segments before cement grouting at the short edges (longitudinal connections of 4 joints constituting a ring) of the segment. It can be assumed that this contact surface is only 1/3 of the segment thickness at the narrowest area, the lateral tractive force of $T = 17$–22 t/m will be generated in segments under short-term load conditions. Therefore, sufficient reinforcement shall be equipped at the long edges (circumferential joints) of the segment to withstand the thrust from TBM, and stirrups shall be provided at the short edges (longitudinal joints) to withstand the lateral tractive force generated in segments under short-term load. Special attention shall be paid to the strength and rigidity at segment edges.
b. When continuous and uniform contact is always maintained between the lining and rocks and the lining only withstands uniform circumferential rock pressure, the maximum load is 0.7 MPa and the axial pressure of the lining is $N = 190$ t/m^2 (1.9 MPa). The average compressive stress generated in 30 cm-thick concrete lining ring is $\sigma = 6.3$ MPa and the concrete lining can withstand this load.

c. If the deformation of rocks in each direction has a different nature, the rock pressure acting on the lining vary in vertical and horizontal direction. However, the possibility of this load condition is low.
d. Other possible concentrated loads during excavation

After taking the above factors into account, the internal forces taken for light and heavy segments are determined through calculation (Table 11.25). The internal force of "light" segments is calculated as 60% of "heavy" segments.

According to the internal force calculated, Lombardi submitted the reinforcement drawings of prefabricated segments mapped in February 1989, July 1989 and June 1990 successively. See Tables 11.26 and 11.27 for previous reinforcement allocated.

② The *Technical Report for Prefabricated Segment Lining of International #2 Lot 30 A and #38 Tunnels of the Datong River-Qinwangchuan Diversion Project* and two reinforcement drawings of segments submitted by Lombardi in May 1989 have been reviewed by Gansu Water Conservancy and Hydropower Survey and Design Research Institute for the second time. The comments proposed are as follows:

a. The third reinforcement rate of load-bearing reinforcement in drawings SELI 5050 and 5052/A adopted finally is 0.192–0.207%, which is in line with the

Table 11.25 Bending moment and axial force on unit length of segments

Segment type	Load condition 1		Load condition 2
	Axial force N (kN)	Bending moment M (kN-m)	Bending moment M (kN-m)
Heavy segments	1200	90	35
Light segments	700	55	20

Table 11.26 Reinforcement allocated for prefabricated segments

Item		First time	Second time	Third time
Drawing No		SELI 5050 5052	SELI 5050 5052	SELI 5050/A 5052/A
Reinforcement type		$FeB_{44}K$	$FeB_{44}K$	$FeB_{44}K$
Heavy segments	Inner-layer reinforcement rate (%)	0.213	0.229	0.286
	Outer-layer reinforcement rate (%)	0.147	0.164	0.207
	Steel content (kg/m^3)	56.0	63.0	77.2
Light segments	Inner-and outer-layer reinforcement rate (%)	0.126	0.143	0.192
	Steel content (kg/m^3)	46.4	53.3	67.5

Table 11.27 Comparison on calculation results of tunnel rock pressure

Grade of rocks	Buried depth (m)	This report		Lombardi report	
		Steady pressure under short-term load (10^{-2} MPa)	Steady pressure under long-term load (10^{-2} MPa)	Steady pressure under short-term load (10^{-2} MPa)	Steady pressure under long-term load (10^{-2} MPa)
1	350	62.0	78.5	59.5	70.0
2	350	38.5	53.0	41.5	53.0
2	200	22.5	37.5	20.8	31.0
3	350	22.0	33.3	22.0	31.0

minimum reinforcement rate of U = 0.15% specified in *Design Code for Hydraulic Concrete Structures* (SDJ 20—1978) in China.

b. If calculating as per domestic regulations such as *Design Code for Hydraulic Concrete Structures* (SDJ 20—1978) and *Code for Design of Concrete Structures* (TJ 10—1974), the reinforcement allocated for both heavy and light segments in the construction drawings can meet the requirements of safely withstanding the strength under different load conditions in the reports above.

4. Study on segment lining structure

Through prudent consideration, to further study on the applicability of the segment lining structure designed and submitted by Italy CMC under specific conditions of the 30 A and #38 Tunnels for reference in construction and to provide design theories, calculation methods and corresponding calculation procedures for reference in future independent design of circular water tunnel segment linings by our hydropower sectors, the construction headquarters of the Datong River-Qinwangchuan Diversion Project, Gansu Water Conservancy and Hydropower Survey and Design Research Institute and Lanzhou Jiaotong University jointly applied to Gansu Academy of Sciences for such scientific research project as *Research on Segment Lining Structure of Water Tunnels of Datong River-Qinwangchuan Diversion Project*, and 3 different calculation methods and review results have been obtained.

(1) *Load Structure Method of Circular Tunnel Segment Lining*: Based on general principles of load structure, this method has established the basic equation for plane circular curved beam and obtained the analytical solution for displacement and internal force of such beam. Based on the load schema in the *Specification for Design of Hydraulic Tunnel* (SDJ 34—1984) (Trial), the vertical pressure of grade-IV rocks is $q_v = 0.0717$ MPa and the horizontal lateral pressure coefficient is $\alpha_q = 0.1$–0.2. The vertical pressure of grade-V rocks is $q_v = 0.1433$ MPa, the lateral pressure coefficient is $\alpha_q = 0.3$–0.35 and the elastic reactance coefficient is $k = 10{,}000$ t/ m^3. The displacement and internal force of lining structures are calculated based on this, and the concrete stress and reinforcement stress of lining structures are calculated as per the strength formula for reinforced

concrete compression-flexural members. The prefabricated segments are made of C25, and the concrete feature adopted by Lombardi is the bending compressive design strength of $Rw = 180$ kgf/cm^2 (18 MPa). The reinforcement used is $FeB_{44}K$ in Switzerland, with a design strength of $Rg = 4400$ kgf/cm^2 (440 MPa). According to the calculation results, the concrete compressive stress and the reinforcement tensile stress are less than the design strength of the material (except for several cases). However, if domestic materials {design strength of $[\sigma_h] = 140$ kgf/cm^2 (14 MPa) for C25 concrete and the design strength of $[\sigma_g] = 2400$ kgf/cm^2 (240 MPa) for reinforcement} are used, the concrete compressive stress can meet requirements, but the reinforcement tensile stress goes beyond the design strength in grade-V rocks, and some of the requirements are not met in grade-IV rocks.

(2) *Principle of Characteristic Curve Method and Its Application in 30 A Tunnel Design*: The characteristic curve (i.e. the convergence-confinement method) is a tunnel design method based on theories and predominated by site survey with experience for reference. Main difference between the calculation methods adopted in this research report and those adopted by Lombardi: The 3D effect of the excavation process is reflected with the semi-rock core theory by Lombardi, but this effect is reflected by the displacement released on the construction surface in this research report. Once the plastic zone of rocks is developed to certain degree, the tunnel top will withstand certain broken-rock pressure, which is not covered in the characteristic curve of Lombardi. However, this pressure is considered in this research report. According to comparison between calculation results in this research report and those of Lombardi, except that the calculation results of grade-2 rocks and the overburden thickness of 350 m are the same, the calculation results in this report are 0–19% higher than those proposed by Lombardi, which is within a reasonable range.

(3) Finite Element Analysis on Elastoplasticity of Circular Tunnels with Segment Linings: In this method, the tunnel axis length is taken as 1 m, and finite element analysis on linear elasticity as well as that under common effect of lining and rocks are carried out as per plane strain. Before lining supporting, 70% load is released from rocks, and 30% released will be withstood after lining supporting. According to analysis results of elastoplasticity, the lining deformed is elliptical, and the settlement of rocks at tunnel top is 43.3126 mm, with a lining displacement of 2.3439 mm at tunnel top and an outward horizontal displacement of 0.6472 mm at waist. The maximum stress of lining is seen on the tunnel bottom section, with the maximum compressive stress of 13.1 MPa for concrete and the maximum tensile stress of 182.6 MPa for reinforcement, which are all below the design strength of the material.

The followings can be concluded based on the review results of 3 different calculation methods above and another 3 calculation methods:

(1) The concrete prefabricated segment linings adopted for the 30 A and #38 Tunnels can guarantee the stability of the chamber for grade-II and grade-III rocks and general Tertiary stratum. For the Tertiary stratum (mainly the sandstone/layers

intercalated with clay), fault fracture zone and loess tunnel sections with groundwater, according to the calculation results, it is recommended to increase the circumferential reinforcement amount used for segments from the original 57 kg/m^3 designed by CMC to 80 kg/m^3. The C25 concrete design strength is taken as $[\sigma_h] = 140$ kgf/cm^2 (14 MPa) and the reinforcement design strength as $[\sigma_g] = 2400$ kgf/cm^2 (240 MPa), to meet the requirements for long-term safe operation of the tunnel. Since the 25MnSi reinforcement with a design strength of $[\sigma_g] = 3800$ kgf/cm^2 (380 MPa) is used for construction, the reinforcement amount of light and heavy segments preliminarily designed is changed from 43 kg/m^3 and 57 kg/m^3 to 67.5 kg/m^3 and 77.2 kg/m^3 by CMC.

(2) The circular erected segment lining is a conditional stable structure (geometrically invariant system). According to calculation results, when the elastic reactance coefficient of rocks is less than 50,000 kN/m^3, circumferential segments may slide in a relative manner at the joints. Therefore, under the adverse geological conditions in which rocks collapse in case of water, the lining is in the risk of losing stability. For this, special anti-seepage measures are taken for the loess rock sections at tunnel opening.

The results of another 3 analytical calculations are basically identical to the review results above.

5. Technical parameters and characteristics of TBM

One Robbins double shield TBM is used for single-working-face tunneling for the 30 A Tunnel, which has a circular section, with an excavation diameter of 5530 mm (5540 mm for a new cutter) and an inner diameter of 4800 mm. Four prefabricated reinforced concrete segment linings are used for each ring, and the gap between the segment and the tunnel is backfilled with crushed gravel and grouted with mortar.

The double shield TBM arrived at the 30 A Tunnel entrance in July 1990, and was assembled in the end of November, and then was delivered to the excavated preparatory tunnel 18 m long. Trial tunneling was started on December 5, and 125 m was tunneled in current month, with a continuous breakthrough of 1000 m since March 1991 and the maximum daily advance of 65.6 m. The 30 A Tunnel was completed on January 20, 1992.

(1) Structural Features

The double shield TBM used for the 30 A Tunnel has a shield made of 40 mm-thick steel plates welded and an outer diameter of 5530 mm. The maximum stroke of extension and retraction of the front and rear shields is 910 mm and the maximum total length of the double shield during extension is 13,610 mm.

The front shield is installed with a cutter head, a main bearing and a cutter head drive device. The cutter head is driven by six 160 kW motors, and each motor is linked to one reducer through a hydraulic clutch, driving the cutter head to rotate under the effect of bull gear. The cutter head is installed with 37 disc cutters with a diameter of 394 mm, with clockwise rotation for cutting. Four radial jacks are installed inside the front shield to extend the 2 lateral grippers outside the shield to support the stratum, with a total support force of 3000 kN.

Located inside the rear shield, the segment erector is of an annular disc structure and is operated through electrical remote control, with full hydraulic transmission. Eight jacks are installed axially along the circumference inside the rear shield. Just like the front shield, 2 lateral grippers are also installed inside the rear shield, and the total support force is 12,900 kN.

In total, 12 hydraulic jacks are installed between the front and rear shields, to push the front shield to move forward. The 12 jacks are divided into 6 groups, with two jacks as a group and spheroidal hinging. Such jacks are arranged in a lambdoidal manner, to avoid relative rotary displacement of the front and rear shields and to accurately control the forward direction of the front shield.

The back-up gantry is about 130 m long and is pulled by the rear shield. Hydraulic oil pumps, hydraulic oil tanks, pipelines, control valves, transformers, electrical control cabinets, cable reels and grouting machines are installed on both sides of the back-up gantry, while the belt conveyor, hoists and gravel jet machines are installed at the top and the TBM cab is located at the front end of the frame.

(2) Operating Principle

The lateral support of the rear shield shall be extended first to support the stratum during tunneling of the double shield TBM, to fix the rear shield, with cutter head rotating for tunneling. Meanwhile, the front shield moves forward by pushing the jack. The bucket around the cutter head collects the soil muck during rotation of the cutter head, and the muck is unloaded to the belt conveyor in the center via a muck chute while rotating to the top, and then to the mining vehicle via the belt conveyor and finally to outside the tunnel in rows. Segments of the rear shield are installed during forward tunneling of the front shield.

Once the cutter head is pushed forward for 800 mm, the front shield will stop pushing and the lateral support of the front shield will extend. The front shield is fixed in the tunnel and the lateral support of the rear shield is retracted. The rear shield moves forward for 800 mm by pushing the jack, and then the lateral support of the rear shield is extended. The rear shield is fixed and the lateral support of the front shield is retracted. At this moment, tunneling of the next half ring can be started, with regripping once tunneling for half a ring.

When the rock strength is less than 3 MPa, lateral supports shall not be used. TBM is supported on the segment through the rear shield jack for pushing. At this time, tunneling and installation of segments shall not be carried out simultaneously.

(3) Technical parameters

TBM model: TBM-188–227;
Total mass of main machine: 331 t;
Total length of TBM: 145.5 m;
Minimum curvature radius: 300 m;
Cutter head diameter (new cutter): 5540 mm;
Outer diameter of shield: 5530 mm;
Length of front shield: 5400 mm;

Length of rear shield: 7300 mm;
Front shield jack: 12 sets × 650 kN × 910 mm (total thrust of 7800 kN);
Rear shield jack: 8 sets × 1950 kN × 2950 mm (total thrust of 15,600 kN);
Cutter head torque: 1610/3210 kN m;
Cutter head speed: 5.72 r/min, 2.875 r/min;
Cutter head drive power: 160 kW × 6 (960 kW);
Disc cutter: 37;
Disc cutter diameter: 394 mm;
Power supply system: 1350 kVA (primary 10 kV/secondary 660 V);
Cutter-drive hydraulic system: 17.5 MPa;
Thrust cylinder hydraulic system: 34.5 MPa;
Segment erector: annular turnplate type longitudinal translation stroke of 1000 mm;

Belt conveyor capacity: 6 m^3/min.

6. Engineering Construction.

(1) Tunneling

The 30 A Tunnel is of a circular section, with an excavation diameter of 5.53 m, a lining inner diameter of 4.8 m and a longitudinal slope of 1/1000. The double shield TBM is used for excavation, with prefabricated reinforced concrete segment lining. Pea gravels shall be spray-filled behind the segment, with neat cement grout being injected. The tunneling cycle is mainly made up of excavation, ventilation, dust prevention and removal, muck discharge and transportation, transportation and installation of prefabricated segments, filling of pea gravels and grouting, guidance, extension of wind, water and power pipelines and tracks, machinery maintenance and service, etc. The rocks are crushed through squeezing and rotation of 37 disc cutters (15.5°) on the TBM cutter head, and the rock muck excavated will be collected constantly via the bucket around, and be unloaded to the belt conveyor 24 m long via a hopper and a chute and finally to the mining vehicle.

Two tunneling methods are adopted in construction based on the geological conditions. In case of hard geology, the reaction force of thrust is provided from lateral supports on the tunnel walls, and the front shield pushes forward the jack to provide thrust. Excavation and installation of segments shall be carried out in a synchronous manner, and 2 segments shall be installed by the rear shield segment erector per tunneling of 800 mm by the front shield cutter head. Afterwards, regripping shall be made for supporting and then the next 800 mm cycle shall be started. In case of soft geology or faults, the rear shield shall support the segment to provide the thrust by pushing forward the jack. The segment shall withstand the reaction force without the use of lateral supports. At this moment, the double shield is retracted, which is equivalent to one shield. Tunneling and installation of segments shall not be carried out simultaneously, and tunneling shall be stopped per tunneling of 800 mm. Tunneling shall be consumed after installation of 2 segments and then the segment shall be installed.

(2) Segment lining

Hexagonal prefabricated reinforced concrete segments are used for lining of the 30 A Tunnel, and 4 segments shall be used for each ring and be erected in a mosaic manner. Two segments shall be installed in a symmetric manner in the sequence of "up-down" or "left–right" per pushing of 800 mm. The entire segment shall be installed per pushing of 1600 mm. The segment shall be erected alternately in the sequence of 2 for "up-down" and 2 for "left–right". Bolts are not used for connection among segments, and the end surface of each segment has a protruding arc and a bending groove tenon, forming hinged connection. A curved water-stop seal groove is provided around the segment, with the rubber sealing strips immersed into asphalt being installed inside the groove in advance. The segment has an outer diameter of 5400 mm, an inner diameter of 4800 mm, a thickness of 300 mm and a width of 1600 mm.

Two lifting holes with a diameter of 120 mm are prefabricated in the middle of each segment, and such holes can also be used for grouting. During installation of segments, two jacks on the lifting plier of the segment erector shall be inserted into the holes simultaneously to clamp the segment firmly for installation. The gap between the segment and the tunnel stratum shall be backfilled timely in two steps. First, the gravel with a grain size of 5–10 mm shall be sprayed via a gravel jet machine at a pressure of 0.25–0.30 MPa and then 1:1 cement mortar shall be grouted at the pressure of 0.2 MPa.

(3) Segment fabrication

"Heavy" and "light" segments are adopted for the 30 A Tunnel based on geological conditions, and the shape and dimensions of two such segments are completely the same. The only difference is the amount of reinforcement (77.2 kg/m^3 and 67.5 kg/m^3 respectively) and C25 concrete is used.

The segment prefabrication plant is set at the area about 300 m away from the tunnel exit, with workshops for reinforcement fabrication, concrete mixing, segment shaping, segment steam curing and temporary storage being set. The ordinary grade-525 cement made by Gansu Yongdeng Cement Factory is used, and the mix proportion of cement, sand and gravel is 1:1.25:4.28. The maximum grain size of gravel is 40 mm.

Manufacturing process of segments: The cement, aggregate, water and additives shall be mixed for about 3 min via a forced mixer after weighing on the batching table, and then be delivered to the delivery drum via a belt conveyor for injection into the segment mold (a reinforcement framework shall be placed inside it in advance for a steel mold), and then be poured and vibrated via a HF vibrator, until the segment is formed. The poured segments will be pushed into the pre-heating room (the temperature shall be controlled at 40–50 °C) for about 25 min with the mold equipped with a roller, and then be delivered to the steam curing room (80–85 °C) for 2 h, and be de-molded and then be cooled naturally for 15 min for numbering. Afterwards, the segment shall be delivered to the outside after indoor storage for 12 h and then is available for use after storage for 72 h in the open air.

(4) Mucking Transport

Muck discharge and tunneling are carried out simultaneously, and the muck discharged is transported through tracks, with a gauge of 1000 mm. Rails are fixed onto the lined bottom arch, and each train has 7 sections with 77 m^3 muck being loaded, which is exactly the volume of muck generated during tunneling for one ring. Once the train departs, both the cutter head and the belt conveyor will stop running and wait for the next train. This time interval can be used for TBM regripping.

(5) Tunnel ventilation

The TBM dedusting system is controlled by the dustproof shield, which is installed around the muck chute close to the TBM conveyor belt to isolate the dust generated from tunneling on the working face from the tunnel space. Clean air will flow back to the tunnel once the dust contained in the air is separated via a dustproof shield purifier.

Forced ventilation is adopted for the tunnel, and 4 axial flow fans (diameter: 1220 mm; power: 4×35 kW; air supply capacity: 24 m^3/s) in series are installed at the area 25 m away from the tunnel portal. The air ducts are plastic nylon hoses made in Italy with a diameter of 1400 mm. Each duct is 100 m long, and air ducts are interconnected with steel rings and are hung on vault. One ventilator is installed at the back-up tail of the TBM for secondary ventilation to supply fresh air to various parts of TBM.

(6) Construction power supply

The power for TBM is 660 V and 50 Hz and the lighting power is 220 V, and 10 kV cables are used for transformation inside the tunnel. Each cable reel is 250 m long and is installed on follow-up trains and will extend with the trains.

(7) Water supply

Water supply pump stations and reservoirs are set around the tunnel opening, and the water will be delivered to inside the tunnel with steel pipes with a diameter of 3 in at a pressure not less than 0.3 MPa mainly for machinery cooling, dustproof, backfill grouting, the water consumption of which is not large.

(8) Laser guidance

Construction survey of tunnel is realized by one laser alignment telescope of the ZED tunnel guidance system after introduction of the horizontal control and elevation control network outside the tunnel in the tunnel. This is a computer-controlled laser beam and inclinometer. The laser beam will display light spots on the fluorescent screen on the left at the TBM head and transmit them to the operator cabinet, to provide the operator with the plane orientation of the cutter head and the front shield. In this way, the axis control during tunneling can always be accurate to the millimeter.

(9) Cycle operation

The working system of 24 h 3-shift is adopted for TBM, with 2 shifts for continuous tunneling and 1 for machinery maintenance and service. One segment lining shall be installed per tunneling of 2×0.8 m with the rock chips of 1 train being delivered.

11.2.2 Shanxi Wanjiazhai Yellow River Diversion Project

1. Project Overview

The Shanxi Wanjiazhai Yellow River Diversion Project is a life project which can fundamentally solve the problem concerning water shortage, promote industrial and agricultural development in Shanxi, improve the people's living standards and maintain the development of the national energy heavy chemical base. It consists of the Wanjiazhai hydro-junction, main line, southern line, connecting sections, northern line, etc. See Fig. 11.100 for details.

The key project is a 896,000,000 m^3 concrete gravity dam 90 m high built on the western Yellow River of the Wanjiazhai Village, Pianguan County, Shanxi. One peak-regulation HPP is built behind the dam, with an installed capacity of 1,080,000 kW and an annual energy output of 2,750,000,000 kW·h.

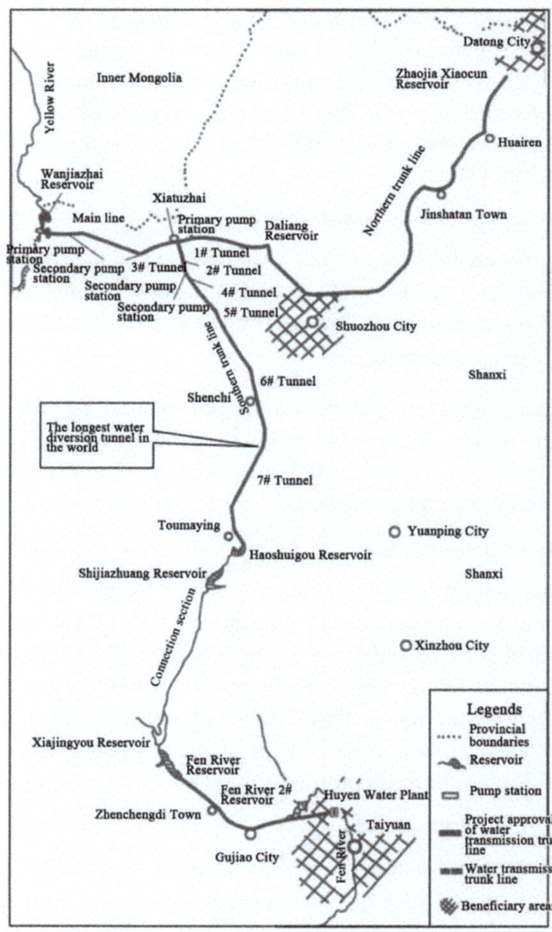

Fig. 11.100 Shanxi Wanjiazhai Yellow River Diversion Project

The Shanxi Wanjiazhai Yellow River Diversion Project takes water from the reservoir area of Wanjiazhai hydro-junction. The annual water diversion amount is 1,200,000,000 m^3. The main line is from eastern Wanjiazhai to Xiatuzhai Village in Pianguan County, with a total length of 44.35 km and a diversion discharge of 48 m^3/s. The southern line is from the Xiatuzhai (crossing the Pianguan River and the Guancen Mountain by diverting water in the south) Village to the Toumaying Village in Ningwu County, with a total length of 102.4 km and a diversion discharge of 25.8 m^3/s, and 640,000,000 m^3 water can be supplied to Taiyuan each year. The northern line is from the Xiatuzhai (crossing Shuozhou and Shentou in the east) Village to the reservoir in Zhaojiaxiao Village in Datong to the north, with a total length of 166.88 km and a diversion discharge of 22.2 m^3/s, and 560,000,000 m^3 water can be supplied to Shuozhou and Datong each year. The connecting section is from the Nangan Tunnel exit in Toumaying Village in Ningwu County to the water plant in Huyan Village of Taiyuan, with a total length of 138.60 km, including 81.20 km natural river course and 57.40 km waterlines.

The Shanxi Wanjiazhai Yellow River Diversion Project is completed in two phases. Construction of the main line, the southern line, the connecting section and installation of some units are involved in the phase-I project, to solve the problem concerning water consumption in Taiyuan in a centralized manner. The estimate of phase-I project is CNY 11.297 billion, including USD 0.4 billion loaned from the World Bank, and the rest is from the water resource compensation. Construction of the northern line and installation of remaining units at the pump stations of the southern line are involved in the phase-II project. The international tendering and bidding system is adopted for the project. The project completed can basically meet the water needs for industry and living in Taiyuan, Datong, Shuozhou and other cities in Shanxi before 2020 or longer.

The Employer of the project is the head office of Shanxi Wanjiazhai Yellow River Diversion Project, the Construction Contractor includes above 40 domestic contractors and 3 international contractors and the Supervisor is YREC. The Shanxi Wanjiazhai Yellow River Diversion Project is a World Bank Loan Project. Environmental supervision required by the World Bank is undertaken by Henan Kehua Engineering Consulting Co., Ltd. subordinated to the North China University of Water Resources and Electric Power.

2. Usage of TBM

In total, 5 Robbins and 1 NFM double shield TBMs (Fig. 11.101) have been used successively in the Shanxi Wanjiazhai Yellow River Diversion Project, and the tunnel 125.25 km long has been excavated.

The Shanxi Wanjiazhai Yellow River Diversion Project combines the double shield TBM tunneling technology with the hexagonal honeycombed segment lining and pea gravel backfill grouting technologies, setting records for daily tunneling of 113 m and monthly tunneling of 1637 m. Specifically, 1 Robbins double shield TBM (Fig. 11.102) is used for the #6, #7 and #8 tunnels of the main line, with an excavation diameter of ϕ6.125 m and a tunneling length of 21.42 km. Four double shield TBMs are used for the #4, #5, #6 and #7 tunnels of the southern line, including 3 from

Fig. 11.101 NFM ϕ4.94 m double shield TBM

Robbins and 1 from NFM (Fig. 11.103), with a diameter of ϕ4.82–ϕ4.94 m and a tunneling length of 90.31 km. One Robbins double shield TBM is used for the #7 tunnel at the connecting section, with a diameter of ϕ4.819 m and a tunneling length of 13.52 km.

See Table 11.28 for main technical parameters of the NFM double shield TBM (ϕ4.94 m).

3. TBM construction

The geology along the #6, #7 and #8 tunnels of the main line is predominated by limestone stratum with N_2 red soil layers locally. Both the entrance and exit of the tunnels are covered by Q_2 and Q_3 loess, without groundwater or large geological structures. One Robbins double shield TBM is used for construction. The Contractor is Italy CMC. The excavation diameter is ϕ6.125 m and the advance diameter is ϕ5.46 m. The segments are 25 cm thick and 1.6 m wide, with 4 segments as one

Fig. 11.102 Robbins double shield TBM

Fig. 11.103 NFM double shield TBM

Table 11.28 Main technical parameters of NFM double shield TBM (ϕ4.94 m)

Item	Unit	Technical Parameters
Cutter head diameter	m	4.94
Total length of main machine + back-up	m	245
Outer diameter of front shield	m	4.94
Outer diameter of rear shield	m	4.854
Number of disc cutter (ϕ432 mm, 17 in)	Nr	33
Main thrust cylinder thrust	kN	8 × 2000 = 16,000
Auxiliary thrust cylinder thrust	kN	14 × 1850 = 25,900
Cutter head drive power	kW	6 × 250 = 1500
Cutter head rotation speed	r/min	0–9
Cutter head maximum torque	kN·m	2600(4.5 r/min)
Cutter head torque at maximum speed	kN·m	1300(9 r/min)
Cutter head breakout torque	kN·m	3900
Thrust cylinder stroke	m	0.70
Auxiliary thrust cylinder stroke	m	2.30
Total support force	kN	30,000
Main thrust system power	kW	37
Auxiliary thrust system power	kW	110
Support system power	kW	110
HV	kV	15
LV	V	690/400/50 Hz
Transformer power	kV·A	1 × 2000;1 × 800
Belt conveyor conveying capacity	t/h	488

ring. The segments can be classified into A, B and C types, with a design strength of segment concrete being C30. The grain size of the pea gravel backfilled is 5–10 mm and the grouting pressure is 0.2 MPa. Tunneling of the #6, #7 and #8 tunnels was started in July 1994, and was completed in September 1997 (lasting for 38 months), with a tunneling length of 21.42 km.

The geology along the #4, #5, #6 and #7 tunnels of the southern line is predominated by limestone (the first 57 km) and sandstone and shale interbedding (the last 33 km). Karst caves, groundwater and local soft layers are developed along the #6 tunnel, and groundwater, coal bed, swelling rocks and Motianling Large Fault (the influence zone is about 300 m long) are developed along the #7 tunnel. Four double shield TBMs are used, including 3 from Robbins and 1 from NFM. Specifically, the #7 tunnel is excavated for 41 km via TBM. The Contractor is the Wan Long Joint Venture made up of Italy Impregilo, Italy CMC and Sinohydro Bureau 43 Co., Ltd. The excavation diameter is ϕ4.82–ϕ4.94 m and the advance diameter is 4.20–4.30 m. The segments are 22 cm and 25 m thick, and are 1.4 m wide, with 4 segments as one ring. The segments can be classified into A, B, C, D and E types, and C55 concrete is used. The #4, #5, #6 and #7 tunnels of the southern line were completed in April 2001 and lasted for 44 months since issuance of the commencement order by the chief supervision engineer in September 1997, with a tunneling length of 90.31 km.

One Robbins double shield TBM is used for the #7 tunnel at the connecting section, with an excavation diameter of ϕ4.819 m, an advance diameter of ϕ4.14 m and a tunneling length of 13.52 km. The bid winner is the Italy CMC. Dolomite limestone and marlstone are developed along the tunnel, and the groundwater level is below the tunnel line. The segments are 25 cm thick and 1.2 m wide with 4 segments as one ring. The segments can be classified into A, B and C types, and C45 concrete is used, setting records for the maximum daily advance of 113 m and the maximum monthly advance of 1637 m. TBM tunneling started on December 2, 2000 and this tunnel was completed on September 27, 2001, lasting for 10 months.

The double shield TBM includes the main machine and the back-up system. The former is made up of the front shield, the rear shield and the telescopic casing linking the front and rear shields. The front shield supports the cutter head, cutter head drive motor and the gearbox, and the main support system and the auxiliary thrust system are provided inside the rear shield. The telescopic casing between the front and rear shields is linked by the horizontal thrust cylinder, and the back-up system is made up of 1 back-up gantry containing various kinds of equipment.

The rear shield is firmly fixed onto the tunnel wall through grippers, and the thrust cylinder inside the telescopic casing acts on the rear shield, providing thrust for the cutter head. Meanwhile, the cutter head drive motor drives the cutter head to rotate, and the disc cutters at different positions of the cutter head form the excavation trajectory on the tunnel face with rotation of the cutter head. Once one excavation stroke is completed, i.e. the stroke preset for the thrust cylinder, the stabilizer inside the front shield will extend and support the tunnel wall, and the gripper of the rear shield will retract to its original position. The thrust cylinder acts on the front shield, and starts retracting to the original position. With the return of the thrust cylinder, the rear shield and all back-up systems linked to it will move forward for 1 stroke, to complete 1 tunneling stroke.

(1) Excavation muck

A bucket and a muck inlet are provided around the cutter head, and the rock muck excavated falls under the effect of gravity. The bucket rotates with the cutter head, and collects rock muck from the lower tunnel. The rock muck will fall into the muck hopper under the effect of gravity once rotating to the upper tunnel, and then be unloaded to the belt conveyor for muck discharge via a muck chute, and then to the train via the belt conveyor. The train departs once the rock muck of 1 excavation cycle is loaded.

(2) Installation of segments and backfill grouting of pea gravel

The double shield TBM is characterized by excavation and lining once and synchronous excavation and lining. A segment erector is installed at the tail of the rear shield, with 6 degrees of freedom, to guarantee accurate in-place of the segment.

Installation of segments and TBM tunneling are free from mutual interference, so they can be made simultaneously. Once the segments are installed, the gap between the outer side and rocks shall be backfilled with pea gravel (grain size: 5–10 mm) first and then with cement mortar, making it become pre-loaded aggregate concrete. In this way, stability of segments during construction is guaranteed, and the segments can closely fit with the rocks, so that they can withstand the external force as a whole. The pea gravel shall be delivered to inside the tunnel via a special tank vehicle, and be pumped-in through the holes reserved on the hose and segments. The segments on the bottom arch shall be backfilled first, and then those of the arch on both sides and finally those on the top arch. Grouting shall be made at the area about 150–200 m away from the back-up tail at a pressure of around 0.2 MPa.

Since the bottom of the TBM rear shield of the International #V Lot #7 tunnel at the connecting section of the Shanxi Wanjiazhai Yellow River Diversion Project is open, the segments on the bottom arch can sit on rocks directly. To guarantee certain gap between the segments on the bottom arch and rocks, 4 square supporting legs (3.5 cm high; side length: 33 cm) are designed for such segments. In this way, the stability after installation of segments is guaranteed and pea gravel can be backfilled and cement can be grouted between the segments on the bottom arch and rocks, forming an integral part.

(3) Tunneling guidance

A laser target is installed on the front shield, and a laser machine is installed at the area 100–250 m away from the laser target. The surveyor shall determine the position of the laser target and the laser machine through construction guide lines. Current direction parameters related to TBM are measured through the laser from the laser machine on the laser target, and delivered to the computer of the TBM cab. The computer will then compare such parameters with the target orientation preset by the surveyor, so that correction parameters for TBM tunneling direction can be provided for the TBM operator at any moment. According to the tunneling speed and advance, the surveyor may measure the construction guide lines and move the laser machine per day or once a day.

(4) Extension of tunnel wind, water, power and track

Air duct boxes, water hose reels and cable reels are installed at the back-up tail, and about 250 m air ducts, 150 m water hoses and 250 m cables can be saved respectively, extending forward with the back-up system. The track is extended at the connections between the rear shield and the back-up system.

Moreover, a gas detector is installed on TBM to monitor possible gas and can shut down TBM automatically, with a warning sound being given.

(5) Construction quality control

After 7d grouting of the tunnel, 5 holes shall be taken for water pressure tests per 100 m, with a diameter of 38 mm. The depth of the holes shall reach half of the pea gravel concrete layer, with a steady pressure of 0.2 MPa. The pressure shall be read per 10 min when the pressure is stable. If the grout take is less than 0.1L, the tunnel 10 m in front of and behind such hole is qualified. On the other hand, coring test shall be made after 28d grouting and 3 cores shall be taken per 100 m with a diameter of 50–75 mm. Once the compressive strength of the core sample reaches 20 MPa, the tunnel 15 m in front of and behind such hole is qualified.

(6) Water-stop measures

Firstly, backfill grouting of pea gravels, and then swelling water-stops and finally the jointing mortar at dovetail slots. The water-stop is made of chloroprene rubber (hardness: 45 ± 5; tensile strength: more than 3.5 MPa; the maximum elongation: above 500%), which will swell for 100%–200% in case of water. The Handan grade-525 cement and 0–1.25 mm fine sands are used for jointing mortar, with 4–5% Master swelling agent being mixed. The joint surface shall be applied with Master emulsion adhesive binder before jointing. Both the slots reserved for water-stop and the dovetail slots are located within 1/3 (68 mm) at the upstream side on the contact surface between segments. Such slot is 22.5 mm wide and 8.5 mm deep, and the dovetail slots are 35 mm wide. The spacing between the slot reserved for the water-stop and the dovetail slots is 8 mm.

(7) Segment fabrication

Concrete prefabricated segments shall be hexagonal (Fig. 11.104), with 4 segments as one ring. The segments are 22 cm and 25 cm thick and are 1.2 m, 1.4 m and 1.6 m wide respectively depending on the hole diameter. They are classified into A, B and C types as well as A, B, C, D and E types based on rock grade. Concrete segments are manufactured at the prefabrication plant.

The error in side length and diagonal of segment molds shall not exceed ± 1 mm. The production scale and operation shifts of the prefabrication plant shall be determined as per the tunneling speed and the requirements for construction period. The rapid steam curing process is adopted for the segment plant, including reinforcement processing, concrete pouring and curing. The concrete entering the silo

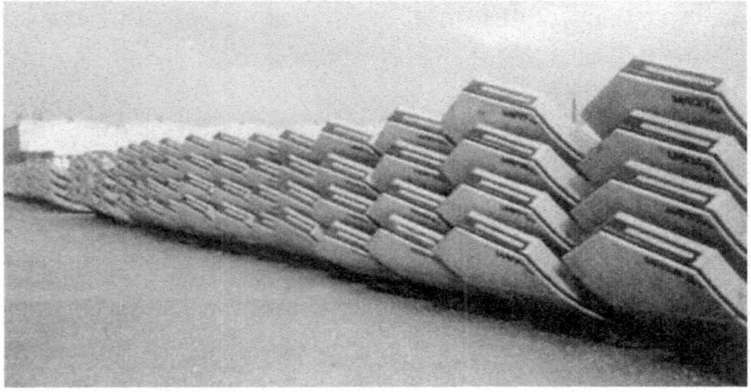

Fig. 11.104 Hexagonal segments

shall be vibrated via a hydraulic vibration bench with manually vibration, and one segment poured shall be pushed into the pre-heating kiln (45 °C, 0.5 h), and then be transferred to the HT kiln (80 °C) for steam curing immediately for 2–3 h. The segments de-molded shall be lifted to inside the plant building and be covered for curing for about 48 h and then be stored in the open air. Each segment shall be subject to visual inspection during pre-cooling, and quality problems such as honeycomb, spongy surface and fallen corners shall be repaired immediately. If such repair is impossible or defects still exist after repair or in case of other quality problems, such segment shall be delivered to the waste treatment plant for disposal or other purpose.

4. Problems encountered in TBM tunneling

(1) Karst cave

Two large karst caves (30–50 m^3) have been encountered when TBM tunneling is progressed to the #6 tunnel of the main line in the Project. The following methods are adopted: Shut down first and then observe through the manhole on the head, and then backfill the bottom with pea gravels or concrete and compact it as per the inspection results of karst cave. When the bottom is filled to the elevation of excavation diameter, start TBM and install segments while moving forward. Manholes are digged on two segments for filling of aggregate, grouting or filling of concrete at other positions of the karst cave on both sides and at top arch, so that the karst cave is filled densely with concrete, and form an integral part with the segments installed, playing a role of complete rocks.

To prevent sinking of the head due to karst, the TBM used for the sections where karst is developed is equipped with a probe drilling machine. The depth of probe drilling shall be larger than daily tunneling length, to guarantee the safety of TBM tunneling. Probe drilling may be carried out during daily maintenance of TBM.

For some small karst caves, pea gravels can be backfilled between the lined segments and rocks after TBM tunneling, and then be grouted. Karst caves are also encountered during tunneling of the #6 tunnel of the southern line and are treated in the same manner.

Large karst caves are encountered during tunneling of the #7 tunnel at the connecting section, since lacking of sufficient support force from rocks and hard sealing of segment joints, the segments installed are connected with steel plates, and pea gravels are filled in the karst cave as much as possible.

(2) Fault fracture zone

The Motianling Large Fault, a tectonic breccia, is a regional large fault in the #7 tunnel of the southern line and its influence zone is about 300 m long. On May 28, 1999, TBM was stuck when tunneling towards the influence zone of the Motianling Fault and tunneling had to be stopped.

Once TBM was stuck, the probe drilling machine was used firstly for chemical and cement grouting, and then the upper pilot tunnel was excavated for chemical and cement grouting in front of and above TBM, with cement grouting for the collapsed body ahead. On August 8, a large cavity was identified in front of the upper pilot tunnel with cement backfill grouting. The loose rock mass around the TBM cutter head was excavated during consolidation grouting of the collapsed body, guaranteeing successful launching of TBM on August 31.

The rear shield was deformed due to excessive pressure of the TBM rear shield from loose rock stratum, and half of the connecting bolts of the shields installed for segments were broken. Besides, steel plates were welded, so that TBM can pass stably during tunneling and treatment. After above 100d day-and-night hardwork, TBM passed through the large fault successfully on September 7.

Dolomite limestones and marlstones are mainly developed along the #7 tunnel at the connecting section. The TBM was stuck at the near-core part of an arrested anticline where extension fracture is developed. Karst caves and interlayer shear zones are developed between the dolomite limestones and marlstones, leading to interlayer staggering due to tectonism and crushed rocks. This tunnel has a buried depth of 300 m, with a high ground stress and the maximum main stress of 10 MPa, exceeding the uniaxial compressive strength (R_c < 5 MPa) of marlstones. Large concentrated stress is generated from TBM excavation, leading to rapid interlayer shear action, with a deformation rate of 3–4 cm/h. TBM was stuck in the midnight on June 18, 2001.

Handling methods: Two geological observation windows were opened on the telescopic shield, and one 500 mm × 500 mm (side length) window was cut each in the middle of both sides of the front and rear shields. Overcutting was made towards up, down, front and behind of the head via an air pick through these 6 side windows, so that the outer wall of the shield can be disengaged from the rocks as soon as possible, to prevent damage to the machine. The depth of overcutting was 400–500 mm and the rocks were supported with 100 mm channel steel and L70 mm × 70 mm angle steel during overcutting. Within the deformation zone of rocks, adjacent segments were subject to reinforced connection with 20 mm-thick steel plates and setscrews. TBM resumed tunneling after 9d and then passed through this section rapidly without shutdown. Heavy segments were modified, to enlarge the load-bearing capacity of segments, with backfilling of pea gravels inside the cavity immediately and further grouting to guarantee the tunnel quality.

(3) Swelling rocks

Considering series of unfavorable engineering characteristics such as swelling, shrinkage, collapse and softening of swelling rocks, the following measures were taken for TBM while passing by the medium and strong swelling-rock stratum 2.5 km long in the #7 tunnel of the southern line:

① Reinforced lining supporting: The saturated extreme swelling pressure of swelling rocks in the #7 tunnel can reach above 3.0 MPa, so the possible load on segments from swelling force of rocks is fully considered during structural design, to guarantee the safety of lining structure.
② Anti-seepage: Special attention is paid to control over joint width of lining segments and installation quality of water-stops during construction. The loss in water content of swelling rocks shall be as less as possible, to prevent TBM sinking due to rock collapse and softening.
③ Increasing excavation section: To reserve certain swelling deformation of rocks, the protruding amount of the disc cutter at the edge is increased during construction or relieved teeth are welded at the edge of the TBM cutter head to realize overcutting, the amount of which shall be adjusted as per the engineering property of the rock stratum through which TBM passes and the stability monitoring data of rocks and tunnels.

(4) Soil layer

A long N_2 red soil layer was encountered during tunneling of the #6 tunnel of the main line, and such layer is plastic due to its high water contents, making the adhesive mud cut unable to be discharged from the muck hopper smoothly. Therefore, it can only be digged out slowly from the muck hopper.

Head sinking has also been encountered in soil layers, and this phenomenon has been seen in the Q_2 and Q_3 loess layers of the #7 tunnel of the main line. The maximum sinking reaches 30–50 cm at one location, forming a low-lying section at the tunnel bottom. The main reason is that the TBM operator fails to lift the head in advance, making it upslope gradually to offset the sinking. This is triggered by underestimate on whether the load-bearing capacity of this stratum can meet such a large pressure of the TBM head.

(5) Segment dislocation

Dislocation is a common problem in installation of segments, and each ring of the lining of the #6, #7 and #8 tunnels of the main line and the #4, #5, #6 and #7 tunnels of the southern line is made up of 4 segments. Each block and each ring shall be assembled in strict accordance with design requirements. However, considering the gap between the segments and rocks, the joints shall be aligned carefully and pea gravels shall be backfilled with grouting immediately to seal such gap, making the segment firm enough. The jointing evenness required in the Contract is not more than 5 mm (20–30 mm at some locations actually). The reasons for dislocation mainly include the followings: Firstly, The TBM excavation accuracy is too poor due to geological reasons. Secondly, the accuracy of segments is not high enough, and the installation personnel of segments are not skilled or careful enough. The surface of dislocation shall be jointed with mortar mixed with swelling agent. Links such as segment design, segment installation, pea gravel backfilling and grouting are improved through cooperation among the Construction Contractor, the Employer, the Designer and the Supervisor in construction with TBM of the #7 tunnel at the connecting section, making above 90% joints qualified (dislocation: less than 5 mm).

5. Construction experience

Six double shield TBMs have been used successively for the headrace tunnel of the Shanxi Wanjiazhai Yellow River Diversion Project, which is the project using the largest number of TBMs in China up to now. The following experience and lessons can be concluded:

The double shield TBM is used for construction of long tunnels under coordination from hexagonal honeycombed prefabricated segments, bringing the advantages such as fast, safe and simultaneous completion of tunneling and lining of this technology into full play.

Even though the construction quality of the #6, #7 and #8 tunnels of the main line is unsatisfactory, experience and lessons are concluded through cooperation among the Employer, the Designer, the Construction Contractor and the Supervisor, and the following satisfactory indexes are reached in construction of the #7 tunnel at the connecting section:

(1) The excavation error is controlled at $\pm$ 150 mm horizontally and $\pm$ 50 mm longitudinally (vertically);
(2) In total, 90% of segment dislocation is controlled within 5 mm and the segment joints are basically qualified;
(3) The pea gravel backfilling and cement grouting are basically full and completed once;
(4) The segment joints are smooth and uniform and are firmly bonded without tiny cracks;
(5) No obvious damage or other quality defects are identified in segment production and installation;
(6) The construction records for the maximum daily advance of 113 m and the maximum monthly advance of 1637 m of TBM are set in China.

TBM shall be used for long tunnels. If the geology ahead may block TBM tunneling or make TBM tunneling out-of-control, advance prediction shall be made for geology. It shall be shut down for pre-treatment of the geology ahead if necessary, or otherwise unfavorable geology will not only postpone construction progress, but also will make the project in a passive position.

If TBM is used for long tunnels and the tunnel diameter is not large enough, great attention shall be paid to traffic safety of light rail vehicles inside the tunnel, to avoid casualties. The traffic accidents in the tunnel using TBM on the southern line of the Shanxi Wanjiazhai Yellow River Diversion Project have triggered 5 casualties and 2 injuries in total.

If the double shield TBM is used, the installation personnel of segments must be trained first, or the experienced operators of segment installation shall be hired, or otherwise the installation quality and progress of segments cannot be guaranteed in initial phases.

The rock powder generated from the double shield TBM tunneling is deposited within the 120° range at the bottom of the tunnel, and the rock powder is so compacted by the main machine under its dead weight that it is difficult for cement grouting to penetrate into the rock powder layer, forming a weak layer with lower strength than the pea gravel layer after grouting.

Since all 6 TBMs used for the Shanxi Wanjiazhai Yellow River Diversion Project are double shield ones, the rocks can only be observed from the small windows at the bottom or sides of the shield, making geological mapping difficult, which is also a disadvantage of the double shield TBM.

11.2.3 Headrace Tunnel of Qinghai Datong River-Huangshui Diversion Project

1. Project Overview

Datong River-Huangshui Diversion Project is a long-distance water diversion project in Qinghai, and it aims at diverting the water in the Datong River to the Huangshui River in water shortage by crossing the Daban Mountain. It is made up of the Shitouxia hydro-junction, main canal, Heiquan reservoir, Huangshui north canal and the dike irrgation works on the south bank of the Huangshui River. As an integral part of the main canal, the headrace tunnel is located in Menyuan and Datong Counties of Qinghai, with the exit in Nala (91 km distant from Xining) from the cattle farm of the Baoku River to Kongjialiang in Datong County. Th exit is located at the Baoku River basin with an elevation of around 2941 m. The Daban Mountain and the mid-and high-mountains on both sides are located at the middle sections of the tunnel, and the elevation is generally 3300–4200 m, belonging to tectonic eroded high mountain areas with a relative height difference up to 1000 m. A ventilation shaft is set at the wide and gentle platform on the left bank of the Moshangou and the ground elevation is about 3282 m. The maximum buried depth of the headrace tunnel is 1028 m and the average buried depth is about 480 m. The tunnels with a buried depth more than 600 m are about 10 km long.

The headrace tunnel has a total length of 24,165.83 m, and the lower proterozoic to the Tertiary stratum are distributed along the line, with complicated lithology such as metamorphic rocks, igneous rocks and sedimentary rocks. The grade-II rocks of the entire tunnel are 7017.4 m, accounting for 33.2% of the total length; the grade-III rocks are 4923 m, accounting for 23.3% of the total length; the grade-IV rocks are 3841.52 m, accounting for 18.2% of the total length; the grade-V rocks are 5383.08 m, accounting for 25.4% of the total length. The construction section adopting TBM is 19,940 m, including 6587.65 m for grade-II rocks, 4152.75 m for grade-III rocks, 3841.52 m for grade-IV rocks and 5358.08 m for grade-V rocks. The main canal has key and difficult points such as ventilation and transportation of long-distance tunnels, construction of fault fracture zones, construction of HP water-rich stratum, construction of coal-derived gas stratum, construction under high ground stress, environmental protection, engineering survey and re-survey, etc.

The headrace tunnel is built by one Wirth ϕ5.93 m double shield TBM introduced by China Railway Group Co., Ltd. Prefabricated concrete segment lining, with an outer diameter of ϕ5.7 m and an inner diameter of ϕ5 m, is adopted. The segments are 35 cm thick with a ring width of 1.5 m, and 6 segments serve as one ring (including key segments). The maximum mass of a single segment is about 4.5 t.

2. TBM technical parameters

The TBM technical parameters are shown in Table 11.29.

3. Handling of TBM shield jamming

(1) Problems and Descriptions

TBM has experienced 10 large-scale and long-term jammings (Table 11.30) since entry in the fault zones (fault groups F5, F4 and F19) at the south edge of the Daban Mountain in April 2008, making the project unable to be progressed.

(2) Ten TBM jammings and breakout

① First TBM jamming and breakout

A. Process of TBM jamming

At 00:17 on April 3, 2008, the thrust was increased from 5500 to 9000 kN suddenly when TBM tunneling was progressed to K17 + 140.416. At the same time, the TBM driver found that the penetration was dropped to 0, and the cutter head was retracted immediately, but the TBM cannot move backward. Afterwards, the single shield tunneling mode was adopted, but the penetration was still 0, so it was judged that the TBM front shield was jammed.

B. Analysis on cause of TBM jamming

The distance between the TBM cutter head and the front shield tail was 4 m. If calculating as per the most adverse conditions, the distance between the worst jamming location and the cutter head was 4 m. At 19:50 on April 2, 2008, when the TBM tunneling was progressed to the chainage K17 + 144.153, the thrust was 4169 kN, the cutter head rotation speed was 3.6 rpm and the penetration was 14.6 mm/r, and the TBM tunneling was normal. However, when the TBM tunneling was progressed to K17 + 140.416 at 00:17, the thrust was increased to 9000 kN and the penetration also dropped to 0, and TBM was judged to be jammed. The TBM has excavated for 3.737 m in this process, lasting for 4 h and 27 min.

Even though the appearance of this rock is complete, swelling intercalated mud is still developed locally with microfissure of a compacted structure. The tunnel has a large buried depth, and rock convergence is fast after excavation since affected by the high ground stress. The excavation diameter is 593 cm, and the cutter head is lifted for

Table 11.29 Main technical parameters of double shield TBM (ϕ 5.93 m)

S/N	Item	Parameter	S/N	Item	Parameter
1	Excavation diameter	5930	18	Gripper support force (kN)	45,000
2	Maximum torque (3.9 r/min) (kN·m)	4000	19	Maximum ground pressure of gripper (MPa)	4
3	Rated torque (7.4 r/min)(kN·m)	2000	20	Belt width (mm)	800
4	Cutter head breakout torque (kN·m)	5200	21	Conveying capacity of TBM belt conveyor (m^3/h)	600
5	Cutter head drive power (kW)	6 × 286 = 1716	22	Belt conveyor speed (m/s)	0–2.5
6	Cutter head rotation speed (variable frequency) (r/min)	0–7.8	23	Belt conveyor (kW)	55
7	Disc cutter (pcs.)	50	24	Exploration drilling rig (kW)	55
8	Maximum bearing capacity of a single cutter (kN)	250	25	Segment backfill mortar (kW)	15
9	Stroke of main thrust cylinder (mm)	1800	26	Deduster (dry-type) (kW)	53
10	Total thrust of main thrust cylinder (kN)	25,000	27	Secondary ventilator (kW)	45
11	Stroke of auxiliary thrust cylinder (mm)	2350	28	Air compressor (kW)	220
12	Total thrust of auxiliary cylinder (kN)	34,000	29	Water supply system (kW)	65
13	Total mass of TBM (t)	About 450	30	Waste water pump (kW)	16
14	Back-up mass (t)	About 260	31	Pea gravel pump (kW)	30
15	TBM length (m)	12.5	32	Total installed power (KW)	2810
16	Total length of TBM (m)	160	33	Transformer total capacity (kVA)	3200
17	Number of telescopic cylinders (pcs.)	14	34	Primary voltage/secondary voltage	20 kV/690 V

Table 11.30 Statistics of TBM jamming

Number of jamming	Date	Chainage	Reasons
First TBM jamming	April 3, 2008	K17 + 140.416	Convergence of rocks
Second TBM jamming	April 30, 2008	K17 + 135.284	Collapse + convergence
Third TBM jamming	June 29, 2008	K17 + 133.868	Collapse
Fourth TBM jamming	September 3, 2008	K17 + 117.984	TBM attitude out-of-control + convergence
Fifth TBM jamming	November 3, 2008	K17 + 102.586	Collapse + convergence
Sixth TBM jamming	December 24, 2008	K17 + 086.993	Collapse + convergence
Seventh TBM jamming	September 18, 2009	K16 + 905	Collapse
Eighth TBM jamming	October 18, 2009	K16 + 981	Collapse + convergence
Ninth TBM jamming	December 25, 2009	K16 + 876	Collapse + convergence
Tenth TBM jamming	July 23, 2010	K16 + 775	Tunnel face instability + convergence

20 cm. The TBM front shield has a diameter of 586 cm and the spacing between the shield top and the chamber is 12 cm. In other words, the 4.5 h convergence reaches above 12 cm, and the convergence speed reaches 27 mm/h.

Therefore, this TBM jamming was caused by too fast rock convergence.

C. TBM breakout

General ideal of this breakout: Clear surrounding rocks above the large span of the shield, and use steel mesh + planks + square timber + I-beam for temporary supporting, forming a space 1.2 m high, to release the pressure of the shield, thus helping TBM to break out. An air pick was used for excavation under coordination from weak blasting, and an annular chamber 1.2 m high was excavated outside the shield. Anchor rods, meshes, steel frames and timbering supports were used timely during excavation, with manual muck discharge in strict accordance with the established scheme and cycle advance.

An impact drill was used, and the bit diameter was 24 mm with a hole depth of 40 cm. In total, 4 holes were drilled per cycle, and 0.5 explosive was loaded in each hole. The millisecond detonator was detonated by flash detonators, and the emulsion explosives were detonated by the detonator. Eight holes were drilled at the complete part, and 1 and 5 ms detonators were used for blasting in sections. The millisecond detonator was detonated by one flash detonator. The muck was delivered to the observation window manually and then be delivered to the mining vehicle via the #1 and #2 belt conveyors, and finally be delivered to outside the tunnel.

D. Conclusion on breakout schemes

The breakout schemes of "manual excavation + temporary supporting" can effectively release the pressure on the shield from collapse and convergence of rocks, realizing TBM breakout, to meet the conditions of TBM tunneling. However, it has no prevention role for the next TBM tunneling. Temporary supporting uses the TBM shield structure as supports, so all supporting structures will be damaged after TBM tunneling.

Tunneling was resumed after TBM breakout on April 30, 2008, but it was jammed again while tunneling towards K17 + 135.284. This breakout lasted for 27d, and the TBM advance was only 5.132 m. Meanwhile, the methods of "manual excavation + temporary supporting" have great hidden danger.

If analyzing as per current conditions, the geological conditions did not turn good, and geological disasters such as collapse, fast convergence and groundwater accompanied by mud inrush may occur on the rocks on the tunnel face. Moreover, since the TBM operation was not stable, continuous and rapid passing cannot be guaranteed.

Therefore, this scheme has numerous disadvantages as for current construction of the project.

② Second TBM jamming and breakout

A. Process of TBM jamming

Tunneling was resumed after the first TBM breakout on April 30. 2008 and the single shield mode was adopted for TBM tunneling at that moment. Tunneling was stopped when the TBM tunneling was progressed to K17 + 135.280 for replacement of the muck car. However, the cutter head was found jammed and cannot be started when tunneling was resumed. The cutter head still cannot be rotated after the double shield tunneling mode was adopted, i.e. the clutch tripped when the cutter head motor was started manually. It was found through inspection that the front of the cutter head was filled with rock muck. The cutter head cannot be started first, and then the rocks of the entire front shield telescopic shield collapsed, leading to jamming of this shield. The tail shield was also jammed due to fast convergence of rocks.

B. Analysis on cause of TBM jamming

According to the conclusions reached on the expert meeting on May 10 after TBM jamming, this collapse occurred in the F5 fault, which is a boundary strip featured by large buried depth, high ground stress, poor rock integrity and changed lithology. The rocks on the tunnel face lost stability during TBM tunneling at this section under the single shield mode, leading to collapse and consequential cutter head jamming. The original supporting structure formed during breakout of the telescopic shield section became ineffective under the effect of high ground stress, leading to rock collapse and consequential TBM jamming.

C. TBM breakout

Once TBM was jammed, the project management office decided to handle with the muck hole in which the cutter head was jammed, attempting to discharge some of the muck in front of the cutter head, so that the cutter head can be started. Therefore, the muck at the cutter head was digged out. One face disc cutter was taken from the top right corner of the cutter head, to coordinate with opening the muck hole on the right of the cutter head. Up to the afternoon on May 6, 2008, 450 m^3 muck (not compacted) had been discharged. The collapsed body still pressed onto the cutter head and the shield according to the muck discharge known and the collapsed body observed at the muck discharge outlet. The method of muck discharge cannot help TBM break out.

According to the comments on the expert meeting, it was decided to enter outside the shield from the opening at the vault of the second last segment arch of the tail shield, for manual excavation of the rocks converged and the collapsed body from backward to forward, to help TBM break out. Once entering outside the shield, the rocks above the large span were excavated manually.

The aquifuge was damaged by the collapse at the cutter head and the front shield during TBM breakout, forming a channel linking the construction section and the aquifer. Before excavating towards the collapsed body, linear sprinkling occurred on the working face, which hindered construction to a large extent. Once excavation was progressed to the collapsed section, the collapsed body became liquid muddy structure after immersion into water. Numerous mud inrushes occurred during excavation, which greatly affected construction safety and progress. Ductules were inserted densely by the project management office during construction of this section, to shorten the excavation advance for forced supporting. The tunnel face was excavated to K17 + 131.28 in this breakout.

D. Conclusion on breakout schemes

The breakout scheme of "manual excavation + steel frames + bolt-shotcrete supporting" was adopted, which can effectively release the pressure on the shield from rock collapse and convergence. Meanwhile, the problems concerning cutter head jamming and control over further collapse of the tunnel face can be solved, realizing TBM breakout, so that the conditions of TBM tunneling can be met. However, it has no prevention role for the next TBM tunneling, and TBM tunneling safety cannot be guaranteed. Tunneling was resumed after TBM breakout on June 29, 2008, but it was jammed again when tunneling was progressed to K17 + 133.868. This breakout lasted for 60d, and the TBM advance was only 1.416 m.

There was no significant difference in the purpose and effect between the first breakout methods and this time. However, the construction safety was somewhat improved, but the section selected and the supporting structure have poor resistance to disturbance during TBM tunneling. Presently, the geological conditions did not turn good, and the geological disasters such as collapse, fast convergence and groundwater accompanied by mud inrush may occur on the rocks on the tunnel face. Moreover, since the TBM operation was not stable, continuous and rapid passing cannot be guaranteed. Therefore, this scheme has numerous disadvantages as for current construction of the project.

③ Third TBM jamming and breakout

A. Process of TBM jamming

The ring beam was replaced as per the approved scheme in the morning on June 27, 2008, and a long square timber was used to support the shield locally. The vault sinkage monitored was 5 mm, and the maximum elastic deformation of the ring beam support was 2 cm. The convergence at 14 ring beams was 4 mm and the height of 5 cm was cut at the ring beam bottom. TBM started tunneling at 5:30 by rotating the cutter head in a trial manner.

The double shield mode was adopted for the cutter head, with grippers for local supporting. The supporting shield and the rear shield were used to provide the supporting reaction force. The double shield tunneling mode was proven successful, and the front shield was broken out. The front shield was pushed forward successfully for 600 mm (the original groove excavated for cutter head breakout), and TBM was broken out successfully and moved forward normally. However, when the cutter head moved to K16 + 133.868, the right side of the tunnel face collapsed under the disturbance from TBM. Even though stability can be guaranteed for the rocks ahead during manual excavation, continuous collapse occurred from the tunnel face under the TBM tunneling mode. Finally, the collapsed body linked to the original collapsed chamber at the ring beam at around 8:30. The height of the collapsed chamber was unknown. The collapsed body in front of the cutter head and the 13 ring beams buried the cutter head shield. Since the area above the supporting shield of the 19–22 ring beams was empty, the ring beam dropped in front of the cutter head, and linked to the original collapsed chamber, making the cutter head unable to move forward. Therefore, it was decided to stop tunneling, and the equipment was jammed again.

B. Reason for TBM jamming

The direction reason for this TBM jamming was poor stability of the rocks at the cutter head and on the tunnel face. The front shield was located within the range of the original collapsed body, and the tunnel face collapsed during tunneling, making the rocks at the ring beam in front of the cutter head loose, thus damaging the supporting structure under the common effect of rock pressure and arch foot instability. The unstable support linked to the original collapsed body, leading to larger collapse. Both the unstable supporting structure and the collapsed rock mass pressed onto the cutter head, making it jammed, leading to TBM jamming.

C. TBM breakout

Before breakout, the ring beam was adopted for the existing undamaged supporting structure first for reinforcement, including 1–14 original ring beams. Specifically, 1–9 ring beams were reinforced with square timber radially, with 3 supporting points (1 in the center and 1 each on both sides symmetrically) for each ring beam. Shotcreting was made again for the ring beam after supporting. Ring beams were inserted inside the 7–14 ring beams, with 120 cm C20 concrete strip foundation being applied at the arch foot.

Once the original support was reinforced, conventional excavation methods were adopted for the collapsed body and the area in front of the cutter head. The excavation surface from the cutter head surface to the rear shield was 50 cm above the shield bottom. Full face excavation was adopted in front of the cutter head, and the excavation progress was 7.5 m advance for the cutter head, with 50 cm-thick shotcreting for invert. The TBM passed by through stepping. In excavation at non-collapsed sections, since the rocks were of a compacted structure and the mother rocks had a high hardness, there was great difficulty in manual excavation, so blasting was made locally.

After full face excavation for 7.5 m in front of the cutter head (to K17 + 126.368) and construction of the invert, according to the results of probe drilling on the tunnel face, the rocks ahead were under a unstable–extremely unstable state. To prevent further spreading of collapse during TBM tunneling and re-jamming due to failure in rear support, the long canopy tube (8 m long; external angle: 10°–15°) grouting must be adopted for the rocks ahead for advance pre-reinforcement before TBM tunneling.

D. Conclusion on breakout schemes

The methods resuming TBM tunneling were basically identical to those of the second breakout. The only difference was that the original supporting structure was reinforced. The methods of "advance full face excavation + advance pre-reinforcement" were adopted this time, which played a prevention role for the next TBM tunneling. However, since the total length of "advance excavation + advance pre-reinforcement" was only 15.5 m, so the TBM tunneling beyond this range in front of the cutter head (the area in front of K17 + 118.368) still lacked of preventive measures. On September 3, 2008, when TBM tunneling was progressed to K17 + 117.984, TBM was jammed due to attitude out-of-control and rock convergence.

This breakout lasted for 64d, and the TBM advance was 15.884 m, which was still low. The main reasons were: The quantities of full face excavation were large, with manual excavation and muck discharge only. Moreover, the efficiency of auxiliary procedures was remarkably lowered due to space limit.

④ Fourth TBM jamming and breakout

A. Process of TBM jamming

The third TBM breakout was completed on September 3, 2008. After 7.5 m in front of the cutter head was excavated, TBM started stepping forward. In this process, since the bottom invert had lost bearing capacity after immersion into water and the rocks around the shield had been cleared, the gripper cannot play its role effectively, so the cutter head cannot be lifted. TBM plummeted towards the lower right direction during tunneling. After TBM tunneling was progressed to rocks, since TBM severely deviated, the cutter head cannot be lifted and adjusted leftwards effectively. When TBM tunneling was progressed to K17 + 117.984, the TBM attitude had severely deviated from the design position (deviated for 600 mm rightwards horizontally and for 770 mm downwards vertically on the cutter head surface), which had triggered difficulty in segment installation and poor installation quality. In this case, TBM tunneling cannot be continued, and the rocks converged, leading to TBM jamming.

B. Analysis on cause of TBM jamming

The direct reason for this TBM jamming was fast convergence of rocks under high ground stress, leading to jamming of the TBM shield by rocks and consequential jamming of this equipment. The indirect reason was that the TBM attitude was out-of-control due to insufficient bearing capacity at the bottom during stepping at the full face excavation section, leading to great deviation in tunneling direction and great difficulty in segment installation, and forced shutdown.

C. TBM breakout

Since the TBM attitude was out-of-control during stepping and tunneling and deviated severely towards the lower right direction, the gap at the tail of the upper left shield was too small during subsequent installation of segments with severe out-of-standard dislocation of segments. Moreover, since outside the segment was an original empty chamber manually excavated, the waist of the segment installed was bulged and longitudinal joints at the vault of segments were opened due to untimely horizontal confinement after segment installation. It was decided through research that the single shield tunneling be adopted for TBM the next to facilitate steering. To avoid damage to the existing segment structure by the thrust of the segment cylinder during single shield tunneling, it was decided to reinforce the segment first. Segment connection was reinforced with U160 channel steel and bolts, to enhance the structural strength of segment rings. Channel steel binder plates were added symmetrically on the inner and outer sides of the segment for reinforcement, with tie rods for connection through grouting holes. The binder plates were arranged horizontally and longitudinally, and connection among B, C and D blocks was mainly reinforced.

There was great deviation between tunnel axes and design during this TBM tunneling, and the tunneling was severely deviated, so conventional methods cannot be adopted to adjust the TBM attitude. It was decided through discussion by the parties involved in construction that the TBM be moved backwards, with a reinforced concrete guide table being poured in front of the cutter head for TBM steering.

The converged rocks at the shield section were excavated for this TBM breakout, to help TBM break out by releasing the pressure on the shield. Overcutting was adopted for the original supporting section from K17 + 128.586, with excavation to K116.586 on the same section. After the sections in front of the cutter head were excavated to the full face, concrete reaction frames were placed on the tunnel face, and square timber blocks were used as thrust frames for TBM retraction. TBM was retracted for 8 m and 6 segments were removed. C25 reinforced concrete invert was cast-in-place from the cutter head to the tunnel face and then TBM stepped forward, with guide rails, the first ring and guide tables being set.

The following data were obtained through actual measurement of TBM and the segments installed before TBM stepping: The last segment (4508 rings) was 354 mm higher than the design elevation. The shield tail was upward, and was 660 mm higher than the design elevation. The cutter head was sank and 90 mm lower than the design elevation. The guide table cutter head was 90 mm lower than the design elevation. In combination with practical conditions on site, through calculation, if building as per the existing segment elevation and horizontal and vertical attitude of the shield, the back-up system will be unable to pass through after segment installation vertically. To guarantee smooth TBM stepping and tunneling, through measurement of practical conditions on site, it was decided by the project management office that the last four

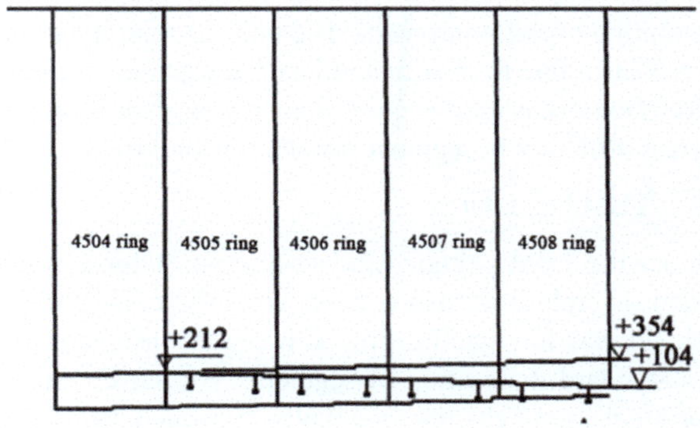

Fig. 11.105 Adjustment of TBM attitude (elevation Unit: mm)

segments be chiseled for 15 cm downwards (moving the four tracks downwards for 15 cm, to guide downward movement of attitude for 15 cm when TBM passed by), with the concrete accessible of the TBM at the lower part of the four segments being chiselled away.

The first four segments were installed outside the shield during TBM stepping, and the upper surface at the lower edge of the segment was aligned with the installed segments chiseled for 15 cm downwards, to guarantee smooth tracks. The C block of the segments (4513, 4514, 4515, 4516, 4517 and 4518 rings) installed after TBM stepping was replaced with steel segments, and D block of these six segments was centered through straight jointing, to guarantee that the C block was always at the top right corner. The back-up system of TBM passed through the section adjusting TBM attitude successfully after lowering TBM as a whole and increasing the spacing between TBM and segments with steel segments. See Fig. 11.105 for the adjustment of TBM attitude.

⑤ Fifth TBM jamming and breakout

A. TBM jamming

After TBM steering and treatment of the bottom of the ultra-high segments at the rear part of the tail shield, TBM started tunneling on November 2, 2008 at the chainage K17 + 116.586. It was found that TBM was unable to regrip after tunneling to K17 + 102.588 (13.998 m). Severe collapse of the rocks at vault was identified through the observation holes of telescopic shields and observation of the cutter head, and the waist rocks were close to the shield, and TBM jamming was confirmed.

B. Analysis on cause of TBM jamming

Tunneling was normal before shutdown (3:00 a.m. on November 4, 2008), and the parameters were: cutter head rotation speed of 1.3 r/min; torque of 740 kN·m; penetration of 20.2 mm/r; thrust of 4732 kN). Afterwards, failure in regripping occurred, indicating that fast convergence and collapse occurred at the shield top from excavation to installation of segments due to broken geology, leading to jamming of the shield by the collapsed body and TBM jamming.

The chainage was K17 + 102.588 when TBM was jammed, and TBM had entered rocks wholly. Severe collapse of the rocks at vault was identified through observation via the cutter head, observation ports and observation outside the segment with steel segments. The rocks at waist were converged under high ground stress and stayed close to the shield. It was concluded through analysis that this TBM jamming was triggered by the collapsed body at the vault of the shield and squeezing of the shield due to convergence of rocks around.

C. TBM breakout

The overall scheme of this breakout: The collapsed body at the upper shield was excavated and supported from the tail shield to release the pressure on the shield (excavation range: the arch rings above 50 cm below the large span), to help TBM break out.

At the exposed section of the collapsed chamber, the wall of which was firmly supported with an arch being erected and shotcreting for supporting. At the position with thick collapsed body, it passed through from inside the collapsed body without clearing the collapsed body. After completion of excavation and supporting, an advance large canopy tube was drilled on the tunnel face at the excavation section formed before TBM tunneling. See Fig. 11.106 for the breakout scheme.

The collapsed chamber was further expanding under the effect of crushed rocks after TBM jamming. The highest area of such chamber was above 10 m through observation. During excavation of the collapsed body, muck slip and secondary collapse occurred frequently, and the muck slipped damaged the advance small canopy tube for numerous times, making construction extremely hard. In actual construction, the construction scheme of excavation with methods similar to CRD (advance excavation of a single ring beam, with 20 cm-thick shotcreting of each ring beam for supporting), and advance excavation of the small pilot tunnel on the right side to the cutter head and reverse excavation of the collapsed body after overcutting was adopted.

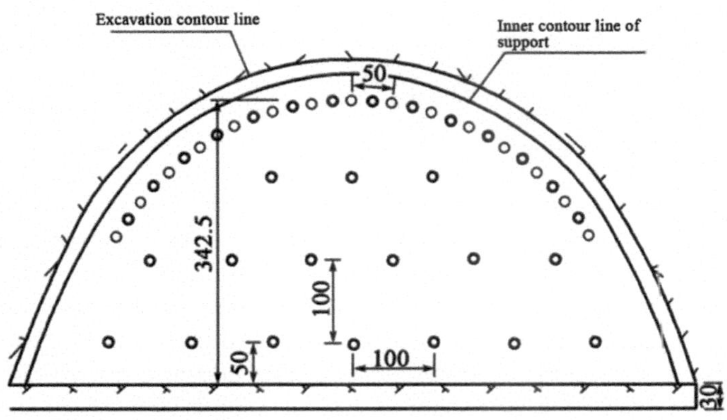

Fig. 11.106 Breakout scheme (unit: mm)

Once the first half section of the cutter head was excavated to K17 + 096.471, according to the requirements of the Employer, the Designer and relevant regulations, the project management office started advance large segment grouting on December 12 to reinforce construction of the rocks ahead. Once the tunnel face was sealed by 40 cm-thick C25 shotcreting, orifice pipes, of quincuncial layout, were used for grouting at K17 + 092–K17 + 097, with a hole depth of 5 m, a longitudinal and horizontal spacing of 1 m. At K17 + 082–K17 + 097, ϕ60 mm advance long canopy tubes, arranged at the arch, were used for reinforced grouting, with $L = 15$ m and a spacing of 50 cm. Orifice pipes were used for curtain grouting first, and advance long canopy tubes were used for advance pre-grouting after curtain grouting.

D. Conclusion on breakout schemes

Considering the construction scheme of "advance full face excavation + advance pre-reinforcement" adopted for the third breakout and the role played in the next TBM tunneling, the construction scheme of "half section excavation + advance pre-reinforcement" was adopted for this breakout. The length of advance pre-reinforcement was increased to 15 m. However, since the rocks ahead were of a compact structure if not disturbed, the grouting amount was too small. In total, 1324L grout was injected into 13 holes of large canopy tubes, and 101.8L grout was injected per hole averagely. The maximum grouting amount per unit hole was 250L (grout carry-over) and the minimum grouting amount was 20L (grout in orifice pipes after slurry carry-over). The maximum grouting ending pressure was 7 MPa for grouting of large canopy tubes, and the minimum grouting ending pressure was 2 MPa (see the construction report for advance consolidation grouting of the Qinghai Datong River-Huangshui Diversion Project).

On December 23, 2008, when TBM tunneling was progressed to K17 + 086.993, TBM was jammed due to rock convergence ahead and collapse of the rear canopy tube as a whole, leading to TBM jamming.

This breakout lasted for 50d and the TBM advance was only 15.593 m. The main reasons were: Manual excavation and muck discharge were adopted only. Moreover, the efficiency of auxiliary procedures was remarkably lowered due to space limit.

⑥ Sixth TBM jamming and breakout

A. Process of TBM jamming

TBM tunneling started on December 23, 2008, and tunneling and final equipment commissioning were carried out in a synchronous manner. Commissioning of equipment systems was completed after tunneling for 3.754 m, and all such systems were under a normal working state. The TBM tunneling was progressed to the tunnel face at 3:40 a.m. on December 25, and TBM tunneling had been normal after that, and the muck quantity was normal. However, when TBM tunneling reached 14.388 m at 12:09 on December 25, the front shield of TBM was found unable to move forward. At this moment, the double shield mode was adopted, and the rear shield moved backwards for 5 cm. The TBM front shield was confirmed jammed, and the cutter head chainage was K17 + 086.993.

B. Analysis on cause of TBM jamming

Once TBM was jammed, the rocks at the front shield were converged as a whole through observation of the cutter head and the front shield, leading to jamming of the front shield and consequential TBM jamming. There was 0.8 m empty chamber between the cutter head and the tunnel face during TBM jamming, and the rocks around the cutter head were complete. There were traces of rocks around the edge cutter, and there was no gap between the front shield and rocks. Four advance canopy tubes at the vault of the tail shield fell onto the tail shield directly due to collapse, and other canopy tubes acted on the tail shield with rocks around, and the rocks stayed close to the tail shield. This jamming occurred under the common effect of rock convergence and collapse.

C. TBM breakout

The overall scheme of this breakout: Enter the outside of the shield from the opening (left side of the tunnel entrance) of the existing steel segments, and excavate and support towards the cutter head from backwards. Excavation started from the original cutter head chainage (K17 + 101.381), and overcutting was required for the supporting section of the original six ring beams in front of the cutter head with excavation towards K17 + 086.993. The excavation section was a portal section with an vault clearance of 120 cm and a large span clearance of 80 cm. Steel supports + anchor rods + advance ductule + steel meshes + shotcreting were adopted for supporting.

D. Construction of test sections for half section excavation

After completion of the sixth breakout, it was required in the document issued by the Employer to compare half section construction with the full face drill and blast of bypass tunnel. For this, according to the excavation section and supporting parameters issued by the Designer, the test sections were built before drill and blast of bypass tunnel. The upper pilot tunnel was excavated to K17 + 059.7 at this section, and 27.293 m was excavated in total, lasting for 37d. The half section TBM tunneling had reached 22.353 m since drill and blast of bypass tunnel on February 22. It was hard to control the TBM attitude in half section tunneling.

In half section construction, since it was hard to control the TBM half section tunneling attitude and the progress was slow, the rocks at the arch footing were much more crushed under the effect of changing rocks during excavation of the upper guide pit. Therefore, there was risk of arch falling during TBM half section tunneling, and the upper guide pit was too close to the TBM during excavation and blasting. Considering the adverse impact of impact wave on TBM, it was decided not to adopt the half section construction scheme after taking the followings into account:

E. Full face drill and blast of bypass tunnel

Full face drill and blast (hereinafter referred to as explosion) of the bypass pilot tunnel was started on February 22, 2009 and completed on September 14. Auxiliary

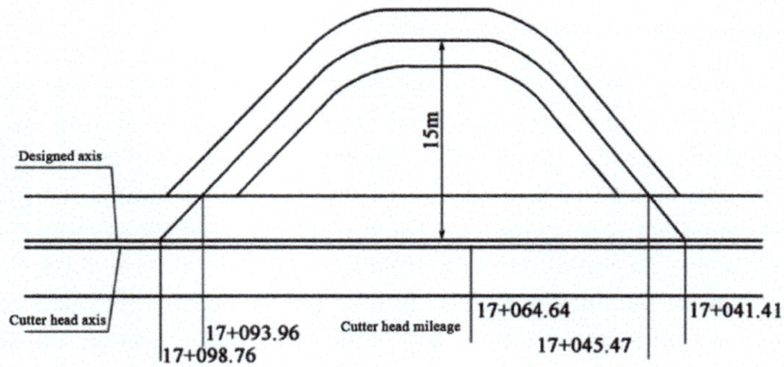

Fig. 11.107 Schematic diagram of project

works such as downward excavation for TBM breakout on both sides of the shield, backfilling of mortar rubbles at side walls in the main tunnel, construction of the first ring in the main tunnel and entrance and exit plugging of bypass tunnel were completed in addition to main works such as full face drill and blast (58.84 m for the bypass tunnel and 144.39 m for the main tunnel) of the bypass tunnel and the main tunnel and concrete pouring (145.39 m) of the invert in the main tunnel within this period. All the above works lasted for 234d. The bypass tunnel had been excavated for 58.84 m and the main tunnel had been excavated for 144.39 m. In total, 145.6 m of the main tunnel invert had been lined, and TBM stepped for 144.64 m, with 96 segments being installed.

a. Construction of the bypass tunnel

The opening chainage of the bypass tunnel entrance was K17 + 093.96, while that of the exit was K17 + 045.47. The intersection angle between the entrance and exit of the bypass tunnel and the main tunnel was about 40°, and the axis distance of the bypass tunnel and the main tunnel was 15 m (Fig. 11.107).

b. Construction of entrance/exit locking of bypass tunnel

Construction of the bypass tunnel entrance opening was started on February 22, 2009, and four segments were removed with construction of the bypass tunnel entrance opening. The two segments in the middle were removed first and then the segments on both sides of the opening after the tunnel opening locking was firmly supported. Roof ripping was not adopted between the bypass tunnel exit and the main tunnel, and excavation was stopped when the bypass tunnel exit surface was excavated to the side walls of the main tunnel for locking. The locking sections at the bypass tunnel entrance and exit were reinforced with I20 steelworks arranged in a fanshaped manner.

c. Construction of bypass tunnel body

The bypass tunnel was of a portal section, and the upper and lower sections were excavated into short steps based on geology. The steps were 2–3 m long, and the first sections were excavated manually under coordination from an air pick. Blasting was adopted for excavation after entering 20 m of the bypass tunnel. A small excavator was adopted for muck discharge during excavation of the bypass tunnel under coordination from mining vehicles since limited by section space, with a low working efficiency and slow construction progress.

d. Secondary lining of bypass tunnel

The 30 cm-thick C35 reinforced concrete was adopted for secondary lining of the bypass tunnel, with ϕ22 mm main bars of a longitudinal spacing of 25 cm being arranged and two bars being spaced for reinforcement; ϕ18 mm longitudinal bars with a spacing of 30 cm were arranged, and ϕ8 mm tie bars were spaced in a quincuncial manner. The bottom plate of secondary lining was poured in the total length of the bypass tunnel. Arch wall lining was arranged at the bypass tunnel entrance and exit as per deformation after initial support, i.e. K0 + 0–K0 + 26 and K0 + 50.114–K0 + 59.114 respectively.

e. Construction in the main tunnel

Construction in the main tunnel was started on April 5, 2009 and completed on August 21. Specifically, 144.39 m initial supports were excavated, 145.6 m invert was poured and 144.6 m guide rails (double-rail) were installed.

f. Excavation in the main tunnel

Excavation in the main tunnel was started after construction of the bypass tunnel. The main tunnel excavated can be classified into two sections, i.e. egg-shaped section for K17 + 064.39–K17 + 035 due to mutual influence with the bypass tunnel and horseshoe-shaped section for K17 + 035–K16 + 920, to reduce the excavation section and to lower the backfill amount behind after segment installation. To avoid roof ripping while entering the main tunnel from the bypass tunnel, the main tunnel was excavated from the upper section in front of the cutter head before the bypass tunnel exit chainage. The lower section of the main tunnel was excavated from the bypass tunnel once the upper section passed by the bypass tunnel exit and the intersection point chainage of the main tunnel. Once the lower section of the intersection was excavated, construction of the two tunnel faces was carried out simultaneously. Short steps were excavated towards the tunnel entrance for initial support, and the steps were 3–7 m long. For excavation in the main tunnel, an excavator was used for muck discharge under coordination from a crawler loader. However, since lacking of supporting equipment, the muck discharge efficiency was low, and the time required for muck discharge per cycle was too long, affecting construction progress.

F. Conclusion on construction scheme

This drill and blast construction was successful as a whole. TBM passed by the unfavorable geological sections successfully through stepping through full face excavation, and the TBM attitude was adjusted through invert guiding. However, with regard to progress, the expected indicators were not reached and the reasons mainly include the followings:

a. Manual excavation and muck discharge were adopted only from the bypass tunnel entrance opening and the cutter head on the upper section of the main tunnel to the bypass tunnel exit, leading to a slow progress.
b. Since reinforced supporting was required for construction at the bypass tunnel exit and the main tunnel intersection, weak blasting was adopted with numerous procedures, leading to a long time consumption since affected by construction safety.
c. The bypass tunnel was severely deformed after initial support, and normal procedures were disturbed by ring beam reinforcement and secondary lining.
d. Local collapse occurred for numerous times under the effect of unfavorable geological conditions, leading to repetitive re-work.
e. A small amount of construction equipment was available for choice due to space limit of the excavation section and influence from ventilation in the tunnel and TBM, leading to a low construction efficiency.

⑦ Seventh TBM jamming and breakout

A. Process of TBM jamming

After TBM stepping at the drill and blast section with system maintenance, TBM tunneling was recovered in the single shield mode at 8:30 a.m. on September 17, 2009. In total, 7 rings were tunnelled in the morning on September 18, 2009 and the tunnel face chainage was K16 + 909. Only small-scale chipping occurred during tunneling at this section, with subtle rock convergence. Sudden increase in tunneling pressure did not occur. To guarantee passing of the K16 + 900–K16 + 887 secondary fault zone detected by TSP for the second time once, the project management office decided to stop tunneling for equipment maintenance. TBM tunneling was recovered in the noon on September 18, 2009. When TBM tunneling was progressed to K16 + 907, the muck quantity was increased suddenly. At K16 + 906.6– + 906.1, the muck quantity reached 144 m^3, which was far beyond the normal muck quantity (21 m^3). At this moment, the deformation of the rocks around the rear shield and the telescopic shield was not severe in addition to chipping by observing in the observation window. However, the front shield top already touched with rocks, but there was still space between both sides and rocks. To prevent the front shield from being jammed, tunneling was recovered at 22:00 on September 18, 2009, and the cutter head was found jammed. The cutter head was then rotated clockwise and counterclockwise manually, but the muck quantity was still large. The muck quantity reached 32 m^3 when there was no advance and start. The cutter head still was not started normally after endeavoring for numerous times. The cutter head top was completely pressed by large collapsed body by observing in the cutter head from the #1 belt conveyor, and the area between the cutter head and the tunnel face was full of collapsed body. According to the muck quantity and the results of advance geological forecast, continued tunneling will further expand the collapse zone. To prevent shield jamming due to pressing by the collapsed body, we decided to adopt manual breakout other than TBM breakout by itself.

B. Analysis on cause of TBM jamming

The direct reason of this jamming was collapse of the rocks from the cutter head to the front shield due to instability, which triggered cutter head jamming first and then the consequence that the entire front shield was pressed by the collapsed body. The excavation of the TBM construction revealed that the surrounding rock at the front shield and the surrounding rock at the end of the telescopic shield had obvious changes in lithology, the rock changed from grayish white quartz diorite to light green gabbro, the cutter and the front shield were in the contact zone of the two different surrounding rocks, the stability of the surrounding rock was very poor and collapsed under the disturbance of the cutter, which led to the trapping of the front shield under the combined effect of high ground stress convergence.

C. TBM breakout

This breakout can be divided into two phases.

Phase I: rapid excavation of the small pilot tunnel on both sides to the cutter head surface.

The project management office immediately organized personnel and equipment for TBM breakout after TBM jamming. Since this jamming was triggered by collapse of the rocks at the cutter head in early phase, by borrowing the experience in double shield breakout in Xinjiang, the project management office decided to excavate the small pilot tunnel to the cutter head on both sides of the shield. The collapsed body was consolidated through chemical grouting. Once the collapsed body at the cutter head was cleared, TBM breakout was realized rapidly. The small pilot tunnel was excavated from the observation port on both sides of the shield manually via an air pick, with square timber + planks + steel meshes (10 cm × 10 cm) for supporting. In this process, the collapsed chamber was further expanding, and the collapsed body and the converged rock mass pressed onto the entire front shield. Besides, since the cementing body of chemical grouting was made of flexible materials whose strength is not enough to resist the high ground stress, rapid breakout of the small pilot tunnel was not realized.

Phase II: conventional excavation above large span of front shield and telescopic shield.

Once the scheme of the small pilot tunnel failed, according to the reasons concluded and the practical conditions (the rocks at the front shield were collapsed, and those at the telescopic shield converged rapidly and pressed onto the shield with certain clearance between the rock excavation surface at the supporting shield and the rear shield and the shield) on site, by borrowing the experience in double shield breakout in Xinjiang, we decided to adjust the TBM breakout scheme and to adopt full face excavation above the large span of the shield immediately, with steel supports + steel meshes (10 cm × 10 cm) + advance ductule + square timber for supporting. This enlarged excavation was carried out manually via an air pick based on the small guide tunnel on both sides. There was a loose channel at the arch of the excavated section since the rocks had been disturbed during excavation of the small pilot tunnel. Local collapse of the rocks on site was severe during secondary enlarged excavation, so this excavation was featured by high risk and great difficulty. There was one personal injury due to collapse of crushed rock mass during construction. After this breakout, according to the longitudinal geological profile and the results of TSP advance geological forecast, the project management office decided to pass through the rest unfavorable geological sections by manually excavating the upper pilot tunnel and the lower TBM section.

⑧ Eighth TBM jamming and breakout

A. Brief description of TBM jamming

At 7:00 a.m. on October 18, 2009, TBM excavation on the lower section was started after the upper pilot tunnel was excavated to K16 + 891.826. Since the single shield mode failed to provide sufficient thrust when TBM was started, the double shield mode was started. Since the stress on both sides of the supporting shield was unbalanced when the gripper was opened under this mode, the TBM tunneling deviated rightwards horizontally. After tunneling for 70 cm in current shift, the cutter head deviated rightwards, leading to blockage of the PPS2 prism, so automatic guidance failed. The TBM tunneling was adjusted leftwards instantly, but the cutter head failed to be adjusted successfully after tunneling for one ring. The survey team moved the guide system forward in the tunnel in the afternoon that day, but the PPS guide system did not be recovered under the existing conditions. Therefore, a single prism was measured manually, and it was determined that the #1 prism 3.1 m away from the cutter head surface deviated rightwards for 20 cm. After that, manual guidance measuring had always been made during TBM tunneling, which was always adjusted leftwards. However, TBM tunneling attitude failed to be adjusted leftwards until TBM was jammed again on October 19.

B. Analysis on cause of TBM jamming

a. Analysis on reasons for TBM tunneling attitude out-of-control and jamming

Reason 1: unbalanced stress of gripper when TBM was started.

When TBM tunneling was started after breakout, since the thrust under the single shield mode was insufficient, the double shield mode was started. However, the TBM deviated rightwards horizontally during startup due to unbalanced stress of the gripper on both sides of the supporting shield.

Reason 2: TBM tunneling stress on the upper section.

Since the upper pilot tunnel was empty above the semicircle during TBM tunneling, the cutter head rotated clockwise during TBM tunneling, and right was always pointed to horizontally with the counter-acting force from counterclockwise rotation of rocks. Therefore, TBM tunneling was always affected by the horizontal force rightwards, which was one reason why the TBM attitude was hard to be adjusted leftwards.

b. Reason for TBM jamming

The direct reason for this TBM jamming was that the tail shield was jammed by unstable supporting system and rocks. Above the tail shield was the collapsed body, and the lithology of the rocks at this section was changed, with distributed water being developed. After the square timber supporting system became invalid due to TBM tunneling, the supporting system lost stability rapidly under the dead weight of the collapsed body and the ground stress on both sides, and pressed onto the tail shield, leading to TBM jamming.

C. TBM breakout

This TBM breakout can be divided into two phases.

Phase I: replacement of primary supports.

According to the reasons for TBM jamming and practical conditions on site, construction of TBM breakout in phase I was made from the following three aspects:

Support replacement due to invasion of the upper pilot tunnel: The convergence of the section was severe after supporting due to high ground stress of this tunnel section. The existing supporting structure of the remaining upper pilot tunnel sections was converged continuously after TBM jamming, leading to invasion finally. Therefore, the project management office decided to replace the supports adopted at this section. The deformed sections were excavated manually with I20 A profile steel supports + ϕ22 mm longitudinal connecting reinforcement + ϕ42 mm advance ductule + + ϕ8 mm steel meshes (10 cm × 10 cm) + ϕ22 mm feet-lock bolts + C25 shotcreting for supporting. In total, 4.2 m had been excavated for replacement at this section, and 7 ring beams were replaced.

Replacement of unstable supports at the front shield and the telescopic shield: Since the attitude of TBM tunneling deviated rightwards, the supports on the right side of the original upper pilot tunnel section was damaged partially during TBM tunneling. The supporting system deformed rapidly under the effect of high ground stress, and stayed close to the right side of the shield. Manual excavation was adopted at this section, with I 20 A profile steel supports + ϕ22 mm longitudinal connecting reinforcement + ϕ42 mm advance ductule + + ϕ8 mm steel meshes (10 cm × 10 cm) + C25 shotcreting for supporting. In total, 5.4 m had been excavated for replacement for primary supports at this section, and 10 ring beams were replaced.

During excavation on the right side of the supporting shield and the tail shield, pressing of the tail shield was the direct reason for this TBM jamming. The supports were replaced towards the tail shield outside the shield from the observation port on the right side of the shield. Manual excavation via an air pick was adopted at this section, with I 16 profile steel supports + ϕ22 mm longitudinal connecting replacement + ϕ42 mm advance ductule + + ϕ8 mm steel meshes (10 cm × 10 cm) for supporting. Shotcreting was not carried out due to narrow space and groundwater. The groundwater channel was expanded due to secondary collapse at the shield top during excavation at this section. Since the amount of groundwater was remarkably increased than before, the operators can only wear raincoats for operation. Since collapsed body was distributed from the arch waist on the right side to the vault, the loose rock mass was softened after immersion and rushed out in a fluidal plastic manner during excavation. Advance ductules were inserted densely during excavation for each cycle. The construction difficulty was extremely large due to narrow space. In total, 4.5 m had been excavated for replacement for supports at this section, and 10 ring beams were replaced.

Phase II: replacement of secondary supports.

The phase I support replacement for breakout lasted to November 3, and 7 ring beams (I20 profile steel) in front of the cutter head were replaced. In total, 6 ring beams (I20 profile steel) were replaced for the front shield, and the right side of the supporting shield and the tail shield was excavated, with 9 ring beams (I16 profile steel) being installed. The upper pilot tunnel was excavated to K16 + 887.9 (after excavation of the upper pilot tunnel for supporting, the ring beam at the waist on the right side pointed to the tunnel face soon) for supporting. On November 4, the convergence rate of the supporting structure replaced in front of the cutter head and at the shield was increased suddenly, and the ring beam at vault was distorted. The waist on the right side of the ring beam installed after replacement in front of the cutter head and at the shield section had invaded again up to November 7, and stayed close to the shield (the original clearance of 60 cm). The ring beam at the vault was severely distorted, and broken locally. The square timber at the shield was deformed and broken, so TBM failed to break out as scheduled.

If analyzing as per this TBM breakout failure, the original supporting strength failed to resist the high ground stress, and the supporting structure not replaced at the upper pilot tunnel and the shield was subjected to continuous convergence under the effect of high ground stress. The ring beams replaced on the right side of the supporting shield and the tail shield also deformed again. Based on this, the project management office decided to re-formulate excavation and supporting parameters: The excavation section was expanded to 100 cm at vault, with a large span of 120 cm (reference spacing with the cutter head). The supporting parameters were reinforced, with a ring beam spacing of 50 cm. I-beam was adopted as the longitudinal connecting system of the ring beam, to enhance the longitudinal bending resistance. The grouting behind the initial support and inside the collapsed chamber was enhanced, to limit the loose range of rocks.

Second Stage The upper pilot tunnel section of phase II support replacement was completed completely, and 7.4 m had been excavated in total, with 14 ring beams being replaced. After the deformed supports of the upper pilot tunnel were replaced, according to the results of TSP advance geological forecast, the project management office decided to keep excavating forward manually to K16 + 880 temporarily.

⑨ Ninth TBM jamming and breakout

A. Brief description of TBM jamming

TBM half section tunneling (cutter head chainage: K16 + 895.3) was carried out on December 20, 2009, and tunneling (cutter head chainage: K16 + 884.8) was stopped on December 22, 2009, and then the shield and the cutter head damaged under the ground stress were handled with. TBM tunneling was recovered in the noon on December 24, 2009, after entering the full face, since located in the F4 fault featured by crushed rocks, with the forward tunneling of the cutter head, the vault was subject to continuous collapse. In the morning on December 25, when TBM full face tunneling was progressed for 5.4 m, there was no penetration with the gradual increase in tunneling thrust. Afterwards, the tunneling thrust was increased to 11,000 kN, but TBM tunneling still failed. The maximum height of the collapsed chamber at the front shield and the cutter head was found more than 5 m by observing from the observation port to the cutter head by manual dig-out of the shield observation port, with substantial rock mass falling onto the shield cutter head. Finally, the entire shield of TBM was jammed.

B. Analysis on cause of TBM jamming

The tunneling section was located in the F4 fault featured by crushed rocks, with red substances among rocks and clogged fillers locally. With the disturbance from the cutter head, the rock mass lost stability, leading to collapse. Afterwards, the ground stress at tunnel top was released, and both sides of the front shield were squeezed under the common effect of continuously acting horizontal ground stress and the dead weight of the collapsed body at top. Finally, TBM was jammed.

C. TBM breakout

This TBM breakout was divided into two parts, i.e. breakout of the front shield and replacement of the deformed supports from the supporting shield to the rear shield.

a. Breakout of front shield

Breakout of the front shield was carried out from the middle to both sides. After the ring beams at vault were installed as per the design section under the protection from the arch of the original collapsed chamber, a square timber was used to support the shield temporarily. Afterwards, the arch foot was excavated on the left and right, with timely supporting.

b. Replacement of the deformed supports from the supporting shield to the rear shield

The excavation section for replacement of ring beams this time was identical to that of front shield breakout. The ring beams were replaced from the last ring beam of the original upper pilot tunnel to the tail shield. The ring beams were stripped in sections one by one from right to left of the entrance direction. If the rock surface was loose and chipping during stripping, initial shotcreting may be made to stabilize the rock surface, or advance ductule may be arranged as per practical conditions on site for stripping of the existing supports.

The ring beams shall be inserted among the existing ring beams in principle. If the spacing of ring beams is too large, it may be reduced properly, with the number of ring beams being increased. The specific supporting parameters were identical to those of the front shield breakout.

⑩ Tenth TBM jamming and breakout

After the ninth breakout, the parties involved in construction decided to adopt the construction method of “excavation of the upper semi-section + TBM tunnelling on the lower semi-section” at this unfavorable geological section based on the experience in the previous 9 TBM jammings and breakouts. It can help avoid jamming when TBM passes through this unfavorable geological section, so this method is still adopted currently. However, the semi-section construction also has risks and difficulties. Accidents such as debris mud flow and deformation occurred for numerous times after the tunnel surface turned into mylonite intercalated with fault gouge on May 2, 2010.

On July 23, 2010, during manual excavation of the upper semi-section, the tunnel face was unstable, and the entire tunnel face was moved backward for 10 cm before supporting. Meanwhile, the rear supporting structure completed was deformed successively due to an extremely high stress. At night of July 23, it was temporarily decided by the Employer, the Supervisor, the Designer and us to plug the tunnel face on site, to handle with the rear deformed supporting structure. In total, 3 ring beams were replaced during handling of the deformed ring beams, and the size of excavation section was increased relatively the last time, with I22 two-layer ring beams + three-layer meshes + 50 cm-thick shotcreting + 5 m self-advancing anchor bolts for supporting. Finally, resisting to the damage by this special unfavorable geological section failed, and construction had to be stopped.

(3) Experience concluded in previous TBM jammings and breakouts

① Countermeasures must be taken for the soft rock tunnel with a high ground stress and the tunnel sections with a long fault fracture zone in advance
The soft rock tunnel with a high ground stress is featured by high ground stress and fast convergence, so the possibility of TBM jamming is extremely high, with great difficulties and danger for handling. Therefore, the double shield TBM construction shall not be adopted as much as possible under this geological condition.
The unfavorable geological section shall be handled with through shafts or inclined shafts in advance as much as possible if possible, and TBM shall pass by through stepping. Countermeasures must be adopted for the tunnel sections with a long fault fracture zone in advance.

② Poor effect in anti-jamming through chemical grouting
When TBM passes through short fault fracture zones, the use state of TBM, the attitude of TBM tunnelling and its smooth pass-by of main anti-jamming measures shall be guaranteed. In the Project, the effect of anti-jamming through chemical grouting is poor at the deformed soft rock tunnel sections with high ground stress. The plastic body which cannot withstand the dead weight of the collapsed body is used for chemical grouting, and will be squeezed rapidly under the high ground stress, so it cannot realize anti-jamming effectively.

③ Ring beams shall be adopted as much as possible for reinforcement in case of deformed initial supports, and the shield + square timber for supporting shall be avoided as much as possible, to prevent damage to the TBM main machine.

④ Breakout through small pilot tunnel and small face excavation is infeasible in the deformed soft rock stratum with high ground stress. Large face excavation can only be adopted for breakout, with supports being reinforced, to reserve sufficient deformation so that sufficient space can be provided for adding ring beams in case of severe deformation of initial supports.
⑤ The breakout scheme of "bypass tunnel + full face drill and blast + TBM stepping" is preferred.

According to the previous breakout methods, TBM breakout can be classified into the followings:

A. Excavation at the location of TBM jamming to release pressure, to realize full face tunnelling rapidly after TBM breakout;
B. Full face manual advance excavation;
C. Half face manual excavation + advance large canopy tube grouting for pre-reinforcement + TBM full face tunnelling;
D. Bypass tunnel + full face drill and blast + TBM stepping;
E. Rapid excavation of the small pilot tunnels on both sides to release pressure;
F. Excavation on the upper half face + TBM tunnelling on the lower half section.

According to practice, as for the non-foreseeability, complexity and particularity of the geology, Measure D and Measure F of the six construction schemes above are feasible. If analyzing from the perspective of construction progress, construction quality control and construction safety, the scheme of "bypass tunnel + full face drill and blast + TBM stepping" is preferred.

11.2.4 Headrace Tunnel of Hongyan River-Shitou River Water Diversion Project

1. Project Overview

As an interbasin water diversion project in Shaanxi, the Hongyan River-Shitou River Water Diversion Project diverts the water in the Hongyan River, a tributary of the Baohe River on the north bank of the Hanjiang River to Shitou River, a tributary on the south bank of the Weihe River under the effect of gravity. The headrace tunnel has a total length of 19.8 km with a total investment of CNY 1.068 billion. The maximum diversion discharge designed for the Project is 13.5 m^3/s and the annual diversion amount designed is 92,100,000 m^3, including 46,960,000 m^3 for ecological water replenishment to Weihe River with a total construction period of 66 months planned. The drill and blast method is adopted at the Lot I, Lot II and Lot III at the tunnel entrance (upstream). At the Lot IV of the Hongyan River-Shitou River Water Diversion Project undertaken by China Railway Tunnel Stock Co., Ltd., the double shield TBM is used for tunnelling from exit to the upstream for about 11 km to the boundary. The Hongyan River-Shitou River Water Diversion Project is featured by complex geological conditions, with 18 faults of different sizes and a 1300 m water-rich moraine. The maximum risks of the Project are collapse of rock mass, mud burst and water burst. The rocks are loose and susceptible to collapse, with frequent geological disasters such as collapse, mud burst, water burst, etc. The TBM main machine has been jammed for numerous times during tunnelling due to collapse and convergence arising from fault fracture zones, non-uniform hardness of rocks and large water burst. Tunnelling was started on October 8, 2008 and TBM was transformed in the midway. The jamming has been handled with for above 50 times during construction. The performance of monthly advance of 435 m and daily advance of 29.7 m has been achieved when the rocks are under good conditions. The double shield TBM used for the Project has tunnelled for 7 km at the Shanggong Tunnel in Yunnan, and is a piece of old equipment removed due to hard handling of the jamming arising from mud burst and water burst. Even though it is produced by Robbins, it lacks of intelligent design due to too early design and production (designed in the 1980s and produced in the 1990s), and the cutter head will be jammed easily in case of mud burst and water burst since the cutter head torque is small.

The headrace tunnel starts from the Guan Mountain in Taibai County and ends at the eastern Wulipo, and crosses the evaded Zhongshan Mountain and the Taibai Basin from west to east. The chamber has a buried depth of 150–300 m (420 m at most), and is located in the bedrock, which includes the Qinling Group Upper Subgroup gneiss, Central Asian Group marble, Lower Subgroup gneiss and Yanshan Stage granite, and all of them are hard rocks. Specifically, the marble section is about 3.45 km long, the gneiss section is about 6.63 km long and the granite section is about 1 km long. The elevation on bedrock surface is 1623–1,750 m, which is 160–300 m higher than the tunnel top. The rocks are weakly–slightly weathered, and the rock mass is relatively complete. The grade-V rocks are 1047.2 m, which accounts for 9.44% of the total length; the grade-IV rocks are 559.8 m, which accounts for 5.05%; the grade-III–grade-IV rocks are 6657 m, which accounts for 60.03%; the grade-III rocks are 2825 m, which accounts for 25.48%.

The headrace tunnel line is located at the south wing of the Taibai-Taochuan syncline, where compression-torsional faults of the NEE and NE groups are relatively developed. The marble and fault fracture zones (15) in the chamber are featured by rich groundwater, with certain influence on rock stability and construction. The tunnel has crushed rock mass under the influence from F1 and F21 branch fractures, with poor tunnel conditions. It is worth mentioning in particular that the rock mass at the small-angle crossing section with F1 is much more crushed. In total, 7 boreholes are distributed within this lot of the headrace tunnel, and confined water is distributed at the 6 (1 for Quaternary phreatic/confined water) boreholes in addition to the ZK8 borehole along the Shitou River at the exit. The confined water includes the bedrock fissure confined water and the marble karst confined water. Crushed rock mass is developed on main aquifers. The confined water aquifers at deep bedrocks are distributed in a non-uniform manner with poor continuity, belonging to distal high-mountain weakly replenished confined water, with poor hydraulic connection with the upper phreatic water. The distribution space of aquifers is mainly controlled by the rock stratum surface and the fracture occurrence. However, the regional large-scale fault zone may be linked to the upper phreatic water aquifers with rich water.

A double shield TBM is used for tunnelling for 11,095 m at the Lot IV of the Hongyan River-Shitou River Water Diversion Project, and the chainage is K8 + 700–K19 + 795. The stratum lithology revealed in construction is basically in line with design drawings, and the difference between the location of the rock stratum boundary and design is within 50 m. The tunnelling section mainly crosses the hornblende schist as well as the hornblende schist intercalated with gneiss, plagioclase gneiss, mica quartz schist, chlorite schist, gneiss, etc. Medium-hard rocks are mainly developed here, and the uniaxial saturated compressive strength of coring tests is within 45 MPa, and the rock core cannot be taken out effectively at crushed rock sections. Since the rocks revealed in double shield construction are limited, the observation and statistics on the angle between rock stratum strike and the tunnel axes are not systematic. The geological compass cannot work effectively under magnetic field disturbance from the equipment in the tunnel, so the rock stratum strike cannot be measured accurately. The angle between the rock stratum strike and the tunnel is within 40° through observation, while that between the rock stratum strike and the tunnel is within 20° during shield breakouts, and some is nearly parallel to the tunnel axes. Rock joint fissures are developed, with a small TBM thrust. The uniform laminar slags which are unique after cutting are not identified during tunnelling. The structures within the observable range are almost micro-tensional, flat and rough and filled with mud and rocky debris both. The groundwater here is rich, and the grade-III rocks are seeping–dripping; the grade-IV/V rocks are under linear flowing and water burst state, with tremendous influence on rock stability. A rectangular weir is provided for measuring at the tunnel opening, and the cumulative water flow in the total length of 3992 m reaches 200 m^3/h.

2. TBM breakout technology

One of the most important geological characteristics of the Project is hard rocks intercalated with soft rocks, with great phase change and fast change. The spacing of soft and hard rock strata is short, with great rock deformation at the interbedding and soft rocks, leading to frequent jamming, so that the TBM direction is hard to be controlled.

(1) Practical conditions of jamming

① Cutter head jamming: On June 28, 2009, when the TBM tunnelling was progressed to the chainage K16 + 814, the crushed rocks on the tunnel face lost stability and triggered collapse. Substantial crushed rocks and rock ballast entered the cutter head with muddy water, with remarkable increase in the muck quantity of the belt conveyor and sharp increase in the cutter head torque and motor current. Finally, the cutter head failed to rotate and the belt conveyor failed to operate.

② Jamming of front shield: On April 12, 2009, the friction generated on the shield under the effect of rock stress was larger than the maximum thrust of the main thrust cylinder due to great deformation of rocks. When the pressure of the main thrust system was increased to 345 bar, TBM still failed to move forward even under the maximum thrust pressure, leading to jamming of the front shield. The connecting bolts between the front shield and the external telescopic shield were broken under the severest conditions. This phenomenon was frequently seen on May 3, 2009, September 22 and October 17.

③ Jamming of supporting shield: At the unfavorable geological sections, the supporting shield was squeezed due to great deformation of rocks. The pressure of the auxiliary thrust cylinder was constantly increasing during regripping of TBM, and the friction from rocks on the supporting shield still failed to be overcome even under the maximum limit value of 700 bar. The supporting shield failed to move forward, so regripping cannot be realized. This phenomenon was frequently seen during construction, and the supporting shield was jammed on February 1, 2009, February 15, March 1, May 19, May 26 and June 2.

④ Jamming of tail shield: The rocks were subjected to continuous convergence under the effect of ground stress, forcing the tail shield to shrink inwards. There was no gap between the tail shield and the segment, and even the edge block of the segment cannot be installed in-place. Under fast convergence of rocks, the tail shield was jammed firmly under the squeezing effect from rocks and segments, leading to jamming of the auxiliary thrust cylinder since the pressure reached the limit value during regripping of the supporting shield. The connecting bolts between the tail shield and the supporting shield were broken under the severest conditions. This phenomenon was frequently seen on May 19, 2009, September 22 and October 17.

(2) Reasons for jamming.

① Geology

The stratum through which the tunnel of the Project passed was featured by complicated geological structure, i.e. hard rocks intercalated with soft rocks, with a fast change frequency of lithology. TBM entered the unproven tensional fault zones, the rock mass of which was extremely crushed after numerous cutting and crumple by tectonic movement and was rich in bedrock fissure water. After tunnelling, the initial stress field of rocks lost balance on the excavation interface, and the strength and deformability of crushed rock mass were far below the secondary stress field of rocks after stress redistribution. The tunnel face lost stability within a short time period after excavation, leading to collapse. The collapsed chamber was further expanding under the effect of rich groundwater in the fault zone, with increasing collapsed slag body above the cutter head. A continuous increase in cutter head torque and substantial slags inside the cutter head were the main reasons for cutter head jamming.

After numerous geological tectonic movements, the project area was featured by large tectonic stress, and the direction of the maximum main stress was nearly perpendicular to the tunnel strike, and one of the characteristics of the double shield TBM was that shotcreting cannot be made for supporting after tunnelling. After redistribution of tunnel excavation stress, the rock deformation was nearly in a linear relationship with time growth within several days after tunnelling, and its convergence was larger than the gap between the tunnelling outline and the shield. This deformation was restrained by primary supporting structure, and the friction was formed due to residual secondary rock stress acting on the shield, increasing tunnelling resistance. When the friction arising from residual stress was increased to the content that it cannot be overcome by the main thrust system or the auxiliary thrust system, the front shield or the supporting shield was jammed.

The tail shield is the farthest part from the cutter head (12.5 m distant from the cutter head), and the exposure of the corresponding rocks is longer after excavation. The squeezing mechanism of rock convergence to the tail shield is similar to that to the supporting shield. However, the residual stress of rocks on the tail shield is larger than that on the supporting shield. Since the elastic deformation of the tail shield has eliminated the gap between the tail shield and the segment, the segment resists the rock convergence with the tail shield beside it. The part of the tail shield not disconnected from the segment is hard to be retracted under the squeezing effect from the rocks and the segment, which not only increases the regripping resistance of the supporting shield, but also leads to jamming of the tail shield.

② Inefficient early-stage exploration

The double shield TBM is mainly applied to Shanxi Wanjiazhai Yellow River Diversion Project, Datong River-Qinwangchuan Diversion Project, Xinjiang Ili River Diversion Project and Qinghai Datong River-Huangshui Diversion Project, and jamming has been identified for numerous times during tunnelling of the projects above. According to practice, the double shield TBM is extremely sensitive to weak rocks and fault fracture zones. The great discrepancy between the practical geology revealed during the double shield TBM tunnelling of the Hongyan River-Shitou River Water Diversion Project and the design geology is the main reason for frequent TBM jamming.

If a double shield TBM is used for construction, in case of severe rock collapse, the collapsed rock mass will invade the internal telescopic shield "groove" between the external telescopic shield and the supporting shield. When the gripper is retracted for regripping, the collapsed rock mass will enter the supporting shield with the gripper, and the front end surface of the supporting shield and the gripper hole have a shear effect on the rock mass during regripping. This unique structural design of the double shield TBM further increases the regripping resistance, which is one of the reasons for jamming of the supporting shield. Besides, the tail shield is 3.5 m long, which increases the overlapping length between the tail shield and the segment. The tail shield will be squeezed to a large extent by the rocks and the segment under a fast rock convergence, which increases the friction resistance during regripping and the possibility of jamming.

(3) Breakout measures.

① Ultrahigh-pressure regripping

The Robbins double shield TBM is equipped with a ultrahigh-pressure pump station whose maximum pressure can reach 700 bar, and the auxiliary thrust cylinder can withstand the pressure above 700 bar. In case of jamming of the supporting shield and the tail shield, the ultrahigh-pressure pump station and the auxiliary thrust cylinder can be used for ultrahigh-pressure regripping first. This breakout measure is generally applied to slight jamming of the supporting shield and the tail shield.

② Manual overcutting

In case of jamming of the front shield or if the supporting shield and the tail shield still cannot be pushed forward under ultrahigh-pressure, the surroundings of the shield can be emptied through manual overcutting to release the pressure of rocks on the shield. The specific methods are as follows: With 3 observation windows of the telescopic shield and the square holes cut temporarily at the tail shield top as a channel, conduct manual overcutting for the areas squeezed by rocks towards the supporting shield, the front and the tail shields via an air pick and a crowbar, with square timber, planks and clinchers to erecting a temporary support. The size of overcutting is 60–80 cm, until the position above the shield semicircle is empty.

③ Chemical grouting for consolidation

Since the geological conditions of the Hongyan River-Shitou River Water Diversion Project are generally poor featuring crushed tunnel rocks and rich groundwater, when the TBM cutter head is jammed, consolidation grouting must be adopted for the loose rocks collapsed outside the cutter head, and the construction personnel can only handle with the collapsed rocks in front of the cutter head after certain strength is reached. Conventional grouting materials such as neat cement grout or the mixture of cement and water glass used for traditional consolidation grouting have disadvantages such as easy dilution by groundwater or flush-away after injection of grout due to inefficient sealing of the tunnel face and low early-stage strength of grout due to a long consolidation time and a narrow operation space inside the cutter head. To improve consolidation strength and to save the time required for handling with collapse, PUR and Silicate Modified PUR chemical grouting materials are used. Such materials are made up of component A and component B, the mix proportion of which is 1:1 (fixed volume ratio), with 1 package for component A and component B each. There is no need for weighing on site. The personnel can enter the cutter head to handle with the collapsed loose rocks 1 h after grouting, and the cutter head can break out successfully within 2d after grouting. Loose rocks at the top will be consolidated into arch rings around 2 m thick, to resist the disturbance from thrust. If the front shield, the telescopic shield or the tail shield is jammed and the rocks are extremely crushed for overcutting, chemical grouting can be adopted first and then overcutting to guarantee the safety of the construction personnel.

(4) Preventive measures for jamming.

① Increasing excavation diameter of cutter head

In total, 25 cutters are designed for the double shield TBM cutter head of the Hongyan River-Shitou River Water Diversion Project. When the rocks have a low uniaxial saturated compressive strength but a fast convergence, the #21–#25 cutters of the cutter head may be paved with cushion blocks successively for 3 mm, 6 mm, 9 mm, 12 mm and 15 mm, and the #25 edge cutter can be extended for 15 mm. Paving of the rest 4 cutters is to guarantee smooth transition among adjacent cutters. More convergence space can be provided for rocks if increasing the excavation diameter of the cutter head for 30 mm, so as to reduce the possibility of shield jamming.

② Lowering the friction resistance between the shield and rocks

Several grease injection holes shall be drilled on the upper semicircle of the shield on the front end of the external telescopic shield and the supporting shield, with 1 pneumatic grease pump and distribution valve system being connected. Afterwards, cheap grease or other lubrication materials shall be injected between the shield and the rocks, to lower the friction coefficient between the shield and rocks, thus lowering the friction resistance during thrusting and regripping.

③ Tunnelling in the single shield mode

The single shield tunnelling mode should be adopted for the weak rock fracture zone, to reduce the disturbance from squeezing of the gripper on rocks, so as to reduce the amount of slags to be cleared at the tail shield and to improve the installation speed of the segment, so that TBM can pass through weak rocks rapidly.

④ Other auxiliary measures

A. Overcutting in front of the supporting shield

Even though the rocks are complete, the convergence is great. After the regripping pressure of the auxiliary thrust cylinder is increased to above 400 bar, the internal telescopic shield is retracted to inside the external telescopic shield. Afterwards, the rocks in front of the supporting shield shall be subject to overcutting for 5–10 cm via an air pick (Fig. 11.108) during regripping. Certain space for convergence shall be reserved above the supporting shield after regripping.

B. Shortening tail shield and enhancing lubrication

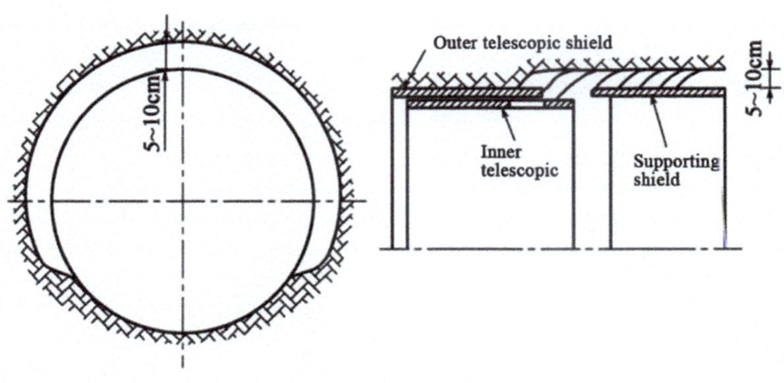

Fig. 11.108 Schematic diagram for overcutting in front of supporting shield

The original length designed for the tail shield is 3.45 m, with too long overlapping length with segments, so that it will be jammed by segments and rocks easily, increasing regripping resistance. The connecting bolts between the tail shield and the supporting shield will be broken under severe conditions. According to practical conditions on site, the tail shield is cut for 600 mm during overcutting of the tail shield. Meanwhile, grease is applied to the inner side of the tail shield during installation of segments.

C. Enhancing geological accuracy

The geology ahead shall be accurately forecast as much as possible through surface exploration and advance geological forecast in the tunnel. Advance chemical grouting shall be made in the cutter head for the rocks ahead.

(5) Summary

The root problems of tunnelling generally depend on the inhomogeneity of physical and geotechnical properties of the stratum through which tunnelling passes. For the full face and mechanical TBM tunnelling, since this excavation method is not flexible, the inhomogeneity of the strength of excavation materials is much more important. Mastering the geological and geotechnical conditions of the project in advance through appropriate methods is of great significance for TBM tunnelling. So far, generally not much fund is spent in early-stage survey. These facts prove that the capital used for early-stage survey will be compensated to a large extent due to lowering of construction expenses and shortening of the construction period. The double shield TBM is generally used for excavating long tunnels featuring complicated stratum, and is generally applicable to the tunnels with a mid-and-large buried depth, mid-and-high strength and basically favorable and stable geology. Since the shield TBM is sensitive to fast tunnel convergence, it will usually be jammed by a converged stratum. The change in lithology on tunnel walls can be observed directly during open TBM tunnelling, and auxiliary equipment such as ring beam erectors and shotcreting can be provided to adapt to geological change. When the hydrological and geological data mastered are not sufficient, open TBM can bring the advantages for timely supporting through the New Austrian Tunnelling Method into full play.

The Employer and the Designer shall do the work concerning geological exploration design in the early phase of construction in a more accurate and meticulous manner, with careful assessment on the adaptation of the double shield TBM to geology, to guarantee smooth TBM tunnelling. Meanwhile, to make the double shield TBM much more flexible in tunnelling in unfavorable geology and to avoid or reduce jamming, the equipment manufacturer is recommended to minimize the length of the main machine during design. A lubricant injection system shall be designed on the shield, with appropriate advance geological forecast and probe drilling machine being equipped.

3. TBM refurbishment technology

(1) Background of refurbishment

The double shield TBM of the project was assembled and commissioned on the construction site on October 8, 2008, and has stepped for 466.5 m at the drill and blast sections of the exit on November 21, and tunnelling was started officially.

On March 23, 2011, the Employer issued a construction suspension notice to the project management office out of the purpose of preventing bury of TBM at the imminent F27 fault and protecting equipment with the construction schemes of remaining works undertermined. During the suspension period, the Employer decided to assign the construction task of remaining 5.1 km (K10 + 200–K15 + 300) tunnels to the joint venture of the SELI of Italy and China Railway 17th Bureau Group Co., Ltd., and this joint venture was also entrusted with equipment refurbishment.

(2) Introduction to equipment refurbishment

SELI of Italy has made the following refurbishment to equipment:

Increase the excavation diameter of the cutter head;
Increase the outer diameter of the front shield;
Reduce the length of the supporting shield;
Reduce the diameter of the gripper cylinder and increase the stroke;
Eliminate the internal steps between the telescopic inner shield and the supporting shield;
Redesign and manufacture the tail shield;
Change the main drive as frequency-variable;
Add the probe drilling machine system, etc.

① Cutter head

The original excavation diameter of the cutter head is 3.655 m and the #18–#25 disc cutter housings are moved outwards based on reserving the original design. See Fig. 11.109 for the refurbishment.

The excavation diameter of the cutter head is increased from 3.655 to 3.755 m after refurbishment, with HARDOX500 anti-wear materials being welded on the cutter head surface, increasing the abrasion resistance of the cutter head. Besides, the layout of the bucket and the bucket teeth is adjusted, to improve its capability of collecting fine residues at the bottom.

See Table 11.31 for the outward movement parameters of the cutter seat, and Fig. 11.110 for the cutter trajectory section after refurbishment of the cutter head.

② Front shield

The original diameter designed for the front shield is 3.58 m, and one 40 mm steel plate is welded based on the original shield structure, to increase its outer diameter from 3.58 to 3.68 m. The bottom is welded in layers (one 40 mm and one 22 mm anti-wear arc steel plates). Welding of the steel plate is extended for 150 mm towards the cutter head, to efficiently protect the cutter head and to reduce the amount of muck entered. See Fig. 11.111 for the refurbishment of the front shield.

③ Supporting shield

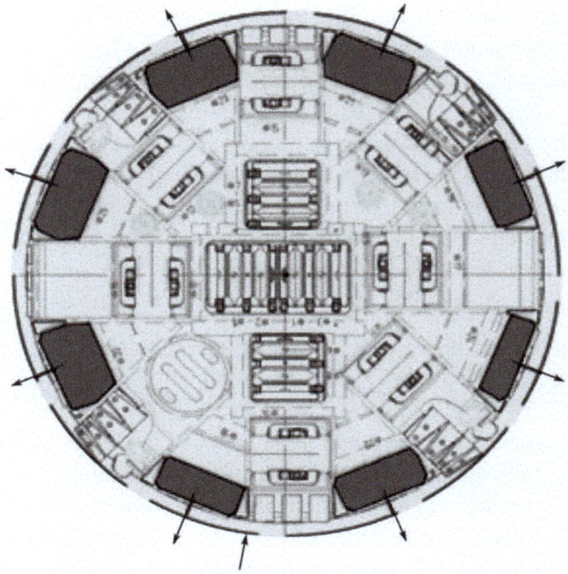

Fig. 11.109 Outward movement of cutter housing for cutter head refurbishment

Table 11.31 Outward movement parameters of cutter housing

S/N	No. of cutter seat	Outward movement of cutter seat (mm)
1	18	10
2	19	16
3	20	23
4	21	31
5	22	38
6	23	45
7	24	51
8	25	53

The original diameter and length designed for the supporting shield are 3580 mm and 3450 mm respectively. The length is reduced to 2400 mm after refurbishment, but the diameter and the original design remain unchanged. With the refurbishment of the supporting shield design, the diameter of the gripper cylinder is changed from ϕ330/230 mm to ϕ280/240 mm, and the stroke is changed from 254 to 354 mm. After refurbishment of the supporting shield, the overall length of TBM is shortened, and the space in the chamber is larger, with 18 holes for probe drilling and advance grouting being reserved around. See Figs. 11.112 and 11.113 for refurbishment of the supporting shield.

④ Telescopic shield

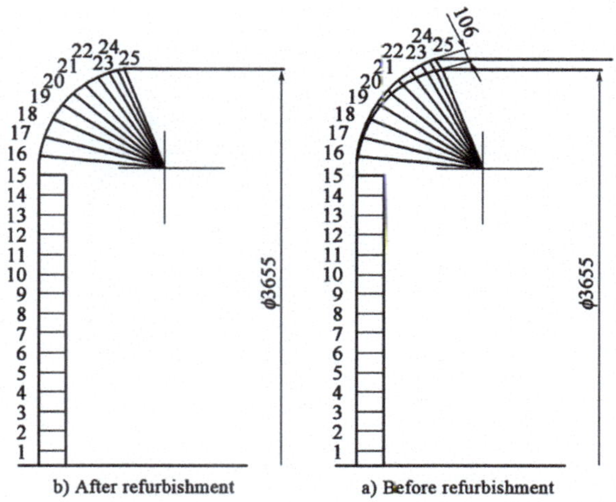

Fig. 11.110 Trajectory section of cutter head cutter

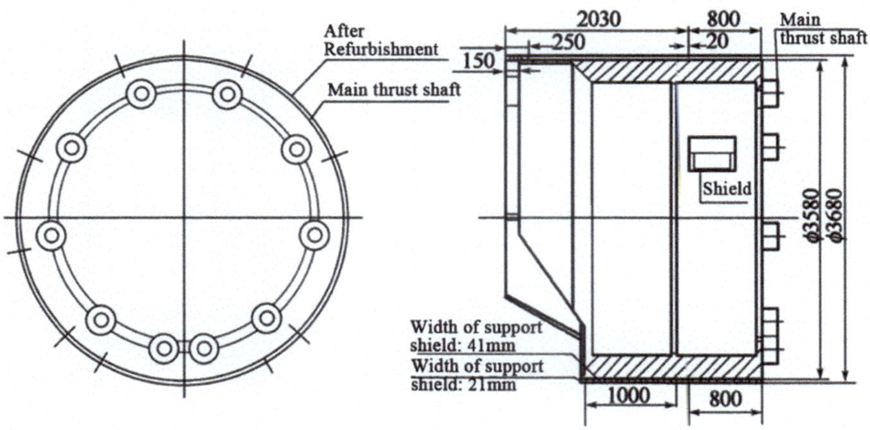

Fig. 11.111 Schematic diagram of the front shield refurbished (unit: mm)

The telescopic shield links the front shield and the supporting shield. The inner shield diameter of the telescopic shield refurbished is identical to that of the supporting shield, while the outer shield diameter is identical to that of the front shield. However, the original inner steps between the telescopic inner shield and the supporting shield designed are eliminated. The telescopic inner shield refurbished no longer has an opening, and the muck will no longer be cleared during TBM tunnelling. However, since there is only a gap of 10 mm between the telescopic outer shield and the telescopic inner shield, the direction adjustment during TBM tunnelling should not be too large. See Fig. 11.114 for the telescopic shield refurbished.

⑤ Re-manufacturing of tail shield

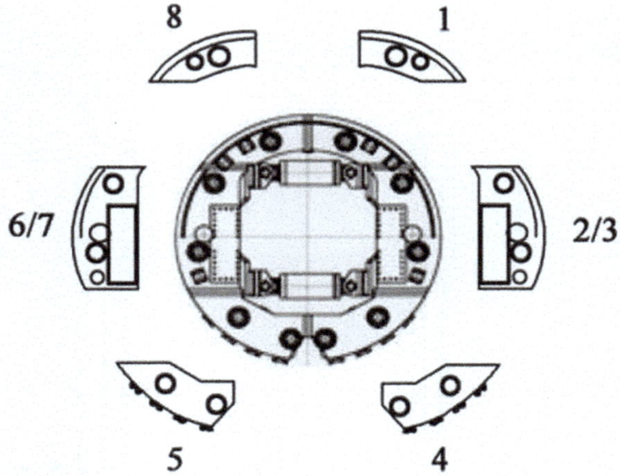

Fig. 11.112 Refurbishment of supporting shield

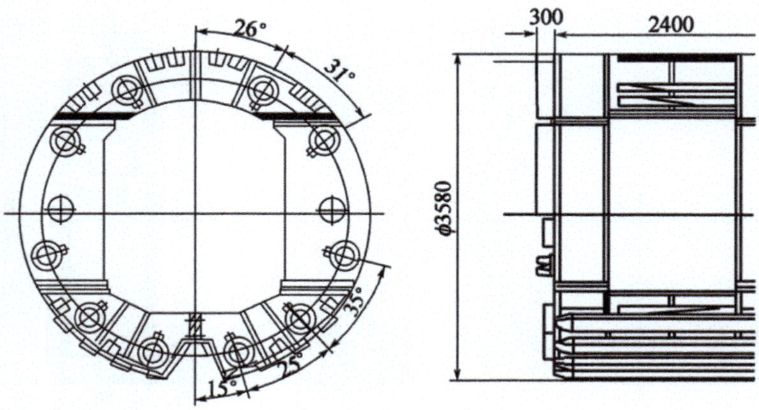

Fig. 11.113 Refurbishment of supporting shield (unit: mm)

Fig. 11.114 Schematic diagram for telescopic shield refurbished (unit: mm)

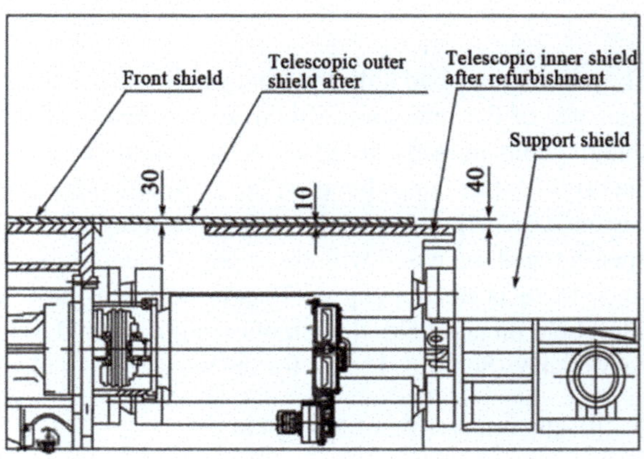

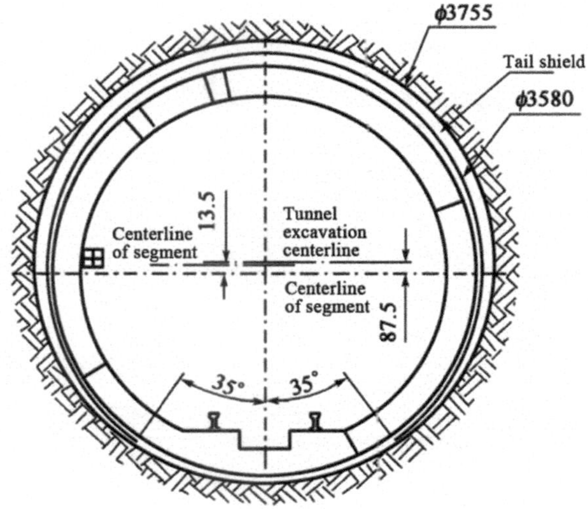

Fig. 11.115 Schematic diagram for tail shield refurbished (unit: mm)

The original length designed for the tail shield is 3450 mm, with a 120° opening at the bottom. The original tail shield is replaced in refurbishment, and a new tail shield 2500 mm long is manufactured, with a 70° opening at the bottom. Besides, spring plate seals and spring steel sheets are installed on the inner and outer sides of the tail shield respectively, to prevent flowing of backfill in the tunnel and the area in front of TBM. The diameter of the tail shield remains unchanged after refurbishment of TBM, but the excavation diameter is increased to 3.755 m. In this way, the center axis of the segment lining ring is moved downwards for 50 mm relatively, and the gap between the lining ring at the top segment and the tail shield is increased for 50 mm, i.e. from 34 to 84 mm. See Fig. 11.115 for the tail shield refurbished.

After refurbishment, the total length of the TBM main machine is 10.33 m, and the excavation diameter of the cutter head is 3.755 m. The diameter of the front shield is 3.68 m, while that of the supporting shield and the tail shield is 3.58 m. See Fig. 11.116 for the structure of the main machine before and after refurbishment, and Fig. 11.117 for the size comparison.

⑥ Main drive system

The original design mode of the cutter head is non-variable frequency drive, with five 260 kW/130 kW double-speed motors being equipped. The cutter head only has two speeds, i.e. 11.4 and 4.7 r/min, with a rated torque of 1089 kN m and a breakout torque of 4118 kN·m, so it cannot adapt to different rocks better.

One variable-frequency drive system is added for the cutter head refurbished based on reserving the original drive mode, and this system can realize stepless speed change within the cutter head rotation speed, to tackle poor geological conditions or pass through the fault by lowering the TBM cutter head rotation speed. After refurbishment, the normal working torque under the frequency converter drive mode is 1089 kN m, and the maximum torque is 2150 kN·m. See Fig. 11.118 for the main

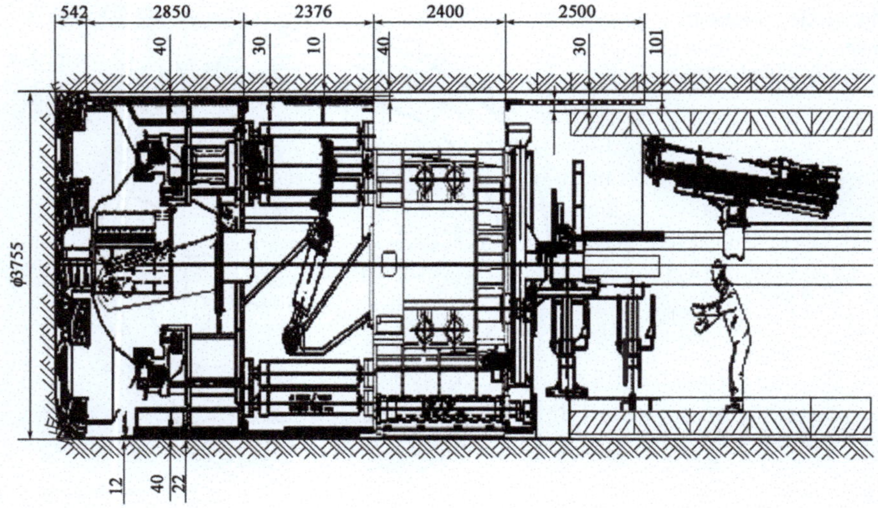

Fig. 11.116 Schematic diagram for main machine refurbished (unit: mm)

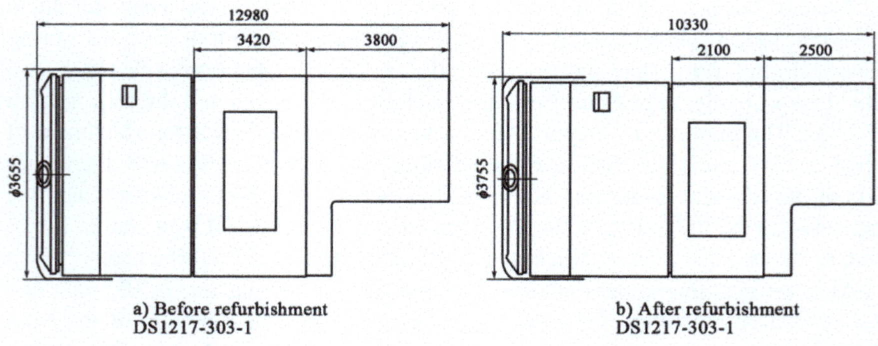

Fig. 11.117 Comparison on overall dimensions before and after refurbishment (unit: mm)

drive torque curve. Switch the variable frequency drive to the existing clutch mode if necessary, and then back to the variable frequency drive after cutter head breakout. The breakout torque remains at 4118 kN m.

The power for each motor is supplied from corresponding frequency converter, and each system is independent. In case of fault in any motor or frequency converter, the rest motors can still work normally, without influence on normal operation of TBM.

⑦ Adding probe drilling machine system

One probe drilling machine (model: ATLAS COPCO 1238ME) is added behind the tail shield in this equipment refurbishment, to realize probe drilling for TBM and advance treatment for unfavorable geological conditions, which is very important

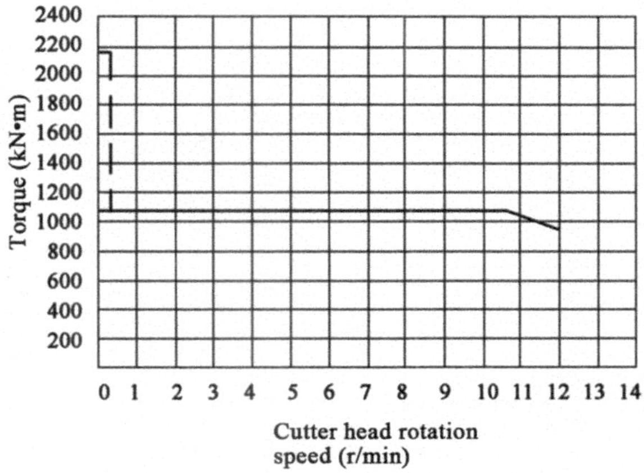

Fig. 11.118 Main drive torque curve

for TBM tunnelling at unfavorable geological sections. In total, 18 advance guide holes are arranged in a circumferential manner for the supporting shield structure. See Fig. 11.119 for the layout.

The probe drilling system is fixed onto the belt bridge, and the probe drilling machine can realize 360° rotation. Drill towards the area ahead of the TBM cutter head at an external angle of 7°through the advance guide holes reserved for the supporting shield. The drilling depth can reach 30 m, to realize probe drilling at the unfavorable geological sections and rock reinforcement. See Fig. 11.120 for the layout of the probe drilling machine.

⑧ Maintenance in other aspects

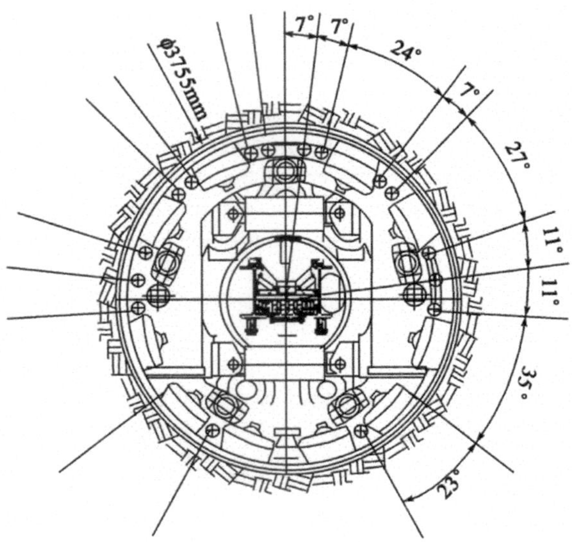

Fig. 11.119 Layout of advance guide holes of supporting shield

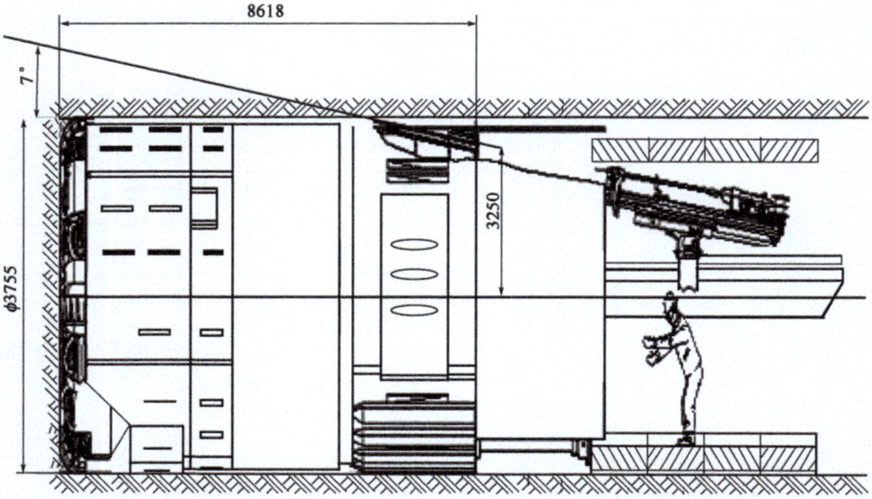

Fig. 11.120 Layout of probe drilling machine (unit: mm)

In equipment refurbishment from August 2011 to May 2012, maintenance and repair have been made for the followings based on the original design of TBM equipment: segment erectors, TBM belt conveyors, back-up belt conveyors, hydraulic system, electrical system, grease injection and lubricating system, pea gravel filling system, hoist and material transportation system, air compressor system, water supply and drainage system, secondary ventilation system, fire extinguishing system, communication system and grouting system.

(3) Comparison on technical parameters before and after refurbishment of double shield TBM (Table 11.32)

(4) Use description of SELI equipment refurbished

The TBM equipment refurbished by SELI of Italy has been shut down and placed still for half a year, without maintenance. However, the systems have numerous left-overs.

After "static handover" of the project, series of inspection and commissioning have been made to meet the conditions and requirements for trial tunnelling on March 21 first. The following problems have been solved in a centralized manner in this process:

Left-overs include connection of cutter head bolts and the shield structure, repair welding of tail shield notch, reinforcement of back-up trailer pillars, connection inspection of hydraulic system pipelines, pressure and flow detection of hydraulic system, cleaning of hydraulic/gear oil tank and replacement of oil, on-line no-load commissioning of control system, maintenance of back-up accessories, etc.

① Problems identified and handled with during trial tunnelling.

After trial TBM tunneling, the problems identified in this process are solved immediately to solve the problems of the system.

Table 11.32 Comparison on main technical parameters before and after refurbishment of double shield TBM

Serial number	Equipment component	Technical parameters	
		Before refurbishment	After refurbishment
1	Excavation diameter	ϕ3.655 m	ϕ3.755 m
2	Cutter head drive	Double-speed motor (5 × 260/130 kW)	Variable-frequency and double-speed motors (5 × 260/130 kW)
3	Cutter head rotation speed	11.4/5.7 r/min	0–11.4 and 11.4 r/min
4	Breakout torque	4118 kN m	2117/4118 kN·m
5	Auxiliary thrust cylinder stroke	2.4 m	1.8 m
6	Outer diameter of front shield	3.58 m	3.68 m
7	Length of supporting shield	3450 mm	2400 mm
8	Gripper cylinder	ϕ330/230 mm	ϕ280/240 mm
9	Stroke of gripper cylinder	254 mm	354 mm
10	Telescopic shield	Internal steps between the telescopic inner shield and the supporting shield	Eliminate
11	Length of tail shield	3450 mm, with a 120° opening at the bottom	2500 mm, with a 70° opening at the bottom
12	Probe drilling machine system	N/A	Add one probe drilling machine system

A. Problems in hydraulic system.

a. It is hard to control the elevation attitude of TBM during trial tunneling, and it is always under a bowing attitude, which cannot be adjusted effectively. It is found through inspection that the balance valves of the #1 and #10 main thrust cylinders of the main thrust hydraulic system fail, which is solved by replacing a balance valve and resetting the backpressure.
b. The indications on the pressure gauge of the main thrust cylinder and the auxiliary thrust cylinder are abnormal sometimes during tunneling, so the working state of the hydraulic system cannot be assessed effectively. Besides, substantial grease is identified in the pressure measuring pipes through detection of actual pressure and pipeline cleaning, leading to blockage, which is solved by cleaning and dredging and replacing some unreliable control group valves.
c. Substantial oil seepage at the pipeline joints of the hydraulic system leads to excessive consumption of hydraulic oil, and one barrel of hydraulic oil needs to be added per day averagely. The machine is shut down for 4d to replace the

hydraulic pipeline joints and the pipelines subjected to oil leakage in batches, with re-layout of the hydraulic pipelines of each system and signs being marked.

d. The degree of freedom of the segment erector is insufficient, so the installation efficiency is low and the operation is not reliable. Missing components such as the cylinder and the seat are re-made, and the original design functions are recovered, realizing 6 degrees of freedom.
e. The hydraulic pump is subject to severe aging, and the main thrust pump is damaged during operation. One damaged main thrust pump is recovered through repair on site, and can be used temporarily. The damaged piston pumps and vane pumps of the #7 pump station have been replaced.
f. The same hydraulic source is shared for the #1 belt conveyor and the oil return pump, and is refurbished during use. An independent drive system is provided separately for the oil return pump, to control both separately and to avoid mutual interference.
g. The clutch system is subject to severe aging, and the fault in the clutch hydraulic system during trial tunneling leads to severe damage to parts such as clutch thrust cylinder seals, control group valves and rotary joints, so the parts above cannot be used normally. The clutch cylinder thrust device is lengthened for the moment, to realize direct engagement for the clutch. The original design is recovered after arrival of the parts purchased such as seals, control group valves and rotary joints at site.

B. Problems in electrical control system

a. The multi-core control cables are adopted for signal transmission of the main machine control system originally, and such cables are usually broken during regripping, leading to frequent faults in the main machine control system. Besides, troubleshooting is extremely hard due to a long line and a narrow space, which affects construction efficiency greatly. As for this problem, the PLC distributed bus control method is adopted at sole discretion, and the original multi-core control cables are replaced by a single communication bus, so this problem is solved fundamentally.
b. The main motor of the frequency converter drive system is subject to severe asynchronization, which is solved by changing the original analog given signal as the digital signal under the bus mode. Meanwhile, the original human machine interface in the operator cabinet is replaced, with one frequency converter being added for monitoring on the human machine interface. In this way, relevant parameters of the frequency converter can be looked up in a real-time manner.
c. A multi-point communication system is added, with the functions of whole machine loudspeaker, building interphone system, video monitoring, traveling interphone, etc.

C. Problems in back-up accessories.

a. The pea gravel shotcrete system and the grouting system are repaired, and 3 air compressors are recovered.

b. Since affected by large water burst, it is difficult for drainage of TBM equipment. For this, a drainage system is added for the TBM equipment, and 6 drainage pipelines are arranged in total, with two 5.5 kW, two 20 kW and two 37 kW water pumps for forced drainage.

D. Refurbishment of probe drilling machine

Atlas 1238 probe drilling machine is equipped for SELI. According to the drilling effect at the tunnel sections with unfavorable geological conditions for numerous times, the maximum drilling depth is 10 m, which is far below the depth required for construction.

Through site guidance by the expert group, the Atlas 1840 drilling machine is recommended. Parts such as drill rods and adapter assemblies have not been arrived yet after system refurbishment.

② The defects of TBM are revealed during trial tunneling

A. Problems in main drive torque

a. Although the variable frequency drive is recovered, the anticipated requirements cannot be met. When the tunnel face is collapsed, the requirements for a low rotation speed and a high torque of the cutter head cannot be realized. Besides, according to the current hardware configurations of the cutter head drive system, the performance of the frequency converter drive system cannot be further improved technically, so the problem in cutter head drive torque cannot be solved fundamentally. The root cause for insufficient torque of the frequency converter drive system is that the main drive motor is an ordinary three-phase asynchronous motor, which cannot reliably match with the control mode of the frequency converter, which can only be that simple "frequency control" does not comply with the control requirements for "a low rotation speed and a large torque".
b. Since affected by water burst, the #2 main drive motor is damaged during operation of the project after handover, and is damaged again after using for a whole after replacement. The root cause is that the #2 and #3 main drive motors are soaked into water for a long term, and the insulation grade of the motor is IP55, with a low waterproof performance. The cutter head torque is remarkably lowered than the design value when 4 main drive motors or 3 main drive motors are used, and substantial crushed rocks will collapse at the tunnel sections with unfavorable geological conditions, so the cutter head cannot be rotated, leading to jamming.

B. Problems in belt conveyor

Allowing for insufficient transfer capability of the belt conveyor, the belt conveyor is always "pressed" in case of excessive muck. The optimal start time of the cutter head is postponed during startup, leading to jamming by flowing muck.

The reasons are:

a. In case of slight excessive muck, the width of the belt conveyor is small, so the friction with the drive pulley is insufficient, leading to "slip";
b. In case of severe excessive muck, the belt conveyor is "pressed" by substantial muck, and the speed slows down gradually. Slip does not occur between the drive pulley and the belt conveyor, but the belt conveyor is stopped directly.

C. Problems in controlling left and right TBM attitudes

The TBM attitude has deviated rightwards as a whole horizontally since tunneling in 2013, and it is extremely hard for direction adjustment. Through numerous discussions by the project management office, no problems have been identified in inspection on the steering control system. Numerous methods have been attempted and the effect is still not obvious even though only the right cylinder is adopted for thrust in the steering mode. The root cause has not been figured out up to now.

D. Problems in main bearing seals

A new main bearing seal is used during refurbishment of TBM by SELI, but the model, quality and installation process are still unknown. Besides, only one simple rod-type pneumatic grease pump is used for the grease lubricating system installed by SELI, without special grease distribution valves, flow and pressure monitoring devices, so it cannot be used normally. To meet the tunneling requirements as soon as possible, the sealing mode is recovered as the original gear oil lubrication designed.

The water content has been found off-standard severely through detection and tests on gear oil since trial TBM tunneling, and such problem still exists through sampling after tunneling for a certain time period after replacement of the gear oil for numerous times. No problems have been found through inspection on the system pipelines, and no effective solutions have been proposed so far.

Through analysis on the cause, the oil channel at the lower half of the main bearing has always been soaked into water due to substantial water burst on the tunnel face, so the possibility of flooding from the channel is high.

E. Problems in fresh blower at tunnel opening

The original fresh blower configured is GIA 4×25 kW variable frequency fans, one of which has been removed by SELI, and the motor and the blade cannot be repaired. At present, only 3 of them are under full speed operation, and all frequency converters have been damaged, so the ventilation requirements of subsequent construction cannot be met.

(5) Summary

① Even though the double shield TBM is featured by high safety, high quality and high efficiency, it has harsh requirements for geological conditions. At the early stage of construction, geological survey must be detailed and objective enough, which is of great significance in formulating construction methods.

② The adaptation of TBM can be enhanced through equipment refurbishment, to enhance the adaptation of equipment. Besides, advance geological forecast shall be adopted as much as possible during construction, to know the geology in front of the tunnel face, with reasonable advance measures being taken.

③ Since hard rocks play a predominant role, the pressure from the construction period is huge. For the tunnel works requiring synchronous lining, the double shield TBM tunneling can bring its advantages into full play, but the construction cost is relatively high.
④ Since limited by numerous conditions, the double shield TBM will be jammed easily during construction of weak crushed rocks, so it must be used carefully.

11.2.5 Taiwan Hsuehshan Tunnel

1. Project Overview

Taipei-Lian Expressway links Taipei and Lian County in eastern coastal areas. In this way, the traveling period between these two cities will be remarkably shortened, thus promoting and accelerating economic development of eastern coastal areas. With a total length of 31 km, this expressway starts from Nankang in Taipei and crosses Shiting and Pinglin via the Lanyang Plain to the southeast. This expressway crosses the middle mountains in Taiwan in the eastern Pinglin. In total, 5 tunnels shall be built for this line, with a total length of 20.1 km. The largest tunnel is the 12.9 km Taiwan Hsuehshan Tunnel, the construction of which is the key to the success of the project. As the largest highway tunnel in Southeast Asia, the Taiwan Hsuehshan Tunnel is the third largest highway tunnel in this world. The entire project will need 8-year construction, and the tunnel is built from east to west.

The construction contract of the tunnel above and its auxiliary structures is signed by and between TANEEB, the Owner of the project, and the agency of RET-SER. The project is managed by Sinotech Engineering Consultants, Inc. The subcontracts for construction of the main tunnel in the east line and for technical assistance of the main tunnel in the west line are signed by and between the Contractor and Spie Batignolles of France. Besides, the contract for 2 TBMs is signed by and between Spie Batignolles and Wirth for construction of the two tunnels.

The main tunnel of the Taiwan Hsuehshan Tunnel is designed as a two-lane highway tunnel, and these two main tunnels are linked to one service tunnel in the middle. There are 3 ventilation shafts along the tunnel, and the diameter of the service tunnel is 4.8 m.

The purpose of the service tunnel is as follows:

(1) Detect the geology along the line;
(2) Pre-treat the unfavorable stratum;
(3) Serve as an auxiliary tunnel during and after construction of the main tunnel.

To avoid one fault zone, about 800 m of the main tunnel is excavated through traditional methods. TBM used for the main tunnel is assembled at the tunnel opening, and then pushed forward to the tunnel face along the invert paved. At this moment, the pilot tunnel shall be excavated in advance for treatment of the fault zone in front of the TBM moving forward.

2. Geological overview

According to the tendering documents, the rock stratum along the Taiwan Hsuehshan Tunnel is mainly made up of the siltstone, sandstone, shale, argillite, mudstones and quartzite. The maximum compressive strength designed for TBM is 300 MPa. The tunnel overburden is generally 300–700 m thick. The tunnel must traverse a fault zone 50 m wide. These fault zones contain clay and lumpish and locally weathered rocks, mud and water. Such geological conditions challenge construction of any type of TBM.

The primary lining of the Taiwan Hsuehshan Tunnel is designed as precast concrete segment lining, and a ring beam at tunnel bottom is erected behind TBM, with waterproof film being paved. Afterwards, the secondary lining is completed through in-place concreting. See Fig. 11.121 for the typical section of the tunnel.

3. Technical parameters and characteristics of TBM

(1) Technical parameters and characteristics of TBM

Two full face double shield TBMs (Fig. 11.122) are used for the Taiwan Hsuehshan Tunnel and the excavation diameter is ϕ11.74 m. See Table 11.33 for main technical parameters.

If proper supporting of the gripper on the tunnel walls is allowed by rocks, synchronous progress of installation and segment installation can be realized for the double shield TBM. During TBM tunneling, the front shield (including the cutter head) pushes forward, and the rear shield is fixed onto the rocks via a gripper. In this way, the thrust and torque will not be delivered to the tunnel lining layer. The reinforced concrete precast segments are installed via a segment erector behind the

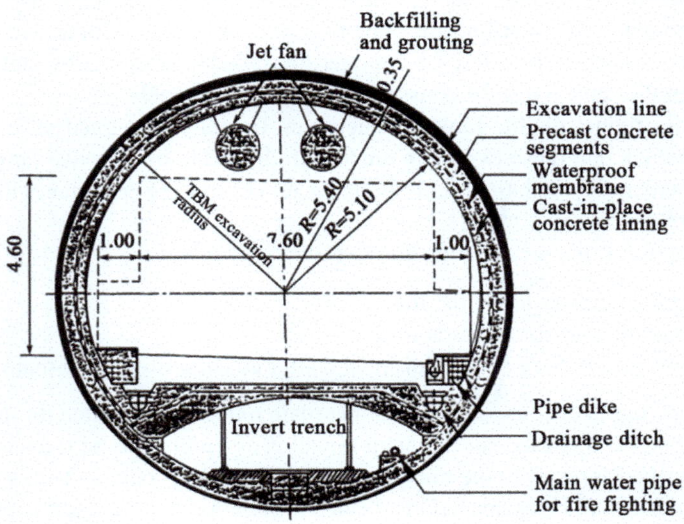

Fig. 11.121 Typical Section of Tunnel (Unit: m)

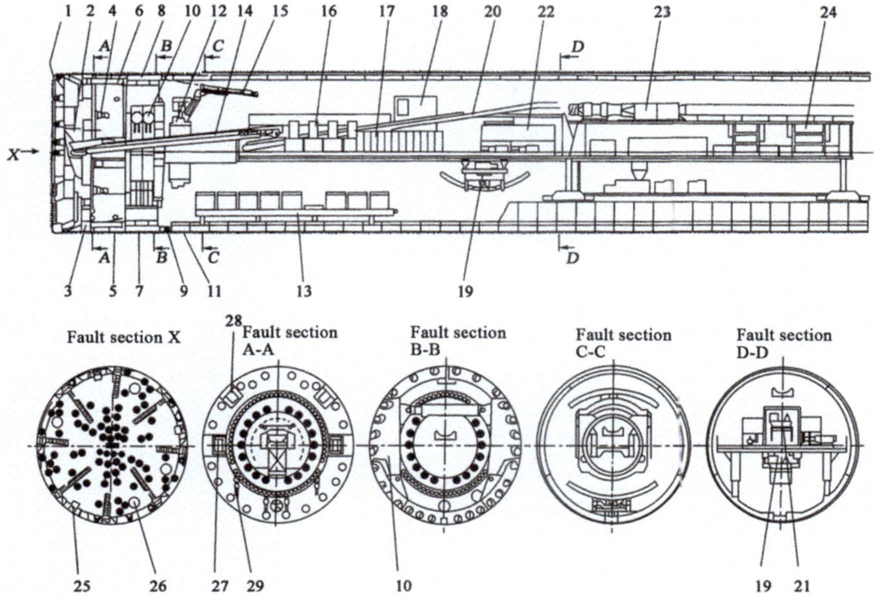

Fig. 11.122 Double Shield TBM (ϕ11.74 m) for Construction of Pinglin Main Tunnel. 1-cutter head; 2-ballast bucket with gate; 3-front shield; 4-cutter head shield; 5-telescopic shield; 6-telescopic shield cylinder; 7-rear shield; 8-auxiliary thrust cylinder; 9-thrust ring; 10-gripper; 11-tail shield; 12-segment erector; 13-segment storage table; 14-conveyor; 15-drilling machine; 16-hydraulic power box; 17-switch cabinet; 18-operating station; 19-segment crane; 20-bridge conveyor; 21-segment crane; 22-hydraulic power box; 23-deduster; 24-transformer; 25-overcutter; 26-manhole; 27-torque beam; 28-stabilizer; 29-cutter head cylinder

gripper shield under the protection from the tail shield. Only when the gripper re-supports can installation of the annular segments and TBM tunneling be mutually dependent. The segment cylinder at the tail shield pushes the gripper shield of TBM away from the installed annular segments. In this way, a cylindrical section is formed. At the same time, the supporting system at the tail is also pushed forward.

Tunnel supporting may not be made at the sections with favorable rocks. At the sections with unstable rocks, if TBM cannot acquire sufficient support force, segment installation and tunneling cannot be made synchronously for TBM just like the shield.

To pass through the unfavorable rock stratum and squeezing rock stratum smoothly, the following characteristics are designed for TBM:

① When TBM pass through the possible squeezing rock stratum, to acquire certain overcutting, the excavation diameter is increased for TBM via an overcutter;
② An eccentric cutter head is designed, and the offset on the shield axis of the cutter head realizes vertical adjustment of the cutter head center according to the front shield;
③ The cutter head shield is as close to the tunnel face as possible;

Table 11.33 Main technical parameters of double shield TBM (ϕ11.74 m)

System name	Item	Unit	Technical parameters
Cutter Head	Cutter head diameter	m	11.74
	Disc cutter (ϕ432 mm)	Piece	77
	Bucket lip	Piece	92
	Overcutter	Piece	3
	Drive mode		Hydraulic motor
	Drive power	kW	4000
	Rotation speed	r/min	0–4
	Torque (4 r/min)	kN m	7200
	Breakout torque (0.95 r/min)	kN m	30,000
Thrust system	Maximum thrust of main thrust cylinder	kN	50,600 (18 cylinders)
	Stroke of main thrust cylinder	mm	1850
	Maximum thrust of auxiliary thrust cylinder	kN	78,700 (28 cylinders)
	Auxiliary thrust cylinder stroke	mm	2000
Support system	Gripper	Pcs	4
	Support force	kN	65,000
Segment Erector	Working load	kN	1100
	Rotation range	(°)	± 220
Belt conveyor	Belt width	mm	1400
	Belt velocity	m/s	0–2 (hydraulic drive)
Electrical system	HV	kV	22.8
	LV	V	690/440
	Transformer	kV A	3 × 3150(690 V);1 × 1250(440 V)
	Emergency generator power	kW	240
	Total installed power	kW	5540
Main machine + back-up	Total mass	t	1800

④ The hydraulic drive of the cutter head can be reversed, with a high breakout torque;

⑤ To enable smooth pass of TBM at the complicated and hard rock stratum, TBM is equipped with a disc cutter and a picker, and these cutters can be replaced under the protection from the cutter head.

⑥ The opening of the cutter head scraper and the bucket can be partially sealed to limit falling of rock muck.

⑦ The opening of the edge bucket can be sealed.

⑧ The muck hopper can be sealed with one hydraulic drive door, to prevent entry of water/muck in the shield.

⑨ In the entire tunneling process, the torque support beam can deliver the reaction force of torque to the supporting shield through the telescopic shield.
⑩ A multi-functional drilling system is installed behind the tail shield for secondary detection and grouting drilling.

(2) Shield device

The shield is about 11 m long and is mainly made up of 4 parts, i.e. front shield, rear shield (also called gripper shield or supporting shield), telescopic shield and tail shield.

The front shield supports the drive device and the cutter head of TBM. The front shield is linked to the rear shield through the telescopic cylinder, which is arranged in 4 groups. In this way, the front shield and the cutter head can rotate to any required direction. In case of water burst, the conveyor door will be closed, and the water pump arranged at the shield bottom can drain the water inside the cutter head.

The telescopic part of the shield links the front shield to the rear shield. The purpose is to guarantee synchronous progress of excavation and installation of precast concrete segments. The telescopic part is designed in such a manner that the curve radius for tunneling correction is 200 m. The jack of the telescopic part links the front shield to the rear shield, and can deliver thrust and pulling force. This design characteristics can prevent downward tilting of the front shield when unstable rock stratum with a high buried depth is encountered.

The torque supporting beam of the guide pad with 2 hydraulic cylinders delivers the cutter head torque from the front shield to the rear shield. The rolling adjustment between the front shield and the rear shield can be made through the torque reaction cylinder.

The 2 hydraulically actuated stabilizers at the upper front shield play a role in stabilizing the shield in the tunneling cycle, and can be used to fix the front shield and help the rear shield pull forward.

The gap between two shields of the telescopic part can be checked and cleaned. To facilitate inspection, numerous windows are set. The sealing device between the inner and outer layers can help clean the gap between these two layers with water or bentonite. The 120° scraper at the top of the outer layer installed can help maintain this gap clean.

The rear gripper shield covers the thrust jack and the gripper, the geometrical shape of which can enable acting of the support force on the tunnel wall and tunnel bottom. In this way, three points of the shield are firmly fixed onto the tunnel. The gripper shield withstands all the top thrusts of the front shield, and can make the front shield move backwards. The gripper liners are arranged as per size, so that the surface pressure is within 4 MPa.

The thrust jack is used to re-locate the gripper when it is unlocked, or push forward the shield when TBM is under the single shield mode. Such jack is arranged at different pressure zones, to facilitate operation of TBM when TBM tunneling is made at unfavorable rock sections and the gripper and the telescopic shield are not used. When TBM operates in a simple shield form, its cutter head torque must be supplemented by the thrust jack of the shield. Such jack is linked to 1 common thrust ring, which can rotate under the effect of the hydraulic cylinder, to deviate all jacks, so as to form 1 circumferential force acting on the cutter head torque. Another feature of the equipment is that 3 thrust jacks (one at vault and one each for the upper quadrant) are equipped with a single jack foot pad, so that 3 top segments can be supported when the thrust ring moves backwards.

A complete tail shield can realize smooth installation of the concrete lining segment 1.5 m wide, including the caulking. Since pea gravel is backfilled at the rear part of the segment, seals constituted by narrowly overlapped spring steel members are arranged at 270° at the top on the inner and outer sides of the tail, to prevent entry of the pea gravel in the tail and flow to the surroundings of the shield. When the segment departs from the tail for mortar backfilling, the tow-bar attached to the tail can support the segment.

(3) Cutter head drive and main bearing

Anticipated unfavorable rock stratum shall be considered for design and manufacturing of the cutter head (Fig. 11.123). Based on this, the cutter head is designed as a tremendous welded steel structure, with 77 disc cutters for hard rock excavation and 8 muck buckets with protruding teeth for weak rock stratum excavation. These protruding teeth cut at the rear part of the disc cutter. Both the disc cutter and the protruding teeth can be inspected and replaced from inside the cutter head. The disc cutter seat is concave and is a part of the cutter head. Only some part of the disc cutter ring protrudes from the cutter head. This kind of cutter can prevent large rocks from blocking the cutter head in the fault fracture zone. A smooth contour surface of the cutter head can stabilize the tunnel face. The muck bucket and the bucket lip are shallow, so that the distance between the shield cutting edge and the tunnel face can be minimized.

The lattice can be inserted into the hole on the cutter head surface to limit the size of the muck entered. The opening of the muck bucket around is opened and closed (Fig. 11.124) by a hydraulically actuated muck bucket. The rock muck is cleared through 8 scrapers and muck buckets around the cutter head and the muck holes installed radially ahead. The guide plate inside the cutter head delivers rock muck to the conveyor in the center of TBM via a muck hopper, which is located at top of the muck cylinder. To avoid entry of water in the shield during tunneling at the fault zone, TBM is equipped with one hydraulically actuated door to close the hopper.

With new cutters, the cutter head can overcut for 55 mm above the shield under standard conditions. The cutter head is provided with 3 extendable cutters/scrapers, which can increase the excavation diameter to 200 mm. The cutter is hydraulically extended, and mechanically locked for extended excavation. The device is also equipped with an additional scraper, to guarantee that rock muck can be effectively cleared during overcutting. To determine the entire overcutting position above the shield, the cutter head center (including the main bearing and the drive) can be adjusted vertically. After concentric overcutting, the cutter head is moved to the eccentric position 100 mm above the shield to keep overcutting. In this way, the

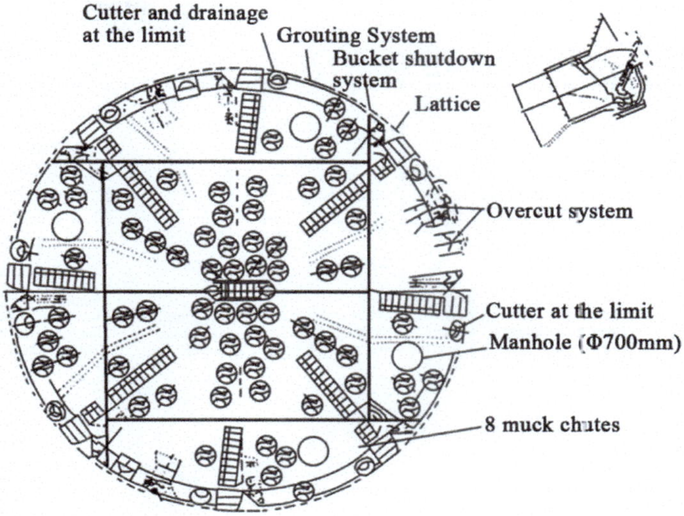

Fig. 11.123 Cutter head

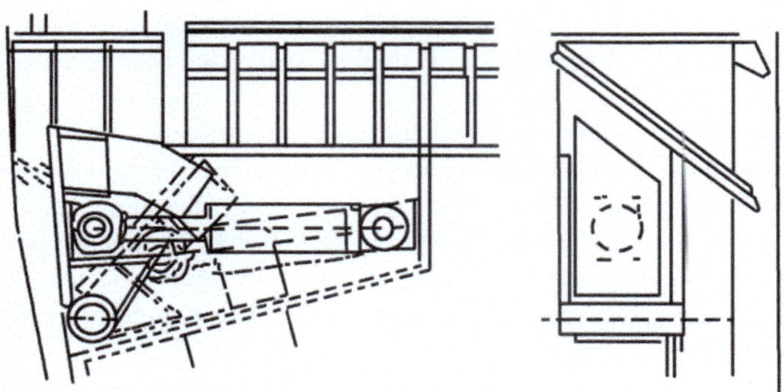

Fig. 11.124 Muck bucket closing device

cutter head can be flush with the invert for overcutting, and the overcutting at the top can reach 255 mm. See Fig. 11.125 for the cutter head overcutting.

The hydraulic drive device of the cutter head can realize constant change in rotation speed control of the cutter head, as well as locating and two-way start. The cutter head manufactured can only be used for one-way excavation. In total, 18 drive devices are provided, including the variable capacity hydraulic motor, the planetary gear box and the pinion. The cutter head is powered from the inner ring gear installed inside the large bearing.

The advantage of the drive device is that the cutter head rotation speed can be adjusted in the most economical manner, to adapt to geology. In this way, the cutter head can be adjusted to the highest possible rotation speed, to reach the optimal

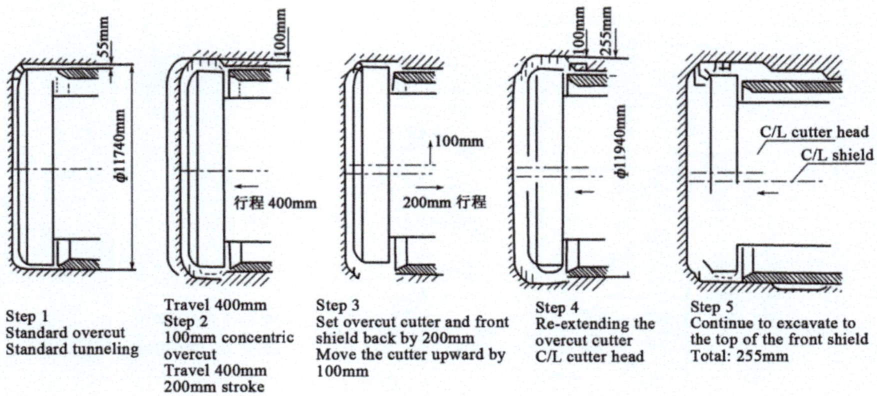

Fig. 11.125 Sequence of cutter head overcutting

penetration rate on the hard and stable rock stratum. In fault zones such as earthy and crushed rock stratum, the cutter head may operate at a low speed to acquire the largest possible torque. The capacity of the oil pump and the motor can be changing, so that the cutter head drive can operate in two forms. In general, the motor is under partial capacity, and is under the highest capacity if a high torque is required.

The main bearing is designed as a 3D double-shaft radial roller bearing. Considering the maximum thrust load and overturn to be encountered while passing by the curved sections and fault zones with unstable rock stratum, 2 main drive sealing devices are designed (1 on the inner diameter side of the bearing and the other on the outer diameter side). These 2 sealing devices can protect the main bearing and the gear drive device from being contaminated by dirt. To further enhance protection, a labyrinth seal is installed on the side facing the dirt of the sealing device. The lip seal is cleaned with lubrication oil to prevent entry of dirt. The main bearing is lubricated through a sprinkling and brake oil system with a pump and a filter. The pressure, flow, temperature and contamination of oil are monitored. The lubrication system and the drive system are mutually interlocked, and will be disconnected automatically in case of any error.

(4) Installation and delivery of segments

The hydraulically actuated segment erection system installs segments at the tail shield. The segments are delivered to the erector head via a mechanical clamping device, which has extendable bolt pins. These bolt pins can be inserted into the grouting holes of segments to clamp segments. The segment erector can rotate for 220° in any direction.

The head of the segment erector has 6° of freedom, and the segment erector is installed on the suspension beam of the gripper shield structure, and can move longitudinally.

The segment ring is 1.5 m wide. The segment ring is made up of 5 segments and 1 caulking block K (Fig. 11.126). The heaviest segment is about 11 t and the

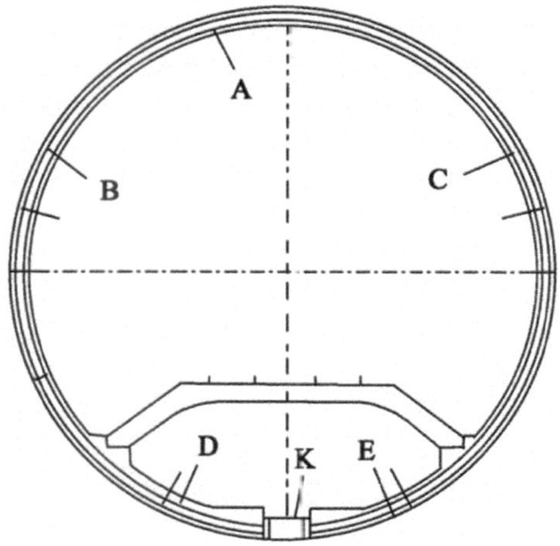

Fig. 11.126 Segment ring and ring beams at tunnel bottom

installation sequence of segments is as follows: First, install two invert segments at the bottom of the tail shield, and then the side wall segments. Afterwards, locate and hold the vault segments via a segment erector. The side wall segments are reinforced with the hydraulically actuated arm attached to the thrust ring to prevent overturn. The next installation steps of segments are: Separate 2 invert segments under the help from the hydraulic cylinder attached to the bottom of the thrust ring, and then insert the caulking block into this gap under the help from the hydraulic crane of the segment erector. Move the side wall segment upwards while separating the invert segments, and link it to the vault segments supported by the segment erector.

In this way, a circular ring can be formed for the segment, and the void above the tail shield can be minimized. During and after TBM resetting and the next cycle, the segment ring shall be grouted and be backfilled with pea gravel. The stroke of the segment erector is long, so that the vault segment can be supported when the erector is disconnected from the tail shield, until the segment ring backfilled is completely stable.

The lining segment is lifted via the mechanical clamping device of the segment erector from the segment storage table. Such device can move axially along the supporting structure of the erector. The stroke of the segment erector is long, so that the erector can support the vault segment, and the erector can disengage from the last segment installed. The segments are delivered from the tunnel opening to the front end of the secondary back-up system via the rail transportation system.

The #1 crane delivers segments from the unloading area to the primary back-up system, and stores segments on the deck at the bottom of the primary back-up trailer. The #2 crane lifts the segments from the deck at the bottom, and delivers them to the segment storage table. The #2 crane can move within the length of the primary back-up system, and enter the TBM area, so that segments can be delivered to the tail of TBM, without interference on the segments stored on the segment storage table.

Located on the invert behind TBM, the segment storage table can deliver segments and caulking blocks to a gripping range of the erector. The segment storage table can accommodate 9 segments, and can push forward along the invert segment automatically via a travelling beam. In this way, when another entrance is required at the invert behind the tail shield, the segment storage table can move backwards.

(5) Detection and grouting borehole

A Montabert HC.80 multi-functional drilling system is set behind the segment erector. General detection boreholes can be drilled in front of the cutter head through this system. For drilling of the detection boreholes above TBM, the drilling jumbo is fixed onto the primary back-up bridge behind the segment erector. Detection can be made through the guide pipes at top of the gripper shield. The same drilling jumbo can be erected easily on the segment erector, and can shake along the entire peripherals of the shield. Therefore, grouting boreholes can be drilled at 24 locations at the unstable rock stratum, so that a fanshaped protective layer is formed in front of and around TBM. The drilling jumbo can also be used for installing the anchor rods at the top if necessary.

The hydraulic power for drilling is supplied from the power box of the primary back-up system. This equipment is used for secondary grouting only. The grouting pumps are designed as a two-piston type. The grouting pumps are used for backfilling of invert lining segment and HP grouting of grouting holes. The dry mixture required for grouting and the mortar required for invert segment backfilling are delivered to inside the tunnel via a container, and then to the primary back-up system via the #1 segment crane.

(6) Conveyor system

Substantial rock muck will be generated from excavation, so an efficient conveyor system is required. The cutter head loads rock muck to the TBM conveyor, and then to the tail of the primary back-up system via a secondary conveyor. Afterwards, the rock muck is delivered to the shuttle-type conveyor via the next conveyor (crossing the ring beam at tunnel bottom and the track laying area), and then to the muck car on-track, which can stop on three rail lines. The rail line on the side is used for muck loading generally, and that in the middle can also be used for muck loading.

The TBM conveyor has a 1400 mm wide belt, and is basically located in the center of the cutter head in the muck ring. The conveyor is linked to the front shield, and moves with the front shield. The conveyor is hydraulically actuated, and can realize stepless speed regulation. The direction of the conveyor can be reversed if necessary, so that construction materials can be delivered to the shield. The conveyor can move backwards hydraulically for maintenance.

The shuttle-type conveyor loading muck to the train can move along the full length of the muck car, which requires that only one end of the conveyor can be used for muck unloading. The train keeps still during muck loading. The direction of the short-span muck conveyor at the tail of the shuttle-type conveyor can be reversed, and can load muck for the muck car on the outer rail. If the muck needs to be loaded for the muck car on the middle rail, the short-span conveyor will be retracted. The rock muck will be directly loaded to the train via a muck chute.

(7) Data acquisition

The control console is located in one sound insulation room at the upper primary back-up system. The control console contains the power and hydraulic elements required for operation and control of TBM. All operation data can be displayed in the operator cabinet. Meanwhile, the parameters much more important such as cutter head torque, rotation speed, tunneling speed and thrust, extension of telescopic shield and thrust jack and the information acquired through monitoring can be delivered to ground. These data can be displayed on the screen during TBM tunneling, and can be recorded by the digital recording system. The construction management department can print these data recorded for storage or analysis by the operator. In case of faults, the warning light will warn the operator and display the type and location of the fault.

TBM is equipped with a methane detector to detect the concentration of methane. The reading of such concentration can be displayed in the operation room. Once the concentration reaches a low limit of the allowed value, the warning light will illuminate, and will sound if reaching a high limit. At this moment, the operator shall shut it down.

(8) Back-up system

The back-up system of TBM of the Taiwan Hsuehshan Tunnel is mainly made up of two parts, i.e. primary bench-type back-up system and secondary back-up system on-track. The former provides the main power box and other equipment for TBM and the latter is used for delivering construction materials and rock muck. The primary bench-type back-up system operates on the supporting wall of the ring beam at tunnel bottom, and the secondary back-up system operates on the double-track laid in the tunnel.

The primary back-up system is linked to the supporting structure of the segment erector, and is dragged forward by the gripper shield. The connection between the primary back-up system and the segment erector enables correction of any deviation of the shield through the hydraulic cylinder, so that the workbench is located at a correct location in the tunnel. All hydraulic power boxes, switching devices, transformers, dedusters, detection and grouting drills with pumps and aggregate containers are installed on the primary back-up system. The segment storage table is located on the bottom layer of the primary back-up system.

The primary back-up system is connected with the secondary back-up system with connection bridges, which cross the ring beam at tunnel bottom and the track laying area. Once the gripper shield moves forward, the connection bridge and the secondary back-up system are pushed forward by 2 hydraulic cylinders. The stroke of TBM is equal to a segment width of 1.5 m. A rock muck conveyor and a segment crane are equipped inside the superstructure of the connection bridge, and a crane for installing ring beams at tunnel bottom and track laying is provided on the lower layer.

Substantial muck is generated from construction of the Taiwan Hsuehshan Tunnel, so the transportation volume is also high, which requires an efficient back-up system. This can only be realized by a rail back-up system. Segments, ring beams at tunnel bottom, backfilling and grouting materials can be delivered to the working face through this three-track back-up system. In this way, more time can be spared for the crane for unloading. Generally, muck loading of the train is only carried out on the outer track, and the conveyor can also be used for muck loading of the muck car on the middle track.

A typical transportation train is made up of 2 material cars, 6 muck cars, 1 locomotive and 1 man car. The 2 material cars can be used to deliver segments, or 1 container is provided to deliver pea gravel and mortar.

The transportation train enters the back-up system from the double-track through a California switch, and enters one track of the three-track back-up system through the switch shunting system.

Power cable reels, data transmission and telephone line reels, water supply and drainage hose reels are arranged at the tail of the back-up system.

4. Treatment of instability of excavation surface

Taiwan Hsuehshan Tunnel is an educational case of great significance in tackling instability of tunnel face. Three double shield TBMs are used for the tunnel, including 1 Robbins ϕ4.8 m double shield TBM for parallel pilot tunnel and two ϕ11.74 m Wirth TBMs for main tunnel excavation. For the main tunnel, the average monthly progress of TBM tunneling is predicted to be 360 m, while that adopting the drill and blast method is 50 m. The construction period of the 12.9 km Taiwan Hsuehshan Tunnel is 4 and a half years for TBM tunneling, while that adopting the drill and blast method (15 years) is much shorter.

Main problems encountered during excavation of the Taiwan Hsuehshan Tunnel are related to the extremely complicated geology at the first 2–3 km of the east tunnel opening. Unfortunately, this complicated geology is not fully assessed during design. This geology can be concluded as follows:

(1) As a stratified structure with a thin clay, the rock mass is made up of the extremely hard sandstone (quartz, with a uniaxial compressive strength of 350 MPa and a quartz content of 98%). Such rock mass may be a loose rock stratum or a squeezed stratum since the existence of the clay interlayer;
(2) Unexpected burst of substantial groundwater to the isolated zone, fault zones and fracture zones (underground reservoir) in rock mass;
(3) The ϕ4.8 m TBM is specially used for excavating parallel pilot tunnels in favorable rock mass, and the construction progress is predicted to be high. However, this TBM has encountered severe problems (the tunnel face has collapsed for more than 10 times in excavation of the above 1 km tunnel, leading to TBM jamming) under unfavorable geological conditions;
(4) Considering high strength and abrasion resistance of the rock mass, it has been proven that it is difficult for drilling exploration holes and drainage holes while excavating parallel pilot tunnels and this is both time-consuming and expensive.

Meanwhile, to improve the property of the rock mass and to facilitate excavation of the main tunnel via a large-diameter TBM, drilling grouting holes along the section of the parallel pilot tunnel in the circumferential direction is required but is hard, and this is both time-consuming and expensive;

(5) Based on the predicted unfavorable geological conditions, the two ϕ11.74 m TBMs are manufactured with the most advanced technology at that time. However, the 2 TBMs work in the extremely unfavorable rock mass, on which grouting cannot be made from inside the parallel pilot tunnel for improvement as predicted.

The followings occur in construction due to the factors above:

(1) Due to the extremely high abrasion resistance of the rock, the entire cutter head will be worn quickly. Therefore, sufficient thrust cannot be applied to the cutter head, leading to instability on the tunnel face and difficulty in controlling it;
(2) Since the tunnel vault and side walls are unstable, it is hard to grout the annular gap between the concrete lining and the excavation section as usual;
(3) The shield is always jammed due to possible interaction (friction of rock blocks to the shield) between the shield and rocks. Since it is impossible to apply axial thrust high enough to the thin concrete lining, it is hard for the shield to break out.

Therefore, to overcome the unfavorable geology along the tunnel, a systematic and continuous pre-treatment scheme is proposed for the area in front of the tunnel face. However, this scheme is excluded due to its high expected costs and long time required. Finally, it is decided to pass through this quartz stratum through the drill and blast method, and then 2 large-diameter TBMs are used at the sections with favorable geological conditions.

5. TBM construction analysis

It is expected that this 12.9 km Taiwan Hsuehshan Tunnel will be excavated rapidly. Since the technology adopted for TBM has this advantage, the monthly average tunneling speed expected at that time can reach around 360 m, while that adopting the drill and blast method is only around 50 m. Approximately 4 and a half years need to be taken for TBM tunneling of the 12.9 km Taiwan Hsuehshan Tunnel, while 15 years are expected if adopting the drill and blast method.

Few geological materials are mastered during construction, and the weathered rock stratum and dense vegetations make detection, drilling and exploration extremely hard. Most and other auxiliary investigations made along the 69 rock-core boreholes (2271 m) set out of the 12.9 km tunnel are carried out within the tunnel opening of the east line, and only few rock-core boreholes traverse the sections with a buried depth above 720 m, reaching the tunnel elevation. Six boreholes are drilled horizontally (1560 m). The main tunnel (east line) was commenced on July 23, 1993 and 732 m was excavated through the drill and blast method at the tunnel opening first. TBM was adopted in May 1996 and 3925 m has been excavated by TBM in the east line, with treatment and fault shutdown for 5 times (374d).

(1) Analysis on construction process

The ϕ11.74 m double shield TBM is used for two main tunnels, and the ϕ4.8 m double shield TBM is used for one parallel pilot tunnel parallel to the main tunnel and relative to the area below the center of the main tunnel. Three TBMs work from Tou-Cheng to Pinglin (1.255% uphill).

① Main tunnel (east line)

The main tunnel (east line) was commenced on July 23, 1993 and 732 m was excavated through the drill and blast method at the tunnel opening first. TBM was adopted in May 1996 and 3925 m has been excavated by TBM in the east line, with treatment and fault shutdown for 5 times (374d). The rock mass at the vault and side walls on the tunnel face collapsed continuously for 80 m during TBM tunneling on July 10, 1997 and the depth of such collapse reached above 3.5 m, with increasing water burst, making TBM tunneling impossible. Therefore, TBM tunneling was suspended, and the drill and blast method was taken for the pilot tunnel at top for supporting and reinforcement with treatment of the geology ahead. Afterwards, TBM was used to excavate the lower half section, with installation of segment lining. TBM full face tunneling was recovered only after excavation of nearly 1935 m tunnel by adopting this "mixed" method.

During TBM tunneling in the east line, the shield is always jammed due to deformed rocks around it. When the thrust is increased to move the shield forward and the segment strength cannot provide sufficient reaction force, the cutter head does not have sufficient thrust to support the tunnel face, leading to failure in controlling collapse of the tunnel face. The deformation at the vault and side walls makes grouting of the gap between segments and tunneling impossible, which also disturbs advance supports.

② Main tunnel (west line)

The main tunnel (west line) was commenced on July 23, 1993 and 881 m was excavated first through the drill and blast method at the tunnel opening. In August 1996, TBM was used for construction, and 456 m has been excavated during TBM tunneling of the west line, with treatment and fault shutdown for 10 times (961d). When TBM tunneling was progressed to 456 m in December 1997, the groundwater not detected burst suddenly at the upper left of the tunnel section while passing through the shear zone crisscross region of the upper fault zone, damaging the segment lining not connected by bolts. At this moment, 750L/s and 18 bar water burst was also identified at the area about 30 m behind the TBM cutter head. The installed segments are broken by water burst and rock muck, and the TBM main machine and the 90 m long back-up system behind it are buried by about 7000 m^3 rock blocks above it.

The damage to TBM of the west line is unrecoverable, and the cost of repair is much higher than that of buying a new one. Therefore, this TBM was removed at the end of 1999. The west line of the main tunnel was built through the drill and blast method and completed in April 2003.

③ Pilot tunnel

The pilot tunnel is located at a lower position (Fig. 11.127) between two traveling holes of the main tunnel. The pilot tunnel is excavated not only to detect the structure of bedrock, but also to drain the water burst of the mountain to lower the groundwater level and water pressure. Besides, the pilot tunnel can handle with the rock stratum in front of two TBMs of the main tunnel when necessary. The pilot tunnel completed can serve as the refuge tunnel and the drain tunnel. In November 1991, one ϕ4.8 m double shield TBM was purchased, and the pilot tunnel works 12.9 km long was completed in March 1995.

The pilot tunnel was commenced in July 1991 and 525 m was excavated through the drill and blast method, and TBM tunneling was adopted on December 22, 1992. In total, 5169 m has been excavated when this line was completed on October 20, 2003, and TBM has been shut down for 1228d due to jamming. Specifically, 1.8 months have been consumed to tackle the 8 jammings, i.e. 10 months for the 9th jamming, 7.5 months for the 10th jamming, 4.3 months for the 11th jamming, 2.5 months for the 12th jamming, 2.4 months for the 13th jamming and 2.8 months for this replacement of the bull gear.

In the process of TBM tunneling in the pilot tunnel, the cutter head has been jammed for 10 times due to collapse of the tunnel face. Another excavation was required from the tail to bypass the pilot tunnel after each collapse, and a detour was made to the area in front of the TBM head. Grouting was also required during excavation, and the muck collapsed at the head was cleared. The damaged components of TBM were checked and repaired.

The main reason for TBM jamming of the pilot tunnel is that the shear zone is almost crushed and some of them contain "shear gouge" under the effect of geological structure. Once TBM tunneling is progressed to this weak geological zone, the rocks on the excavation surface will collapse easily, and substantial soil-rock muck will enter the cutter head and the shield structure, leading to clogging of the TBM cutter head and jamming due to untimely handling by the muck delivery system. The location where water burst occurs is a relatively permeable fracture zone, which collects substantial water. The excavation surface will collapse first when excavation is progressed to this weak zone, and water will burst from the loose fracture zone instantly.

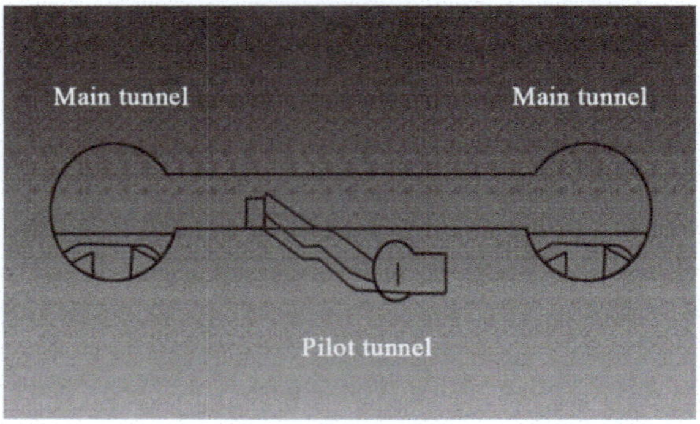

Fig. 11.127 Location of pilot tunnel

(2) Analysis on cause of TBM failure

① Great difference between geological materials and practical geology

Only 69 rock-core boreholes (cumulative depth: 2271 m) are drilled along the 12.9 km Taiwan Hsuehshan Tunnel, and most and other auxiliary investigations of these boreholes are made within the tunnel opening of the east line, and only few rock-core boreholes traverse the overburden and reach the tunnel elevation (general buried depth: less than 300 m; maximum buried depth: 720 m). Since the period of the exploration is short with a low input, the geological exploration materials provided finally are remarkably different from the practical geology, and the severe fault and shear zones and the tremendous water burst bring disasters to construction.

② The auxiliary pilot tunnel fails to play a role in escort

The pilot tunnel of the Taiwan Hsuehshan Tunnel is excavated not only to detect the structure of bedrock, but also to drain the water burst of the mountain to lower the groundwater level and water pressure. Besides, the pilot tunnel can handle with the rock stratum in front of two TBMs of the main tunnel when necessary. The pilot tunnel completed can serve as the refuge tunnel and the drain tunnel.

In fact, the pilot tunnel is not adopted for geological detection, drainage or reinforcement due to its severe lagging of construction, so it does not play a role in escort.

③ Geological adaptation is not considered for TBM design.

The double shield TBM of the main tunnel has been improved in numerous aspects:

a. The reaming cutter is installed, and the tunneling diameter can be increased in the squeezing stratum;
b. The cutter head can be lifted relative to the shield, to expand the gap above the shield;
c. The cutter head can rotate clockwise and anticlockwise, to acquire the maximum starting torque;
d. The muck inlet on the cutter head can be closed partially to control entry of rock muck;
e. The muck chute is closed to prevent entry of water and sposh in the shield;
f. A probe drilling machine is installed behind the tail shield, to facilitate supporting.

④ Using a double shield TBM for the pilot tunnel is wrong.

a. It is hard to detect the lithology of rocks with a double shield TBM;
b. It is hard to support rocks with a double shield, and the weak rocks in the main tunnel cannot be treated;
c. The double shield itself lacks of sufficient capability and tunneling speed while passing by the fault and weak fracture zones, so it cannot play a role in advancing.
d. An open type TBM shall be used for the pilot tunnel, and it has strong advantages in rock observation, probe drilling, passing of faults, weak fracture zones and rocks with a high ground stress, drainage, support of the main tunnel, tunneling speed, etc.

⑤ Lacking of experience in tackling weak rocks.

Damage to initial stability, excavation methods and speed, supporting methods and type all affect re-stability of rocks during rock excavation. Artificial disturbance may also be a reason for disasters of TBM.

One of outstanding problems on the construction site of TBM of the Taiwan Hsuehshan Tunnel is lacking of experienced construction technicians. Once the Sibeibadinuoer in France is demobilized, the engineering personnel in Taiwan must handle with TBM independently. The hired TBM operators from Russia are more accustomed to receiving orders other than playing actively.

11.2.6 Headrace Tunnel of Lesotho Project

1. Project Overview

The Lesotho Project aims at diverting the river water to the southwest in the rainy Lesotho Highland to the dry industrial areas around the Johannesburg and Pretoria in South Africa. The total length of the headrace tunnel exceeds 200 km and 75% of the tunnel and reservoirs are built in Lesotho.

The phase-I project can be classified into phase-IA and phase-IB.

I The total length of the phase-IA tunnel works is 82 km. The arch dam built will retain the Malilamato river water to the southwest, and divert the water to the natural river system in South Africa through the headrace tunnel and the water tunnel. The headrace tunnel from the water inlet of the Kaze Reservoir to the underground powerhouse of the Mola HPP has a total length of about 45 km, and the Lesotho basalt is mainly developed here. The south section of the water tunnel is about 15 km and Clarence block sandstones are mainly developed here. The north section of the water tunnel is 22 km with complicated geological conditions such as weak mudstones, sandstones, clay and granulite dikes.

I TBMs made by different manufacturers such as Atlas Copco, Robbins and Wirth are used for the phase-IA tunnel works. In total, the following 5 full face TBMs are used:

(1) One Atlas Copco JARVA MK-15 TBM;
(2) One Robbins 167–266 TBM and one Robbins 167–267 TBM;
(3) One Robbins 186–206 TBM;
(4) One Wirth TBM.

The first 4 TBMs are open type and are applicable to simple geological conditions and are equipped with Atlas copco 1238 probe drilling machines, which can realize probe drilling at the same time of tunneling. The last TBM is a double shield type, and is applicable to complicated geological conditions.

See Table 11.34 for the construction sections of 5 TBMs.

The northernmost section in South Africa is called the north water tunnel with a total length of 22 km and it has the most complicated geological conditions of the

Table 11.34 Brief introduction to phase-I tunneling of Lesotho Project

Project section	Tunneling works	Tunneling length (km)	TBM model	Construction contractor
Headrace tunnel (ϕ5.03 m)	North of Katze water inlet	10.7	Atlas Copco JARVA MK-15	LHPC joint venture
	In front of the Hlotse parallel pilot tunnel	17.4	Robbins 167–267	
	In front of the Muela parallel pilot tunnel	17.5	Robbins 167–266	
South section of water tunnel (ϕ5.03 m)	Water inlet from Holoio to Muela	2.1	Robbins 186-206	
	Ngoajane to Holoio	5.2		
	Ngoajane to #5 ventilation shaft	5.7		
North section of water tunnel (ϕ5.39 m)	Caledon tunnel section	8.15	Wirth	HMC joint venture
	Ash tunnel section	11.0		

phase-IA diversion project, and about 3 km of the section is built through the drill and blast method. Other construction sections, 8.15 km Caledon tunnel and 11 km Ash tunnel, are built by one ϕ5.39 m Wirth double shield TBM, and the tunnels above are the pressurized water tunnels lined with precast concrete segments for the first time in this world.

I The Maheier headrace tunnel of the phase-IB tunnel works has a total length of 32 km, and Lesotho malpais is mainly developed along the tunnel. According to durability study results, this rock stratum is susceptible to disintegration in case of gas and water. Two double shield TBMs are used for excavation from both ends of the tunnel to the center. One Wirth ϕ5.39 m double shield TBM is used for the project after excavation of the north section of the water tunnel of the phase-IA tunnel works, and the other is NFM/Mitsubishi-Bortec ϕ4.88 m double shield TBM made in 1984, and it has excavated for 20 km in Spain, Ecuador, etc.

Construction of TBM used for the north section of the water tunnel with complicated geological conditions will mainly be introduced in this section.

2. Geological overview

The geology at the north section of the water tunnel is the most complicated in the entire project. The tunnel traverses crushed settled layers such as sandstones, clay and mudstones respectively, and the rock stratum itself has strong convergence. It is worth mentioning in particular that the fault rocks in the dyke contain substantial groundwater, and the dyke is vertical and slightly inclined and diorite magma sedimentary rocks are developed here. The largest fault reaches 20 m and the rocks are severely crushed and some of them are lump and are widely weathered. The compressive strength of hard diorite is up to 300 MPa and the soft sandstone sections have a high abrasiveness since containing of large sharp-edge quartz particles.

3. Construction overview

The 11 km Ash Tunnel was tunneled at the end of June 1992 and was completed on November 12, 1993. The 8.15 km Caledon Tunnel was tunneled On March 12, 1994 and completed on March 3, 1995.

About 200 construction personnel are staffed, and the 12 h two-shift system is adopted for tunneling. The maximum team advance is 36.4 m, the maximum daily advance is 70 m, the maximum weekly advance is 306.6 m and the maximum monthly advance is 1076.6 m.

In early stages of tunneling, the consumption of cutters in the sandstones with a high abrasiveness is: About 15 mm of the edge cutter will be worn per tunneling for 120 m, and the service life of the edge cutter is 1.5–2 km in the weak mudstones with a low abrasiveness.

At special geological conditions, a probe drilling machine is used to detect the presence of methane, the position and width of the dyke and other geological characteristics in front of the tunnel face. Grouting is made when necessary to prevent burst of substantial groundwater. During continuous rock drilling, each probe borehole is 115 m deep with a diameter of 65 mm. In total, 48 drill rods 2.4 m long are used, and the overlapping length of such borehole is at least 10 m. Each drilling lasts for about 4.5 h averagely. TBM will stop working during probe drilling, and the downtime therefrom accounts for about 5%–10% of that in normal operation.

The segments are 1.4 m wide and 5 + 1 segments are erected. The heaviest segment is 2.5 t and the segments are 240 mm thick. The side seals of segments are driven by rotating the segment erector to make the tension reach 35 kN/m. Ring seals are pre-tightened by the segment cylinder with a constant force being maintained, until segments are linked with bolts. Once the segment is disengaged from the tail shield, the backfill grouting of about 3.2 m^3 is required for each ring. To tackle possibly high convergence of the rocks in geological forecast, the radial overcutting designed is 145 mm. However, high convergence is not encountered in construction.

Peeled rocks are identified on the tunnel face and at the top of the tunnel sections with a maximum buried depth of 350 m, indicating presence of stress. Most rocks fallen are left outside the TBM shield, and the void is backfilled through annular grouting. In weak rock areas, the rocks are further stabilized through secondary protective grouting from the back-up grouting station of TBM about 200 m behind the tunnel face to the tunnel top and consolidation grouting after TBM tunneling.

4. Technical parameters and characteristics of TBM

The Wirth ϕ5.39 m double shield TBM (Fig. 11.128) is made up of the shield assembly, the cutter head, the cutter head drive and the main bearing, the segment erector and operation devices, the probe drilling machine, the muck discharge system, the operator cabinet and the back-up trailer, etc.

Fig. 11.128 ϕ5.39 m double shield tbm for Lesotho headrace tunnel

The shield is about 12 m long and the tail shield plate is closed in the circumferential direction. The segments are sealed with rubber gaskets at the ends, and the excavated tunnel walls are sealed with multi-layer spring steel gaskets.

Considering anticipated difficult geological conditions, the cutter head is specially designed and manufactured, and 42 single disc cutters for hard rocks are equipped on it. There are 4 radial grooves in front of the cutter head, with 24 scrapers (also called bucket lip) being provided for use in soft stratum. The scraper is 25 mm lower than the disc cutter, and the excavation diameter can be increased for 40 mm through the overcutter. The cutter head, the main bearing and the drive device can be regulated axially and radially relative to the front shield, and overcutting can be made towards the tunnel wall as planned during radial regulation in the X and Y direction, to improve the operability of the shield system. Axial displacement is mainly used for monitoring the thrust applied to the cutter head and the cutter during forward movement of the front shield.

The probe drilling machine is installed on the working platform behind the segment erection bridge, and a probe drilling hole can be drilled in front of the cutter head. Moreover, the drilling machine can also swing with the segment erector within the range of diameter. Grouting holes are drilled through the shield at 17 locations, so that a sealed protective layer can be formed in unstable rocks. The customized sleeves 5 m long are arranged along the 5.5°probe drilling angle, and are linked to the probe drilling port in the shield and tunnel walls.

The segment erector can move axially along the erection bridge, and the movement distance is long enough to install enclosed wedges on one end and to remove the second last segment on the other end.

The segments of each bottom plate are linked with bolts and 2 sleepers, so that rails can be pulled out from the rail storage and be laid. The 38 platform trailers about 300 m long operate on these rails. The platform trailer is arranged with double-track, so that the second mining vehicle can operate in parallel during muck loading of the first mining vehicle. Hydraulic platform trailer pulling devices are installed in the muck loading area, and the operator can move the mining vehicle freely as per the muck loading amount. There is 1 short swingable muck bucket at the end of the #3 curved belt conveyor about 230 m long, so that the rock muck can be loaded to the empty mining vehicle parked on the back-up left or right rail.

A California switch is installed at the area about 4 km away from the tunnel opening behind TBM, so that the vehicle can give another the right of way in the tunnel. A double-track system is used on the back-up trailer behind TBM and 3 mining vehicles are used. Each vehicle has eight 10 m^3 rotary mining vehicles pulled by a 25 t diesel locomotive.

See Table 11.35 for main technical parameters of TBM.

11.3 Engineering Case of Single Shield TBM

11.3.1 Headrace Tunnel of Yao River Water Diversion Project

1. Project Overview

With the Yao River, an important tributary of the Yellow River, as the water source, the Yao River Water Diversion Project in Gansu is a large interbasin water diversion project solving problems regarding drought and water shortage in the central part of Gansu Province. The water supply range of the phase-I project mainly involves 6 counties (districts) such as Yuzhong, Weiyuan, Linyao, An'ding, Longxi and Huining under the jurisdiction of Lanzhou, Ding'xi and Baiyin. The construction contents mainly include the 109.73 km main canal, 4 canals with a total length of 148.40 km and 26 branch distribution pipes and branch canals with a total length of about 210.34 km.

The main canal of the Yao River Water Diversion Project takes water from the right bank in the reservoir area of the Jiudianxia Hydro-junction at the middle reaches of the Yao River and the water supply area is located in Dingxi and its east. It needs to traverse primary tributary basin such as the Yao River, the Weihe River and the Zuli River and has crossed the watershed of the basin for numerous times. To shorten the line and to facilitate water diversion, the main canal is arranged in the central part of the water supply area in a WE strike as a whole. Canals at all levels are arranged on the south and north sides, constituting a water canal network covering the entire water supply area.

The total length of the phase-I main canal line is 109.73 km, including 18 tunnels 94.43 km long, accounting for 86.1% of the total length of the main canal; 9 flumes 1.53 km long; 11 covered channels 2.96 km long; 10.81 km open channels. The #7 and #9 tunnels (17.24 km and 18.25 km long respectively) of the main canal are 2 longer tunnels of the main canal, and a single shield TBM is used. The total length of the two tunnels is 35.49 km, accounting for 32% of that of the main canal of the

Table 11.35 Main technical parameters of ϕ 5.39 m double shield TBM

System name	Item	Unit	Technical Parameters
Cutter head	Excavation diameter	m	5.39
	Drive power	kW	1500
	Drive mode		Hydraulic
	Rotation speed	r/min	0–8.6
	Torque (4.0 r/min)	kN m	2700
	Torque (8.6 r/min)	kN m	1250
	Breakout torque	kN m	4500
	Disc cutter (ϕ416 mm)	Piece	42
	Scraper	Piece	24
	Reaming cutter	Piece	2
Cutter head	Reaming range	mm	+ 20
	Horizontal displacement of cutter head	mm	300
	Radial displacement in X and *Y* directions	mm	± 30
Thrust system	Main thrust cylinder thrust	kN	2500 × 12 = 30,000
	Stroke of main thrust cylinder	mm	1530
	Articulation cylinder thrust	kN	2500 × 4 = 10,000
	Articulation cylinder stroke	mm	110
	Thrust of auxiliary thrust cylinder	kN	1250 × 20 = 25,000
	Auxiliary thrust cylinder stroke	mm	1900
Gripper	Support force	kN	9500 × 4 = 38,000
Segment erector	Working load	kN	30
	Swing range	(°)	± 210
Belt conveyor	Belt width	mm	800
	Belt speed (hydraulic)	m/s	0–2.4
Electrical system	Main voltage	kV	11
	Working voltage	V	380
	Transformer	kV·A	3 × 1000 = 3000
	Installed power	kW	1800

phase-I project, and the tunnels above are characterized by great length, complicated geological conditions, high cost, great construction difficulties and long construction period.

The Cretaceous and the upper Tertiary strata are developed at the #7 tunnel built by China Railway Group Co., Ltd. Extremely soft rocks are mainly developed, and the section with grade-IV rocks is 2.50 km long, accounting for 14.5%, and the Cretaceous sandstones, muddy fine siltstones and sandy mudstones are mainly developed. The section with grade-V rocks is 14.74 km long, accounting for 85.5%. The attitude of rocks is gentle, and is slightly affected by structure, and fractures are not developed here, and only the gentle brachy-axis folds are developed, so the water abundance is poor as a whole. The groundwater is mainly replenished by atmospheric precipitation. Since the precipitation here is low and is non-uniform within a year, the amount of groundwater (the maximum flow of spring water measured is less than 5 L/min) is relatively small due to weak permeability of the stratum. According to drilling, test and hydrogeological survey, the porosity of glutenite and sandstones is around 20%, belonging to aqueous permeable strata. Groundwater is generally revealed by drilling, and spring water is developed at sandstone and glutenite. As a relative aquiclude, the argillaceous siltstone and silty mudstone have non-uniform groundwater, which is generally laminar and load-bearing locally. The mountain is a weak laminar aqueous mountain.

2. Engineering design

The slope designed for the tunnel is 1/1500 and the water conveyance discharge designed is 32 m^3/s (36 m^3/s for enlarged discharge). As a mid-section and mid-span long tunnel, this tunnel is a grade-2 hydraulic tunnel.

The net inner diameter of the tunnel is 4.96 m if adopting the TBM scheme, and C30 precast reinforced concrete lining segments are used, with a thickness of 280 mm and an excavated tunnel diameter of 5.75 m. The lining segments are divided into 4 in the circumferential direction, constituting one ring. Each segment is a non-equilateral rhombic hexagon, and is subject to stagger-jointed erection in the circumferential and longitudinal direction, forming a stable embedded honeycomb structure. The longitudinal width is 1.60 m and the bottom arch serves as the track bed. The hole between segments and rocks is filled with pea gravels with a diameter of 30–50 mm and is subject to LP backfill grouting with cement mortar or cement slurry to stabilize the structure and to improve the integral stress of segments. Based on the difference in grade of rocks, heavy segments with a higher reinforcement rate are used at the fault fracture zone and the tunnel sections with unfavorable rock mass, and light segments can be used at general sections. Segments can be classified into light, heavy and extra heavy precast C30 reinforced concrete ones based on the reinforcement rate. The tunnel entrance and exit are the chamber sections where the mechanical equipment of TBM is assembled, enters and leaves the tunnel, and are also the reinforced section of the tunnel structure. According to technical requirements of TBM, the reinforced section at tunnel extrance and exit is determined as 50 m, and is excavated through the drill and blast method, with primary shotcreting for reinforcement. Once TBM enters or leaves the tunnel, the segment track bed of the bottom arch paved during TBM tunneling shall be removed, with secondary cast-in-place of C20 reinforced concrete lining as a whole. TBM works from the tunnel exit to the entrance at the upstream.

3. Scheme adjustment

The single shield TBM used for the #7 tunnel of the Yao River Water Diversion Project started tunneling from the exit to the entrance on December 29, 2009. The monthly advances are 245.0 m, 493.3 m, 666.2 m and 961.0 m respectively from January to April 2010 and TBM can work normally at the tunnel sections where stability can be realized for geology. Since entering the water-containing loose sandstones on April 20, 2010, TBM has been subjected to jamming, mud and sand burst and severe bowing successively, leading to jamming for three times. It was determined on January 20, 2011 that TBM was not applicable to this stratum, and tunneling was suspended. The total advance from May to November 2010 was 454.5 m and the cumulative advance of TBM was 2820 m. The bowing of TBM still cannot be controlled through breakout measures such as increasing the pressure of the upper and lower main thrust cylinder by adding HP pumps and steel gaskets. After pushing forward for 8 m in a forced manner, the bowing was further worsened, and reached 66.3 cm at most. Meanwhile, the segments were damaged in this process, with severe dislocation (15 cm at most), leading to severe mud burst the tail shield. Consequently, 3/4 shield of the TBM was buried, so tunneling was suspended at the chainage K60 + 931.

To prevent severer consequences due to instability of segments, through demonstration by experts, it was believed that TBM was not applicable to the water-containing loose fine sandstone stratum and the construction scheme was adjusted. TBM was no longer pushed forward in a forced manner, and was disassembled at 60 + 931. The original TBM jammed at the exit of the #7 tunnel was disassembled, and the cutter head and the shield were re-manufactured and assembled at the tunnel entrance for excavating the tunnel not on the water-containing loose fine sandstone stratum from the entrance.

TBM tunneling was started at the entrance on August 15, 2011 and the monthly advances were 1515 m, 1718 m, 1868 m, 1297 m, 1094 m, 1010 m, 1104 m and 1632 m respectively from September 2011 to April 2012. TBM maintained under an efficient and stable tunneling state at the tunnel sections where stability can be realized for geology. On April 30, 2012, TBM tunneling was suspended at the tunnel sections featuring water-containing loose sand layers (chainage: K57 + 835). The cumulative advance of TBM at the entrance was 11,120 m and the chainage at the shutdown area was K57 + 835 and the distance from K59 + 762 on the completed tunnel face (disassembling tunnel) of the #4 inclined shaft was 1927 m.

It was decided through demonstration to add the #1 and #2 shafts for TBM breakout and treatment of remaining tunnel sections with unfavorable geological conditions through active freezing. TBM was pushed forward under no-load thrust for segment installation to the disassembling tunnel. Besides, the #5 and #6 on-track inclined shafts were added, to realize synchronous artificial excavation in combination with numerous working faces of the #3 and #4 inclined shafts. However, since the freezing of the #1 and #2 shafts severely lagged behind and the cost of recovering TBM was high, the construction period was relatively long. It was decided finally to disassemble TBM locally, and segment installation under no-load thrust of TBM was changed as cast-in-place lining at remaining tunnel sections. The TBM back-up was removed smoothly in November 2013. On October 7, 2014, the freezing sections of the #6 inclined shaft and the #2 shaft of the #7 tunnel were completed, marking completion of the main canal of the Yao River Water Diversion Project in Gansu.

4. Single shield TBM

The single shield TBM used for the #7 tunnel of the Yao River Water Diversion Project has an excavation diameter of 5.75 m and is mainly made up of the cutter head, front shield, middle field, tail shield, segment erector, hydraulic pump station, electrical system, back-up system, etc. Two continuous belt conveyors are used for muck discharge. See Table 11.36 for the overall configuration parameters.

5. Refurbishment of single shield TBM

When TBM tunneling was progressed to the chainage K60 + 931 at the entrance, thick laminar sandstones of a particle structure were encountered with caliche nodules locally. The strength is about 1 MPa (boulder in the silty fine sands) and the rocks will be softened and collapsed easily in case of water. Since the stability of rocks was poor, the excavation surface was subject to severe collapse, and the muck quantity was larger than that during normal tunneling. The groundwater in the stratum rushed out in the form of a fountain on the free face, forming sand burst with sands, which was severer at the tail shield opening. Consequently, the thrust of TBM was increased, leading to TBM shutdown and numerous jammings of the cutter head and the shield, interrupting construction and severely affecting the construction period.

As for the severe geological problems encountered during TBM tunneling of the #7 tunnel, through demonstration by experts, it was believed that TBM was not applicable to the water-containing loose fine sandstone stratum of the #7 tunnel, and the TBM jammed was disassembled. The single shield TBM was refurbished to adapt to this geology by China Railway Engineering Equipment Group Co., Ltd. The shield and the cutter head were re-designed and re-manufactured and assembled at the tunnel entrance for tunneling.

The single shield TBM (Fig. 11.129) refurbished of the Yao River Water Diversion Project started working on August 15, 2011 and the records of monthly advances of 1515 m, 1718.6 m and 1868 m were set by China Railway Group Co., Ltd. in September, October and November 2011 respectively.

(1) Background of TBM refurbishment

The geological conditions of the tunnel are extremely complicated, with intense phase change of rocks. The tunnel stability is poor, and soft rocks and extremely soft rocks are mainly developed, with substantial groundwater at local tunnel sections. The groundwater has numerous structures, and is load-bearing. Since controlled by the structure and the stratum lithology, the groundwater has great change in distribution and enrichment.

Firstly, TBM tunneling was made from the tunnel exit to the entrance. When TBM tunneling was progressed for 2800 m, water-containing loose sandstones were encountered, and the TBM shield was buried by the sediment due to mud and sand burst. The shield bowing was severe after the machine was recovered, and the shield was wrapped by the sdiment, leading to severe damage to the installed segments due to a high thrust. Consequently, TBM tunneling was interrupted. See Fig. 11.130 for the original TBM cutter head.

When tunneling from the exit to the entrance was impossible, it was decided to disassemble TBM and its back-up at the entrance for tunneling from the entrance to

Table 11.36 Technical parameters of single shield TBM of the Yao River water diversion project

Main components	Sub-components	Parameter
Complete machine	TBM length	10.3 m
	Back-up length	170 m
	Switch	200 m
	Total weight of the main machine and the back-up	About 1700 t
	Minimum turning radius	500 m
Cutter Head	Material of cutter head	Q345C
	Cutter	17 in disc cutter
	Number of blocks	2
	Excavation diameter	5.75 m
	Eccentricity	20 mm
	Overcutter (maximum overcutting capacity)	1/90 mm
	Number/diameter of face disc cutter	31/432 mm
	Number/diameter of center disc cutter	4/432 mm
	Number/diameter of gauge disc cutter	3/432 mm
	Other cutters	27 scrapers, 6 bucket lips, 3 edge protection cutters
	Cutter load	250 kN
	Opening ratio	12%
	Muck discharge outlet	6%
	Cutter change method	Wedge-lock, back-mounted
	Nozzle	16
	Foam nozzle	4
	Manhole	1
Cutter Head drive	Drive type	VFD variable-frequency electric
	Rotation speed	0–8r/min
	Rated torque	4000 kN·m
	Breakout torque	6312 kN·m
	Service life of main bearing	> 15000 h
	Working pressure of seal	10 bar
	Seal type/quantity of main bearing	Lip type 2 × 4
Shield	Form	Single-shield TBM
	Front shield diameter/length/thickness/mass/number of blocks	5.68 m/3.38 m/85 t/60 mm/3

(continued)

Table 11.36 (continued)

Main components	Sub-components	Parameter
	Middle shield diameter/length/thickness/mass/number of blocks	5.66 m/4.40 m/80 t/40 mm/4
Shield	Tail shield diameter/length/mass/number of blocks	5.66 m/2.98 m/16 t/1
	Thickness of tail shield	50 mm/40 mm (bottom 120°)
	Tail shield opening	Bottom 60°
	Installation range of grout excluder plate	270°
	Tail shield clearance	60 mm
Stabilizer	Number of stabilizers	2
	Cylinder	ϕ300 mm × ϕ280 mm × 185 mm
	Maximum support force of a single cylinder	740 kN
	Maximum contact pressure	4 MPa
Anti-rolling cylinder	Qty	4
	Stroke	105 mm
Thrust system	Maximum total thrust	29733 kN
	Quantity of cylinder	16
	Stroke of cylinder	2600 mm
	Maximum advance rate	120 mm/min
	Maximum retraction speed	3000 mm/min
	Number of displacement sensors	4
	Quantity of thrust cylinder zones	4
Articulation system	Number of articulation cylinders	14
	Stroke of cylinder	150 mm
	Maximum pulling force of a single cylinder	25860 kN@300 bar
	Number of displacement sensors	4
C1 belt conveyor	Operating speed of belt conveyor	0–3 m/s
	Belt transfer capability	375 m^3/h
	Belt width	800 mm
	Length of belt conveyor	25 m
	Drive type	Hydraulic
	Pump station power	75 kW
Back-up	Connection Bridge	2
	Trailer	17

(continued)

Table 11.36 (continued)

Main components	Sub-components	Parameter
	Structure	Enclosed platform
	Rolling stock (L × W × H)	130 m × 1.6 m × 2.1 m
	Rail storage capacity	10 pcs (12.5 m/pcs)
	Storage of HV cables	400 m
C2 belt conveyor	Operating speed of belt conveyor	0–3.0 m/s
	Transportation capability	375 m^3/h
	Belt width	800 mm
	Length of belt conveyor	130 mm
	Drive mode	Hydraulic
	Pump station power	132 kW
	Maximum gradient	7°
Bottom mortar unit	Specifications and model of grouting pump	Two-piston type
	Number of grouting pumps	2 sets
	Capacity	9 m^3/h (each)
	Capacity of slurry storage tank	0.8 m^3
	Agitator	7.5 m^3/h
Pea gravel system	Specifications and model of pea gravel pumps	Pneumatic
	Quantity of pea gravel pumps	2 sets
	Theoretical capacity	2 × 15 m^3/h
Pea gravel system	Actual capacity	2 × 10 m^3/h
	Capacity of pea gravel storage tank	6.5 m^3
Pea gravel grouting	Grouting pump	Screw
	Theoretical capacity	12 m^3/h
Segment Erector	Rated pick-up capacity	55 kN
	Pick-up mode	Mechanical type
	Drive mode	Hydraulic drive
	Pump station power	45 kW
	Degree of freedom (DOF)	6
	Torque	180 kN·m
	Movement stroke	2750 mm
	Rotation angle	± 220°
	Control mode	Wireless control with cord connectors
	Rotating speed	0–1.5 r/min
	Lateral force of segment installation	15 kN

(continued)

Table 11.36 (continued)

Main components	Sub-components	Parameter
Segment hoist	Form	Electric double-track
	Lifting capacity	55 kN
	Working range	Trolley to segment feeder
	Gradient	7°
	Lifting speed	40 m/min
Segment feeder	Form	Hydraulic automatic
	Stroke	4.7 m
	Feeding capacity	1 pcs
Probe Drilling Machine	Specifications and model	B1000R
	Impact energy	225–520 J
	Power	90 kW
	Borehole	6 × ϕ110 mm (ID) ± 60° for front shield
		11 × ϕ110 mm (ID) ± 110° for middle shield
	Accuracy	2 s
Monitoring system	Number of cameras	4
	Number of display screens	1
Cooling water system	Form	Closed circulation
Cooling water system	Specifications of water hose reel	ϕ80 mm
	Storage length of water hose reel	60 m
Air compressor	Quantity of air compressor	2 sets
	Power	110 kW/set
	Displacement	17.1 m^3/min
	Rated pressure	7.5 bar
	Quantity of air tanks	2
	Capacity of air tanks	2 × 4 m^3
Secondary Ventilation System	Flow of secondary ventilator	12 m^3/s
	Fan power	37 kW
	Duct diameter	710 mm
	Duct storage	100 m
Dedusting System	Duct diameter	490 mm
	Form	Dry-type dedusting
	Filter accuracy	0.5 μm
	Capacity	240 m^3/min
Power system	Primary voltage	20 kV
	Secondary voltage	420 V/690 V

(continued)

Table 11.36 (continued)

Main components	Sub-components	Parameter
	Transformer capacity	2500 kV·A/1250 kV·A
	Transformer protection grade	IP55
Emergency Generator	Specifications and model	Diesel type
	Power	182.5 kW
Monitoring and alarm system for toxic and harmful gas	Type of monitored gas	CH_4, CO_2, CO, O_2
	Quantity of detectors	4 for CH_4, 2 for CO, 2 for CO_2, 1 for O_2

Fig. 11.129 Single shield TBM refurbished of the Yao River water diversion project

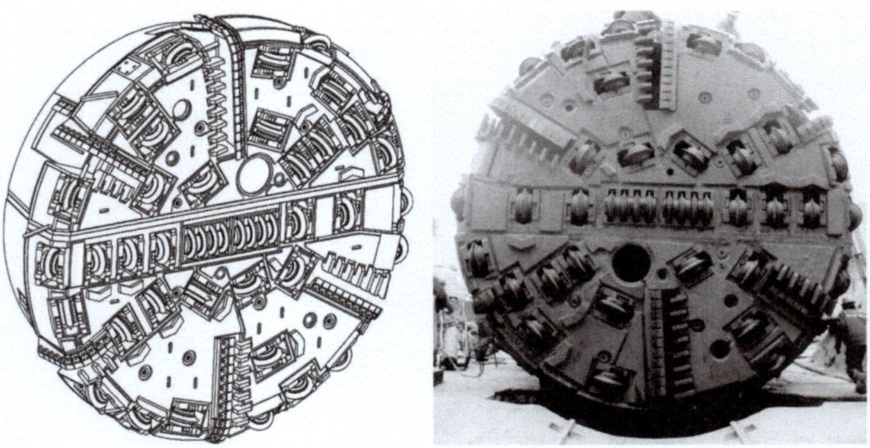

Fig. 11.130 Structure of original TBM cutter head

the exit. Since the TBM cutter head and the shield cannot be disassembled from the tunnel, a new TBM cutter head and a new shield were used.

In combination with the construction of TBM tunneling at the exit and geological conditions at the entrance, the design of the new cutter head and the shield was improved. The cutter head and the shield improved can adapt to engineering geology better, realizing continuously stable and efficient tunneling.

(2) Refurbishment of cutter head

① Two-way rotation

Passing of special geology is not fully considered in the original one-way rotation design of the cutter head, and the original anti-rolling cylinder preventing rolling of the shield does not have enough reaction force to resist the reversed rolling of the shield arising from cutter head rotation.

The shield will roll under the effect of reaction force during tunneling. The original cutter head can only rotate clockwise during tunneling, and the shield always rotates counterclockwise. To restrain shield rolling, the angle of the thrust cylinder must be adjusted. In this way, a force acting on the segment will be generated in the circumferential direction, driving the segment to roll directly.

Only two-way rotation of the cutter head during tunneling can solve the problem concerning segment rolling properly.

A two-way muck hole is designed for the new cutter head (Fig. 11.131).

The improved TBM can realize two-way rotation during tunneling, so it is unnecessary to adjust the angle of the thrust cylinder. Rolling of the shield is adjusted through two-way rotation of the cutter head.

② Re-layout of edge cutter

Fig. 11.131 Improved cutter head

The original cutter head is designed with 42 cutters, including 4 double disc cutters (center disc cutters), 31 face disc cutters and 3 edge cutters. See Fig. 11.132 for the trajectory of the original cutter head cutters.

The new cutter head is designed with 39 cutters, including 4 double disc cutters (center disc cutters), 30 face disc cutters and 1 edge cutter. See Fig. 11.133 for the layout of the refurbished cutter head edge cutters.

③ Cushion blocks are used for overcutting via an edge cutter.

The M39, M41and M42 cutter seats of the original cutter head are designed as a movable one (Fig. 11.134). The cutter seat is linked to the cutter head with bolts. If overcutting is required, remove the bolts and move the cutter seat to the excavation radius required.

Cutters are not installed in the M39 and M42 cutter holes during normal tunneling, and such holes are sealed with steel plates. If overcutting is required, install cutters in these holes.

Cushion blocks are used for overcutting designed for the new cutter head (Fig. 11.135). The overcutting radius can be increased by 15, 25 and 50 mm. Five cutters shall be heightened per heightening spacing, i.e. M35–M39 cutters. As for different heightening radii, the cushion blocks of different heights shall be installed for the M35–M39 cutters, to guarantee a smooth excavation contour line in the arc area of the cutter head.

A movable cutter seat is originally designed for overcutting, and the operation is complicated. When the single shield cutter head cannot move backwards, it is quite

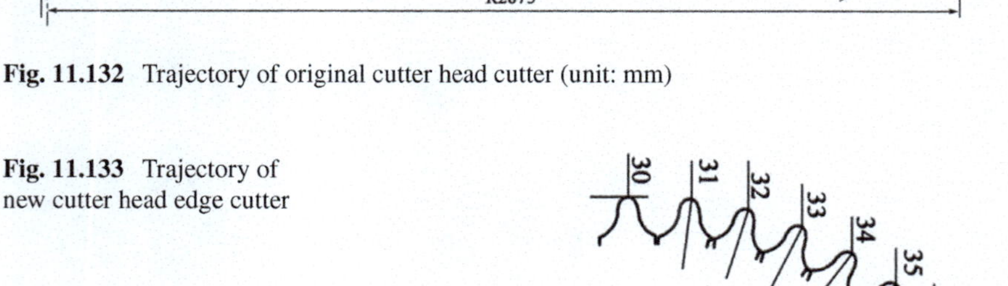

Fig. 11.132 Trajectory of original cutter head cutter (unit: mm)

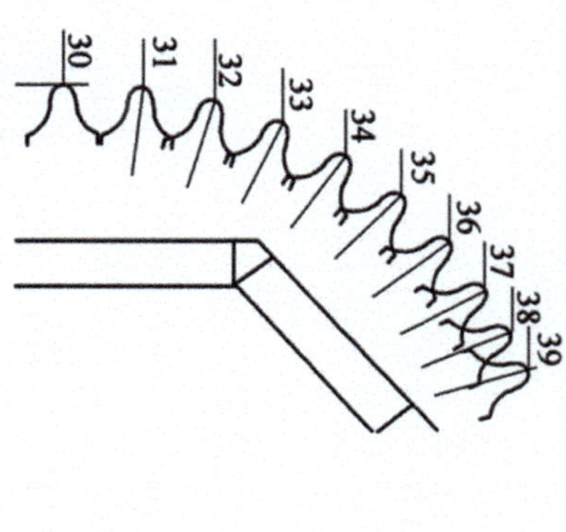

Fig. 11.133 Trajectory of new cutter head edge cutter

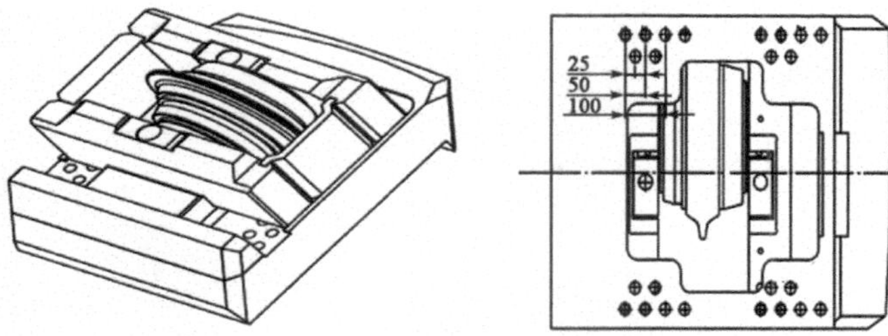

Fig. 11.134 Movable cutter seat designed for the original cutter head (unit: mm)

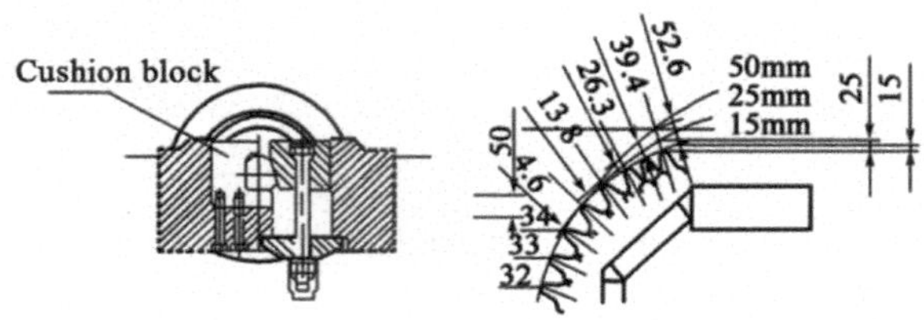

Fig. 11.135 Overcutting designed for new cutter head (unit: mm)

difficult to move the cutter seat inside the cutter head. Moving cutter seat also cannot guarantee a smooth transition of the contour line at the cutter tip in the arc area.

Cushion blocks are newly designed, and the operation is simple and convenient, and a smooth transition can be realized for the cutter trajectory and spacing in the arc area.

④ Optimization of cutter head structure

The original cutter head is designed for hard rocks, so the cutter head structure is too heavy, and is not applicable to loose rocks. The thickness (the length from the flange surface to the disc cutter tip) of the new cutter head is optimized, and the original length designed is reduced from 1995 to 1818 mm. The distance from the front edge of the front shield to the cutter tip of the cutter head is also reduced from 960 to 765 mm.

The resistance from rocks to the cutter head is remarkably lowered when the cutter head is used for tunneling in loose rock stratum since the cutter head structure is optimized. The mass of the cutter head is reduced from 80 to 60 t, which greatly mitigates the "bowing" of the cutter head and the front shield since the shield is too long.

⑤ Optimization of hard facing on cutter head surface

The hard facing on the surface of the original cutter head adopts the 10 mm-thick hardox400 steel plates, and the cutter head panel is greatly worn in sandstones (Fig. 11.136).

The new cutter head adopts SA1750CR chrominum carbide composite panels to increase its abrasion resistance. The SA1750CR chrominum carbide composite panel adopts special production process and chemical formulation, and the main chemical

Fig. 11.136 Hardox wear-resistant plate designed for original cutter head

components of the hard facing are carbon (4–5%) and chromium (30–40%). The thickness of the base plate is the practical thickness, and the dilution rate of the base plate and the hard facing is around 1 mm. The hardness gradient from the hard facing to the base plate is around 700 HV (Fig. 11.137).

(3) Refurbishment of shield.

① Improved design of front shield

The upper part of the front shield improved is 400 mm longer than the lower part (Fig. 11.138). The muck inlet capacity of the muck hole in the circumferential direction of the cutter head is improved, thus increasing the TBM tunneling speed. The amount of muck stacked at the horn of the front shield is lowered under a high penetration, thus reducing abrasion of the edge cutter.

② Drilling injection holes around the shield

Fig. 11.137 Wear of composite steel plates of new cutter head

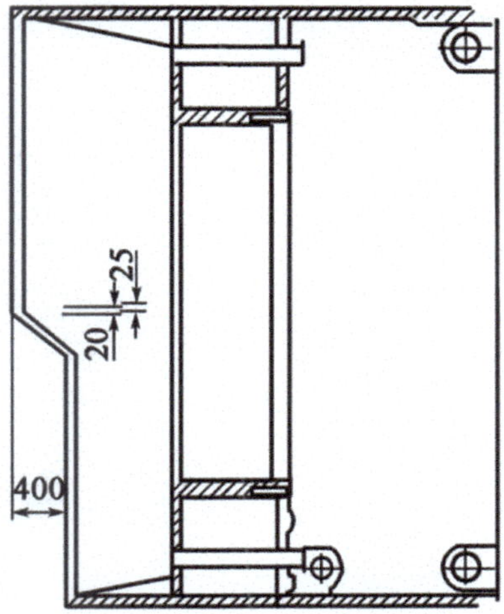

Fig. 11.138 Schematic diagram of the front shield (unit: mm)

Injection holes are drilled around the shield, so that waste oil or bentonite can be injected to the outer wall of the shield in case of collapse or convergence of the rocks, to reduce the friction resistance between the shield and rocks.

③ Optimized layout of thrust cylinder

The working area of the original thrust cylinder can be classified into 4, i.e. up, down, right and left (4 cylinders each; 16 in total) (Fig. 11.139). The newly designed thrust cylinder still has 4 working areas, but one thrust cylinder is added at the bottom, so there are 17 thrust cylinders in total (Fig. 11.140).

The oil lines of the added thrust cylinder are connected to one thrust cylinder at the bottom in parallel, and ball valves are installed at the oil inlet/outlet of two chambers. Such valves can be opened if it needs to be used.

The added thrust cylinder shall not be used during tunneling at the tunnel sections with normal geological conditions, and shall be used under the following conditions:

The thrust needs to be increased in case of collapse or rock convergence of TBM; the thrust difference between the cylinder at the bottom and that at the top needs to be increased while adjusting the TBM tunneling attitude.

④ Improved structure of tail shield

Since buttress is designed at 45° of the segments at the bottom, the tail shield is designed as a 60° opening. The 40 mm steel plates are designed for the 120° bottom of the original tail shield, and the upper 240° is 50 mm (Fig. 11.141).

For the tail shield refurbished, 25 mm steel plates are designed in the 60°–90° area, 40 mm steel plates are designed in the 90°–120° area and the rest is 50 mm (Fig. 11.142). The spacing of the tail shields is increased after refurbishment, which effectively improves installation quality of segments.

⑤ Setting mounting holes for extending shield for the front shield

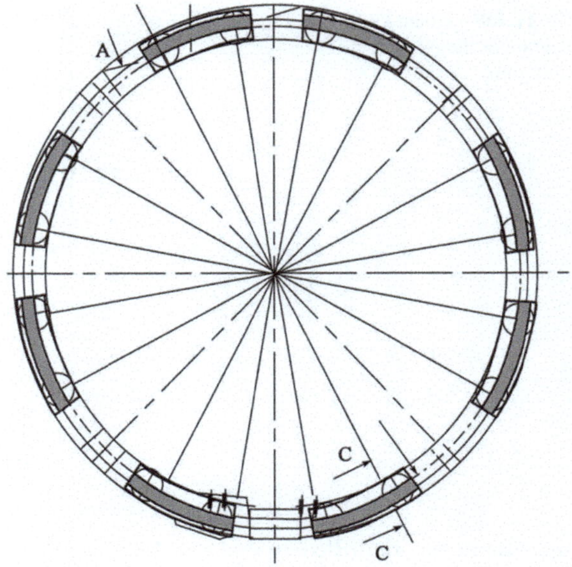

Fig. 11.139 Distribution of original thrust cylinder

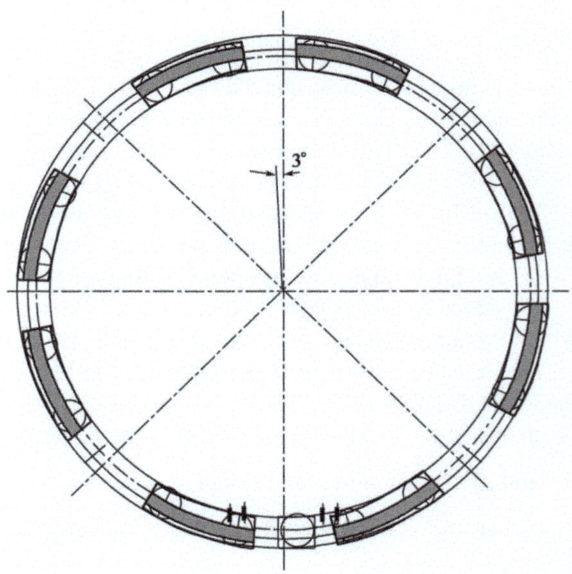

Fig. 11.140 Distribution of refurbished thrust cylinder

Two rows of holes with a diameter of 33 mm are drilled in the upper 150° area at the front edge of the front shield. Extension of the front shield can effectively reduce the resistance from rocks to the cutter head, and TBM can work smoothly in the loose sand stratum with a low water content.

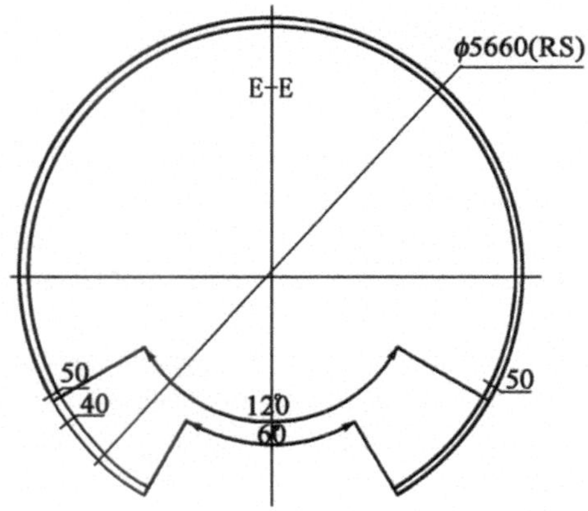

Fig. 11.141 Original tail shield designed (unit: mm)

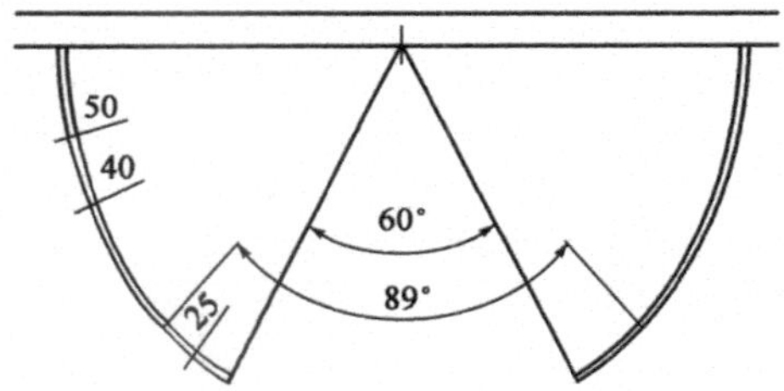

Fig. 11.142 Refurbished tail shield (unit: mm)

Since the geology of the project is featured by a low strength and a low abrasiveness, the cutter head can realize two-way muck discharge, without influence on normal use due to excessive and too rapid abrasion of the scraper and the scraper seat, which is special.

6. Construction technology of single shield TBM

(1) Single shield TBM tunneling at the tunnel sections featuring water-containing loose sand layers.

① Geological characteristics of water-containing loose sand layers

The lithology of rocks is mainly composed of the upper Tertiary N2L3 and the Cretaceous K1hk4 rock strata. The muddy cemented fine siltstones of K1hk4, locally intercalated with thin mudstone, are weak, with a uniaxial saturated compressive strength of $R_b = 5–6$ MPa, and are relatively soft, easy to soften and swell in case of water, and disintegrate in case of full water with plastic flow and other engineering characteristics. The sandy mudstones and argillaceous siltstones of the upper Tertiary N2L3, with sandstones and glutenite locally, are mainly muddy and cemented, with a poor cementing degree and a uniaxial saturated compressive strength of $R_b = 1.5–2.0$ MPa. As extremely soft rocks, they will swell easily in case of water, and disintegrate in case of full water, and dry and shrink in case of water loss and is weakly swelling.

② Design of construction countermeasures and schemes

A. Control over muck quantity

The engineer on-duty shall make statistics on the muck quantity generated per cycle of advance. The theoretical actual muck quantity is 41.5 m^3, and the loosening coefficient of muck is 1.79. The theoretical muck quantity is 37 m^3/advance and 74 m^3/ring. Each muck car is 16 m^3 and 10 muck cars are designed for one row normally. The muck quantity is around 9.4 cars per tunneling of two rings. The rotation speed and thrust are lowered to reduce the disturbance from rocks. Observe for collapse through the tunnel face, observation windows and grouting holes. The engineer on-duty shall record backfilling of pea gravel, and judge on whether collapse occurs based on the backfilling quantity.

B. Optimization of tunneling parameters

Tunneling parameters must be selected as per rock conditions. At normal rock sections, when the attitude of TBM well matches with that of segments, the thrust and the cutter head rotation speed may be increased properly. At the sections with unfavorable geological conditions, continuous TBM tunneling shall be guaranteed as appropriate under the preconditions of controlling the muck quantity and reducing the downtime.

C. Lowering friction between the shield and rocks

The bentonite and waste grease shall be injected from outside the shield, and tunneling shall be made continuously to reduce the TBM downtime. The tunneling parameters shall be optimized, and the thrust and the cutter head rotation speed shall be lowered to reduce the disturbance from rocks.

(2) Breakout technology for cutter head jamming of single shield TBM.

① Construction overview

Since the water-containing loose sand layers basically cannot realize stability, the TBM cutter head is jammed due to severe collapse and convergence, large mass and great length of the cutter head.

② Design of construction countermeasures and schemes

As for the cutter head jamming, it is considered to mitigate the soil pressure and dead weight of the cutter head. Since the dead weight of the cutter head cannot be changed, it is considered to reduce the soil pressure born by the cutter head by extending the front shield, with overcutting around the shield, so as to realize TBM breakout.

③ Construction process and methods

A. Construction process flow

See Fig. 11.143 for the flow chart of construction process.

B. Construction methods

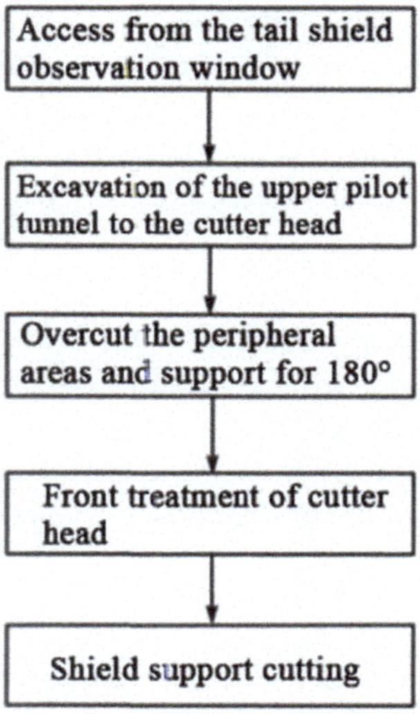

Fig. 11.143 Flow chart of construction process

a. See Fig. 11.144 for excavation and support of the small pilot tunnel right above the shield.

Excavation shall be made manually via an air pick or a pneumatic drill from the observation window at the tail shield and the excavation dimensions of the small pilot tunnel are 100 cm (W) × 120 cm (H). Since limited by space, the muck is poured to the #1 belt conveyor via a small ash bin through the observation window, then to the muck car via the #2 belt conveyor and finally to outside the tunnel via rolling stock. The I16 I-beam ring beams are used for supporting at a spacing of 45 cm, with ϕ18 mm thread steel for longitudinal connection. In total, 7 longitudinal connecting bars are arranged in parallel, and the arch foot of ring beams is welded to the shield. The ring beams are supported with square timber temporarily in the period from installation to welding, and shall be removed after welding. Steel breasting plates or

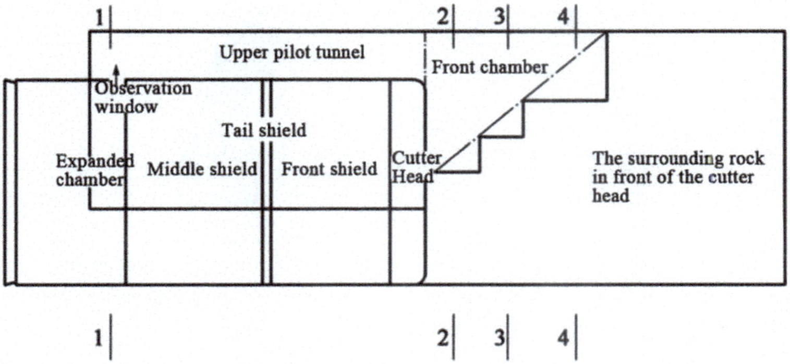

Fig. 11.144 Excavation of small pilot tunnel

planks shall be installed behind the ring beam after completion of the ring beams and the longitudinal connecting bars, and be fixed onto the ring beam with steel wires (Fig. 11.145). The exposed excavation surface behind the tail shield is plugged with planks, which shall also be used for plugging the excavation surface after excavation and support are progressed to in front of the cutter head.

b. See Fig. 11.146 for overcutting support from shield top to 250°on both sides.

After excavation of the upper pilot tunnel, overcutting on both sides of the shield is made from the cutter head to the tail shield manually via an air pick or a pneumatic drill. The space of the next ring beam can only be excavated after supporting per excavation of the space of one ring beam. The welds between the ring beam and the shield shall be cut when excavation and support are progressed to 250° and then

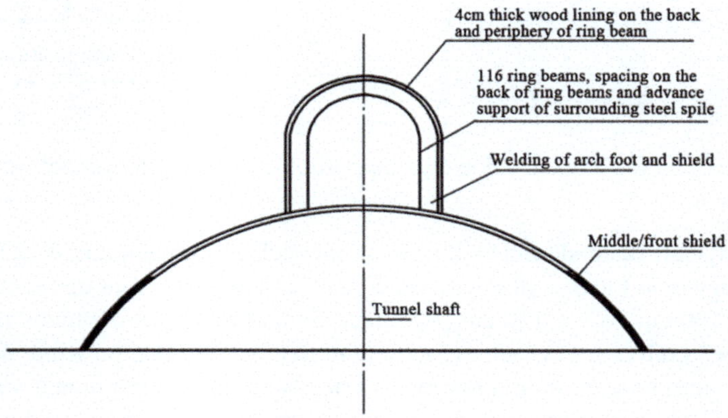

Fig. 11.145 Support of pilot tunnel

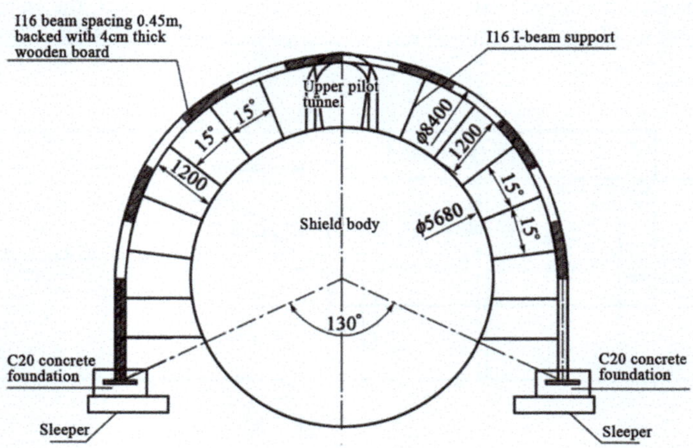

Fig. 11.146 Overcutting support (unit: mm)

be supported with square timber, to guarantee construction safety. Once the welds between the I-beam supports and the shield are cut, the pilot tunnel shall be backfilled with pea gravel.

c. Treatment in front of the cutter head

Support: The I-beam supports shall be extended to in front of the cutter head, and the I-beam ends shall be supported with square timber, with steel breasting plates being installed behind the I-beam. The exposed tunnel face shall be plugged with planks.

Excavation: The upper cutter head is excavated in the pilot tunnel. The middle and bottom of the cutter head are excavated through the cutter holes in the cutter silo to guarantee safety. Great attention must be paid to safety during excavation, and the principle of "excavation after support" shall be followed.

C. Implementing effect of schemes

Excavation and support of the pilot tunnel are completed smoothly, and the spacing of I-beam supports is uniform, with a tidy and aesthetic outlook. No safety accidents have occurred during excavation and support, and extension and refurbishment of the cutter head are completed successfully, so TBM breaks out smoothly.

The exposed excavation surface close to the tail shield shall be plugged during excavation of the pilot tunnel in loose sand layers, and the principle of "short advance and support first" must be followed. Technical clarification must be strictly followed, and the spacing of ring beams and the overlapping position of connecting bars shall be controlled. Since rocks are collapsed during original TBM tunneling, the collapsed area must be handled with by chiseling out the protruding rock mass, and be plugged with planks. After completion of the TBM breakout chamber, the steel supports welded to the shield shall be separated from the shield, and the breakout chamber shall be backfilled with pea gravel.

(3) Control technology for steps between segments

A circular section is designed for the #7 tunnel of the phase-I Yao River Water Diversion Project and the excavation diameter designed is 5.75 m. The ϕ5.75 m single shield TBM designed and manufactured by NFM and Northern Heavy Industries Group Co., Ltd. jointly is used, and the segment has an inner diameter of 4.96 m and a thickness of 280 mm. The upper 270° behind the segment is subject to backfill grouting with pea gravel (5–10 mm), and the strength is required to be C15 after completion, and the lower 90° is backfilled with M15 cement mortar.

Hexagonal segments are designed for the project, with a ball socket self-locking structure of a male and female face longitudinally. The outer diameter designed is 5520 mm and the inner diameter is 4960 mm. The segments are 280 mm thick and 1600 mm wide, and each ring has 4 segments (1 bottom segment, 2 side segments and 1 top segment). The maximum mass of a single segment is about 5.2 t. See Fig. 11.147 for erection of segments.

① Construction of single shield TBM segments

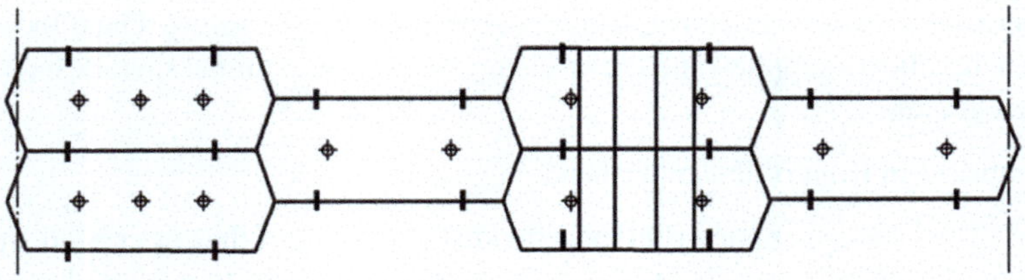

Fig. 11.147 Erection of Segments

Construction procedures of the single shield TBM mainly include slag removing during tunneling, segment installation, jointing, sealing of working holes, filling voids behind segments, etc. See Fig. 11.148 for details.

The single shield TBM is similar to the shield, and tunneling and segment installation cannot be made synchronously. The segment shall be installed per tunneling for 800 mm, and then tunneling shall be continued. Under the precondition of guaranteeing installation quality of segments, the installation speed directly affects TBM tunneling progress.

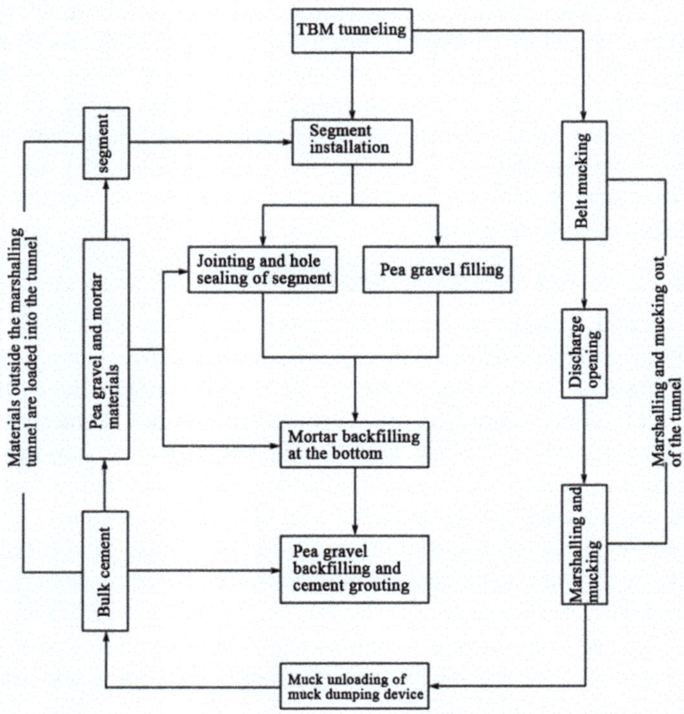

Fig. 11.148 Construction procedures of TBM tunneling

Large steps often occur during segment installation in the early phase of the single shield TBM tunneling, which directly affects normal construction of follow-up supporting procedures of TBM, leading to lowering in waterproof performance of segment lining and bringing certain hidden danger to the tunnel.

② Analysis on cause of steps between segments

A. "Upwarp" of the TBM tail shield during tunneling

To avoid "bowing" of the cutter head, during TBM tunneling, the upper and lower articulation cylinders in the shield shall be adjusted into the form of "short at the upper and long at the lower part", leading to "upwarp" of the tail shield, so that the bottom segments cannot fall onto the bedrock surface completely during installation. It sinks after pulling out from the tail shield, leading to large steps at the joints with the side segments. See Figs. 11.149 and 11.150 for the working state of the articulation cylinders.

When the stroke of the upper and lower articulation cylinders in Fig. 11.149 is the same, the trend values of the front shield and the middle and tail shields are identical, so that the segments can fall onto the excavation rock surface smoothly.

"Bowing" of the cutter head can be prevented by adjusting the upper and lower articulation cylinders in Fig. 11.150 into the form of "short at the upper and long at the lower part". The bottom segments cannot fall onto the bedrock surface during installation, leading to sinking after pulling out from the tail shield.

B. Failure in adjusting TBM tunneling attitude and segment installation attitude synchronously

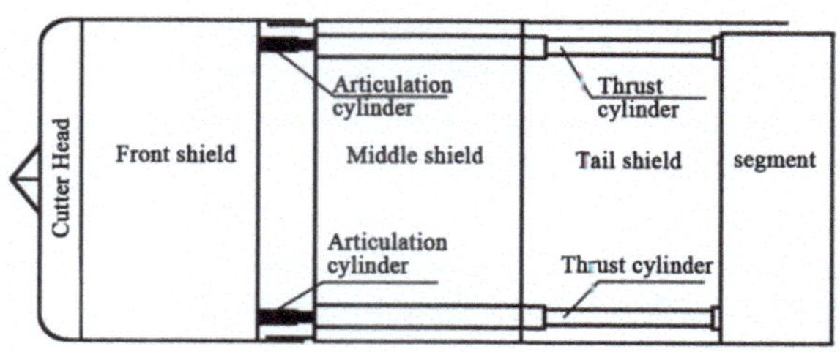

Fig. 11.149 TBM articulation cylinder (without stroke difference)

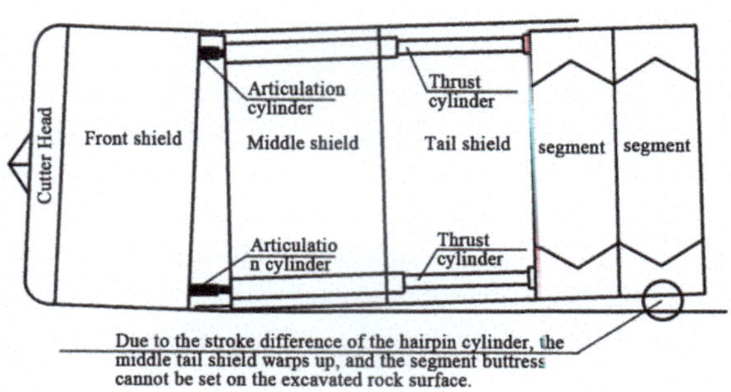

Fig. 11.150 TBM articulation cylinder (with stroke difference)

During the single shield TBM tunneling, the thrust reaction force is provided from the installed segments behind the shield, the inner diameter of which is larger than the outer diameter of segments. Under normal construction, there is certain clearance (tail shield clearance) between both, to facilitate attitude adjustment during segment installation. During TBM tunneling, in case of too fast steering, or sudden change in TBM attitude on the weak stratum, the tail shield will stay closely to the segments on one side or local area. It will be hard to adjust the attitude smoothly during segment installation if without tail shield clearance, leading to steps finally.

C. Low bearing capacity of bedrock at water-rich rock sections

The #7 tunnel along the main canal of the phase-I Yao River Water Diversion Project has complicated geological conditions, and soft and extremely soft rocks are mainly developed, with groundwater at local tunnel sections. The groundwater is load-bearing on multiple layers, and the distribution and volume of groundwater are greatly affected by the structure and stratum lithology. When TBM tunneling is progressed to water-rich rock tunnel sections, the water leaked is gathered at the tail shield since downhill is adopted for TBM, softening the rocks in such area continuously, thus lowering the bearing capacity of bedrock. When the bottom segments around 5 t are installed and are pulled out from the tail shield, the acting force between bottom segments and side segments is less than their dead weight, and the support force of the bottom rocks of the bottom segments is lowered, leading to sinking of bottom segments, forming large steps with side segments.

D. Poor adaptation of segment erection and connecting methods at the tunnel sections with unfavorable geological conditions

The segments of the project are of a ball socket structure of a male and female face longitudinally, and a positioning pin is used for connection during installation of two adjacent segments, to control segment erection quality. When it is hard to control the TBM tunneling attitude at the tunnel sections with unfavorable geological conditions, there is a high possibility of damage to the positioning pin during segment installation, so the positioning pin is hard to play its role designed. Controlling the segment erection quality relying on the ball socket self-locking structure of a male and female face alone is hard to control steps between segments.

③ Control measures for steps between segments

A. Refurbishment to bottom of tail shield

Since the segments are hard to fall onto the excavation rocks, the opening range at the bottom of the tail shield is refurbished by cutting the tail shield steel plates within 146° of the bottom, re-welding a thin steel plate and increasing the angle to the rock surface around, to guarantee that the buttress of the bottom segments can fall onto the excavation rock surface smoothly. The clearance between bottom segments and the excavation rock surface is reduced after refurbishment, lowering the falling range of bottom segments (Figs. 11.151 and 11.152).

B. Formulating standards for TBM attitude control

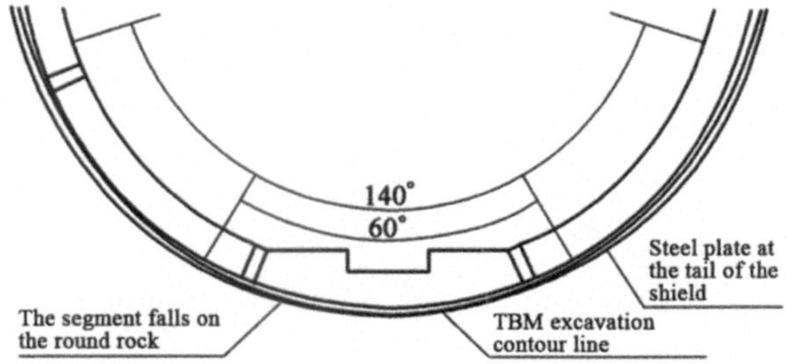

Fig. 11.151 Segments falling onto rocks after refurbishment of tail shield

Fig. 11.152 Diagram for segments falling onto rocks after refurbishment of tail shield

As for too fast steering by the TBM operator, or sudden change in TBM attitude on weak stratum, steering principles are formulated. Segment attitude (rotating and adjusting tail shield clearance) shall be adjusted frequently and slowly, with mandatory standards for TBM tunneling attitude control being formulated (Table 11.37).

C. Replacing softened rocks with timely backfilling of mortar at the bottom

For softened rocks, loose rock muck shall be replaced by dry mixture of cement and sands of certain proportion to increase the bearing capacity of rocks on one hand. Appropriate additives such as accelerator or water glass shall be added to the dry mixture when necessary, to narrow the setting time of mortar. On the other hand, once the segments are installed, M15 mortar shall be backfilled within the 90° range at the bottom of the segments timely, to fill the gap between bottom segments and rocks. Appropriate accelerator shall be added to the M15 mortar under poor geological conditions.

D. Adding limit devices on segments

When TBM tunneling is progressed to weak rock sections, the bottom and top segments will sink, and the side segments will be subjected to severe backward leaning. To tackle this tunnel section with unfavorable geological conditions, limit devices are added for segments, and positioning steel plates are added on the inner

Table 11.37 Standards for TBM tunneling attitude control

S/N	Item	Control standards
1	Horizontal attitude	Prewarning value for deviation: ± 100 mm; trend value: ± 0.5%
2	Vertical attitude	Prewarning value for deviation: + 20– + 80 mm; trend value: + 0.2%– + 0.6%
3	Steering value	The horizontal and elevation steering value shall not be more than ± 5 mm per tunneling for 800 mm; the change in trend value shall be less than 0.1%
4	Stroke difference of thrust cylinder	The change in cylinder stroke difference shall not be more than 15 mm per tunneling for a single ring; the cumulative cylinder stroke difference shall not be more than 30 mm
5	Stroke difference of articulation cylinder	Stroke difference of the upper and lower articulation cylinders: not more than 15 mm during normal tunneling and 20 mm if steering is required. Stroke difference of the left and right articulation cylinders: not used as much as possible during normal tunneling and not more than 10 mm if steering is required
6	Elevation difference between middle shield and tail shield	Not more than 20 mm
7	Rolling value of TBM front shield	0–1.5%

arc surface (upstream face) at the upper segments, to mutually restrain its backward leaning or sinking. Once the void between segments and rocks is filled by pea gravel and cement grouting and the design strength is reached, the added limiting steel plates will then be cut, to avoid any influence on operation of the tunnel.

(4) Construction technology for passing by the sections with unfavorable geological conditions

The single shield TBM has remarkable characteristics such as fast tunneling, synchronous lining and forming once under the mid-and hard-rock geological conditions featuring single lithology. However, jamming and bowing of the cutter head and the shield will occur easily if tunneling at the sections with unfavorable geological conditions such as the soft rocks with a high ground stress, crushed faults and without stability capability, especially in the weak water-containing loose sandstones with a poor foundation bearing capacity, thus severely affecting normal TBM tunneling. This section analyzes the cause of the jamming arising from geological disasters such as rock convergence, deformation, collapse, water burst and quicksand during tunneling via a single shield TBM at the sections with unfavorable geological sections of the #7 tunnel along the main canal of the phase-I Yao River Water Diversion Project in Gansu in a systematic manner and mainly introduces and proposes some of the problems during TBM tunneling at the sections with unfavorable geological conditions and countermeasures.

① Lithology of unfavorable geology

The tunnel rocks are mainly made up of the upper Tertiary (N_2L^3) argillaceous siltstones, silty mudstones, glutenite, (pebbly) sandstones, loose sandstones which are soft and interbedded with intense phase change. The attitude of rocks is gentle, and is slightly affected by structure, and fractures are not developed here, and only the gentle brachy-axis folds are developed, so the water abundance is poor as a whole. The groundwater is mainly replenished by atmospheric precipitation. Since the precipitation here is low and is non-uniform within a year, the amount of groundwater (the maximum flow of spring water measured is less than 5 L/min) is relatively small due to weak permeability of the stratum. According to drilling, test and hydrogeological survey, the porosity of glutenite and sandstones is around 20%, belonging to aqueous permeable strata. Groundwater is generally revealed by drilling, and spring water is developed at sandstone and glutenite. As a relative aquiclude, the argillaceous siltstone and silty mudstone have non-uniform groundwater, which is generally laminar and load-bearing locally. The mountain is a weak laminar aqueous mountain.

② Engineering characteristics and difficulties

A. The tunnel traverses loess weak stratum sections, including the water-containing loose sand layers without stability capability. TBM tunneling in this stratum has great geological risks.
B. The strength of rocks at some tunnel sections is relatively low, with a poor foundation bearing capacity, which may lead to "bowing" during TBM tunneling.
C. Groundwater is distributed at local tunnel sections. The groundwater is load-bearing on multiple layers, and the distribution and volume of groundwater are greatly affected by the structure and stratum lithology, and will affect mechanical, electrical, hydraulic, and sensing equipment of TBM.
D. Hexagonal segments of a ball socket self-locking structure of a male and female face are used for the project, without rigid bolted connection. Appearance defects such as rotation and steps will be triggered easily under the effect of external force after installation.
E. The erection structure of segment rings are susceptible to rock collapse and convergence, and non-uniform water burst and quicksand at the sections with unfavorable geological conditions may lead to bias pressure, leading to instability of segment structure.
F. Since the TBM tunnel is an extra long hydraulic tunnel featuring blind advancing, the difficulty in organizing ventilation, drainage, communication, dispatching and transportation will be extremely great.
G. Since the tunneling equipment used for the project is the first single shield TBM nationwide, little construction experience can be borrowed since the TBMs used before are double shield or open type, tunneling cannot be forecast to a large extent.

③ Main problems in construction at the sections with unfavorable geological conditions and countermeasures

Since the rocks on the tunnel face basically cannot realize stability, the collapse of rocks is severe. Large crushed rocks cannot be discharged smoothly, leading to clogging of the TBM cutter head and constantly increasing torque. The cutter head cannot be started normally if beyond the breakout torque.

As for this kind of tunnel sections with unfavorable geological conditions, TBM breakout is mainly considered in terms of rock reinforcement and lowering tunneling torque of the cutter head. The countermeasures adopted mainly include:

A. Consolidation grouting: To prevent severe influence on the TBM cutter head due to severe collapse of the rocks in front of the cutter head and to guarantee full filling of the void behind the segment, consolidation grouting is made for some of the weak rocks in front of the TBM cutter head. Ruimijiefulai U-type and Malisan N-type polyurethanes chemical grouting materials are proposed.
B. The starting torque of the cutter is reduced, and the abrasion resistance of the cutter head is enhanced to lower the frequency of abnormal damage. A cutter ring with favorable abrasion resistance is used to extend the service life of the cutter.
C. Service of the TBM muck discharge system is intensified, to guarantee that the cutter head and the belt conveyor can reach a normal breakout capacity. On the other hand, the cleaning frequency of the muck control door is increased, to prevent clogging of the door by large rocks, leading to jamming of the cutter head.
D. Proper tunneling parameters are selected, and the tunneling torque, penetration, rotation speed and thrust are strictly controlled, to guarantee that the muck quantity matches with the cutter head muck hole and the capacity of the belt conveyor and to prevent too fast much discharge, leading to increasing resistance in muck scraping and insufficient muck discharge capacity of the belt conveyor.
E. When the cutter head is jammed and cannot be started normally, the TBM front shield can be extended for refurbishment. The steel plates of the front shield can be extended to the heightened overcutter. Some of the 212° range of the upper cutter head shall be covered to reduce the rotating torque of the cutter head, so as to realize breakout.

Since the downtime of TBM is too long, the rocks behind the shield are collapsed severely or converged obviously, so that the rock pressure acts on the shield. When TBM tunneling is recovered, the thrust of TBM tunneling may increase remarkably, leading to severe damage to the segment providing reaction force and intensified rotation, resulting in tunneling suspension due to shield jamming in the worst case.

As for this stratum, TBM breakout shall mainly be considered by lowering the rock pressure acting on the shield, and the countermeasures adopted mainly include:

A. The advance geological forecast at the tunnel sections with unfavorable geological conditions shall be intensified. The Beam geological forecast system and probe geological drilling equipped for TBM shall be used for forecast based on the geological exploration data and achievements of material detection. In routine construction, geological clarification shall be made for the first-tier construction management and technical personnel in tunnel sections.

B. According to analysis based on geological exploration data, when TBM tunneling is progressed to the tunnel sections where rocks are converged or collapsed, the work concerning emergency preparation shall be done well. Meanwhile, the work concerning service of TBM equipment and reserve of parts shall be done well in an overall manner. The TBM abnormal downtime shall be shortened as much as possible.
C. Before long-term shutdown of TBM, lubrication materials (such as the bentonite, waste grease, etc.) shall be injected between the shield and rocks through the grease hole reserved at the upper shield, to reduce the friction coefficient and the friction resistance, thus avoiding TBM jamming after construction is recovered.
D. The overcutting radius of the cutter head shall be increased properly as per the rock convergence rate, to guarantee that the shield has enough time to pass by without jamming due to rock convergence after tunneling.
E. The articulation cylinder of the front shield and the back-up towing cylinder shall be put in a free state during breakout, and the maximum thrust of the main thrust cylinder shall be concentrated for breakout in parts.
F. In case tunneling is suspended due to jamming of the shield by rocks, the upper rocks close to the shield can be excavated, digged out and supported through the observation window reserved at the TBM tail shield, to release some rock stress acting on the shield. Meanwhile, lubrication materials shall be injected between the shield and rocks to lower the friction resistance, so as to guarantee normal TBM tunneling and realize breakout. See Fig. 11.153 for construction of the TBM breakout chamber.

When the rocks with a high water content are encountered during TBM tunneling, the rocks at tunnel bottom are severely argillized, leading to remarkable lowering of the bearing capacity, so that the stability needs of the TBM structure and segment lining are hard to be met. Consequently, unfavorable phenomena such as cutter head bowing

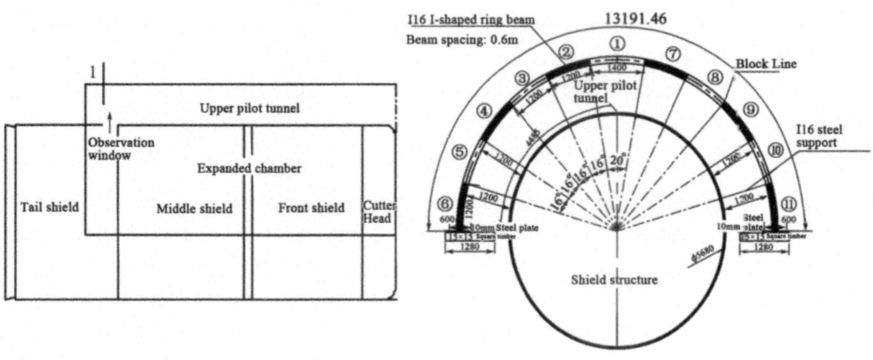

Fig. 11.153 Construction of TBM breakout chamber (unit: mm)

and shield sinking occur during TBM tunneling, and severe consequences such as hard control over TBM tunneling state, hard installation of segments, severe damage to segment tips, large steps between segments or obvious segment convergence will occur in serious cases.

TBM jamming shall be avoided mainly by improving bearing capacity of rocks, controlling attitude and segment reinforcement on this stratum, and the countermeasures adopted mainly include:

A. The bottom plate shall be reinforced by injecting cement mortar or two-fluid slurry by arranging grouting holes at a proper position of the shield bottom, to improve the bearing capacity of the stratum, so as to prevent bowing of the TBM cutter head and sinking of the shield. However, the grouting amount shall be strictly controlled in this process, to avoid difficult thrust due to gluing of the shield.
B. It is considered to access the main thrust cylinder at the bottom to the ultrahigh-pressure oil pump, to increase the thrust difference of the top and bottom thrust cylinders and to prevent bowing of the TBM cutter head and sinking of the front shield.
C. The tunneling parameters shall be controlled and a correct shield attitude shall be guaranteed. The stroke difference of the front articulation cylinder shall be increased, and the shield shall be maintained under a slightly head-up trend, with constant-speed tunneling and slow adjustment of the attitude.
D. At the tunnel sections with unfavorable geological conditions, to control the joints and steps generated from segment installation and convergence, limiting steel plates are added on the segment or circumferential steel plate rings are added in the segment, so as to control or reduce steps between segments and convergence.
E. For the failure in water-stop due to steps between segments, the segment joints can be filled with GBW self-adhesive water-stop which will swell in case of water, to guarantee the waterproof performance of the segment lining.

When TBM tunneling is progressed to the tunnel sections featuring water-containing loose sand layers, since the water content of rocks is high and the rocks are subject to severe collapse without stability, the water burst and quicksand in the stratum will affect the electrical equipment in the shield to certain extent.

TBM breakout shall mainly be considered by equipment protection, segment reinforcement, pump drainage of sand water and lowering the pressure of rocks on the shield on this stratum, and the countermeasures adopted mainly include:

A. Close attention shall be paid to muck properties and tunnel face. The observation frequency for muck properties and rocks on the tunnel face shall be increased during tunneling, especially the change in groundwater. The tunnel face may be observed in the cutter head when appropriate, and the tunneling parameters shall always be adjusted as per the geological conditions on the tunnel face.
B. During routine service of equipment, some of the equipment (such as thrust cylinder, main motor, stabilizer, observation window, grease fitting, boreholes,

distribution cabinet) in the shield shall be sealed to prevent damage from groundwater to the electrical and hydraulic equipment in the shield at the tunnel sections with unfavorable geological conditions.

C. Sediment cleaning in the tail shield (at the opening) shall be intensified during tunneling and before segment installation, and sufficient personnel and tools shall be provided, to guarantee fast installation of segments.
D. Emergency plans shall be formulated before TBM tunneling is progressed to the tunnel sections featuring water-containing loose sand layers, and the work concerning service and maintenance of TBM equipment shall be done well in an overall manner, to avoid abnormal shutdown of TBM as much as possible. Meanwhile, the duration of normal shutdown for service of TBM shall be minimized, to guarantee fast tunneling.
E. Lubrication materials shall be injected between the shield and rocks through the grease hole reserved on the shield during TBM tunneling, to reduce the friction coefficient and the friction resistance, so as to avoid TBM jamming.
F. The cutter ring with a favorable abrasion resistance shall be used, to enhance the abrasion capability between the cutter and rocks, to lower the frequency of abnormal damage to cutters and to avoid replacement of cutters due to abnormal shutdown.
G. Drainage in the tunnel shall be enhanced. During construction, a high-power waste water pump shall be provided around the TBM tail shield, the TBM room and the back-up, and the waste water shall be drained to the back-up waste water tank directly. Moreover, one blowdown pipeline shall be laid to the tail of the TBM california switch, to guarantee timely drainage of the groundwater.
H. Considering severe collapse of rocks at the tunnel sections featuring water-containing loose sand layers, pea gravel backfill grouting and mortar filling cannot be made as per design requirements. Therefore, it is considered to drill a ϕ48 mm floral ductule through the segment working hole for deep consolidation grouting, to guarantee the stability of the lining structure erected and the water-stop effect.
I. If TBM is jammed due to severe water burst and quicksand and a high water content of rocks at the tunnel sections featuring water-containing loose sand layers, and breakout cannot be realized smoothly through conventional methods, such sections shall be excavated manually immediately, mainly including adding inclined shafts and shafts or excavation and support of bypass tunnel, with thrust and installation of segment lining for pass-by when TBM tunneling is progressed to this position after the integral sliding track bed is applied.

(5) Other construction measures of single shield TBM.

① Measures for tackling "bowing"

Increase the thrust of the bottom cylinder and decrease the thrust of top cylinder, forming a pressure difference of the upper and lower cylinders. Adjust the stroke of the articulation cylinder, forming a "V-shaped" angle. Maintain upward tunneling of TBM and avoid idling of the cutter head. Match segment attitude with the TBM attitude and avoid excessive adjustment of a single attitude.

② Measures for tackling geological disasters such as mud and sand burst

Know and be familiar with distribution of unfavorable geological conditions, with planning and discussion in advance. Formulate tunneling parameters at the sections with unfavorable geological conditions and reasonably optimize tunneling organization. Formulate a feedback mechanism followed by the person in charge of M&E, civil work and production management. For possible problems, prepare personnel and materials in advance by specially-assigned personnel. Study on, discuss on and formulate opinions for possible emergency, and divide responsibilities to specific persons and zones. Service and maintain the machinery and equipment in advance, to guarantee smooth tunneling at the tunnel sections with unfavorable geological conditions. Optimize procedures and reasonably organize transportation, to avoid influence of TBM tunneling on procedures and transportation. Plug the muck hole with steel plates, to reduce the muck quantity. The engineer on-duty shall pay close attention to the change in geology and monitor the change in water content and the test personnel shall measure the mud content. Record the backfilling of pea gravel and mortar, to facilitate quality analysis on backfill grouting of pea gravel and tunnel waterproof.

③ Reinforcement of lined segments at the tunnel sections with unfavorable geological sections

The hexagonal segments of a self-locking structure are used for the project, and the integral structural stability can be guaranteed through the structure of the segment itself. According to the convergence monitored of the structure, the sections severely deformed are reinforced with inner sleeve I16 I-beam steelwork for the entire circle at a spacing of 80 cm. Such steelwork is placed at the middle of the longitudinal joints of the segment, and limiting steel plates are installed at the contact area of longitudinal joints of side and top segments.

④ Measures for TBM tunneling to pass by collapsed and converged tunnel sections

Formulate the clarification scheme in advance, and enhance construction organization. Prepare sufficient materials to avoid postpone of the progress and minimize the downtime to guarantee continuous tunneling. Lower the rotation speed and thrust to reduce the disturbance from rocks, and reduce the amplitude of attitude adjustment as much as possible. Inject bentonite or waste grease from outside the tail shield to reduce the friction of rocks.

11.3.2 Chongqing Rail Transit Line 5

1. Project Overview

(1) Overview

The Chongqing Rail Transit Line 5 has a total length of 39.75 km, including 31.18 km (accounting for 78.44% of line length) for underground lines, 8.12 km (accounting for 20.43% of line length) for elevated lines, 0.45 km (accounting for 1.13% of line length) for ground lines and transition sections. This line has 25 stations, including 5 elevated stations and 20 underground stations, with an average spacing of 1.62 km. It has 3 overtaking stations and 9 transfer stations. The phase-I project of the Line 5 starts at the EXPO Garden Center station and ends at Tiaodeng.

(2) Lot and single shield TBM sections

With a total length of 10.42 km, the construction line of the lot starts at the ventilation shaft behind the Chongguang Station and ends at the receiving shaft in front of the Dashiba Station, including 4 stations, 6 sections (including 2 refurbished stations) and 1 release line. The stations include Hemulu Station, Renhe Station, Ranjiaba Station (refurbished) and Dalongshan Station (refurbished). The sections include Chongguang Station—Hemulu Station—Renhe Station—Dazhulin Station and Xingfu Square Station—Ranjiaba Station—Dalongshan Station—Dashiba Station. The release line is the Dazhulin release line.

The section of the single shield TBM is located in the Chongqing New North Zone, mainly include the ventilation shaft—Hemulu Station—Renhe Station—Dazhulin Station section. Two pieces of equipment start tunneling from Dazhulin and has tunneled for 6501 m after passing by the Renhe Station and the Hemulu Station, and is lifted out from the ventilation shaft at this section. Two composite EPB shields are used for tunneling from the launch shaft behind the Chongguang Station at the Chongguang Station—ventilation shaft section, and are lifted out from the ventilation shaft at this section.

(3) Tunnel design

The minimum turning radius of this section is R = 350 m and the maximum slope is 44‰. Reinforced concrete segment lining is adopted for the tunnel. The segment has an outer diameter of 6.6 m, an inner diameter of 5.9 m and a breadth of 1.5 m. A general ring structure (segment wedge: 39.6 mm) is adopted for stagger-jointed erection. Each segment is made up of 3 standard segments + 2 adjacent segments + 1 key segment, and each ring is connected with 10 longitudinal and 12 circumferential M27 bolts. Concrete self-waterproof (C50 P12) + EPDM rubber water-stop are used for the tunnel.

(4) Geology and hydrology

The buried depth at tunnel vault of the single shield TBM is 15–100.3 m at this section. The tunnel traverses the Middle Jurassic Shaximiao Formation (J2S) sedimentary rock layers, where thick moderately weathered sandy mudstones play a predoninant role with thin sandstones being intercalated. Moderately weathered thick sandstones are mainly developed locally. Grade-IV rocks are developed on the stratum through which the tunnel traverses (Fig. 11.154).

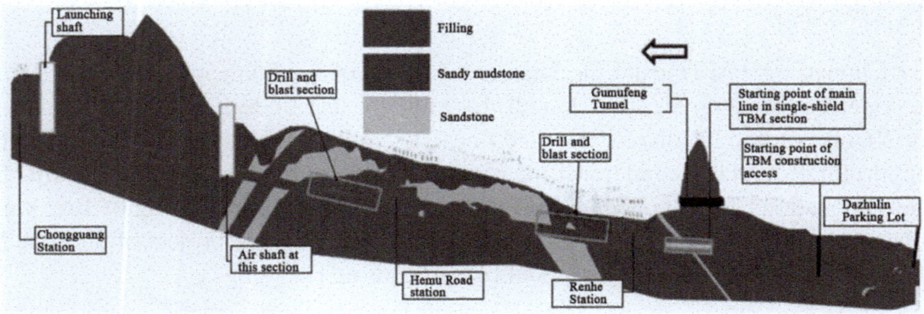

Fig. 11.154 Geological Longitudinal Section

The natural compressive strength of moderately weathered sandstones within the tunnel body is 37–53 MPa, while that of moderately weathered sandy mudstones is 8–15 MPa. See Table 11.38 for the recommended design parameters for geological exploration.

The surface water mainly includes the hardened water from urban roads and utilities, including production and household waste water. The groundwater mainly includes the Quaternary loose pore water, which is mainly distributed at gentle slopes and groove center and is mainly replenished from atmospheric precipitation and the water leaked from underground pipelines.

Since this line traverses hills and gullies, there is no uniform groundwater level in the soil stratum. The pore-fissure water of clastic rocks includes weathering fissure water and structural fissure water.

2. Single shield TBM

The single shield TBM (Fig. 11.155) used for this lot is refurbished based on the original Robbins open type TBM used for the phase-I project of the Chongqing Line 6. The cutter head has an excavation diameter of 6.83 m, a maximum thrust of 39,000 kN, a maximum excavation speed of 80 mm/min, a breakout torque of 6080 kN·m and a rated torque of 4054 kN m. The entire machine is 137 m long and the mass is about 750 t.

(1) Cutter head design

The cutter head of the single shield TBM is optimized based on the construction experience of the Yao River Water Diversion Project. See Table 11.39 for the characteristic parameters of the cutter head.

Two-way muck scraping is designed for the cutter head, and a device preventing gathering of muck is installed at the cone plate. It is easy to adjust the rolling value, which effectively improves the TBM tunneling efficiency.

(2) Main parameters of equipment

See Table 11.40 for main parameters of equipment.

(3) Other equipment

Table 11.38 Recommended design parameters for geological exploration

Lithology	Weathering degree	Unit weight (kN/m^3)	Standard value of uniaxial limit compressive strength of rocks		Foundation bearing capacity characteristic value (kPa)	Rock/soil shear strength		Rock				
			Saturability (MPa)	Natural value (MPa)		Internal friction angle φ(°)	Internal cohesion *C*(kPa)	Modulus of deformation (MPa)	Elastic modulus (MPa)	Poisson's ratio μ	Elastic wave velocity V_p(m/s)	Tensile strength (kPa)
Silty clay		20.4			150	12	40					
Sandy mudstone	Moderately weathered	25.9	7.74	12.63	1000	34	612	1316	1720	0.36	4203	168
Sandstone	Moderately weathered	25.2	27.49	37.16	2500	42	1818	3634	4254	0.16	3550	520

Fig. 11.155 Single shield TBM

Table 11.39 Characteristic parameters of cutter head

Item	Cutter head designed for yao river water diversion project	Cutter head designed for this lot
Excavation diameter	5750 mm	6830 mm
Length of front shield cut	1120 mm	995 mm
Length of the roadheader whose cutter head is extended from the front shield cut	960 mm	910 mm
Opening ratio	11%	12.7%
Angle of cone plate of cutter head	15°	35°
Cutter head mass	80 t	130 t

① Supporting equipment

Vertical lifting equipment: Two 45 t gantry cranes with a span of 25.8 m are used, and segments, air duct, water pipes, grease and oil and muck can be lifted.

Ventilation system: Forced ventilation via a axial flow fan, and the canvas air duct 1200 mm in diameter is used.

Mortar mixing system: A JS-1000L mortar mixer is used, and the slurry is cement paste.

Pea gravel pouring: One ZL50D loader is used.

Cutter repair workshops and oil water detection stations.

② Horizontal transportation equipment

Track laying: H175 profile steel is used for the sleeper double-track, and 160 U-steel is used for a single line; 43 kg/m rails are used, with a gauge of 0.97 m.

Rolling stock: 55 t locomotives + 6 muck cars (20 m^3) + 1 (6.5 m^3) pea gravel car + 1 mortar car (8 m^3) + 2 (15 t) segment cars, with three rolling stocks for each equipment.

Table 11.40 Main parameters of equipment

S/N	Item	Technical parameters	Unit	Remarks	S/N	Item	Technical parameters	Unit	Remarks
1	Cutter head rotation speed	0–5.4–11.9	r/min		5	Longitudinal climbing ability	± 60	‰	
2	Maximum advance rate	80	mm/min		6	Cutter head specifications	ϕ6830 × 1845	mm	
3	Maximum thrust	39,000	kN		7	Rotation direction	Clockwise / counterclockwise		
4	Horizontal turning radius	250	m		8	Number of muck holes	5	Group	
9	Number of water ports	7	Pcs		19	Tail shield grouting excluder	2	Row	
10	Quantity of central twin disc cutter	4	Piece		20	Tail shield installation clearance	50	mm	
11	Single disc cutter	35	Piece		21	Cylinder	21	Pcs	
12	Disc cutter extension	135	mm		22	Advancing stroke	2100	mm	
13	Quantity of scraper	178	Piece		23	Two-piston grouting pump for primary grouting	1	Pcs	
14	Height of scraper	25–35	mm		24	Grouting capacity	10	m^3/h	
15	Total drive power	7 × 330	kW	Variable frequency drive	25	Piston pump of secondary grouting system	1	Pcs	
16	Main bearing diameter	3285	mm		26	Quantity of pea gravel pumps	2	Pcs	

(continued)

Table 11.40 (continued)

S/N	Item	Technical parameters	Unit	Remarks	S/N	Item	Technical parameters	Unit	Remarks
17	Number of inner and outer lip seals	3	Nr	3 each	27	Pea gravel pump capacity	2×15	m^3/h	
18	Quantity of pre-grouting pipes	6	Pcs						

3. Construction problems and countermeasures

(1) Long-distance and large-gradient horizontal transportation of locomotives

Cause analysis: The single shield TBM uphills (4.4 and 3.4%) at a long distance from Dazhulin, and downhills under a heavy load. The longest transportation distance reaches 4.7 km. The risk of sliding exists during horizontal transportation in the construction process.

Countermeasures mainly include:

① A high-power 55 t battery locomotive is used, to guarantee that parameters such as power and braking performance can meet the requirements for safe transportation.
② Special transportation schemes and emergency plans are formulated, with eyecatching warning signs at important sections such as large-gradient sections and cross operation areas.
③ Anti-sliding devices are provided at the towing and large-gradient sections of running tunnels.
④ The dual-gas-circuit braking is used, to effectively reduce the brake delay.
⑤ Specially-assigned personnel shall be staffed for regular inspection and service per day, to guarantee that the machine is under good conditions.
⑥ A professional rail service team shall be set up for patrol and service of the rails per day.
⑦ The operation personnel shall be trained on safety education regularly, to improve their safety awareness.

(2) Construction of the overcross sections of running railway

Cause analysis: The running tunnel overcrosses the Chongqing-Huaihua Railway, Shanghai-Chongqing-Chengdu High-speed Railway and inclined shafts, and the minimum clearance with the Shanghai-Chongqing-Chengdu High-speed Railway is 4.5 m. Tunneling will undoubtedly bring certain influence to the tunnel structure of the existing railway under service.

Countermeasures:

① The design scheme is optimized by adjusting the slope from 34 to 44‰ and the clearance from 2.5 to 4.5 m.
② A professional assessment institution is entrusted with checking calculation and analysis on the railway tunnel as requested by the railway sector through communication with it, with a special construction protection scheme being formulated.
③ The railway tunnel is monitored in a real-time manner through automatic methods so that tunneling parameters can be adjusted and optimized timely as per monitoring results, to guarantee structural safety of the railway tunnel.
④ The railway inclined shaft is backfilled with plain concrete before tunneling.
⑤ Backfill grouting shall be made behind the segment timely after tunneling, to guarantee compact backfilling.

(3) Stepping at this section

Cause analysis: After being erected at the Dazhulin Station, the two pieces of equipment shall pass through the station through stepping in turns and the sections subjected to concealed excavation is 2148 m long. The integral construction progress is affected by frequent switch of procedures.

Countermeasures:

① The stepping scheme and the scheme of concealed excavation shall be planned before construction, and a quick stepping trolley is designed in combination with the stepping scheme of the open type TBM and shield.
② The structural form of the enlarging end (receiving tunnel and launch tunnel) of the station shall be optimized in advance, to guarantee stepping of the TBM and the trailer as a whole.
③ Setting-out shall be made before stepping to simulate the stepping route to avoid shield jamming.

(4) Vault backfilling

Cause analysis:

There is a 0–3 m backfilling section at the 61 m vault behind the Renhe Station, and the original design of drill and blast first and then stepping of the single shield TBM is changed as TBM tunneling to minimize procedure switch and to improve construction efficiency. There are risks of land subsidence and pipeline damage during tunneling.

Countermeasures:

① Supplementary geological exploration is made for such section, and the borehole spacing is decreased from 25 to 5 m, to explore the geology at such section in a detailed manner.
② Setting-out for 1 m shall be made along the tunnel contour on the ground, and two-fluid slurry shall be injected with sleeve valve pipes at a spacing of 1 m, to consolidate such area into a tank.
③ Cement paste shall be injected in a quincunx manner with sleeve valve pipes at a spacing of 1.5 m × 1.5 m within the consolidation contour of the two-fluid slurry and the reinforcement effect shall be verified.
④ The preparation for advance reinforcement shall be made in the tunnel.
⑤ The cutter shall be replaced in advance, with cleaning of the excavation chamber, equipment maintenance and service, and replacing cutters at this section shall be avoided, to guarantee balanced and continuous pass-by.
⑥ Ground monitoring shall be enhanced, with timely feedback of monitoring data.

(5) Sinking of tail shield segments at the opening

Cause analysis:

An opening is designed at 60° of the tail shield bottom, and the segments will sink easily before effective filling of pea gravel after segment erection, leading to steps between segments and water seepage.

Countermeasures:

① A batch of concrete cushion blocks shall be precast, and four precast concrete cushion blocks shall be paved within the 60° opening before erection of bottom segments, and then the bottom segments, side segments and top segments shall be erected successively.
② Hydraulic reclamation of the pea gravel at bottom and on both sides shall be made timely once the segment is retracted from the tail shield.

(6) Tunneling parameters and attitude

According to previous tunneling experience, construction parameters such as set thrust, tunneling speed, cutter head rotation speed and torque are as follows:

① Tunneling thrust: The rated thrust designed for the single shield TBM is 39,000 kN and the thrust set at normal sections is 9000–12,000 kN.
② Tunneling speed: The maximum tunneling speed is 80 mm/min and the normal speed set is 40–60 mm/min.
③ Cutter head rotation speed: The rated rotation speed of the cutter head is 5–6 r/min and the rotation speed set for the cutter head at normal tunneling sections is 5.0–5.4 r/min and the penetration is 9–13 mm/s.
④ Cutter head torque: The rated torque of the cutter head is 4054 kN·m and the torque set for the cutter head at normal tunneling sections is 800–1400 kN m.

Since the form of pea gravel + primary grouting + secondary grouting is adopted for filling behind the segments of the single shield TBM, timely and effective filling cannot be realized like synchronous grouting of the shield. The formed attitude will sink for 4–6 cm after segment erection, and the vertical attitude shall be maintained at around 6 cm during tunneling.

Questions

1. Briefly introduce the reasons why the superiority of TBM is not fully played.
2. What's the basic sequence for erection of the TBM main machine?
3. Briefly introduce the handling schemes of large, medium and small-sized collapse.
4. Briefly introduce the preparations and steps of replacing inner seals and wear-resistant strip of the main bearing.
5. Briefly introduce the breakout technology of the double shield TBM.
6. Briefly introduce handling of instability of the excavation surface.
7. Briefly introduce the breakout process for the cutter head of the single shield TBM.
8. Briefly introduce the methods tackling jamming in front of the cutter head.

9. Briefly introduce the measures tackling “bowing” of the single shield TBM.
10. Briefly introduce the measures tackling geological disasters such as mud and sand burst of the single shield TBM.
11. Briefly introduce the measures reinforcing lined segments at the tunnel sections with unfavorable geological conditions

Bibliography

1. Chen K, Hong K (2016) Construction case of shield tunnels at home and abroad. China Communications Press Co., Ltd., Beijing
2. Chen K, Hong K (2016) In: Shield construction technique. Rev. 2; Beijing, China Communications Press Co., Ltd.
3. Chen K, Zhou B (2017) Shield in China. Yilin Press, Nanjing
4. Chen K (2006) Application prospect of TBM in construction of railway tunnels. Construct Mach 8:14–17
5. Chen K, Feng H (2013) Risks in tunneling by TBM and their countermeasures. Tunnel Construct 33(2):91–97
6. Chen K, Yang Y (2016) Discussion on design of TBM with high adaptability to Gaoligongshan tunnel. Tunnel Construct 36(12):1523–1530
7. Wang M, Li D (2004) Tunneling with rock TBM and related case histories. China Railway Publishing House Co., Ltd., Beijing
8. Li J, Chen K (2006) Technical characteristics and engineering application of double shield TBM. Construct Machanization 3:46–49
9. Gong Q (2014) Introduction to TBM tunneling. Science Press, Beijing
10. Amberg R (1994) Die Bewältigung der Serpentinzone in der Zweispurstrecke-Vereina-Tunnel Nord. Schweiz.Ing.Architekt 44:920–923
11. Wei W, Wang M, Yu L (2015) Application technology of open type TBM in tunnel construction. Chengdu, Southwest Jiaotong University Press
12. Hilar M, Srb M (2009) Long railway tunnels-comparison of major projects. Planning 53:2020
13. Perras MA, Diederichs MS, Besaw D (2014) Geological and geotechnical observations from the Niagara tunnel project. Bullet Eng Geol Environ 73(4):1303–1323
14. Zuo Z, Song S (2012) Construction difficulties of ND pumped storage power station in Switzerland. Express Water Resour Hydropower Inform 33(6):30–31
15. Lumetzberger M, Kitzler C (2013) Detailed surveys of the existing condition of the Bärenwerk pressure tunnel for plant renewal/Detaillierte Bestandserhebung für den Druckstollen Bärenwerk zur Anlagenerneuerung. Geomech Tunnell 6(5):434–445
16. Long W, Yin J, Lin Y et al (2005) The development and application of TBM, and discussion of correlated engineering geological problems. Explor Eng (Rock & Soil Drilling and Tunneling) 2:55–59
17. Wang J, Huang Z, Yang X (2014) Assemble of open TBM main drive. Sci-Technol Enterprise 6:256–256
18. Li X (2016) Construction technology of single shield TBM. Eng Technol (English Version) 12:131–132
19. Wang W, Fan D, Wang Q et al (2016) Application and improvement of single shield TBM in metro tunnel construction. Construct Machanization 37(8):51–54

K. Chen et al., *TBM Design and Construction*,
https://doi.org/10.1007/978-981-99-0059-6

20. Wan Q, Yao X, Liu X (2014) Brief analysis on structural form and working principle of double shield TBM. Technol Innov Appl 2:76
21. Xu D, Jialin H, Li Y et al. (2014) Study on design of TBM cutter head. Public Commun Sci Technol (1)
22. Liu Z, Teng H, Shi Y et al. (2008) Cutterhead design key issues of a full face rock tunnel boring machine(TBM). China Mech Eng
23. Sun W, Ling J, Huo J et al. (2013) Dynamic characteristics study with multidegree-of-freedom coupling in TBM cutterhead system based on complex factors. Mathem Problems in Eng (3):657–675
24. Zhou S (2008) Study on drive system of the full-face rock tunnel boring machine. Zhejiang University, Hangzhou
25. Li Y (2014) Study on installation process of main drive of double shield TBM for hard rocks. China Sci Technol Panorama Magazine 7:103–104
26. Rao Y, Gong G, Yang H (2017) Study on thrust coordination control of single horizontally supported TBM. J Central South University (Science and Technology) 48(3):666–674
27. Zhang L, Zhang G, Wang Y (2005) Study on design methods of TBM back-up system. Railway Construct Technol 37(S1):16–17
28. Xu W (2014) Study on adaptability of urban public transport and urban development. Beijing Jiaotong University, Beijing
29. Zhang L (2006) Application of AHP-fuzzy comprehensive evaluation in assessment of eco-industrial park. Dalian, Dalian University of Technology
30. Han W (2007) Application of AHP-based fuzzy comprehensive evaluation in concrete bridge assessment. Guangxi University, Nanning
31. Zhang N (1981) Subordinate and propability characteristics of random phenomena. J Wuhan Univer Technol 1981(1)
32. Yuan X, Li W, Zhao Y (2009) Brief analysis on TBM tunneling speed and influence factors. Water Conserv Construct Managem 4:75–77
33. Yan Y (2016) Study on influence factors of double shield TBM tunneling speed. Gansu Agricultural University, Lanzhou
34. Li Z (2014) Analysis on TBM tunneling speed for hard rocks and influence on project economy. High Speed Railway Technol 5(1):11–14
35. Zhao W (1999) Study on matching of main working parameters of full face TBM for Qinling tunnel. Beijing Jiaotong University, Beijing
36. Zhu Q, Liu X, Pan C (2006) Analytical calculation of main thrust of full face TBM cutter head. Mining Process Equipm 34(11):20–21
37. Li J, Chen K (2006) Geology adaptability of double-shield TBMs and relevant calculations. Tunnel Construct 26(2):76–78
38. Zhang L (2013) Study on main thrust system of double shield TBM. Yanshan University, Qinhuangdao
39. Administration Bureau for Shanxi Wanjiazhai Yellow River Diversion Project (2011) Application and study on double shield TBM. Beijing, China Water and Power Press
40. Song K, Yuan D, Wang M (2005) Discussion on cutter head rotation speed of tunnel boring machine (TBM). Construct Mach 8:63–65
41. Dong S (2016) Breakout technology of open type TBM in fault fracture zone. Tunnel Construct 36(3):326–330
42. Wen S, Kong Q (2014) Risk assessment on the accident of TBM cutterhead jamming caused by tunnel collapse. J Yangtze River Scient Res Instit 31(4):59–62
43. Li P, Chai Y (1998) TB880E TBM and supporting equipment for Qinling tunnel. World Tunnel 4:6–14
44. Du Y, Du L (2011) Full face hard rock tunnel boring machine: system principles and integrated design. Wuhan, Huazhong University of Science and Technology Press
45. Wang S (1998) Summary on relationship between TBM and geological factors. World Tunnel 2:39–43

46. He F, Gu M, Wang S (2002) Study on grading methods of rocks in TBM tunnel. Chinese J Rock Mech Eng 21(9):1350—1354
47. Li C, Peng Y (2006) Discussion on rock classification in TBM construction tunnel. J China and Foreign Highway 26(3):235–237
48. Qi S, Wu F (2011) Study on rock mass grading of TBM based on fuzzy mathematics. Chinese J Rock Mech Eng 30(6):1225−1229
49. Benardos AG, Kaliampakos Dc (2004) A methodology for assessing geotechnical hazards for TBM tunnelling—illustrated by the Athens Metro, Greece. Int J Rock Mech Mining Sci 41(6):987–999
50. Mikaeil R, Naghadehi MZ, Sereshki F (2009) Multifactorial fuzzy approach to the penetrability classification of TBM in hard rock conditions. Tunnell Underground Space Technol 24(5):500–505
51. Hamidi JK, Shahriar K, Rezai B et al (2010) Application of fuzzy set theory to rock engineering classification systems:an illustration of the rock mass excavability index. Rock Mech Rock Eng 43(3):335–350
52. Wang P, Guo W, Zhu D (2014) Prediction model for tunnelability grading of tbm rocks based on fuzzy clustering theory. Modern Tunnell Technol 51(6):58–65
53. Huang J (2016) Analysis on geological adaptation of double shield TBM in Qingdao metro running tunnel. Modern Tunnell Technol 53(3):42–46
54. Wu YY, Xiangbin W, Juntao Y (2006) Study on rock grading methods of TBM tunnels. Hydrogeol Eng Geol 33(5):120−122
55. Barton NR (2000) In: TBM tunnelling in jointed and faulted rock. Boca Raton, CRC Press
56. Xuechao W, Zhenhong L, Qingbo W (2007) Discussion on rock grading of the west line of South-to-North water diversion project based on TBM. In: Xuechao W, Faquan W (eds) Exploring rock mechanics and engineering geology of the west line of the South-to-North water transfer project, Beijing, Science Press
57. Yan C, Yan S, Liu Z (2013) Change in quartz content of rocks and influence on TBM in the west line of the South-to-North water diversion project. J Eng Geol 21(4):657–663
58. Zhang J, Fu B (2008) Rock burst and criteria and prevention. Chinese J Rock Mech Eng 27(10):2034−2042
59. Wang S (2011) Construction countermeasures for TBM of jinping headrace tunnel under high ground stress. Railway Construct Technol 4:28–33
60. Wang JA, Park HD (2001) Comprehensive prediction of rockburst based on analysis of strain energy in rocks. Tunnell Underground Space Technol 16(1):49–57
61. Azimi Y, Osanloo M, Aakbaepour-Shirazi M et al. (2010) Prediction of the blastability designation of rock masses using fuzzy sets. Int J Rock Mechan Mining Sci 47(7):1126–1140
62. Dudt JP, Delisio A (2016) The "penalty factors" method for the prediction of TBM performances in changing grounds. Tunnell Underground Space Technol 57:195–200
63. Chen F, He C, Deng J (2015) Definition of high ground stress and qualitative and quantitative criteria. Rock and Soil Mechan 36(4):971–980
64. Yin J, Shang Y, Fu B et al. (2005) Development of TBM tunneling technologies and analysis on engineering geological problems and countermeasures. J Eng Geol 13(3):389−397
65. Li Q, Zhang D, Zhang C et al (2013) Experimental study on damage to metro tunnel rocks under different buried depths. Modern Tunnell Technol 50(6):85–93
66. Xu Z, Huang R, Wang S (2000) Division of tunnel buried depth. The Chinese J Geol Hazard and Control 11(4):5–10
67. Xu L (1999) Research on rockburst character and prevention measure of Erlang mountain highway tunnel. Chengdu University of Technology, Chengdu
68. Zhang Y, Wang P, Du Y et al. (2015) Study on geological factors affecting TBM and discussion on construction countermeasures. Chinese J Underground Space and Eng 11(S2):651−656
69. Liu Z, Wang X, Wang Q et al (2005) Analysis on tunnel rock grading and deformation of the west line of the South-to-North water diversion project. Chin J Rock Mechan Eng 24(20):3625–3630

70. Yin J (2005) Study on main engineering geological problems related to TBM. Central South University, Changsha
71. Liu Y, Tang M (2008) Rock mechanics. Chemical Industry Press Co., Ltd., Beijing
72. Liu X, Ma X (2009) Fuzzy safety evaluation for water bursting and mud surging of tunnels through the fault fracture zones. Chinese J Underground Space and Eng 5(S2):1786–1790
73. Shang Y, Wang S, Xue J et al (2002) Analysis on engineering geology of TBM jamming at marlstone sections of Shanxi Wanjiazhai yellow river diversion project and accident handling. J Eng Geol 10(3):293–298
74. Song T, Xiao Z, Su H et al. (2004) Analysis on engineering geology of 2·22 TBM jamming of Shangongshan tunnel. Chinese J Rock Mech Eng 23(S1):4544−4546
75. Wang J (2011) Analysis on jamming of double shield TBM in construction of water diversion tunnel and jamming releasing technology. Tunnel Construct 31(3):364–368
76. Liu Q, Huang X, Shi K et al (2013) Principle of full face TBM jamming above 1000m. J China Coal Soc 38(1):78–84
77. Liu Q, Huang X, Shi K et al (2014) The mechanism of TBM shield jamming disaster tunnelling through deep squeezing ground. J China Coal Soc 39(S1):75–82
78. Wen S, Yang S, Dong Z et al (2015) Study on mechanism of TBM jamming at deep tunnels and control measures. Chinese J Geotech Eng 37(7):1271–1277
79. Liang J (1999) Discussion on problems related to TBM construction at sections with unfavorable geological conditions of the south line of the Shanxi Wanjiazhai yellow river diversion project. Shanxi Hydrotech 2:10–12
80. Li C, Peng D (1998) Characteristics of water burst at exit of Qinling tunnel and influence on TBM construction. World Tunnel 4:18–21
81. Hamidi JK, Shahriar K, Rezai B et al. (2010) Application of fuzzy set theory to rock engineering classification systems: an illustration of the rock mass excavability index. Rock Mechanics and Rock Eng 43(3):335–350
82. Yang Y, Jiang G, Sun W (2012) Preliminary discussion on the relationship of permeability rate of fractured rock and its intactness index and BQ value. Geotech Invest Surv 40(1):35–38
83. Code of Water Pressure Test in Borehole for Water Resources and Hydropower Engineering (SL 31—2003). Beijing, China Water & Power Press
84. Code for Water Resources and Hydropower Engineering Geological Investigation (GB 50487—2008). Beijing, China Planning Press
85. Song Y, Jia G (2004) Main engineering geological problems and exploration of deep and long tunnels. J Eng Geol S1:218–222
86. Karadogan A, Kahriman A, Ozer U (2008) Application of fuzzy set theory in the selection of underground mining method. J Southern African Institute of Mining and Metallurgy 108(2):73–79
87. Ren G, Fu W (2002) Preliminary exploration on geological conditions triggering harmful gas at deep and long tunnels. Mountain Res 20(1):122–125
88. Guan B, Mai C (1988) Japan Society of Civil Engineering. Japanese Tunnel Standard Specification (Mountain Chapter) and Interpretation. Trans. Chengdu, Southwest Jiaotong University Press
89. Yuan Y (2012) New argumentation for grade repartition of gas tunnel. J Railway Eng Soc 29(9):65–70
90. Wang L, Li H, Lu Y et al. (2009) Study on grading of highway gas tunnel. Southwest Highway 4:154—157
91. Kang X, Xu M, Ding R (2010) Preliminary exploration on risk evaluation of tunnel gas disasters. J Railway Eng Soc 27(5):39–42
92. Technical Code for Railway Tunnel with Gas (TB 10120—2002). Beijing, China Railway Publishing House
93. Zhang G (2012) Discussion on grading of railway tunnel gas. J Railway Eng Soc 29(10):75–79
94. Mo Z (2011) Study on rock-breaking mechanism and efficiency of TBM disc cutter. University of Chinese Academy of Sciences, Wuhan

95. Wang C (2013) Analysis on stress of TBM cutter head and study on rock-breaking mechanism of disc cutter. Shenyang Jianzhu University, Shenyang
96. Xiao N, Zhou X, Zhai S (2015) Numerical study on rock-breaking mechanism of TBM edge disc cutter. J China Coal Soc 40(6):1263–1269
97. Xia YM, Wu Y, Guo JC et al. (2014) Numerical simulation of rock-breaking mechanism by gage disc cutter of TBM. J China Coal Soc 39(1):172–178
98. Zhang P (2009) Study on layout of TBM disc cutter and optimal design of structural parameters of cutter head. Dalian University of Technology, Dalian
99. Su P, Wang W, Huo J et al (2010) Study on optimal design for layout of TBM disc cutter. J Northeastern Univer 31(6):877–881
100. Zhag Z, Qiao Y (2011) Study on layout rules of disc cutter of full face TBM. Eng Mech 28(5):172–177
101. Xu X, Yu J (1984) Rock fragmentation. Beijing, China Coal Industry Publishing House
102. Lai H, Zhu CZ, Li XB et al (1991) Mechanical rock fragmentation. Central South University Press, Changsha
103. Mao C (1998) General of full face TBM: one of technical lectures of full face TBM. Construct Mach 8:30–34
104. Mao C, Liu C (1996) Indentation test of disc cutter of full face TBM. Construct Mach Equipm 13(1)
105. Yang J (2008) Brief analysis on relationship of TBM construction and operation system of Xinjiang Daban tunnel. Water Conserv Construct Managem 28(3):24–26
106. Jiang X, Wang C (2005) Analysis on optimal configuration of muck discharge system through TBM tunneling. Mining Process Equipm 6:8–9
107. Wang H (2014) Management of muck discharge system of long-distance headrace tunnel belt conveyor. Water and Water Technol
108. Zhang J, Mei Z, Tang Y et al (2010) Study on selection of TBM and optimization of support system of extra long tunnel of Dahuofang water diversion project. J Railway Eng Soc 27(3):61–65
109. Guo M (2016) Scheme design of tunnel transportation system of lot TBM2 of the Wanjiazhai yellow river diversion project in the middle part of Shanxi. Shanxi Hydrotech 2:55–57
110. Liu G (2016) Quick erection and commissioning technology of TBM under narrow site conditions. Equipm Managem Maintenan 1:83–85
111. Zhao J, Li E, Feng Z et al. (2009) Site erection of full face TBM. West China Developm 11:81
112. Su H, Wang J (2004) Quality control and management of TBM construction. Chin J Rock Mechan Eng 23(11):1930
113. Wang Z, Zhang Y (2013) Construction quality management and control of TBM. Shanxi Hydrotechn 3:37–39
114. Yang J (2015) TBM construction safety and management measures of some hydraulic tunnel in Xinjiang. Water Resour Planning and Design 7:101–103
115. Wei Z (2002) Mobilization transportation of TBM. Tunnel Construct 2 6–7
116. Yao X (2007) Transportation of large parts of TBMs. Tunnel Construct 27(6):112–113
117. Jia Q (2012) Problems in highway transportation of large parts of TBM and countermeasures. Traffic Eng Technol National Defence 10(3):54–57
118. Wang Y (2009) Assembling and commissioning technology of TB880E TBM under narrow sites. Construct Machanization 30(12):56–58
119. Liu G (2016) Quick assembling and commissioning technology of TBM under narrow site conditions. Equipm Managem Maintenan 1:83–85
120. Tang H (2015) Tunneling attitude control of single shield TBM. Shandong Indus Technol 24:104
121. Wu G, Wang Y, Wang Q (2015) Risk and control over TBM tunneling of extra long tunnel. China Sci Technol Panorama Magazine 16:121
122. Cao S (2009) Control and survey of one-way TBM tunneling of long tunnels. Shanxi Hydrotechnics 3:7–8

123. Wang G, Liu L (2014) Selection of tunneling parameters of TBM under different geological conditions. Water and Water Technol
124. Zhao W (2003) Selection of tunneling parameters of TB880E TBM in various rocks. Railway Construct Technol 5:16–18
125. Song J, Yang H, Luan C (2016) Quality control over initial support of open type TBM. Yunnan Water Power 32(1):104–106
126. Zhang J (2013) Application of construction technologies of open type TBM in ultrahard and soft rock strata. Urban Construct Theory Res (9)
127. Wang M (2005) Construction technique of open TBM for long railway tunnels in very hard or soft rock strata. Chin Civil Eng J 38(5):54–58
128. Zhang J (2017) Study on construction technology of open type TBM at Karst and soft rock deformed sections of Gaoligongshan tunnel. Urban Construct Theory Res (1)
129. Wen B (2015) Study on construction technology of TBM of long railway tunnels on hard rock stratum. Architect Eng Technol Design (3)
130. Zhang B (2017) Discussion on construction technology of TBM on the stratum at strong water-rich sandy cobble gully sections. Techn Supervision in Water Resour 25(1):121–123
131. Wang W (2012) Construction technology of TBM for passing by crushed weak water-rich stratum. Construct Machan 33(10):93–96
132. Zhao H et al (2017) Study on construction technology for TBM to pass by Limestones continuously of lot 4 of the Songhuajiang river diversion water supply project. Tunnel Construct 37(3):354–362
133. Wang M, Wang Z (2001) Construction technology for TBM to pass by fault fracture zones. Tunnel Construct 3:1–4
134. Yu X (2012) Construction technology for TBM to pass by existing Karst cave of the headrace tunnel of the Jinping cascade-II HPP. Urban Construct Theory Res (12)
135. Hu Y, Gu H, Zhang G (2002) Failure analysis of TBM disc cutter. Heat Treatment of Metals 27(1):55–56
136. Yang H (2007) Brief discussion on TBM disc cutter failure and cutter management in tunnel construction. Construct Machanization 28(12):60–62
137. Wang S (2012) Abrasion mechanism of TBM disc cutter and numerical simulation. Jiangsu University, Nanjing
138. Zhang Z, Pu L, Zhao H et al (2013) Study on material performance of TBM disc cutter ring. J Iron and Steel Res 41(1):18–21
139. Ji G, Zhang X (2015) Influence of geological factors on abrasion of TBM disc cutter. National Drill Steel/Tool Annual Meeting
140. Yang Y, Chen K, Guo L et al (2016) Analysis of influencing factors of wear of disc cutter of full face hard rock tunnel boring machine. Tunnel Construct 36(11):1394–1400
141. Tan Q, Sun X, Xia Y et al (2017) Prediction model for wear of TBM disc cutter. J Central South Univer (Science and Technology) 48(1):54–60
142. Zhou Y (2006) Quality control and detection and maintenance technology of TBM disc cutter. Construct Machin Technol Managem 44(10):85–92
143. Li X, Zhao L, Xu A (2015) Emergency handling scheme for failure of TBM main bearing seal. Techn Supervision in Water Resour 23(2):64–66
144. Dong R (2015) Fault diagnosis and handling of main bearing seal of TBM. Equipm Managem Maintenan 4:32–34
145. Xue J, Ji Z, Xu M (2015) Study on TBM equipment fault and rock relevance. Railway Construct Technol (11):82—85
146. He L (2009) Brief discussion on TBM environment-friendly management, maintenance and service technologies. Railway Construct Technol 11:43–45
147. Liang Z, Wang Y, Liao Z et al (2014) Current situations and development trend of study on cutter rock-breaking mechanism. Adv Sci Technol Water Resour 34(2):85–94
148. Jia F (2013) Repair of sealing system of TBM main bearing in tunnel. Tunnel Construct 33(3):252–258

149. Wang Y (2011) Analysis on fault cause of TBM main bearing and maintenance and service technologies. Traffic Eng Technol for National Defence 9(2):46–49
150. Yin S (2009) Fault analysis of lubrication system of TBM-319 TBM main bearing. Railway Construct Technol 11:46–48
151. Zhou Y (2007) Quality control and detection and maintenance technology of TBM disc cutter. Modern Tunnell Technol 44(3):85–92
152. Wei N, Sha M (1999) The analysis of TBM cutter wear characteristic in the construction of Qinling tunnel. J Shijiazhuang Railway Institute (Natural Science) 2:86–89
153. Wen S, He D, Yang S (2014) Analysis on comprehensive risks of TBM construction accidents arising from tunnel deformation. Rock and Soil Mechan 35(6):1727–1734
154. Du S, Chuai L (2011) Application of open type TBM. China Water & Power Press, Beijing
155. Wallis S (1993) Lesotho highlands water project. 2. Laserline, Surrey(UK) 40
156. Wu S, Zhou J, Chen B et al (2015) Study on influence of TBM excavation scheme of headrace tunnel of Jinping cascade-II HPP on rock burst risk. Rock Mechan Eng 4:728–734
157. Lu C (2013) Risk analysis and prevention of tunnel construction by TBM. Henan Water Resour South-to-North Water Divers 10:59–60
158. Dong Y (2013) Risk analysis and safety management of TBM for construction of Chongqing metro. Urban Rapid Rail Transit 3:107–111
159. Fang D, Liu N, Zhang C et al (2013) Rockburst risk control for large diameter TBM boring in high geostress region. Chin J Rock Mechan Eng 32(10):2100–2107
160. Huang P (2009) Theoretical methods and application of analysis on TBM construction risks of headrace tunnel. Tianjin University Hydraulic Structure Engineering Department, Tianjin
161. Du S, Yang S, Ji H (2008) Determination of axis for concrete lining of tunnel excavated by TBM. Water Resour Hydropower Eng 39(2):36–39
162. Du S (2008) Demonstration of comparison on changed equipment selection for TBM and support lining. Water Resour and Hydropower Eng 39(1):52–56
163. Yao Z, Du S, Chuai L (2007) Turnaround of TBM in extra long tunnel construction. Water Resour Hydropower of Northeast China 35(7):21–22
164. Yao Z, Du S, Du Y (2007) Assembling of TBM outside tunnel. Water Resour Hydropower of Northeast China 35(6):22–23
165. Du S (2006) Superiority of TBM in tunneling with large-diameter disc cutter. Water Resour Hydropower Eng 37(10):40–42
166. Zhang J, Shu L, Zhang C et al. (2010) Application of synthetical index method based on entropy weight in evaluation on groundwater quality. Water Resour Power 28(8):36–38
167. Yao Z, Du S, Yang J (2006) Necessity of setting middle construction adits at TBM construction sections of extra long tunnel. Water Resour Hydropower Eng 37(4):30–31
168. Wei Y, Du S (2006) Construction characteristics of large-section extra long water tunnel. Water Resour Hydropower Eng 37(3):8—11
169. Lu S, Gu T, Li Z et al. (2010) Comprehensive evaluation on advance geological forecast at sections with complicated geology of headrace tunnel. Chinese J Eng Geophys 7(5):630–633
170. He X (2010) Brief analysis on advantages of TBM construction technologies. Equipm Geotech Eng 11(2):35–37
171. Liang J, Huang N (2008) Application study of transient electromagnetic method in collapse columns in Shanxi. Chinese J Eng Geophys 5(5):579–583
172. Zhao B (2008) Geological characteristics and countermeasures of Hongyan River-Shitou river water diversion project. J Water Resour Architect Eng 6(2):62–66
173. TBM Application and Relevant Problems and Prospect (1999) Chinese J Rock Mechan Eng 18(3):363–363
174. Liu X, Shen W, Liu G et al (2010) Analysis on deformation reserved for TBM construction of red-bed soft rock tunnels of Yao River water diversion project. Chinese J Underground Space and Eng 6(6):1207–1214
175. Wang Y, Li H, Li J (2010) Countermeasures of TBM construction under complicated geology of Zhongtianshan tunnel. Traffic Eng Technol National Defence 8(6):45–47

176. Wang R, Chen Z, Yang X (2010) Brief analysis on risk management of TBM construction items of some large headrace tunnel. Project Managem Technol 8(9):32–35
177. Wen S, Xu W (2008) Risk analysis on TBM jamming accidents of deep tunnel. J Yangtze River Scient Res Institute 25(5):135–138
178. Meng X (2008) Discussion on construction of long-distance and double-shield TBM. Tunnel Construct 28(4):429–433
179. Hou B, Yu Z, Liu J (2007) Potential risks and countermeasures for TBM construction of long tunnel projects. Water Resour Hydropower of Northeast China 25(11):19–20
180. Yin J, Shang Y, Fu B et al. (2005) Development of TBM tunneling technologies and analysis on relevant engineering geological problems and countermeasures. J Eng Geol 13(3):389—397

The manufacturer's authorised representative in the EU is Springer Nature Customer Service Centre GmbH, Europaplatz 3, 69115 Heidelberg, Germany. If you have any concerns regarding our products, please contact ProductSafety@springernature.com

Printed and bound by CPI Group (UK) Ltd, Croydon, CR0 4YY
30/09/2026
02225249-0006